MOLECULAR MECHANISMS OF NEURONAL RESPONSIVENESS

ADVANCES IN EXPERIMENTAL MEDICINE AND BIOLOGY

Recent Volumes in this Series

MOLECULAR MECHANISMS OF NEURONAL RESPONSIVENESS

Edited by

Yigal H. Ehrlich
Robert H. Lenox and
Elizabeth Kornecki

The University of Vermont College of Medicine
Burlington, Vermont

and

William O. Berry

U.S. Air Force Office of Scientific Research
Bolling A.F. Base
Washington, D.C.

PLENUM PRESS • NEW YORK AND LONDON

Library of Congress Cataloging in Publication Data

Molecular mechanisms of neuronal responsiveness.

(Advances in experimental medicine and biology; v. 221)
"Proceedings of a symposium of a national conference on molecular mechanism of neuronal responsiveness, held March 21–23, 1986, at the University of Vermont College of Medicine, Burlington, Vermont."
Bibliography; p.
Includes index.
1. Neural transmission – Regulation – Congresses. 2. Neural receptors – Congresses. 3.Neuroplasticity – Congresses. I. Ehrlich, Y. H. II. Series.
QP364.5.M65   1987   599/.0188   87-22074

ISBN 978-1-4684-7620-0   ISBN 978-1-4684-7618-7 (eBook)
DOI 10.1007/ 978-1-4684-7618-7

Proceedings of a symposium of a National Conference on Molecular Mechanisms of Neuronal Responsiveness, held March 21-23, 1986, at The University of Vermont College of Medicine, Burlington, Vermont

© 1987 Plenum Press, New York
Softcover reprint of the hardcover 1st edition 1987
A Division of Plenum Publishing Corporation
233 Spring Street, New York, N.Y. 10013

PREFACE

The interaction of neurotransmitters, neuromodulators and neuroactive drugs with receptors localized at the cell surface initiates a chain of molecular events leading to integrated neuronal responses to the triggering stimuli. Major advancements in the characterization and isolation of receptor molecules have answered many questions regarding the nature of the elements that determine the specificity in these interactions. At the same time, recent studies have provided evidence that delicate regulation by intracellular enzymatic systems determines the efficiency of the stimulus-response coupling process, mediates the interaction between receptors, operates in feedback control mechanisms and transduces signals from the receptors to various effector sites in a highly coordinated fashion. These studies are at the focus of the present volume, which is an outcome of a symposium held at the University of Vermont College of Medicine on March 21-23, 1986, in conjunction with the seventeenth annual meeting of the American Society for Neurochemistry. The symposium has demonstrated clearly that the concerted efforts of investigators in neurophysiology, biochemistry, pharmacology, cell-biology, molecular genetics, neurology, and psychiatry are required to achieve better understanding of the processes underlying neuronal responsiveness. This volume includes contributions provided by prominent investigators in all these research areas. We hope that the readers will find here a useful source of information and ideas for stimulating further studies which may serve to narrow the gap between basic neuroscience research and its clinical implications.

Whereas many of the processes under discussion operate similarly in all eukaryotic cells, this volume emphasizes those features believed to be unique to neurons. In particular, cells in the nervous system have the capability of undergoing extremely long-lasting alterations in response to hormonal, pharmacological and environmental stimulations. These adaptive processes can result in behavioral changes. Elucidation of the molecular mechanisms involved will undoubtedly yield novel strategies for the treatment of neurological and neuropsychiatric disorders. Accordingly, this volume is arranged in three sections: I. Signal Transduction and Stimulus-Response Coupling; II. Neuronal Adaptation and Synaptic Plasticity; and III. Behavioral and Clinical Implications of recent findings on the mechanisms of neuromodulation. The chapters included in each section provide up-to-date reviews and summaries of recent developments in the field, as well as descriptions of specific studies which may offer potential new directions in this rapidly growing area of research.

Section I of this volume focuses on basic mechanisms of intracellular communication and includes chapters on the regulation of stimulus-secretion coupling, and on the role of ion-channels, mobilization of calcium ions, metabolism of fatty acids and polyphosphoinositides, cyclic nucleotides and protein phosphorylation systems--in receptor-mediated stimulation. A diversity of experimental approaches is represented: electrophysiological studies, biochemical investigations conducted on several levels of organization (cell-free assays, intact cultured cells, in situ and in-vivo studies) and the use of model systems and novel procedures of molecular biology for shedding new light on molecular mechanisms underlying neuronal responsiveness. In section II different approaches to molecular studies of adaptive processes are demonstrated, including mechanisms of receptor desensitization, long-term regulation of ion channels, synaptic-potentiation, development of tolerance, the kindling process, genetic influences on chemoreception and the expression of genomic changes induced by receptor blockade. The behavioral and clinical implications of recent advances in neurochemical studies of neuronal responsiveness are highlighted in section III, with chapters spanning from grooming behavior in the rat to neurological and neuropsychiatric disorders, and including studies on seizure activity and epilepsy, stress, neurodegeneration and regeneration, mechanism of action of benzodiazepine drugs and antidepressants, as well as recent studies implicating specific biochemical systems in the etiology and treatment of major affective disorders.

The editors wish to express their sincere appreciation to all the individuals who helped us organize the symposium and generate this volume. Special thanks are due to the staff of the Continuing Medical Education Office of the University of Vermont and its director Maureen E. Hanagan. We are indebted to Diane Carbonneau who typed all the chapters and helped us overcome many unforeseen obstacles. Finally, we acknowledge the generous support by the U. S. Air Force Office of Scientific Research without which the symposium could not have taken place.

May, 1987

Y. H. Ehrlich
R. H. Lenox
E. Kornecki
W. O. Berry

CONTENTS

SECTION II

Neuronal Adaptation and Synaptic Plasticity

SECTION III

Behavioral and Clinical Implications

FURTHER STUDIES ON DEPOLARIZATION RELEASE COUPLING

IN SQUID GIANT SYNAPSE

R. Llinás, M. Sugimori and K. Walton

Department of Physiology and Biophysics, New York
University Medical Center, 550 First Avenue
New York, NY 10016

INTRODUCTION

During the past few years significant information has been accumulated
concerning the physiological and biochemical properties by which presynaptic
membrane depolarization leads to transmitter release in chemical synapses.
The physiological aspects of this property of pre-terminals is often
referred to as "depolarization-release coupling" and is known to be a function of intracellular calcium concentration at the appropriate cytosolic
compartment (cf. Llinas, 1982). Most of the information relating to this
aspect of synaptic transmission has come from recent studies in the squid
stellate ganglia (Llinas, Steinberg and Walton, 1976-1981; Llinas, Sugimori
and Simon, 1982; Charlton, Smith and Zucker, 1982; Augustine and Eckert,
1984; Augustine, Charlton and Smith, 1985a,b; Simon and Llinas, 1985; Llinas, McGuiniess, Leonard, Sugimori and Greengard, 1985) and frog and crayfish
neuromuscular junctions (Mallart and Martin, 1967; Dudel 1983 a,b; Parnas
et al., 1984). Among these the squid giant synapse continues to be a most
useful preparation, particularly when studying those aspects of the release
function which require injections of materials into the presynaptic terminal
or the direct measurement of the presynaptic Ca. In this paper we would
like to address three main issues, (i) the effects of direct injection of
Ca ions into the presynaptic terminal on transmitter release, (ii) voltage
dependence of Ca-dependent release, and (iii) the temperature dependence of
transmitter release.

Presynaptic Calcium Injection and Transmitter Release

The first demonstration that Ca ions trigger the release of synaptic
transmitters when injected presynaptically was carried out in the squid giant
synapse by Miledi in 1973. This particular design consisted of relating Ca
injected into the pre-terminal to subsequent release (as determined by an
appropriate postsynaptic membrane potential change). The results demonstrated unambiguously that Ca injection can trigger release directly. Recently
we have repeated some of these initial experiments and confirmed the results
of Miledi and of Charlton, Zucker and Smith, 1980 (Llinas, Sugimori and
Bower, 1983; Llinas, Sugimori and Leonard, 1984). In addition, we found
that if Ca was pressure-injected, rather than introduced iontophoretically
(Miledi, 1973; Charlton, Zucker and Smith, 1980), a more robust transmitter
release occurred. Thus, as shown in Fig. 1, a single short pressure injection of Ca into the presynaptic terminal from an electrode containing 0.5 M

1

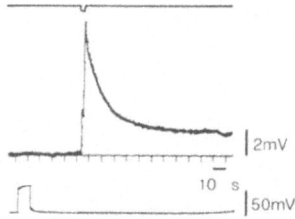

FIG. 1: Injection of Ca into the presynaptic terminal. Top trace: injec-
         tion of Ca into pre-terminal from 0.5M Ca acetate electrode.
         Middle trace: postsynaptic response to Ca injection. Bottom
         trace: Pre-voltage pulse prior to Ca injection failed to elicit
         post-response. (Bath saline contained TTX, 4-AP, 0.2 mM $CaCl_2$
         and 0.5 mM $CoCl_2$).

Ca acetate (top trace), was followed by a postsynaptic response of 11 mV
lasting 50 ms (middle trace); indicating the release of a substantial amount
of neurotransmitter. This type of recording was obtained after the Ca in
the bathing saline was reduced from 10 mM to 0.2 mM and 0.5 mM $CoCl_2$ was
added such that direct depolarization of the presynaptic terminal (bottom
trace) produced no release.

     While such results indicate that under the proper experimental condi-
tions transmitter is released following presynaptic Ca injection, a large
variability in synaptic release was found among synapses, although in all
cases the injecting microelectrode impaled the actual terminal digits (the
Ca injection was monitored by direct observation). For example, Fig. 2
illustrates the time course and amplitude of postsynaptic potentials
observed after single injections. In A, Ca injection produced a postsynaptic potential having a 5-mV peak amplitude and a 55-sec duration, with
a close-to-exponential decay. In B, in another synapse, the postsynaptic

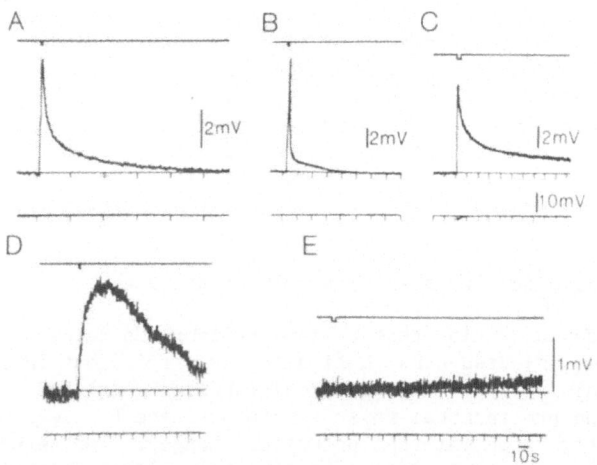

FIG. 2: Variation in post-response to pre-Ca injection. A-C, Note rapid
         onset of post-potential with close-to-exponential decay in A and
         two time courses in decay of response in B and C. D and E, Slower
         post-onset and lower amplitude responses. (Same recording condi-
         tions as for Fig. 1).

2

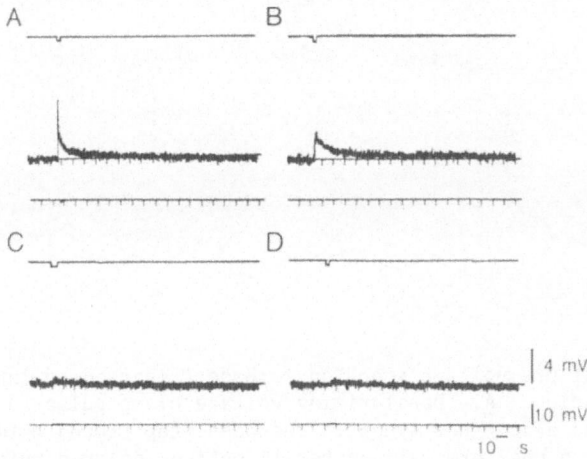

FIG. 3: Effect of repeated Ca injection into the presynaptic terminal. Ca
was injected at 10-min intervals. Note that response amplitude and
rate of rise progressively decreased until in D response is just
above the noise level.

response declined with two different time courses, a fast initial decay followed by a second slower component lasting 35 sec. In a third synapse (C),
the postsynaptic potential is shown with a fast rising phase but a rather
slow fall, again demonstrating two time courses. Two other examples in
which smaller amounts of transmitter were released are shown in D and E.
In D, release was slow and very prolonged, reaching a peak after 20 sec and
not having quite returned to baseline after 90 sec. In E the transmitter
release increased even more slowly. Occasionally, the injection was followed by a prolonged depolarization of the preterminal indicating damage to
the terminal digit by the pressure pulse.

These results suggest that the distribution of the injected Ca varied
quite widely from one synapse to the next and that the transmitter release
itself depended very particularly on the distance of the microelectrode tip
from the presynaptic active zone on the internal plasmalemmal surface. The
most parsimonious explanation for the variability is that the injection
triggered fast transmitter release if Ca becomes immediately available to
the release sites; the slower component probably reflecting Ca diffusing to
other release sites. Finally, recordings such as those shown in Fig. 2 D
and E suggest that large Ca injections could produce long-term effects if
the initial release of transmitter was very small. The reason for this
supposition is shown in Fig. 3 where a series of four Ca injections were
applied at different intervals. The initial injection produced a postsynaptic potential with a rapid rise followed by a slow depolarization (lasting
for 20 sec) (A). Ten minutes after the response shown in A, a second injection elicited a smaller response with a diminished initial component leaving
a slow potential (B). At 20 min (C) and 30 min (D) after the injection in
A, similar injections produced almost no release. The results of this type
of experiment suggest that following local intraterminal Ca injection, there
is a large initial release of transmitter which exhausts the immediately
available supply; thus, subsequent Ca injections do not trigger further
release. An alternative hypothesis would be that sufficient Ca has been
injected to block the release mechanism (Kusano, 1970); however, this explanation does not seem attractive when considering that continuous release may
occur for a minute or longer after a single Ca injection (e.g. Fig. 1).

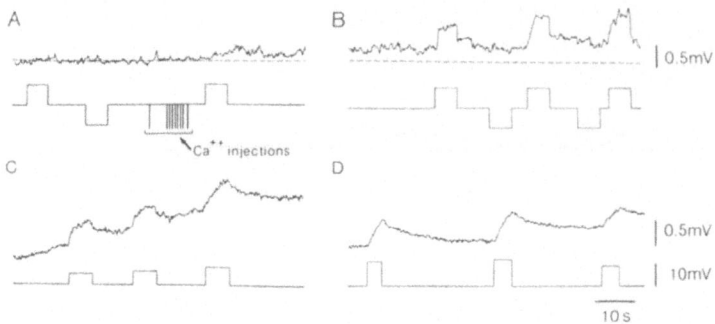

FIG. 4: Effect of pre-voltage steps on release triggered by intraterminal
Ca injection. <u>A</u>. Depolarizing voltage clamp pulse (bottom trace)
failed to elicit postsynaptic response (top trace) before injection
of Ca into pre-terminal. After injection, delayed release is seen.
<u>B</u>. Same synapse as in A about 1 min after Ca injection (broken
line giving postmembrane potential prior to Ca injection). <u>C</u> and
<u>D</u>. Modulation of Ca-triggered release by voltage in two other
synapses 7 min (C) and 4 min (D) after Ca injection. Note in C
that voltage pulses of 4.4, 5.5, and 7.6 mV elicit different amplitude responses. (Bathing saline contained TTX, 4-AP, 0 CaCl$_2$ and
0.5 mM CoCl$_2$.)

Voltage Dependence of Transmitter Release

A set of experiments was conducted in which the presynaptic terminal
was voltage clamped with prolonged voltage steps at various times before and
after intracellular injection of Ca [the bath containing tetrodotoxin (TTX),
4-aminopyridine (4-AP), zero Ca and 0.5 mM CoCl$_2$]. The results of five such
experiments are shown in Figs. 4 and 5 where the top traces show the postsynaptic voltage, which reflects the amplitude and time course of transmitter release, and the lower traces the presynaptic voltage clamp steps from
a holding potential of -70 mV. In Fig. 4A a 4.3-mV, 6-sec depolarizing
voltage clamp pulse delivered before the Ca injection did not trigger any
transmitter release. Following Ca injection, after a delay, an identical
depolarizing pulse elicited an increase in the amount of transmitter release.
About one minute after Ca injection (Fig. 4B), three voltage steps clearly
modulated the slow postsynaptic response elicited by the Ca injection.
Similar increases in transmitter release in other synapse are shown in Fig.

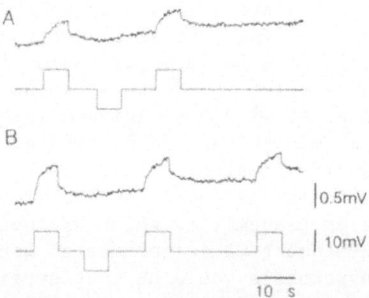

FIG. 5: Depolarizing but not hyperpolarizing pulses increase postsynaptic
response. Modulation of transmitter release by depolarizing pulses
in two other synapses. (Same recording conditions as in Fig. 4).

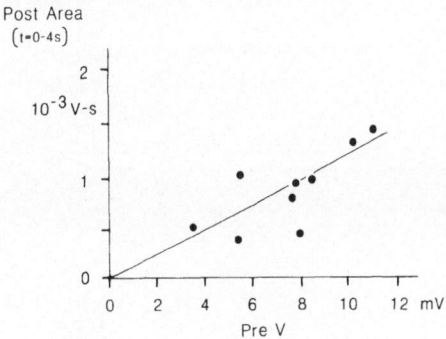

FIG. 6: Presynaptic voltage and transient postsynaptic response. The area
under the postsynaptic response during the first 4 sec after the
onset of the presynaptic depolarization is plotted as a function
of pre-amplitude. Note the clear dependence of post-response on
the pre-voltage step amplitude.

4C and D. In C voltage pulses of three amplitudes elicited different amplitude voltage transients on the slow postsynaptic response due to the Ca
injection itself. Fig. 5 illustrates two examples in other synapses, showing increases in the post-response similar to those in Fig. 4. In addition,
they demonstrate that hyperpolarizing voltage clamp steps fail to produce
any postsynaptic voltage change (Fig. 5). Thus, only depolarizing voltage
steps occurring after the Ca injection could modulate the amount of transmitter release by Ca injection, excluding electrotonic coupling artifacts as
well as the possibility that the postsynaptic response was due to increased
extracellular K concentration produced by the presynaptic voltage step.
Indeed, even prior to addition of 4-AP, when g_k would be large, presynaptic
depolarization did not generate changes in the postsynaptic membrane potential prior to Ca injection.

While the responses usually occurred promptly after the onset of the
increased voltage pulse, the release could take as long as 12 sec to reach
its peak value; also a certain variability was observed in the amplitude and
duration of the voltage-modulated release. Nevertheless, transmitter release

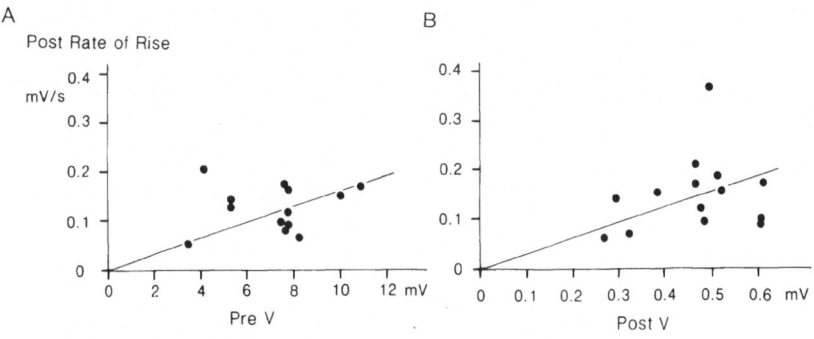

FIG. 7: Presynaptic and postsynaptic voltage and rate of rise of post-
response. The rate of rise of the transient postsynaptic response
elicited by presynaptic depolarization is plotted against the pre-
amplitude in A and against the amplitude of the transient post-
response in B. Note that despite the scatter shown in B, reflecting the variability inherent in this type of experiment, there is
a clear relationship between pre-voltage and post-response in A.

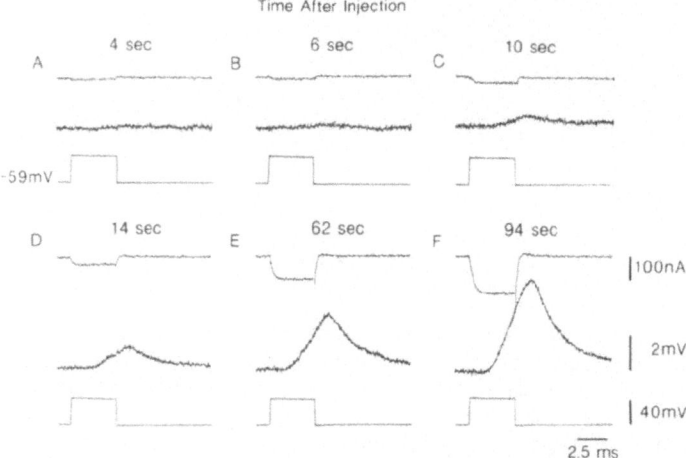

FIG. 8: I_{Ca} and EPSP following extracellular focal addition of single injection of Ca to the pre-terminal. $CaCl_2$ bolus was delivered in the immediate vicinity of the presynaptic terminal at t=0 sec, the ganglion was bathed in TTX, 4-AP, 0.2 mM $CaCl_2$ and 50 mM $MgCl_2$. As the extracellular Ca concentration increased, a Ca current was recorded and a post-response was first seen after 4 sec, both I_{Ca} and the EPSP increasing and reaching a peak after 94 sec.

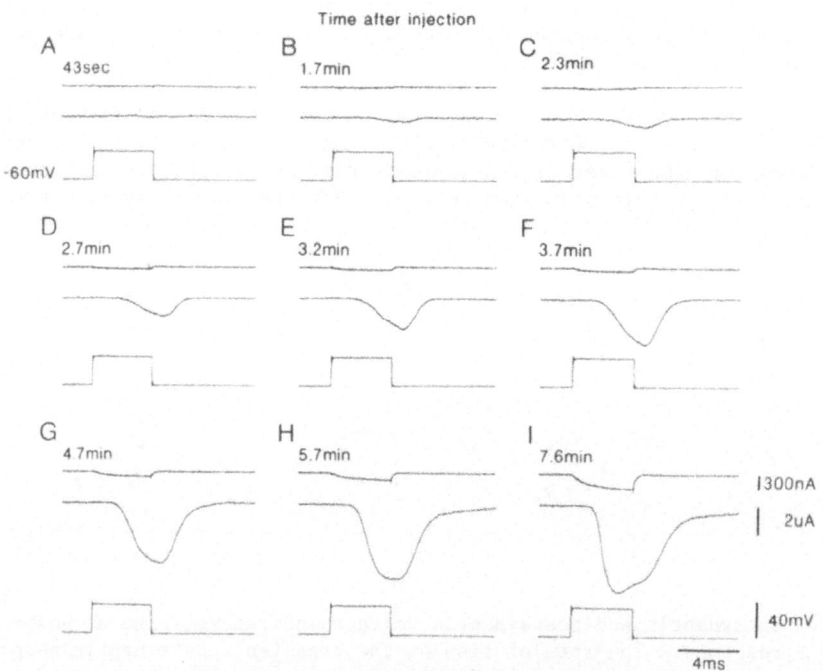

FIG. 9: I_{Ca} and I_{post} following single focal addition of Ca. 43 sec after addition of Ca at the presynaptic terminal both presynaptic and postsynaptic currents were recorded. Both increased, reaching peak values after 7.6 minutes. (Same recording conditions as for Fig. 8.)

was found to be modulatable by presynaptic voltage. This can be seen in Fig. 6 where the transmitter released (estimated from the area under the transient postsynaptic potential)(in the first four seconds t=0 - t=4 sec) after presynaptic depolarization is plotted as a function of presynaptic voltage. A clear dependence of postsynaptic response on presynaptic voltage can also be seen in Fig. 7A where the rate of rise of the transient increase in postsynaptic response is plotted as a function of presynaptic voltage. The variability in the postsynaptic response itself is illustrated in Fig. 7B where the rate of rise of the postsynaptic transient is plotted as a function of the post amplitude at 4 sec after the onset of the presynaptic depolarization. Results similar to those illustrated here were observed in 13 of 15 synapses that showed postsynaptic responses to presynaptic Ca injection. The reason for failure of voltage-dependent modulation of the postsynaptic response in the two other synapses is not clear.

These experiments suggest, therefore, that when sufficient intracellular Ca is present at release sites to produce prolonged transmitter release (for instance, after Ca injection), direct depolarization of the terminal, which does not elicit transmitter release prior to Ca injection, is capable of modulating the release process.

Effects of Local Extracellular Ca Perfusion on Transmitter Release

Recent experiments by Smith, Augustine and Charlton (1985) have indicated that if synaptic release in the squid giant synapse is elicited during a well localized continuous extracellular perfusion of the preterminal region with saline containing Ca, as opposed to that produced when Ca is in the bath, the more localized Ca entry produced with their paradigm may give a more accurate measure of the relationship between Ca and release. The argument presented by Augustine et al. indicates that with well localized extracellular Ca perfusion the distribution of Ca entry would be restricted to the area of the presynaptic terminal under voltage control and thus the Ca current measured would be limited to the actual site of superfusion. A similar set of experiments was performed by us with an extracellular Ca concentration of 0.2 mM in which, rather than applying a constant flow of extracellular Ca, a single bolus of Ca was delivered to the bath at the presynaptic terminal (Llinas, Sugimori and Leonard, 1984). The records shown in Figs. 8-10 demonstrate the relationship between the I_{Ca} and transmitter release following this method of perfusion. Here we have both restricted the site of Ca entry and tested the effect of a continuously increasing $[Ca]_o$ on transmitter release. Fig. 8 shows the effect of a presynaptic depolarizing pulse delivered (from a holding potential of -50

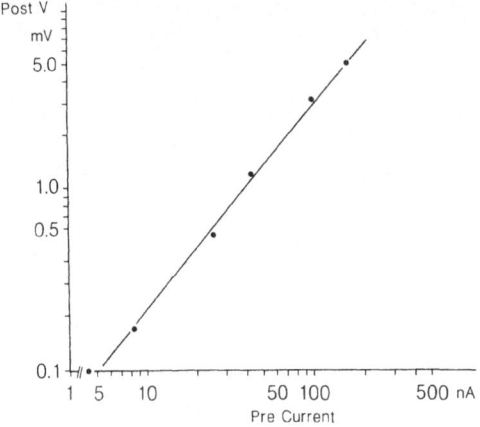

FIG. 10: Double logarithmic plot of EPSP and I_{Ca} after focal addition of Ca extracellularly.

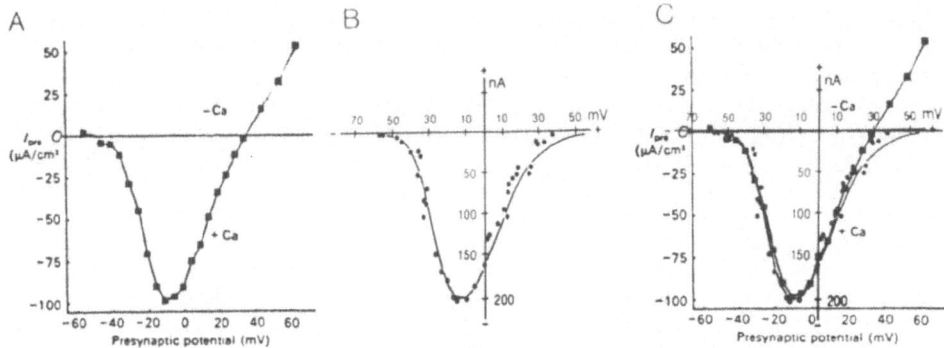

FIG. 11: Comparison of I_{Ca} amplitude as a function of presynaptic potential. <u>A</u>. Data from Augustine et al. (1985a). <u>B</u>. Data from
Llinas et al. (1981a) replotted with the I_{Ca} in the lower qua-
drants of the graph. <u>C</u>. Superposition of plots in A and B.
Note that the plots are close to identical and well within the
variation seen among individual synapses.

mV) at six different times after a single extracellular Ca injection. The
top trace gives the presynaptic current, the middle the postsynaptic
response, and the bottom the presynaptic voltage pulse. Note that the
amplitude of the postsynaptic response increased with time after the Ca
injection, reaching a maximum at 94 sec; after this the synaptic potential
quickly returned to the unresponsive level shown in Fig. 8A. Similar
results can be seen in Fig. 9 from another synapse where a large extracellular Ca injection was applied and a post synaptic current, rather than the
potential, was used to monitor transmitter release. These types of experiments indicate that the relationship between Ca current and transmitter
release has a stoichiometry just over 1 as shown in the double
logarithmic plot of the data in Fig. 8. This is an important consideration in that it points out that differences in the relationship between Ca
entry and transmitter release seen between experimental preparations may
be due to the manner in which the data are measured, rather than to the
results themselves or to the methodology utilized. This issue is further
addressed in Fig. 11 where the Ca currents obtained by Augustine et al.
(1985a, their Fig. 8B) using presynaptic local bathing are compared with
the Ca currents published 4 years previously, with the usual Ca in the
bathing solution (Llinas et al., 1981a). Because these two sets of Ca
currents are indistinguishable, the argument that the differences between
our results and those of Augustine et al. are due to differences in voltage
clamp control is untenable.

Temperature Effects on Synaptic Transmission

The effects of changes in temperature on spontaneous and evoked synaptic transmission has long been used as a method to study transmitter
release properties and, for the most part, has been carried out in the
neuromuscular junction (nmj) (Katz and Miledi, 1965; Hubbard, Jones and
Landau, 1971; Barrett et al. 1978).

Early studies of the temperature dependence of transmission at the
squid giant synapse were similar to nmj work in that the major finding was
a marked decrease in synaptic delay and an increase in EPSP amplitude with
increased temperature (Lester, 1970; Weight and Erulkar, 1976; Charlton
and Atwood, 1979). Lester (1970) demonstrated that synaptic delay was a
logarithmic function of temperature while Weight and Erulkar (1976) showed

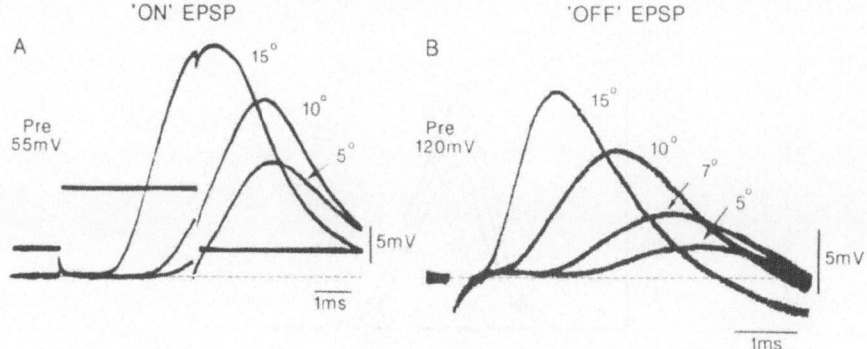

'ON' EPSP   'OFF' EPSP

FIG. 12: Effect of temperature on EPSP. The postsynaptic response to a 55 mV presynaptic depolarization is shown in A. As the temperature was lowered from 15° to 5°C, the synaptic delay increased and the rates of rise, fall and amplitude of the EPSP decreased. Similar temperature effects are shown in B for the 'off' EPSP following a voltage pulse to the supression potential.

that the increased delay with low temperature was independent of spike amplitude. These findings were confirmed by Charlton and Atwood (1979) who also found that low temperature decreased the EPSP amplitude, even though the presynaptic action potential amplitude and duration increased. In preliminary voltage clamp studies they found a decrease in presynaptic I_{Ca} amplitude and rate of rise at low temperature, and that the increase in I_{Ca} onset was small compared to the change in synaptic delay. They also reported a good correlation between the rate of rise of the EPSP and of the I_{Ca}, the latter being most marked at high levels of presynaptic depolarization. The results of our voltage clamp studies of temperature effects (Llinas, Walton and Sugimori, 1978) showed a less marked correlation between changes in I_{Ca} and the EPSP.

In the set of experiments reported here the effect of temperature (5 to 20°C) was studied using either the presynaptic voltage clamp technique of Adrian et al. (1970), or a double voltage clamp of both pre- and post-synaptic terminals as described by Llinas and Sugimori (1978). G_{Na} and G_{k} were blocked respectively by 10^{-6} TTX, by intracellularly injected tetraethylammonium (TEA) and by bath application of 4-AP.

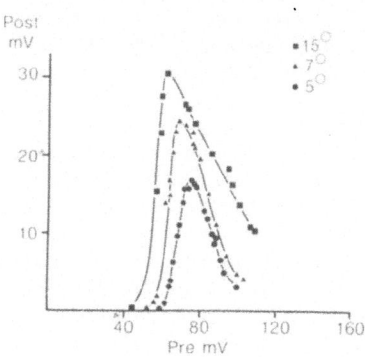

FIG. 13: Presynaptic voltage and postsynaptic response at three temperatures. EPSP amplitude is plotted as a function of pre-voltage for 15°, 7° and 5°C. Note that the 'threshold' voltage for the EPSP moved to the right as the temperature decreased and the peak EPSP amplitude decreased.

9

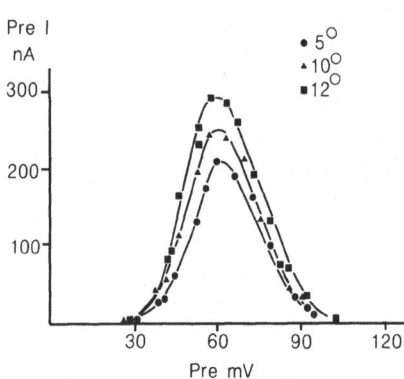

FIG. 14: Presynaptic voltage and I_{Ca} at three temperatures. I_{Ca} is plotted
as a function of pre-voltage for 15°, 10° and 12°C. Note that the
relationship is bell-shaped at all three temperatures and that
temperature has little effect on the voltage at which I_{Ca} was first
seen. The major effect of temperature is on I_{Ca} amplitude at all
levels of pre-voltage.

Synaptic delay and postsynaptic response amplitude. The postsynaptic
response elicited by two levels of presynaptic depolarization in two
synapses are shown in Fig. 12. Two main effects of decreasing the temperature
from 15° to 5° C are evident: an increased delay for the postsynaptic
potential and a decrease in its peak amplitude. The temperature dependence
of these two paramets is reflected in their Q_{10}s, being 4 for the delay and
close to 3.5 for the EPSP amplitude (15° to 5° C). Voltage clamp experiments demonstrated a similar temperature dependence for these parameters.
Indeed, the synaptic delay between the onset of the presynaptic voltage
clamp pulse and the EPSP had a Q_{10} of 3.8 (10°-20°C) and 2.3 (5°-15°C)
while in double voltage clamp studies the Q_{10} was 2.9 for the delay to the
onset of the postsynaptic current (see Fig. 15).

Fig. 13 plots EPSP amplitude as a function of presynaptic depolarization for 5°, 7° and 15° C. As in the current clamps studies, EPSP amplitude was markedly temperature-dependent, with Q_{10}s of 3.1 and 3.6, while
in double voltage clamp studies, the Q_{10} for the postsynaptic current
amplitude was 4.6. The rate of rise of the EPSP was also quite sensitive
to temperature (Q_{10} of 3.8; 5°-15°C), while other parameters were less so.
For example, the threshold for the postsynaptic response as a function of
presynaptic voltage had a Q_{10} just under 2 in three different experiments
(1.56, 1.94, 1.85; 5-15°C), and the amplitude of the presynaptic depolarization eliciting the maximum post response had a Q_{10} just over 1 (1.13,
1.17, 1.18; 5-15°C).

As in all previous studies we found that two parameters were profoundly
affected by temperature, synaptic delay and postsynaptic potential or current
amplitude. The effects of temperature on these parameters, however, differ
in one important respect. While the amplitude of the EPSP recorded at low
temperatures could be increased by increasing the presynaptic depolarization
amplitude or duration, the synaptic delay was less sensitive to such manipulation. For example, a 39-mV depolarization at 7° C elicited a 2.5 mV EPSP
with a delay of 2.37 msec. If the depolarization step was increased to 53
mV, the EPSP amplitude became 6.2 mV (an increase of 248%) but the delay
decreased only by 8% to 2.17 msec. Thus, the effect of low temperature on
transmitter release can be ameliorated by changing the potential across the
presynaptic membrane; however, the effect of temperature on the onset of

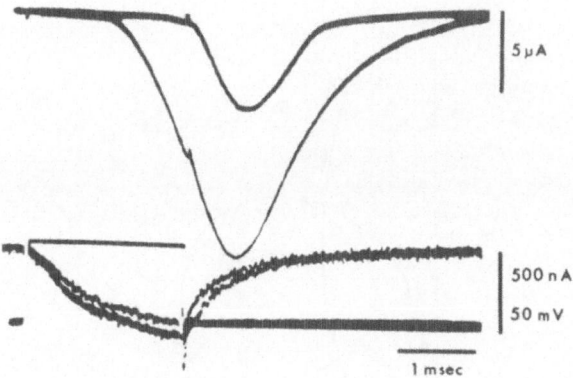

5 µA

500 n A

50 m V

1 msec

Fig. 15: Effect of temperature on I_{Ca} and I_{post}. The presynaptic terminal was depolarized with a 50 mV square pulse at 7° (smaller responses) and 10°C. The postsynaptic current (top traces) shows a marked increased synaptic delay and decreased response amplitude with the 3° decrease in temperature while the presynaptic current (middle traces) showed no significant change in onset and only a small decrease in amplitude.

the release event is not significantly affected by the amplitude of the pre-synaptic voltage pulse.

The amplitude and time course of I_{Ca}. The effect of temperature on I_{Ca} is shown in Fig. 14 where I_{Ca} amplitude is plotted as a function of presynaptic depolarization for 5°, 10° and 12°C. The figure illustrates that the 'threshold' voltage for I_{Ca} did not change significantly with temperature, while the current amplitude did. The Q_{10} for the latter was voltage-dependent, I_{Ca} amplitude being more sensitive to temperature at higher presynaptic depolarizations than at lower ones. The values for the Q_{10} in the experiment shown in Fig. 14 were 1.28 for a 40-mV depolarization and 2.5 for a 60-mV voltage pulse. The voltage at which the maximum current was elicited was less dependent on temperature, having Q_{10}'s of 1.06 to 1.13 (5°-15°C) in two synapses. Finally, the suppression potential was not significantly affected by temperature.

The temperature dependence of Ca channel kinetics shows a low Q_{10} value. The current onset was delayed up to 500 usec for a decrease from 10° to 5°C and the time required for the current to reach one-half its maximum value was slightly increased under these conditions. The Q_{10} for the latter was 1.14 (5°-15°C) for a 40-nA current but could be as high as 4 for a 210-nA current. Thus, as with I_{Ca} amplitude, the Q_{10}'s for the rate of rise of I_{Ca} depends on the amplitude of the presynaptic voltage pulse. The latter may, in fact, be the dominant factor, but has not yet been determined. Mea-surements of tail currents indicate that channel closing had a Q_{10} near 1.4 (1.32, 1.53; 5°-15° C), this value was not dependent on the voltage clamp step amplitude. These results are consistent with those obtained for the temperature dependence of other voltage-dependent conductances in the squid giant axon (Hodgkin and Huxley, 1952). It is of interest to note two points, (1) that the temperature effects on I_{Ca} are dependent on the level of pre-synaptic depolarization and (2) that with one exception the Q_{10}'s for the various properties of I_{Ca} range between 1.1 and 2.5 while those for EPSP latency, amplitude and rate of rise are 3 and higher.

Calcium current and postsynaptic response. While the I_{Ca} is certainly

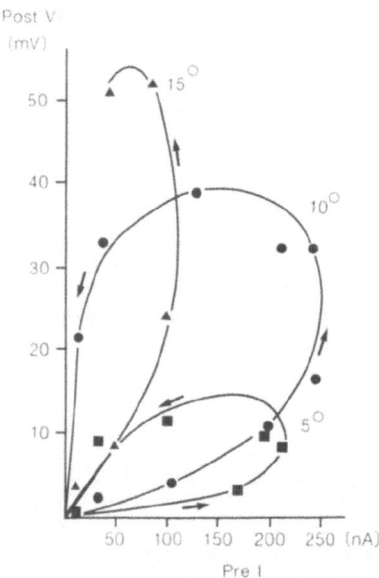

FIG. 16: Relationship between presynaptic Ca current and postsynaptic
response at three temperatures. EPSP amplitude is plotted against
presynaptic current for 15°, 10° and 5° C. Right-going arrows
correspond to voltage clamp pulses up to 60 mV, left-going arrows
to voltage clamp pulses above 60 mV. Note that asymmetrical relationship holds at all three temperatures but that the relationship
becomes less steep with decreasing temperature.

affected by temperature, changes in this parameter do not seem to be directly
responsible for the dramatic effects observed on synaptic delay and do not
account entirely for effect on EPSP amplitude. This is best illustrated by
results from the double voltage clamp experiments. In the experiment shown
in Fig. 15 a square voltage clamp pulse was used to briefly depolarize the
presynaptic terminal by 50 mV from rest (lowest traces), while the postsynaptic terminal region was clamped at the resting potential. By superimposing
the results obtained at 7° C and at 10° C a marked change in synaptic delay
and postsynaptic current amplitude (top traces) are evident, while there is
only a small change in I_{Ca} amplitude and time course (middle traces).

Since the relationship between presynaptic depolarization and I_{Ca} at
the squid synapse is bell-shaped (Figs. 11 and 14) (i.e. two levels of
depolarization below and above 60 mV elicit the same amplitude I_{Ca}), the
plot of postsynaptic potential as a function of I_{Ca} shows asymmetry. The
same amplitude of I_{Ca} triggers more or less transmitter release, depending
on whether the I_{Ca} was elicited by a depolarization larger or smaller than
60 mV (see Simon and Llinas, 1985). This asymmetry was maintained when
the temperature was changed, the plot being shifted downward with decreasing temperature and upward with increasing temperature as shown for 15°,
10° and 5°C in Fig. 16. To further evaluate the effect of temperature on
the relationship between I_{Ca} and postsynaptic response, these parameters
were also plotted using double logarithmic coordinates (not shown here)
(Llinas et al., 1981b). At each temperature studied the points were found
to fall on a straight line and the slope of the line changed directly with
temperature, being 0.76 at 5° C, 1 at 10° C, and 1.84 at 15° C.

12

DISCUSSION

The results of the above experiments indicate that depolarization-release coupling, from the viewpoint of the secretory event triggered by Ca current, is far from being understood. Several problems may now be posed regarding the areas addressed by the experiments presented in this paper. The first relates to intracellular Ca injection.

Injection of Ca into the Presynaptic Terminal

The experimental results described in this paper indicate that, as shown by Miledi (1973), Charlton et al. (1980), Llinas, Sugimori and Bower (1983), Llinas, Sugimori and Leonard (1984), synaptic transmission can occur in the absence of transmembrane inward Ca current. Furthermore, synaptic transmission was observed in the presence of extracellular Cd at a concentration sufficiently high to block I_{Ca} (Llinas, Steinberg and Walton, 1981a), demonstrating that this ion does not interfere significantly with the release phenomenon. Perhaps the most surprising finding was that release was not always achieved following intracellular Ca injection, even in a synapse which showed healthy synaptic transmission just prior to injection. In all cases injection was monitored visually, and in several experiments aequorin was pre-injected into the terminal to monitor the distribution of Ca during and after the injection.

The absence of release observed in many synapses following Ca injection was explained by the aequorin experiments as they showed that the Ca concentration increase following injection was generally limited to the vicinity of the microelectrode tip (also see Martin and Miledi, 1978), indicating a very avid and efficient intracellular Ca buffering system in the pre-terminal.

In those experiments where intracellular Ca did release transmitter, variations were often seen in the characteristics of transmitter release. Indeed, as shown in Fig. 2 some synapses responded with a rather fast and massive release which peaked rapidly and decayed rather slowly. In some, the falling phase consisted of two distinct components and the postsynaptic response could last minutes (Fig. 2C). This probably indicates that the Ca injection reached the plasmalemmal membrane and acted directly at the release site to produce transmitter liberation and then diffused along the surface of the membrane to evoke further slow release.

A rough calculation of the amount of transmitter released by such an injection may be attempted. Assuming that miniature potentials have an amplitude of 10 μV and a time constant of 2 msec (Miledi, 1973; Mann and Joyner, 1978), a quantal content of 2×10^6, with a maximum rate of release of 7 mV/10 msec or 7×10^5 quanta/sec, may be calculated for the postsynaptic potential in Fig. 1. The quantal content of this synaptic potential corresponds to about one-third to one-tenth of the total number of vesicles available in the preterminal (Pumplin, personal communication; Martin and Miledi, 1986), suggesting that the Ca concentration increase after injection may be restricted to about one-fifth of the presynaptic active zone once it reaches the internal plasmalemmal surface. It was not surprising then that in many of the synapses only one transmission event was elicited for each Ca injection. The observation of two time constants in the decline of the response further suggests that the presynaptic area reached by the injected Ca was probably depleted of available transmitter quite rapidly and that the second component of the release probably relates, as suggested above, to mobilization of Ca along the membrane to reach other active zones or to continuous release from a certain area by increased availability of transmitter accompanying increased intracellular Ca concentration.

In short, following an injection of Ca, vesicular release is probably

maximal over a given active membrane area. Repeated Ca injections, by depleting the vesicle population at that point, affords less and less release.
This is in fact demonstrated in Fig. 3 which shows that the fast rising
response seen after the first injection was completely absent after the
second injection, suggesting that the area immediately adjacent to site of
Ca injection was completely depleted of vesicles (also see Martin and Miledi,
1978). The fact that the third and fourth injections released very little
transmitter is consistent with the view that the initial massive release of
transmitter exhausted the available transmitter substance. This phenomenon
was reported in this synapse following Ca spikes by Katz and Miledi (1971)
and by Llinas and Nicholson (1975). The latter investigators showed that,
although Ca currents continued to flow during the action potential, the
amplitude of the postsynaptic response decreased very quickly, even during
the plateau phase of the presynaptic Ca spike.

Finally, since under certain conditions synaptic transmission can occur
for a protracted period (Charlton and Atwood, 1977), it may be suggested
that, if sufficient Ca enters the terminal, the release process may be
maintained for a very long time. Thus, the transmitter release system may
be less affected by intracellular Ca than the kinetic properties of the Ca
channels, as suggested by experiments in which increased intracellular Ca
concentration blocked the voltage-dependent Ca conductance (cf. Eckert and
Chad, 1984). While these kinds of experiments indicate that intraterminal
accumulation of Ca may produce transmitter release for a prolonged period,
it must be remembered that the total Ca influx following injection from an
electrode containing 0.5 M Ca acetate must be orders of magnitude higher
than that obtained even by tetanic activation of the terminal under normal
extracellular Ca concentration.

Voltage Dependence of Transmitter Release

The second issue raised in this paper concerns the question of direct
voltage effect on transmitter release. In our initial voltage clamp study
of the squid giant synapse we suggested that, although transmitter release
was clearly dependent on an influx of Ca ions into the presynaptic terminal,
it could also have a voltage-dependent component (Llinas, Steinberg and Walton, 1981b). This hypothesis was considered in further detail in the crayfish neuromuscular junction (Parnas, Dudel and Parnas, 1984). Indeed, while
the voltage-dependence hypothesis has met with criticism (Zucker and Lando,
1986; but see Parnas and Parnas, 1986), the fact remains (as shown here)
that a certain degree of voltage modulation for synaptic release can be
demonstrated after presynaptic Ca injection in the squid synapse. Note,
however, that the voltage dependence found in our experiments is rather
small compared to that expected from our initial results.

With respect to the squid giant synapse, a second interpretation of
our original results relates to the fact that local intracellular Ca concentration is not linearly related to the Ca current amplitude (Simon,
Sugimori and Llinas, 1984; Simon and Llinas, 1985). This is so because the
total Ca current observed at higher values of membrane depolarization is
more evenly distributed over the active release sites in the pre-terminal
than the total Ca current at low depolarization levels. Indeed, at high
depolarization, many channels are open and each allows low Ca influx
because of the decreased driving force for this ion during depolarization.
The same level of I_{Ca} can be attained at low levels of depolarization, but
in this case more Ca is flowing across <u>fewer channels</u> and thus the probability of vesicular release is higher at higher levels of depolarization where
more Ca channels are open (Simon and Llinas, 1985). The above does not,
however, explain the voltage dependence of release demonstrated in this
paper. It simply states that several mechanisms are probably at work in
this voltage-dependent release. Among other explanations, an iontophoretic

redistribution of Ca concentration against the internal surface of the plasmalemmal membrane could explain to a certain extent our present findings. Clearly this problem must be studied further.

Synaptic Release Following Local Changes in Extracellular Ca Concentration

The efficiency of release by extracellular Ca when its concentration in the extracellular medium is abruptly increased was examined in the next set of experiments. Here a transient Ca concentration increase was introduced at the presynaptic site by a rapid injection of Ca into the bath containing 0.2 mM Ca and 50 mM Mg (i.e. insufficient Ca to produce transmitter release following presynaptic activation of the ganglion). If under these circumstances a voltage step was given, transmitter release did not occur. On the other hand, transmitter release was seen to occur as the Ca concentration near the presynaptic terminal was rapidly increased. In many of these experiments a single extracellular pulse of Ca onto the pre-terminal was sufficient to produce an increase in release within a short period (30 sec) and to support maximum release within 5 min. This clearly indicates that the mobility of Ca in the immediate extracellular environment of the pre-terminal membrane is not as impaired by the surrounding tissue as suggested by the morphology (Young, 1973; Pumplin and Reese, 1978; Martin and Miledi, 1986). More importantly, under these circumstances excellent voltage clamp experiments can be carried out (Augustine et al., 1985a). In fact, a good correlation was found between Ca current amplitude and postsynaptic response with this protocol. This type of experiment yielded a stoichiometry of Ca current to postsynaptic potential close to 2. The significant point of the experiment, however, is that transmitter release seems to be well related to the Ca current amplitude whether the extracellular Ca is introduced abruptly into the extracellular medium or is present prior to presynaptic depolarization.

Effect of Temperature on Transmitter Release

Probably the most intriguing parameter studied with respect to modulation of release is temperature. Our results indicate that when the temperature of the bath is lowered, the kinetics of Ca channel activation and the amount of transmitter released for a given Ca current are decreased. However, the more striking finding is the effect on the synaptic delay. Thus, while the decrease in Ca channel kinetics and of the postsynaptic channels are involved in the increased delay, most of this increase must be explained by other mechanisms as the two do not demonstrate a sufficiently large temperature dependence to account for the increase in synaptic delay reported. The target of choice for maximum temperature dependence would be the mechanism of synaptic release (i.e. Ca to vesicular fusion) or the fusion mechanism itself altered by the change in membrane fluidity. The first possibility may involve direct action of Ca on release or on a possible biochemical step of importance in either synaptic fusion or vesicular availability (as discussed in a chapter by Greengard et al. in this volume).

Because the techniques developed recently allow a rather direct study of the coupling of depolarization to transmitter release, we feel that further utilization of this approach will contribute significantly to our understanding of this fundamental issue. The phenomena that follow Ca entry and that ultimately regulate the release of transmitter in chemical synapses are clearly a nontrivial set of events involving multiple biophysical and biochemical compartments. Much is yet to be understood.

REFERENCES

Adrian, R. H., Chandler, W. K. and Hodgkin, A. L., Voltage clamp experiments in striated muscle fibres, J. Physiol. (Lond.) 208:607-644 (1970).

Augustine, G. J. and Eckert, R., Divalent cations differentially support transmitter release at the squid giant synapse, J. Physiol. (Lond.) 346:257-271 (1984).

Augustine, G. J., Charlton, M. P. and Smith, S., Calcium entry into voltage-clamped presynaptic terminals of squid, J. Physiol. (Lond.) 367:143-162 (1985a).

Augustine, G. J., Charlton, M. P. and Smith, S., Calcium entry and transmitter release at voltage-clamped nerve terminals of squid, J. Physiol. (Lond.) 367:163-181 (1985b).

Barrett, E. F., Barrett, J. N., Botz, D., Chang, D. B. and Mahaffey, D., Temperature-sensitive aspects of evoked and spontaneous transmitter release at the frog neuromuscular junction, J. Physiol. (Lond.) 279:253-273 (1978).

Charlton, M. P. and Atwood, H. L., Modulation of transmitter release by intracellular sodium in squid giant synapse, Brain Res. 134:367-371 (1977).

Charlton, M. P. and Atwood, H. L., Synaptic transmission: Temperature sensitivity of calcium entry in presynaptic terminals, Brain Res. 170:543-546 (1979).

Charlton, M. P., Zucker, R. and Smith, S. J., Presynaptic injection of calcium facilitates transmitter release in the squid giant synapse, Biol. Bull. 159:481 (1980).

Charlton, M. P., Smith, S. J. and Zucker, R., Role of presynaptic calcium ions and channels in synaptic facilitation and depression at the squid giant synapse, J. Physiol. (Lond.) 323:173-193 (1982).

Dudel, J., Transmitter release triggered by a local depolarization in motor nerve terminals of the frog: role of calcium entry and of depolarization, Neurosci. Lett. 41:133-138 (1983a).

Dudel, J., Graded or all-or-nothing release of transmitter quanta by local depolarizations of nerve terminals on crayfish muscle? Pflugers Arch. 398:155-164 (1983b).

Eckert, E. and Chad, J. E., Inactivation of Ca channels, Prog. Biophys. Molec. Biol. 44:215-267 (1984).

Greengard, P., Browning, M. D., McGuinness, T. L. and Llinas, R., Synapsin I, a phosphoprotein associated with synaptic vesicles: Possible role in regulation of neurotransmitter release, in "Molecular Mechanisms of Neuronal Responsiveness", Adv. Exptl. Med. & Biol., edited by Y. H. Ehrlich, R. H. Lenox, E. Kornecki and W. Berry, Plenum, New York (1987).

Hodgkin, A. L. and Huxley, A. F., A quantitative description of membrane current and its application to conduction and excitation in nerve, J. Physiol. (Lond.) 117:500-544 (1952).

Hubbard, J. I., Jones, S. F. and Landau, E. M., The effect of temperature change upon transmitter release, facilitation and post-tetanic potentiation, J. Physiol. (Lond.) 216:591-609 (1971).

Katz, B. and Miledi, R., The effect of temperature on the synaptic delay at the neuromuscular junction, J. Physiol. (Lond.) 181:656-670 (1965).

Katz, B. and Miledi, R., The effect of prolonged depolarization on synaptic transfer in the stellate ganglion of the squid, J. Physiol. (Lond.) 216:503-512 (1971).

Ksano, K., Influence of ionic environment on the relationship between pre- and postsynaptic potentials, J. Neurobiol. 1:435-457 (1970).

Lester, H., Transmitter released by presynaptic impulses in the squid stellate ganglion, Nature 227:493-496 (1970).

Llinas, R., Calcium in synaptic transmission, Sci. Amer. 247:56-65 (1982).

Llinas, R., McGuiness, T. L., Leonard, C. S., Sugimori, M. and Greengard, P., Intraterminal injection of synapsin I or calcium calmodulin-dependent protein kinase II alters neurotransmitter release at the squid giant synapse, Proc. Natl. Acad. Sci. (USA) 82:3035-3039 (1985).

Llinas, R. and Nicholson, C., Calcium role in depolarization-release coupling. An aequorin study in squid giant synapse, Proc. Natl. Acad. Sci. (USA) 72:187-190 (1975).

Llinas, R., Steinberg, I. Z. and Walton, K., Presynaptic calcium currents and their relation to synaptic transmission: A voltage clamp study in squid giant synapse and a theoretical model for the calcium gate, Proc. Natl. Acad. Sci. (USA) 73:2918-2922 (1976).

Llinas, R., Steinberg, I. Z. and Walton, K., Presynaptic calcium currents in squid giant synapse, Biophys. J. 33:289-322 (1981a).

Llinas, R., Steinberg, I. Z. and Walton, K., Relationship between presynaptic calcium current and postsynaptic potential in squid giant synapse, Biophys. J. 33:323-352 (1981b).

Llinas, R. and Sugimori, M., Double voltage clamp study in squid giant synapse, Biol. Bull. 155:454 (1978).

Llinas, R., Sugimori, M. and Bower, J. M., Visualization of depolarization-evoked presynaptic calcium entry and voltage dependence of transmitter release in squid giant synapse, Biol. Bull. 165:529 (1983).

Llinas, R., Sugimori, M. and Leonard, C. S., Transmitter release by local increase in $[Ca^{++}]_o$ and by prolonged presynaptic voltage steps in the squid giant synapse, Biol. Bull. 167:529 (1984).

Llinas, R., Sugimori, M. and Simon, S. M., Transmission by presynaptic spike-like depolarization in the squid giant synapse, Proc. Natl. Acad. Sci. (USA) 79:2415-2419 (1982).

Llinas, R., Walton, K. and Sugimori, M., Voltage-clamp study of the effects of temperature on synaptic transmission in the squid, Biol. Bull. 155:454 (1978).

Mallart, A. and Martin, A. R., An analysis of facilitation of transmitter release at the neuromuscular junction of the frog, J. Physiol. (Lond.) 193:679-694 (1967).

Mann, D. M. and Joyner, R. W., Miniature synaptic potentials at the squid giant synapse, J. Neurobiol. 9:329-335 (1978).

Martin, R. and Miledi, R., A structural study of the squid synapse after intraaxonal injection of calcium, Proc. R. Soc. London B 201:317-333 (1978).

Martin, R. and Miledi, R., The form and dimensions of the giant synapse of squids, Phil. Trans. R. Soc. London B 312:355-377 (1986).

Miledi, R., Transmitter release induced by injection of calcium ions into nerve terminals, Proc. R. Soc. B 183:421-425 (1973).

Parnas, I., Dudel, J. and Parnas, H., Depolarization dependence of the kinetics of phasic transmitter release at the crayfish neuromuscular junction, Neurosci. Lett. 50 (1984).

Parnas, I. and Parnas, I., Calcium is essential but insufficient for neurotransmitter release: The calcium-voltage hypothesis. Gif Lectures in Neurobiology (Regulation and Synaptic Plasticity) (1986).

Pumplin, D. W. and Reese, T. S., Membrane ultrastructure of the giant synapse of the squid Loligo pealei, Neuroscience 3:685 (1978).

Simon, S. M. and Llinas, R., Compartmentalization of the submembrane calcium activity during calcium influx and its significance in transmitter release, Biophys. J. 48:485-498 (1985).

Simon, S. M., Sugimori, M. and Llinas, R., Modelling of submembranous calcium concentration changes and their relation to rate of presynaptic transmitter release in the squid giant synapse, Biophys. J. 45:264a (1984).

Smith, S., Augustine, G. and Charlton, M. P., Transmission at voltage-clamped giant synapse of the squid. Evidence for cooperativity of presynaptic calcium action, Proc. Natl. Acad. Sci. (USA) 82:622-625 (1985).

Weight, F. F. and Erulkar, S. C., Synaptic transmission and effects of temperature at the squid giant synapse, Nature 261:720-723 (1976).

Young, J. Z., The giant fibre synapse of Loligo, Brain Res. 57:457-460 (1973).

Zucker, R. S. and Lando, L., Mechanism of transmitter release: voltage hypothesis and calcium hypothesis, Science 231:574-579 (1986).

Zucker, R. S. and Stockbridge, N., Presynaptic calcium diffusion and the time course of transmitter release and synaptic facilitation at the squid giant synapse, J. Neurosci. 3:1263-1269 (1983).

TEMPORAL AND SPATIAL EVENTS IN THE CALCIUM MESSENGER SYSTEM

Howard Rasmussen and Paula Barrett

Departments of Cell Biology and Internal Medicine, Yale
University School of Medicine, New Haven, CT 06510

INTRODUCTION

The great complexity of the central nervous system makes it a difficult
object of biochemical study. Yet, some of the most important biochemical
discoveries having implications for the field of cell regulation have been
made in CNS tissue. A case in point is the discovery by Nishizuka and
coworkers of a new kind of protein kinase, the so-called phospholipid-dependent, calcium-activated protein kinase or C-kinase, in brain tissue (Takai
et al.). This kinase was found to be distinct from either the classic cAMP-
dependent or Ca^{++}-CaM-dependent protein kinases. It was, however, found to
be activated by Ca^{++}, phospholipids, and diacylglycerols (Takai et al.,
1977; Kishimoto et al., 1980). After its discovery in the brain, where it
exists in very large amounts, it was found to be widely distributed in animal tissues (Nishizuka and Takai, 1981). Its discovery coincided in time
with a significant breakthrough in our understanding of the role of inositol
polyphosphatase in the transducing events which occur in the calcium messenger system (Michell, 1975; Berridge, 1984) (see chapters by Agranoff and
by Lapetina in this volume).

Although a link between Ca^{++} and inositol lipids had been postulated
for some time (Michel, 1975), work by Downes and Michell (1981, 1982a,b)
and later Berridge (1984) and Irvine (Irvine et al., 1984a,b) led to the
recognition that the polyphosphoinositides, phosphatidylinositol-4 phosphate
and -4,5 bis phosphate (PIP and PIP_2) are of key importance. Agonist-recep-
tor interaction in a number of systems has been shown to lead to the rapid
hydrolysis of PIP and PIP_2 leading to the generation of 1,4,5-inositol-tris-
phosphate (IP_3), 1,4-inositol bisphosphate ($InsP_2$) and 1,2-diacylglycerol
(DG). Further, it was found that InsP induces a release of Ca^{++} from the
endoplasmic reticulum (Streb et al., 1983). Studies of the platelet release
reaction by Nishizuka and coworkers (Kaibuchi et al., 1983; Nishizuka, 1983),
then showed that Ca^{++} and diacylglycerol acted synergistically in prompting
this release reaction. These results provided a direct link between PI turnover, DG production, and C-kinase activation.

Using this evidence as a starting point, we and our colleagues have
attempted to contruct a model of how the calcium messenger system operates
to bring about sustained cellular responses in such specific cases as the
control of aldosterone secretion from adrenal glomerulosa cells (Barrett et
al., 1986a,b; Foster et al., 1981, 1982; Kojima et al., 1984a,b, 1985a,b;
Rasmussen and Barrett, 1984), insulin secretion from pancreatic islet cells
(Zawalich et al., 1983, 1984), contraction of tracheal and vascular smooth
muscle (Park and Rasmussen, 1985; Forder et al., 1985), and prolactin secre-

tion from pituitary cells (Delbeke et al., 1984a,b). None of these endocrine systems has an exact counterpart in specific neural systems, but each has the virtue of being a relatively uniform cell population with a well defined physiological response, and each was thought to be regulated by the intracellular messenger Ca^{++}.

As this work progressed, a collaborative effort was initiated with Dr. Daniel Alkon to explore the possibility that the model developed form studies in these endocrine systems could help explain long term (hrs to days) changes in neural cell function (Alkon, 1984). The results of these endeavors are discussed by Dr. Alkon in his chapter of this volume. Our purpose is to describe the model of cell activation developed from our studies in endocrine cells, and to discuss its implications to the field of cell regulation.

Two Branch Model of Calcium Messenger System Function

To put this discussion into an appropriate context, it is worth briefly discussing the models of Ca^{++} messenger system function which predated our present one, and which are still widely accepted. One can do this most easily by considering a specific example, the carbacholamine-induced contraction of tracheal smooth muscle (Adelstein and Eisenberg, 1980; Bolton, 1973; Blumenthal and Stull, 1980; Silver and Stull, 1984). When carbachol is applied to this tissue, there is a rapid contraction which is sustained as long as carbachol is present. The accepted model of how this contraction is produced is as follows: carbachol-receptor interaction leads both to a mobilization of Ca^{++} from an intracellular pool and an increase in rate of Ca^{++} influx across the plasma membrane. These two events lead to a rise in the intracellular $[Ca^{++}]$. This rise in $[Ca^{++}]$ is detected by a specific Ca^{++} receptor protein, calmodulin (CaM). The binding of 3 or 4 Ca^{++} to CaM leads to its association with a specific enzyme, myosin light chain kinase. Activation of this enzyme leads to the phosphorylation of myosin light chains (MLC). The phosphorylation of MLC initiates actin-myosin interactions which lead to a contraction. The assumptions in this model are that: a) as long as carbachol is present and the muscle is contracted, the $[Ca^{++}]$ remains high; b) as long as carbachol is present the content of MLC·P remains high; c) the sustained increase in Ca^{++} influx rate maintains $[Ca^{++}]$ at its high value; and d) that the flow of information from cell surface to cell interior is qualitatively the same throughout the duration of the cellular response.

Recent experiments show that not one of these assumptions is valid. When an agonist acts in smooth muscle there is a very rapid but brief (1-2 min) rise in $[Ca^{++}]$ followed by a fall to a value very close to its basal value (Morgan and Morgan, 1982, 1984, 1986). The $[Ca^{++}]$ then remains at this near basal value as the contractile response develops and is maintained. Likewise, the content of MLC·P increases rapidly, reaches a peak in 3-5 min and then declines slowly to reach pretreated values in 15-30 min even though the response is maintained (Aksoy, 1983; Silver and Stull, 1984). Finally, as will be discussed below, there is a changing pattern of information flow from cell surface to cell interior as a function of time after carbachol addition (Park and Rasmussen, unpublished).

The basic problem we set out to investigate was how Ca^{++} serves it second messenger function in a simple, highly specialized endocrine cell which produces a single product. The system we chose was the adrenal glomerulosa cell (Abano et al., 1974; Fakunding et al., 1979; Fakunding and Catt, 1980; Foster et al., 1981, 1982; Fujita et al., 1979; Fujita et al., 1981; Hyath et al., 1986; Kojima et al., 1984a,b, 1985a-c; Sala et al., 1979; Williams et al., 1981). This cell produces the steroid hormone, aldosterone, in response to increases in either angiotensin II, adrenocor-

ticotropin (ACTH), or extracellular K^+ concentration. Each of these agonists activates aldosterone secretion by a somewhat different mechanism, but
each has one effect on adrenal cells which it shares with the others. Each
induces a sustained increase in the Ca^{++} influx rate, and in each instance
this input is important to the mechanism by which the agent acts.

The response most thoroughly studied is angiotensin II-induced aldosterone secretion. Addition of AII to these cells induces a slowly developing
(5-8 min) increase in aldosterone secretory rate which rises to a sustained
level and which persists for several hours (as long as angiotensin is present). Angiotensin II does not activate adenylate cyclase, but does alter
cellular Ca^{++} metabolism. When angiotensin II acts, there is a transient
increase in cytosolic $[Ca^{++}]$ which reaches a peak in 1 min and falls to
nearly its basal value within 2-4 min as the aldosterone secretory response
is initiated (5-8 min). A slow increase to a sustained elevated rate of
secretion occurs subsequent to this redistribution of intracellular calcium
(15-20 min) (Apfeldorf and Rasmussen, 1986). The addition of AII also
induces an immediate and sustained increase in Ca^{++} influx rate across the
plasma membrane. Thus, there is an initial release of Ca^{++} from an internal
pool which leads both to a transient rise in intracellular $[Ca^{++}]$ and a net
loss of Ca^{++} from the cell. At 5 min or greater, the intracellular $[Ca^{++}]$
is at nearly its basal value, total cell Ca^{++} is relatively constant, yet
influx rate is high. These observations indicate that because there is no
net gain of Ca^{++} by the cell during this sustained phase of the response,
Ca efflux rate must be high enough to balance the elevated rate of Ca
influx. In other words, during the sustained phase of the response there
is an increase in Ca^{++} cycling across the plasma membrane.

These effects on Ca^{++} metabolism occur as a consequence of hormone
receptor interaction. Addition of angiotensin II leads to an hydrolysis of
PIP_2 and PIP with the generation of $InsP_3$, $InsP_2$ and DG. The $InsP_3$ in turn
is responsible for the mobilization of intracellular Ca^{++} and hence the
transient rise in $[Ca^{++}]$ and net efflux of Ca^{++} from the cell. The nature
of the coupling between the hormone-receptor interaction and the sustained
increase in Ca^{++} influx rate is not yet understood. On the one hand, it
may be a direct result of hormone-receptor interaction, or, on the other,
may be a consequence of PIP_2 hydrolysis.

That simply a transient rise in $[Ca^{++}]$ and a sustained increase in
Ca^{++} influx rate are not, in and of themselves, sufficient to induce a sustained cellular response, is shown by the fact that addition of low concentrations of the divalent cation ionophore, A23187, causes similar changes in
Ca^{++} metabolism but induces only a transient (lasting 25-30 min) increase in
the aldosterone secretory response. However, if the A23187 is combined with
either oleolylacetylglycerol (OAG) or 12-0-tetradecanoylphorbol-13-acetate
(TPA), activators of protein kinase C, a sustained rather than a transient
increase in aldosterone secretory rate is observed despite identical changes
in cellular Ca^{++} metabolism. The concentrations of OAG and TPA employed
lead to no significant change in cellular Ca^{++} metabolism or aldosterone
secretion if employed alone.

These results led to the development of a two branch model of angiotensin II action in which the initial phase of the response is mediated by
the transient $InsP_3$- induced increase in intracellular $[Ca^{++}]$ acting via
CaM-dependent systems; and the sustained phase by the sustained activation
of C-kinase (Rasmussen and Barrett, 1984). Further work has led to the following refinements of this model. First, it was found that the size of the
initial $[Ca^{++}]$ transient is related to the magnitude of the sustained
response. This is accounted for by the fact that the rise in $[Ca^{++}]$ along
with the increase in the DG content of the plasma membrane acts to bring
about the conversion of the C-kinase from its Ca^{++}-insensitive to its Ca^{++}-

sensitive, membrane-bound form (Kojima et al., 1985a; Wolf et al., 1985a,b). Second, it was found that once the C-kinase is in its Ca^{++} sensitive form, the expression of its activity is controlled by the rate of Ca^{++} influx (or cycling) across the plasma membrane. Thus, the magnitude of the sustained cellular response is determined by two factors: a) the amount of the Ca^{++}-sensitive form of the C-kinase associated with the plasma membrane; and b) the rate of Ca^{++} cycling across the plasma membrane.

The importance of the second factor can be most easily illustrated with the use of the Ca^{++} channel blocker, nitrendipine (NT). When adrenal cells are pretreated with 1 μM NT, there is practically no change in the basal Ca^{++} influx rate, but when AII is then added, there is no increase in Ca^{++} influx rate: NT blocks the angiotensin II-stimulated but not basal Ca^{++} influx rate. Addition of angiotensin II under this circumstance leads to a normal increase in PIP_2 hydrolysis, $InsP_3$ production, DG production, and a transient rise in intracellular $[Ca^{++}]$, but induces a transient rather than sustained aldosterone secretory response.

These observations have been the basis for a model of calcium messenger system function in which there are two temporally distinct phases. In each phase, Ca^{++} serves a specific messenger function, but both the cellular and molecular sites of Ca^{++} action differ. During the initial phase, it is the increase in $[Ca^{++}]$ in the bulk cytosol which serves as messenger, and the major targets of Ca^{++} action are CaM-dependent enzymes including CaM-dependent protein kinases. During the sustained phase it is the increase in Ca influx or cycling across the plasma membrane which serves as messenger, and its major target is the membrane-associated C-kinase.

Validation of this model has been obtained by studying the temporal patterns of protein phosphorylation (Barrett et al., 1986a,b). These results are summarized in Figure 1. When angiotensin II acts, the patterns of protein phosphorylation seen at 1 min and at 30 min differ. At 1 min six proteins show an increase in their extent of phosphorylation; and at 30 min only four proteins display an increase. One of these is a protein which showed an increase at 1 min; the other three are newly phosphorylated. This means that five of the six proteins displaying an increase in extent of phosphorylation at 1 min no longer display this increase at 30 min. These data show that different subsets of proteins are phosphorylated early (1 min) and late (30 min). When A23187 acts there is at 1 min an increase in the extent of phosphorylation of five of the same six proteins, seen 1 min after angiotensin II addition, but no proteins displaying an increase in extent of phosphorylation are seen at 30 min. This difference in temporal patterns of protein phosphorylation after angiotensin II and A23187 correlates with a difference in their effects on aldosterone secretion. Angiotensin II induces a sustained response, A23187 a transient response which is over in 30 min. When TPA alone acts there is neither a measurable increase in protein phosphorylation nor in aldosterone secretion during a 30 min incubation period (data not shown). When A23187 is combined with TPA, a sustained aldosterone secretory response is evoked, and the identical patterns of protein phosphorylation are observed as seen after AII action: six proteins at 1 min, four at 30 min with only one of the proteins being common to the 1 min and 30 min subsets. These data have been interpreted to signify that the initial Ca^{++} transient (or that induced by A23187) leads to the immediate phosphorylation of five proteins which are substrates for CaM-dependent protein kinases, and that the appearance of these phosphoproteins is transient. The activation of the C-kinase branch of this system leads to the delayed phosphorylation of four others, all of which persist during the sustained phase of cellular response.

Validation that NT blocks the events in the C-kinase branch is provided by the demonstration (Fig. 1) that exposure of NT-pretreated adrenal cells

22

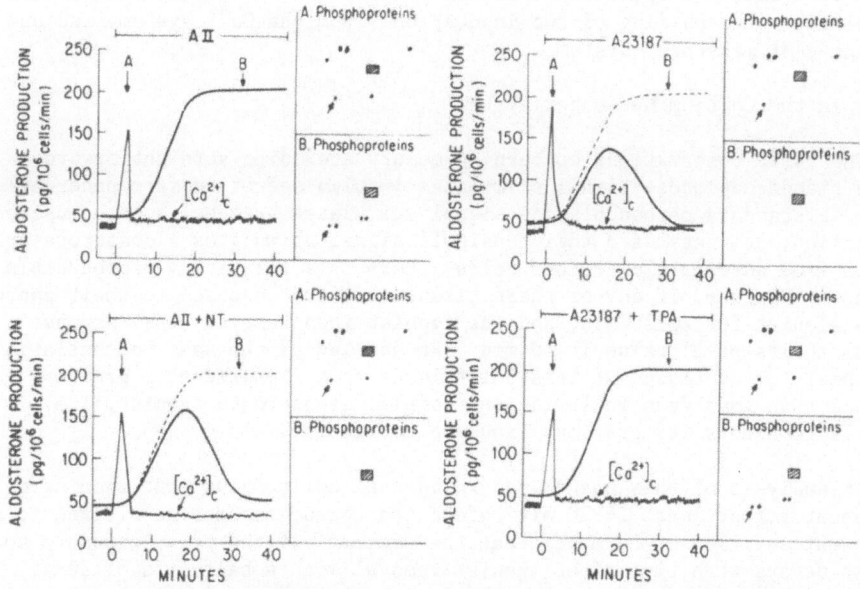

FIG. 1: A schematic representation of the responses of adrenal glomerula
cells to the continuous exposure to 1 nM angiotensin II (AII) (upper
left); 50 nM A23187 (upper right); 1 μM nitrendipine (NT) and 10 nM
angiotensin II (lower left); and 50 nM A23187 and 10 nM 12-0-tetra-
decanoylphorbol-13-acetate (TPA) (lower right). The solid lines
(———) in the separate figure represent the rates of aldosterone
production produced by the particular agonist or combination of
agonists. The dashed lines (-----) in the upper right and lower
left figures represent the expected response from angiotensin II
perifusion. The wavy line marked, $[Ca^{2+}]$, represents the intracellular free Ca^{++} concentration as estimated with the photoprotein aequorin. The enclosures labeled A and B to the right hand
side of each separate figure represent the patterns of protein
phosphorylation seen at 1 min (A) and 30 min (B) after agonist
addition. The solid black dot (·) represents the location of specific phosphoproteins; the shaded square (▨) represents the location of a reference protein whose state of phosphorylation does not
change. (→) represents the one protein thought to be a putative
substrate of both calmodulin-dependent protein kinases and protein
kinase C.

to angiotensin II leads to a normal $[Ca^{++}]$ transient, the phosphorylation
of all six early phase proteins, but a transient rather than sustained cellular response associated with a lack of enhanced phosphorylation of any of
the late phase proteins.

These data provide unequivocal evidence for a model in which there is a
temporal sequence of events in the Ca^{++} messenger system with a distinctly
different pathway of information flow from cell surface receptor to the cell
interior as a function of time after agonist action.

Before considering the broader implications of these findings, there
are two other aspects of Ca^{++} messenger system to be discussed: 1) evidence
that a type of memory develops in this system; and 2) that activation of the
C-kinase branch of this system is not the only way that a transient Ca^{++}

signal can lead to a sustained cellular response. There is an alternative possibility: the pairing of two inputs; one from the Ca^{++} system, and one from the cAMP messenger system.

Memory in the Calcium Messenger System

The basic observations concerning memory are quite straightforward. When a standard glucose signal stimulates insulin secretion from pancreatic islets, a standard carbacholamine signal stimulates tracheal smooth muscle contraction; or a standard angiotensin II signal stimulates aldosterone production from adrenal glomerulosa cells, there is a relatively reproducible response. However, if any of these tissues is first exposed to their appropriate agonist for 20-30 min, and the agonist then removed, the response returns to its basal value in 10 min. Readdition of the same concentration of agonist to the tissue at this point leads to a significantly greater response than that seen following the initial exposure to agonist, i.e., the cell remembers its previous exposure to agonist.

An analysis of this phenomenon in adrenal cells shows that agonist must be present for at least 15-20 min before the tissue becomes sensitized to a subsequent addition of hormone; that the 'memory' of the prior exposure to agonist decays with time after agonist removal with a half-life of 30-45 min; and that during this memory phase the cells are responsive to the calcium channel agonist, BAY K 8644 whereas control cells are not responsible (in terms of aldosterone secretion) even though BAY K 8644 induces the same 2-fold stimulation of CA^{++} influx rate in both control and primed cells. This latter feature has proven of great value in dissecting out the possible molecular basis of this memory.

When 10 nM BAY K 8644 is added to normal adrenal cells, it induces a nearly 2-fold increase in Ca^{++} influx rate, but this change alone is insufficient to bring about a significant change in aldosterone secretory rate. On the other hand, if BAY K 8644 is combined with an activator of C-kinase such as OAG, which also by itself has no effect on aldosterone production, then a sustained aldosterone secretory response is induced. Parenthetically, this result provides additional support for the concept that both C-kinase activation and an increase in Ca^{++} influx rate are necessary to induce the sustained phase of the aldosterone secretory response.

When adrenal cells are exposed for a period of 20 min to angiotensin II and then the peptide hormone removed, or its actions inhibited, and the cells then treated with BAY K 8644, there is a prompt increase in aldosterone secretory rate: the cells previously exposed to agonist behave as though they had been treated with OAG, i.e., as though their C-kinase was still in its CA^{++} sensitive form. Support for this concept comes from the observations that in control cells addition of BAY K 8644 leads to no discernible change in the extent of protein phosphorylation, but in agonist-preexposed cells, the extent of phosphorylation of the four late proteins increases. Thus, when hormone is removed, the hormonally-induced increase in Ca^{++} influx rate rapidly declines ($t\frac{1}{2}$ = 1 min), and as a consequence the rate of aldosterone production falls ($t\frac{1}{2}$ = 4-6 min), but the C-kinase persists in its Ca^{++}-sensitive form for a considerably longer period ($t\frac{1}{2}$ = 30-40 min). During this period of slow relaxation (the C-kinase returning to its Ca^{++} insensitive form), readdition of an agent which increases Ca influx rate is sufficient to reactivate the cellular response. The magnitude of this response will depend on the amount of the C-kinase which remains in its Ca^{++}-sensitive form.

This discovery emphasizes in another way the fact that there is a temporal sequence of biochemical events at the plasma membrane when the calcium messenger system is activated.

24

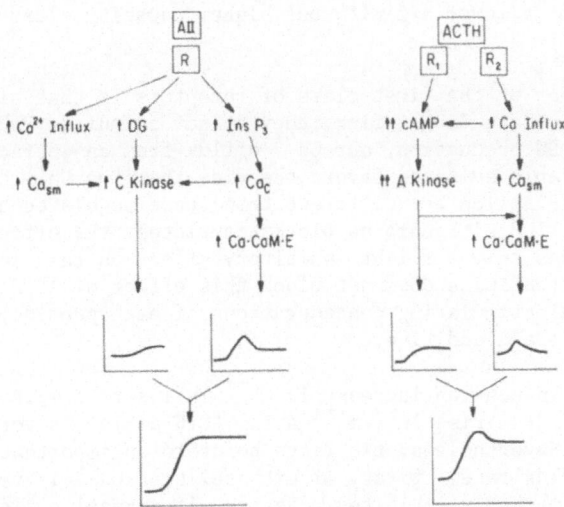

FIG. 2: A representation of the differing patterns of information flow when angiotensin II (AII) (left) and ACTH (right) act to induce a sustained increase in aldosterone production rate (lower line drawings) from adrenal glomerulosa cells. In the case of AII action, three different messengers are produced as a consequence of the interaction of hormone with a single class of receptor (R): an increase in Ca^{++} influx rate which causes an increase in the Ca^{++} concentration in a submembrane domain at the plasma membrane (Ca_{sm}); an increase in the DG content of the plasma membrane; and an increase in the cytosolic concentration of InsP3. This rise in InsP3 leads to a transient increase in the cytosolic free Ca^{++} (Ca_c) which activates, transiently, CaM-dependent enzymes (Ca·CaM·E) leading to the initial phase of response. The increase in Ca_{sm} leads to an increased flow of information through the C-kinase branch which mediates the sustained phase of the response. In the case of ACTH, there are two classes of receptor (R_1 and R_2) linked, respectively to adenylate cyclase and Ca^{++} influx. The increase in Ca^{++} influx resulting from the activation of R_2 acts synergistically with the activated R_1 to stimulate adenylate cyclase and cause an increase in cAMP concentration. This, in turn, activates the cAMP-dependent protein kinase (A-kinase). The rise in the Ca influx rate leads to an increase in Ca_{sm} which acts synergistically with cAMP to bring about a sustained cellular response.

An Alternative Mechanism of Sustained Cellular Response in the Calcium Messenger System

As noted at the outset of our discussion of adrenal glomerulosa cells, an increase in aldosterone production can be induced by either of two peptide hormones, angiotensin II or ACTH. The mechanism of action of these two peptides is quite different (Fig. 2), but each induces a sustained increase in aldosterone production rate; each increases the Ca^{++} influx rate; and in each case, this increase is an important signal in activating the cell. However, in other respects their actions are quite different: angiotensin II but not ACTH stimulates PI turnover; ACTH but not angiotensin II activates adenylate cyclase. The combination of an increase in CA^{++} influx rate and adenylate cyclase activation are responsbile for mediating

25

the actions of ACTH (Fig. 2). These two effects are mediated by different ACTH receptors: a high affinity, low capacity class of receptors is linked to the Ca^{++} system; a lower affinity but higher capacity class is linked to the cAMP system.

A unique aspect of the first class of receptors is that although, when occupied, they stimulate Ca^{++} influx, they do not stimulate PIP_2 hydrolysis, $InsP_3$ production, DG production, nor Ca^{++} efflux from an intracellular pool. Further, the available evidence favors the view that the Ca^{++} channels opened as a result of ACTH action are different than those regulated by angiotensin II. For example, 1 μM nitrendipine blocks completely the effect of AII on Ca^{++} influx, but has only a slight inhibitory effect on this action of ACTH, and even 10 μM nitrendipine does not block this effect of ACTH completely. Further, at maximal stimulating concentrations of each peptide, their effects on Ca^{++} influx rate are additive.

Because ACTH induces an increase in Ca^{++} influx rate without mobilizing intracellular Ca^{++}, the rise in $[Ca^{++}]$ after ACTH action is very small and quite transient. Nevertheless, two facts point to an important role of this change in Ca^{++} influx rate. First, if extracellular Ca^{++} is removed, the action of ACTH is markedly inhibited. Second, if adrenal cells are incubated in low Ca^{++} media, the effect of ACTH is blunted, and only a transient aldosterone secretory response observed. This result is explained by the fact that when ACTH is applied to cells incubated in a low Ca^{++} medium, the peptide hormone no longer exerts its characteristic effect on Ca^{++} influx. Third, if one compares the effects of ACTH with those of forskolin, a direct activator of adenylate cyclase, on both adolsterone production and adenylate cyclase activation there is a clear difference. The dose response curve for aldosterone production is superimposable on that for [cAMP] when forskolin is the activator; but the curve for aldosterone production is shifted to the left of the [cAMP] curve when ACTH is the activator. Clearly, some input other than cAMP, i.e., Ca^{++} influx, plays a role in ACTH action.

The manner in which these two inputs interact is not yet completely determined. The Ca^{++} signal has at least two effects: it helps activate adenylate cyclase (Fig. 2), and it also interacts with cAMP in some manner to determine cellular response. It seems quite likely that these two signals will be found to interact at the plasma membrane level and in some way provide a new pathway of information flow from cell surface to cell interior; not the classical view in which a sustained increase in the cAMP content of the cytosol mediates response. In other words, Ca^{++} cycling across the plasma membrane provides in this system as well as in the system operated by angiotensin II, a signal at or just beneath the plasma membrane which activates some transducing system (analogous to the C-kinase) which provides the means for information to flow from cell surface to cell interior during the sustained phase of the response.

Implications of Two Branch Model

If the two branch model of Ca^{++} messenger system function were confined to the actions of angiotensin II on aldosterone production, it would not necessarily be of great general interest. However, this does not appear to be the case. Less complete evidence suggest that the Ca^{++} messenger system operates in much the same way in a variety of systems: glucose-induced Insulin secretion (Montague et al., 1985; Zawalich et al., 1983, 1984); TRH-induced prolactin secretion (Albert and Tashjian, 1984a,b; Delbeke et al., 1984a,b; Durst and Martin, 1982a,b, 1984; Fearon and Tashjian, 1985; Gershengorn, 1981; Gershengorn and Thaw, 1983, 1985; Martin, 1984; Martin and Kowalchyk, 1984); carbacholamine-induced tracheal smooth muscle contraction (Park and Rasmussen, 1985); norepinephrine-induced vascular smooth muscle contraction (Danthulun et al., 1984; Forder et al., 1985; Rasmussen et al.,

1984); and carbacholamine-induced pancreatic exocrine secretion (Pandol et al., 1985). Undoubtedly, this list will grow as a variety of other systems are investigated. It seems safe to conclude that this is one general mechanism by which the Ca^{++} messenger system operates to mediate sustained cellular responses. Even so, it is necessary to point out that the patterns of response in these various systems are not identical. Thus, for example, when angiotensin II acts there is a monotonic rise in aldosterone secretory rate to a sustained plateau; when glucose acts there is a biphasic pattern of insulin secretion with a very rapid but brief phase followed by a fall then a slowly progressive rise to a high sustained rate; and when TRH acts there is also a biphasic pattern but in this case, the initial transient peak is quite large, and the sustained phase relatively small. Hence, a variety of response patterns can be mediated by this two branch system, but the cellular and molecular details underlying these different patterns is not yet known.

The most important features of this new model are first, the introduction of the element of time into the consideration of second messenger action; and, second, the emphasis on the role of different plasma membrane events during this temporal sequence. There is simply more than an activation of the PLC-mediated hydrolysis of PIP_2 occurring at this site. There follows a Ca^{++}-flux mediated control of C-kinase activity. Although this particular change in plasma membrane transducing function is the one emphasized in the preceding discussion, it is not the only one either in a temporal sense, nor in the spatial sense of changing plasma membrane function. At least two others are: a) changes in Na^+/H^+ exchange; and b) generation of the active metabolites of arachidonic acid.

One of the common consequences of an increased C-kinase activity is the phosphorylation of plasma membrane bound proteins which, in turn, influence cell behavior. A specific example is the activation via C-kinase of the plasma membrane Na^+/H^+ exchanger so that intracellular pH rises (H^+ falls) and intracellular Na^+ increases (Busa and Nuccitelli, 1983; Macara, 1985). These are global changes in intracellular ionic composition that are communicated throughout the cell thereby influencing: a) activities of enzymes-nearly all of which are pH dependent; b) altering the association of Ca^{++} with its receptor proteins, e.g., CaM; and c) altering the fluxes of other ions across both the plasma and subcellular membranes. They, too, clearly play an informational role during long term stimulation of cellular response.

An additional consequence of PIP_2 hydrolysis is the generation of free arachidonic acid (AA) derived from the DG (and/or phosphatidic acid) produced as a result of PIP_2 hydrolysis (Ballow and Cheung, 1985; Feinstein et al., 1985; Holmsen, 1985; Sha'afi and Naccache, 1985). The AA in turn gives rise to prostacyclins, prostaglandins, leukotrienes, and/or thromboxanes. These provide an additional messenger input arising from the plasma membrane during the evolving temporal sequence of changes in plasma membrane activities. These metabolites act as either autocrine or paracrine agents to modify cellular response by acting via either the Ca^{++} or cAMP messenger systems.

This emphasis on a temporal sequence of information flow also helps one to address another common feature of cellular response. This feature is that of hypertrophy and hyperplasia of particular cell types in response to their continued stimulation by an agonist. If we take the specific example of the adrenal glomerulosa, we can recognize a temporal sequence of hypertrophy and hyperplasia, stretching over days or weeks, when this organ is exposed to continued high levels of AII. The initial events occurring in minutes and hours, are mediated respectively, by the two branches of the Ca^{++} messenger system as discussed above. Within hours to days these cells have synthesized new proteins and increased their individual capacities to produce aldosterone--the first step in their hypertrophy. Over days and weeks, they undergo a proliferative response and increase in number further increasing the organ's capacity to produce aldosterone. We do not yet know the signals necessary to invoke these more long term responses but they surely must involve a further temporal sequence of information flow from cell surface to cell interior. One can predict that there is a continually evolving pattern of information flow from cell surface to cell interior which allows the cell to recognize temporal information and react accordingly. Finally, the observation that at least a short term type of biochemical memory can reside in a membrane-associated enzyme focus attention on this cellular site as a logical one at which longer term memory may be also be stored.

ACKNOWLEDGMENTS

Supported by grants from the National Institutes of Health (AM 19813, HL 35849, and AM 33001) and the Muscular Dystrophy Association of America.

REFERENCES

Adelstein, R. S. and Eisenberg, E., Regulation and kinetics of the actin-myosin-ATP interaction, Ann. Rev. Biochem. 49:921-956 (1980).

Aksoy, M. O., Mras, S., Kamm, K. E., Murphy, R. A., Ca^{++}, cAMP, and changes in myosin phosphorylation during contraction of smooth muscle, Am. J. Physiol. 245:C255-C270 (1983).

Albano, J. D., Brown, B. L., Ekins, R. P., Tait, S. A. and Tait, J. R., The effects of potassium, 5-hydroxytryptamine, adrenocorticotrophin and angiotensin II on the concentration of adenosine 3':5'-cyclic monophosphate in suspensions of dispersed rat adrenal zona glomerulosa and zona fasiculata cells, Biochem. J. 142:391-400 (1974).

Albert, P. R. and Tashjian, A. H., Thyrotropin-releasing hormone-induced spike and plateau in cytosolic free Ca^{++} concentrations in pituitary cells, J. Biol. Chem. 259:5827-5832 (1984a).

Albert, P. R. and Tashjian, A. H., Relationship of thyrotropin-releasing hormone-induced spike and plateau phases in cytosolic free Ca^{++} concentrations to hormone secretion: selective blockade using ionomycin and nifedipine, J. Biol. Chem. 259:15350-15363 (1984b).

Alkon, D. L., Calcium-mediated reduction of ionic currents: a biophysical memory trace, Science 226:1037-1045 (1984).

Apfeldorf, W. F. and Rasmussen, H., Angiotensin II induces a transient rise in cytosolic calcium in adrenal zona glomerulosa. (In preparation).

Ballocu, L. R. and Cheung, W. Y., The role of calcium in prostaglandin and thromboxane biosynthesis, in: "Calcium and Cell Physiology," D. Marme, ed., Springer-Verlag, Berlin (1985).

Barrett, P., Kojima, I., Kojima, K., Zawalich, K., Isales, C. and Rasmussen, H., Temporal patterns of protein phosphorylation of angiotensin II, A23187 and/or TPA in adrenal glomerulosa cells, Biochem. J. (Submitted), (1986a).

Barrett, P., Kojima, I., Kojima, K., Zawalich, K., Isales, C. and Rasmussen, H., Short term memory in the calcium messenger system: evidence for a sustained activation of C-kinase in adrenal glomerulosa cells, Biochem. J. (Submitted) (1986b).

Berridge, M. J., Inositol triphosphate and diacylglycerol as second messengers, Biochem. J. 220:345-360 (1984).

Berridge, M. J. and Irvine, R. F., Inositol triphosphate, a novel second messenger in cellular signal transduction, Nature 312:315-321 (1984).

Blumenthal, D. K. and Stull, J. T., Activation of skeletal muscle myosin light chain kinase by calcium and calmodulin, Biochemistry 19:5608-5624 (1980).

Bolton, T. B., Mechanism of action of transmitters and other substances on smooth muscle, Physiol. Rev. 59:606-718 (1973).

Busa, W. B. and Nuccitelli, R., Metabolic regulation via intracellular pH, Am. J. Physiol. 246 (Regulatory Integrative Comp. Physiol. 15):R409-R438 (1984).

Danthulun, N. R. and Deth, R. C., Phorbol ester-induced contraction of arterial smooth muscle and inhibition of α-adrenergic response, Biochem. Biophys. Res. Commun. 125:1103-1109 (1984).

Delbeke, D., Kojima, I., Dannies, P. and Rasmussen, H., Synergistic stimulation of prolactin release by phorbol ester, A23187, and forskolin, Biochem. Biophys. Res. Commun. 123:735-741 (1984a).

Delbeke, D., Scammell, J. G., Dannies, P. S., Difference in calcium requirements for forskolin-induced release of prolactin from normal pituitary cells and GH_4C_1 cells in culture, Endocrinology 114:1433-1440 (1984b).

Downes, C. P. and Michell, R. H., The polyphosphoinositide phosphodiesterase of erythrocyte membranes, Biochem. J. 198:133-140 (1981).

Downes, C. P. and Michell, R. H., The control of Ca^{++} of the polyphosphoinositide phosphodiesterase and the Ca pump ATPase in human erythrocytes, Biochem. J. 202:53-58 (1982a).

Downes, C. P. and Michell, R. H., Phosphatidylinositol-4 phosphate and phosphatidylinositol-4,5 bis-phosphate: lipids in search of a function, Cell Calcium 3:467-502 (1982b).

Drust, D. S. and Martin, T. F., Thyrotropin-releasing hormone rapidly and transiently stimulates cytosolic calcium-dependent protein phosphorylation in GH_3 pituitary cells, J. Biol. Chem. 257:7566-7573 (1982a).

Drust, D. S., Sutton, C. A. and Martin, T. F., Thyrotropin-releasing hormone and cyclic AMP activate distinctive pathways of proteins phosphorylation in GH pituitary cells, J. Biol. Chem. 257:3306-3312 (1982b).

Drust, D. S. and Martin, T. F., Thyrotropin-releasing hormone rapidly activates protein phosphorylation in GH_3 pituitary cells by a lipid-linked, protein kinase C-mediated pathway, J. Biol. Chem. 259:14520-14530 (1984).

Fakunding, J. L., Chow, R. and Catt, K. J., The role of calcium in the stimulation of aldosterone production by adrenocorticotropin, angiotensin II, and potassium in isolated glomerulosa cells, Endocrinology 105:327-333 (1979).

Fakunding, J. L. and Catt, K. J., Dependence of aldosterone-stimulation in adrenal glomerulosa cells on calcium uptake: effects of lanthanum and verapamil, Endocrinology 107:1345-1353 (1980).

Fearon, C. W. and Tashjian, A. H., Jr., Thyrotropin-releasing-hormone induces redistribution of protein kinase C in GH_3C_1 rat pituitary cells, J. Biol. Chem. 260 (in press) (1986).

Feinstein, M. B., Halenda, S. P. and Zavocio, G. B., Calcium and platelet function, in: "Calcium and Cell Physiology," D. Marme, ed., Springer-Verlag, Berlin (1985).

Forder, J., Scriabine, A. and Rasmussen, H., Plasma membrane calcium flux, protein kinase C activation and smooth muscle contraction, J. Pharm. Exp. Ther. 235(2):267-273 (1985).

Foster, R., Lobo, M. V., Rasmussen, H. and Marusic, E. T., Calcium: its role in the mechanism of action of angiotensin II and potassium in aldosterone production, Endocrinology 109:2196-2201 (1981).

Foster, R., Lobo, M. V., Rasmussen, H. and Marusic, E. T., The effect of calcium in the potassium induced depolarization in adrenal glomerulosa cells, FEBS Lett. 149:253-257 (1982).

Fujita, K., Aguilera, G. and Catt, K. J., The role of cyclic AMP in aldosterone production by isolated zona glomerulosa cells, J. Biol. Chem. 254:8567-8574 (1979).

Gershengorn, M. C., Thyrotropin-releasing hormone stimulation of prolactin release from clonal rat pituitary cells. Evidence for action independent of extracellular calcium, J. Clin. Invest. 67:1769-1776 (1981).

Gershengorn, M. C., Calcium influx is not required for TRH to elevate free
cytoplasmic calcium in GH$_3$ cells, Endocrinology 113:1522-1524 (1983).

Gershengorn, M. C. and Thaw, M., Thyrotropin-releasing hormone (TRH) stimu-
lates biphasic elevation of cytoplasmic free calcium in GH$_3$ cells:
Further evidence that TRH mobilizes cellular and extracellular Ca^{++},
Endocrinology 116:591-596 (1985).

Holmsen, H., Receptor-controlled phosphatidate synthesis during acid hydro-
lase secretion from platelets, in: "Calcium in Biological Systems,"
R. P. Rubin, G. B. Weiss, J. W. Putney, Jr., (eds.), Plenum Press, New
York (1985).

Hyatt, P. J., Tait, J. F. and Tait, A. S., The mechanism of the effect of
K$^+$ on the steroidogenesis of rat zona glomerulosa cells of the adrenal
cortex: role of cyclic AMP, Proc. R. Soc. Lond. B266:21-42 (1986).

Kishimoto, K., Takai, Y., Mori, T., Kikkawa, U. and Nishizuka, Y., Activation of calcium and phospholipid-dependent protein kinase by diacylglycerol, its possible relation to phosphatidylinositol turnover, J. Biol.
Chem. 255:2273-2276 (1980).

Kojima, I., Kojima, K., Kreutter, D. and Rasmussen, H., The temporal integration of the aldosterone secretory response to angiotensin occurs via
two intracellular pathways, J. Biol. Chem. 259:14448-14457 (1984a).

Kojima, I., Kojima, K. and Rasmussen, H., Role of calcium fluxes in the sustained phase of angiotensin II mediated aldosterone secretion from adrenal glomerulosa cells, J. Biol. Chem. 260:9177-9184 (1985a).

Kojima, I., Kojima, K. and Rasmussen, H., Effect of angiotensin II and K$^+$
on Ca^{++}-efflux and aldosterone production in adrenal glomerulosa cells,
Am. J. Physiol. 248:E36-E43 (1985b).

Kojima, I., Kojima, K. and Rasmussen, H., Role of calcium and cAMP in the
action of adrenocorticotropin on aldosterone secretion, J. Biol. Chem.
260:4248-4256 (1985c).

Kojima, I., Kojima, K. and Rasmussen, H., Characteristics of angiotensin
II-, K$^+$, and ACTH-induced calcium influx in adrenal glomerulosa cells,
J. Biol. Chem. 260:9171-9176 (1985d).

Kojima, I., Kojima, K. and Rasmussen, H., Intracellular calcium and adenosine 3;,5;-cyclic monophosphate as mediators of potassium-induced
aldosterone secretion, Biochem. J. 228:69-76 (1985e).

Kojima, K., Kojima, I. and Rasmussen, H., Dihydropyridine calcium agonist
and antagonists effects on aldosterone secretion Am. J. Physiol. 245:
E645-E650 (1984a).

Macara, I. G., Oncogenes, ions and phospholipids, Am. J. Physiol. 248:C3-
C11 (1985).

Martin, T. F., Thyrotropin-releasing hormone rapidly activates the phospho-
diester hydrolysis of polyphospholinositides in GH$_3$ pituitary cells.
Evidence for the role of a polyphosphoinositide-specific phospholipase
C in hormone action, J. Biol. Chem. 258:14816-14822 (1983).

Martin, T. F. and Kowalchyk, J. A., Evidence for the role of calcium and
diacylglycerol as dual second messengers in thyrotropin-releasing hormone action: involvement of Ca^{++}, Endocrinology 115:1527-1536 (1984).

Montague, W., Morgan, N. G., Rumford, G. M. and Prince, C. A., Effect of
glucose on polyphosphoinositide metabolism in isolated rat islets of
Langerhans, Biochem. J. 227:483-489 (1985).

Morgan, J. P. and Morgan, K. G., Vascular smooth muscle: the first recorded
Ca^{2+} transients, Pflgers Arch 395:75-77 (1982).

Morgan, J. P. and Morgan, K. G., Stimulus-specific patterns of intracellular
calcium levels in smooth muscle of ferret portal vein, J. Physiol. 351:
155-167 (1984).

Morgan, J. P. and Morgan, K. G., Alteration of cytoplasmic ionized calcium
levels in smooth muscle by vasodilators in the ferret, J. Physiol.
(Lond) 357:539-551 (1986).

Nishizuka, Y. and Takai, Y., Calcium and phospholipid turnover in a new
receptor function for protein phosphorylation, in: O. M. Rosen and E.
G. Krebs, eds, "Protein Phosphorylation," New York: Cold Spring Harbor
Laboratory (1981).

Nishizuka, Y., Calcium, phospholipid turnover and transmembrane signalling, Phil. Trans. Roy. Soc. (Lond) B302:101-112 (1983).

Pandol, S. J., Schoeffield, M. S., Sachs, G. and Muallem, S., Role of free cytosolic calcium in secretagogue-stimulated amylase release from dispersed acini from guinea pig pancreas, J. Biol. Chem. 260(18):10081-10086 (1985).

Parks, S. and Rasmussen, H., Activation of tracheal smooth muscle contraction: synergism between Ca^{++} and activators of protein kinase C, Proc. Natl. Acad. Sci. USA 82:8835-8839 (1985).

Rasmussen, H. and Barrett, P. Q., Calcium messenger system: an integrated view, Physiol. Rev. 64:938-984 (1984).

Rasmussen, H., Forder, J., Kojima, I. and Scriabine, A., TPA-induced contraction of isolated rabbit vascular smooth muscle, Biochem. Biophys. Res. Commun. 122:776-784 (1984).

Sala, G. B., Hyahi, K., Catt, K. J. and Dufan, M. L., Adrenocorticotropin action in isolated adrenal cells, J. Biol. Chem. 254:3861-3865 (1979).

Sha'afi, R. I. and Naccache, P. H., Relationship between calcium, arachidonic acid metabolites and neutrophil activation, in: "Calcium in Biological Systems," R. P. Rubin, G. B. Weiss, J. W. Putney, Jr., eds, Plenum Press, New York (1985).

Silver, P. J. and Stull, J. T., Phosphorylation of myosin light chain and phosphorylase in tracheal smooth muscle in response to KCl and carbachol, Mol. Pharmacol. 25:267-274 (1984).

Streb, H., Irvine, R. F., Berridge, M. J. and Schulz, I., Release of Ca^{++} from a non-mitochondrial store in pancreatic acinar cell by inositol-1-4,5-trisphophate, Nature (Lond) 306:67-69 (1983).

Takai, Y., Kishimoto, A., Inoue, M. and Nishizuka, Y., Studies on a cyclic nucleotide-independent protein kinase and its pro-enzyme in mammalian tissues, J. Biol. Chem. 252:7603-7609 (1977).

Takai, Y., Kishimoto, A., Kikkawa, U., Mori, T. and Nishizuka, Y., Unsaturated diacylglycerol as a possible messenger for the activation of calcium-activated, phospholipid-dependent protein kinase system, Biochem. Biophys. Res. Commun. 91:1218-1224 (1979).

Williams, B. C., McDougall, J. G., Tait, J. F. and Tait, S. A., Calcium efflux and steroid output from superfused rat adrenal cells: effects of potassium, adrenocorticotropic hormone, 5-hydroxytryptamine, adenosine-3'5'-cyclic monophosphate and angiotensin II and III, Biochim. Biophys. Acta. 639:243-295 (1981).

Wolf, M., Cuatrecasas, P. and Sahyoun, N., Interaction of protein kinase C with membranes is regulated by Ca^{++}, phorbol esters, and ATP, J. Biol. Chem. 2260:15718-15722 (1985a).

Wolf, M., Levine, H., III, May, W. S., Jr., Cuatrecasas, P. and Sahyoun, N., A model for intracellular translocation of protein kinase C involving synergism between Ca^{++} and phorbol esters, Nature 317:546-549 (1985b).

Zawalich, W., Brown, C. and Rasmussen, H., Insulin secretion: combined effects of phorbol ester and A23187, Biochem. Biophys. Res. Commun. 117:448-455 (1983).

Zawalich, W., Zawalich, K. and Rasmussen, H., Insulin secretion: combined tolbutamide, forskolin and TPA mimic action of glucose, Cell Calcium 5:551-558 (1984).

POTASSIUM CHANNELS IN MOUSE SPINAL CORD CELLS

J. E. Kimura and B. McIver

Department of Physiology and Biophysics, University of
Vermont, Given Building, Burlington, Vermont 05405

Activation of voltage gated potassium channels is important for the
repolarization of the action potential in a number of systems (for example,
the squid giant axon: Hodgkin and Huxley, 1952b; and the frog node of Ranvier: Stampfli and Hille, 1976), though it may not be necessary in others
(mammalian node of Ranvier: Chiu, Ritchie, Rogart and Stagg, 1979). Barrett, Barrett and Crill (1980) reported the existence of both fast and slow
voltage-activated outward currents in cat motor neurones in vivo, and suggested that the fast component may be involved in repolarization of the
fast action potential. However, iontophoretic application of tetra-ethyl
ammonium (TEA), sufficient to reduce the amplitude of this fast outward
current, caused only a minor increase in the duration of the action potential (Schwindt and Crill, 1980; Schwindt and Crill, 1981).

Detailed quantitative analysis of conductance systems in cells of the
central nervous system is impeded by: 1) Difficulties in adequately controlling the voltage over the whole cell surface in these geometrically
complex cells (Rall and Segev, 1985). The dendritic tree of mammalian
motor neurones may have as much as twenty to thirty times the surface area
of the neuronal soma (Rall, 1959); 2) The problem of pharmacologically
separating the several conductance pathways to characterize the current
flow through a single channel type (Schwindt and Crill, 1980); and 3)
Limitations in the bandwidth of both voltage recording and voltage control,
caused by the necessarily high impedences of electrodes used to voltage
clamp these cells.

Application of the patch-clamp technique (Hamill, Marty, Neher, Sakmann and Sigworth, 1981) eliminates these problems. It allows investigation of a single channel type, under conditions that should permit almost
perfect voltage control of the area of the membrane under study. This
chapter describes some of the results of a patch clamp study of a fast
activating outward potassium current in cultured mouse spinal cord neurones. The electrical response of this culture system (Ransom and Holz,
1977; Heyer and MacDonald, 1982; Biltucci, McIver and Kimura, unpublished)
resembles that reported for in vivo cat motor neurones (Schwindt and Crill,
1982), and rat spinal cord slices (Murase and Randic, 1983).

The aim of these experiments was to obtain a series of descriptive
equations with which to perform modelling of the electrical response of
these cell types. The analysis has been performed with this in mind.

METHODS

Briefly, the bathing ('external') solution contained 140 mM NaCl, 1 mM KCl, 2 mM CaCl, 10 mM gucose, 10 mM N-2-hydroxyethylpiperazine-N'-2-ethanesulphonic acid (HEPES), and 300 nM Tetrodotoxin (TTX). The electrode ('internal') solution contained 135 mM KCl, 10 mM ethyleneglycol-bis-(-ami-noethylether)-N,N'-tetra-acetic acid (EGTA), 10 mM Glucose, and 10 mM HEPES. Both solutions were adjusted to pH 7.3 to 7.4 with 1 M sodium hydroxide. Membrane patches were held at a potential of -90 mV ('inside' surface of the membrane patch negative), and depolarizing voltage clamp steps were applied for 160 to 200 msec. Successive voltage clamp steps were separated by an interval of 2 seconds, to permit complete recovery to steady state. All experiments were performed at a temperature of between 8 and 10°C.

In all traces, an upward deflection of the trace represents an outward current, while a downward deflection represents an inward current. All potentials are given as inside potential minus outside potential.

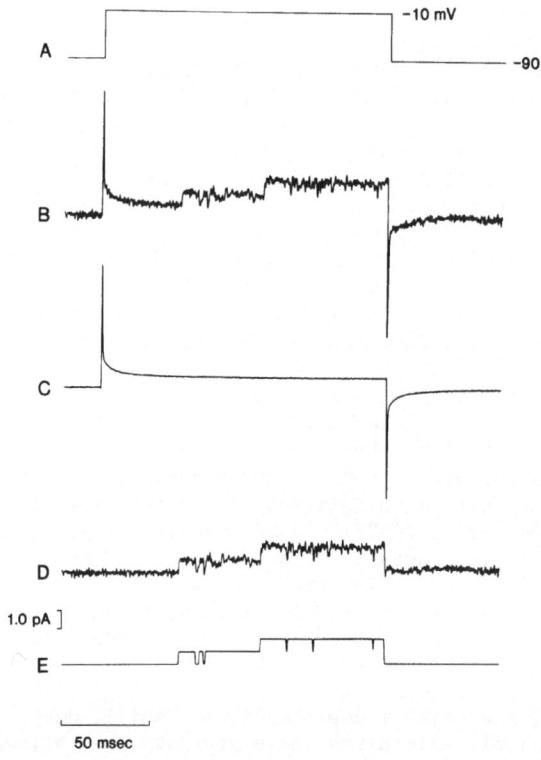

FIG. 1: Single channel records elicited in response to a depolarizing voltage clamp step from -90 mV to -10 mV. Outward currents are shown as upward deflections of the current trace. Potentials are given as if the membrane was still attached to the cell, inside relative to outside. A, the applied voltage clamp step. B, the current response elicited. C, the mean response of the membrane and electrode, in the absence of single channel activity. D, the result of subtracting C from B, resulting in single channel currents, with a baseline at 0 pA. E, results of the partially automated analysis procedure. The right column of traces shows two channels open simultaneously, towards the end of the voltage clamp step. Data were filtered at 1.0 kHz, and sampled at 16.7 kHz (1 point every 60 usec).

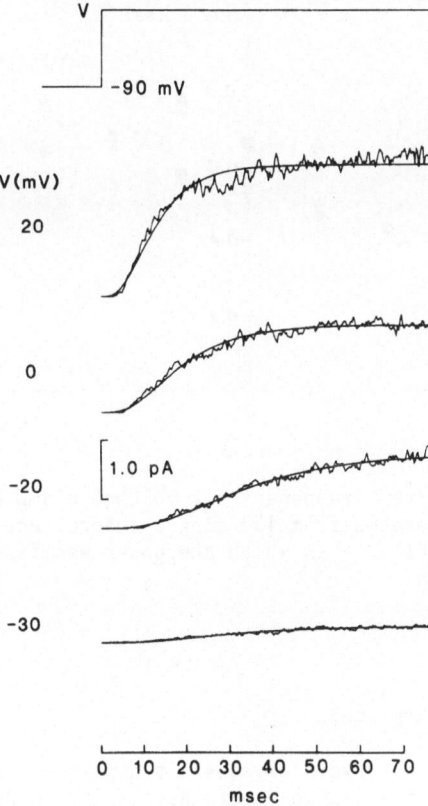

FIG. 2: The current voltage relationship for the open channel. The membrane potential was stepped from -90 mV to the indicated test potential for 200 msec. The filter was set to a corner frequency of 1.5 kHz, and data was acquired at a rate of 16.7 kHz. Null responses have been subtracted from each of the current traces. The slope of the linear part of the curve (up to +20 mV) is 8.7 pS.

RESULTS

Figure 1 shows a voltage clamp step to -10 mV (Fig. 1A), and the resulting current traces (Fig. 1B). The voltage clamp steps induce currents through the membrane patch, consisting of a rapid capacity charging transient, an ohmic leak pedestal, and single channel events resulting from channel activation (Fig. 1B). A 'null response' was generated shown in figure 1C. The null response was subtracted from each record, providing Fig. 1D. This also removed any baseline offset from zero, simplifying the analysis by eliminating the need to calculate both the baseline and the channel amplitude levels. Addition of noise to the raw data was minimal by this procedure, if a sufficiently large number of null traces were averaged together (10 to 20), although this was difficult to achieve in patches with several channels under strongly activating conditions, where the proportion of such traces was small.

The potassium single channel currents appeared as small rectangular (outward) current steps late in the trace. The patch shown in figure 1 contained at least 2 channels, since 2 superimposed outward current steps are visible in the right panel of figure 1. In general, patches contained between 2 and 5 single channels, as estimated by counting the maximal number of superimpositions observed.

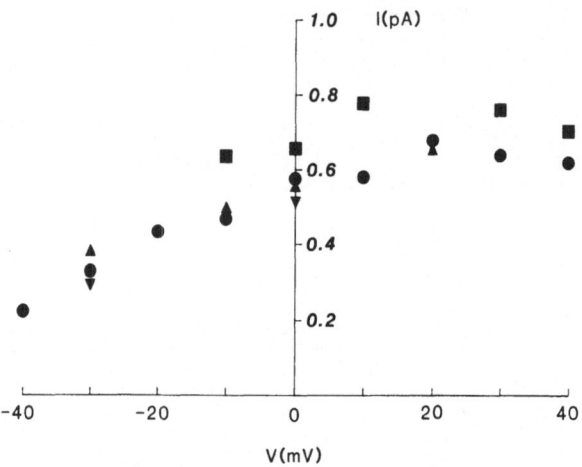

FIG. 3: The average current response to a voltage clamp step from -90 mV
to -20 mV, generated from 150 single channel current traces, and
fitted to equation 1, in which the power was fixed to values
between 2 and 6.

Single Channel Current Amplitude

Single channel currents were of constant amplitude at any one potential. The amplitude of the single event increased with depolarization
(fig. 2) up to around +10 to +20 mV, beyond which the current-voltage plot
saturated. Figure 2 plots the current-voltage relationships for 5 separate
experiments. The slope conductance of the linear portion of the current-
voltage curve, negative to +10 mV, was 8.7 pS, according to a least squares
fitting procedure. The saturation that was visible at more depolarized
potentials (most noticable positive to +30 mV) may be due to an increased
rate of channel 'flickering' at these potentials, causing a reduction in
the apparent single channel amplitude. The linearly extrapolated reversal
potential (-68 mV) is not the same as E calculated by the Nernst equation
(-120 mV). This could be caused by the channel being permeant to other
ionic species (sodium, for example, which had a calculated reversal potential of +121 mV in this system). Constant field predicts a non-linear current-voltage relationship at voltages close to the reversal potential, that
could also explain the difference between calculated and extrapolated
reversal potentials.

Averaged Currents

Averaging the single channel current flow of a large number of traces,
recorded under identical conditions from the same patch, produces a 'macroscopic' current trace, as it would appear under perfect voltage clamp conditions (see figure 3). Idealised reconstructions of the single channel
data were used to generate the average current response, improving the sig-
nal-to-noise ratio of the current trace. Typically, between 100 and 200
single channel traces were used to produce a single averaged current.

Averaged currents showed a distinct inflection in the rising phase
attributable to the delay to opening of single channels. This rules out a
first-order description of the activation process, and requires a higher
order description.

Figure 3 shows averaged currents, that were fitted to the Hodgkin-Huxley (1952b) equation:

EQUATION 1:

$$i_k = A \cdot (1-e^{-t/t_n})^x \quad \text{where} \quad A = \bar{g}_k \cdot n_\infty^x \cdot (v_m - E_k)$$

using a fourth power (x=4). The fits are shown superimposed in fig. 3. The final steady state current increased with depolarization, while the rate of rise of the current became shorter. The curve fit provided values for both τ_n, and the steady state current value 'A'.

By assuming that the activation variable ('n') reaches a value of 1 under strong depolarizing voltage-clamp steps, the maximal potassium conductance $(\bar{g}_k)$ can be calculated from the equation

EQUATION 2:

$$\bar{g}_k = \frac{A'}{V_m - E_k}$$

A' is A taken from curve fits at large depolarizations. The value of E (-68mV) used to calculate this value was obtained by linear extrapolation of the current-voltage curve, for voltages less than +40 mV. This permits the calculation of values for n at other potentials, by

EQUATION 3:

$$M_\infty = \sqrt[x]{\frac{A}{\bar{g}_k \cdot (V_m - E_k)}}$$

Although the I-V plot appears to saturate at strong depolarizations, this may be due to the increase in the degree of 'flickering', apparent at these potentials, that will cause a reduction in the measured channel amplitude

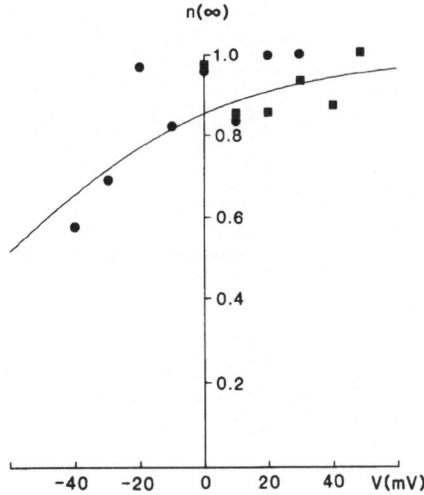

FIG. 4: The steady-state fraction of 'n' particles in the activating site plotted against the applied voltage.

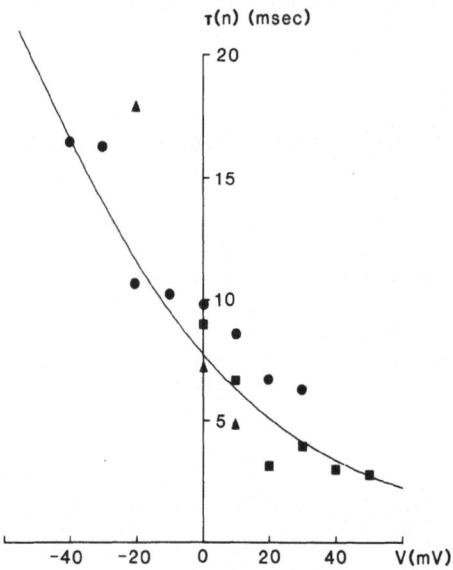

FIG. 5: The time constant (τ_n) plotted as a function of the applied voltage. The solid curve is that predicted from α_n and β_n , as described in Results.

because of the bandwidth limitations of the recording system. Figure 4 shows a plot of the value of n as a function of voltage. The mid-point of the n curve was at a voltage of approximately -40 mV. The full predicted sigmoidal shape of this curve was not represented, since we were unable to record these single channel currents at potentials negative to -50 mV, owing to their very small amplitude at these voltages.

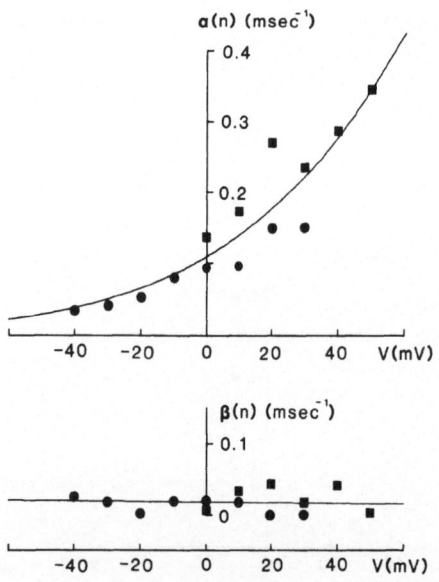

FIG. 6: The forward- and back- rate constants (α_n and β_n respectively) for movement of the imaginary 'n' particle, required for channel activation, calculated from 2 different patches.

The time constant of relaxation of 'n' (τ_n) is plotted as a function of voltage in figure 5. τ_n fell with depolarization, from 16 msec at -30 mV to 5 msec at +30 mV.

Figure 6 plots values for the forward and back rate constants (α_n and β_n), calculated using equations.

EQUATION 4:

$$\alpha n = \frac{n_\infty}{\tau n}$$

EQUATION 5:

$$\beta n = \frac{1 - n_\infty}{\tau n}$$

The values of alpha were superimposed with the expression:

EQUATION 6:

$$\alpha n = \frac{0.9\gamma}{e\gamma - 1} \qquad \gamma = \frac{0.1 - Vm}{0.003}$$

and beta with

EQUATION 7:

$$\beta n = \frac{0.45^\delta}{e^5 - 1} \qquad \delta = \frac{0.4 + Vm}{0.3}$$

in both of which Vm is the membrane potential, expressed in volts.

The forward rate constant (α_n) was voltage-dependent, increasing steeply with depolarization, while the back rate constant (β_n) was slower, and less voltage-dependent over this potential range. The solid curves in figures 5 and 6 were calculated from the values of α_n and β_n obtained from equations 6 and 7.

DISCUSSION

This paper describes experiments that examined some of the properties of a potassium channel in spinal cord neurones. The current flow through single potassium channels was recorded from excised, outside-out membrane patches (Hamill and Sakmann, 1981). This permitted accurate voltage control, without requiring an intracellular electrode to monitor the cell membrane potential. The potassium concentrations were maintained at approximately physiological levels, to mimic cellular conditions. TTX was added to the bathing solution to block sodium channels, that were often seen in the same patches in its absence. Since the calcium concentration in the electrode solution ('internal' surface of the membrane) was buffered to extremely low levels (with 10mM EGTA), no contamination was seen from the large conductance calcium-activated potassium channels that exist in this preparation (Kimura, McIver, Pun and Wong, 1983) and activate at calcium concentrations above about 10μM.

The single channel slope conductance reported here (approximately 9 pS) is similar to potassium-selective, inwardly rectifying voltage-gated channels recorded from Guinea-pig ventricular cells (13 pS with 40 mM external potassium, and 5.5 pS with 11 mM potassium: Sakmann and Trube, 1984a), and to the potassium channels recorded from the Human T-lymphocyte (10-15 pS) (Cahalan, Chandy, DeCoursey and Gupta, 1985). A so-called 'delayed rectifier' potassium channel, recorded from an embryonal carcinoma cell line, has a single channel conductance of 10 pS (Ebihara and Speers, 1984), and the anomalous rectifier channel, recorded in Tunicate egg membrane, that activates with hyperpolarizing pulses, also has a small unitary conductance (5-12 pS) (Fukushima, 1982). These potassium channels represent a quite heterogeneous group of channel types, however, and do not resemble each other kinetically. Larger voltage-gated potassium channels have also been reported from some preparations: 35 pS from cultured Purkinje neurones, (Gruol, 1984); 52 pS from clonal cells of the anterior pituitary (Wong and Lecar, 1982); and 62 pS for similar channels in Chick heart (Clapham and Defelice, 1984). Calcium-activated potassium channels tend to have a much larger single channel amplitude (see, for example, Pallota, Magleby and Barrett, 1981).

Our value for the single channel conductance (8.7 pS), together with estimates of the number of channels in each patch, yielded a preliminary estimate of the membrane potassium conductance of between 4.5 and 11.1 mS/cm^2. However, we are unable to determine the degree to which channels are damaged during seal formation, and patch excision, nor can we determine the geometry of the patch of membrane within the electrode, though it has been suggested that the patch assumes an omega shape in the pipette tip (Hamill et al., 1980). This may result in a larger area of voltage-clamped membrane, that therefore reduces the estimates of membrane conductance. A semi-spherical membrane geometry, of 0.7 μm diameter has an area of 1.5 μm^2, and the conductance estimate would then be one fourth of that already calculated. These figures for maximal conductance should therefore be treated as first approximations only.

Nevertheless, the estimated membrane conductance was similar to that measured by Barrett et al (1980) for the 'fast outward current' in cat motor neurones. Extrapolation of one of their tail currents (their Fig. 6), recorded from an in vivo motor neurone, under a two-electrode-voltage-clamp, provides a current flow of 30 nA at the end of a depolarizing step to approximately 0 mV. The calculated driving potential was 35 mV, resulting in a membrane conductance of 11 mS/cm (activated at 0 mV), assuming a spherical cell of 50 μm diameter.

The delayed rectifier in squid giant axon has a maximal conductance of between 26 and 49 mS/cm (Hodgkin and Huxley, 1952a), though it may be as high as 70 mS/cm (Conti, DeFelice and Wanke, 1975). These values are considerably higher than our estimates for the spinal cord cells. However, noise analysis (Conti et al., 1975) gave an estimated single channel density at 60 per μm^2, in the squid giant axon, resulting in an estimated single channel conductance of 12 pS, similar to the 9 pS observed in this study. Direct measurement of unitary current steps in cut open squid axons, over a limited voltage range, suggested a slightly higher single channel conductance, of 16 pS (Liano and Bezanilla, 1983).

The current-voltage relationship of the potassium channel was non-linear at positive potentials, showing saturation at a current density of around 0.85 pA per channel. However, it is unclear whether our estimates of single channel current flow at very depolarized potentials are true estimates of the open channel amplitude. The noise on the open channel, that may indicate a 'flicker' closed state, appears to become more pronounced at positive voltages, and will tend to reduce estimates of the single channel current flow at those potentials. Although rectifying current-voltage relationships have been reported for single channel potassium currents in Guinea-pig heart (Sakmann and Trube, 1984a; Trube and Hescheler, 1984), and

Rabbit atrio-ventricular node cells (Kakei and Noma, 1984), most relation-ships are approximately linear.

The classical Hodgkin-Huxley formalism (1952b) was used to describe the averaged currents. This employed a fourth order activation scheme, with no inactivation. An implicit assumption of the method of calculation used here for n is that the single channel current-voltage relationship is linear. Although a degree of rectification was observed at depolarized potentials, the remainder of the curve (from -40 mV to +30 mV) is described adequately as a straight line, of conductance 8.7 pS.

Values of n at voltages negative to -40 mV could not be calculated because we were unable to record averaged currents at potentials negative to around -40 mV. The single channel current amplitude became extremely small at those voltages, making identification difficult. In addition, the probability of eliciting an event was reduced at these voltages, requiring a greater number of traces to be averaged to produce an adequate current. The midpoint of the n curve was close to a potential of -40 mV, that is close to that of the squid delayed rectifier (mid-point around -50 mV: Palto, 1971). Barrett et al (1980) report that the mid-point of the 'steady state activation curve' for the fast outward current of cat motor neurones lies at approximately +10 mV (70 mV depolarized from the resting potential). This steady state activation curve is equivalent to the fourth power of our n_∞ curve (since they were forced to assume a first order kinetic scheme, because of technical limitations of two-electrode voltage-clamping these cells). The mid-point of the fourth power of our n_∞ curve lies at +5 mV. The 'open probability curve' (equivalent to our n_∞) for single voltage-gated channels in Guinea-pig heart has a mid-point at -50 to -60 mv (Sakmann and Trube 1984b), which is considerably to the left of ours, although Chick heart cell potassium channels are 50% activated at approximately -15 mV (Clapham and Defelice, 1984).

The time constant of relaxation of 'n' (τ_n) was dependent on voltage, becoming shorter with depolarization. The τ_n-V plot for the squid delayed rectifier shows a distinct peak at about -70 mV (Palti, 1971), followed by an approximately exponential fall with depolarization. Extrapolation of our calculated τ_n-V relationship, to negative potentials, predicts a peak in this curve at -120 mV, although this potential is well beyond the range of voltages in which we are able to record this current. τ_n was considerably slower for this conductance than for the squid delayed rectifier, at similar temperatures, ranging from 16 msec to 3 msec at 10°C in the spinal cord, but from 6 msec to 1 msec at 6°C in the squid (Palti, 1971). Further experiments are required to determine the effect of temperature on this channel, since these cells normally operate at 36°C.

Over the voltage range studied here, β_n showed only a very small voltage-dependence, becoming slightly smaller with depolarization. On the other hand, α_n increased steeply with depolarization, rising from 0.02 msec at -40 mV to 0.3 msec at +40 mV. These rate constants are considerably smaller than those of the squid axon delayed rectifier current, in which α_n rises from 0.3 $msec^{-1}$ at -40 mV, to 1.1 msec at +40 mV, at a similar temperature (6°C) (Hodgkin and Huxley, 1952b).

In conclusion, the channel described here has some of the properties required of a fast outward charge carrier, perhaps partially responsible for the repolarization of fast action potentials (Barrett et al., 1980; Schwindt and Crill, 1980). It is similar, in many respects, to the delayed rectifier of the squid axon, and is likely to be the same current as the 'fast outward current' already partially described in the cat motor neurone (Barrett et al., 1980). Incorporation of the information described here, and similar information from other voltage-gated channels in these neurones, into a computer model should allow determination of how these conductance systems participate in the genesis of action potentials in the cultured spinal cord neurone.

ACKNOWLEDGMENTS

Much of the work described in this Chapter was carried out in collaboration with Dr. Chris Huang, of the Physiological Laboratory, Cambridge, England. The authors would like to thank Mr. A. Krawitt, and Mr. S. Biltucci for skilled assistance. Dr. R. Y. K. Pun provided valuable instruction on the growth and maintenance of the culture system. Bryan McIver is grateful to the University of Edinburgh for leave of absence to do this work. This work was supported by grant number BNS-8217484, from the National Science Foundation, and by grants from the Whittaker Foundation, and the Epilepsy Foundation of America, all to John E. Kimura at the University of Vermont.

REFERENCES

Barrett, E. F., Barrett, J. N., and Crill, W. A., Voltage sensitive outward currents in cat motorneurones, Journal of Physiology 304:251-276 (1980).

Calahan, M. D., Chandy, K. G., DeCoursey, T. E. and Gupta, S., A voltage gated potassium channel in Human T-lymphocytes, Journal of Physiology 358:197-237 (1985).

Chiu, S. Y., Ritchie, J. M., Rogart, R. B. and Stagg, D., A quantitative description of membrane currents in rabbit myelinated nerve, Journal of Physiology 292:149-166 (1979).

Clapham, D. E. and DeFelice, L. J., Voltage activated K channels in embryonic chick heart, Biophysical Journal 45:40-42 (1984).

Conti, F., DeFelice, L. J. and Wanke, E., Potassium and sodium current noise in the membrane of the Squid giant axon, Journal of Physiology 248:45-82 (1975).

Ebihara, L. and Speers, W. C., Ionic channels in a line of embryonal carcinoma cells induced to undergo neuronal differentiation, Biophysical Journal 46:827-830 (1984).

Fukushima, Y., Blocking kinetics of the anomalous potassium rectifier of Tunicate egg studied by single channel recording, Journal of Physiology 331:311-331 (1982).

Gruol, D. L., Single channel analysis of voltage sensitive K channels in cultured purkinje neurons, Biophysical Journal 45:53-55 (1984).

Hamill, O. P., Marty, A., Neher, E., Sakmann, B. and Sigworth, F. J., Improved patch clamp techniques for high resolution current recording from cells and cell-free membrane patches, Pflugers Archiv 391:85-100 (1981).

Hamill, O. P. and Sakmann, B., Multiple conductance states of single acetylcholine receptor channels in embryonic muscle cells, Nature 294:462-464 (1981).

Heyer, E. J. and Macdonald, R. L., Calcium- and sodium-dependent action potentials of mouse spinal cord and dorsal root ganglion neurons in cell culture, Journal of Neurophysiology 47:641-655 (1982).

Hodgkin, A. L. and Huxley, A. F., Current carried by sodium and potassium ions through the membrane of the giant axon of Loligo, Journal of Physiology 116:449-472 (1952a).

Hodgkin, A. L. and Huxley, A. F., A quantitative description of membrane current and its application to conduction and excitation in nerve, Journal of Physiology 117:500-544 (1952b).

Kakei, M. and Noma, A., Adenosine-5'-triphosphate sensitive single potassium channel in the atrioventricular node cell of the Rabbit heart, Journal of Physiology 352:265-284 (1984).

Kimura, J. E. and Meves, H., The effect of temperature on the assymetrical charge movement in Squid giant axons, Journal of Physiology 289:479-500 (1979).

Liano, I. and Bezanilla, F., Bursting activity of potassium channels in the cut-open axon, Biophysical Journal 41:38a (1983).

Murase, K. and Randic, M., Electrophysiological properties of rat spinal dorsal horn neurones in vitro: calcium dependent action potentials, Journal of Physiology 334:141-153 (1983).

Pallota, B. S., Magleby, K. L. and Barrett, J. N., Single channel recordings of Ca-activated K currents in rat muscle cell culture, Nature 293:471-474 (1981).

Palti, Y., Description of axon membrane ionic conductances and currents, p. 168-182. In: Biophysics and Physiology of Excitable Membranes. W. J. Adelman (ed). Van Nostrand Reinhold, New York (1971).

Rall, W., Branching dendritic trees and motoneuron membrane resistivity, Experimental Neurology 1:491-527 (1959).

Rall, W. and Segev, I., Space-clamp problems when voltage clamping branched neurons with intracellular microelectrodes, p. 191-215. In: Voltage and Patch clamping with microelectrodes. T. G. Smith, H. Lecar, S. J. Redman, and P. W.. Gage (eds). American Physiological Society (1985).

Ransom, B. R. and Holz, R. W., Ionic determinants of excitability in cultured mouse dorsal root ganglion and spinal cord cells, Brain Research 136:445-453 (1977).

Sakmann, B. and Trube, G., Conductance properties of single inwardly rectifying potassium channels in ventricular cells from Guinea-pig heart, Journal of Physiology 347:641-657 (1984a).

Sakmann, B. and Trube, G., Voltage dependent inactivation of inward rectifying single channel currents in the Guinea-pig heart cell membrane, Journal of Physiology 347:659-683 (1984b).

Schwindt, P. C. and Crill, W. E., Effects of barium on cat spinal motoneurons studied by voltage clamp, Journal of Neurophysiology 44:827-846 (1980).

Schwindt, P. C. and Crill, W. E., Differential effects of TEA and cations on outward ionic currents of cat motoneurons, Journal of Neurophysiology 46:1-16 (1981).

Schwindt, P. C. and Crill, W. E., Factors influencing motoneuron rhythmic firing: results from a voltage clamp study, Journal of Neurophysiology 48:875-890 (1982).

Stampfli, R. and Hille, B., Electrophysiology of the peripheral myelinated nerve. In: Frog Neurobiology, R. Llinas and W. Precht (eds). Springer-Verlag, Berlin (1976).

Trube, G. and Hescheler, J., Inward rectifying channels in isolated patches of the heart cell membrane: ATP-dependence and comparison with cell-attached patches, Pflugers Archiv 401:178-184 (1984).

Wong, B. S. and Lecar, H., Differentiation of the two potassium channels in clonal anterior pituitary cells by the patch clamp technique, Biophysical Journal 37:325a (1982).

POLYUNSATURATED FATTY ACIDS AND INOSITOL PHOSPHOLIPIDS AT THE SYNAPSE IN NEURONAL RESPONSIVENESS

Nicolas G. Bazan and Dale L. Birkle

Louisiana State University Medical School, LSU Eye Center
2020 Gravier Street, Suite B, New Orleans, Louisiana 70112

INTRODUCTION

Neuronal communication involves synthesis, release, interaction with receptors, and degradation of a wide variety of chemical mediators at synapses. These mediators comprise neurotransmitters and other neuroactive substances and in some instances, two or more of these substances coexist in a given nerve ending. However, only one-tenth or less of the chemical mediators of the nervous system have been identified to date, and our understanding of how the processing of information involves the components of excitable membranes is just beginning.

Excitable membranes, with specialized presynaptic and postsynaptic domains, represent a large surface area in nervous tissue. The role of membrane components in neurotransmitter release (e.g., control of the release of different neurotransmitters from the same synaptic ending), is not understood. In the postsynaptic membrane, target responses are specified initially by interaction with receptors at the outer leaflet, coupled to transducer proteins that provide a link with a chain of chemical events leading to physiological actions on the postsynaptic neuron and/or on a given neuronal circuitry. Off signals, not well understood as yet, modulate the chemical events back to resting levels, allowing the transduction of subsequent extracellular messages.

To gain a better understanding of chemical events that take place within excitable membranes and constitute neuronal responsiveness, an experimental approach must be followed that preserves the integrity of the neuronal circuitry and precludes the introduction of artifacts, particularly during sampling. Simple cellular systems and preparations of isolated cells or subcellular fractions are often used. However when the intact tissue, i.e., brain, is studied, complex interactions between various cell types can make interpretations of data difficult. The use of various forms of stimulation (e.g., convulsions) is a useful strategy for studying biochemical changes that result from neuronal activity, because under basal conditions, many of these changes remain below the limits of detection by current procedures. One such form of stimulation, electroconvulsive shock, is of particular interest because it can be used to produce a short-lived tonic-clonic seizure that is rapidly reversible. Since this type of stimulus is transient, it is possible to follow the time course of increasing alterations in cerebral lipid metabolism and subsequent reversion to resting values. The profile of alterations of brain lipids during convulsions can

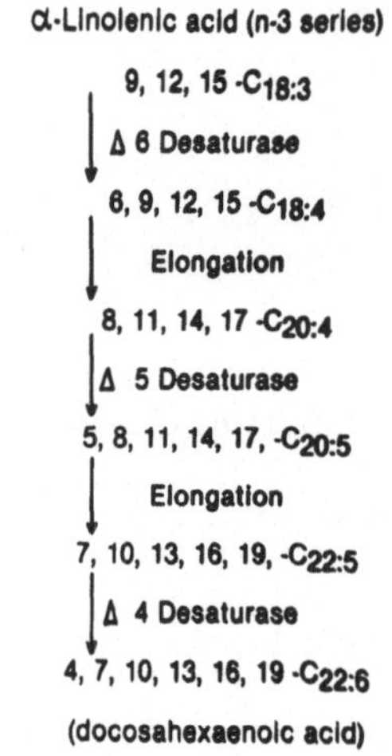

α-Linolenic acid (n-3 series)

$$9, 12, 15 \text{-} C_{18:3}$$

Δ 6 Desaturase

$$6, 9, 12, 15 \text{-} C_{18:4}$$

Elongation

$$8, 11, 14, 17 \text{-} C_{20:4}$$

Δ 5 Desaturase

$$5, 8, 11, 14, 17, \text{-} C_{20:5}$$

Elongation

$$7, 10, 13, 16, 19, \text{-} C_{22:5}$$

Δ 4 Desaturase

$$4, 7, 10, 13, 16, 19 \text{-} C_{22:6}$$

(docosahexaenoic acid)

FIG. 1:   Synthesis of docosahexaenoic acid (22:6, n-3) from the dietary
precursor and essential fatty acid, linolenic acid (18:3, n-3).
(Reprinted with permission from Birkle and Bazan, 1986.)

be used to understand membrane events during enhanced neurotransmitter
release and membrane depolarization in intact nervous tissue under in vivo
conditions. Methodologically, this can be achieved by the rapid inactivation of brain enzymes using, for instance, head-focused high powered microwave irradiation as discussed below. The electroconvulsive shock model is
also useful because it reveals aspects of the neurochemical basis of seizures; such information is of value for the understanding of epilepsy. One
additional feature that makes electroconvulsive shock a model of interest
is its effectiveness in the treatment of bipolar depression. Currently, the
neurochemical basis of both depressive disorders and the therapeutic action
of electroconvulsive shock is lacking.

A single seizure (e.g., electroconvulsive shock) transiently alters
lipids of excitable membranes; however, repeated convulsions, as in experimental status epilepticus, result in sustained alterations of brain lipid
metabolism. During convulsions there is a rapid accumulation of endogenous
free polyunsaturated fatty acids (arachidonic and docosahexaenoic acids) in
brain (for recent review, see Bazan et al, 1986). In fact, the rate of
accumulation of free fatty acids is comparable to that observed in adipose
tissue undergoing maximal hormonally-mediated lipolysis of triacylglycerols
(Aveldano and Bazan, 1975a; Bazan et al., 1982a). Because there is a very
small pool of triacylglycerols in the central nervous system, loss of fatty
acids from this pool is minimal (Bazan, 1970). Therefore rapid deacylation
of phospholipids is probably the major mechanism for fatty acid accumulation.
It has been difficult to pinpoint the sources of the released polyunsaturated
fatty acids because the mass of phospholipids is very large compared with
the small pool size of the accumulated free fatty acids. Loss of arachidonate from some phospholipids has been observed in ischemia (Ikeda et al,
1986), as well as in the brain of animals undergoing convulsions (see below).
Phospholipids of excitable membranes are a heterogeneous group of fatty
acyl-glycerophosphate derivatives divided into classes on the basis of their
polar head groups (phosphatidylcholine, phosphatidylethanolamine, phosphatidylserine, phosphatidylinositol, phosphatidylinositol 4-phosphate, phosphatidylinositol 4,5-bisphosphate, etc.). These classes comprise many molecular species which are defined individually by the fatty acyl substituents
present at C_1 and C_2 of the glycerol backbone. Most of the fatty acyl chains
at C_1 are saturated or lower unsaturated, and most at C_2 are unsaturated.

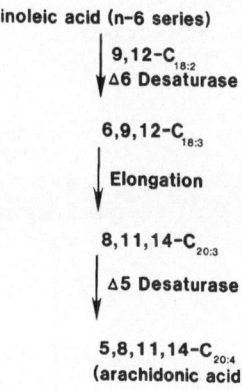

Linoleic acid (n-6 series)

$9,12-C_{18:2}$
$\Delta 6$ Desaturase

$6,9,12-C_{18:3}$

Elongation

$8,11,14-C_{20:3}$

$\Delta 5$ Desaturase

$5,8,11,14-C_{20:4}$
(arachidonic acid)

FIG. 2: Synthesis of arachidonic acid (20:4, n-6) from the dietary precursor and essential fatty acid, linoleic acid (18:2, n-6). (Reprinted with permission from Birkle and Bazan, 1986.)

Thus, palmitic (16:0), stearic (18:0) and oleic (18:1) acids are found most frequently at C_1. Docosahexaenoic (22:6, n-3) and arachidonic (20:4, n-6) acids prevail at C_2. Like other membranes, excitable membranes contain different proportions of each class of phospholipid. However, excitable membranes are unique in their very high content of polyunsaturated acyl groups in phospholipids, particularly decosahexaenoic acid. Another unique feature is that in the central nervous system, there are molecular species of phospholipids that contain two docosahexaenoate chains esterified to C_1 and C_2 (Aveldano and Bazan, 1977).

The physiological significance of excitable membrane phospholipids is not well understood, and knowledge regarding their role in neuronal communication is limited. One exception to this is the profusion of information being generated currently on inositol phospholipids and cell signaling. Until recently, it was believed that the major function of phospholipids was to form anular rings around receptors and/or ionic channels, providing the appropriate hydrophilic or hydrophobic milieus, and that the highly unsaturated molecular species provided increased membrane fluidity. However now it is known that phospholipids are the source of several crucial messenger-type compounds and play an important and direct role in modulation and mediation of cellular responses.

This article summarizes current work on polyunsaturated fatty acids and inositol phospholipids in neuronal responsiveness. Specifically the role of membrane lipid composition on development of synaptic circuitry, and alterations in excitable membrane lipids as a result of short-term and long-term neuronal stimulation are discussed. The focus is on inositol phospholipids as key lipids, involved in the generation of messengers in the central nervous system. The fate of accumulated polyunsaturated fatty acids, e.g. further metabolism through oxygenases to produce eicosanoids and docosanoids, and the possible physiological and pathological roles of disruption or modification in membrane lipid turnover are reviewed briefly. Finally a summary of potentially fruitful avenues of research on excitable membrane lipids is provided.

Essential Fatty Acids of Cellular Membranes in Nervous Tissue

The highly unsaturated nature of phospholipids of excitable membranes is unique to the nervous system and retinal photoreceptor membranes. This enrichment in polyunsaturated fatty acids is relatively stable even when

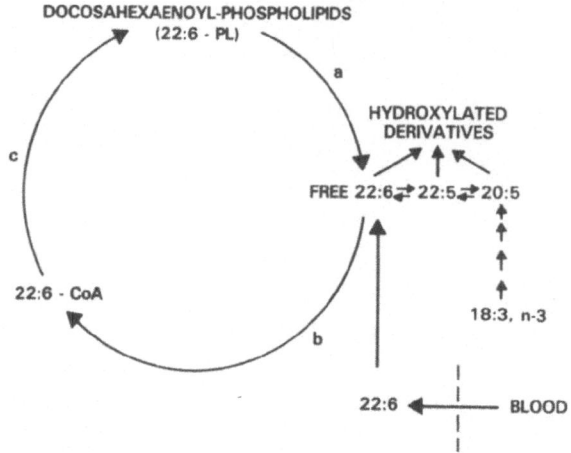

FIG. 3: Docosahexaenoic acid (22:6, n-3) metabolism, a = phospholipase A_2, giving rise to lysophospholipid or monoacylphosphoglyceride (not shown) and free 22:6; b = docosahexaenoyl-coenzyme A synthetase; c = acyltransferase. 18:3, n-3, the precursor of docosahexaenoic acid, is connected to 20:5 with arrows that represent elongation and desaturation reactions. Hydroxylated derivatives (docosanoids) are products of lipoxygenase activity (Bazan et al, 1984). (Reprinted with permission from HEP Bazan et al, 1986.)

animals are deprived of dietary essential fatty acids (for recent review, see Bazan et al, 1985a). The biochemical basis of this avid retention is not well understood. However, some unique enzymes, receptors and/or other proteins must be expressed that assure a high level of polyunsaturated fatty acids in brain and retina. Functional failure, including decreased visual acuity in primates (Neuringer et al, 1984), results from prolonged dietary deprivation of linolenic acid. Through a sequence of elongaton and desaturation steps, linolenic acid is converted to docosahexaenoic acid (Fig. 1). A similar sequence yields arachidonic acid from linoleic acid (Fig. 2). There are no metabolic steps linking the linoleic acid family with the linolenic acid family. However, the fatty acids compete with each other for the delta 6- and 5-desaturases. Once docosahexaenoic or arachidonic acid has been synthesized, esterification to phospholipids occurs. The fact that very low amounts of essential fatty acids or elongation and desaturation intermediates are found in the brain or in retina, implies that elongation and desaturation occur rapidly after dietary intake of linoleic or linolenic acid. Subsequently, the elongated and desaturated fatty acids are sequestered by tissues and esterfied into phospholipids. Nervous tissue under resting conditions contains negligible amounts of free (unesterified) arachidonic and docosahexaenoic acids (see below). It is likely that elongation and desaturation of dietary fatty acids takes place outside the nervous tissue, and that long-chain polyunsaturated fatty acids are distributed to tissues via systemic circulation. The liver is probably the major organ contributing to the elongation and desaturation of essential fatty acids. It should be noted that all tissues contain sizable amounts of arachidonic acid; however, docosahexaenoic acid is concentrated selectively in the synapse. Hence, uptake mechanisms for arachidonic acid may be distributed more widely than those for docosahexaenoic acid, which may be highly specific for nervous tissue.

In the perinatal period of life when the bulk of synaptogenesis occurs, large amounts of docosahexaenoic acid are required, which are supplied first

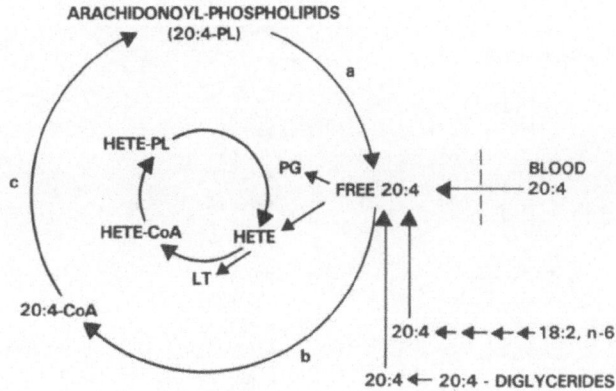

FIG. 4: Arachidonic acid metabolism (20:4, n-6), a = phospholipase A_2, that gives rise to lysophospholipid or monoacylphosphoglyceride (not shown) and free 20:4; b = arachidonoyl - CoA synthetase; c = acyltransferase. Linoleic acid (18:2, n-6), the precursor of arachidonic acid is shown as well as arachidonoyl diacylglycerols giving rise to free arachidonic acid through the action of diacylglycerol lipases. In the center of the cycle free arachidonic acid is oxygenated to prostaglandins (PG) and hydroxyeicosatetraenoic acids (HETE). HETE can be esterified in phospholipids via a CoA intermediate in a manner analogous to esterification of other fatty acids. LT indicates another branching of the arachidonic acid cascade towards sulfopeptido leukotrienes.

by the placenta and then maternal milk. When linolenic acid is provided, the liver may elongate and desaturate this fatty acid (Fig. 1). Certain blood lipoproteins may be the vehicle for redistribution of docosahexaenoic acid to the brain and retina. Later in life, docosahexaenoic acid content remains high in synaptic membranes, and in spite of the fact that stimulation releases some of the esterified fatty acid, there is a conservation mechanism in operation within the tissue, and the dietary essential fatty acids remain a requirement. We hypothesize that there are two basic processes involved in maintaining high levels of docosahexaenoic acid in the central nervous system. First, there may be some kind of hormonal mechanism whereby the nervous system signals the liver to send docosahexaenoic acid when needed. At the time that liver secretes a lipoprotein with docosahexaenoic acid into the blood, all organs are exposed to this. However, only the central nervous system takes up the majority of this fatty acid. Therefore, our second hypothesis is that the nervous system has a receptor-mediated recognition mechanism which mediates the sequestration of docosahexaenoic acid-containing lipoproteins from the blood. Other tissues may express only a fraction of such receptors, and hence are capable of retaining a relatively small amount of docosahexaenoic acid.

The differentiation of the retina also involves active synaptogenesis and membranogenesis. Visual cells develop photoreceptor outer segments very rich in docosahexaenoic acid. Throughout life, a renewal process takes place, characterized by photoreceptor membrane biogenesis at the base of the outer segment of the visual cell, and shedding of the distal tip with subsequent phagocytosis by retinal pigment epithelial cells. In spite of all these dramatic changes in photoreceptor membranes, docosahexaenoic acid is retained avidly in the visual cell outer segment. Here, in addition to the operation of a liver-blood-tissue transport system, a similar system may exist between the pigment epithelial cell and the visual cell. The space

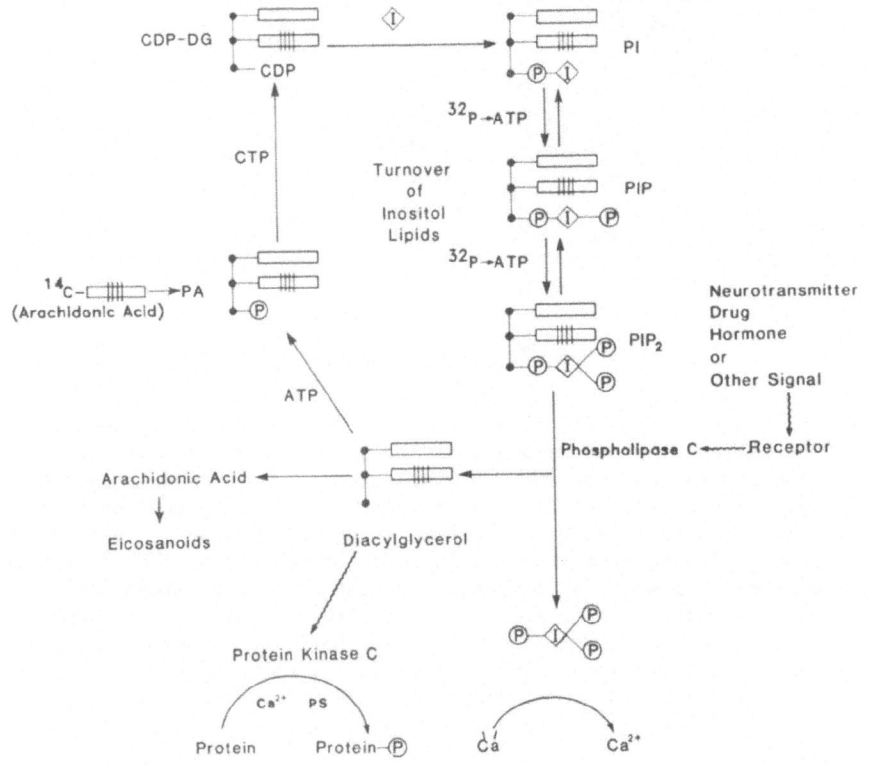

FIG. 5: Phosphoinositide cycle involved in cell signaling. Routes for introduction of radiolabeled precursors are shown. Eicosanoids indicates cyclooxygenase and lipoxygenase reaction products: PA, phosphatidic acid; I, inositol; PI, phosphatidylinositol, PIP, PI-4-phosphate; PIP_2, PI-4,5-bisphosphate; CDP-DG, cytidine diphosphate diacylglycerol; PS, phosphatidylserine.

between these two cells is called the interphotoreceptor matrix, where the prevalent protein is interphotoreceptor retinoid binding protein. This protein and other as yet unpurified proteins contain docosahexaenoic acid, implicating them as candidates for transport proteins for docosahexaenoic acid, (Bazan et al, 1985b). Hence, intercellular retrieval mechanisms may occur, in addition to the interorgan redistribution involving the blood stream. We do not have evidence as yet as to whether analogous intercellular mechanisms exist in the brain, but it is conceivable that glial cells could play a supportive role in the maintenance of high levels of docosahexaenoic acid in neurons.

Once the polyunsaturated fatty acid enters the cell it is converted to a thiol ester of coenzyme A (CoA). These conversion steps may be complementary and crucial in the retention mechanism of the nervous system; the kinetic properties of acyl CoA synthetase indicate that the synthesis of long-chain fatty acyl CoA could serve as an intracellular trapping mechanism for the fatty acid (Reddy and Bazan, 1983, 1985a, 1985b; Reddy et al, 1984), in a manner analogous to the intracellular trapping of glucose by phosphorylation. Figs. 3 and 4 outline the synthesis of docosahexaenoyl- and arachidonoyl-CoA and the further acylation to glycerol to synthesize phospholipids.

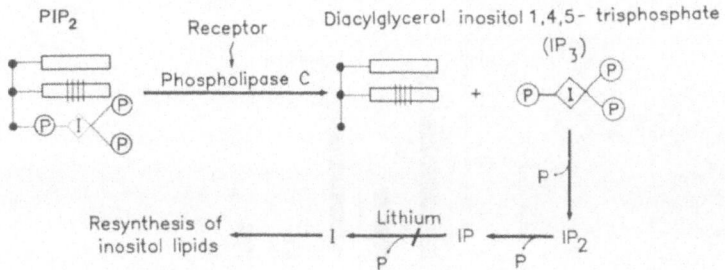

FIG. 6: Working hypothesis for the seizure-induced accumulation of diacy-
glycerols and inositol-1,4,5-trisphosphate via receptor-linked
degradation of phosphatidyl-inositol-4,5-bisphosphate (PIP$_2$),
mediated by phospholipase C. The resynthesis of phosphoinositides
is interrupted by lithium, which inhibits catabolism of inositol
phosphate by inositol phosphatase.

Inositol Phospholipids in Cell Signaling: Biochemical Events

     The phosphodiesteratic cleavage (mediated by phospholipase C) of ino-
sitol phospholipids is activated by neurotransmitters, hormones, growth
factors or other agonists that lead to intracellular ionization of Ca^{2+}.
The products of this reaction are diacylglycerol and water-soluble inositol
phosphates (Fig. 5). Because increased formation of diacylglycerol and
inositol 1,4,5-trisphosphate takes place in the brain during convulsions
(see below, Bazan, 1976; Bazan et al, 1982b; Van Rooijen et al, 1986), an
enhanced phosphodiesteratic cleavage of polyphosphoinositides probably
occurs. Since ischemia also triggers similar neurochemical changes (Aveldano and Bazan, 1975a; Bazan, 1970; Bazan and Rodriguez de Turco, 1980;
Ikeda et al, 1986), the vulnerability of the brain at synapses may involve
inositol phospholipid degradation. Careful study of these neurochemical
changes in membrane lipids may uncover different reactions that occur concurrently in various cells or in different neuronal circuits. One compli-
cation in interpreting observed changes is the variety and heterogeneity
of pools of a single lipid.

     Diacylglycerol, in addition to being a product of phospholipase C-
mediated degradation of phosphoinositides, is also a key intermediate in the
biosynthesis of other phospholipids and neutral lipids, and is a substrate
for lipases that release arachidonic acid for the synthesis of eicosanoids.
Convulsions may activate phospholipase C (Fig. 6) to produce diacylglycerols, which can be deacylated by diacylglycerol lipase, giving rise to free
arachidonic and stearic acids. Diacylglycerol is also a product of triacylglycerol degradation. Although there is a low content of triacylglycerol
in the central nervous system, it displays relatively high turnover (Cook et
al, 1982). Diacylglycerol can also be generated by the back reaction of the
choline- and/or ethanolamine phosphotransferases. Recently diacylglycerol
has been defined as an intracellular messenger for the activation of a cal-
cium- and phospholipid-sensitive protein kinase (protein kinase C, Kikkawa
et al, 1982). Also the water-soluble moiety of phosphatidylinositol 4,5-
bisphosphate hydrolysis, inositol 1,4,5-trisphosphate, acts as a second
messenger for the release of calcium from non-mitochondrial intracellular
stores through a specific receptor (Streb et al, 1984; Berridge, 1984;
Berridge and Irvine, 1984). Other phosphorylated derivatives (e.g., inositol tetraphosphate) have been described recently in brain (Irvine et al,
1986); their role is still unknown.

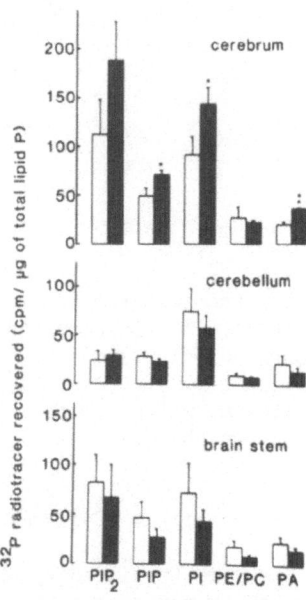

FIG. 7: Effect of bicuculline on phospholipid labeling in brain areas.
Rats were treated with ^{32}P and bicuculline (black bars) or vehicle
(open bars). Rats were sacrificed and different brain areas were
collected and phospholipids analyzed for incorporation of radiotracer. Data are from one experiment (n = 4, mean $\pm$ S.E.M.) and
are expressed as cpm recovered/mg of total lipid phosphorus.
Bicuculline-treated animals yielded data significantly different
from control, *p $\leq$ 0.04 and **p $\leq$ 0.0001. (Reprinted with permission
from Van Rooijen el al, 1986.)

Experimental Status Epilepticus Enhances Polyphosphoinositide Metabolism in
Rat Brain

    We have studied the turnover of phosphoinoisitides in brain during
bicuculline-induced seizures (Van Rooijen et al, 1986). In these studies,
rat brain phosphoinositides were radiolabeled in vivo by intraventricular
injection of either ^{3}H-myo-inositol or ^{32}P. Bicuculline was administered
(10 mg/kg, ip.) and rats were killed by head-focused high power microwave
irradiation five minutes later, during status epilepticus. Cerebrum was
removed and phosphoinositides and water-soluble inositol phosphates were
extracted and analyzed. Bicuculline-induced seizures resulted in a 50%
increase in ^{32}P-labeling of phosphatidic acid and phosphatidylinositol, as
compared to vehicle controls. ^{32}P-labeling of phosphatidylinositol 4-phos-
phate and phosphatidylinositol 4,5-bisphosphate also increased 24% and 36%,
respectively (Fig. 7). Other lipids were unchanged. Water-soluble inositol phosphates were analyzed by high voltage paper electrophoresis. Bicuculline treatment caused a 24% increase in the amount of radiotracer (^{3}H-
myo-inositol) recovered in inositol 1,4-bisphosphate, and a 50% increase
in the labeling of inositol 1,4,5-trisphosphate. Most of the effects of
bicuculline on phosphoinositide metabolism were reduced by pretreatment with
atropine, a muscarinic cholinergic antagonist. These data indicate that the
inositide cycle in brain is accelerated during status epilepticus, which may
account for increases in cerebral diacylglcyerols and arachidonic acid. The
effects of bicuculline on the inositide cycle may be mediated in part through
activation of muscarinic receptors.

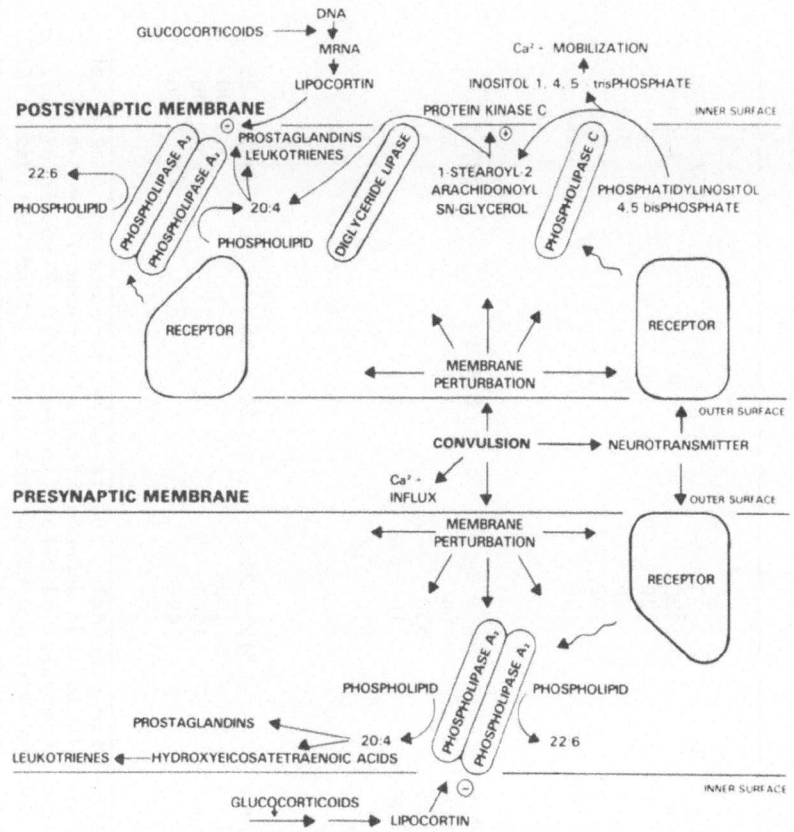

FIG. 8: Working models of reactions at the pre- and post-synaptic membrane affected by convulsions. In each membrane, biochemical events at each leaflet (inner and outer) are depicted.

These data support the hypothesis that seizures cause the activation of phospholipase C. This mechanism implies that upon receptor activation intracellular messengers are generated. One messenger is diacylglycerol, a regulator of protein kinase C. The other messenger is inositol-1,4,5-tris-phosphate, which ionizes intracellular calcium. This cell signaling system (Fig. 8) seems to be postsynaptic (Agranoff and Fisher, 1982), since the enhanced phosphoinositide turnover during bicuculline-induced status epilepticus is antagonized by atropine. Increases in brain diacylglycerols, enriched in arachidonate and stearate, accumulate in brain during electroconvulsive shock (Aveldano and Bazan, 1979) and in bicuculline-induced seizures (Rodriguez de Turco et al. 1983). Because phosphoinositides are predominately the stearoyl-arachidonoyl species (Baker and Thompson, 1972), their degradation could account for the accumulation of diacylglycerols. A sizable portion of the free fatty acids may arise by deacylation of diacylglycerols.

The coupling of receptors to phosphoinositide hydrolysis may be altered during sustained convulsions. Overstimulation of the phosphoinositide hydrolysis may result in a down-regulation of muscarinic receptors as an adaptive response. Muscarinic cholinergic receptors in brain are complex and unevenly distributed among different regions (Birdsall et al, 1980; Birdsell and Hulme, 1983; Brown and Brown, 1984). In kindled seizures,

TABLE 1: Apparent Rates of Accumulation of Free Fatty Acids and Diacylglycerols and Loss of Phosphatidylinositol 4,5-Bisphosphate in Rat Brain after a Single Electroconvulsive Shock

	16:0	18:0	18:1	20:4	22:6	TOTAL
Free Fatty Acids						
Control[a]	37 ± 2,4[b]	49 ± 3.5	28 ± 2.8	33 ± 3.2	10 ± 1	157 ± 9.3
2[c]	990[d]	1230	480	1380	150	4260
2-5[c]	-[f]	240	60	220	100	380
5-10	-	36	-	96	100	380
10-20	-	162	30	192	30	390
20-30	-	156	48	156	6	348
Diacylglycerols						
Control	56 ± 4	103 ± 6	52 ± 3	96 ± 2	8 ± 1	322 ± 17
2	80	240	60	80	60	520
2-5	180	360	150	630	60	1320
5-10	48	276	108	216	12	576
10-30	-	-	-	-	-	-
Phosphatidylinositol 4,5-bisphosphate						
Control	46 ± 6	446 ± 14	nd	310 ± 21	22 ± 34	822 ± 25
2	40	870	-	150	-	1590
2-5	240	1520	-	1300	-	2590
5-10	-	996	-	996	-	816
10-30	-	-	-	-	-	-

[a]Represents control animals killed by microwave irradiation (two seconds). [b]Units of measurement are nmol/g wet tissue weight. [c]Represents animals killed by microwave irradiation two seconds after electroconvulsive shock. [d]Values are nmol/min/g wet weight measured between the time intervals indicated (in seconds). [e]Time intervals, in seconds, used for calculation of apparent rates. [f]Dash indicates return to control levels. Abbreviations: 16:0, palmitic acid, 18:0, stearic acid, 18:1, oleic acid, 20:4, arachidonic acid, 22:6, docosahexaenoic acid, nd, not detectable.

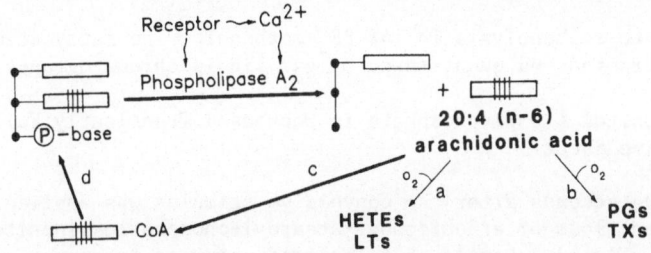

FIG. 9: Working hypothesis for the mechanism of release of arachidonic
acid due to seizures, via calcium-sensitive, receptor-linked phospholipase A_2. The source of arachidonic acid is not known, but is
likely to be a glycerolipid with a saturated fatty acid (open bar)
in the 1-position, arachidonic acid (hatched bar) in the 2-position
and a phosphate-base group in the 3-position of the glycerol backbone. This lipid may be a pool of phosphatidylinositol not directly engaged in the inositol cycle. The fate of free arachidonic
acid is designated by a) lipoxygenase; b) cyclooxygenase; c) acyl
CoA synthetase; d) reacylation. PG, prostaglandins; TXs, throm-
boxanes, HETEs, hydroxyeicosatetraenoic acids; LTs, leukotrienes.

down-regulation of muscarinic cholinergic receptors which is not a result of
increased levels of acetylcholine, occurs as a consequence of repeated seizures, but not as a consequence of kindling per se (Dasheiff et al, 1981,
1982; Dasheiff and McNamara, 1980; Savage and McNamara, 1982). Moreover,
activation of cholinergic receptors during sustained seizures damages brain
by unknown mechanisms (Olney et al, 1983). Cholinergic agents induce seizures and brain damage in lithium-treated rats (Honchar et al, 1983), further supporting a role for the phosphoinositides in brain damage, since
lithium inhibits inositol-1-phosphatase (Fig. 6) and may have other effects
on the phosphoinositide cycle (Berridge et al, 1982).

Electroconvulsive Shock Stimulates Phosphoinositide Metabolism

Although the diacylglycerol pool arising from phosphoinositide degradation has been proposed to give rise to free arachidonic acid, it is not
clear that in vivo polyphosphoinositides selectively generate diacylglycerols as a result of cell stimulation and that, in turn, the deacylation
of this lipid results in accumulation of free arachidonic acid. The aim
of recent experiments was to quantitate the fatty acid content of phosphoinositides, diacylglycerols and free fatty acids in the rat cerebrum as a
function of time after electroconvulsive shock. This design allows direct
testing of the hypothesis that accumulation of stearic and arachidonic acids
and diacylglycerols in stimulated brain is the result of enhanced operation
of the phosphoinositide cycle.

For administration of electroshock, platinum needle subdermal electrodes were implanted in anesthetized rats and the animals were placed in
plastic restrainers. Electroshock was given at a stimulation rate of 150
pps at 130 V and 750 msec of train duration. Rats were killed by microwave
irradiation. To sample the early time points after electroconvulsive shock,
the animals were given the stimulus after being placed in the microwave
instrument. For all other time points, the shock was given within the plastic restrainer outside of the microwave, and then the animals were killed.
Control animals were handled in an identical manner, except that no current
was applied to the electrodes. Cerebrum was removed, and lipids were
extracted by standard procedures (for recent review of lipid methodology,
see Bazan et al, 1987). Individual lipid bands isolated from TLC plates

were subjected to methanolysis in 14% BF_3 methanol. The fatty acid methyl esters were extracted and quantitated by gas liquid chromatography.

Phosphatidylinositol 4,5-Bisphosphate is Decreased Transiently Following Electroconvulsive Shock

Within five seconds after the convulsive stimulus was applied, there was a significant loss of arachidonoyl-stearoyl-phosphatidylinositol 4,5-bisphosphate (Table 1). Levels of this lipid returned to control levels by one minute. The loss of palmitic and docosahexaenoic acid from phosphatidylinositol 4,5-bisphosphate was negligible; oleic acid was not detectable in this lipid pool. The other phosphoinositides (data not shown) tended to decrease after electroconvulsive shock, although these decreases were not statistically significant.

Diacylglycerols Accumulate Concurrently with Loss of Phosphoinositides

Although at two seconds after electroconvulsive shock there was an accumulation of diacylglycerol, it was statistically significant only after five seconds, with the largest change in the stearoyl-arachidonoyl species (Table 1). The levels of diacylglcyerols reached a maximum by 30 seconds and returned to control values by one minute. This time course was parallel to the loss of arachidonoyl-stearoyl-phosphatidylinositol 4,5-bisphosphate and could indicate that stimulation of the phospholipase C-mediated degradation of phosphoinositide results in the accumulation of diacylglycerols.

Time Course of Fatty Acid Accumulation after Electroconvulsive Shock

There was a significant accumulation of free fatty acids within two seconds after a single electroconvulsive shock, increasing over the first 30 seconds, starting to decrease by one minute and reaching control values by five minutes (Table 1). Stearic and arachidonic acids were accumulated more rapidly than other fatty acids with the fastest apparent rates of accumulation taking place during the first two seconds, thus slightly preceding the maximum period of phosphoinositide degradation and diacylglycerol accumulation.

This study shows that within five seconds after electroconvulsive shock, phosphoinositides, as measured by fatty acid content, decrease in rat cerebrum. Most of this change is due to the loss of arachidonic and stearic acids. At the same time, arachidonoyl-stearoyl-diacylglycerols and free arachidonic and stearic acids accumulate. The accumulation of free fatty acids occurs slightly earlier than the accumulation of diacylglycerols, suggesting that at least part of the free fatty acid pool arises via the action of phospholipases A_1 and A_2 on phospholipids other than the phosphoinositides (Fig. 9). The accumulation of docosahexaenoic and palmitic acids, which cannot be accounted for by phosphoinositide breakdown, supports this mechanism. The accumulation of stearic and arachidonic acids in the free fatty acid and diacylglycerol pool is concurrent with the loss of phosphatidylinositol 4,5-bisphosphate, five seconds after electroshock. At this point, accumulation of these fatty acids may be via phospholipase C-mediated cleavage, and further deacylation of diacylglycerols by diacylglycerol lipase and monoacylglycerol lipase. Phospholipase A_1 and A_2 activation may occur at the presynaptic membrane in an event involved in neurotransmitter release, and phospholipase C activation may occur at the postsynaptic membrane as a consequence of the released neurotransmitters (Fig. 8).

Polyunsaturated Fatty Acids in the Stimulated Brain

Seizures occur in the epilepsies, are commonly the result of head

TABLE 2: Effect of Bicuculline on Free Fatty Acids in Synaptosomes or
Microsomes from Rat Cerebrum.

Fraction	Free arachidonic acid (% control)	Total free fatty acids (% control)
Synaptosomes	139.9	118.9
Microsomes	98.6	93.9

Ventilated rats were treated with bicuculline (10 mg/kg, i.p.), sacrificed
and the subcellular fractions were isolated. Free fatty acids were separated from total lipid extracts by gradient-thickness thin-layer chromatography and quantitated by gas chromatography, using nonodecanoic acid as an
internal standard (Reprinted with permission from Bazan et al, 1986).

trauma, injury at birth, and stroke, and are sometimes associated with
tumors. They are a symptom of damage to the brain, and can cause neuronal
damage, as when status epilepticus occurs. Recent studies show changes in
membrane lipids associated with the development of epileptic seizures, as
well as the pathogenesis of epileptic brain damage. For example, seizures
associated with cortical injections of iron (a model of post-traumatic epilepsy) can be prevented with antioxidants which block peroxidation of lipids
(Willmore et al, 1978; Willmore and Rubin, 1981). Seizures cause a prompt
and dramatic increase in free fatty acids, particularly arachidonic acid
(Bazan, 1970, 1971a; Bazan and Rakowski, 1970; Seisjo et al., 1982). Arachidonic acid, in addition to being the precursor of potent metabolites
(prostaglandins, hydroxyeicosatetraenoic acids and leukotrienes), can also
give rise to oxygen radicals through lipid peroxidation; these changes may
play a direct role in neuronal pathology (Marion and Wolfe, 1978; Bazan,
1976; Bazan et al, 1982a, 1982b, 1983; Rodriguez de Turco et al, 1983;
Collins and Olney, 1982; Bazan and Rodriguez de Turco, 1980).

The seizure-induced changes in membrane lipids have been correlated
in time with the events of a single seizure where there is no neuropathology (electroconvulsive shock model), and with sustained seizures that lead
to damage (bicuculline-induced status epilepticus model). The observed
changes in free fatty acids provide insight into the relationship between
membrane lipids and the susceptibility of brain regions to neuropathology.
Changes in synaptic lipids during convulsions may contribute causally to
the phase of postictal depression, where synaptic function is depressed.
Moreover, the synapse, particularly the postsynaptic dendrite, is the
structure most sensitive to seizure-induced damage (Collins and Olney, 1982;
Collins et al, 1983a, 1983b; Wong and Prince, 1979; Traub and Llinas, 1979;
Ribak and Reiffenstein, 1982; Schwartzkroin and Pedley, 1979).

Seizure-induced influx of calcium directly due to depolarization (pre-
and postsynaptic), or coupled to postsynaptic effects of neurotransmitters,
activates phospholipase A_2, which leads to release of arachidonic acid (Fig.
9). The changes in free fatty acids due to seizures may reflect an overstimulation of normal retailoring of fatty acids in phospholipids of excitable
membranes mediated by deacylation-reacylation mechanisms. Recent studies
in our laboratory have shown that bicuculline-induced accumulation of free
arachidonic acid occurs specifically in the synaptosomal fraction of rat
brain, and not in the microsomes, a fraction enriched in nonexcitable mem-

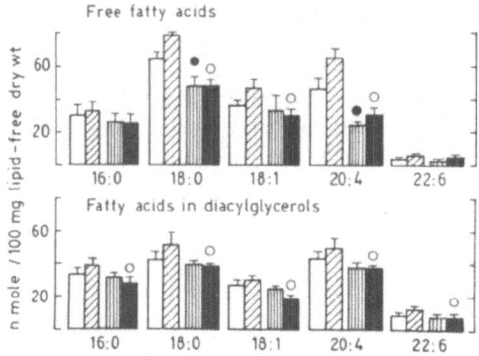

FIG. 10: Effects of electroconvulsive shock on the content of free fatty
acids and diacylglycerols in mouse brain. Actions of α-methyl-p-
tyrosine. A closed circle depicts significant difference between
basal levels of controls (⌐) and drug-treated (Ⅷ) animals. An
open circle indicates significant differences between groups of
animals subjected to electroconvulsive shock with (■) or without
(▨) drug-treatment. Drug-treated animals received 80 mg/kg of
α-methyl-p-tyrosine ip. six hours prior to the experiment (Reprinted with permission from Bazan et al, 1980).

branes (Table 2). Differences in free fatty acids among gross brain regions
(Ginobili et al, 1986) and rates of release that are of the same order of
magnitude as those observed during maximal activation of lipolytic hormones
support this hypothesis. Also the antagonism of electroshock-induced accumulation of free fatty acids by pretreatment with α-methyl-p-tyrosine provides strong evidence that turnover of fatty acids is linked to neurotransmission (Fig. 10, Aveldano and Bazan, 1979). The exact source and subcellular origin of free fatty acids are difficult to determine, because the
mass of acyl groups remaining in brain lipids is considerably larger.
Phospholipases A_1 and A_2 have been studied in subcellular fractions from
cerebral cortex or hypothalamus by adding radioactive substrates (Bazan,
1971b), but this ex vivo method has not been useful for the study of seizure-induced changes in lipids. Using [^{14}C]arachidonic acid administered
in vivo to label metabolically active pools, the relationship of alterations
in phospholipid and fatty acid metabolism can be studied. Radiolabeled
fatty acids are incorporated readily into brain lipids, and into pools that
are affected by convulsive stimuli and ischemia. Fig. 11 (Pediconi et al,
1986) shows the effects of electroconvulsive shock and post-decapitation
ischemia on the removal of intracerebrally injectd [^{14}C]arachidonic acid
from the free fatty acid pool in rat brain. Both decreased acylation and
increased deacylation caused by the stimuli could account for the observed
effects.

Role of Phospholipase A_2 in Polyunsaturated Fatty Acid Accumulation

Many enzymes, including diacylglycerol lipases, acyl-CoA synthetase
and in particular, phospholipase A_2 may be involved in the accumulation of
polyunsaturated fatty acids following cerebral stimulation or trauma. All
of these enzymes occur in brain (Bazan, 1971b; Woelk and Porcellati, 1973;
Cabot and Gatt, 1978; Reddy and Bazan, 1983). Recently, an endogenous protein, lipomodulin, which is an inhibitor for phospholipase A_2, has been
isolated from rabbit neutrophils (Hirata et al, 1980). The synthesis of
lipomodulin is induced by glucocorticoids. Previously, it has been reported
that anti-inflammatory steroids such as dexamethasone, can induce the bio-

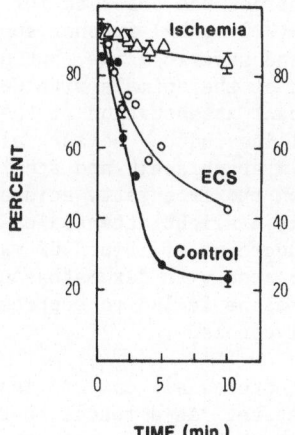

FIG. 11: Effect of a single electroconvulsive shock and postdecapitation
ischemia on the uptake of intracerebrally injected $[1-{}^{14}C]$-ara-
chidonic acid. Values represent the percent of radiolabel in the
free fatty acid fraction as a function of time. (Reprinted with
permission from Bazan et al, 1983a).

synthesis of a phospholipase A_2 inhibitor which prevents the synthesis of
prostaglandins, metabolic products of arachidonic acids (Flower and Blackwell, 1979). Furthermore, dexamethasone is effective in the treatment of
brain edema (Fishman, 1982). Thus, we have used dexamethasone as a tool to
elucidate a possible mechanism for free fatty acid release during stimulation of the brain by seizures, and damage to the brain by cryogenic edema.
Our working hypothesis is that pretreatment with dexamethasone can inhibit
phospholipase A_2 activity, which would in turn decrease the quantity of free
fatty acids released by stimulation.

Rats were pretreated with either saline or dexamethasone at various
doses and times prior to sacrifice. At the time of the experiment, the
experimental groups were injected with bicuculline (10 mg/kg ip.) and sacrificed by decapitation four minutes after the injection when status epilepticus had developed. Control animals were given saline. Alternatively,
a cryogenic lesion was made by applying a liquid N_2-cooled steel rod to the
exposed right parietal bone for one minute. Edema was estimated by determining % dry weight (Politi et al, 1985). The heads were immediately frozen
in liquid N_2, the cerebrum was removed, and tissue was weighed, pulverized
and the lipids were extracted. In some experiments, animals were killed by
head-focused microwave irradiation. Free fatty acids and diacylglycerols
were analyzed by gas liquid chromatography.

Dexamethasone pretreatment caused a slight attenuation of the basal
free fatty acid level, particularly arachidonic acid. Bicuculline-induced
seizure caused a dramatic increase in the total free fatty acids, and a more
than four-fold increase in arachidonic and stearic acids. Pretreatment with
dexamethasone caused a significant attenuation of the seizure-induced rise
in free fatty acids. The total amount of free fatty acid increased from 100
to 266 pmol/ug lipid P upon seizure, while pretreatment with dexamethasone
suppressed the elevation of free fatty acids to 144 pmol/ug lipid P. The
degree of suppression was dose-dependent and also exhibited a certain amount
of specificity towards the polyunsaturated fatty acids and stearic acid.
When the animals were pretreated with dexamethasone (1.25 mg/kg for two
days), there was a 30% decrease in the seizure-induced elevation of arachidonic acid. Arachidonic acid was lowered from 81 pmol/ug lipid P to 50

pmol/ug lipid P. A similar effect was observed for stearic and docosahex-aenoic acids. A higher dose (2.0 mg/kg) further suppressed the seizure-induced rise in arachidonic and stearic acids, but not in docosahexaenoic acid. Prolonged pretreatment of the animals with dexamethasone (3 days) did not show any further significant attentuation of the seizure-induced change in free fatty acids. Bicuculline-induced seizure also caused an increase in diacylglycerols enriched in arachidonic and stearic acids. However, in contrast with the results from the free fatty acid analysis, pretreatment with dexamethasone caused only a slight attentuaion of the seizure-induced rise in diacylglycerols. A decrease of about 40% was observed only with those animals that were pretreated with dexamethasone at a dose of 1.25 mg/kg. Higher doses of dexamethasone failed to suppress further the seizure-induced elevation in diacylglycerols.

Microwave irradiation denatures all enzymes instantaneously, resulting in sealing of the metabolic state. As a result, post mortem changes can be minimized. The basal level of the total free fatty acids in microwaved brain was about 30 pmol/ug lipid P with 29.7%, 9.7% and 10% of stearic, ara-chidonic and docosahexaenoic acids, respectively. Bicuculline-induced sei-zures caused a three-fold increase in total free fatty acids, which included a 12-fold increase in arachidonic acid. Dexamethasone pretreatment (1.25 mg/kg) resulted in a slight attenuation of the basal free fatty acid level, and after seizure there was about a 30% decrease in both arachidonic and stearic acids. The degree of suppression of the seizure-induced rise in free fatty acids was in direct agreement with that obtained from liquid N_2-fixed brains. Dexamethasone caused no significant change in diacylgly-cerols.

Cryogenic lesion also caused an increase in free fatty acids and poly-unsaturated diacylglycerols. Twenty-four hours after the lesion, maximal changes in both water content and cerebral lipid metabolism were observed. Total free fatty acids were increased nine-fold and diacylglycerols were increased three-fold in the injured hemisphere. In the contralateral hemi-sphere, where edema was also present, only diacylglycerols accumulated (50% increase). Within 48 hours, prior to the resolution of the edema, lipid levels had returned to normal. Dexamethasone pretreatment (1.25 mg/kg, every 12 hours for two days prior to the experiment) completely inhibited fatty acid accumulation in the lesioned hemisphere and decreased diacylgly-cerol accumulation by 30% in both hemispheres. Dexamethasone also attenu-ated edema formation. Interestingly, the degree of edema correlated more closely with changes in diacylglycerols than with changes in free fatty acids (Politi et al, 1985).

Dexamethasone is a glucocorticoid which possesses potent inhibitory effects on phospholipase A_2. Hirata et al (1980) have proposed that gluco-corticoids bind with surface receptors that can initiate the synthesis of mRNA. In turn, a signal is directed for the synthesis of a protein inhibi-tor, lipomodulin, with molecular weight of about 40 K. Lipomodulin binds with the enzyme matrix of phospholipase A_2 and renders the enzyme inactive. Our results indicate that pretreatment with dexamethasone causes a signifi-cant decrease in the amount of free fatty acids generated due to seizure or cryogenic edema. The effect was more pronounced with the arachidonate spe-cies. These results further substantiate the hypothesis that phospholipase A_2 is involved in releasing free fatty acids upon cerebral stimulation (Bazan, 1970). It is interesting that dexamethasone cannot completely block the free fatty acid release upon seizure. This is in contrast with cryogenic injury where dexamethasone causes a complete blockage of the fatty acid release (Politi et al, 1985). This indicates that during seizures, enzyma-tic processes other than phospholipase A_2 are involved in fatty acid release. One possible process could be the inhibition of the acyl CoA syn-thetase. Active arachidonoyl-CoA synthetase is present in the brain (Reddy and Bazan, 1983).

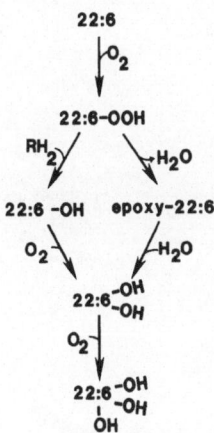

FIG. 12: Possible pathways involved in the synthesis of docosanoids
through lipoxygenation of docosahexaenoic acid (22:6). Hydro-
peroxy (22:6-OOH) and epoxy-22:6 intermediates are proposed,
based on the mechanism for the lipoxygenation of arachidonic
acid. (Reprinted with permission from Birkle and Bazan, 1986.)

Microwave irradiation inactivates all enzymes instantaneously, thus
eliminating any post-mortem metabolic changes. Because post mortem ischemic conditions can generate tremendous amount of free fatty acids in a very
short interval (Bazan, 1970), the large quantity of free fatty acids accumulated by ischemia could mask an effect of dexamethasone. In the experiment where the brain tissues were fixed by microwave irradiation, there was
no significant difference in the degree of suppression by dexamethasone of
the seizure-induced free fatty acid accumulation, as compared to liquid
N_2-fixed brain.

Dexamethasone showed only a slight modification of the increase in
diacylglycerols induced by seizures or cryogenic edema. This implies that
the enzymatic route through which diacylglycerides are released, the action
of phospholipase C on phosphoinositides, was not affected by dexamethasone
pretreatment.

Fate of Free Arachidonic Acid: Synthesis of Prostaglandins, Leukotrienes,
Hydroxyeicosatetraenoic Acids and Related Compounds

Enhanced neuronal activity or pathological insult triggers the accumulation of endogenous free arachidonic acid in the brain. It has been difficult to correlate these findings with the further oxidative metabolism of
arachidonic acid, although it has been found that eicosanoids are increased
in the brain during convulsions. Enhanced synthesis of prostaglandins
occurs after electroshock (Zatz and Roth, 1975) or pentylenetetrazol-induced
seizures (Marion and Wolfe, 1978) and in synaptosomes during bicuculline-
induced status epilepticus (Bazan et al, 1986). Little is known about the
regulation of prostaglandin formation during convulsions, although it has
been suggested that prostaglandins may be endogenous anticonvulsants (Forstermann et al, 1982). Data are lacking about prostaglandins and lipoxygenase reaction products during convulsions in brain regions. Prostaglandin
synthesis is altered in ischemia (Crockard et al, 1982; Ianotti et al, 1981;
Gaudet and Levine, 1979), hypoxia (Spagnuolo et al, 1979: Gaudet et al,
1980), and head injury (Ellis et al, 1981), and leukotrienes are increased
during recirculation after ischemia (Moskowitz et al, 1983). The lipoxy-

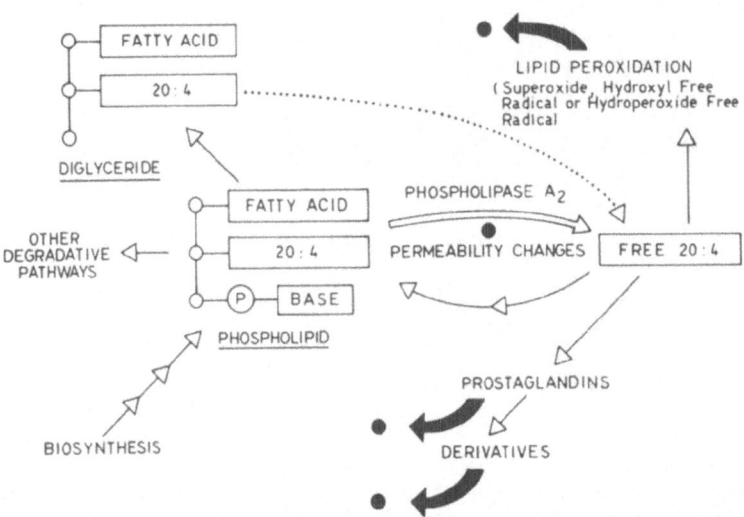

FIG. 13:   Possible pathways involved in the metabolism and fate of arachidonic acid in the central nervous system. A phospholipid is
represented as a source of free arachidonic acid. The double
arrow indicates the possible rate-limiting step catalyzed by
phospholipase A_2. Arachidonoyl-diacylglycerols may also be produced from phospholipids and in turn, might release free arachidonic acid. Black circle indicate possible sites of effects
(Reprinted with permission from Bazan and Rodriguez de Turco,
1980.)

genase pathway is very active in the central nervous sytem (Birkle and
Bazan, 1984a, 1984b; Bazan et al, 1984; Adesuyi et al, 1986; Lindgren et al,
1984) and is altered in synaptosomal fractions during convulsions (see
below, Bazan et al, 1986). Convulsion-induced accumulation of arachidonic
acid and prostaglandins occurs in ventilated and well-oxygenated rats (Forstermann et al, 1983; Steinhauer et al, 1979), hence, these changes are
caused by increased neuronal activity, rather than complications such as
hypoxia or hypertension.

    We have used the retina as a brain slice model, and intravitreal injec-
tion of radiolabeled arachidonic acid to label lipid pools in vivo for the
study of eicosanoid synthesis. We have obtained evidence of a K^+-sensitive
pathway that leads to an increased accumulation of lipoxygenase reaction
products in the retina (Birkle and Bazan, 1984b). To confirm the association of this pathway with the synapse, synaptosomes were isolated from the
brain of rats that had been labeled in vivo by intraventricular injection of
radioactive arachidonic acid. Upon induction of status epilepticus by administration of bicuculline, accumulation of lipoxygenase reaction products
occured mainly in synaptosomal fractions and not in microsomal fractions
(Bazan et al, 1986). It is possible that sustained stimulation (e.g., in
status epilepticus or prolonged depolarization by K^+) overactivates remodel-
ing reactions that normally regulate the physicochemical properties of excitable membranes, and results in increased levels of both prostaglandins and
lipoxygenase reaction products.

Fate of Free Docosahexaenoic Acid: Synthesis of Docosanoids

    Docosahexaenoic acid also accumulates following cerebral ischemia and stimulation of neuronal activity. Recently we have reported that oxygenation of docosahexaenoic acid to docosanoids (Fig. 12, hydroxy-docosahexaenoic acids, HDHE) occurs in the retina (Bazan et al, 1984). It is of importance to determine whether oxygenation of docosahexaenoic acid occurs in brain, because of the high concentration of this fatty acid in synaptic membranes and the large accumulation that occurs after pathological stimulation. To study this question, rats were injected intraventricularly with $[1-^{14}C]$docosahexaenoic acid, treated 60 minutes later with 10 mg/kg (ip.) bicuculline, and killed by high power microwave irradiation during status epilepticus. Approximately 6% of the total radioactivity of the lipid extract was oxygenated metabolites of docosahexaenoic acid with properties of HDHEs. There was active production (20-40% of total labeling) of compounds somewhat more nonpolar that monohydroxy derivatives, which have been identified tentatively as hydroperoxides and may be intermediates in the conversion of 22:6 to hydroxy derivatives (Fig. 12). Preliminary evidence suggests an enhanced production of HDHEs in cerebrum from bicuculline-treated rats, as compared to vehicle controls. These studies suggest that the production of docosanoids, like the lipoxygenation of arachidonic acid, is under neuronal control.

CONCLUDING REMARKS

Our knowledge of the role of excitable membrane lipids in neuronal responsiveness is still in the early stages. There are two broad areas in which to focus research efforts. First is the role of alterations in membrane lipids in brain damage. Polyunsaturated fatty acids are extremely susceptible to lipid peroxidation, caused by the unregulated generation of free radicals. Free radicals arise in the brain from the extravasation of blood, uncoupling of electron transport, and as side products of cyclooxygenase and lipoxygenase reactions. Peroxidation of polyunsaturated fatty acids causes degradation of the structural integrity of cellular membranes. Furthermore, activation of lipases destroys membrane lipids directly (Fig. 13). The interesting observation that neonatal mammals and adult poikilotherms are less susceptible to ischemia-induced brain damage and are resistant to ischemia-reduced free fatty acid accumulation (Aveldano and Bazan, 1975b; Bazan, 1971c) points to a strong connection between alterations in lipid metabolism and cerebral damage. The second major area of research is the roles of membrane lipids as messengers and precursors of messengers in cell signaling systems. The use of neurotransmitter antagonists and agonists, calcium channel blockers, phospholipase inhibitors, antioxidants such as Vitamin E, pro-oxidants like heme and Fe^{2+}, and inhibitors of cyclooxygenase and lipoxygenase should provide important new insights into the physiological role of lipid-derived messengers.

ACKNOWLEDGMENTS

The authors gratefully acknowledge the support of National Institutes of Health research grant NS23002.

REFERENCES

Adesuyi, S. A., Cockrell, C. S., Gamache, D. A. and Ellis, E. F., Lipoxygenase metabolism of arachidonic acid in brain, J. Neurochem. 34:1331 (1986).

Agranoff, B. W. and Fisher, S. K., Stimulated phospholipid labeling in nerve ending preparations: Studies on localization and biochemical mechanism, in: "Phospholipids in the Nervous System", L. Horrocks, ed., Raven Press, New York (1982).

Aveldano, M. I. and Bazan, N. G., Rapid production of diacylglycerols enriched in arachidonate and stearate during early brain ischemia, J. Neurochem. 25:919 (1975a).

Aveldano, M. I. and Bazan, N. G., Differential lipid deacylation during brain ischemia in a homeotherm and a poikilotherm. Content and composition of free fatty acids and triacylglycerols, Brain Res. 100:99 (1975b).

Aveldano, M. I. and Bazan, N. G., Acyl groups, molecular species and labeling by ^{14}C-glycerol and ^{3}H-arachidonic acid of vertebrate retina glycerolipids, Adv. Exp. Med. Biol. 83:397 (1977).

Aveldano, M. I. and Bazan, N. G. Alpha-methyl-p-Tyrosine inhibits the production of free arachidonic acid and diacylglycerols in brain after a single electroconvulsive shock, Neurochem. Res. 4:213 (1979).

Baker, R. and Thompson, W., Positional distribution and turnover of fatty acids in phosphatidylcholine and phosphatidylethanolamine in rat brain in vivo, Biochim. Biophys. Acta 270:489 (1972).

Bazan, H. E. P., Ridenour, B., Birkle, D. L. and Bazan, N. G., Unique metabolic features of docosahexaenoate metabolism related to functional roles in brain and retina, in: "Molecular and Biochemical Pharmacology of Phospholipids in the Nervous System", L. Horrocks, L. Freysz, and G. Toffano, eds., Liviana Press, Italy (in press) (1986).

Bazan, N. G., Effects of ischemia and electroconvulsive shock on free fatty acid pool in the brain, Biochim. Biophys. Acta 218:1 (1970).

Bazan, N. G., Changes in free fatty acids of brain by drug-induced convulsions, electroshock and anesthesia, J. Neurochem. 18:1379 (1971a).

Bazan, N. G., Phospholipases A_1 and A_2 in brain subcellular fractions, Acta Physiol. Latino. Amer. 21:101 (1971b).

Bazan, N. G., Modifications in the free fatty acids of developing rat brain, Acta Physiol. Latino. Amer. 21:15 (1971c).

Bazan, N. G., 1976, Free arachidonic acid and other lipids in the nervous system during early ischemia and after electroshock, Adv. Exp. Med. Biol. 72:317 (1976).

Bazan, N. G. and Rakowski, H., Increased levels of brain free fatty acids after electroconvulsive shock, Life Sci. 9:501 (1970).

Bazan, N. G. and Rodriguez de Turco, E. B. Membrane lipids in the pathogenesis of brain edema: Phospholipids and arachidonic acid, the earliest membrane components changed at the onset of ischemia, Adv. Neurol. 28:197 (1980).

Bazan, N. G., Aveldano de Caldironi, M. I., Cascone de Suarez, G. D. and Rodriguez de Turco, E. B., Transient modifications in brain free arachidonic acid in experimental animals during convulsions, in: "Neurochemical and Clinical Neurology", L. Batistin, G. Hashim, and A. Lajtha, eds., Alan R. Liss, New York (1980b).

Bazan, N. G., Aveldano de Caldironi, M. I. and Rodriguez de Turco, E. B., Rapid release of free arachidonic acid in the central nervous system due to stimulation, Prog. Lipid Res. 20:523 (1982a).

Bazan, N. G., Morelli de Liberti, S. M. and Rodriguez de Turco, E. B., Arachidonic acid and arachidonoyl-diglycerides increase in rat cerebrum during bicuculline-induced status epilepticus, Neurochem. Res. 7:839 (1982b).

Bazan, N. G., Rodriguez de Turco, E. B. and Morelli de Liberti, S. G., Free arachidonic acid and membrane lipids in the central nervous system during bicuculline-induced status epilepticus, Adv. Neurol. 34:305 (1983).

Bazan, N. G., Morelli de Liberti, S. G., Rodriguez de Turco, E. B. and Pediconi, M. F., Free arachidonic and docosahexaenoic acid accumulation in the central nervous system during stimulation, in: "Neural Membranes", G. Y. Sun, N. G. Bazan, J. Y. Wu, G. Porcellati, A. Y. Sun, eds., Humana Press, New Jersey (1983a).

Bazan, N. G., Birkle, D. L. and Reddy, T. S., Docosahexaenoic acid (22:6, n-3) is metabolized to lipoxygenase reaction products in the retina,

Biochem. Biophys. Res. Comm. 125:741 (1984).

Bazan, N. G., Birkle, D. L. and Reddy, T. S., Biochemical and nutritional aspects of the metabolism of polyunsaturated fatty acids and phospholipids in experimental models of retinal degradation, in: "Retinal Degeneration: Contemporary Experimental and Clinical Studies," M. M. LaVail, G. Anderson, J. Hollyfield, eds., Alan R. Liss, Inc., New York (1985a).

Bazan, N. G., Reddy, T. S., Redmond, T. M., Wiggert B. and Chader, G. J., Endogenous fatty acids are covalently and non-covalently bound to interphotoreceptor retinoid-binding protein in the monkey retina, J. Biol. Chem. 260:13677 (1985b).

Bazan, N. G., Birkle, D. L., Wilson, T. and Reddy, T. S., The accumulation of free arachidonic acid, diacylglycerols, prostaglandins, and lipoxygenase reaction products in the brain during experimental epilepsy, Adv. Neurol. 44:879 (1986).

Bazan, N. G., Reddy, T. S. and Scott, B. L., Quantitative analysis of acyl group composition of brain phospholipids and neutral lipids, in: "Neuromethods," A. Boulton, ed., Humana Press, New Jersey, (in press) (1987).

Berridge, M. J. Inositol trisphosphate and diacylglycerols as second messengers, Biochem. J. 220:345 (1984).

Berridge, M. J. and Irvine, R. F., Inositol trisphosphate, a novel second messenger in cellular signal transduction, Nature 312:315 (1984).

Berrige, M. J., Downes, C. P. and Hanley, M. R., Lithium amplifies agonist-dependent phosphatidylinositol responses in brain and salivary glands, Biochem. J. 206:587 (1982).

Birdsall, N. J. M. and Hulme, E. C., Muscarinic receptor subclasses, Trends Pharm. Sci. 4:459 (1983).

Birdsall, N. J. M., Hulme, E. C. and Burgen, A., The character of muscarinic receptors in different regions of the rat brain, Proc. Roy Soc. Lond B. Biol. Sci. 207:1 (1980).

Birkle, D. L. and Bazan, N. G. Lipoxygenase and cyclooxygenase reaction products and incorporation into glycerolipids of radiolabeled arachidonic acid in the bovine retina, Prostaglandins 27:203 (1984a).

Birkle D. L. and Bazan, N. G., Effects of K^+ depolarization on the synthesis of prostaglandins and hydroxyeicosatetra(5,6,11,14)enoic acids (HETE) in the rat retina. Evidence for esterification of 12-HETE in lipids, BIochim. Biophys. Acta 795:564 (1984b).

Birkle, D. L. and Bazan, N. G., The arachidonic acid cascade and phospholipid and docosahexaenoic acid metabolism in the retina, Prog. Retinal Res. 5:309 (1986).

Brown, J. H. and Brown, S. L., Agonists differentiate muscarinic receptors that inhibit cyclic AMP formation from those that stimulate phosphoinositide metabolism, J. Biol. Chem. 259:3777 (1984).

Cabot, M. C. and Gatt, S., The hydrolysis of triacylglycerol and diacylglycerol by a rat brain microsomal lipase with an acidic pH, Biochim. Biophys. Acta 530:508 (1978).

Collins, R. C. and Olney, J. W., Focal cortical seizures cause distant thalamic lesions, Science 218:117 (1982).

Collins, R. C., Olney, J. W. and Lothman, E. W., Metabolic and pathologic consequences of focal seizures, in: "Epilepsy," A. A. Ward, J. K. Penry, eds., Raven Press, New York, (1983a).

Collins, R. C., Lothman, E. W. and Olney, J. W., Status epilepticus in the limbic system: Biochemical and pathological changes, Adv. Neurol. 34: 277 (1983b).

Cook, H. W., Clarke, J. T. R. and Spence, M. W., Involvement of triacylglycerols in the metabolism of fatty acids by cultured neuroblastoma and glioma cells, J. Lipid Res. 23:1292 (1982).

Crockard, H. A., Bhakoo, K. K. and Lascelles, P. T., Regional prostaglandin levels in experimental ischemia, J. Neurochem. 38:1311 (1982).

Dashieff, R. M. and McNamara, J. O., Evidence for an agonist independent down regulation of hippocampal muscarinic receptors in kindling, <u>Brain Res</u>. 195:345 (1980).

Dasheiff, R. M., Byrne, M. D., Patrone, V. and McNamara, J. O., Biochemical evidence of decreased muscarinic cholinergic neuronal communication following amygdala kindled seizures, <u>Brain Res</u>. 206:233 (1981).

Dashieff, R. M., Savage, D. D. and McNamara, J. O., Seizures down-regulate muscarinic cholinergic receptors in hippocampal formation, <u>Brain Res</u>. 235:327 (1982).

Ellis, E. F., Wright, K. F., Wei, E. P. and Kontos, H. A., Cyclooxygenase products of arachidonic acid metabolism in cat cerebral cortex after experimental concussive brain injury, <u>J. Neurochem</u>. 37:892 (1981).

Fishman, R. A., Steroids in the treatment of brain edema, <u>New England J. Med</u>. 306:359 (1982).

Flower, R. J. and Blackwell, G. J., Anti-inflammatory steroids induce bio-synthesis of a phospholipase A_2 inhibitor which prevents prostaglandin generation, <u>Nature</u> 278:456 (1979).

Forstermann, U., Heldt, R., Friedhelm, K. and Hertting, G., Potential anti-convulsive properties of endogenous prostaglandins formed in mouse brain, <u>Brain Res</u>. 240:303 (1982).

Forstermann, U., Heldt, R. and Hertting, G., Increase in brain prostaglan-dins during convulsions is due to increased neuronal activity and not to hypoxia, <u>Arch. Int. Pharmacodyn. Ther</u>. 263:180 (1983).

Gaudet, R. J. and Levine, L., Transient cerebral ischemia and brain prosta-glandins, <u>Biochem. Biophys. Res. Comm</u>. 86:839 (1979).

Gaudet, R. J., Alam, I. and Levine, L., Accumulation of cyclooxygenase pro-ducts of arachidonic acid metabolism in gerbil brain during reperfusion after bilateral common carotid artery occulsion, <u>J. Neurochem</u>. 35:653 (1980).

Ginobile, M. S., Rodriguez de Turco, E. B. and Barrantes, F. J., Asymmetry of diacylglycerol metabolism in rat cerebral hemispheres, <u>J. Neurochem</u>. 46:1382 (1986).

Hirata, J., Schiffmann, E., Venkatasubramanian, K., Soloman, D. and Axelrod, J., Phospholipase A_2 inhibitory protein in rabbit neutrophils induced by glucocorticoid, <u>Proc. Natl. Acad. Sci. USA</u>, 77:2583 (1980).

Honchar, M. P., Olney, J. W. and Sherman, W. R., Systemic cholinergic agents induce injury by cholinergic neuronal communication following amygdala kindled seizures, <u>Brain Res</u>. 206:233 (1983).

Iannotti, F., Crockard, A., Ladds, G. and Symon, L., Are prostaglandin levels altered in experimental ischemic edema in gerbils, <u>Stroke</u> 12:301 (1981).

Ikeda, M., Yoshida, S., Busto, R., Santigo, M. and Ginsberg, M., Polyphospho-inositides as a probable source of brain free fatty acid accumulated at the onset of ischemia, <u>J. Neurochem</u>. 47:123 (1986).

Irvine, R. F., Letcher, A. J., Heslop, J. P. and Berridge, M. J., The inosi-tol tris/tetrakisphosphate pathway-demonstration of Ins(1,4,5)P_3 3-kinase activity in animal tissues, <u>Nature</u> 320:631 (1986).

Kikkawa, U., Takai, Y., Tanaka, Y., Miyake, R. and Nishizuka, Y., Protein kinase C as a possible receptor protein of tumor-promoting phorbol esters, <u>J. Biol. Chem</u>. 257:7841 (1982).

Lindgren, J. A., Hokfelt, T., Dahlen, S. E, Patrono, C. and Samuelsson, B., Leukotrienes in the rat central nervous system, <u>Proc. Natl. Acad. Sci. USA</u> 81:6212 (1984).

Marion, J. and Wolfe, L. S., Increase in vivo of unesterified fatty acids, prostaglandin $F_{2\alpha}$ but not thromboxane B_2 in rat brain during drug induced convulsions, <u>Prostaglandins</u> 16:99 (1978).

Moskowitz, M. A., Kiwak, K. J., Hekimian, K. and Levin, L., Synthesis of compounds with properties of leukotrienes C_4 and D_4 in gerbil brains after ischemia and reperfusion, <u>Science</u> 224:886 (1983).

Neuringer, M., Connor, W. E., Van Petten, C. and Barstad, L., Dietary omega-3 fatty acid deficiency and visual loss in infant rhesus monkeys, <u>J. Clin. Invest</u>. 73:272 (1984).

Olney, J. W., Gubareff, T. and Labruyere, J., Seizure-related brain damage in lithium-treated rats, Science 220:323 (1983).

Pediconi, M. F., Rodriguez de Turco, E. B. and Bazan, N. G., Reduced labeling of brain phosphatidylinositol, triacylglycerols and diacylglcyerols
by $[1-{}^{14}C]$ arachidonic acid after electroconvulsive shock. Potentiation of the effect by adrenergic drugs and comparison with palmitic
acid labeling, Neurochem. Res. 2:217 (1986).

Politi, L. E., Rodriguez de Turco, E. B. and Bazan, N. G., Dexamethasone
effect on free fatty acid and diacylglycerol accumulation during expe-
rimentally-induced vasogenic brain edema, Neurochem. Pathol. 3:249
(1985).

Reddy, T. S. and Bazan, N. G., Kinetic properties of arachidonoyl-coenzyme A
synthetase in rat brain microsomes, Arch. Biochem. Biophys. 226:125
(1983).

Reddy, T. S. and Bazan, N. G., Synthesis of docosahexaenoyl-, arachidonoyl-
and palmitoyl-coenzyme A in ocular tissues, Exp. Eye Res. 41:87 (1985a).

Reddy, T. S. and Bazan, N. G., Synthesis of arachidonoyl coenzyme A and doco-
sahexaenoyl coenzyme A in synaptic plasma membranes of cerebrum, cerebellum and brain stem of rat brain, J. Neurosci. Res. 13:381 (1985b).

Reddy, T. S., Sprecher, T. and Bazan, N. G., Long-chain acyl coenzyme A synthetase from rat brain microsomes: Kinetic studies using $[1-{}^{14}C]$doso-
sahexaenoic acid substrate, Eur. J. Biochem. 145:21 (1984).

Ribak, C. E. and Reiffenstein, R. J., Selective inhibitory synapse loss in
chronic cortical slabs: A morphological basis for epileptic susceptibility, Can. J. Physiol. Pharmacol. 50:864 (1982).

Rodriguez de Turco, E. B., Morelli de Liberti, S. and Bazan, N. G., Stimulation of free fatty acid and diacylglycerol accumulation in cerebrum and
cerebellum during bicuculline-induced status epilepticus, J. Neurochem.
40:252 (1983).

Savage, D. D. and McNamara, J. O., Kindled seizures selectively reduce a
subpopulation of $[{}^{3}H]$quinuclidinyl benzilate binding sites in rat dentate gyrus, J. Pharm. Exp. Therap. 222:670 (1982).

Seisjo, B. K., Ingvar, M. and Westerberg, E., The influence of bicuculline-
induced seizures on free fatty acid concentrations in cerebral cortex,
hippocampus, and cerebellum, J. Neurochem. 39:796 (1982).

Schwartzkroin, P. A. and Pedley, T. A., Slow depolarizing potentials in
"Epileptic" neurons, Epilepsia 20:267 (1979).

Spanguolo, C., Sautebin, L., Galli, G., Racagni, G., Galli, C., Mazzari, S.
and Finesso, M., $PGF_{2\alpha}$, thromboxane B_2 and HETE levels in gerbil brain
cortex after ligation of common cartoid arteries and decapitation, Prostaglandins 18:53 (1979).

Steinhauer, H., Anhut, H. and Hertting, G., The synthesis of prostaglandins
and thromboxane in the mouse brain in vivo: Influence of drug induced
convulsions, hypoxia and the anticonvulsants trimethadione and diazepam, Naunyn-Schmiedebergs Arch. Pharmacol. 310:53 (1979).

Streb, H., Bayerdorffer, E., Haase, W., Irvine, R. F. and Schulz, I., Effect
of inositol-1,4,5-trisphosphate on isolated subcellular fractions of
rat pancreas, J. Membrane Biol. 81:241 (1984).

Traub, R. D. and Llinas, R., Hippocampal pyramidal cells: Significance of
dendritic ion conductances for neuronal function and epileptogenesis,
J. Neurophysiol. 42:476 (1979).

Van Rooijen, L. A. A., Vadnal, R., Dobard, P. and Bazan, N. G., Enhanced
inositide turnover in brain during bicuculline-induced status epilepticus, Biochem. Biophys. Res. Comm. 136:827 (1986).

Willmore, J. L. and Rubin, J. J., Antiperoxidant pretreatment and iron-
induced epileptiform discharges in the rat: EEG and histopathologic
studies, Neurology 31:62 (1981).

Willmore, J. L., Sypert, G. W. and Munson, J. B., Recurrent seizures induced
by cortical iron injection: A model of post-traumatic epilepsy, Ann.
Neurol. 4:329 (1978).

Woelk, H. and Porcellati, G., Subcellular distribution and kinetics properties of rat brain phospholipase A_1 and A_2, Hoppe-Seyler's Z. Physiol.
Chem. 354:90 (1973).

Wong, R. K. S. and Prince, D. A., Dendritic mechanisms underlying pencillin-induced epileptiform activity, <u>Science</u> 204:1228 (1979).

Zatz, M. and Roth, R., Electroconvulsive shock raises prostaglandins F in rat cerebral cortex, <u>Biochem. Pharmacol</u>. 24:2101 (1975).

RECEPTOR-MEDIATED PHOSPHOINOSITIDE METABOLISM

Bernard W. Agranoff

Department of Biological Chemistry and Neuroscience
Laboratory Building, University of Michigan, Ann Arbor
MI 48104-1687

INTRODUCTION

In a scant few years, the ligand-stimulated turnover of phosphatidy-linositol (PI) has progressed from a curious observation to a major biochemical and pharmacological enterprise. In the 1940's, de Hevesy (see de Hevesy, 1964) introduced the use of beta-emitting radioisotopes to biochemistry, and demonstrated that the addition of ^{32}P-labeled inorganic phosphate to tissue preparations led to highly radioactive phospholipids. In 1953, Hokin and Hokin published their seminal observation that two minor phospholipids, phosphatidylinositol (PI) and phosphatidate (PA) were selectively labeled to high specific activities in such preparations (Hokin and Hokin, 1953). Furthermore, the labeling was greatly intensified by the presence of cholinergic ligands in the incubation medium, and the stimulated labeling could be reduced to the basal level by the addition of atropine, a known antagonist of muscarinic cholinergic receptors. This latter observation was important, since it demonstrated that the stimulated labeling is indeed receptor-linked. Hence their observation constituted a biochemical "handle" into the transduction process whereby a receptor-ligand interaction on the outer leaflet of the plasma membrane is converted to an intracellular response. Over the intervening years, it became apparent that a large number of receptor-ligand interactions could be coupled to stimulated PA and PI labeling, and that they formed a class distinct from receptor-ligand interactions linked to cyclic AMP. In 1975, Michell summarized the then known inositol lipid-linked neurotransmitters and hormones, noting that a common thread in all these studies was the induction of increased intracellular Ca^{2+} (Michell, 1975). It was initially thought that this Ca^{2+} came from extracellular sources. As indicated in the scheme shown in Fig. 1, it is presently held that the receptor-mediated stimulation of phospholipid turnover involves the initial breakdown of phosphatidylinositol bisphosphate (PIP_2) to generate two moieties, each of which is believed to have second messenger characteristics: diacylglycerol (DG), which activates protein kinase C (Nishizuka, 1985) and inositol 1,4,5-trisphosphate (IP_3), which mobilizes Ca^{2+} from the endoplasmic reticulum (Berridge, 1984). This review will emphasize chemical and biochemical aspects of two intermeshed metabolic cycles whereby PIP_2 is regenerated from DG on the one hand, and inositol recovered from inositol phosphates on the other. In addition, it will consider the possible physiological consequences of stimulated phosphoinositide turnover in the nervous system.

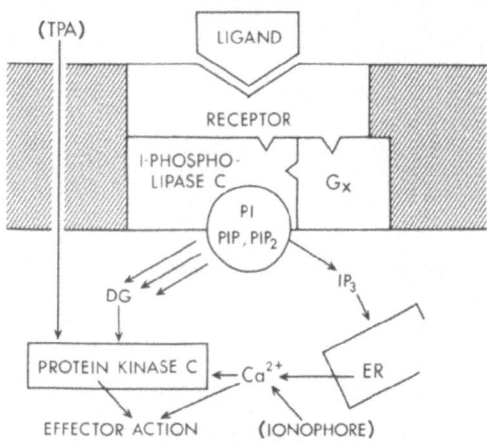

FIG. 1: Signal transduction of ligands that stimulate turnover of inositol
lipids. The ligand-receptor complex, presumably through the intervention of an integral membrane protein (G_x) stimulates the breakdown of the inositol lipids to DG and inositol phosphates. DG
then activates protein kinase C. In the case of PIP_2 breakdown,
IP_3 is also released. IP_3 releases Ca^{2+} from the endoplasmic
reticulum (ER). Phorbol esters such as TPA penetrate the cell
membrane and activate protein kinase C at the DG binding site.
Ionophore increases cytosolic Ca^{2+} from intra- and extracellular
sources.

Chemistry and Biochemistry of the Inositol Lipids

The structure of PI is shown in Fig. 2A. The DG moiety of isolated
PI is characterized by an enrichment in the 1-stearoyl, 2-arachidonoyl
sn-glycerol species. This enrichment holds for the other inositol lipids,
phosphatidylinositol phosphate (PIP), and PIP_2 as well. The sn-3 position
is phosphodiesteratically linked to myo-inositol. Note that in its chair
configuration, myo-inositol has 1 axial and 5 equatorial hydroxyl groups.
The molecule can be conveniently visualized as a turtle (2B) in which the
axial hydroxyl is the head. The right fore leg, labeled "1", is bound to
PA (i.e., DG·P). Looking down on the turtle from above and proceeding
counterclockwise, the 2 position is the axial hydroxyl represented by the
turtle's head, while the left fore leg, left hind leg, tail and right hind
leg become, respectively, the 3,4,5 and 6 positions. The myo-inositol
molecule is represented diagrammatically in Fig. 2C, wherein the open triangle represents the axial hydroxyl (the turtle's head) which projects out
of the plane of the paper toward the reader. There is an alternate chair
(2D) conformation, in which each equatorial hydroxyl becomes axial and conversely, the axial becomes equatorial. Filled triangles represent axial
hydroxyls that extend below the plane of the paper. In general, the favored
conformation of a cyclic polyol is the one that maximizes the number of
equatorial hydroxyls. Thus, while the two chair forms of myo-inositol are
freely interconvertible, the preponderant conformation is the one represented in Fig. 2C. The numbering system used here refers to the "D" convention. With L-numbering, the left front leg is 1, the axial hydroxyl
remains 2, the right fore leg is 3, etc.

There are nine hexahydroxycyclohexanes, and they were originally
described collectively as "inositols." To avoid confusion of this generic
term with that for the isomer depicted in Fig. 2, the myo-prefix was added

and the various isomers shown in Fig. 3 have been termed "cyclitols." In
those cyclitols in which there are 3 equatorial and 3 axial hydroxyls, the
alternate chairs are identical, i.e., the two structures can be superimposed.
An exception to this is allo-inositol, in which the alternate chairs form a
stereoisomer pair, as indicated in Fig. 3. Up to the present, only myo-inositol, by far the most commonly occurring cyclitol, has been found to occur
in the phosphoinositides (Posternak, 1965; Cosgrove, 1980).

There are still other conformations of the cyclitols, including "boat"
forms, skewed chairs, etc. Substitutions on the various hydroxyl positions
can be expected to alter the conformation, as might interaction of myo-inositol and its derivatives with receptors, enzyme sites, etc. In this
regard, it is of interest that the silkworm larvae has a sensillum that is
responsive to myo-inositol (Ishikawa, 1967). When we performed an electrophysiological structure-activity study with the nine cyclitols, only myo-
inositol and epi-inositol (and to a lesser extent, allo-inositol) were
active (Jakinovich and Agranoff, 1971). The results are suggestive that
the receptor binding site is complex and may bind to a strained conformation of the cyclitol.

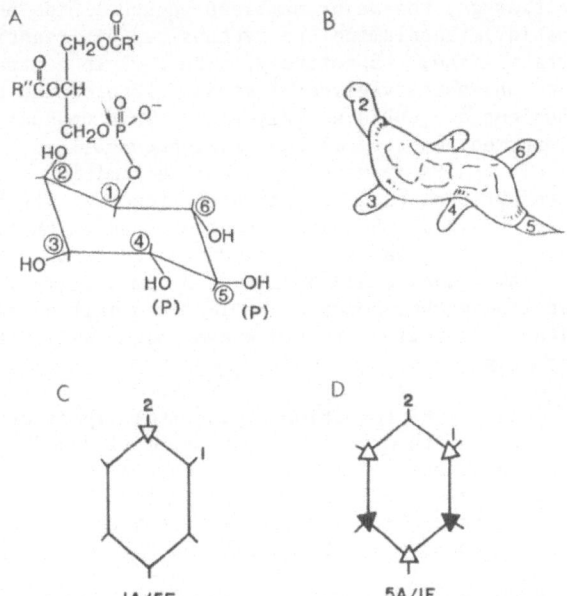

FIG. 2:   Structure of phosphatidylinositol (PI) and related lipids. PI is
          1,2-diacyl-sn-3-glycerophosphoryl D-1 myo-inositol. Additional
          phosphorylation at the 4' position, or at the 4' and 5' positions,
          yields phosphatidylinositol phosphate (PIP) and phosphatidylinosi-
          tol bisphosphate (PIP$_2$), respectively. Cleavage by a phospholipase
          C type phosphodiesterase (see arrow) results in formation of dia-
          cylglycerol (DG) and D-myo-inositol 1,4,5-trisphosphate. Myo-ino-
          sitol is conveniently visualized as a turtle (B), and the various
          inositol phospholipids can be considered to be derivatives in which
          DG is phosphodiesteratically linked to the right fore leg. Myo-ino-
          sitol is seen from above in C. The open triangle represents the
          axial hydroxyl at position 2. Fig. 2D shows the alternate chair
          form of myo-inositol in which there are 1 equatorial and 5 axial
          hydroxyls (see text). Filled triangles are directed below the
          plane of the paper.

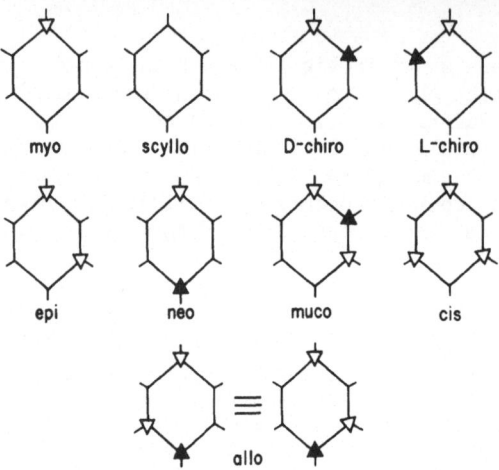

FIG. 3: The isomers of hexahydroxycyclohexane. Myo-inositol and its 8 isomers are represented diagramatically.(see Fig. 2).

Biosynthesis of Inositol Lipids

        In eukaryotic tissues, the major membrane phospholipids phosphatidyl-
choline and phosphatidylethanolamine are synthesized via reaction of CDP-
choline and CDP-ethanolamine, respectively, with DG. In contrast, the phosphatidylinositol and phosphatidylglycerol series are produced by reaction
of PA with a liponucleotide, CDP-diacylglycerol. This unusual molecule is
also a metabolic intermediate for all bacterial phospholipid syntheses.
Its formation from CTP and PA appears to be a rate-limiting step in PI biosynthesis, and since pyrophosphate is generated from CTP, its formation is
considered to be irreversible. However, the hydrolase which cleaves CDP-DG
into PA and CMP is present in bacterial and mammalian tissues (Raetz et al.,
1972; Rittenhouse et al., 1981), and 5'-AMP is a competitive inhibitor of
the enzyme obtained from either source. While the significance of the
inhibition is unclear, regulation of this enzyme might well play a role in
inositol lipid synthesis.

        Biosynthetic steps in the formation of inositol lipids are summarized
in Fig. 4. CDP-DG reacts with myo-inositol to form PI and CMP. The other
cyclitols and various available derivatives do not serve as phosphatidyl-
transferase acceptors. Galactinol, an inositol galactoside, inhibits the
reaction (Benjamins and Agranoff, 1969). Two kinase reactions employing
ATP next sequentially phosphorylate PI to PIP and thence to PIP_2.

        Myo-inositol can be derived from exogenous sources, since it is ubiqui-
tous in the diet. It can alternatively be biosynthesized by cyclization of
glucose 6-phosphate to D-myo-inositol 3-phosphate, followed by dephosphorylation to free inositol. Of the various body tissues, brain and testes have
high activities of the cyclizing enzyme (Eisenberg, 1967). This may reflect
the relative unavailability of inositol from the plasma to cells of these
organs because of impermeability.

        What then is the relationship of these biosynthetic mechanisms to sti-
mulated labeling of inositol lipids? It has been demonstrated that if the
incorporation of biosynthetic intermediates such as glycerol into phospholipid are measured under conditions in which ^{32}P incorporation is stimulated
by ligands such as carbachol, a concomitant increase in labeled glycerol
incorporation is not seen (Fisher and Agranoff, 1981). The stimulated
labeling is thus not due to increased de novo synthesis, but rather the
accelerated recycling of existing lipids. It was initially proposed that
PI breakdown was stimulated by ligand-receptor interaction, giving rise to

inositol monophosphate (IP_1) and DG (Durell et al., 1968). DG could then be converted to PA via the enzyme DG kinase, completing a loop. This idea was not quite correct, and it took many years to demonstrate that in fact it was PIP_2 breakdown that was initially stimulated (Agranoff et al., 1983; Berridge et al., 1983). PIP and PIP_2 have long been known as membrane components and in retrospect, it was probably for technical reasons that stimulated polyphosphoinositide breakdown was not discovered sooner. First, unless an acidic extraction of lipids is made, PIP and PIP_2 are not recovered in lipid solvents. Secondly, until about 10 years ago, separation techniques for PIP and PIP_2 were rather cumbersome. Thirdly, the monoester phosphates on the 4' and 5' position rapidly exchange via their kinases and phospholipid monoesterases. They thus quickly become highly labeled and this may obscure stimulated breakdown via the inositol-specific phosphodiesterase. In addition, stimulated breakdown with ^{32}P requires prelabeling of tissue preparations prior to addition of ligand. Under these conditions there remains a high specific activity of ATP in cells. Since the labeled lipids are rapidly regenerated, one must look at very brief time points, or transient decreases in PIP_2 will not be detected (Agranoff, et al., 1983). The phospholipid labeling cycle as currently envisioned is demonstrated in Fig. 5. While a common Ca^{2+}-activated phosphodiesterase can break down PI, PIP and PIP_2 (Wilson et al., 1984), it is generally believed that PIP_2 breakdown initiates the stimulated labeling cycle. It thus becomes apparent that experiments that use ^{32}P for stimulated labeling are measuring the restorative, rather than the initial phase of stimulated inositol lipid turnover.

A major advance in the elucidation of the stimulated cycling and related metabolism was the development by Berridge and associates of a method for the use of Li^+ and ^{3}H-inositol for lipid labeling (Berridge et al., 1982). It had been known for some time that Li^+ blocked the breakdown of inositol monophosphates to free inositol (Allison et al., 1976). By adding Li^+ in the presence of agonists such as carbachol, 5HT, etc., to tissues containing appropriate receptors, the accumulation of labeled inositol monophosphate could be used as a measure of the rate of cycling. As shown in Fig. 5, IP_3 formed by the cleavage of PIP_2 is converted to IP_2 by the action of a 5'-phosphohydrolase. IP_2 is in turn degraded by a phosphatase to IP_1, and in the presence of Li^+, is not further cleaved to free inositol. While the IP_3 released upon PIP_2 activation leads to

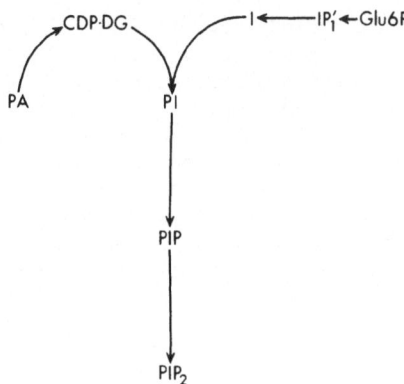

FIG. 4: Enzymatic steps involved in the biosynthesis of inositol lipids (see text).

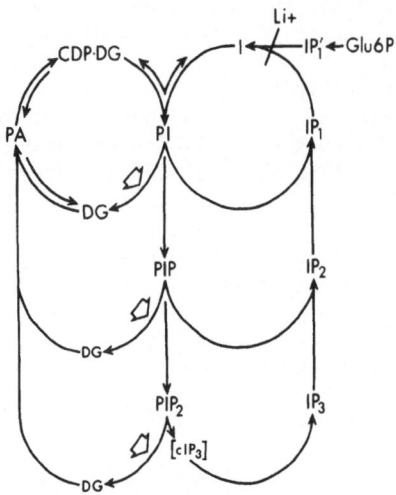

FIG. 5: Turnover of inositol lipids and of its hydrolysis products. Note
that the biosynthetic reactions in Fig. 3 are embedded in the left
hand cycle. DG released by phosphodiesteratic cleavage (arrows) of
either PI, PIP or PIP_2 gives rise to DG which can be converted via
DG kinase to PA. There is considerable evidence that in most, if
not all cells, ligand-activated cleavage is at the level of PIP_2.
On the right hand side, the water-soluble products of the phosphodiesteratic cleavage are shown. 1,4,5-IP_3 may form through the
intermediate formation of 1,2 cyclic IP_3. Sequential dephosphorylation of IP_3 to IP_2 and IP_1 and eventually to myo-inositol, followed by its reincorporation in PI completes this cycle. Note that
IP_2 and IP_1 can also form via direct phosphodiesteratic cleavage of
PIP or PI, respectively. In this diagram IP_1 represents D-myo-
inositol 1-phosphate and IP_1' is the 3-phosphate. The same enzyme
cleaves both isomers and is blocked by Li^+.

increased intracellular Ca^{2+}, the PIP_2 phosphodiesterase itself is a Ca^{2+}-
requiring enzyme. This is not a "chick and egg" problem: it appears that
sufficient Ca^{2+} for phosphodiesterase activation is present at the membrane.
Thus, in most instances, Ca^{2+} appears to be required for, but does not
regulate PIP_2 breakdown.

Diacylglycerol Metabolism

The left hand of the "bicycle" shown in Fig. 5 represents the enzymatic
steps whereby PIP_2 hydrolysis activated by receptor-ligand interaction leads
to DG, and the restorative steps that reconvert DG to PIP_2, constituting a
closed cycle. While PIP_2 breakdown appears to be an initial event, it is
possible in some tissues at least, that once intracellular Ca^{2+} has been
elevated as a result of IP_3 action, PIP and PI breakdown also occurs. Since
PI is present in membranes at significantly higher levels than PIP or PIP_2,
its breakdown could contribute significantly to DG production. While DG is
a common product of inositide breakdown, one might hope that the nature of
the inositol phosphate released would permit one to judge which of the three
inositides has been cleaved. However, IP_3 breaks down rapidly, so that it
is often difficult to assess the relative rates of PI, PIP and PIP_2 breakdown. Thus, an increase in IP_1 and DG following cell activation could
reflect increased breakdown of either of the three inositol lipids. In any
case, the DG produced will be enriched in 1-stearoyl, 2-arachidonoyl (ST/
AR)-DG. The DG released upon breakdown of inositides is generally held to
be linked to the activation of protein kinase C by DG (Nishizuka, 1984).

One might then anticipate selective binding of ST/AR-DG to the enzyme binding site relative to other DG species. Up to the present, however, there has been no evidence of such specificity, and in most in vitro systems tested, a wide variety of diacylglycerols and analogs appear to be effective in activating the enzyme. If if eventually proves to be the case that there is no selectivity for ST/AR-DG, the inferred link between stimulated inositol lipid turnover and protein kinase C should be questioned. Alternative sources of DG in cells are of the non-ST/AR variety. They include DG released upon sphingomyelin synthesis from phosphatidylcholine, DG produced by the action of phosphatidate phosphohydrolase on biosynthetic phosphatidate, breakdown of triglyceride, etc.

It is also frequently proposed that arachidonate in ST/AR-DG is released in the inositol lipid labeling cycle, and is thus a reservoir for eicosanoid synthesis. If this proves to be the case, we are left to explain how the lipid cycle in Fig. 5 can regenerate itself, unless there is reacylation at the sn-2 position of the glycerol backbone of these lipids after arachidonate release.

Phorbol esters have proven useful tools for the study of DG action, since they penetrate the cell membrane and occupy the DG binding site on protein kinase C. Experiments indicate that in addition to stimulating protein kinase C phosphorylation, phorbol esters also slow the stimulated labeling cycle, presumably via protein kinase C-catalyzed phosphorylations. The possible substrates for these regulatory phosphorylations include the receptor itself, an intervening G-protein, and the inositol lipid-specific phosphodiesterase. Still another possible regulatory site is at the level of PIP kinase. Gispen et al. (1985) report that a membrane-bound enzyme termed B50-kinase is identical with protein kinase C. It catalyzes a phosphorylation which inhibits PIP_2 synthesis. ACTH fragments block the stimulating effect of B50-kinase with the net result that PIP turnover is increased. It has also been proposed that the substrate is identical to a protein also known as GAP43, and as F_1 (see chapter by Routtenberg et al., in this volume).

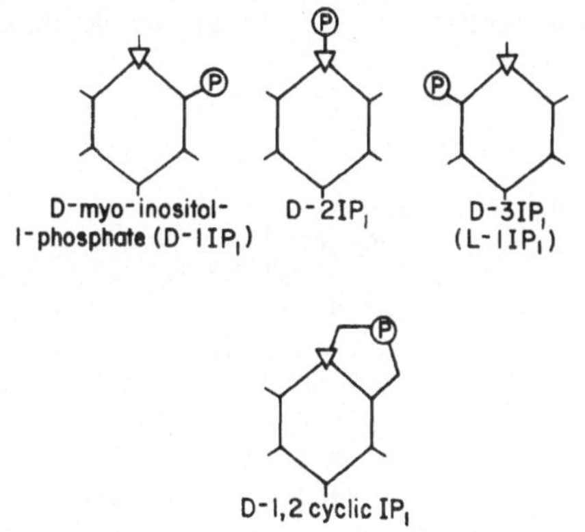

FIG. 6: Some inositol monophosphates. Note that D-1 IP_1 and its cyclic derivative are formed from enzymatic cleavage of PI. The stereoisomer of D-1 IP_1, D-3 IP_1 (or L-1 IP_1), is formed from glucose via cyclization of glucose 6-phosphate and subsequent phosphatase action. Also shown is the D-2 isomer, in which the phosphate is on the axial hydroxyl. It has been isolated as a chemical hydrolysis product of phytic acid (IP_6; see Fig. 7).

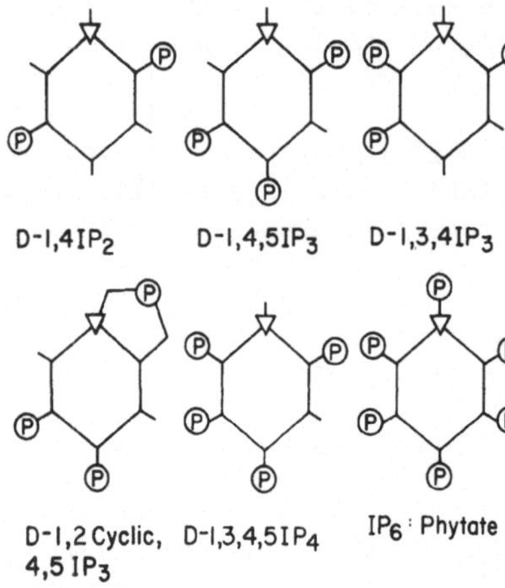

FIG. 7: Higher phosphates of D-myo-inositol. $1,4\text{-}IP_2$ is formed either by the cleavage of PIP, or following the action of a 5'-phosphohydrolase on $1,4,5\text{-}IP_3$. $D\text{-}1,3,4\text{-}IP_3$ does not mobilize Ca^{2+}, in contrast with the 1,4,5 isomer. The 1,2 cyclic 4,5-trisphosphate may also be formed during enzymatic cleavage of PIP_2 and is believed to be active in Ca^{2+} mobilization. $D\text{-}1,3,4\text{-}IP_3$ has been postulated to arise from $D\text{-}1,3,4,5\text{-}IP_4$. Also shown is phytate (IP_6), commonly found in plant seeds.

Inositol Phosphates

When PI is cleaved by a purified inositol lipid-specific phosphodiesterase, two products can be identified in addition to DG: D-myo-inositol 1-phosphate, and a cyclic derivative in which the 1- and 2-hydroxyls of inositol form a 5-member ring via a cyclic phosphodiester linkage (Fig. 6). While it has been speculated that the cyclic derivative is the initial product of PI breakdown and that the ring is then cleaved enzymatically to yield the D1-phosphate, such a sequence has not been clearly established. We are left then with two derivatives, each of which appears to be a partial product of the cleavage of PI.

Inositol Trisphosphates

D-myo-inositol 1,4,5-trisphosphate was originally identified as a hydrolysis product of PIP_2 (Grado and Ballou, 1961). It was proposed, by analogy with the identified products of PI hydrolysis, that the 1,2 cyclic 4,5-trisphosphate might also be formed (Agranoff and Seguin, 1974; Fig. 7). It has not up to the present been possible to demonstrate the formation of this substance in tissues, but evidence that it is a product of the action of purified inositol-specific phosphodiesterase on PIP_2 has been reported (Connolly et al., 1986). Both the 1,4,5 and 1,2 cyclic $4,5\text{-}IP_3$ mobilize intracellular Ca^{2+} (Wilson et al., 1985). Structure-activity studies suggest that it is highly likely that the 4,5-vicinal phosphates are required for this action. An active 5'-phosphohydrolase cleaves IP_3 to produce $1,4\text{-}IP_2$, the same IP_2 that is produced by phosphodiesteratic cleavage of PIP. The 5'-phosphohydrolase is considered to be an "off" enzyme, terminating the Ca^{2+}-mobilizing effect of IP_3. The 1,3,4 isomer of IP_3 has been found to accumulate in cells, and unlike the 1,4,5 isomer, to be inactive in mobilizing Ca^{2+}. Questions have arisen regarding both its origin and function (Irvine et al., 1984). Most recently, evidence has been reported for the presence of IP_4 (Batty et al., 1985). If the IP_4 is the $1,3,4,5\text{-}IP_4$ isomer, cleavage by a 3'-specific phosphohydrolase would then yield the $1,3,4\text{-}IP_3$ (Batty et al., 1985). As to the origin of IP_4, it may arise from

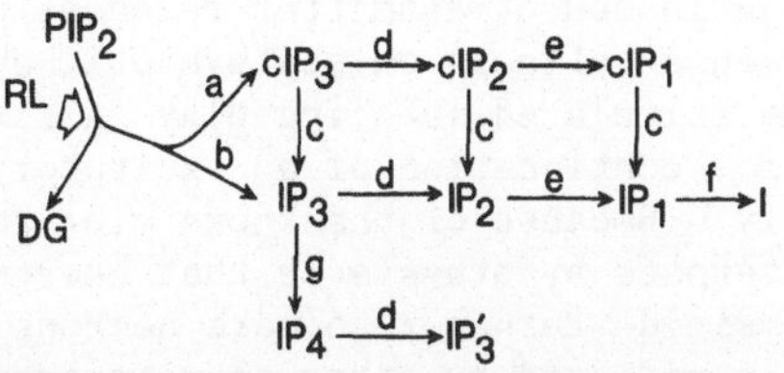

FIG. 8: Interrelationship of water-soluble products of PIP_2 cleavage.
Phosphodiesteratic cleavage may give rise to both cIP_3 and 1,4,5-
IP_3 (a and b). Alternatively, cIP_3 may be an obligatory intermediate which is converted to IP_3 via action of a cyclic phosphodiesterase (c). An active 5'-phosphohydrolase (d) converts cIP_3 or
IP_3 to cIP_2 or IP_2, respectively. The 4'-phosphate is then cleaved
(e), followed by hydrolysis of the remaining 1-phosphate to yield
inositol. Postulated here is a 3'-kinase which converts IP_3 to
$1,3,4,5-IP_4$. The 5'-phosphohydrolase would then produce 1,3,4-IP3
(see text).

phosphorylation of $1,4,5-IP_3$ (Irvine et al., 1986). Possible interrela-
tions of these products is shown in Fig. 8.

Evidence for the presence of an IP_3 kinase in our laboratory comes
from experiments with ^{32}P-labeled IP_3 and a brain high-speed supernatant
fraction in the presence of ATP. Material co-migrating with an IP_4 standard from phytate hydrolysis is formed in a time- and enzyme-dependent
fashion. The significance of the phosphorylation of IP_3 rests on whether
or not IP_4 proves to mobilize Ca^{2+}. If it does, then the question arises
of how IP_4's action differs from IP_3. If it does not mobilize Ca^{2+}, IP_4
could be considered to be a storage state which can be converted to an active
or inactive IP_3. Depending on the relative activities of 5'- and 3'-phos-
phohydrolases, they would produce the 1,3,4- and $1,4,5-IP_3$'s, respectively.
It should be noted that a further phosphorylation of IP_4 at the 6 position
would yield the 1,3,4,5,6-pentakisphosphate, in which all of the equatorial
positions are phosphorylated (Fig. 7). This IP_5 is known to be present in
chick erythrocytes, where it binds to hemoglobin, as does 2,3-disphospho-
glyceric acid in mammalian erythrocytes. A further question is whether any
of these derivatives occur as the 1,2-cyclic phosphodiester. Depending on
relative activities of a cyclic IP_3 phosphodiesterase, phosphomonoesterases
and IP_3 kinases, a variety of pathways are possible, as illustrated in Fig.
8. Not shown is the further possibility that cIP_3 can be phosphorylated to
cIP_4. Also, there may be alternate pathways for the degradation of IP_2 via
phosphatases specific for the 1 and 4 positions (Michell, 1986).

Physiological Significance of Stimulated PI Turnover in the Nervous System

Stimulated phospholipid turnover has been demonstrated in the nervous
system with muscarinic cholinergic, as well as α-adrenergic, H_1-histami-
nergic, serotonergic ($5HT_2$) and glutamatergic agents. Numerous neurohumoral peptides are also effective activators of stimulated lipid turnover.
Of these various agents, the best studied is the muscarinic receptor. In
a series of lesion experiments, we were able to demonstrate that in the
guinea pig hippocampus, most if not all, of the cholinergic stimulation of
labeling can be accounted for by postsynaptic receptor sites (Fisher et al.,
1980, 1981). This means that the particles active in nerve ending preparations which support stimulated labeling are either pinched-off postsynaptic terminals ("dendrosomes") or are presynaptic terminals which are also
postsynaptic to incoming cholinergic fibers. In addition, since in none of
the other systems investigated is there any evidence as yet for the role of

stimulated lipid labeling in neurotransmitter release, we should for the present seek second messenger effects on postsynaptic events. These include the possibility that the stimulated labeling plays a direct causal role in signal transmission, i.e., continuation of an excitatory depolarization wave initiated presynaptically. Because of the known slow response of the muscarinic receptor, an alternate hypothesis is that neuronal activity is modulated via stimulated inositide turnover to make neurons more or less excitable to incoming signals mediated by other neurotransmitters. This could be accomplished via phosphorylation of ion channels or by phosphorylation of other proteins involved in signal transduction, etc. It is also possible that the products of PIP_2 cleavage play a role in neuroplasticity (see chapter by Routtenberg et al., in this volume).

It is clear from the foregoing that a new set of biochemical tools has become available and that a great deal of progress and understanding of its operation has been achieved in a short period of time. We can expect in the foreseeable future to see the integration of the new findings into the known body of neuropharmacology, which must now be reinterpreted and extended.

REFERENCES

Agranoff, B. W. and Seguin, E. B., Preparation of inositol triphosphate from brain: GLC of trimethylsilyl derivative, Prep. Biochem. 4:359-366 (1974).

Agranoff, B. W., Murthy, P. and Seguin, E. B., Thrombin-induced phosphodiesteratic cleavage of phosphatidylinositol bisphosphate in human platelets, J. Biol. Chem. 258:2076-2078 (1983).

Allison, J. H., Blisner, M. E., Holland, W. H., Hipps, P. P. and Sherman, W. R., Increased brain myo-inositol 1-phosphate in lithium-treated rats, Biochem. Biophys. Res. Commun. 71:664-670 (1976).

Batty, I. R., Nahorski, S. R. and Irvine, R. F., Rapid formation of inositol 1,3,4,5-tetrakisphosphate following muscarinic receptor stimulation of rat cerebral cortical slices, Biochem. J. 232:211-215 (1985).

Benjamins, J. A. and Agranoff, B. W., Distribution and properties of CDP-diglyceride: Inositol transferase from brain, J. Neurochem. 16:513-527 (1969).

Berridge, M. J., Inositol trisphosphate and diacylglycerol as second messengers, Biochem. J. 220:345-360 (1984).

Berridge, M. J., Downes, C. P. and Hanley, M. R., Lithium amplifies agonist-dependent phosphatidylinositol responses in brain and salivary glands, Biochem. J. 206:587-595 (1982).

Berridge, M. J., Dawson, R. M. C., Downes, C. P., Heslop, J. P. and Irvine, R. F., Changes in the levels of inositol phosphates after agonist-dependent hydrolysis of membrane phosphoinositides, Biochem. J. 212: 473-482 (1983).

Connolly, T. M., Wilson, D. B., Bross, T. E. and Majerus, P. W., Isolation and characterization of the inositol cyclic phosphate products of phosphoinositide cleavage by phospholipase C, J. Biol. Chem. 261:122-126 (1986).

Cosgrove, D. J., Inositol Phosphates. Their Chemistry, Biochemistry and Physiology. Elsevier Scientific Publishing Co., New York (1980).

de Hevesy, G., Some applications of isotope indicators, in Nobel Lectures. Chemistry, 1942-1946, pp. 9-41, Elsevier Publishing Company, Amsterdam (1964).

Durell, J., Sodd, M. A. and Friedel, R. O., Acetylcholine stimulation of the phosphodiesteratic cleavage of guinea pig brain phosphoinositides, Life Sci. 7:363-368 (1968).

Eisenberg Jr., F., D-myoinositol 1-phosphate as product of cyclization of glucose 6-phosphate and substrate for a specific phosphatase in rat testis, J. Biol. Chem. 242:1375-1382 (1967).

Fisher, S. K. and Agranoff, B. W., Enhancement of the muscarinic synaptoso-
    mal phospholipid labeling effect by the ionophore A23187, J. Neurochem.
    37:968-977 (1981).
Fisher, S. K., Boast, C. A. and Agranoff, B. W., The muscarinic stimulation
    of phospholipid labeling is independent of its cholinergic input, Brain
    Res. 189:284-288 (1980).
Fisher, S. K., Frey, K. A. and Agranoff, B. W., Loss of muscarinic receptors
    and of stimulated phospholipid labeling in ibotenate-treated hippocam-
    pus, J. Neurosci. 1:1407-1413 (1981).
Gispen, W. H., Leunissen, J. L. M., Oestreicher, A. B., Verkleij, A. J. and
    Zwiers, H., Presynaptic localization of B-50 phosphoprotein: The
    (ACTH)-sensitive protein kinase substrate involved in rat brain poly-
    phosphoinositide metabolism, Brain Res. 328:381-385 (1985).
Grado, C. and Ballou, C. E., Myo-inositol phosphates obtained by alkaline
    hydrolysis of beef brain phosphoinositide, J. Biol. Chem. 236:54-60
    (1961).
Hokin, M. R. and Hokin, L. E., Effects of acetylcholine on phospholipides
    in the pancreas, J. Biol. Chem. 209:549-558 (1953).
Irvine, R. F., Letcher, A. J., Lander, D. J. and Downes, C. P., Inositol
    trisphosphates in carbachol-stimulated rat parotid glands, Biochem. J.
    223:237-243 (1984).
Irvine, R. F., Letcher, A. J., Heslop, J. P. and Berridge, M. J., The inosi-
    tol tris/tetrakis phosphate pathway--demonstration of inositol (1,4,5)-
    trisphosphate-3-kinase activity in animal tissues, Nature, in press
    (1986).
Ishikawa, S., Maxillary chemoreceptors in the silkworm, in Olfaction and
    Taste 2. Wenner-Gren Center International Symposium Series, Vol. 8
    (Hayashi T., ed), pp. 761-777. Pergamon Press, New York (1967).
Jakinovich Jr., W. and Agranoff, B. W., The stereospecificity of the inosi-
    tol receptor of the silkworm bombyx mori, Brain Res. 33:173-180 (1971).
Michell, R. H., Inositol phospholipids and cell surface receptor function,
    Biochim. Biophys. Acta 415:81-147 (1975).
Michell, R. H., Profusion and confusion, Nature 319:176-177 (1986).
Nishizuka, Y., Turnover of inositol phospholipids and signal transduction,
    Science 225:1365-1370 (1984).
Posternak, T., The Cyclitols (Lederer E., ed), Holden-Day, Inc., San Fran-
    cisco, CA (1965).
Raetz, C. R. H., Hirschberg, C. B., Dowhan, W., Wickner, W. T. and Kennedy,
    E. P., A membrane-bound pyrophosphatase in Escherichia coli catalyzing
    the hydrolysis of cytidine diphosphate-diglyceride, J. Biol. Chem. 247:
    2245-2247 (1972).
Rittenhouse, H. G., Seguin, E. B., Fisher, S. K. and Agranoff, B. W., Pro-
    perties of a CDP-diglyceride hydrolase from guinea-pig brain, J. Neuro-
    chem. 36:991-999 (1981).
Wilson, D. B., Bross, T. E., Hoffman, S. L. and Majerus, P. W., Hydrolysis
    of polyphosphoinositides by purified sheep seminal vesicle phospholi-
    pase C enzymes, J. Biol. Chem. 259:11718-11724 (1984).
Wilson, D. B., Connolly, T. M., Bross, T. E., Majerus, P. W., Sherman, W.
    R., Tyler, A., Rubin, L. J. and Brown, J. E., Isolation and characte-
    rization of the inositol cyclic phosphate products of polyphosphoino-
    sitide cleavage by phospholipase C. Physiological effects in permea-
    bilized platelets and limulus photoreceptor cells, J. Biol. Chem. 260:
    13496-13501 (1985).

MUSCARINIC ACETYLCHOLINE RECEPTOR-LINKED INOSITIDE CYCLE

IN THE CENTRAL NERVOUS SYSTEM

Lucio A. A. van Rooijen and Jörg Traber

Neurobiology Department, Troponwerke GmbH and Co. KG
Neurather Ring 1, 5000 Cologne 80, Federal Republic
of Germany

The Inositide Cascade

The inositide cycle (Fig .1) constitutes a second messenger system mediating the cellular response to activation of particular receptors (for reviews: Abdel-Latif, 1983; Berridge, 1984; Fisher et al., 1984a; Nishizuka, 1984). It is generally accepted that the initial event following receptor activation is phosphodiesteratic degradation of phosphatidylinositol 4,5-bisphosphate (PIP_2) to diacylglycerol (DG) and D-myo-inositol 1,4,5-trisphosphate (IP_3). The latter is degraded by specific phosphatases eventually to myo-inositol. DG is readily phosphorylated to phosphatidate (PA), which is converted through a liponucleotide-intermediate to phosphatidylinositol (PI). In two sequential phosphorylation steps, PI is converted to PIP_2, via phosphatidylinositol 4-phosphate (PIP), thus closing the inositide cycle. One way to analyze receptor-sensitive operation of the inositide cycle is to measure the appearance of labeled inositol phosphates in [^{3}H]-inositol (pre)incubated cells. The use of Li^+, which inhibits the last step of the degradation of IP_3 to myo-inositol (Hallcher and Sherman, 1980), has proven valuable in this analysis (Berridge et al., 1982). Alternatively one can measure concurrent labeling of phospholipids from $^{32}P_i$. Receptor activation will then enhance incorporation of radiotracer into PA and PI.

As is depicted in Fig. 1, inositide turnover can be involved in the release of free fatty acids (i.e. arachidonate), intracellular mobilization of Ca^{2+} by IP_3, and activation of protein kinase C by diacylglycerol (for reviews: Irvine, 1982; and refs. above).

Control of the Inositide Cycle

Various studies have led to the notion that phosphodiesteratic degradation of PIP_2 may be the initial event in the inositide cycle following appropriate receptor-activation. For instance, Berridge (1983) showed a sequential transient increase in IP_3 and IP_2 followed by a rise in IP_1 when blowfly salivary glands were stimulated with serotonin. In thrombin stimulated platelets, an increase in IP_3 was observed (Agranoff et al., 1983). With time, the loss of radiotracer from prelabeled precursor lipid followed the sequence PIP_2, PIP and PI (Shukla et al., 1983; Thomas et al., 1983). This loss of radiotracer from the inositides is transient. Since the loss from PI is continued longer than that from PIP_2 and PIP, a replenishment of

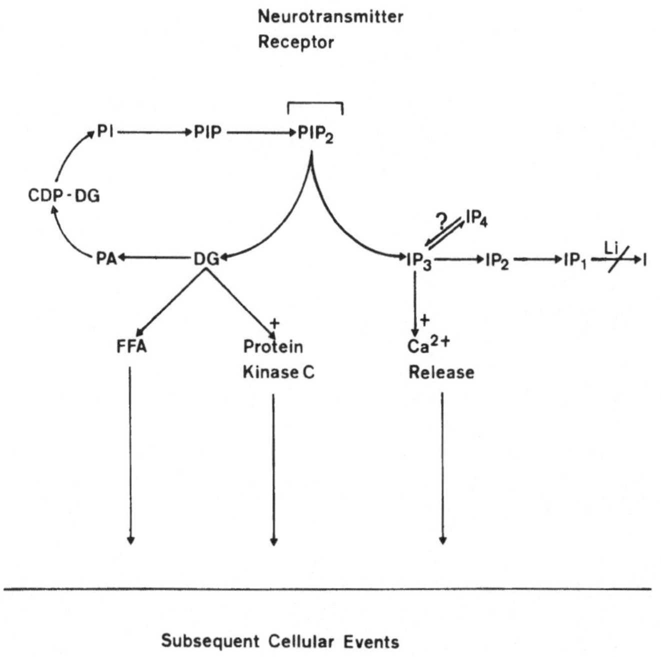

FIG. 1: The inositide cascade. Receptor sensitive operation of the inositide cycle and subsequent events have been outlined as discussed in the text.

the polyphosphoinositides from a much larger pool of PI is envisioned. Downes and Wusteman (1983) further investigated this in muscarinic cholinergically stimulated parotid gland. ATP-depletion with dinitrophenol abolished agonist-evoked increases in IP_3, IP_2 and IP_1. Although the possible involvement of another ATP-dependent event can not be excluded, it was concluded that, upon receptor activation, polyphosphoinositides are degraded and then replenished from PI (Downes and Wusteman, 1983).

The conclusion that PIP_2 is first degraded and then replenished, through PIP from PI implies a possiblity to control the inositide response. Two possible mechanisms have been proposed sofar. In the first there is product feed-back inhibition of the inositide-kinases: PIP_2 has recently been shown to inhibit PIP-kinase to even 50% at a concentration equimolar to the substrate (Van Rooijen et al., 1985b). Phosphodiesteratic degradation of PIP_2 following receptor activation may thus indirectly stimulate synthesis of this lipid for replenishment.

The second mechanism to control polyphosphoinositide-levels consists of a feed-back loop involving protein phosphorylation (Gispen, 1986). In neuronal membranes, a protein named B-50 has been identified by Gispen and colleagues as a substrate for Ca^{2+}-dependent, DG-modulated protein kinase C (Zwiers et al., 1980; Aloyo et al., 1983). Phosphorylated B-50 appears to inhibit the activity of PIP-kinase (Jolles et al., 1980). It was proposed (Gispen, 1986) that B-50 serves a function in the control of the levels of PIP_2 which are subject to alteration during receptor activation: Enhanced phosphodiesteratic degradation of PIP_2 will generate IP_3 (then mobilizing intracellular Ca^{2+}) and DG, both of which stimulate phosphorylation of B-50 by protein kinase C. Phosphorylated B-50 will then inhibit PIP-kinase-activity, resulting in an impairment of the replenishment of PIP_2, discussed above. This feed-back mechanism may serve an "off" signal to receptor mediated enhancement of inositide turnover. It is to be expected that a protein with a function similar to B-50 will also be found in non-neuronal systems. The proposed loop as a control mechanism is further supported by the observation that the protein kinase C activator, 12-0-tetra-decanoylphorbol-13-acetate (TPA) inhibits carbamylcholine (CCh)-induced production of inositol phosphates in hippocampal slices (Labarca et al., 1984; Schrama et al., 1986). Furthermore, $ACTH_{1-24}$, known to inhibit phosphorylation of B-50 (Jolles et al., 1980), has recently been found to overcome the TPA-inhibition of CCh-induced production of inositol-phosphates in hippocampal slices (Schrama et al., 1986).

Calcium Dependence

The inositide response is often observed in conjunction with Ca^{2+}-dependent cellular responses, such as smooth muscle contraction or secretion. As omission of Ca^{2+} from the external medium can greatly reduce or even abolish the physiological response, in the presence of an intact inositide response, it was postulated that enhanced inositide turnover mediates influx of Ca^{2+} which then evokes the subsequent physiological response (Michell, 1975).

Alternatively, since phosphodiesteratic degradation of the inositides is Ca^{2+}-dependent, (Downes and Michell, 1981; Van Rooijen et al., 1983), the mobilization of Ca^{2+} could also stimulate inositide turnover. Two lines of evidence have been considered to support this suggestion for the inositide response elicited by muscarinic acetylcholine receptor (mAChR) activation. First, EGTA has been found to inhibit the mAChR-mediated response in nerve endings and cerebral slices (Griffin et al., 1979; Fisher and Agranoff, 1980; Kendall and Nahorski, 1984). Since Ca^{2+}-flux across the plasma membrane should be considered as a very dynamic process, chelation of extracellular Ca^{2+} by EGTA most likely will have its effect on the intracellular availability of the cation as well (Beaven et al., 1984). An inhibition by EGTA of Ca^{2+}-dependent intracellular events, such as the inositide phosphodiesterase, is hence to be expected. The data from studies with EGTA may therefore not be taken as evidence for a Ca^{2+}-flux to mediate the stimulated inositide turnover to receptor-activation, but may only indicate that the cation is involved in the response. Second, the Ca^{2+}-ionophore A23187 was found to enhance basal and CCh-stimulated inositide turnover in nerve endings (Fisher and Agranoff, 1980; 1981). It appeared however that the stimulatory effects of A23187 were only observed when Ca^{2+} was omitted from the incubation medium (Van Rooijen and Traber, 1986). The effects of A23187 could be particular for this compound, since elevation of intracellular Ca^{2+} with veratridine (Miller and Kowall, 1983; Van Rooijen and Traber, 1986) and the dihydropyridine BAY K 8644 (Van Rooijen and Traber, 1986) did not mimic or enhance the inositide response to CCh in nerve endings.

Ca^{2+}-channel antagonists can be classified as: dihydropyridines, diphenylalkylamines and other (Spedding, 1985). An example from each group, being nimodipine, flunarizine and verapamil respectively, was unable to inhibit the inositide response to CCh in nerve endings (Van Rooijen and Traber, 1986). In addition, muscarinic stimulation of nerve endings did not alter $^{45}Ca^{2+}$-uptake (Van Rooijen, 1984). In ileum smooth muscle, CCh-stimulated PI-labeling was also not affected by various Ca^{2+} antagonists (Jafferji and Michell, 1976). If these channels were activated in the course of the muscarinic acetylcholine receptor-mediated enhancement of PA- and PI-labeling an inhibition by the Ca^{2+}-channel antagonists would most likely have been observed.

The inositide response to mAChR-activation is therefore Ca^{2+}-dependent but not mediated by Ca^{2+}-influx. This conclusion is in consensus with the notion that the inositide response mediates intracellular mobilization of Ca^{2+}, since IP_3 can stimulate release of the cation from the endoplasmic

TABLE 1: Neurotransmitter receptors reported to be linked to the inositide
cycle in preparations of the central nervous system.

receptor	reference
α_1-adrenergic	Brown et al., 1984; Janowsky, 1984a; 1984b; Kendall et al., 1984; Minneman and Johnson, 1984; Schoepp et al., 1984
H_1-histaminergic	Daum et al., 1983; Brown et al., 1984; Daum et al., 1984
muscarine-acetylcholinergic	Hokin and Hokin, 1953; Durell et al., 1968; Schacht and Agranoff, 1972; Abdel-Latif et al., 1977; Miller, 1977; Griffin et al., 1979; Yandrasitz and Segal, 1979; Fisher and Agranoff, 1980; 1981; Aly and Abdel-Latif, 1982; Berridge et al., 1982, Downes, 1982; Fisher et al., 1983; Brown et al., 1984; Fisher et al., 1984b; Gonzales and Crews, 1984; Janowsky et al., 1984b; Labarca et al., 1984; Brown et al., 1985; Fisher and Bartus, 1985; Gil and Wolfe, 1985; Jacobsen et al., 1985; Lazareno et al., 1985; Van Rooijen et al., 1985; Schrama et al., 1986; Van Rooijen et al., 1986
serotonergic	Berridge et al., 1982; Brown et al., 1984; Conn and Sanders-Bush, 1984; Janowsky et al., 1984b; Godfrey et al., 1985; Kendall and Nahorski, 1985.

reticulum. Since the original observation by Strebb et al. (1983) IP_3-induced liberation of Ca^{2+} has now been demonstrated in a variety of tissues (Burgess et al., 1984; Irvine et al., 1984; Joseph et al., 1984; Prentky and Wollheim, 1984; Epstein et al., 1985; Somlyo et al., 1985; Yamamoto and Van Breemen, 1985). To our knowledge no evidence on this event in a central nervous system preparation has been reported yet.

Muscarinic Acetylcholine Receptor Mediated Inositide Response

In the central nervous system, several neurotransmitter receptors have been investigated in regard to their linkage to the inositide cycle (Table 1). For the adrenergic and histaminergic receptors, the α_1- and H_1-subtypes appear to mediate the inositide response has been defined (for refs.: Table 1). The inositide response to acetylcholine receptor activation is muscarinic and not nicotinic. Although of all neurotransmitter receptors the muscarinic acetylcholine receptor has been the most intensively studied in this respect, a particular subtype is not readily identified for its linkage to the inositide cycle.

mAChR-subtypes

The mAChR does not form a homogenous population. Binding-competition experiments with mAChR-ligands revealed up to three different agonist-affinities of the mAChR, which have been termed superhigh, high and low affinity states (Birdsall and Hulme, 1976; Birdsall et al., 1980; 1983). For CCh, the affinity constants are 0.08, 1.6 and 125 µM respectively (Birdsall et al., 1980). Pharmacological studies on the physiological response of the opossum lower esophageal spincter (Goyal and Rattan, 1978; Gilbert et al., 1984) gave rise to definition of M_1- and M_2 AChR-subtypes. The mAChR-affinities for ligands which preferentially bind to the M_1 AChR (pirenzepine) and to the M_2AChR (oxotremorine-M) differ across various tissues and even regionally within the brain (Hammer et al., 1980; Watson et al., 1983; Birdsall et al., 1983; Garvey et al., 1984; Wamsley et al., 1984; Spencer et al., 1985, 1986).

Compounds, like GTP or the sulfhydryl alkylating N-ethylmaleimide (NEM), can decrease the affinity of the M_2AChR while the M_1AChR seems unaffected (Aronstam and Eldefrawi, 1979; Watson et al., 1983; Flynn and Potter, 1985). Autoradiographic analysis of binding of ^3H-PZ and ^3H-OXO-M to brain slices revealed that NEM (Horvath et al., 1986) and GTP (unpublished data) decreased labeling only by ^3H-OXO-M, while the regional differences of binding by both radioligands was not altered. GTP abolished the high affinity component of ^3H-OXO-M binding to cerebral cortex particulate material, while the total number of binding sites was unaltered (Baven, 1984). The total number of mAChR binding sites and the affinity as determined with the nonselective antagonist ^3H-QNB (quinuclidinylbenzilate) are not altered by NEM or the non-hydrolyzable GTP-analog GppNHp (Matsumoto et al., 1983; Aronstam et al., 1978). Although, in various protein separation systems (polyacrylamide gel electrophoresis; affinity chromatography), mAChR migrates as a single band or peak (Amitai et al., 1982; Haga and Haga, 1985), the mAChR solubilized from different brain areas retains its heterogeneity of affinity (Wenger et al., 1985). Hence a single mAChR protein may have different binding sites and/or conformational states with different agonist affinities, at least part of which are interconvertible.

The two models of mAChR subtypes (superhigh, high and low states; and M_1- and M_2AChR are sofar not readily compatible. It is obvious that this unclarity has not facilitated elucidation of which mAChR-subtype, if any in particular, is linked to the inositide cycle.

TABLE 2: Similar potencies of CCh to elicit an inositide response in different cerebral regions.

preparation	measure	CCh EC_{50}(M)	n	ref.
rat cerebrum nerve endings	^{32}P -PA	7.6×10^{-5}	0.93	1
guinea pig cerebral cortex nerve endings	^{32}P -PA	$\sim 7 \times 10^{-5}$	n.d.	2
guinea pig cerebral cortex slices	^3H -IP	2×10^{-4}	n.d.	3
rat cerebral cortex slices	^3H -IP	4.9×10^{-5}	n.d.	4
		1.2×10^{-4}	0.93	5
		6.8×10^{-5}	0.98	6
rat hippocampus nerve endings	^{32}P -PA	5.3×10^{-5}	0.99	7
rat hippocampus slices	^3H -IP	$\sim 5 \times 10^{-5}$	n.d.	8
rat medulla pons slices	^3H -IP	1.5×10^{-4}	1.14	5
rat midbrain nerve endings	^{32}P -PA	2.4×10^{-5}	0.89	7

The reported potencies (and Hill coefficients: n) of CCh to enhance production of ^{32}P-PA or ^3H-IP in slices or nerve endings from various brain structures are listed. n.d.: not determined; $\sim$: estimated from figure. References: 1) Van Rooijen et al., 1986; 2) Fisher et al., 1983; 3) Fisher and Bartus, 1985; 4) Gonzales and Crews, 1984; 5) Lazareno et al., 1985; 6) Jacobson et al., 1985; 7) Van Rooijen et al., manuscript in preparation; 8) Janowsky et al., 1984.

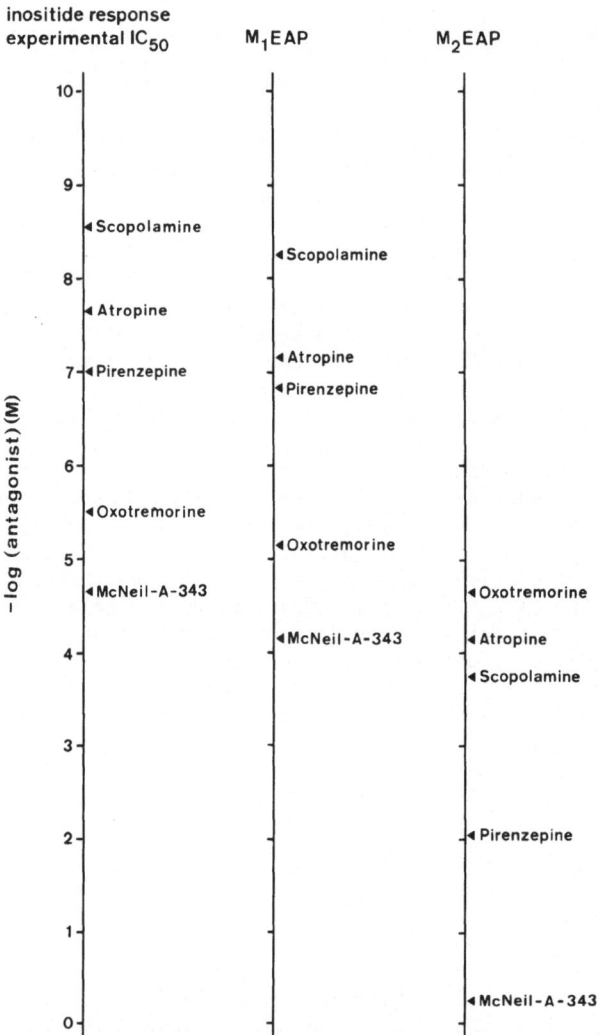

FIG. 2: Equi-affinity prediction of potency (EAP) of mAChR-antagonists to inhibit the inositide response to CCh. The equi-affinity prediction values were calculated from binding data reported for atropine and oxotremorine to M_1AChR (Yamamura et al., 1985) or to M_2AChR (Baven, 1984). The experimental values are derived from data reported by Lazareno et al. (1985; atropine), Fisher et al. (1984b; oxotremorine). All values for scopolamine, PZ and McN-A-343 are calculated from data in Van Rooijen et al. (1986).

mAChR-subtype-mediated Inositide Response

Linkage of the inositide cycle to either M_1- or M_2AChR-subtypes is not readily defined, since both the M_1AChR-selective antagonist pirenzepine (PZ) can inhibit (Gonzales and Crews, 1984; Brown et al., 1985; Gil and Wolfe, 1985; Lazareno et al., 1985; Smith and Yamamura, 1985; Van Rooijen et al., 1986) and the M_2AChR-selective agonist oxotremorine-M (OXO-M) can evoke (Fisher et al., 1984b; Fisher and Bartus, 1985; Jacobson et al., 1985; Van Rooijen et al., 1986) an inositide response. It therefore appears necessary to closely compare the potencies of several agonists and antagonists on inositide turnover with their affinities for mAChR subtypes.

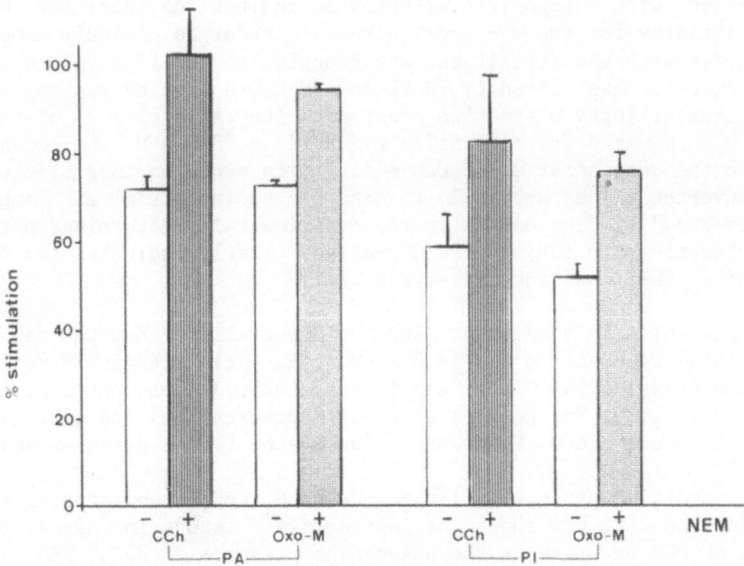

FIG. 3: Enhancement by NEM of CCh and OXO-M-stimulated labeling of PA and
PI. Nerve endings, prepared from male Wistar rat (Winkelmann, Bor-
chen, FRG) cerebri, by differential and sucrose gradient centrifugation, were incubated in a final volume of 0.5 ml medium (30 mM
HEPES, pH 7.4; 142 mM NaCl; 5.6 mM KCl; 1.0 mM $MgCl_2$; 2.2 mM $CaCl_2$;
3.6 mM $NaHCO_3$; 5.6 mM glucose) with various drugs and 10-25 µCi of
$^{32}P_i$ for 30 min at 37°C in a water bath with shaking. After termination of the incubation by the addition of 1.5 ml of chloroform:
methanol (1:2, by vol), phospholipids were extracted by two phase
separation under acidic conditions, separated by thin layer chromatography on K-oxalate impregnated TLC plates (Merck, Darmstadt,
FRG, silica gel 60, 20 x 20 x 0.025 cm), localized, scraped off,
and radiotracer was quantitated by scintillation counting (Van
Rooijen et al., 1983; Van Rooijen and Traber, 1986). The presence
of 1 mM CCh or OXO-M and of 30 µM NEM is indicated.

CCh is commonly used to elicit an inositide response and binds with
higher affinity to the M_2- than to the M_1 AChR (Van Rooijen et al., 1986).
If the inositide cascade were equally linked to the M_1- and M_2 AChR, varying
potencies of CCh to enhance operation of the inositide cycle with Hill coefficients of less than 1 will be expected across different brain areas. As
is listed in Table II, in nerve endings or slices from various brain structures, CCh is equally potent to stimulate the production of ^{3}H-inositol
phosphates or ^{32}P-PA. In all cases, the Hill-coefficient was determined to
be close to 1 (Table II). Most likely therefore only one subtype of the
mAChR is linked to the inositide cycle.

The EC_{50} values of CCh-stimulated inositide turnover (Table II) are
very similar to the affinity constant for CCh-binding to the ^{3}H-PZ site
(Luthin and Wolfe, 1984; Yamamura et al., 1983; Van Rooijen et al., 1986),
as well as to the low affinity constant derived from binding competition
experiments with CCh and a tritium labeled nonselective antagonist such as
QNB (Fisher et al., 1983). This is also the case for OXO-M (Fisher et al.,
1984b; Van Rooijen et al., 1986).

Another approach is to compare the potency of mAChR antagonists (and partial agonists with antagonists effects) to inhibit the inositide response with their affinity for the M_1- and M_2AChR. In order to directly compare these potencies with the affinities, a correction of the IC_{50} values obtained for displacement of high affinity ^3H-PZ and ^3H-OXO-M binding was performed through an equi-affinity prediction equation: The ratio of IC_{50} of the antagonist vs. CCh required for high affinity ^3H-PZ or ^3H-OXO-M binding was multiplied with the concentration of CCh employed to evoke an inositide response. The thus converted values available to date are shown in fig. 2. Comparison of the "corrected" binding data with the experimentally determined potency of these antagonists to inhibit the CCh-effect clearly indicates the M_1AChR to be linked to the inositide cycle (Fig. 2).

This conclusion is consistent with the observations that the magnitude of the inositide response to CCh is different in various brain structures (Gonzales and Crews, 1984; Fisher and Bartus, 1985; Jacobson et al., 1985; Lazareno et al., 1985; Van Rooijen et al., in preparation) and qualitatively related to the occurence of M_1- rather than M_2AChR (refs. given above).

If the inositide cycle were linked to the M_1AChR, then receptor affinity modification with NEM should be ineffective. Incubation of nerve endings with 1 mM CCh or OXO-M in the absence or presence of 30 µM NEM, disclosed however an enhancement of both the CCh- and the OXO-M stimulated labeling of PA and PI by NEM (Fig. 3). Labeling of the polyphosphoinositides was unaltered (not shown). Although it should not be excluded that NEM may exert its enhancing effect through another mechanism, the observation that the agonist-enhancement of PI-labeling was affected by NEM similar to that of PA indicates that NEM does not exert its effect by inhibition of further metabolism of PA. If indeed these effects of NEM reflect conversion of the mAChR to a low agonist affinity state, they could indicate that the M_1- and M_2AChR classification may not be sufficient in regard to its linkage to the inositide response. It is expected that the approach to relate ligand properties in regard to mAChR affinities with a biochemically defined second messenger system, may not only identify possible receptor subtype linkage, but also may provide more information on the mAChR itself.

ACKNOWLEDGMENTS

We wish to thank Drs. W. U. Dompert, E. Horvath and D. G. Spencer for their helpful discussions in the course of the work described here.

REFERENCES

Abdel-Latif, A., Metabolism of phosphoinositides, in: Handbook of Neurochemistry, Vol. 3, pp. 91-131, Lajtha, A. (ed.), Plenum Press, New York (1983).

Abdel-Latif, A. A., Akhtar, R. A. and Hawthorne, J. N., Acetylcholine increases the breakdown of triphosphoinositide of rabbit iris muscle prelabeled with ^{32}P-phosphate, Biochem. J. 161:61 (1977).

Agranoff, B. W., Murthy, P. and Seguin, E. B., Thrombin-induced phosphodiesteratic cleavage of phosphatidylinositol bisphosphate in human platelets, J. Biol. Chem. 258:2076 (1983).

Aloyo, V. J., Zwiers, H. and Gispen, W. H., Phosphorylation of B-50 protein kinase and B-50 kinase, J. Neurochem. 41:649 (1983).

Aly, M. I. and Abdel-Latif, A. A., Studies of the effects of acetylcholine and antiepileptic drugs on $^{32}P_i$ incorporation into phospholipids of rat brain synaptosomes, Neurochem. Res. 7:159 (1982).

Amitai, G., Aissar, S., Balderman, D. and Sokolovsky, M., Affinity labeling of muscarinic receptors in rat cerebral cortex with a photolabile antagonist, Proc. Natl. Acad. Sci. 70:243 (1982).

Aronstam, R. S., Abood, L. G. and Hoss, W., Influence of sulfhydryl reagents and heavy metals on the functional state of the muscarinic acetylcholine receptor in rat brain, Molec. Pharmacol. 14:575 (1978).

Aronstam, R. S. and Eldefrawi, M. E., Reversible conversion between affinity states for agonists of the muscarinic acetylcholine receptor from rat brain, Biochem. Pharmacol. 28:701 (1979).

Baven, P., ^{3}H-Oxotremorine-M binding to membranes prepared from rat brain and heart: evidence for subtypes of muscarinic receptors, Eur. J. Pharmacol. 101:101 (1984).

Beaven, M. A., Rogers, J., Moore, J. P., Hesketh, T. R., Smith, G. A. and Metcalfe, J. C., The mechanism of the calcium signal and correlation with histamine release in 2H3 cells, J. Biol. Chem. 259:7129 (1984).

Berridge, M. J., Rapid accumulation of inositol trisphosphate reveals that agonists hydrolyze polyphosphoinositides instead of phosphatidylinositol, Biochem. J. 212:849 (1983).

Berridge, M. J., Inositol trisphosphate and diacylglycerol as second messenger, Biochem. J. 220:345 (1984).

Berridge, M. J., Downes, C. P. and Hanley, M. R., Lithium amplifies agonist-dependent phosphatidylinositol responses in brain and salivary glands, Biochem. J. 206:587 (1982).

Birdsall, N. J. M. and Hulme, E. C., Biochemical studies on muscarinic acetylcholine receptors, J. Neurochem. 27:7 (1976).

Birdsall, N. J. M., The character of the muscarinic receptors in different regions of the rat brain, Proc. R. Soc. Lond. 207:1 (1980).

Birdsall, N. J. M., Hulme, E. C., Stockton, J., Burgen, A. S. V., Berrie, C. P., Hammer, R., Wong, E. H. F. and Zigmond, M. J., Muscarinic receptor subclasses: evidence from binding studies, in: CNS Receptors - From Molecular Pharmacology in Behavior, P. Mandel and F. V. DeFeudis (eds.), Raven Press, New York (1983).

Brown, E., Kendall, D. A. and Nahorski, S. R., Inositol phospholipid hydrolysis in rat cerebral cortical slices: I. Receptor characterization, J. Neurochem. 42:1379 (1984).

Brown, J. H., Goldstein, D. and Brown-Masters, S., The putative M_1 muscarinic receptor does not regulate phosphoinositide hydrolysis, Molec. Pharmacol. 27:525 (1985).

Burgess, G. M., Irvine, R. F., Berridge, M. J., McKinney, J. S. and Putney, J. W., Actions of inositol phosphates on Ca^{2+} pools in guinea pig hepatocytes, Biochem. J. 224:741 (1984).

Conn, P. J and Sanders-Bush, E., Selective 5HT-2 antagonists inhibit serotonin stimulated phosphatidylinositol metabolism in cerebral cortex, Neuropharmacol. 23:993 (1984).

Daum, P. R., Downes, C. P. and Young, J. M., Histamine-induced inositol phospholipid breakdown mirrors H_1-receptor density in brain, Eur. J. Pharmacol. 87:497 (1983).

Daum, P. R., Downes, C. P. and Young, J. M., Histamine stimulation of inositol 1-phosphate accumulation in lithium-treated slices from regions of guinea pig brain, J. Neurochem. 43:25 (1984).

Downes, C. P., Receptor-stimulated inositol phospholipid metabolism in the central nervous system, Cell Calcium 3:413 (1982).

Downes, C. P. and Michell, R. H., The polyphosphoinositide phosphodiesterase, Biochem. J. 198:133 (1981).

Downes, C. P. and Wusteman, M. M., Breakdown of polyphosphoinositides and not phosphatidylinositol accounts for muscarinic agonist-stimulated inositol phospholipid metabolism in rat parotid glands, Biochem. J. 216:633 (1983).

Durell, J., Sodd, M. A. and Friedel, R. O., Acetylcholine stimulation of the phosphodiesteratic cleavage of guinea pig brain phosphoinositides, Life Sci. 7:363 (1968).

Epstein, P. A., Prentki, M. and Attie, M. F., Modulation of intracellular Ca^{2+} in the parathyroid cell: release of Ca^{2+} from non mitochondrial pools by inositol trisphosphate, FEBS Lett. 188:141 (1985).

Fisher, S. K. and Agranoff, B. W., Calcium and the muscarinic phospholipid labeling effect, J. Neurochem. 34:1231 (1980).

Fisher, S. K. and Agranoff, B. W., Enhancement of the muscarinic synaptosomal phospholipid labeling effect by the ionophor A23187, J. Neurochem. 37:968 (1981).

Fisher, S. K. and Bartus, R. T., Regional differences in the coupling of muscarinic receptors to inositol phospholipid hydrolysis in guinea pig brain, J. Neurochem. 40:1085 (1985).

Fisher, S. K., Klinger, P. D. and Agranoff, B. W., Muscarinic agonist binding and phospholipid turnover in brain, J. Biol. Chem. 258:7358 (1983).

Fisher, S. K., Van Rooijen, L. A. A. and Agranoff, B. W., Renewed interest in the polyphosphoinositides, Trends Biochem. Sci. 9:53 (1984a).

Fisher, S. K., Figueiredo, J. C. and Bartus, R. T., Differential stimulation of inositol phospholipid turnover in brain by analogs of oxotremorine, J. Neurochem. 43:1171 (1984b).

Flynn, D. D. and Potter, L. T., Different effects of N-ethylmaleimide on M1 and M2 muscarinic receptors in rat brain, Proc. Natl. Acad. Sci. 82:580 (1985).

Garvey, J. M., Rossor, M. and Iversen, L. L., Evidence for multiple muscarinic receptor subtypes in human brain, J. Neurochem. 43:299 (1984).

Gil, D. W. and Wolfe, B. B., Pirenzepine distinguishes between muscarinic receptor-mediated phosphoinositide breakdown and inhibition of adenylate cyclase, J. Pharmacol. Exp. Therap. 232:608 (1985).

Gilbert, R., Rattan, S. and Goyal, R. K., Pharmacological identification activation and antagonism of two muscarinic receptors in the lower esophageal sphincter, J. Pharmacol. Exp. Therap. 230:284 (1984).

Gispen, W. H., Phosphoprotein B-50 and phosphoinositides in brain synaptic plasma membranes: a possible feedback relationship, Biochem. Soc. Transact. 14:163 (1986).

Godfrey, P. P., McClue, S. J., Minchin, M. C. W. and Young, M., Ru 24969, a $5-HT_1$ agonist, stimulates inositol phospholipid breakdown in rat brain slices, Br. J. Pharmacol. 84:112 (1985).

Gonzales, R. A. and Crews, F. T., Characterization of the cholinergic stimulation of phosphoinositide hydrolysis in rat brain slices, J. Neurosci. 4:3120 (1984).

Goyal, R. K. and Rattan, S., Neurohumoral, hormonal and drug receptors for the lower esophageal spincter, Gastroent. 74:598 (1978).

Griffin, H. D., Hawthorne, J. N. and Sykes, M., A calcium requirement for the phosphatidylinositol response following activation of presynaptic muscarinic receptors, Biochem. Pharmacol. 28:1143 (1979).

Haga, K. and Haga, T., Purification of the muscarinic acetylcholine receptor from procine brain, J. Biol. Chem. 260:7927 (1985).

Hallcher, L. M. and Sherman, W. R., The effects of lithium ion and other agents on the activity of myo-inositol-1-phosphatase from bovine brain, J. Biol. Chem. 255:10896 (1980).

Hammer, R., Berrie, C. P., Birdsall, N. J. M., Burgen, A. S. V. and Hulme, E. C., Pirenzepine distinguishes between different subclasses of muscarinic receptors, Nature 283:90 (1980).

Hokin, M. R. and Hokin, L. E., Enzyme secretion and the incorporation of P^{32} into phospholipids of pancreas slices, J. Biol. Chem. 203:967 (1953).

Horvath, E., Van Rooijen, L. A. A., Traber, J. and Spencer, D. G., Effects of N-ethylmaleimide on muscarinic acetylcholine receptor subtype autoradiography and inositide response in rat brain, Life Sci. in review.

Irvine, R. F., How is the level of free arachidonic acid controlled in mammalian cells?, Biochem. J. 204:3 (1982).

Irvine, R. F., Brown, K. D. and Berridge, M. J., Specificity of inositol trisphosphate-induced calcium release from permeabilized Swiss-mouse 3T3 cells, Biochem. J. 221:269 (1984).

Jacobson, M. D., Wüsteman, M. and Downes, C. P., Muscarinic receptors and hydrolysis of inositol phospholipids in rat cerebral cortex and parotid gland, J. Neurochem. 44:465 (1985).

Jafferji, S. S. and Michell, R. H., Effects of calcium-antagonistic drugs on the stimulation by carbamylcholine and histamine of phosphatidyli-nositol turnover in longitudinal smooth muscle of guinea pig ileum, Biochem. J. 160:163 (1976).

Janowsky, A., Labarca, R. and Paul, S. M., Noradrenergic denervation increases $_1$-adrenoreceptor-mediated inositol-phosphate accumulation in the hippocampus, Eur. J. Pharmacol. 102:193 (1984a).

Janowsky, A., Labarca, R. and Paul, S. M., Characterization of neurotrans-mitter receptor-mediated phosphatidylinositol hydrolysis in the rat hippocampus, Life Sci. 35:1953 (1984b).

Jolles, J., Zwiers, H., Dekker, A., Wirtz, K. W. A. and Gispen, W. H., Cor-ticotropin (1-24)-tetracosapeptide affects protein phosphorylation and polyphosphoinositide metabolism in rat brain, Biochem J. 194:283 (1981).

Jolles, J., Zwiers, H., Van Dongen, C. J., Schotman, P., Wirtz, K. W. A. and Gispen, W. H., Modulation of brain polyphosphoinositide metabolism by ACTH-sensitive protein phosphorylation, Nature 286:623 (1980).

Joseph, S. K., Thomas, A. P., Williams, R. J., Irvine, R. F. and William-son, J. R., myo-inositol 1,4,5-trisphosphate, a second messenger for the hormonal mobilization of intracellular Ca^{2+} in liver, J. Biol. Chem. 259:3077 (1984).

Kendall, D. A. and Nahorski, S. R., Inositol phospholipid hydrolysis in rat cerebral cortical slices: II. Calcium requirement, J. Neurochem. 42:1388 (1984).

Kendall, D. A. and Nahorski, S. R., 5-Hydroxytryptamine-stimulated inositol phospholipid hydrolysis in rat cerebral cortex slices: pharmacological characterization and effects of antidepressants, J. Pharmacol. Exp. Therap. 233:473 (1985).

Labarca, R., Janowsky, A., Patel, J. and Paul, S. M., Phorbol esters inhibit agonist-induced ^{3}H inositol-1-phosphate accumulation in rat hippocampus slices, Biochem. Biophys. Res. Commun. 123:703 (1984).

Lazareno, S., Kendall, D. A. and Nahorski, S. R., Pirenzepine indicates heterogeneity of muscarinic receptors linked to cerebral inositol phos-pholipid metabolism, Neuropharmacol. 24:593 (1985).

Luthin, G. R. and Wolfe, B. B., Comparison of ^{3}H-pirenzepine and ^{3}H-quinu-clidinylbenzilate binding to muscarinic cholinergic receptors in rat brain, J. Pharmacol. Exp. Therap. 228:648 (1984a).

Matsumoto, K., Uchida, S., Higuchi, H., Mizushima, A. and Yoshida, H., Effect of urea-treatment on agonist binding affinity of the muscarinic receptor, Life Sci. 33:963 (1983).

Michell, R. H., Inositol phospholipids and cell surface receptor function, Biochem. Biophys. Acta 415:81 (1975).

Miller, J. C., A study of the kinetics of the muscarinic effect on phospha-tidylinositol and phosphatidic acid metabolism in rat brain synapto-somes, Biochem. J. 168:549 (1977).

Miller, J. C. and Kowal, C. N., Effects of pentobarbital and veratridine on phosphatidylinositol and phosphatidate metabolism in rat parotid acinar cells, Biochem. Pharmacol. 32:2237 (1983).

Minneman, K. P. and Johnson, R. D., Characterization of alpha-1 adrenergic receptors linked to ^{3}H-inositol metabolism in rat cerebral cortex, J. Pharmacol. Exp. Therap. 230:317 (1984).

Nishizuka, Y., Turnover of inositol phospholipids and signal transduction, Science 225:1365 (1984).

Prentki, M. and Wollheim, C. B., Cytosolic free Ca^{2+} in insulin secreting cells and its regulation by isolated organelles, Experientia 40:1052 (1984).

Schacht, J. and Agranoff, B. W., Effects of acetylcholine on labeling of phosphatidate and phosphoinositides by ^{32}P-orthophosphate in nerve ending fractions of guinea pig cortex, J. Biol. Chem. 247:771 (1972).

Schoepp, D. D., Knepper, S. M. and Rutledge, C. O., Norepinephrine stimulation of phosphoinositide hydrolysis in rat cerebral cortex is associated with the alpha$_1$-adrenoreceptor, J. Neurochem. 43:1758 (1984).

Schrama, L. H., De Graan, P. N. E., Eichberg, J. and Gispen, W. H., Feedback control of the inositol phospholipid response in rat brain is sensitive to ACTH, Eur. J. of Pharmacol. 121:403 (1986).

Shukla, S. D., Buxton, D. B., Olson, M. S. and Hanahan, D. J., Acetylglyceryl ether phosphorylcholine. A potent activator of hepatic phosphoinositide metabolism and glycogenolysis, J. Biol. Chem. 258:10212 (1983).

Smith, T. L. and Yamamura, H. I., Carbachol stimulation of phosphatidic acid synthesis: competitive inhibition by pirenzepine in synaptosomes from rat cerebral cortex, Biochem. Biophys. Res. Commun. 130:282 (1985).

Somlyo, A. V., Bond, M., Somlyo, A. P. and Scarpa, A., Inositol trisphosphate induced calcium release and contraction in vascular smooth muscle, Proc. Natl. Acad. Sci. 82:5231 (1985).

Spedding, M., Calcium antagonist subgroups, Trends Pharmacol. Sci. 6:109 (1985).

Spencer, D. G., Horvath, E., Luiten, P., Schuurman, T. and Traber, J., Novel approaches in the study of brain acetylcholine function: Neuropharmacology, neuroanatomy and behavior, in: Senile dementia of the Alzheimer type, Advances in applied neurological sciences, Vol. 2, pp. 325-354, J. Traber and W. H. Gispen (eds.), Springer Verlag, Berlin, FRG (1985).

Spencer, D. G., Horvath, E. and Traber, J., Direct autoradiographic determination of M$_1$ and M$_2$ muscarinic acetylcholine receptor distribution in the rat brain: relation to cholinergic nuclei and projections, Brain Res. in press (1986).

Strebb, H., Irvine, R. F., Berridge, M. J. and Schulz, I., Release of Ca^{2+} from nonmitrochondrial intracellular store in pancreatic acinar cells by inositol 1,4,5-trisphosphate, Nature 306:67 (1983).

Thomas, A. P., Marks, J. S., Coll, K. E. and Williamson, J. R., Quantitation and early kinetics of inositol lipid changes induced by vasopressin in isolated and cultured hepatocytes, J. Biol. Chem. 258:5716 (1983).

Van Rooijen, L. A. A., Polyphosphoinositide phosphodiesterase: characterization and physiological significance in brain. Dissertation Utrecht. (1984).

Van Rooijen, L. A. A. and Traber, J., Muscarinic cholinergic enhancement of inositide turnover in cerebral nerve endings is not mediated by calcium uptake, Biochem. Pharmacol., in press (1986).

Van Rooijen, L. A. A., Seguin, E. B. and Agranoff, B. W., Phosphodiesteratic breakdown of endogenous polyphosphoinosites in nerve ending membranes, Biochem. Biophys. Res. Commun. 112:919 (1983).

Van Rooijen, L. A. A., Hajra, A. K. and Agranoff, B. W., Tetraenoic species are conserved in muscarinically enhanced inositide turnover, J. Neurochem. 44:540 (1985a).

Van Rooijen, L. A. A., Rossowska, M. and Bazan, N. G., Inhibition of phosphatidylinositol-4-phosphate kinase by its product phosphatidylinositol-4,5-bisphosphate, Biochem. Biophys. Res. Commun. 126:150 (1985b).

Van Rooijen, L. A. A., Dompert, W. U., Horvath, E., Spencer, D. G. and Traber, J., Pharmacological aspects of the inositide response in the central nervous system: the muscarinic acetylchoine receptor, in: Progress in brain research, Gispen, W. H. and Routenberg, A. (eds), Elsevier, Amsterdam, in press (1986a).

Wamsley, J. K., Gehlert, D. R., Roeske, W. R. and Yamamura, H. I., Muscarinic antagonist binding site heterogeneity as evidenced by autoradiography after direct labeling with ^{3}H-QNB and ^{3}H-pirenzepine, Life Science 34:1395 (1984).

Watson, M., Yamamura, H. I. and Roeske, W. R., A unique regulatory profile and regional distribution of ^{3}H-pirenzepine binding in the rat provide evidence for distinct M$_1$ and M$_2$ muscarinic receptor subtypes, Life Sci. 32:3001 (1983).

Wenger, D. A., Parthasarthy, N. and Aronstam, R. S., Regional heterogeneity of muscarinic acetylcholine receptors from rat brain in retained after detergent solubilization, <u>Neurosci. Lett</u>. 54:65 (1985).

Yamamura, H. I., Watson, M. and Roeske, W. R., ^{3}H-Pirenzepine specifically labels a high affinity muscarinic receptor in the rat cerebral cortex, <u>in</u>: CNS receptors - From molecular pharmacology in behavior, pp. 331-336, P. Mandel and F. V. DeFeudis (eds.), Raven Press, New York (1983).

Yamamoto, H. and Van Breemen, C., Inositol-1,4,5-trisphosphate releases calcium from skinned cultured smooth muscle cells, <u>Biochem. Biophys. Res. Commun</u>. 130:270 (1985).

Yandrasitz, J. R. and Segal, S., The effect of $MnCl_2$ on the basal and acetylcholine-stimulated turnover of phosphatidylinositol in synaptosomes, <u>FEBS Lett</u>. 108:270 (1979).

Zwiers, H., Schotman, P. and Gispen, W. H., Purification and some characteristics of an ACTH-sensitive protein kinase and its substrate protein in rat brain membranes, <u>J. Neurochem</u>. 34:1689 (1985).

THE ROLE OF GTP-BINDING PROTEINS IN RECEPTOR ACTIVATION OF

PHOSPHOLIPASE C

Eduardo G. Lapetina

Molecular Biology Department, Burroughs Wellcome Co., 3030
Cornwallis Road, Research Triangle Park, NC 27709

Recent studies indicate that phospholipase C-induced degradation of the inositol phospholipids produces the formation of two molecules that have second messenger properties. The two molecules are 1,2-diacylglycerol, which stimulates protein kinase C, and 1,4,5-inositol trisphosphate, which can mobilize Ca^{2+} from the endoplasmic reticulum to the cytosol. Inositol phospholipid degradation occurs in a wide variety of cells responding to specific stimuli. In platelets, it has been shown that agonists such as thrombin, collagen, platelet-activating factor, vasopressin, serotonin, prostaglandin-endoperoxides and thromboxane A_2 stimulate the metabolism of the inositol phospholipids (Lapetina, 1986a). This receptor-coupled reaction seems to be an early step in the transduction mechanism of those agonists. Recent advances indicate that regulation of inositol phospholipid degradation might be modulated by GTP-binding proteins (Lapetina, 1986a,b; Cockcroft and Gomperts, 1985; Haslam and Davidson, 1984; Lapetina et al., 1986; Litosch et al., 1985; Wallace and Fain, 1985; Ui, 1984), much like the regulation of adenylate cyclase (Katada et al., 1984; Hanski and Gilman, 1982). In this case, two guanine nucleotide-binding regulatory proteins, G_s (N_s) and G_i (N_i), modulate the stimulation and inhibition of adenylate cyclase, respectively. These proteins contain three subunits recognized as α, β and γ. Cholera toxin catalyzes ADP-ribosylation of the α_s-subunit of G_S and stimulates adenylate cyclase. Pertussis toxin, on the other hand, ADP-ribosylates the α_i-subunit of G_i.

Stimulation of Human Platelets with Trypsin or Thrombin Inhibits the Pertussis Toxin-Induced ADP-Ribosylation of α_i-Subunit

Platelets do not have cell surface receptors for pertussis toxin (Ui, 1984), which makes it impossible to study the effects of pertussis toxin on platelet activation. Nevertheless, pertussis toxin can ADP-ribosylate the α_i-subunit in platelet membranes (Katada et al., 1984). Also, thrombin produces activation of G_i in platelet membranes, with subsequent inhibition of adenylate cyclase (Aktories and Jacobs, 1984). We have studied pertussis toxin-induced ADP-ribosylation in platelets that have been permeabilized with saponin (Lapetina et al., 1986). Saponin allows platelet permeabilization to pertussis toxin, which leads to ADP-ribosylation of the α_i-subunit of G_i. Maximal ADP-ribosylation is obtained at 20 min (Lapetina et al., 1986). ADP-ribosylation is not inhibited by prostacyclin, indomethacin, or 1,2-diacylglycerol; it is inhibited by phorbol esters, 5 μg/ml trypsin and 1 mM Ca^{2+}. Another GTP-binding protein, G_o, has recently been isolated from rat brain (Sternweis and Robishaw, 1984). Its α_o-subunit has a molecular weight of 39,000, which is also ADP-ribosylated by pertussis toxin. It is possible that the ADP-ribosylated protein in platelets could be either α_i or α_0 since discrimination of the α_i- and α_0-subunits is difficult. Whether or not the ADP-ribosylated protein in platelets is α_0, α_i, or a mixture of both must be determined. Pertussis toxin-induced ADP-ribosylation in platelets is dramatically decreased if platelets are previously exposed to thrombin (Lapetina et al., 1986). This might reflect an alteration of the α- and β-subunits of the G_i- or G_0-protein. Thrombin is able to completely inhibit ADP-ribosylation, indicating that thrombin can fully activate G-protein (Lapetina et al., 1986). This may represent a unique mechanism for thrombin action and may be related to the proteolytic action of thrombin on the G-protein. Our recent finding that leupeptin inhibits thrombin-induced responses such as secretion, aggregation, protein kinase C and phospholipase C indicates that the activation of platelet proteases by thrombin is essential for platelet activation (Ruggiero and Lapetina, 1985). Trypsin also activates platelet secretion, aggregation, protein kinase C and phospholipase C (Ruggiero and Lapetina, 1985). Since trypsin produces proteolytic cleavage of the α-subunit (Lapetina et al., 1986), it is possible that phospholipase C activation is related to removal of the α_i-subunit of the G_i-protein. Similarly, thrombin might cause modification of the G_i-protein that leads to phospholipase C activation through the involvement of a GTP-binding protein different to G_i or G_0.

Phospholipase C-Induced Degradation of the Inositides by GTPγS in Permeabilized Platelets

Platelet phospholipase C produces the breakdown of the inositol phospholipids (Lapetina, 1986a). This activity can be measured by a decrease of the inositol phospholipids and formation of inositol phosphates, or 1,2-diacylglycerol, or its phosphorylated product, phosphatidic acid (Lapetina, 1986a). GTP or GTP analogs such as GTPγS and GppNHp can stimulate phospholipase C in platelets permeabilized by electrical stimulation (Haslam and Davidson, 1984). In this case, formation of 1,2-diacylglycerol was determined.

We recently determined the effect of GTPγS on phospholipase C in platelets that were prelabeled with [³H]inositol and were permeabilized with saponin (Lapetina, 1986b). The experiments are carried out under the same assay conditions that allow ADP-ribosylation by pertussis toxin of the α_i-subunit of G_i (Lapetina et al., 1986). GTPγS (200 μM) produces formation of [³H]inositol bisphosphate and [³H]inositol trisphosphate after 1 min and is maximal between 15 and 30 min (Lapetina, 1986b). Formation of [³H]inositol polyphosphates is greater than formation of [³H]inositol monophosphate (Lapetina, 1986b). GTPγS at 2 μM stimulates the formation of [³H]inositol polyphosphates; maximal formation is observed at 20-500 μM GTPγS (Lapetina, 1986b). These experiments are performed in the presence of 1 mM EDTA without the addition of Ca^{2+}, which indicates that the effect of GTPγS on phospholipase C is independent of Ca^{2+}. Formation of the [³H]-inositol polyphosphates parallels the disappearance of platelet [³H]phosphatidylinositol monophosphate and [³H]phosphatidylinositol bisphosphate (Lapetina, 1986b). The breakdown of the polyphosphoinositides by GTPγS is not affected by prostacyclin (1 μg/ml) or dibutyril cyclic AMP (1 mM) (Lapetina, 1986b). Pertussis toxin neither affects phospholipase C nor the action of GTPγS on phospholipase C (Lapetina, 1986b).

Thrombin (1 unit/ml) also causes the formation of [³H]inositol polyphosphates in platelets that are permeabilized with saponin (Lapetina, 1986b). In this case, pretreatment with pertussis toxin, which induces ADP-ribosylation of α_i, increases the effect of thrombin on the formation of the [³H]inositol phosphates (Lapetina, 1986b). These results differ from the effect of pertussis toxin in inhibiting the agonist-induced stimulation of phospholipase C in various cell systems (Brandt et al., 1985; Murayama and Ui, 1985; Okajima et al., 1985; Verghese et al., 1985; Volpi et al., 1985) and might indicate a mechanism that is unique for certain agonists. Activation of phospholipase C by thrombin correlates with the decrease of GTP-binding proteins that are susceptible to ADP-ribosylation by pertussis toxin (Lapetina et al., 1986).

This effect of thrombin is similar to that of trypsin. Trypsin stimulates phospholipase C (Ruggiero and Lapetina, 1985) and, at the same time, produces proteolysis of the α_i-subunit that is ADP-ribosylated by pertussis toxin in platelets (Lapetina et al., 1986). These results also suggest that removal of the α_i-subunit of the G_i-protein, and activation of a GTP-binding protein different to G_i or G_o, might correlate with activation of phospholipase C.

Correlations of GTP-Binding Proteins and Phospholipase C in Stimulated Platelets

The effect of GTPγS on platelet phospholipase C indicates that a GTP-binding protein is related to the modulation of this enzyme. The fact that GTPγS stimulates phospholipase C after pretreatment of the cell with pertussis toxin, which dissociates and inactivates G_i-protein, suggests that G_i does not have a stimulatory effect on phospholipase C. Instead, it seems that GTPγS is acting through a GTP-binding protein that is different to G_i or G_o. Pertussis toxin enhances the effect of thrombin on phospholipase C; this might suggest that the removal of G_i or G_o is associated with activation of phospholipase C. Trypsin proteolytically cleaves α_i and in parallel stimulates phospholipase C. The disappearance of G_i or G_o does not seem to be enough for activation of phospholipase C, because pertussis toxin itself does not stimulate phospholipase C. Therefore, the synergistic effect of GTPγS and thrombin indicates that removal of G_i and interaction of GTPγS with another GTP-binding protein might be associated with phospholipase C activation.

Protein Kinase C Stimulates Dephosphorylation of 1,4,5-Inositol Trisphosphate

1,4,5-Inositol trisphosphate is formed in response to specific agonists that cause activation of phospholipase C and degradation of phosphatidylinositol bisphosphate (Lapetina, 1986a). 1,4,5-Inositol trisphosphate is a second messenger that releases Ca^{2+} from the dense tubular system to the cytosol in stimulated platelets (O'Rourke et al., 1985). Our information (Molina y Vedia and Lapetina, 1986) indicates that [3H]inositol trisphosphate is dephosphorylated to [3H]inositol bisphosphate and [3H]inositol monophosphate by human platelets treated with 0.05-0.10% Triton X-100. This dephosphorylation of [3H]inositol trisphosphate to [3H]inositol bisphosphate and [3H]inositol monophosphate is also observed when platelets are permeabilized by electrical stimulation or by 20 μg/ml saponin. These detergents or electropermeabilization allow inositol trisphosphate to access cytosolic inositol trisphosphate-phosphatase. Pretreatment of intact platelets with phorbol dibutyrate and 1-oleyl, 2-acetyl,diacylglycerol for 30 s, at concentrations that maximally activate protein kinase C, stimulates the conversion of inositol trisphosphate to inositol bisphosphate and inositol monophosphate.

This information shows that inositol trisphosphate-phosphatase is stimulated by the action of activators of protein kinase C. The regulation of inositol trisphosphate-phosphatase is important, because the action of this enzyme takes away the signal for the mobilization of intracellular Ca^{2+}. The agonist-induced activation of phospholipase C produces both inositol trisphosphate and 1,2-diacylglycerol (Fig. 1) (Lapetina, 1986a). Inositol

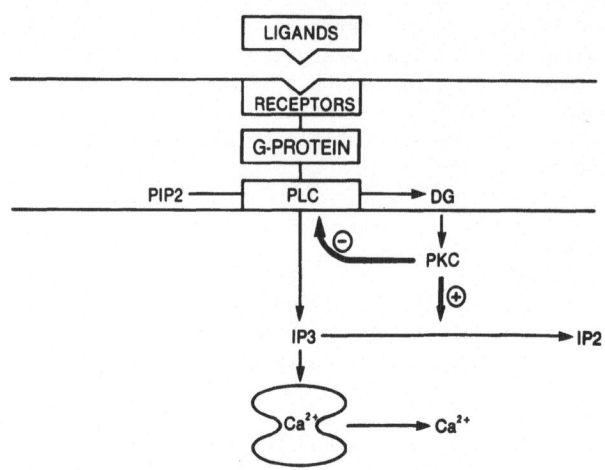

FIG. 1: Receptor-coupled activation of platelets. Different platelet ligands
        interact with specific platelet receptors to stimulate phospholipase
        C (PLC). This activation involves a GTP-binding protein (G-protein).
        PLC degrades phosphatidylinositol 4,5-bisphosphate (PIP2) to produce
        1,2-diacylglycerol (DG) and 1,4,5-inositol-trisphosphate (IP3). DG
        stimulates protein kinase C (PKC) and IP3 causes mobilization of Ca
        from the dense tubular system. The dephosphorylation of IP3 to ino-
        sitol 1,4-bisphosphate is stimulated by PKC. PKC also seems to have
        a negative feedback on PLC.

trisphosphate rapidly mobilizes Ca^{2+} and, with 1,2-diacylglycerol, synergistically increases the activity of protein kinase C (Nishizuka, 1983).
Our results suggest that activation of protein kinase directly or indirectly influences the ability of inositol trisphosphate-phosphatase to stop the
signal for Ca^{2+} mobilization.

     The rapid activation of protein kinase C in stimulated platelets (Ni-
shizuka, 1983; Lapetina et al., 1985) could cause phosphorylation and activation of inositol trisphosphate-phosphatase. In this regard, a recent
communication (Connolly and Majerus, 1986) indicates that inositol trisphosphate-phosphatase is phosphorylated by brain protein kinase C, resulting in a 4-fold increase in inositol trisphosphate-phosphatase activity.
An alternative, indirect effect of the action of protein kinase C on inositol trisphosphate-phosphatase could be mediated through the phosphorylation
of an intermediate that eventually triggers activation of inositol trisphosphate-phosphatase.

        It has been reported that phorbol esters and 1,2-diacylglycerols
inhibit the activation of phospholipase C induced by various platelet agonists (Rittenhouse and Sasson, 1985; Watson and Lapetina, 1985; MacIntyre
et al., 1985; Zavoico et al., 1985; Drummond and MacIntyre, 1985; Naccache
et al. 1985; Lynch et al., 1985). The negative feedback control over receptor-induced hydrolysis of inositol phospholipids and the stimulation of ino-
sitol-trisphosphate are mechanisms directed to stop cellular over-stimula-
tion (Fig. 1).

REFERENCES

Aktories, K. and Jakobs, K. H., Ni-mediated inhibition of human platelet
     adenylate cyclase by thrombin, Eur. J. Biochem. 145:333-338 (1984).

Brandt, S. J., Dougherty, R. W., Lapetina, E. G. and Niedel, J. E., Pertussis toxin inhibits chemotactic peptide-stimulated generation of inositol phosphates and lysosomal enzyme secretion in human leukemic (HL-60) cells, Proc. Natl. Acad. Sci. USA 82:3277-3280 (1985).

Cockcroft, S. and Gomperts, B. D., Role of guanine nucleotide binding protein in the activation of polyphosphoinositide phosphodiesterase, Nature 314:534-536 (1985).

Connolly, T. M. and Majerus, P. W., Protein kinase C phosphorylated human platelet inositol trisphosphate 5-phosphomonoesterase increasing phosphatase activity, Clin. Res. 34:656A (1986).

Drummond, A. H. and MacIntyre, D. E., Protein kinase C as a bidirectional regulator of cell function, Trends Pharmacol. Sci. 6:233-234 (1985).

Hanski, E. and Gilman, A. G., The guanine nucleotide-binding regulatory component of adenylate cyclase in human erythrocytes, J. Cyclic Nucleotide Res. 8:323-336 (1982).

Haslam, R. J. and Davidson, M. M., Receptor-induced diacylglycerol formation in permeabilized platelets; possible role for a GTP-binding protein, J. Recept. Res. 4:605-629 (1984).

Katada, T., Bokoch, G. M., Northup, J. K., Ui, M. and Gilman, A. G., The inhibitory guanine nucleotide-binding regulatory component of adenylate cyclase: Properties and functions of the purified protein, J. Biol. Chem. 259:3568-3577 (1984).

Lapetina, E. G., Inositide-Dependent and -Independent Mechanisms in Platelet Activation, in: "Receptors and Phosphoinositides," J. W. Putney, ed., Alan R. Liss, Inc., New York (in press) (1986a).

Lapetina, E. G., Effect of pertussis toxin on the phosphodiesteratic cleavage of the polyphosphoinositides by guanosine 5'-0-thiotriphosphate and thrombin in permeabilized human platelets, Biochim. Biophys. Acta (in press) (1986b).

Lapetina, E. G., Reep, B. and Chang, K.-J., Stimulation of human platelets with trypsin, thrombin and collagen inhibit the pertussis toxin-induced ADP-ribosylation of a 41,000 dalton protein, Proc. Natl. Acad. Sci. USA (in press) (1986).

Lapetina, E. G., Reep, B., Ganong, B. R. and Bell, R. M., Exogenous sn-1,2-diacylglycerols containing saturated fatty acids function as bioregulators of protein kinase C in human platelets, J. Biol. Chem. 260:1358-1361 (1985).

Litosch, I., Wallis, C. and Fain, J. N., 5-Hydroxytryptamine stimulates inositol phosphate production in a cell-free system from blowfly salivary glands: Evidence for a role of GTP in coupling receptor activation to phosphoinositide breakdown, J. Biol. Chem. 260:5464-5471 (1985).

Lynch, C. J., Charest, R., Bocckino, S. B., Exton, J. H. and Blackmore, P. F., Inhibition of hepatic α_1-adrenergic effects and binding by phorbol myristate acetate, J. Biol. Chem. 260:2844-2851 (1985).

MacIntyre, D. E., McNicol, A. and Drummond, A. H., Tumour-promoting phorbol esters inhibit agonist-induced phosphatidate formation and Ca^{2+} flux in human platelets, FEBS Lett. 180:160-164 (1985).

Molina y Vedia, L. M. and Lapetina, E. G., Phorbol 12,13-dibutyrate and 1-oleyl,-2-acetyl,diacylglycerol stimulate inositol trisphosphate dephosphorylation in human platelets, J. Biol. Chem. (in press) (1986).

Murayama, T. and Ui, M., Receptor-mediated inhibition of adenylate cyclase and stimulation of arachidonic acid release in 3T3 fibroblasts: Selective susceptibility to islet-activating protein, pertussis toxin, J. Biol. Chem. 260:7226-7233 (1985).

Naccache, P. H., Milski, M. M., Volpi, M., Becker, E. L. and Sha'afi, R. I., Unique inhibitory profile of platelet activating factor induced calcium mobilization, polyphosphoinositide turnover and granule enzyme secretion in rabbit neutrophils towards pertussis toxin and phorbol ester, Biochem. Biophys. Res. Commun. 130:677-684 (1985).

Nishizuka, Y., Calcium, phospholipid turnover and transmembrane signalling, Philos. Trans. R. Soc. London [Biol.] 302:101-112 (1983).

Okajima, F., Katada, T. and Ui, M., Coupling of the guanine nucleotide regulatory protein to chemotactic peptide receptors in neutrophil membranes and its uncoupling by islet-activating protein, pertussis toxin: A possible role of the toxin substrate in Ca^{2+}-mobilizing receptor-mediated signal transduction, J. Biol. Chem. 260:6761-6768 (1985).

O'Rourke, F. A., Halenda, S. P., Zavoico, G. B. and Feinstein, M. B., Inositol 1,4,5-trisphosphate releases Ca^{2+} from a Ca^{2+}-transporting membrane vesicle fraction derived from human platelets, J. Biol. Chem. 260:956-962 (1985).

Rittenhouse, S. E. and Sasson, J. P., Mass changes in Myoinositol trisphosphate in human platelets stimulated by thrombin: Inhibitory effects of phorbol ester, J. Biol. Chem. 260:8657-8660 (1985).

Ruggiero, M. and Lapetina, E. G., Leupeptin selectively inhibits human platelet responses induced by thrombin and trypsin; a role for proteolytic activation of phospholipase C, Biochem. Biophys. Res. Commun. 131:1198-1205 (1985).

Sternweis, P. C. and Robishaw, J. D., Isolation of two proteins with high affinity for guanine nucleotides from membranes of bovine brain, J. Biol. Chem. 259:13806-13813 (1984).

Ui, M., Islet-activating protein, pertussis toxin: A probe for function of the inhibitory guanine nucleotide regulatory component of adenylate cyclase, Trends Pharmacol. Sci. 5:277-279 (1984).

Verghese, M. W., Smith, C. D. and Snyderman, R., Potential role for a guanine nucleotide regulatory protein in chemoattractant receptor mediated polyphosphoinositide metabolism, Ca^{++} mobilization and cellular responses by leukocytes, Biochem. Biophys. Res. Commun. 127:450-457 (1985).

Volpi, M., Naccache, P. H., Molski, T. F. P., Shefcyk, J., Huang, C.-K., Marsh, M. L., Munoz, J., Becker, E. L. and Sha'afi, R. I., Pertussis toxin inhibits fMet-Leu-Phe- but not phorbol ester-stimulated changes in rabbit neutrophils: Role of G proteins in excitation response coupling, Proc. Natl. Acad. Sci. USA 82:2708-2712 (1985).

Wallace, M. A. and Fain, J. N., Guanosine 5'-0-thiotriphosphate stimulates phospholipase C activity in plasma membranes of rat hepatocytes, J. Biol. Chem. 260:9527-9530 (1985).

Watson, S. P. and Lapetina, E. G., 1,2-Diacylglycerol and phorbol ester inhibit agonist-induced formation of inositol phosphates in human platelets: Possible implications for negative feedback regulation of inositol phospholipid hydrolysis, Proc. Natl. Acad. Sci. USA 82:2623-2626 (1985).

Zavoico, G. B., Halenda, S. P., Sha'afi, R. I. and Feinstein, M. B., Phorbol myristate acetate inhibits thrombin-stimulated Ca^{2+} mobilization and phosphatidylinositol 4,5-bisphosphate hydrolysis in human platelets, Proc. Natl. Acad. Sci. USA 82:3859-3862 (1985).

MOLECULAR GEOMETRIES AND STERIC ENERGIES OF PHORBOL

10,11-DIACETATE AND 1,2-DIACETYLGLYCEROL MOLECULES

Ubaldo Leli, Mark Froimowitz and George Hauser

Ralph Lowell Laboratories, McLean Hospital, Belmont
Massachusetts 02178 and Departments of Biological
Chemistry and Psychiatry, Harvard Medical School
Boston, Massachusetts 02115

SUMMARY

Protein kinase C, an enzyme that is stimulated physiologically by
diacylglycerol (DAG) and phospholipids in the presence of Ca^{2+}, is involved
in a novel cellular signaling system that is activated by the binding of
appropriate agonists to certain classes of receptors. Phorbol esters are
tumor promoters that can replace DAG in the activation of protein kinase C.
Molecular similarities between the two compounds have been proposed to be
responsible for the capacity to activate the enzyme. We have studied the
molecular geometries and conformational energies of DAGAc and PDAc using
the Molecular Mechanics II program and parameter set developed by Allinger
and Yuh (1980). This was done to establish whether conformers of the two
compounds are geometrically similar and which hydroxyl group of the phorbol molecule corresponds to the C3 hydroxyl of DAG which must be unsubsti-
tuted for activation of protein kinase C.

INTRODUCTION

Diacylglycerol (DAG) is not only an important intermediate in the biosynthesis of phospholipids and triglycerides but also a second messenger
molecule in a recently discovered transmembrane signaling system of general
occurrence in animal cells (Berridge & Irvine, 1984; Hirasawa & Nishizuka,
1985; Hokin, 1985). The source of the DAG that acts as a second messenger
is the phosphodiesteratic cleavage of phosphatidylinositol 4,5-bisphosphate
when certain receptors are activated by appropriate agonists. The physiological role of DAG generated in this manner is now thought to lie in its
capacity to activate a calcium- and phospholipid-dependent protein kinase
(protein kinase C) (Kaibuchi et al., 1981) which can modify proteins by
phosphorylation, thus initiating a series of reactions leading to final
cellular responses possibly involved in neuronal responsiveness and plasticity. Phorbol diesters, a series of diterpenes which are tumor promoters
and potent skin irritants and are found in certain plants of the Euphorbia
genus (Ashendel, 1985), share with DAG the capacity to activate protein
kinase C (Nishizuka, 1984). The reason for this common activity has been
postulated to be the similarity of a certain region of the phorbol ester
molecule and DAG (Nishizuka, 1984) (Fig. 1). We calculated the molecular
geometries and steric energies of two model compounds for DAG and phorbol

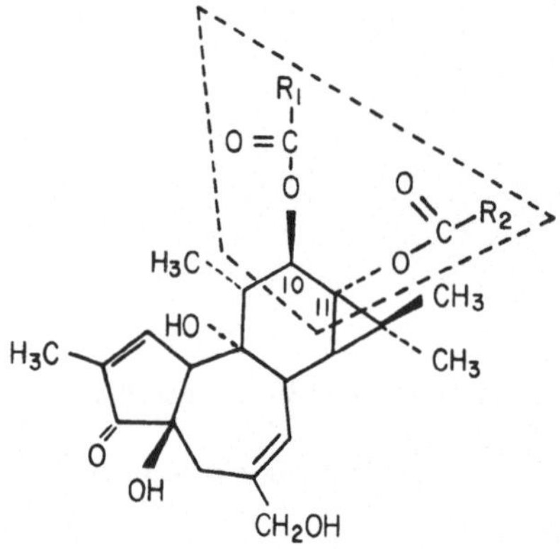

FIG. 1:   Phorbol 10,11-diacetate and diacylglycerol. The region of the
phorbol diester molecule that is similar to DAG is outlined with a
dotted line.

diesters, namely 1,2-diacetylglycerol (DAGAc) and phorbol 10,11-diacetate
(PDAc) in order to document the similarities of these two classes of molecules, and perhaps to explain certain structual features that are necessary
for the biological effect of DAG and phorbol esters.

Carbons 10 and 11 in the system of Pettersen et al. (1968), which is
used in this report, correspond to carbons 12 and 13 in the more commonly
used numbering system.

METHODS

Calculations of conformers of DAGAc and PDAc were performed using the
Molecular Mechanics II program (MM2) and parameter set developed by Allinger and Yuh (1980). Missing parameters for the ester groups in the vicinity of the cyclopropane ring were approximated by using the corresponding
parameters with sp^3 carbons substituted for the cyclopropane carbon atoms.
All starting conformations had their steric energies minimized with respect
to all internal coordinates.

There are seven dihedral angles that are conformationally significant
in DAGAc. Two of these (C6-C5-O1-C1 and C7-C4-O2-C2) were set at 180° due
to the strong preference of the molecule for that conformation. The
remaining five dihedral angles each have three possible conformations (1
trans and 2 gauche). Thus, as many as 3^5 or 243 different conformations
are possible. For PDAc, only C21-O6-C11-C13 and C22-O5-C10-C11 were varied.
These also have three possible conformations for a total of 9 possible conformers. As with DAGAc, the C23-C21-O6-C11 and C24-C22-O5-C10 dihedral
angles were set at their preferred value of 180°.

RESULTS

34 Conformers of DAGAc among the 243 that were submitted for energy
minimization shifted to other conformations and therefore were not considered energetically favorable. The conformers with lower energy were the
ones in which the dihedral angles C2-C3-O3-HO3 and O1-C1-C2-O2 were in a
<u>trans</u> and in a <u>gauche</u> conformation, respectively. The conformer with
lowest energy (10.9 Kcal/mole) is shown in Fig 2A. The dihedral angle of
PDAc O6-C11-C10-O5 is relatively fixed at 279° because C10 and C11 are part

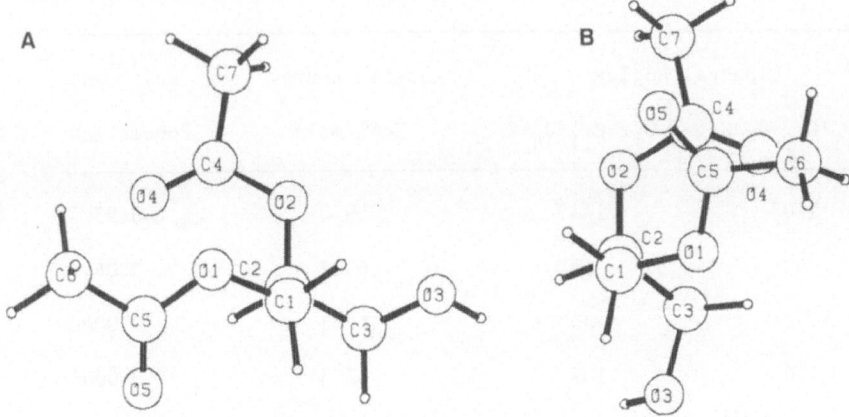

FIG. 2. A. Conformer of DAGAc of lowest steric energy (total steric
         energy = 10.9 Kcal/mole, probability of existence = 0.06). B.
         Conformer probability of DAGAc with geometry corresponding to
         PDAc (total steric energy = 14.1 Kcal/mole, probability of exi-
         stence = 0.002).

of a rigid ring system and the bond between them cannot rotate freely (Fig.
3). The six-membered ring of PDAc can assume a chair or a boat conformation, the former being energetically preferred. Four of the nine conformers of PDAc in the chair form, generated by rotating C11-C10-O5-C22 and
C13-C11-O6-C21, were unstable and shifted to other conformations during the
energy minimizations. The Boltzmann probabilities and the dihedral angles
of the five PDAc conformers in the chair conformation are shown in Table
1. The phorbol diester molecule contains a DAG-like structure (Fig. 1)
(Nishizuka, 1984), so that the dihedral angle O1-C1-C2-C3 in the DAGAc con-

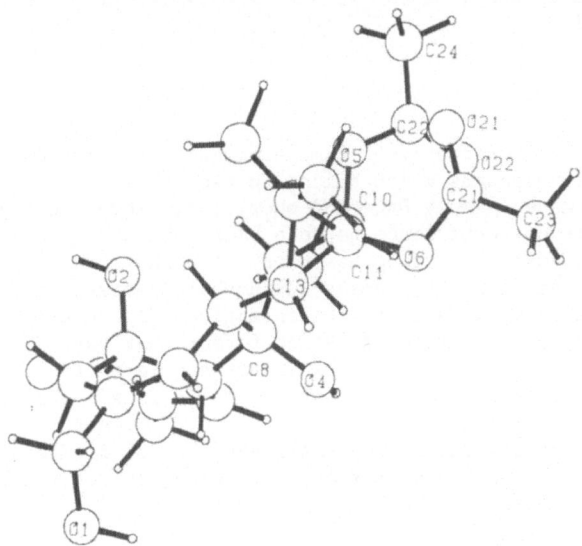

FIG. 3: Phorbol 10,11-diacetate conformer of lowest energy (total steric
         energy = 59.0 Kcal/mole, probability of existence = 0.95).

TABLE 1: Probability of Existence of the Conformers of PDAc (Chair Form)

Dihedral Angles		Steric Energy	Boltzmann
C_{11}-C_{10}-O_5-C_{22}	C_{13}-C_{11}-O_6-C_{21}	Kcal/mole	Population
110°	-141°	59.0	0.95
90°	63°	60.9	0.04
-120°	55°	65.3	0.0000
16	157	63.3	0.0006
-126°	-141°	63.7	0.0003

formers corresponding to PDAc should be about 300°. The DAGAc conformer which best matches the geometry of PDAc is shown in Fig. 2B. Using the Boltzmann equation at room temperature, the relative probability of existence of this conformer (Fig. 2B) is only 0.24%.

The free hydroxyl on C3 of DAG is essential for the activation of protein kinase C (Ganong et al., 1986). If phorbol esters act with the same mechanism, one would expect one of the free hydroxyl groups in the PDAc molecule to correspond to the free hydroxyl of DAGAc. We addressed this problem by calculating the distances between the oxygen of the hydroxyls and the carboxylic carbon of the acyl residues esterified at C10 and C11 of PDAc and comparing them with the distances of the free hydroxyl of DAGAc and the carboxylic carbons esterified at C1 and C2. The calculations revealed that the distance between the oxygen of the hydroxyl at C8 and the carboxylic carbon C21 in PDAc closely approximates the distance between the oxygen of the hydroxyl at C3 and the carboxylic carbon esterified at C1 of DAGAc (Table 2).

DISCUSSION

We have applied for the first time computational methods in order to compare two model compounds which belong to two classes of molecules of biological importance, namely DAGs and phorbol diesters. Our results confirm that there is correspondence between DAGAc and a region of the PDAc molecule. An interesting finding was that the bond between the two carbons which are esterified with fatty acids in phorbol diesters is fixed and the esterified residues are in a gauche conformation. In naturally occurring DAG the predicted concentration of the conformer that has a molecular geometry similar to the PDAc conformer of lowest steric energy is only about 0.2%. This might explain the great difference in potency between DAGs and phorbol diesters in activating protein kinase C. However, these conclusions cannot be directly extended to DAGs and phorbol diesters because of the importance of the nature of the acyl substituents for the biological activity of these molecules.

Our results suggest that the hydroxyl group of the PDAc molecule that corresponds to the free hydroxyl in DAGAc which is required for the activation of protein kinase C (Cabot and Jaken, 1984; Ganong et al., 1986) is the one bound to C8. The fact that the distance between the carboxylic carbon esterified at C11 and the hydroxyl at C8 is very close to that between

TABLE 2: Atomic Distances Between Hydroxyl Oxygen and Carboxylic Carbons
of DAGAc and PDAc

Compound	Distance Between	Angstroms
DAGAc*	03 - C4	4.3
	03 - C5	4.9
PDAc	02 - C22	6.1
	02 - C21	7.2
	04 - C22	5.1
	04 - C21	4.8
	01 - C22	9.9
	01 - C21	9.1

*Conformer with molecular geometry similar to PDAc.

the carboxylic carbon of the acyl residue at C1 and the free hydroxyl group of DAGAc suggests that the absence of a substituent on the C8 hydroxyl may be crucial for the biological activity of phorbol esters. It indicates further that an acyl residue at C1 and C11 on DAGs and phorbol diesters, respectively, may also be required. There is support for these suggestions from the fact that the deoxyphorbol monoesters at C11 are biologically active and the diterpene free alcohol phorbol is inactive as a skin irritant and tumor promoter (Hecker, 1978), properties that are related to the capacity of these compounds to activate protein kinase C (Blumberg et al., 1984). Also, alkyl acetylglycerols in which the ester bond at C1 is replaced by an ether bond are much less effective as activators of protein kinase C (Cabot & Jaken, 1984); further, the substitution of the ester with an amide group at C1 causes total loss of activity, whereas the same substitution at C2 results in a 10-fold decrease in potency (Ganong et al., 1986).

Calculations of molecular geometries and steric energies of analogs of the two classes of molecules under consideration, and experimental testing of the predictions derived by means of these computational methods through measurement of the potency of the same analogs in activating kinase C, represent a novel approach to the determination of the structure of the receptor for these molecules, and to further elucidation of fundamental molecular events related to signal transmission across cell membranes.

ACKNOWLEDGMENTS

This study was supported by research grants NS 06399 and NS 19047 from the National Institutes of Health, U.S.P.H.S.

REFERENCES

Allinger, N. L. and Yuh, Y. H., Quantum Chemistry Program Exchange, Department of Chemistry, Indiana University, Bloomington, IN, Program 395 (1980).

Ashendel, C. L., The phorbol ester receptor: A phospholipid-regulated protein kinase, Biochim. Biophys. Acta 822:219-242 (1985).

Berridge, M. J. and Irvine, R. F., Inositol trisphosphate, a novel second messenger in cellular signal transduction, Nature 312:315-321 (1984).

Blumberg, P. M., Jaken, S., König, B., Sharkey, N. A., Leach, K. L., Jeng, A. Y. and Yeh, E., Mechanism of action of the phorbol ester tumor promoters: Specific receptors for lipophilic ligands, Biochem. Pharmacol. 33:933-940 (1984).

Cabot, M. C. and Jaken, S., Structural and chemical specificity of diradylglycerols for protein kinase C activation, Biochem. Biophys. Res. Commun. 125:163-169 (1984).

Ganong, B. R., Loomis, C. R., Hannun, Y. A. and Bell, R. M., Specificity and mechanism of protein kinase C activation by sn-1,2-diacylglycerols, Proc. Natl. Acad. Sci. USA 83:1184-1188 (1986).

Hecker, E., Structure-activity relationships in diterpene esters irritant and cocarcinogent to mouse skin, in: "Carcinogenesis--A Comprehensive Survey," Vol. 2, "Mechanisms of Tumor Promotion and Carcinogenesis," T. J. Slaga, A. Sivak, and R. K. Boutwell, eds., Raven Press, New York, 11-48 (1978).

Hirasawa, K. and Nishizuka, Y., Phosphatidylinositol turnover in receptor mechanism and signal transduction, Ann. Rev. Pharma col. Toxicol. 25: 147-170 (1985).

Hokin, L. E., Receptors and phosphoinositide-generated second messengers, Ann. Rev. Biochem. 54:205-235 (1985).

Kaibuchi, K., Takai, Y. and Nishizuka, Y., Cooperative role of various membrane phospholipids in the activation of calcium-activated phospholipid-dependent protein kinase, J. Biol. Chem. 256:7146-7149 (1981).

Nishizuka, Y., The role of protein kinase C in cell surface signal transduction and tumor promotion, Nature 308:693-697 (1984).

Pettersen, R. C., Birnbaum, G. I., Ferguson, G., Islam, K. M. S. and Sime, J. G., X-ray investigation of several phorbol ester derivatives. The crystal and molecular structure of phorbol bromofuroate-chloroform solvate at -160°, J. Chem. Soc. (B), 980-984 (1968).

A MODEL OF THE LIGHT DEPENDENT REGULATION OF RETINAL ROD PHOSPHODIESTERASE, GUANYLATE CYCLASE AND THE CATION FLUX

M. W. Bitensky, D. Torney, A. Yamazaki, M. M. Whalen, and
J. S. George

Los Alamos National Laboratory, Los Alamos, New Mexico 87545

Early Observations

In the 1880's Kuhn dissected dark adapted eyes and observed that the retina's red/purple "visual pigment", changed rapidly to a pale yellow upon illumination. Years later, George Wald and Ruth Hubbard chemically characterized the rod visual pigment as consisting of opsin and 11-cis retinal (Wald, 1968). They demonstrated that the photoisomerization of rhodopsin was driven by a photon induced change in the configuration of 11-cis retinal (to all-trans), which was accompanied by conformational changes in the protein opsin. Subsequently, Wald, Yoshizawa and others were able to identify a series of spectral intermediates that appeared in rapid succession following the illumination of rhodopsin. The early intermediates could only be captured by stabilization at low temperatures or ultrafast spectroscopy. The major opsin photoconformers are called batho, hypso, meta I, meta II and meta III rhodopsin (Wald 1968; Birge, 1981). The meta II conformation was subsequently found to be enzymatically active (see below).

In the 60's Tomita observed that vertebrate rods and cones hyperpolarized in response to light (Tomita, 1970). Hagins and colleagues described a longitudinal "dark current" which flowed from inner to outer segment of the retinal rod and was diminished by illumination of rhodopsin (Hagins, Penn and Yoshikami, 1970). In invertebrate species such as Limulus, the photoreceptor membranes depolarize as a consequence of rhabdom illumination (Hartline, Wagner and MacNichol, 1952).

These observations together with the specialized topography of the vertebrate rod outer segment have provoked a fundamental question in the arena of visual science: how is the capture of photons within the disk membrane of the vertebrate rod communicated to the physically separate plasma membrane to produce a striking and rapid fall in sodium current and a consequent membrane hyperpolarization? The changes in the rod outer segment plasma membrane voltage are transmitted by the cable properties of the photoreceptor membrane to the rod synaptic terminus (Fig. 1) where membrane hyperpolarization may be associated with a decrease in the release of an inhibitory neurotransmitter. This ionic/molecular architecture suggests the possibility that rod photoreceptor elements, when at rest, continuously inhibit the retina. Upon illumination of specific photoreceptor elements, visual data are, in this view, communicated via the optic nerve to the occipital cortex as a consequence of the light mediated disinhibition of the corresponding retinal ganglion cells.

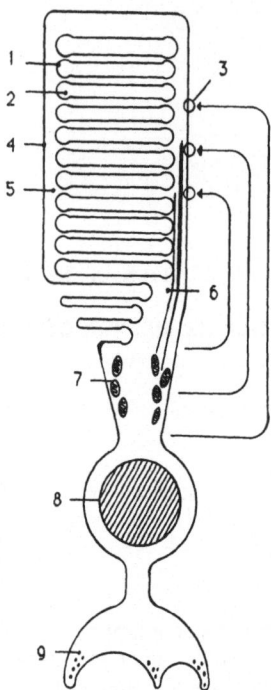

FIG. 1: Vetebrate Rod. 1. Disk Membrane. 2. Intradiskal space (contains Ca^{2+}). 3. Cation channel. 4. Plasma membrane. 5. Rod outer segment cytoplasm. 6. Connecting cilium with microtubules. 7. Mitochondria (inner segment). 8. Nucleus. 9. Synaptic terminus. Na^+ is actively extruded from the inner segment and returns via the cation channels in the outer segment membrane.

Photoreceptor Cell Biology

It was demonstrated in the 1940's on the basis of psychophysical experiments (Hecht, Schlaer and Pirenne, 1942), that the dark adapted vertebrate rod is a true quantum counter able to detect a single photon (Matthews and Baylor, 1981; Baylor, Lamb and K.-W. Yau 1979). Direct electrical coupling of the rod outer segment elements (Gold, 1981) and their extensive synaptic interconnections (via horizontontal, amacrine and bipolar cells to the retinal ganglion cells) suggested that the signals generated by photon capture within photoreceptors undergo significant signal processing and integration within the retina (Sterling, 1983).

This information is further elaborated within the lateral geniculate bodies prior to projection upon the occipital cortical matrix. However, measurements of photoreceptor membrane current in individual rod photoreceptors indicate that a large portion of light adaptation occurs within the photoreceptors themselves (Bastian and Fain, 1982; Lamb, McNaughton, and Yau, 1981); a moment's reflection reveals that if adaptation is to be effective in facilitating receptor function and extending its dynamic range this must be the case (Dawis, 1986).

Morphological studies of the rod indicate that the disks are continuously formed by infolding of the plasma membrane at the point of contiguity between the inner and outer segments. In the early 1970's, Young performed an elegant series of pulse labeling experiments which demonstrated that the insertion of newly synthesized disk proteins (and especially rhodopsin) was

primarily into the newly formed disks. Furthermore he was able to show
that these disks migrated gradually from the base to the apex of the rod
outer segment, where the senescent disks are removed by phagocytosis (Young,
1976). This phagocytic function resides in a pigment epithelium which
forms a (stray) photon absorbing layer at the back of the eye. Phagocytosis of distal disks appears to occur primarily during a burst of activity
following exposure of the rods to first light upon awakening (Burnside,
1978). In animals not exposed to a diurnal rhythm of illumination, there
can be abnormal elongation of the unilluminated rod outer segments.

Papermaster added a novel protein to the (then) modest disk membrane
protein inventory when he identified, by immunohistochemistry, the presence
of a large molecular weight, protein (>280 kd) associated primarily with
the margins of the disks and their radial infoldings, and apparently absent
from the internal plane of the disk membranes (Papermaster, Schneider, Zorn
and Krachenbuhl, 1978). Papermaster also demonstrated that rhodopsin, and
presumably other integral disk membrane proteins, are inserted into lipid
vesicles following translation in the inner segment. This vesicle (the
"schleppersome"), transports newly synthesized disk membrane integral proteins to the base of the outer segment where the vesicle fuses with the
plasma membrane (Papermaster, Schneider, Zorn and Krachenbuhl, 1978).

It is noteworthy that rhodopsin and the closely related cone photopigment molecules evolved at an early period in the development of life on
earth. An extraordinary number of invertebrate and vertebrate species
exhibit photoreceptive organs which utilize rhodopsin or related molecular
species. Recently, Foster and Nakanishi have shown that the visual pigment
of the algae Clamydamonas has spectral characteristics analogous to those
of vertebrate rhodopsin (Foster, Saranak, Patel, Zarilli, Okabe, Kline and
Nakanishi, 1984). Moreover, this photopigment interacts with a series of
synthetic retinal isomers with affinities which are homologous to the
interactions of bovine opsin with the same isomers. These experiments
demonstrate that both mammalian and clamydamonas opsins (the protein moieties of rhodopsin) share a conserved hydrophobic pocket within which
isomers of vitamin A aldehyde are bound.

The Transduction Process

In the early 1970's, Hagins and Yoshikami suggested a model of visual
transduction that became known as the "Ca^{2+} hypothesis" (Hagins, 1972;
Hagins and Yoshikami, 1974). This model suggested that rod disks which are
closed, flattened membrane saccules (see Figure 1) actively acquire and
release Ca^{2+}. It was proposed that Ca^{2+} was released as a result of the
illumination of rhodopsin. This Ca^{2+} was utilized as a cytoplasmic messenger ion to modulate the cation channels of the rod outer segment membrane. This attractive concept was supported by a variety of compelling
experiments. Hagins, and others, found that elevation of Ca^{2+} in the
extracellular milieu of the rod (Yoshikami and Hagins, 1973; Yau, McNaughton, and Hodgkin, 1981), particularly in the presence of ionophore or the
addition of Ca^{2+} to rods by iontophoresis (Brown, Coles and Pinto, 1977)
resulted in a hyperpolarization of the rod plasma membrane that resembled
the light response. Furthermore, lowering external Ca^{2+} (Yoshikami and
Hagins, 1973) or introduction of cytoplasmic Ca^{2+} chelators by vesicle
fusion (Hagins and Yoshikami, 1977) reduced the sensitivity of the photoresponse. Although this model was extremely attractive and was further
enhanced by analogy with the sarcoplasmic reticulum, it faced at least two
problems. First, release of Ca^{2+} upon illumination of isolated disks,
although demonstrated, (Smith and Bauer, 1979; George and Hagins, 1983) was
not observed in quantities and with kinetics appropriate to support light
regulation of the sodium channel. Second, while indirect experiments suggested such Ca^{2+} movements might occur (Yoshikami, George, and Hagins,

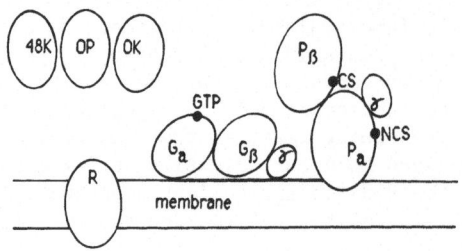

FIG. 2: Light Dependent Enzymes of Vertebrate Rod. R: rhodopsin. 48K: rhodopsin inhibitory protein. O.K.: opsin kinase. O.P.: opsin phosphatase. $G\alpha,\beta,\gamma$: subunits of the GTP binding protein. $P\alpha,\beta,\gamma$: subunits of the PDE. $P\gamma$ is thought to function as an inhibitory subunit. CS: Catalytic site (subunit localization is speculative). NCS: Non-catalytic site(s) (subunit localization is speculative). Capture of photons by R enables $G\alpha$ to bind GTP. GTP·$G\alpha$ activates PDE by changing its relationship with $P\gamma$.

1980; Gold and Korenbrot, 1980) direct measurements of rod cytoplasmic Ca^{2+} during illumination have not been made. Nevertheless, in view of the provocative observations of Hagins and coworkers and the demonstration by Fain and others that disks do indeed contain abundant stores of Ca^{2+} (Schroder and Fain, 1984; Puckett, Aronson, and Goldin, 1985) this hypothesis was accepted for some time as the most compelling explanation of visual excitation in vertebrate rods.

In 1970 Bitensky, Gorman and Miller began a series of experiments that introduced the idea that cyclic nucleotides might play a role in visual transduction. These studies suggested that a light initiated fall in cyclic nucleotide concentrations was mediated by rhodopsin and was closely affiliated with the initial events of visual excitation. In the earliest studies these investigators found a fall in cyclic AMP as a consequence of illumination of photoreceptor membrane suspensions (Bitensky, Gorman, and Miller 1971). The observation that light could lower the concentrations of cyclic nucleotides in suspensions of photoreceptor membranes was valid, but subsequent experiments by Pannbacker (Pannbacker, 1973) and Miki et al. (Miki, Baraban, Keirns, Boyce, and Bitensky, 1975) revealed that the cyclic nucleotide metabolism of the rod was primarily cyclic GMP metabolism. Miki and others also found that light activated cyclic GMP phosophodiesterase (PDE) (rather than inhibiting adenylate cyclase activity) and that PDE activation required the presence of a nucleoside triphosphate. The PDE's preference for cyclic GMP over cyclic AMP was more than 20:1 at substrate concentrations in the 100 µM range.

Light Activated Phosphodiesterase

The early experiments by Miller and others, suggested that rhodopsin was the most likely photopigment associated with the light activation of rod PDE. These findings were later confirmed in two ways. 1) Keirns and others measured a detailed action spectrum for PDE and found that it corresponded perfectly to the action spectrum of rhodopsin (Keirns, Miki, Bitensky and Keirns 1975). Shinozawa, and others, were also able to show that chromatographically purified rhodopsin could activate the photoreceptor PDE complex (Shinozawa, Uchida, Martin, Cafisco, Hubbell, and Bitensky, 1980).

Liebman and coworkers developed a very useful and rapid technique for measuring the enzymatic activity of the phosphodiesterase (Yee and Liebman, 1983). They found that this enzyme's activity could be recorded in real

SIGNAL	RECEPTOR	AMPLIFIER	CATALYTIC MOIETY
LIGHT	RHODOPSIN	$G_{\alpha,\beta,\gamma}$	PDE
CHEMICAL AGONIST	MEMBRANE RECEPTOR	N_S $_{\alpha,\beta,\gamma}$	ADENYLATE CYCLASE

FIG. 3: Functional and Structural Homology Between Light Activated PDE and Hormone/Neurotransmitter Activated Adenylate Cyclase. N_S is the stimulatory guanine nucleotide binding protein of Adenylate cyclase. The inhibitor of PDE is also known to inhibit adenylate cyclase activity.

time by exploiting the generation of protons which follows as a consequence of cGMP hydrolysis. Using this approach they were able to show that the turnover number for PDE was higher than had been estimated with purified enzymes measuring GMP production with radioisotopes. They were also able to obtain succinct kinetics because of the ability to record enzyme rates rather than measuring an isotopically labelled product at a single time (Liebman and Pugh, 1981).

In an interesting series of experiments Cohen, Hall, and Ferendelli (Cohen, Hall, and Ferendelli, 1978) demonstrated that extreme lowering of $[Ca^{2+}]$ in the medium bathing a retina caused a striking increase in cyclic GMP concentrations in the rod outer segments to levels 10-fold greater than observed when the bathing medium contained physiological concentrations of Ca^{2+}. These experiments were among the earliest to suggest the possibility that there could be a coupling between Ca^{2+} concentrations and cyclic nucleotide metabolism.

In the middle 1970's Wheeler and others demonstrated that the illumination of rhodopsin and the activation of PDE were linked by the binding of GTP to a GTP binding protein in the rod (see Fig. 2). These investigators were able to show that this GTP binding protein had light-activated GTPase activity and that the action spectrum for rod GTPase was identical to the action spectra for rhodopsin and PDE. Moreover, Wheeler et al, found that hydrolysis of GTP corresponded to the inactivation of the light-activated PDE (Wheeler and Bitensky, 1977; Wheeler, Matsuo and Bitensky, 1977).

Subsequently, Shinozawa, and others, found that the GTP binding protein was a complex consisting of at least two different components one of which was shown to bind GTP and a second or helper subunit which markedly facilitated both binding and turnover of GTP (Shinozawa, Uchida, Martin, Cafisco, Hubbell and Bitensky, 1980). In subsequent experiments Uchida, and coworkers (Uchida, Wheeler, Yamazaki and Bitensky, 1981) and Stryer et al, (Fung and Stryer, 1980; Stryer, Hurley and Fung, 1981) independently found that the α subunit of the GTP binding complex ($G\alpha$), which weighed about 39 Kd, could activate PDE following the formation of a complex GTP ($G_{\alpha}\cdot GTP$). Uchida et al. identified the subunit as the locus of GTP binding by photoaffinity labeling (Uchida, Wheeler, Yamazaki and Bitensky, 1981). The activation of PDE by the purified subunit bound to GPP(NH)p (a non-hydrolyzable GTP analogue) was no longer dependent on the presence of

111

bleached rhodopsin, notwithstanding the fact that bleached rhodopsin was essential for the initial binding of the GTP analog to the G_α subunit. The helper component was found to consist of two subunits: G_β (Mr 35000) and G_γ (Mr 6500) (Fung and Stryer, 1981).

In the middle 1970's Wheeler, Bitensky, and others, began a series of studies that demonstrated an extraordinary and pervasive homology between the light activated photoreceptor system and the hormone activated adenylate cyclase system (see Fig. 3) (Shinozawa, Sen, Wheeler and Bitensky, 1979). A review describing the homology between the two systems was published by Pober and Bitensky in the late 1970's (Pober and Bitensky, 1979). Bitensky, and others, found interactions between the components of the photoreceptor cascade and components of the adenylate cyclase regulatory cascade in chimeric systems (Bitensky, Wheeler, Rasenick, Yamazaki, Stein, Halliday and Wheeler, 1985). Other examples of such interactions were subsequently reported from a number of laboratories (Kanaho, Tsai, Adamik, Hewlitt, Moss and Vaughan, 1985; Cerrone, Codina, Kilpatrick, Staniszewski, Gershik, Somers, Spiegel, Birnbaumer, Caron and Lefkowitz, 1985). Perhaps one of the most interesting examples of a hybrid system is provided by the observation that the inhibitory moiety of the rod phosphodiesterase could inhibit various forms of adenylate cyclase (rat brain synaptosomal and amphibian erythrocyte) and, furthermore, that this inhibition was relieved by the addition of GTP (Bitensky, Wheeler, Rasenick, Yamazaki, Stein, Halliday and Wheeler, 1982). Recently, Bitensky and Yamazaki found that a soluble (rat testis) adenylate cyclase is fully inhibited by the PDE inhibitor. This inhibition is fully reversed by $G_\alpha \cdot$ GTPγs from the retinal rod (Bitensky and Yamazaki, in preparation). Further examination of the mechanism of the activation of PDE by Yamazaki et al., revealed that the light activation of phosphodiesterase was a consequence of modulation by a heat stable inhibitory moiety, which at that time had not been implicated in the light activated cascade. These workers found that phosphodiesterase isolated in the absence of the inhibitory moiety, was already active and was not further activated by light and GTP (Yamazaki, Stein, Chernoff and Bitensky, 1983). They also found that the addition of the $G_\alpha \cdot$GTP complex to an inactive phosphodiesterase resulted in release of an inhibitory moiety to the supernatant fraction in the amphibian system and concomitant activation of PDE. Moreover, the hydrolysis of GTP bound to G_α was accompanied by return of both the inhibitor and the G_α subunit (now bound to GDP) to the disk membrane surface. The ability of the $G_\alpha \cdot$GTP complex to release the PDE inhibitory moiety has not yet been observed in the bovine system, although an analogous mechanism is thought to operate.

Yamazaki and coworkers also reported the existence of a novel type of non-catalytic cyclic GMP binding site(s) located on one or both of the two larger subunits of the light activated phosphodiesterase (Yamazaki, Sen, Casnelli, Greengard and Bitensky, 1980). These sites were clearly distinguishable from the catalytic site in a number of ways. For example, isobutylme-thylxanthine inhibits the catalytic function of PDE (presumably by competition for the catalytic site) yet stimulates binding of cyclic GMP to the non-catalytic sites. This is also true for the heat stable (endogenous) inhibitory moiety of PDE (Yamazaki, Bartucca, Ting and Bitensky, 1982). While the catalytic site could hydrolyze both cyclic GMP and cyclic AMP (with a marked preference for cyclic GMP), the non-catalytic sites showed strict specificity for cyclic GMP, with a Km below 1 μM. Cyclic GMP could not be displaced from the non-catalytic sites even by a 100-fold excess of cyclic AMP. Furthermore, the non-catalytic sites are rapidly eliminated by trypsin, which does not compromise the catalytic site.

During the late 1970's, Miller and Nicol found that the iontophoretic infusion of cyclic GMP into the rod caused ROS membrane depolarization and an increase in (dark) cation flux (Nicol and Miller, 1978; Miller and Nicol, 1979). Such depolarization of the rod by exogenous cyclic GMP is not observed immediately following a saturating light flash or pulse but does reappear a few tens of seconds after the peak of the light-induced membrane hyperpolarization (Miller and Nicol, 1981). These experiments suggested: 1) that cyclic GMP can somehow regulate the cation channels in the rod; and 2) that after light activation of rod phosphodiesterase, the enzyme's activity persists well after the return of membrane current and voltage to near dark levels for large bleaches. Recent work by Hodgkin, McNaughton and Nunn assays both phosphodiesterase activity and cation current at intervals subsequent to a flash giving 5700 isomerizations/rod (Hodgkin, McNaughton and Nunn, 1985). It is found that the decay of the phosphodiesterase activity to its dark level closely approximates the return of the membrane current to its dark level.

A recent series of innovative experiments which were first reported by Fesenko, et al., (Fesenko, Kolesnikov and Lyubarsky, 1985) and by Nakatani and Yau (Nakatani and Yau, 1985), found that the cation channels of the rod outer segment plasma membrane can be directly studied by excising a membrane patch from the outer segment which is sealed to the tip of a micropipette. The current through such a membrane patch can be readily recorded electronically. The amount of current is clearly sensitive to the nature, abundance and distribution of the different ionic species surrounding the membrane patch. A most striking observation, however, is that cyclic GMP can markedly increase the flow of cations through this membrane patch and that removal of cyclic GMP is associated with a marked reduction in cation flux. The conductance c of the patch is found to obey the following equation with $K = 30$ μM:

$$c = [cGMP]^{1.8} / (K + [cGMP]^{1.8})$$

With micromolar calcium on either side of the patch the fraction of channels open and also the conductance follow this equation with K roughly 40 μM. However, with nanomolar calcium on both sides of the patch the exponent is very close to 3.0 and K is smaller (Haynes, Kay and Yau, 1986; Zimmerman and Baylor, 1986).

Under typical experimental conditions, Ca^{2+} does not significantly influence the flow of cations through this outer segment membrane patch. New experiments with modified experimental architectures have been carried out using intact or broken rod outer segments within suction micropipettes (Yau and Nakatani, 1985). These experiments have strengthened the conclusion that cyclic GMP itself is the direct determinant of the patency of the cation channels in the rod outer segment membrane.

An Alternative View of Ca^{2+} Dynamics

In addition, Yau, and others, have suggested that there is actually a fall in cytoplasmic Ca^{2+} associated with the illumination of outer segments, rather than a rise which had been proposed by Hagins, and others, as a part of the Ca^{2+} hypothesis. The suggestion that light is associated with a fall in cytoplasmic Ca^{2+} follows from the observation that the major influx of Ca^{2+} is through the cation channels in the dark rod. Moreover, Ca^{2+} is constantly extruded from the rod by a sodium/calcium exchanger. The light induced closing of the cation channels would be expected to cause a fall of the intracellular Ca^{2+} as a consequence of the light induced cessation of Ca^{2+} influx and its continuous efflux via the cation exchange mechanism

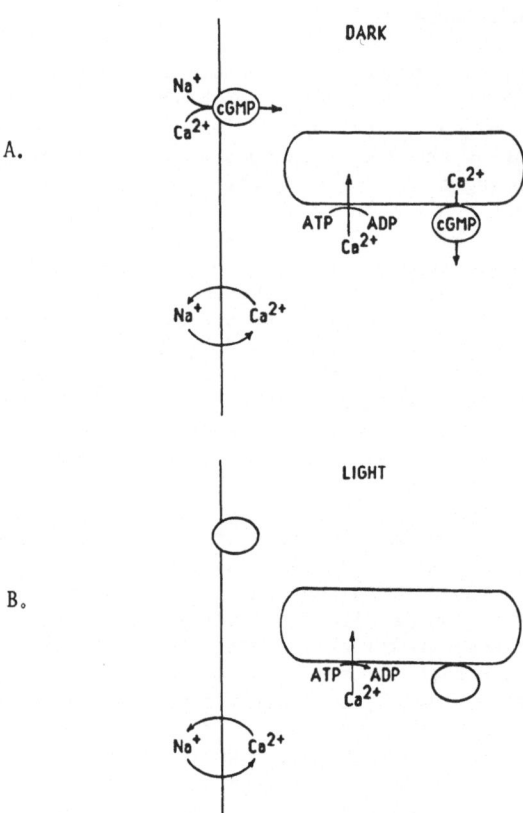

Fig. 4: Plasma and Disk Membrane Influences on Cytoplasmic Ca^{2+}. A. In the dark, extracellular and disk Ca^{2+} enter the cytoplasm via the cGMP dependent cation conductance. Na^+/Ca^{2+} exchange through the plasma membrane and ATP dependent Ca^{2+} uptake by the disks oppose this flux. B. After light exposure, the cGMP dependent conductances in both membrane systems are closed. In the presence of continued Ca^{2+} removal (by exchange and active uptake) cytoplasmic Ca^{2+} levels may fall.

(see Fig. 4). Gold has recently demonstrated this calcium efflux from the retina (Gold, 1986), while the membrane current remains near zero. For seconds after the initial large rate of calcium efflux, there is a sustained rate (approximately 1/10 the initial rate) which terminates at the time of recovery of the membrane current. Recently, Cervetto, et al. using the Ca^{2+} sensitive photoprotein aequorin, found that cytoplasmic Ca^{2+} appeared to rise when the rod dark current was increased by methylisobutyl-xanthine, a PDE inhibitor which increases cytoplasmic cGMP. The Ca^{2+} influx which results, produces light emission from aequorin, followed by cation channel closure and a transient decrease in cytoplasmic Ca^{2+} (Cervetto, McNaughton and Nunn, 1986). Moreover introduction of such Ca^{2+} chelators as quin II (Miller and Korenbrot, 1986) or BAPTA (Mathews, Torre and Lamb, 1986) has little effect on the rising phase of the light response, but it produces an increase in the duration of the light response and an overshoot (a depolarization) upon recovery. One way to interpret this finding is to suggest that initiation of the photoresponse depends upon the activation of PDE and is relatively independent of changes in Ca^{2+} concentration. In this context, the introduction of methylxanthines or cyclic GMP is associated with a prolongation of the initial phase of the light

$$h\nu \rightarrow R^* \rightarrow GTP{\cdot}G_\alpha \rightarrow PDE^*$$

$$PDE^* \longrightarrow \downarrow cGMP \rightarrow Channels\ Close$$

$$Channels\ Close \longrightarrow \downarrow Ca^{2+} \rightarrow G\ Cyclase^*$$

$$G\ Cyclase^* \longrightarrow \uparrow cGMP \rightarrow Channels\ Open \rightarrow \uparrow Ca^{2+}$$

$$\uparrow Ca^{2+} \rightarrow G\ Cyclase_{inactive}$$

FIG. 5:   Coupled Sequence Model of Visual Transduction.

response (Lipton, Rasmussen and Dowling, 1977) also suggesting that the consumption of cyclic GMP by phosphodiesterase provides a major component of this response.

The return of the sodium current and the repolarization of the membrane to dark levels may depend upon a fall in cytoplasmic Ca^{2+} levels. It is this fall in Ca^{2+} which could be delayed by the buffering of calcium and its slow release from Ca^{2+} chelating agents. Clearly, this concept is based upon direct evidence. The precise and direct measurement of cytoplasmic Ca^{2+} during the light response is perhaps more critical than ever.

Toward a New Model of Transduction

To briefly recapitulate this one hundred-odd year history of research into the molecular mechanisms of the visual response, the key elements which have emerged include the following:

1. Light bleaches and activates a macromolecular chromophore rhodopsin.

2. Rhodopsin bleaching triggers activation of a GTP binding protein (i.e., the formation of $G\alpha{\cdot}GTP$) which subsequently interacts with (an inhibitory moiety of) cGMP PDE, to activate the enzyme.

3. The fall in membrane current (the membrane hyperpolarization) appear associated with a phosphodiesterase mediated fall in cytoplasmic cyclic GMP.

4. This reduction in membrane current may be followed by a fall in cytoplasmic calcium.

In the context of these observations it appears reasonable to suggest that the reopening of the sodium channels reflects a combination of both the decrease in activity in PDE and a surge of guanylate cyclase activity responding to the fall in Ca^{2+} that follows the closing of the cation channels by light. This surge in guanylate cyclase activity would restore cyclic GMP toward dark levels. Guanylate cyclase activity must replace the cGMP which has been hydrolyzed by PDE and compensate for PDE activity which persists after the light response. The proposed mechanism permits a relatively rapid return of the rod to a functional status: the stimulated rod is soon prepared to respond again to new photon input.

These, and other observations, can be assembled to support a series of fundamental assumptions about the photoresponse:

1. The cation flux is governed by cyclic GMP.

115

2. Guanylate cyclase activity is governed by calcium.

3. Cytoplasmic calcium is governed by the patency of the cation channels, the activity of the cation exchanger and also by Ca^{2+} dynamics involving uptake by and release from the disk membrane.

The proposed model couples the reestablishment of resting conditions, i.e., the opening of the cation channels, to a fall in cytoplasmic calcium, produced by channel closure (see Fig. 5). This postulated series of coupled responses links the consequences of channel closure (a fall in Ca) to a mechanism which restores membrane voltage to resting levels (a surge in guanylate cyclase activity). The proposed model is supported by a variety of biochemical and electrophysiological observations. For example, isobutylmethylxanthine (Lipton, Rasmussen and Dowling, 1977), might slow the rate of excitation by increasing the amount of cyclic GMP and reducing hydrolytic rates. These data anticipate the data of Fesenko (Fesenko, Kolesnikov and Lyubarsky, 1985), and Yau (Nakatani and Yau, 1985; Yau and Nakatani, 1985), which directly show that cGMP can regulate the cation channels. Calcium buffers could modify the light response in the observed manner by preventing rapid changes in cytoplasmic calcium. Calcium buffers could delay the surge of cyclic GMP that may occur as a consequence of guanylate cyclase activation by falling Ca^{2+} levels and delay the subsequent shutdown of cyclase by rising Ca^{2+} levels, causing a slow recovery of the dark current.

The suggested dual enzyme model also allows the rod to escape from the necessity of rapidly turning off PDE before it can restore the sodium current by permitting falling Ca^{2+} levels to orchestrate a surge in guanylate cyclase activity. This model also is in harmony with the data of Goldberg et al., which demonstrate a striking flux through the cyclic GMP pathway indicating a strong synthetic capability in guanylate cyclase (Goldberg, Ames, Gander and Walseth, 1983).

The sharp fall in cyclic GMP followed by the sharp recovery is reminiscent of a mechanism first suggested by Professor Shiro Kakiuchi (Kakiuchi and Rall, 1968). Professor Kakiuchi proposed that the synthesis of cyclic AMP by adenylate cyclase in response to a neurotransmitter agonist results in an increase in the cytoplasmic Ca^{2+} level, due to the ability of cyclic nucleotides to mobilize calcium both from intra- and extracellular stores. This surge in calcium activates calmodulin dependent neuronal phosphodiesterase which rapidly hydrolizes the newly synthesized cyclic AMP. This produces a sudden rise and fall (a pulse) in cyclic AMP levels as a consequence of neurotransmitter activity. In Shiro Kakiuchi's "square wave" mechanism, adenylate cyclase is responsible for the downward limb. The model proposed here is a mirror image, in which light activated PDE accomplishes the initial reduction in levels of cGMP, and a Ca^{2+} sensitive guanylate cyclase restores the deficit.

The physiological role of the non-catalytic binding sites described by Yamazaki, et al. (Yamazaki, Bartucca, Ting and Bitensky, 1982), is still uncertain. Since the cGMP affinity of these sites is lowered by activation of PDE, they could provide a rapid burst of cyclic GMP hydrolysis as the phosphodiesterase rapidly donates cyclic GMP to its own catalytic sites. However, the small burst of cGMP hydrolysis generated by release of cGMP bound to the non-catalytic sites represents a very modest fraction of the total flux activity observed by Goldberg and collaborators (Goldberg, Ames, Gander and Walseth, 1983). Such data suggest that guanylate cyclase is somehow activated as a consequence of PDE activation. Cyclic GMP binding to the non-catalytic sites on PDE might also influence the Km of cGMP at the catalytic site and thus regulate PDE.

116

Light Adaptation

The implications of the proposed sequential multienzymic mechanism for the mechanism of light adaptation are also not clear. The transient reduction of photoreceptor sensitivity following light exposure requires the creation of a persistent ionic or biochemical consequence which reflects the capture of photons by the rod disk membranes. This might be accomplished by a decrease in the number of rhodopsins which are capable of responding to light since rhodopsin phosphorylation occurs as a consequence of bleaching and is followed by binding of a 48 kD protein (Wilden, Hall and Kuhn, 1986). This seems a rather unsatisfactory explanation for adaptation in view of the great number of rhodopsins in the rod (3×10^9) and given that adaptation can occur even with modest bleaches which would not significantly affect the functional status of so vast an ocean of rhodopsins.

The spectrum of biochemical choices to explain adaptation encompasses other components of the activation sequence including transient functional modification of the GTP binding protein subunits or the phosphodiesterase subunits. The possibility that G_α (following hydrolysis of GTP to GDP) may be temporarily unavailable for further reaction, seems remote since $G_\alpha \cdot GDP$ has a strong affinity for bleached rhodopsin. Perhaps the most intriguing model is that following PDE activation and recovery of the dark current, residual phosphodiesterase activity is balanced by persistent guanylate cyclase activity, thus reducing the effective pool of PDE molecules available for <u>de novo</u> activation by light. One flash followed rapidly by another, would thus require a larger number of photons in the second flash to produce a comparable fall in cGMP levels thereby reducing the intrinsic gain of the system.

The Coupled-Sequence Model

The aggregate of this data is assembled into a speculative model, which synthesizes the light response from a series of sequentially linked reactions: the first event is rhodopsin bleaching which drives the formation of $GTP \cdot G_\alpha$. This provides explosive activation of PDE which catalyzes the hydrolysis of cyclic GMP. A rapid fall in cyclic GMP concentration produces the initial (short latency) phase of the light response a fall in current. Shutdown of the plasma membrane cation conductance allows Na^+/Ca^{2+} exchange to produce a fall in rod cytoplasmic calcium. Guanylate cyclase activity surges as rod cytoplasmic Ca^{2+} falls, thus restoring cGMP and membrane current to resting (dark) levels. Ca^{2+} levels rise as the current is restored producing a shutdown of guanylate cyclase. Other factors which must eventually be included in a more comprehensive model are the disk Ca^{2+} flux, the effects of phosphatidylinositol metabolism (Schmidt, 1983; Waloga and Anderson, 1985), and the functional explanation of the non-catalytic cGMP binding sites on PDE.

The model (still incomplete) is built upon the following assumptions: 1) sodium current is entirely controlled by (cyclic GMP) 2) Ca^{2+} principally regulates the activity of guanylate cyclase (and may also influence the phosphodiesterase activity in a complex way), 3) the model also is based on the idea that at any time, the free [cGMP] is determined by integration of the PDE and guanylate cyclase activities and the affinity of cyclic cGMP for its membrane cytoplasmic binding sites. The following experimental observations are also intergrated into the model: 1) activation of PDE is explosive, easily fast enough to account for the initial transient of the light response; 2) cyclic-GMP increases cation flux through various preparations of rod outer segment plasma membranes; 3) isobutylmethylxanthines lower the membrane voltage and increase the current by raising cGMP; 4) a fall in external calcium results in an increase in cyclic-GMP concentrations and a jump in external Ca^{2+} mimics the light response. These are attributed to an inhibition of guanylate cyclase by calcium; 5) light activated phosphodiesterase activity does not necessarily turn off as quickly as the termination of the light response; 6) steady light produces a tremendous flux through the cyclic-GMP pool, suggesting a futile cycle aspect of the model, specifically the "competition" between PDE and guanylate cyclase; 7) under certain experimental conditions there is a prolongation of the light response e.g., injection of calcium buffers. Computational and Mathematical Models of the Photoresponse:

We have recently generated a 3-dimensional Monte Carlo simulation of the photoresponse (Torney and Bitensky, 1986). The simulation models four components; rhodopsin, $G\alpha$, PDE, and cyclic GMP. The response predicts a rapid decline in the free cyclic GMP pool as a function of time. The model is interesting in its potential to characterize evolutionary constraints which define the stoichiometry of the responding components and the light response time. For example, if one reduces the number of $G\alpha$ molecules to one-half of the observed abundance, what are the consequences for the kinetics of the light response? The model is also useful for examining potential molecular loci which could account for the light desensitization phenomena. Here the model may predict the numbers of $G\alpha$, rhodopsins or PDE molecules which might be "inactivated" by a measured input of photons and how these changes could affect the gain and time course of a subsequent response.

The model described throughout the paper has been translated into a coupled system of six nonlinear ordinary differential equations, which contain eight rate constants. The first choice for the equations giving the (time) rate of change of [cyclic GMP] and $[Ca^{2+}]$ is shown here:

$$d \, [\text{cyclic GMP}]/dt = K_1/(1+[Ca^{2+}]_i/K_{ca}) - K_2[\text{PDE*}][\text{cyclic GMP}],$$

$$d \, [Ca^{2+}]_i/dt = K_3[\text{cyclic GMP}]^{1.8}/(K_{CG}+[\text{cyclic GMP}]^{1.8})$$

$$- \, K_4[Ca^{2+}]_i$$

$[Ca^{2+}]$ is the free cytoplasmic Ca^{2+} concentration, and [PDE*] is the light activated PDE concentration. The term with K_1 describes cyclic GMP synthesis as a function of $[Ca^{2+}]$ and has calcium acting as an inhibitor of guanylate cyclase, the term with K_2 describes enzymatic hydrolysis of cyclic GMP, the term with K_3 describes Ca^{2+} influx through the cGMP dependent cation channel, and the term with K_4 represents Ca^{2+} removal from the cytoplasm of the rod outer segment by Na^+/Ca^{2+} exchange. The membrane current predicted by this model is currently being compared with experimental data to check for consistency. For all bleaches above about 100 isomerizations per rod such an ordinary differential equation formulation should adequately reflect the biochemical kinetics but one may be able to extend the model also to encompass single isomerization events.

REFERENCES

Bastian, B. L. and Fain, G. L., The Effects of Sodium Replacement on the Responses of Toad Roads, J. Physiol. (London) 330:331-347 (1982).

Birge, R. R., Photophysics of Light Transduction in Rhodopsin and Bacteriorhodopsin, Ann. Rev. Biophys. Bioeng. 10:315-354 (1981).

Bitensky, M. W., Gorman, R. E. and Miller, W. H., Adenyl Cyclase as a Link Between Photon Capture and Changes in Membrane Permeability and Frog Photoreceptors, Proc. Natl. Acad. Sci. USA 68:561-562 (1971).

Bitensky, M. W., Wheeler, M. A., Rasenick, M. M., Yamazaki, A., Stein, P. J., Halliday, K. R. and Wheeler, G. L., Functional Exchange of Components Between Light-activated Phosphosiesterase and Hormone-activated Adenylate Cyclase, Proc. Natl. Acad. Sci. USA 79:3408-3412 (1982).

Bitensky, M. W. and Yamazaki, A., Inhibition of Testis Adenylate Cyclase by Rod PDE Inhibitor, in preparation.

Bownds, M. D., Biochemical Steps in Visual Transduction: Roles for Nucleotides and Calcium Ions, Photochem. Photobiol. 32:487-490 (1980).

Brown, J. E., Coles, J. A. and Pinto, L. H., Effect of Injections of Calcium and EGTA into the Outer Segments of Retinal Rods of Bufo Marinus, J. Physiol. (London) 269:707-722 (1977).

Burnside, M. B., Possible Roles of Microtubules and Actin-Filaments in Retinal Pigmented Epithelium, Exp. Eye Res. 23:257-275 (1978).

Cerrone, R. A., Codina, J., Kilpatrick, B. F., Staniszewski, C., Gershik, P., Somers, R. L., Spiegel, A. M., Birnbaumer, L., Caron, M. G. and Lefkowitz, R. J., Transducin and the Inhibitory Nucleotide Regulatory Protein Inhibit the Stimulatory Nucleotide Regulation Protein Mediated Stimulation of Adenylate Cyclase in Phospholipid Vesicle Systems, Biochem. 24:4499-4503 (1985).

Cervetto, L., McNaughton, P. A. and Nunn, B. J., Calcium Current and Aequorin Signals in Isolated Salamander Rods, Biophys. J. 49:281a (1986).

Cohen, A. I., Hall, I. A. and Ferrendelli, J. A., Calcium and Cyclic Nucleotide Regulation in Incubated Mouse Retinas, J. Gen. Physiol. 71:595-610 (1978).

Fesenko, E. E., Kolesnikov, S. S. and Lyubarsky, A. L., Induction by Cyclic GMP of Cationic Conductance in Plasma Membrane of Retinal Rod Outer Segments, Nature (London) 313:310-313 (1985).

Foster, K. W., Saranak, J., Patel, N., Zarilli, G., Okabe, M., Kline, T. and Nakanishi, K. A., Rhodopsin is the Functional Photoreceptor for Photoaxis in the Unicellular Eukaryote Chlamydomonas, Nature 311:756-759 (1984).

Fung, B. K.-K. and Stryer, L., Photolyzed Rhodopsin Catalyzes the Exchange of GTP for Bound GTP in Retinal Rod Outer segments, Proc. Natl. Acad. Sci. USA 77:2500-2504 (1980).

George, J. S. and Hagins, W. A., Control of Ca^{2+} in Rod Outer Segment Disks by Light and Cyclic GMP, Nature (London) 303:344-348 (1983).

Gold, G. H., Photoreceptor Coupling - Its Mechanism and Consequences, Curr. Top. Membr. Transp. 15:59-89 (1981).

Gold, G. H. and Korenbrot, J. I., Light-Induced Calcium Release by Intact Retinal Rods, Proc. Natl. Acad. Sci. USA 77:5557-5561 (1980).

Goldberg, N. D., Ames III, A., Gander, J. E. and Walseth, T. F., Magnitude of Increase in cGMP Metabolic Flux Determined by [18]O Incorporation into Nucleotide -phosphoryls Corresponds with Intensity of Photic Stimulation, J. Biol. Chem. 258:9213-9219 (1983).

Hagins, W. A., The Visual Process: Excitatory Mechanisms in the Primary Receptor Cells, Ann. Rev. Biophys. Bioengr. 1:131-158 (1972).

Hagins, W. A. and Yoshikami, S., A Role for Ca^{2+} in Excitation of Retinal Rods and Cones, Exp. Eye Res. 18:299-305 (1975).

Hagins, W. A. and Yoshikami, S., in: Vertebrate Photoreception, H. B. Barlow and P. Fatt, eds. Academic, New York (1977).

Hagins, W. A., Penn, R. D. and Yoshikami, S., Dark Current and Photocurrent in Retinal Rods, Biophys. J. 10:380-412 (1970).

Hartline, H. K., Wagner, H. G. and MacNichol, E. F., The Peripheral Origin of Nervous Activity in the Visual System, Cold Spring Harbor Symp. Quant. Biol. 17:125-142 (1952).

Hecht, S., Schlaer, S. and Pirenne, M. H., Energy, Quanta and Vision, J. Gen. Physiol. 25:819-840 (1942).

Kakiuchi, S. and Rall, T. W., The Influence of Chemical Agents on the Accumulation of Adenesine, 3',5'-Phosphate in Slices of Rabbit Cerebellum, Mol. Pharmacol. 4:367-388 (1968).

Kanaho, Y., Tasi, S. C., Adamik, R., Hewlitt, E. L., Mass, J. and Vaughan, M., Rhodopsin-Enhanced GTPase Activity of the Inhibitory GTP Binding Protein of Adenylate Cyclase, J. Biol. Chem. 259:7378-7381 (1985).

Keirns, J. J., Miki, N., Bitensky, M. W. and Keirns, M., A Link Between Rhodopsin and Disc Membrane Cyclic Nucleotide Phosphodiesterase Action Spectrum and Sensitivity to Illumination, Biochemistry 4:2760-2766 (1975).

Kuffler, S. W. and Nichols, J. G., From Neuron to Brain (Sinauer Associates, Sunderland, Mass, 1976).

Lamb, T. D., McNaughton, P. A. and Yau, K.-W., Spatial Spread of Activation and Background Desensitization in Toad Rod Outer Segments, J. Physiol. (London) 319:463-496 (1981).

Liebman, P. A. and Pugh, E. N., Current Topics in Membrane and Transport, 15:157-170 (1981).

Lipton, S. A., Rasmussen, H. and Dowling, J. E., Electrical and Adaptive Properties in Rod Photoreceptors in Bufo Marinus II, J. Gen. Physiol. 70:771-791 (1977).

Mathews, H. R., Torre, V. and Lamb, T. D., Effects on the Photoresponse of Calcium Buffers and Cyclic GMP incorporated into Cytoplasm of Retinal Rods, Nature 313:582-585.

Matthews, G. and Baylor, D. A., The Photocurrent and Dark Current of Retinal Rods, Curr. Top. Membr. Transp. 15:3-18 (1981).

Miki, N., Baraban, J. M., Keirns, J. J., Boyce, J. J. and Bitensky, M. W., Purification and Properties of Light-activated Cyclic Nucleotide Phosphodiesterase of Rod Outer Segments, Biol. Chem. 250:6320-6327 (1975).

Miller, D. L. and Korenbrot, J., Light-activated GTPase in Vertebrate Photoreceptors: Regulation of Light-activated Phosphodiesterase, Biophys. J. 49:281a (1986).

Miller, W. H. and Nicol, G. D., Molecular Mechanisms of Visual Transduction, Curr. Top. Membr. Transp. 15:417-437 (1981).

Miller, W. H. and Nicol, G. D., Evidence that Cyclic GMP Regulates Membrane Potential in Photoreceptors, Nature (London) 280:64-66 (1979).

Nakatani, K. and Yau, K.-W., cGMP Opens the Light-sensitive Conductance in Retinal Rods, Nature 313:379-392 (1985).

Nicol, G. D. and Miller, W. H., Cyclic GMP - Injected into Retinal Rod Outer Segments Increases Latency and Amplitude of Response to Illumination, Proc. Natl. Acad. Sci. USA 75:5217-5220 (1978).

Pannbacker, R. G., Control of Guanylate Cyclase Activity in Rod Outer Segments, Science 182:1138-1140 (1973).

Papermaster, D. S., Schneider, B. G., Zorn, M. A. and Kraechenbuhl, J. P., Immunocytochemical Localization of a Large Intrinsic Membrane Protein to the Incisures and Margins of Frog Outer Segment Discs, J. Cell. Biol. 79:415-425 (1978).

Papermaster, D. S., Schneider, B. G., Sorn, M. A. and Kraenchbuhl, J. P., Immunocytochemical Localization of Opsin in Outer Segments and Golgi Zones of Frog Photoreceptor Cells. An Electron Microscopic Analysis of Cross-linked Albumin-embedded Retinas, J. Cell. Biol. 77:196-210 (1978).

Pober, J. S. and Bitensky, M. W., Light-regulated Enzymes of Vertebrate Retinal Rods, Adv. Cyclic. Nucleotides. Res. 11:265-301 (1979).

Puckett, K. L., Aronson, E. T. and Goldin, S. M., ATP-dependent Calcium Uptake Activity Associated with a Disk Membrane Fraction Isolated from Bovine Retinal Rod Outer Segments, Biochemistry 24:390-400 (1985).

Schmidt, S. Y., Phosphatidylinositol Synthesis and Phosphorylation are Enhanced by Light in Rat Retinas, J. Biol. Chem. 256:6863-6868 (1983).

Schroder, W. H. and Fain, G. L., Light-dependent Calcium Release from Photoreceptors Measured by Laser Micromass Analysis, Nature 309:268-270 (1984).

Shinozawa, T., Uchida, S., Martin, E., Cafisco, D., Hubbell, W. and Bitensky, M. W., Additional Component Required for Activity and Reconstitution of Light-activated Vertebrate Photoreceptor GTPase, Proc. Natl. Acad. Sci. USA 77:1408-1411 (1980).

Shinozawa, T., Sen, I., Wheeler, G. L. and Bitsensky, M. W., Predictive Value of the Analogy Between Hormone-sensitive Adenylate Cyclase and Light-sensitive Photoreceptor cGMP Phosphodiesterase: A Specific Role for a Light-sensitive GTPase as a Component in the Activation Sequence, J. Supramolec. Struct. 10:185-190 (1979).

Smith, H. G. and Bauer, P. J., Light-induced Permeability Change in Sonicated Bovine Disks: Aresenazo III and Flow System Measurements, Biochemistry 18:5067-5073 (1979).

Sterling, P., Microcircuitry of the Cat Retina, Ann. Rev. Neurosci. 6:149-185 (1983).

Stryer, L., Hurley, J. B. and Fung, B. K.-K., Transducin: An Amplifier Protein in Vision, Trends in Biochem. Sci. 6:245-247 (1981).

Tomita, T., Electrical Activity of Vertebrate Photoreceptors, Quant. Rev. Biophys. 3:179-222 (1970).

Torney, D. C. and Bitensky, M. W., Computer Simulation of the Light Response of Vertebrate Rods, Biophys. J. 49:31a (1986).

Uchida, S., Wheeler, G. L., Yamazaki, A. and Bitensky, M. W., A GTP-protein Activator of Phosphodiesterase which forms in Response to Bleached Rhodopsin, J. Cyclic Nucleotide Res. 7:95-104 (1981).

Wald, G., Molecular Basis of Visual Excitation, Nature (London) 219:800-807 (1968).

Waloga, G. and Anderson, R. E., Effects of Inositol-1,4,5-trisphosphate Injections into Salamander Rods, Biochem. Biophys. Res. Comm. 126:59-62 (1985).

Wheeler, G. L. and Bitensky, M. W., A Light-activated GTPase in Vertebrate Photoreceptors: Regulation of Light-activated Phosphodiesterase, Proc. Natl. Acad. Sci. USA 74:4238-4242 (1977).

Wheeler, G. L., Matuo, Y. and Bitensky, M. W., Light-activated GTPase in Vertebrate Photoreceptors, Nature (London) 269:822 (1977).

Wilden, U., Hall, S. W. and Kuhn, H., 'PDE Activation by Phosphorylated Rodopsin is Quenched when Rhodopsin is Phosphorylated and Binds Intrinsic 48-kDa Protein of Rod Outer Segment, Proc. Natl. Acad. Sci. USA 83:1174-1178 (1986).

Yamazaki, A., Stein, P. J., Chernoff, N. and Bitensky, M. W., Activation Mechanism of Rod Outer Segment Cyclic GMP Phosphodiesterase: Release of Inhibitor by the GDP/GTP Binding Protein, J. Biol. Chem. 258:8188-8194 (1983).

Yamazaki, A., Sen, I., Casnelli, J., Greengard, P. and Bitensky, M. W., Cyclic GMP-specific, High Affinity, Non-catalytic Binding Sites on Light-Activated Phosphodiesterase, J. Biol. Chem. 255:11619-11624 (1980).

Yamazaki, A., Bartucca, F., Ting, A. and Bitensky, M. W., Reciprocal Effects of an Inhibitory factor on Catalytic Activity and Noncatalytic cGMP Binding Sites of Rod Phosphodiesterase, Proc. Natl. Acad. Sci. USA 79:3702-3706 (1982).

Yau, K.-W. and Nakatani, K., Light-suppressible, Cyclic GMP-sensitive Conductance in the Plasma Membrane of a Truncated Rod Outer Segment, Nature (London) 317:252-255 (1985).

Yau, K.-W. and Nakatani, K., Light-induced Reduction of Cytoplasmic Free Calcium in Retinal Rod Outer Segment, Nature (London) 313:579-582 (1985).

Yau, K.-W., McNaughton, P. A. and Hodgkin, A., Effect of Ions on the Light-sensitive Current in Retinal Rods, Nature (London) 292:502-505 (1981).

Yee, R. and Liebman, P. A., Light-activated Phosphodiesterase of the Rod Outer Segment: Kinetics and Parameters of Activation and Deactivation, J. Biol. Chem. 253:8902-8909 (1983).

Yoshikami, S. and Hagins, W. A., Control of the Dark Current in Vertebrate Rods and Cones in: Biochemistry and Physiology of Visual Pigments, pp 245-255, H. Langer, ed., Springer-Verlag, New York (1973).

Yoshikami, S., George J. S. and Hagins, W. A., Light-induced Calcium Fluxes from Outer Segment Layer of Vertebrate Retinas, Nature (London) 286:395-398 (1980).

Young, R. W., Visual Cells and Concept of Renewal, Invest. Ophthalmol. 15:700-725 (1976).

REGULATION OF NEURONAL ADENYLATE CYCLASE

M. M. Rasenick[1], M. M. Marcus[1], Y. Hatta[1], F. DeLeon-Jones,[1,2] and S. Hatta[1]

[1]Department of Physiology and Biophysics, University of Illinois College of Medicine, [2]Westside VA Hospital P O Box 6998, Chicago, IL 60680

INTRODUCTION

Receptors for a panoply of neurotransmitters and neuromodulators are known to couple to the adenylate cyclase system. In the past few years we have benefitted from an explosion of information concerning the molecular identification of the adenylate cyclase components, and this information has served both to elucidate possible mechanisms for the coupling of adenylate cyclase to receptors as well as the activation cascade of that enzyme. This information has also served to illustrate the extraordinary complexity of the system. Several recent reviews (cf. Birnbaumer et al., 1985; Gilman, 1984) have addresssed the structure and function of adenylate cyclase. The purpose of this manuscript is not to overlap with those, but rather, to illustrate some of the unusual features of adenylate cyclase from neural systems.

Components of the Adenylate Cyclase System

Neuronal adenylate cyclase is responsive to a variety of neurotransmitters which stimulate or inhibit that enzyme. Receptor mediated effects upon the enzyme display an absolute requirement for GTP or an hydrolysis-resistant triphosphate analog. Neurotransmitter effects are exerted through at least two membrane-associated GTP-binding proteins, referred to herein as GNs and GNi, denoting, respectively, the stimulatory and inhibitory proteins (see figure 3). Recently, the alpha, or GTP-binding portion, of GNi has been purified from bovine cerebral cortex and appears to consist of two to three pertussis toxin substrates clustered at about 40 kDa (Neer, et al., 1984; Sternweis and Robishaw, 1984). The alpha subunit of GNs has been purified and appears to be a single polypeptide of about 42-45 kDa (Northup et al., 1982), although, in some tissues, a 48-52 kDa form of GNs also appears (d'Alayer et al., 1983). GNs and GNi regulate the catalytic activity of the adenylate cyclase enzyme through interaction with a distinct catalytic moiety. The mechanism of this regulation is unknown, but it has been suggested that the beta and gamma components (36 kDa and 8 kDa respectively) common to GNs and GNi provide a regulatory function (Codina et al., 1984; Katada et al., 1984).

The coupling of adenylate cyclase refers to the interaction between (or among) neurotransmitter receptor, GNs or GNi and the adenylate cyclase catalytic moiety. Physical interaction among these proteins has been reported (Schlegel et al., 1979; Limbird et al., 1980) and, as suggested below, GNs and GNi may interact directly as well. The coupling between the neurotransmitter receptor and GNs or GNi in nerve cells is diminished or lost upon preparation of subcellular fractions (vide infra). However, as a result of this coupling loss, hydrolysis-resistant guanine nucleotides can activate or inhibit neuronal adenylate cyclase profoundly without benefit of hormone (neurotransmitter). Furthermore, we have demonstrated that coupling between the GN proteins and the adenylate cyclase catalytic moiety is augmented by treatments which alter cytoskeletal or membrane composition (Rasenick et al., 1981; Rasenick et al, 1984) as well as by chronic antidepressant treatment (Menkes et al., 1983). Additionally, we have indicated recently that infection of guinea pig membranes with Creutzfeldt-Jakob agent promotes coupling between GNs and the adenylate cyclase catalytic moiety (Rasenick et al., 1986).

Coupling of Adenylate Cyclase in Neuronal Systems

Intact cell preparations (slices or cubes) of mammalian brain show marked neurotransmitter activation of adenylate cyclase. Membrane preparations from those same tissues, by contrast, show little effect of neurotransmitter in the activation of that same enzyme (Von Hungen and Roberts, 1974). Certain regions such as striatum respond to dopamine, but that response is quite modest compared to the activation of turkey erythrocyte membrane adenylate cyclase by isoproterenol, or some other "classic" hormone-mediated adenylate cyclase system. Inhibitory neurotransmitters remain coupled to adenylate cyclase in several brain regions after the preparation of membranes (Childers and LaRiviere, 1984) under conditions where stimulatory response to adenylate cyclase is uncoupled.

One question which has persisted concerning the loss of neurotransmitter sensitivity as a concommittance of cell disruption is the possibility that some cytosolic component mediates hormone responsiveness in neural cells. Unfortunately, this has never been demonstrated, nonetheless, at least one investigation of this phenomenon led to the discovery of a protein which augmented NaF activated adenylate cyclase (Rasenick and Bitensky, 1980; Nijjar et al., 1980).

One possibility for the dissociation of stimulatory receptor from the adenylate cyclase system in neuronal membranes may reside in the apparent solubility of a portion of the GNs protein. Although convention regarded the GNs protein as membrane bound, this laboratory indicated that some portion of the total GNs protein was released from the synaptic membrane (Rasenick et al., 1984). Neer et al., (1984) suggested that in the brain, the GTP-binding (alpha) subunits of GNi as well as GNs might be soluble when uncomplexed with the $\beta\gamma$ subunits, and both of these observations have been confirmed recently by Sternweis (1986). It is unlikely that GNs release during membrane preparation is responsible for the observed coupling loss, as hydrolysis-resistant GTP analogs and NaF activate the enzyme considerably. Furthermore, although activated transducin, GNs and GNi may be released from the plasma membrane with buffer washing, there is no indication that this release is necessary for enzyme activation. The increased solubility of GN protein may simply reflect the altered conformation expressed by an activated GTP-binding protein (Stein et al., 1985).

Microtubule Disrupting Drugs Increase Adenylate Cyclase Coupling

Microtubule-disrupting drugs have been shown to increase adenylate cyclase activity in synaptic membranes prepared from rat cerebral cortex. Activation of the enzyme by hydrolysis-resistant GTP analogs and NaF is augmented by colchicine or vinblastine ($EC_{50}=5x10^{-7}$ M), while basal and Mn^{2+} stimulated (reflecting catalytic-moiety activation) activities are unchanged (Rasenick et al., 1981; figure 1). These findings suggest that the microtubule-disrupting drugs increase the "coupling" between the GTP-binding pro-

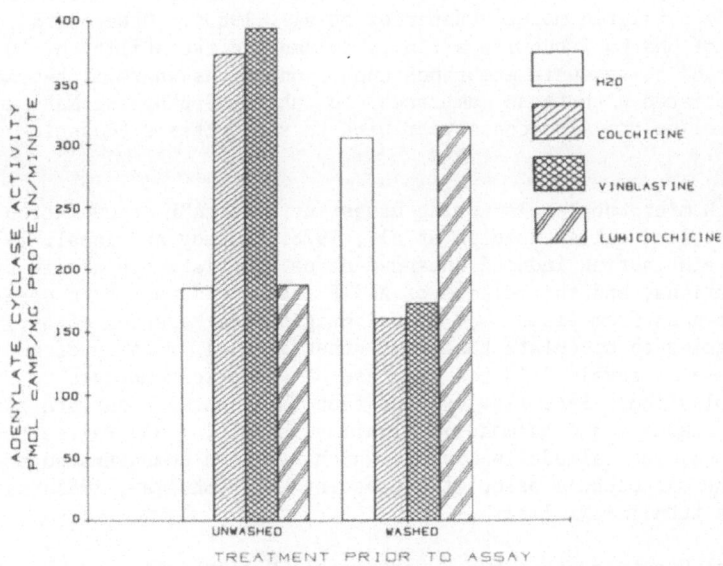

FIG. 1: Effects of Microtubule-Disrupting Drugs on Neuronal Adenylate
Cyclase. (A) UNWASHED. Synaptic membrane enriched fractions from
21-day old male Sprague-Dawley rat cerebral cortex were prepared
as described previously (Rasenick and Bitensky, 1980) and 50 μg
aliquots were incubated with 1 μM colchicine, vinblastine, H_2O or
10 μM lumicolchicine at 30°C for 20 min. Following this, membranes
were incubated for 20 min at 30°C with 2 μM GppNHp and assayed (as
described; Rasenick et al., 1981) for adenylate cyclase activity.
(B) WASHED. Membranes were treated as above except that following
incubation with colchicine, vinblastine, lumicolchicine or H_2O,
membranes were washed twice with buffer. Also, GppNHp and ATP were
added simultaneously in the 10-min assay step, eliminating the
second 20-min preincubation (Data and experimental detail from
Rasenick et al., 1981). Neither colchicine or vinblastine altered
adenylate cyclase activation by $MnSO_4$, indicating that these agents
did not alter the catalytic capacity of the enzyme. (From Rasenick
et al., 1985).

tein which stimulates adenylate cyclase (GNs) and the catalytic moiety of
that enzyme. When these membranes are treated with colchicine or vinblastine
and subsequently washed, activity of the GNs protein is released from the
membranes into the supernatants (figure 1). Photoaffinity labeling studies
(Rasenick et al., 1984) demonstrate that this loss of GNs activity coincides
with the colchicine- or vinblastine-mediated release of GNs from the synaptic
membrane. This release of GNs activity is indicative of facilitated Ns-cata-
lytic moiety coupling and is similar to that observed in the homologous cGMP
phosphodiesterase cascade from retinal rod outer segments (see Stein et al.,
1983).

Purified synaptic membranes have been demonstrated to contain tubulin
as an integral, or at least a tightly associated, membrane component (Zisapel et al., 1980). The ability of microtubule disrupting drugs to promote
release of GNs from the synaptic membrane is similar to the ability of cyto-
chalasin to promote the release of fibronectin from plasma membranes. Since
it appears that membrane actin forms a link for fibronectin (Ali and Hynes,
1977), we have proposed that plasma membrane tubulin is associated with GNs,

and the mitigation of that association augments interaction of the GNs protein with the catalytic moiety (Rasenick et al., 1985). Other data, such as
the binding of GNs to a tubulin affinity column and the ability of 10 nM
tubulin, (added to synaptic membranes under conditions where it becomes
tightly associated with those membranes) to inhibit GppNHp- or NaF-mediated
adenylate cyclase are also consistent with this hypothesis (Rasenick et al.,
1985).

Although microtubule disrupting drugs increase cAMP accumulation in
intact white blood cells (Rudolph et al., 1978; Kennedy and Insel, 1978) colchicine and vinblastine induced enhancement of adenylate cyclase in broken
cell preparations, and the release of AAGTP-labeled GNs has been observed
only in membranes from tissues of neural origin (see Rasenick et al., 1985).
It is intriguing to speculate that some other special property of neural adenylate cyclase is involved in the response of synaptic membranes to colchicine or vinblastine. Specifically, the fact that neural adenylate cyclase is
activated by calcium and calmodulin (Treisman et al., 1983) raises the possibility that calcium calmodulin kinase, which has been demonstrated to phosphorylate the microtubule associated protein, MAP2 (Shulman, 1985) can also
mitigate the tubulin-GNs link.

Physiological Considerations for Cytoskeletal/GN Association

The possibility that cytoskeletal or membrane components participate in
an intracellular regulation of adenylate cyclase is exciting for several reasons. First the prominence of neurotransmitter receptors linked to GNi in
the CNS makes it likely that such receptors act not only subsequent to GNs-
linked neurotransmitters but in a primary fashion as well. Inhibition of
intracellularly activated adenylate cyclase would provide a stronger stimulus than inhibition of a quiescent adenylate cyclase. Clearly, microtubule-
disrupting drugs are not involved in an intracellular process, but enzymes
which modify microtubule-associated components, such as Ca^{2+}/calmodulin
dependent protein kinase (see below), might alter adenylate cyclase activity
in a similar manner. Such a mechanism could account for the observed interaction between agents which result in an elevation of intracellular calcium
and adenylate cyclase (Bell et al., 1985). This is especially relevant in
light of the modulation of neural adenylate cyclase by calmodulin to a much
greater extent than such modulation occurs in non-neural tissues (Treisman et
al., 1983).

Interaction Among GTP-Binding Proteins

In previous studies we have identified several proteins which appear
capable of binding the photoaffinity GTP analog, P^3-(4-azidoanilido)-P^1-5'
GTP (AAGTP) (Rasenick et al., 1984, Hatta et al., in press). These proteins
are of the Mr 52, 48, 42, 40, 36, and 32 kDa. The 40 kDa protein is the
dominant AAGTP-binding protein in membranes which have been washed after
exposure to AAGTP but prior to UV photolysis. The 40 kDa protein is a pertussis toxin substrate and appears to be the inhibitory GTP-binding protein,
GNi (which exists in multiple forms, Neer et al., 1984). The 42 kDa protein
appears to be GNs.

It has been known for some time (cf. Perkins, 1973) that hydrolysis-
resistant GTP analogs can promote the activation of adenylate cyclase in a
manner which persists subsequent to buffer washing of the membranes. Similarly, cerebral cortex (Yamamoto and Shimizu, 1983) and platelet (Katada et
al., 1984) membranes show a stable GTPγS- or GppNHp-induced inhibition of
adenylate cyclase. A general interpretation of these results is that hydrolysis-resistant GTP analogs bind tightly to GNs in the former case and GNi
in the latter.

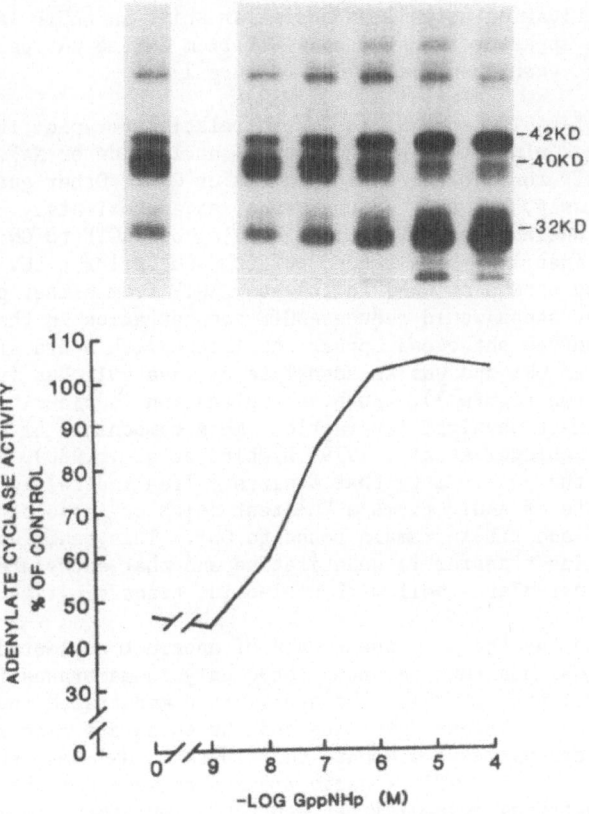

FIG. 2: (Upper). SDS/PAGE analysis of AAGTP-labeled synaptic membranes.
Synaptic membranes were incubated with 1.2×10^{-7} M $[^{32}P]$AAGTP as
above but after 2nd incubation with Gpp-NHp, reactions were subjected to 20 min of UV photolysis on ice. The membranes were then
washed and submitted to SDS/PAGE and autoradiographed. (Lower).
Adenylate cyclase activity of synaptic membranes incubated with
AAGTP. Synaptic membranes prepared from rat cerebral cortex were
incubated with AAGTP (1.2×10^{-4} M) for 3 min at 23°C, subsequently
washed and assayed for adenylate cyclase activity with indicated
concentration of GppNHp. Adenylate cyclase activity is expressed
as a percentage of the control activity of the membranes which were
not incubated with AAGTP. Values are means of 3 experiments and
basal adenylate cyclase activity in control membranes was 50.4
pmol/mg protein/min. Persistent inhibition of adenylate cyclase
was dose-dependent between 10^{-8}-10^{-4} M.

In addition to serving as a photoaffinity GTP probe, AAGTP is an hydrolysis-resistant GTP analog which is capable of supporting sustained stimulation or sustained inhibition of adenylate cyclase. Under conditions where
exposure of AAGTP to rat cerebral cortex synaptic membranes results in stable
inhibition of adenylate cyclase, a subsequent incubation with either GppNHp
or NaF can override this inhibition. When ^{32}P labeled AAGTP is employed
under these conditions, AAGTP is bound, primarily to the 40 kDa GNi. Under
conditions where AAGTP or NaF override the inhibition, $[^{32}P]$AAGTP is increas-
ingly lost from GNi and it appears on GNs. (note: the UV irradiation and
covalent binding of AAGTP occurs only after the incubation with GppNHp or

NaF.) GTPγS or unlabeled AAGTP, which will activate synaptic membrane ade-
nyylate cyclase, will have a similar shift in adenylate cyclase activity from
inhibition to activation as well as a similar shift in AAGTP labeling from
GNi to GNs. The apparent transfer of AAGTP from GNi to GNs persists subsequent to repeated washing of membranes (figure 2).

The shift of AAGTP from GNi to GNs is relatively rapid; it appears to be
completed within 2 min after the addition of nucleotide or NaF. NaF does not
compete with AAGTP for a binding site on GNs or GNi. Other guanyl nucleotides compete slowly, but this process requires approximately 30 min after
the addition of nucleotides. The tight binding of AAGTP to GNi and GNs is
revealed by the inability of a second buffer wash (prior to UV irradiation
but subsequent to GppNHp or NaF) to release AAGTP from either protein.
Since the washing steps would reduce AAGTP concentration in the media to
10^{-12} M, the observed phenomena appear consistent with a transfer of bound
nucleotide between GNi and GNs as adenylate cyclase switches from inhibition
to stimulation (see figure 3). Such an explanation is plausible, considering
likelihood of direct physical interaction among components of the adenylate
cyclase system (Schlegel et al., 1979; Limbird et al., 1980). Whereas we
cannot discount the possibility that a nitrene free radical formed at the
terminal phosphate of AAGTP forms a "nearest neighbor" association with GNs
while the purine and ribose remain bound to GNi. This seems unlikely, in
that the nucleotide transfer is quantitative and that adenylate cyclase
activity change correlates well with nucleotide transfer (figure 2).

It is noteworthy that the phenomenon of apparent nucleotide transfer
between GNi and GNs has thus far been noted only in membranes from neural-
crest derived cell types. This laboratory has demonstrated reversible interactions between cytoskeletal proteins and GNs which are also restricted to
neural cells. Perhaps these elements contribute to GNi-GNs interaction as
well. The existence of a GTP exchange process between GNi and GNs represents a novel switching mechanism between the stimulatory and inhibitory
regulation of neurotransmitter sensitive adenylate cyclase, which might
function at the intracellular level. Such a mechanism would affect, profoundly, neurotransmitter mediated events within the central nervous system.

A Novel Neural GTP-Binding Protein

We have demonstrated recently the presence of a 32 kDa GTP-binding protein (Hatta et al., in press). This protein is not a substrate for ADP ribo-
sylation by pertussis or cholera toxin, and it binds AAGTP loosely yet specifically. Preliminary experiments for the purification of the 32 kDa AAGTP-
binding protein have shown that the 32 kDa protein labeled with [^{32}P]AAGTP
can be extracted by cholate and elute in the 2nd major peak from a DEAE-
sephacel column (figure 4). As shown in figure 4, peak 1 contains GNs as
the predominant AAGTP-labeled protein while peak 2 contains GNi and the 32
kDa protein as well as some GNs. It does not appear that the 32 kDa AAGTP-
binding protein is a proteolytic fragment of GNi or GNs, as Staphylococcus
aureus V8 protease studies yield different fragments for these species
(figure 5). We have observed this protein in membranes from several regions
of rat brain, hamster brain, guinea pig brain, and in rat platelet membranes,
but not in rat liver or kidney membranes. Curiously, the transfer of AAGTP
from GNi to GNs noted above has also been noted only in neural cells. AAGTP
is either ineffective or will support only stimulation of adenylate cyclase
in liver or kidney membranes.

Several recent studies suggest that muscarinic receptor-mediated phospholipase C activation and Ca^{2+} mobilization in 1321N1 astrocytoma cells are
regulated through mechanisms involving an unidentified guanine nucleotide
binding protein(s) other than GNi and GNs (Evans et al., 1984, Hughes et al.,
1985). Attempt to assign function to GTP-binding proteins may yield infor-

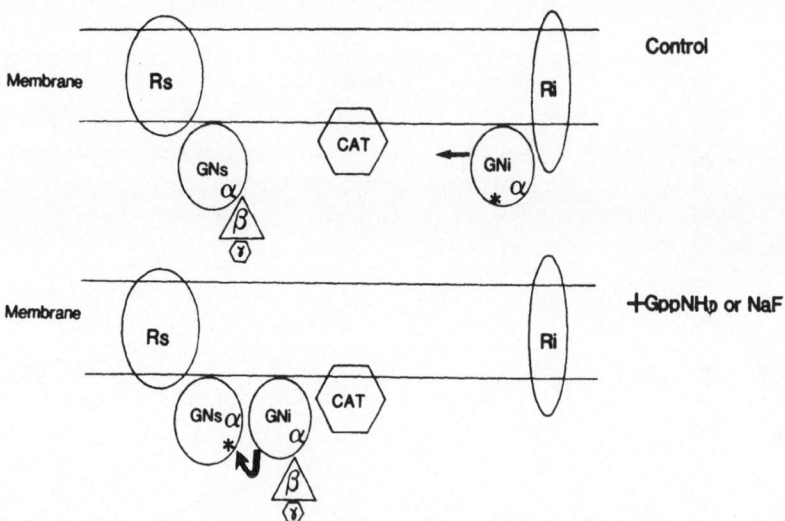

FIG. 3: Model for transfer of guanine nucleotides. Nucleotide is initially
bound to GNi when adenylate cyclase is inhibited. As this inhibition is overriden by GppNHp or NaF, nucleotide [*] is transfered to
GNs from GNi. Although we depict βγ transferring from GNi to GNs
along this process, we have no data which suggest that this occurs
in synaptic membranes.

mation about GTP-binding protein homology without illuminating GN function.
To wit, previous experiments have demonstrated functional interchangeability
between transducin and GNs (Rasenick et al., 1981; Bitensky, et al., 1983)
or GNi (Cerione et al., 1984). This does not mean that transducin normally
functions to stimulate or inhibit adenylate cyclase. Similarly, the 39 kDa
form of GNi, often referred to as GNo, has been seen to alter muscarinic
receptor affinity (Florio and Sternweis, 1985) or alpha 2 receptor-stimulated
GTPase (Cerione et al., 1986). The function of GNo remains unknown. The 32
kDa protein must occupy a similar category.

Antidepressants and Neuronal Adenylate Cyclase

     Chronic treatment of rats with antidepressants has been demonstrated to
decrease beta receptor number and lower isoproterenol stimulated cAMP accumulation (see Sulser, this volume, for review). Conversely, we (Menkes et al.,
1983) as well as Anderson et al. (1984) have observed increased coupling
between GNs and the catalytic moiety of neural adenylate cyclase prepared
from rats treated chronically with tricyclic antidepressants or electroconvulsive shock. Beta receptor down-regulation appears to conflict with augmented GNs-catalytic moiety coupling, but the inability to measure isoproterenol-stimulated adenylate cyclase in rat brain makes it difficult to explore
this apparent inconsistency. One noteworthy factor from our studies is that
the effects of antidepressant treatments in augmenting adenylate cyclase were
brain specific. Membranes prepared from kidney and liver from the treated
animals were indistinguishable from controls in their adenylate cyclase activity.

Homology Among GN Proteins

     Considerable homology exists among several GTP-binding proteins (e.g.
ras p21, EFTu, GNs, GNi, the rod outer segment GTP-binding protein (transducin) and tubulin; see Gilman, 1984; Halliday, 1984). These proteins share a

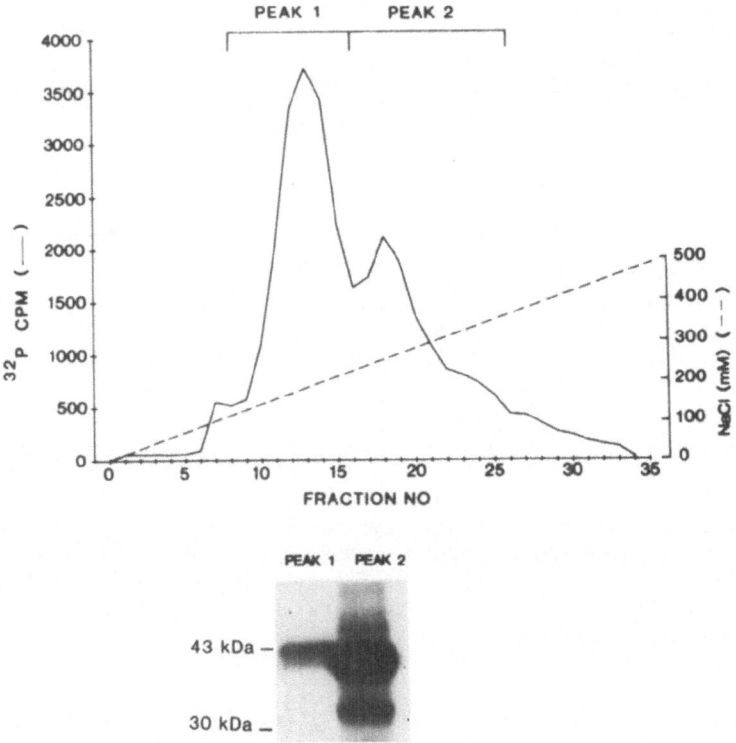

FIG. 4: DEAE-Sephacel chromatography of guanine nucleotide-binding proteins
and SDS PAGE analysis of proteins eluted. A cholate extract of rat
cerebral cortex membranes prelabelled with [^{32}P]AAGTP were applied
to DEAE-Sephacel column (0.5 x 18 cm) pre-equilibrated with 10 mM
Hepes, pH 8.0, 1.0 mM EDTA, 20 mM 2-mercaptoethanol, 30% ethylene-
glycol (Buffer A). The labelled proteins were eluted with a 300-ml
of linear gradient from 0 to 500 mM NaCl in 0.9% cholate in Buffer
A. Fractions of about 2.0 ml were collected in tubes.

common glycine rich sequence which is the likely GTP binding site as well as
other regions of apparent homology. There also appears to be functional
homology among these proteins as the rod outer segment GN and GNs (Rasenick
et al., 1981; Bitensky et al., 1983) or GNi (Cerione et al., 1985) appear
interchangeable. Tubulin has not yet been demonstrated interchangeable with
the GN proteins, but it has GTPase activity and can be ADP-ribosylated by
cholera and pertussis toxins (Lim et al, 1985). Finally, the GTP-binding
protein p21, has been implicated in the alteration of adenylate cyclase
activity (Toda et al., 1985) and we have recently discovered that infection
with the agent causing Creutzfeldt-Jakob disease results in increased coupling between the GNs and catalytic moiety of adenylate cyclase (Rasenick
et al., 1986). In that study we speculated that an infectious agent elicited gene product might act as p21 to increase adenylate cyclase coupling.
Thus, the evolutionary significance of a family of GTP-binding proteins
which interact in a spirit of integrated cooperation toward a common biological end, provides an intriguing source for speculation as well as fertile
ground for research.

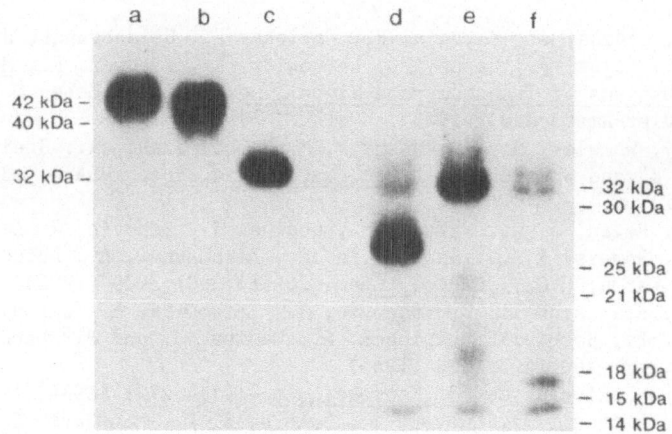

FIG. 5: Autoradiograph of SDS PAGE of Staphylococcus aureus V8 protease digestion products. [^{32}P]AAGTP-labelled bands from SDS PAGE gels were digested without prior elution by packing gel slices containing these bands in the sample wells of a second SDS gel and then overlaying each slice with S. aureus V8 protease (500 ng/lane). Digestion proceeded directly in the stacking gel during the subsequent electrophoresis. Lane a, b, and c, no protease; lane d, e, and f, S. aureus V8 protease (500 ng/lane). Lane a and d, 42 kDa protein; lane b and e, 40 kDa protein; lane c and f, 32 kDa protein.

SUMMARY

It appears that several components function in a spirit of integrated cooperation toward the intracellular regulation of neurotransmitter responsiveness. We have demonstrated that cytoskeletal proteins might interact with GNs and that GNs and GNi might interact with one another. At this juncture, it appears that both of these phenomena might occur only in cells of neural origin. Calmodulin and antidepressants may also affect adenylate cyclase in nervous tissue alone. The effects of AAGTP are different in nervous tissue from other tissues, and experiments with that nucleotide have led to the discovery of a new, 32 kDa GTP-binding protein which appears only in neural crest cells. Appreciation of the intricacies of signal transduction through the adenylate cyclase system are developing along with our understanding of that system. When combined with the complexity of neurotransmitter responsiveness, comprehension of the combined systems remains in its infancy, destined to grow as well as to surprise and delight all who are interested.

ACKNOWLEDGMENT

This work was supported by grants AFOSR 83-0249 and PHS-MH 39595. M. M. Rasenick is a fellow of the Chicago Community Trust.

REFERENCES

Ali, I. and Hynes, R., Biochem. Biophysics Acta 471:16-23 (1977).
Anderson, P., Klysner, R., Geisler, A., Neuropharmacol. 23:445-447 (1984).

Bell, J. O., Buxton, I. and Brunton, L., J. Biol. Chem. 260:2625-2628 (1985).

Birnbaumer, L., Codina, J., Mattera, R., Cerione, R., Hildebrandt, J., Sunyer, T., Rojas, F., Caron, M., Lefkowitz, R., Iyengar, R., Molecular Mechanisms of Transmembrane Signalling. Editors, Cohen & Houslay (Elsevier-Amsterdam) (1985).

Bitensky, M. W., Wheeler, M. A., Rasenick, M. M., Yamazaki, A., Stein, P., Halliday, K. and Wheeler, G., Proc. Natl. Acad. Sci. (USA) 79:3408-3412 (1982).

Cerione, R. A., Regan, J. W., Nakata, H., Codina, J., Benovic, J. L., Gierschik, P., Somers, R. L., Spiegel, A. M., Birnbaumer, L., Lefkowitz, R. J. and Caron, M.G., J. Biol. Chem. 261(8):3901-3909 (1986).

Cerione, R. A., Staniszawski, C., Benovic, J., Lefkowitz, R., Caron, M., Gierschik, P., Somers, R., Spiegel, A., Codina, J. and Birnbaumer, L., J. Biol. Chem. 260:1493-1500 (1985).

Childers, S. and LaRivere, G., J. Neuroscience 4:2764-2771 (1984).

Codina, J., Hildebrandt, J., Sunyer, T., Sekura, R. D., Manclark, C. R., Iyengar, R. and Birnbaumer, L., Adv. Cyclic Nucleotide Protein Phosphorylation Res. 17:111-125 (1984).

D'Alayer, J., Berthilliar, G. and Monneron, A., Biochemistry 22:3948-3953 (1983).

Evans, T., Martin, M., Hughes, A. and Harden, T. K., Mol. Pharm. 27:32-37 (1985).

Florio, V. A. and Sternweis, P. C., J. Biol. Chem. 260(6):3477-83 (1985).

Gilman, A. G., J. Cell 36:577-579 (1984).

Halliday, K., J. Cyclic Nucleo. Prot. Phos. Res. 9:435-448 (1984).

Hatta, S., Marcus, M. M. and Rasenick, M. M., Proc. Natl. Acad. Sci. (USA) 83:5439-5443 (1986).

Hughes, A., Martin, M. and Harden, T. K., Proc. Natl. Acad. Sci. (USA) 81:5680-5684 (1984).

Insel, P. A. and Kennedy, M. S., Nature 273(5662):471-3 (1978).

Katada, T., Bokoch, G. M., Northup, J. K., Ui, M. and Gilman, A. G., J. Biol. Chem. 259:3568-3585 (1984).

Lim, L., Sekura, R. and Kaslow, H. J., J. Biol. Chem. 260:2585-2588 (1985).

Limbird, L. E., Gill, D. M. and Lefkowitz, R. J., Proc. Natl. Acad. Sci. USA 77:775-779 (1980).

Menkes, D., Rasenick, M. M., Wheeler, M. A. and Bitensky, M. W., Science 219:65-67 (1983).

Neer, E. J., Lok, J. M. and Wolf, L. G., J. Biol. Chem. 259:14222-14229 (1984).

Nijjar, M. S. and Ho, J. C., Biochem. Biophys. Acta. 600(1):238-243 (1980).

Northup, J. K., Smigel, M. D. and Gilman, A. G., J. Biol. Chem. 257:11416-11423 (1982).

Perkins, J. P., Adv. Cyclic Nucleo. Res. 3:1-64 (1973).

Rasenick, M. M. and Bitensky, M. W., Proc. Natl. Acad. Sci. (USA) 77:4628-4632 (1980).

Rasenick, M. M., O'Callahan, C. M., Moore, C. A. and Kaplan, R. S., In: Microtubules and Microtubule Inhibitors eds. M. De Brabander and J. DeMey (Elsevier Amsterdam), pp. 313-323 (1985).

Rasenick, M. M., Stein, P. J. and Bitensky, M. W., Nature 294:6560-6562 (1981).

Rasenick, M. M., Valley, S., Manuelidis, E. E. and Manuelidis, L. FEBS Lett. 198:164-168 (1986).

Rasenick, M. M., Wheeler, G. L., Bitensky, M. W., Kosack, C., Maling, R. L. and Stein, P. J., J. Neurochem. 43:1447-1454 (1984).

Rudolph, S. A., Greengard, P. and Malawista, S. E., Proc. Natl. Acad. Sci. (USA) 74:3404-3408 (1977).

Schlegel, W., Kempner, E. and Rodbell, M. D., J. Biol. Chem. 254:5168-5176 (1979).

Schulman, H., In: Microtubules and Microtubule Inhibitors eds. M. De Brabander and J. DeMey (Elsevier Amsterdam) pp. 153-160 (1985).

Stein, P. J., Halliday, K. and Rasenick, M. M., J. Biol. Chem. 260:9081-9084 (1985).

Stein, P. J., Rasenick, M. M. and Bitensky, M. W., Prog. Retinal Res. 1:222-238 (1982).

Sternweis, P. C., J. Biol. Chem. 261:631-637 (1986).

Sternweis, P. C. and Robishaw, J. D., J. Biol. Chem. 259:13806-13813 (1984).

Sulser, F. et al., Adv. Exp. Med. Biol. (this volume) (1987).

Toda, T., Uno, I., Ishikawa, T., Powers, S., Kataoka, T., Brock, D., Cameron, S., Broach, J., Natsomoto, K. and Wigler, M., Cell 40:27-36 (1985).

Treisman, G., Bagley, S. and Gnegy, M., J. Neurochem. 41:1398-1406 (1983).

Von Hungen, K. and Roberts, S., Review of Neuroscience I 231-281 (1974).

Yamamoto, T. and Shimizu, H., J. Neurochem. 40:629-636 (1983).

Zisapel, N., Levi, M. and Gozes, I. J., Neurochem. 34:26-32 (1980).

SYNAPSIN I, A PHOSPHOPROTEIN ASSOCIATED WITH SYNAPTIC VESICLES:

POSSIBLE ROLE IN REGULATION OF NEUROTRANSMITTER RELEASE

Paul Greengard, Michael D. Browning, Teresa L.
McGuinness, and Rodolfo Llinas*

The Rockefeller University, 1230 York Avenue, New
York, NY 10021, and *Department of Physiology and
Biophysics, New York University Medical Center, New
York, NY 10016

INTRODUCTION

Elucidation of the molecular events underlying communication between
neurons is one of the most exciting problems in neuroscience. While it has
been known for many years that the depolarization-dependent influx of calcium into the presynaptic terminal leads to the release of neurotransmitters, little is known about the molecular events which underly this presynaptic process. Similarly, although neurotransmitters are known to produce
graded responses via interaction with specific receptors, little is known
about the molecular events which underlie these postsynaptic responses.
Although it has been generally assumed that protein molecules play important roles in the production of these responses, it is only within the last
15 years that we have come to understand how the activity of proteins could
be regulated with a time course consistent with mediation or modulation of
neuronal communication. What has become apparent during this period is
that protein phosphorylation is a primary mechanism utilized by eukaryotic
cells for post-translational regulation of protein function. Consequently,
protein phosphorylation represents a conceptual framework for analysis of
the molecular mechanisms which underly neuronal communcation. The scheme
shown in Figure 1 summarizes our current concepts about the molecular pathways underlying biological regulation of neuronal communication.

Neuronal Phosphoproteins

There is now considerable direct evidence that protein phosphorylation
plays an important role in neuronal reponses to extracellular signals (see
Nestler and Greengard, 1984; Browning et al., 1985 for pertinent reviews).
Within the last decade there has been a sustained focus on neuronal phosphoproteins and this effort has resulted in the identification of a number of
distinct neuronal phosphoproteins. These studies of brain phosphoproteins
have generally been of two types. In one type of study, investigators have
focused on known proteins and have attempted to determine whether these proteins are phosphorylated (see Table 1). A second type of study involves
the search for previously unknown proteins that become phosphorylated in
response to physiologically relevant stimulation of intact nerve cell preparations (see Table 2 for a restricted list). Implicit in this latter
approach is the assumption that the study of such proteins may lead to the

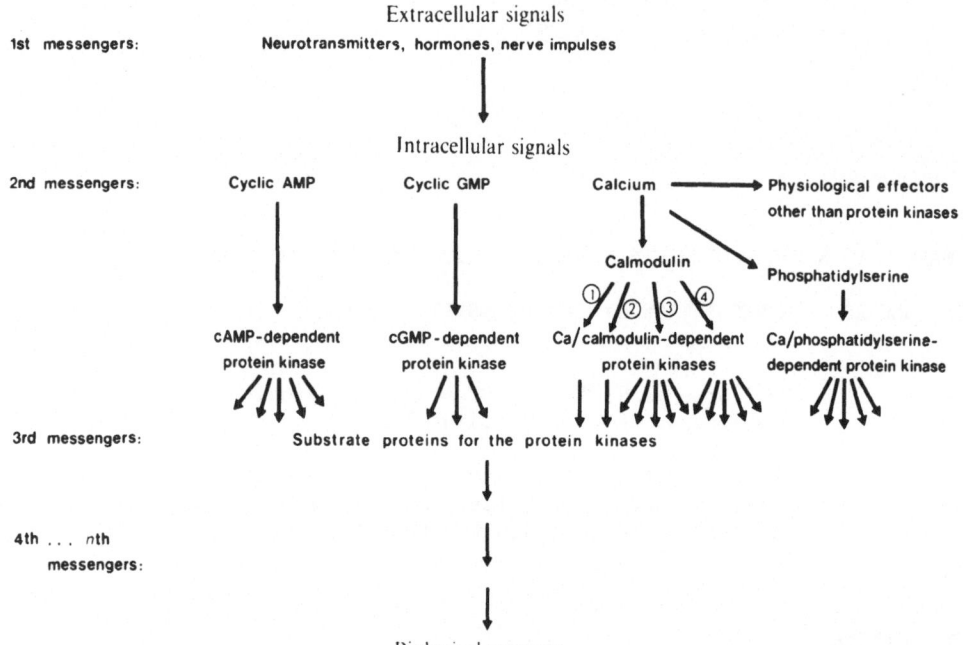

FIG. 1: Signals in the brain. Extracellular signals (first messengers)
produce specific biological responses in target neurons through a
series of intracellular signals (second, third, etc. messengers).
Second messengers in the brain include cyclic AMP, cyclic GMP,
calcium and diacylglycerol. Cyclic AMP and cyclic GMP produce
most of their second-messenger actions through the activation of
virtually one type of cyclic AMP-dependent protein kinase and one
type of cyclic GMP-dependent protein kinase, respectively. The
former enzyme exhibits a broad substrate specificity, and the
latter a more restricted specificity. Calcium exerts many of its
second-messenger actions through the activation of calcium-dependent protein kinases as well as through a variety of physiological
effectors other than protein kinases. Calcium activates protein
kinases in conjunction with calmodulin or with phosphatidylserine/
diacylglycerol. There are at least four types of calcium/calmodu-
lin-dependent protein kinases in brain: a phosphorylase kinase
that phosphorylates only phosphorylase (and possibly glycogen synthase) (Cohen, 1982; Krebs and Beavo, 1979); myosin light chain
kinase, which phosphorylates only myosin light chain (Hathaway et
al., 1981); calcium/calmodulin-dependent protein kinase I (Kennedy
and Greengard, 1981; Kennedy et al., 1983; Nairn et al., 1983);
and calcium/calmodulin-dependent protein kinase II (Kennedy and
Greengard, 1981; Kennedy et al., 1983; McGuinness et al., 1985).
The activation of individual protein kinases causes the phosphorylation of specific substrate proteins in target neurons. In some
cases, these substrate proteins, or third messengers, appear to be
the immediate effectors for the biological response. In other
cases, they appear to produce the biological response indirectly
through fourth, fifth, sixth, etc. messengers. (From Nestler and
Greengard, 1984).

discovery of previously unknown regulatory processes. One such protein,
synapsin I, will be the focus of the remainder of this report.

TABLE 1: Previously Identified Phosphoproteins[a]

A. Enzymes involved in neuro-
    transmitter biosynthesis

Tyrosine hydroxylase	Joh et al., 1978; Edelman et al., 1981.
Tryptophan hydroxylase	Yamauchi and Fujisawa, 1981; Kuhn and Lovenberg, 1982.

B. Neurotransmitter receptors

Nicotinic acetylcholine receptor	Gordon et al., 1977; Teichberg et al., 1977; Huganir et al., 1984.
Muscarinic acetylcholine receptor	Burgoyne, 1983.
β -Adrenergic receptor	Stadel et al., 1983.
GABA receptor (GABA-modulin)	Wise et al., 1983.

C. Ion channels

Sodium channels	Costa et al., 1982; Costa and Catterall, 1984 a,b.
Potassium channel	Kaczmarek et al., 1980; Castellucci et al., 1980; Adams & Levitan, 1982; de Peyer et al., 1982; Alkon et al., 1983.
Calcium channel	Oesterrieder et al., 1982; De Riemer et al., 1985.

D. Enzymes involved in cyclic
    nucleotide metabolism

Adenylate cyclase	Richards et al., 1981; Ehrlich et al., 1982.
Guanylate cyclase	Zwiller et al., 1982.
Phosphodiesterase	Sharma et al., 1980; Marchmount and Houslay, 1980.

E. Autophosphorylated protein
    kinases

Cyclic AMP-dependent protein kinase	Rangel-Aldao and Rosen, 1976.
Cyclic GMP-dependent protein kinase	deJonge and Rosen, 1977.
Calcium/calmodulin-dependent protein kinases	see Nestler and Greengard, 1984.
Calcium/phospholipid-dependent protein kinase	Kikkawa et al., 1982.
Tyrosine-specific protein kinases	Barnekow et al., 1982; Cotton and Brugge, 1983.
Rhodopsin kinase	Lee et al., 1981; Schichi and Somers, 1978.

F. Proteins involved in regulation
    of transcription and translation

RNA polymerase	Hook et al., 1981.
Histones	Langan, 1969; Gurley et al., 1981.
Nonhistone nuclear proteins	Johnson, 1982.
Ribosomal protein S6	Roberts, 1982.

137

Other ribosomal proteins   Roberts, 1982; Thomas, 1982.

G.   Cytoskeletal proteins
        MAP-2   see Nestler and Greengard,
                                        1984.
        Tau   see Nestler and Greengard,
                                        1984.
        Neurofilaments   see Nestler and Greengard, 1984.
        Myosin light chain   Hathaway and Traugh, 1982.
        Actin   DeMaille and Pechere, 1983.
        Tubulin   Goldenring et al., 1982.

[a]Neuronal proteins regulated by phosphorylation are listed with selected references. Some of the proteins listed are specific to neurons while others are present in many cell types including neurons. Not included are many phosphoproteins present in a variety of different tissues (including brain) that play roles in generalized cellular processes and are not thought to play roles in neuron-specific phenomena. (From Browning et al., 1985).

SYNAPSIN I

     Synapsin I (previously called Protein I) was the first identified neuronal substrate for cyclic AMP-dependent protein kinase and was discovered in a synaptic membrane fraction in this laboratory about 15 years ago (Johnson et al., 1972; Ueda et al., 1973). A summary of the salient characteristics of this protein is provided below.

Purification of Synapsin I

     The original purification of synapsin I (Ueda and Greengard, 1977) and more recent modifications (DeGennaro, L. J. & Greengard, P., unpublished results) relied on a step of acid extraction of a lysed crude particulate fraction. This procedure was adequate for bovine brain, but in rat brain poor stoichiometry of phosphorylation of site 1 in synapsin I resulted. We

TABLE 2: Examples of Previously Unknown Neuronal Phosphoproteins that were Detected by Phosphorylation[a]

Synapsin I   Johnson et al., 1971; Ueda and Greengard, 1977.
Protein III   Huang et al., 1982; Browning and Greengard, 1984.
DARPP-32   Walaas and Greengard, 1983; Hemmings et al., 1984 a,b.
87K   Wu et al., 1982; Albert et al., 1984.
B-50   Zwiers et al., 1976, 1982; Oestereicher et al., 1984.
G-substrate   Schlicter et al., 1978; Aswad and Greengard, 1981 a,b; Aitken et al., 1981.
Pyruvate dehydrogenase[b]   Browning et al., 1979; Hoch et al., 1984.

[a]A large number of neuron-specific phosphoproteins have been found (see, for example, Gispen and Routtenberg, 1982; Nestler and Greengard, 1984). Most such proteins have not been included in this table because they have not been extensively characterized. (From Browning et al., 1985).
[b]Pyruvate dehydrogenase was not initially identified in brain by phosphorylation. However, it is included here because of the possibility (Browning et al., 1982) that pyruvate dehydrogenase could have a neuron-specific role.

TABLE 3: Physico-chemical Properties of Synapsin I

	Synapsin Ia	Synapsin Ib
Molar proportion[a]	1	2
Molecular weight	86,000	80,000
Isolectric point	10.3	10.2
Stokes radius	59A	59A
Sedimentation coefficient	2.9 S	2.9 S
Frictional ratio	2.2	2.2
Acid soluble	yes	yes
Amino acid composition	rich in proline and glycine	
Other structural features	a collagenase-insensitive domain and a proline-rich collagenase-sensitive domain	

[a]molar proportion observed in brain homogenate, in purified synapsin I, and in immunoprecipitates from _in vitro_ translation products from brain polysomes.

therefore developed a purification procedure for rat brain synapsin I under non-denaturing conditions using the zwitterionic detergent, 3-[3-chloroamido-propyl-dimethylammonio]-1-propanesulfonate (CHAPS) (Schiebler et al., 1986). Synapsin I, isolated in the native state by use of this detergent, can be stoichiometrically phosphorylated with purified cAMP-or Ca/calmodulin-dependent protein kinases.

Physico-chemical properties

Some of the physico-chemical properties of synapsin I are listed in Table 3. Synapsin I exists as a doublet, synapsin Ia and synapsin Ib, having molecular weights of 86,000 and 80,000, respectively. It is extremely basic and contains two domains: one that is collagenase-insensitive and the other, a proline-rich region, that is degraded by purified collagenase (Huttner et al., 1981). The collagen-like domain accounts for its highly elongated structure. Synapsin I contains three separate phosphoacceptor sites (all serine residues) and these are described in Table 4.

Distribution

Synapsin I is found exclusively in the nervous system (Ueda and Green-gard, 1977; Sieghart et al., 1978; DeCamilli et al., 1979; Fried et al., 1982; DeCamilli et al., 1983a) where it is concentrated in virtually all synaptic terminals (Table 5). In terminals it is primarily localized to synaptic vesicles (DeCamilli et al., 1983b; Navone et al., 1984), very likely binding through the collagenase-sensitive domain to the outer (cytoplasmic) surface of synaptic vesicles (Ueda, 1981; Huttner et al., 1983). Synapsin I was found to be a major component in a highly purified fraction of synaptic vesicles, representing approximately 6% of the total vesicle protein.

Regulation of the state of phosphorylation of Synapsin I

The physiological and pharmacological regulation of the phosphorylation of synapsin I is summarized in Table 6. Of particular note is that agents and conditions that mimic physiological stimuli modulate the state of phosphorylation of synapsin I. These findings were dramatically demonstrated in experiments where cortical brain slices were exposed to successive cycles of depolarization and repolarization with the result that synapsin I was successively phosphorylated and dephosphorylated (Forn and

TABLE 4:   Protein Kinase Specificity of Synapsin I

Synapsin I undergoes multisite phosphorylation.

1.   One serine residue (site 1) in the collagenase-insensitive domain of
     Synapsin I is phosphorylated both by cyclic AMP-dependent protein
     kinase and by calcium/calmodulin-dependent protein kinase I.

2.   Two serine residues (sites 2 and 3) in the collagenase-sensitive
     domain of Synapsin I are phosphorylated by calcium/calmodulin-
     dependent protein kinase II.

3.   Not an effective substrate for cyclic GMP-dependent protein kinase
     or for calcium/diacylglycerol-dependent protein kinase.

Greengard, 1978). The stimulation of synapsin I phosphorylation by depolarizing agents appeared to be due to calcium influx into synaptic terminals.
Experiments with intact synaptosomes also demonstrated calcium-dependent
regulation of synapsin I phosphorylation (Kreuger et al., 1977).

      Synapsin I phosphorylation can also be altered by treatment with cer-
tain neurotransmitters in defined, relatively homogeneous regions of the
nervous system. Norepinephrine was found to increase the state of phosphorylation of synapsin I by 30% in slices of rat frontal cortex (Mobley and
Greengard, 1985). Dopamine increased synapsin I phosphorylation in slice
preparations of rat substantia nigra (Hemmings et al., 1985), striatum (Hemmings et al., 1985), posterior pituitary (Tsou and Greengard, 1982; Treiman
and Greengard, 1985) and superior cervical ganglia (Nestler and Greengard,
1980). Serotonin stimulated synapsin I phosphorylation in rat facial motor
nucleus (Dolphin and Greengard, 1981a,b). In each of these tissue preparations, the particular neurotransmitter appeared to stimulate the phosphorylation of synapsin I in the presynaptic elements present.

      Direct electrical stimulation of presynaptic input in intact prepara-
tions under physiological conditions can increase synapsin I phosphorylation in superior cervical ganglia (Nestler and Greengard, 1982), posterior
pituitary (Tsou and Greengard, 1982) and striatum (D. Haycock, unpublished
results). Interestingly, optimal stimulation of synapsin I phosphorylation
in superior cervical ganglia was obtained with frequencies of nerve stimulation that approximate those found in vivo (Douglas and Ritchie, 1957).

TABLE 5:   Distribution of Synapsin I

1.   Present only in nervous system (both central and peripheral).

2.   Within nervous system, present only in neurons.

3.   Within neurons, concentrated in presynaptic terminals.

4.   Within terminals, associated with synaptic vesicles.

5.   Present at virtually all synapses.

6.   Not present in adrenal chromaffin cells.

7.   Appears simultaneously with synapse formation during development.

TABLE 6: Physiological and Pharmacological Regulation of Synapsin I.

1. In synaptosomes and in slices of nervous tissue, depolarizing agents and cyclic AMP increase state of phosphorylation.

2. In specific anatomical regions of central and peripheral nervous system, the relevant neurotransmitters (serotonin, dopamine, norepinephrine, adenosine) increase state of phosphorylation.

3. In isolated peripheral nervous tissue (superior cervical ganglion), in posterior pituitary, and in striatum, impulse conduction under physiological conditions increases the state of phosphorylation.

4. In whole animals, convulsants increase and depressants decrease state of phosphorylation in cerebrum.

5. In whole animals, neurotransmitters and hormones increase total amount in specific brain regions.

Calcium/calmodulin-dependent protein kinases

As indicated above, Synapsin I was shown to be a major phosphosubstrate for Ca^{2+}-dependent protein kinases activated in brain slice and synaptosomal preparations (Forn and Greengard, 1978; Kreuger et al., 1977). The protein kinase activation by Ca^{2+} in these experiments was later found to be mediated by the action of calmodulin (Schulman and Greengard, 1978a). Subsequently, Ca^{2}/calmodulin-dependent protein kinases I and II (calmodulin kinase I and calmodulin kinase II) were identified as the direct result of the study of the phosphorylation of synapsin I (Schulman and Greengard, 1978a,b; Kennedy and Greengard, 1981). These enzymes have now been purified on the basis of their ability to phosphorylate synapsin I.

Calmodulin kinase II is a relatively abundant brain protein, constituting possibly 0.4% of total brain protein. Some of the physico-chemical properties of calmodulin kinase II are listed in Table 7. Recent immunocytochemical studies have revealed that this kinase is present in several neuronal compartments including dendrites, spines, soma and presynaptic terminals (Fig. 2). A variety of evidence suggests that calmodulin kinase II may be a mediator of the effects of calcium at the presynaptic terminal by

TABLE 7: Properties of Calmodulin Kinase II.

Properties	Calmodulin Kinase II
Native M_r (holoenzyme)	600-650,000 (polymer)
Subunit composition	α: 50,000 M_r β: 61,000 M_r
Ratio of subunits	3:1 (α :β)*
Autophosphorylation	Yes (α and β)

* ratio in brain homogenate; ratio varies amongst different brain regions. The molar ratios in the holoenzyme are as yet unknown.

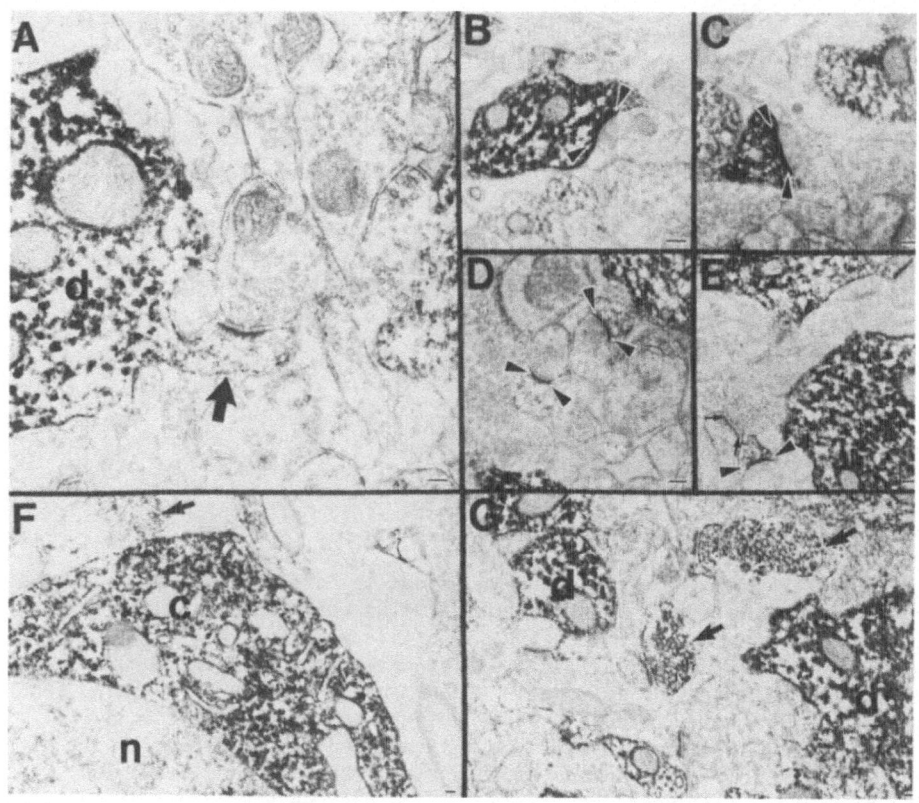

FIG. 2: Electron micrographs of ultrathin sections through the dentate
gyrus. (A) Immunolabeled dendrite (d) giving rise to a spine
(arrow). Immunoreactivity is greater in the dendrite than in its
spine (as can also be seen for two other dendrites and their spines
in the upper right corner of D and E. (B-E) Enhanced immunoreactivity on PSDs (between arrowheads) in sections not counterstained
with uranyl acetate. [In E, compare immunoreactive PSD (arrowheads) to unlabeled PSD (small arrows).] Immunoreactivity on PSDs
is greater in strongly labeled (B, C) than in very lightly labeled
(D, E) processes. (F) Immunoreactivity in the cytoplasm (c) of a
granule cell. By comparison, the nucleus (n) appears to be unlabeled. The arrow points to a lightly labeled nerve terminal. (G)
Immunoreactive dendrites (d) and nerve terminals (arrows). Most
of the immunoreactivity in nerve terminals is associated with
synaptic vesicles. (Bars = 100 nm.) (From Ouimet et al., 1984)

phosphorylating the synaptic vesicle-associated protein synapsin I. Synapsin I is the best substrate found to date for calmodulin kinase II (Table
8). Moreover, calmodulin kinase II has a higher affinity for synapsin I
than other calcium/calmodulin-dependent protein kinases have exhibited for
their physiological substrates.

Possible Role Played by Synapsin I in Neuronal Communication

As mentioned above, synapsin I is a neuron-specific phosphoprotein
localized to presynaptic terminals, where it is associated with synaptic
vesicles. Electron micrographs of ferritin-labeled synapsin I antibody

staining indicated that the protein coats the synaptic vesicles. These studies led naturally to the suggestion that synapsin I modulates some aspect of synaptic vesicle function such as calcium-stimulated exocytosis.

In addition, recent evidence indicates that phosphorylation of site II on the synapsin I molecule by calmodulin kinase II decreases the affinity of synapsin I for synaptic vesicles (Huttner et al., 1983; Schiebler et al., 1986). Taken together, these data have led to a hypothesis concerning the role played by synapsin I in neuronal communication. According to this hypothesis, phosphorylation of synapsin I may enhance neurotransmitter release by stimulating one of the "priming steps" that occur prior to exocytosis. The priming of synaptic vesicles for release might include such steps as translocation of the vesicles from the cytoplasm to the release site and interaction of the vesicle membrane and plasma membrane before fusion occurs. Dephospho-synapsin I could be the active form of the molecule and inhibit one of these steps, and phosphorylation of synapsin I by calmodulin kinase II could facilitate synaptic transmission by removing this inhibition. Alternatively, dephospho-synapsin I could be the inactive form and phosphorylation of synapsin I by calmodulin kinase II could facilitate synaptic transmission by increasing the level of the active phosphorylated form. A series of microinjection studies using the squid giant synapse have distinguished between these possibilities.

Microinjection of Synapsin I and Calmodulin Kinase II into the Presynaptic Terminal of the Squid Giant Synapse

The squid giant synapse is an excellent model system for detailed studies of the components of synaptic transmission. This "giant synapse" was first identified by Young (1938) as the last synapse in a chain of giant nerve cells that comprises the escape system for the animal. The anatomy of this chain of giant neurons has been reviewed in detail elsewhere (Llinas 1982, 1984). At the synapse, the presynaptic terminal digit has a diameter

TABLE 8: Substrate Specificity of Calmodulin Kinase II[a]

Substrate	Concentration (mg/ml)	Relative Rates of Phosphorylation[b] (%)
Synapsin I	0.1	100
Myosin P-light Chain (smooth muscle)	0.8	16
Glycogen Synthase[c]	0.4	13
Histone H1	0.2	3.0
Microtubule-associated Protein 2	0.8[d]	2.2
Acetyl-CoA Carboxylase	0.3	1.0
ATP-citrate Lyase	0.4	0.7
Myosin P-light Chain (skeletal muscle)	0.2	0.1
Tubulin	0.8[d]	0.05
Phosphorylase b	1.5	<0.02
Casein	0.4	<0.02
Phosvitin	2.0	<0.02

[a] From McGuinness et al., 1983. [b] Phosphorylation of substrate proteins was carried out using 60-80 ng of purified kinase and the indicated amounts of substrate proteins. [c] Purified as described (Caudwell et al., 1978). [d] Concentration refers to total amount of 3-times assembly-disassembly purified microtubule protein.

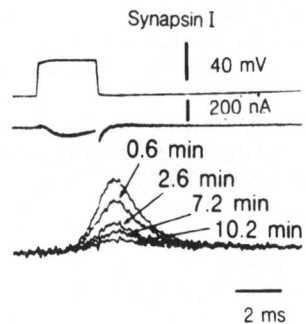

Synapsin I

40 mV

200 nA

0.6 min
2.6 min
7.2 min
10.2 min

2 ms

FIG. 3: Effect of synapsin I injection on the presynaptic current and postsynaptic potential. Presynaptic currents and postsynaptic potentials were generated by 40 mV presynaptic depolarizing voltage
steps delivered immediately before and at the indicated times after
dephosphosynapsin I injection. Upper trace, superimposed presynaptic voltage steps; middle trace, superimposed presynaptic currents; lower trace, superimposed post-synaptic potentials. (From
Llinas et al., 1985)

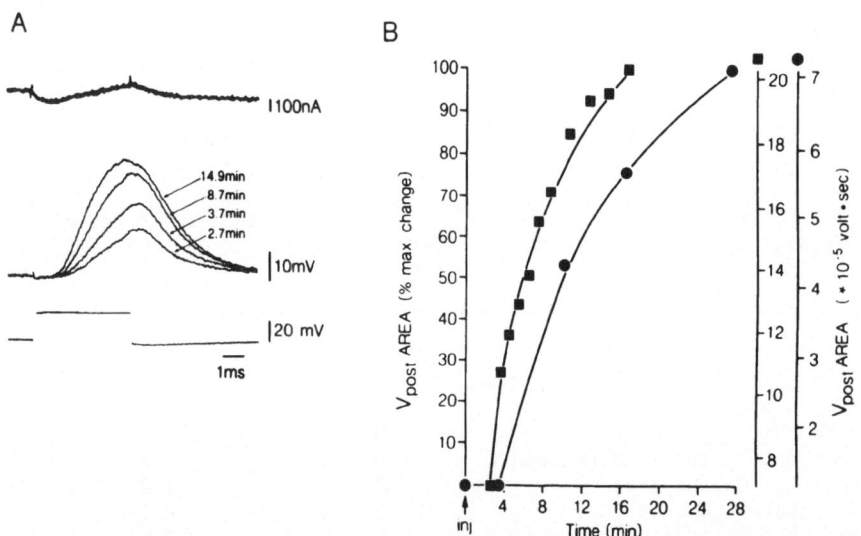

FIG. 4: Presynaptic current and postsynaptic potential before and after
injection of calmodulin kinase II. (A) Presynaptic currents and
postsynaptic potentials were generated by 28-mV presynaptic depolarizing voltage steps delivered at the indicated times after
injection. Upper trace, superimposed presynaptic currents; middle
trace, superimposed postsynaptic potentials; lower trace, presynaptic voltage step. (B) Area of postsynaptic potential (volt sec)
from two different experiments plotted as percentage of maximal
increase (left ordinate) or as the actual values for the area
(right ordinates) as a function of time after injection (inj) of
calmodulin kinase II. (From Llinas et al., 1985)

144

of ≈50 μm and the postsynaptic fiber has a diameter of ≈200 μm (Llinas et al., 1981). These large diameters make it possible for the pre- and post-synaptic fibers to be impaled with electrodes so that synaptic transmission can be studied at the junctional site.

The giant synapse preparation was used to test directly the hypothesis that synapsin I and calmodulin kinase II are involved in regulating or modulating some aspect of synaptic transmission. The technique of intracellular microinjection of proteins into living cells has recently proven to be successful in elucidating the roles of cAMP-dependent protein kinase in neuronal processes (Kaczmarek et al., 1980; Castellucci et al., 1980; also see Levitan et al., 1983). The microinjection experiments were carried out under voltage clamp conditions. The basic design of the injection experiments consisted in analyzing the relationship between calcium entry and transmitter release, as measured by the postsynaptic response, before and after the injection into the presynaptic site of either synapsin I or calmodulin kinase II.

Injection of Synapsin I

Dephospho-synapsin I was pressure-injected into the terminal digit of the squid giant synapse. As shown in Figure 3, injection of synapsin I pro-

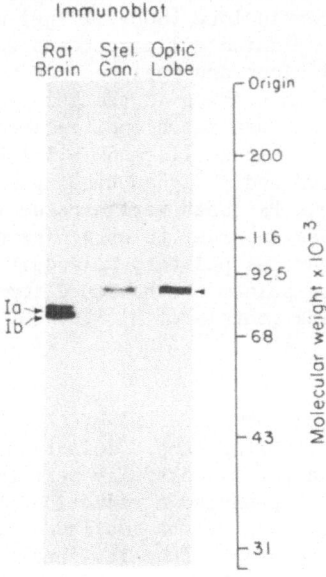

FIG. 5: Immunoblot showing cross-reactivity of squid proteins with antibodies prepared against mammalian brain synapsin I. Homogenates of rat brain (Rat Brain: 5 μg), squid stellate ganglion (Stel. Gan.: 200 μg), and squid optic lobe (Optic Lobe: 200 μg) were subjected to NaDodSO₄/7.5% polyacrylamide gel electrophoresis and the proteins were transferred to nitrocellulose. Immunoblotting was carried out by using rabbit antiserum (raised against bovine brain synapsin I) and ^{125}I-labeled Protein A. Immunoreactive proteins were visualized by autoradiography. Arrows point to the rat brain synapsin Ia/Ib doublet. Arrowhead points to the major squid immunoreactive band. Three minor immunoreactive bands of M_r ≈135,000, ≈56,000 and ≈44,000 were also detected in the squid tissues. The four bands may represent a family of synapsin I-like proteins. (From Llinas et al., 1985)

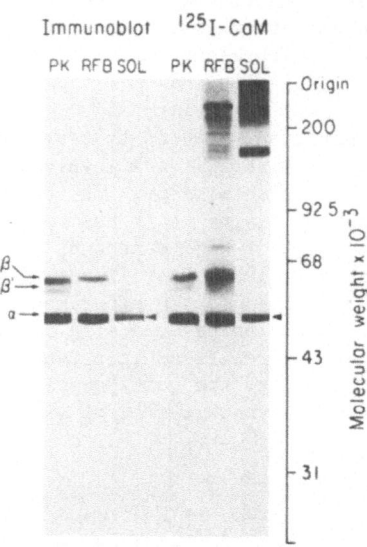

FIG. 6: Immunolabeling and ^{125}I-calmodulin (^{125}I-CaM) binding of purified calmodulin kinase II and of homogenates of rat forebrain and squid optic lobe. Purified kinase (PK: 0.5 μg), rat forebrain (RFB: 50 μg), and squid optic lobe (SOL: 75 μg) were separated by NaDodSO /9% polyacrylamide gel electrophoresis and either trans- ferred to nitrocellulose and analyzed for immunoreactivity or analyzed for ^{125}I-CaM binding in the gel. Immunoblotting was car- ried out by using a mouse monoclonal antibody (raised against rat forebrain calmodulin kinase II) and ^{125}I-labeled goat antimouse IgG. Immunoreactive and ^{125}I-CaM binding proteins were visualized by autoradiography. PK lanes were exposed for shorter times than were the other lanes in order to obtain exposures in the same intensity range. Arrows point to calmodulin kinase IIα and β/β' subunits. Arrowhead points to the squid immunoreactive protein and the ^{125}I-CaM binding protein of $M_r \approx 50,000$. (From Llinas et al., 1985)

duced a decrease in the amount of neurotransmitter released in response to a constant presynaptic depolarizing step. This response was obtained in all cases in which the injection was successfully performed (n = 8). In every case, injection of synapsin I produced a reduction in synaptic transmission without any detectable alteration in the amplitude or the time course of the presynaptic inward calcium current. Thus it appears that dephospho-synapsin I decreased the amount of transmitter released in response to a given amount of calcium influx.

In control experiments, injection of synapsin I that had been heated to 88°C for 2 min had no effect on the postsynaptic response. Similarly, injec- tion of synapsin I that had been phosphorylated by calmodulin kinase II had only a slight effect on synaptic transmission.

Injection of Calmodulin Kinase II

Calmodulin kinase II was pressure-injected into the presynaptic termi- nal. In all cases in which the injection was successfully performed (n = 4), injection of calmodulin kinase II produced an increase in the amount of neu- rotransmitter released in response to a constant presynaptic depolarizing step (Fig. 4). Injection of calmodulin kinase II produced an increase in the amplititude and rate of rise and a decrease in the latency of the post-synaptic response in the absence of any detectable change in amplitude or time course of the presynaptic inward calcium current. Thus these results suggest that calmodulin kinase II increased the amount of transmitter release by a mechanism that was independent of an action on membrane ionic conductances.

Evidence for the Existence of a Synapsin I-Like Protein in Squid Nervous System

Immunoblot analysis, using antibodies specific for synapsin I, revealed that squid optic lobe and stellate ganglion contained a protein immunologically related to mammalian synapsin I but with a slightly higher M_r value than the mammalian synapsin I doublet (Fig. 5). Further studies are still needed to determine the degree of functional similarity between the mammalian synapsin I and the synapsin I-like protein in the squid.

Evidence for the Existence of Calmodulin Kinase II in Squid Nervous System

Immunoblots, using a monoclonal antibody specific for calmodulin kinase II, revealed that both squid optic lobs (Fig. 6) and stellate ganglion (data not shown) contain an immunoreactive protein that migrates in the region of the α subunit (50 kDal) of mammalian calmodulin kinase II. Similarly, ^{125}I-calmodulin binding of proteins in the gel revealed that both squid optic lobe (Fig. 6) and stellate ganglion (data not shown) contain a ≈ 50 kDal calmodulin binding protein.

SUMMARY

The data presented here provide evidence that the study of neuronal phosphoproteins can lead to the identification of previously unknown proteins and that these proteins may play important roles in neuronal communication. Specifically, in the case of synapsin I, direct evidence has been obtained that this phosphoprotein is involved in regulating neurotransmitter release. A tentative explanation of the results obtained in the microinjection studies is as follows: synapsin I, in the dephosphostate, is bound to the cytoplasmic surface of synaptic vesicles and inhibits the ability of the vesicle to interact with the plasma membrane; increases in intracellular calcium activate calmodulin kinase II which in turn phosphorylates synapsin I and the phosphorylated synapsin I dissociates from the synaptic vesicle thus removing a constraint on the release of neurotransmitter.

Clearly, more studies need to be done to rigorously test this hypothesis. Nevertheless these studies of synapsin I suggest that the study of previously unknown phosphoproteins will lead to the elucidation of previously unknown regulatory processes in neurons.

ACKNOWLEDGMENT

This work was supported by U. S. Air Force Office of Scientific Research, Air Force Systems Command, USAF, grant AFOSR 84-0086.

REFERENCES

Adams, W. B. and Levitan, I. B., Intracellular injection of protein kinase inhibitor blocks the serotonin-induced increase of K$^+$ conductance in

Aplysia neuron R 15, Proc. Natl. Acad. Sci. USA 79:3877-3880 (1982).

Aitken, A., Bilham, T., Cohen, P., Aswad, D., and Greengard, P., A specific substrate from rabbit cerebellum for guanosine 3':5' monophosphate-dependent protein kinase. III. Amino acid sequences at the two phosphorylation sites, J. Biol. Chem. 256:3501-3506 (1981).

Albert, K. A., Wu, W. C. S., Nairn, A. C., and Greengard, P., Inhibition by calmodulin of calcium/phospholipid-dependent protein phosphorylation, Proc. Natl. Acad. Sci. USA 81:3622-3625 (1984).

Alkon, D. L., Acosta-Urguidi, J., Olds, J., Kuzma, G., and Neary, J. T., Protein kinase injection reduces voltage-dependent potassium currents, Science 219:303-306 (1983).

Aswad, D. and Greengard, P., A specific substrate from rabbit cerebellum for guanosine 3':5'-monophosphate-dependent protein kinase. I. Purification and characterization, J. Biol. Chem. 256:3487-3493 (1981a).

Aswad, D. and Greengard, P., A specific substrate from rabbit cerebellum for guanosine 3':5'-monophosphate-dependent protein kinase. II. Kinetic studies on its phosphorylation by guanosine 3':5'-monophosphate-dependent and adenosine 3':5'-monophosphate-dependent protein kinase, J. Biol. Chem. 256:3494-3500 (1981b).

Barnekow, A., Schartl, M., Anders, F., and Bauer, H., Identification of a fish protein associated with a kinase activity and related to the Rous sarcoma virus transforming protein, Cancer Research 42:2429-2433 (1982).

Browning, M. D., Dunwiddie, T., Bennett, W., Gispen, W. H., and Lynch G., Synaptic phosphoproteins: Specific changes after repetitive stimulation of the hippocampal slice, Science 203:60-62 (1979).

Browning, M. D. and Greengard, P., A family of synaptic vesicle-associated phosphoproteins: synapsin Ia, synapsin Ib, protein IIIa and protein IIIb, Soc. Neurosci. Abstr. 10:196 (1984).

Browning, M. D., Baudry, M. and Lynch, G., Evidence that high frequency stimulation influences the phosphorylation of pyruvate dehydrogenase and that the activity of this enzyme is linked to mitochondrial calcium sequestration, Prog. Brain. Res. 56:317-338 (1982).

Browning, M. D., Huganir, R., and Greengard, P., Protein phosphorylation and neuronal function, J. Neurochem. 45:11-23 (1985).

Burgoyne, R. D., Regulation of the muscarinic acetylcholine receptor: effects of phosphorylating conditions on agonist and antagonist binding, J. Neurochem. 40:324-331 (1983).

Castellucci, V. F., Kandel, E. R., Schwartz, J. H., Wilson, F. D., Nairn, A. C., and Greengard, P., Intracellular injection of the catalytic subunit of cyclic AMP-dependent protein kinase stimulates facilitation of transmitter release underlying behavioral sensitization in Aplysia, Proc. Natl. Acad. Sci. USA 77:7492-7496 (1980).

Caudwell, B., Antoniw, J. F., and Cohen, P., Calsequestrin, mysosin, and the components of the protein-glycogen complex in rabbit skeletal muscle, Eur. J. Biochem. 86:511-518 (1978).

Cohen, P., The role of protein phosphorylation in neural and hormonal control of cellular activity, Nature 296:613-620 (1982).

Costa, M. R., Casnellie, J. E., and Catterall, W. A., Selective phosphorylation of the alpha subunit of the sodium channel by cAMP-dependent protein kinase, J. Biol. Chem. 257:7918-7921 (1982).

Costa, M. R. C. and Catterall, W. A., Cyclic AMP-dependent phosphorylation of the α-subunit of the sodium channel in synaptic nerve ending particles, J. Biol. Chem. 259:8210-8218 (1984a).

Costa, M. R. C. and Catterall, W. A., Phosphorylation of the α-subunit of the sodium channel by protein kinase C, Cellular and Molecular Neurobiology 4:291-297 (1984b).

Cotton, P. C. and Brugge, J. S., Neural tissues express high levels of the cellular src gene product pp60 [c-src], Mol. and Cell Biol. 3:1157-1162 (1983).

DeCamilli, P., Ueda, T., Bloom, F. E., Battenberg, E., and Greengard, P.,

Widespread distribution of protein I in the central and peripheral nervous system, Proc. Natl. Acad. Sci. USA 76:5977-5981 (1979).

DeCamilli, P., Cameron, R., and Greengard, P., Synapsin I (protein I), a nerve terminal-specific phosphoprotein: I. Its general distribution in synapses of the central and peripheral nervous system demonstrated by immunofluorescence in frozen and plastic sections, J. Cell Biol. 96: 1337-1354 (1983).

DeCamilli, P., Harris, S. M., Huttner, W. B., and Greengard, P., Synapsin I (Protein I), a nerve terminal-specific phosphoprotein: II. Its specific association with synaptic vesicles demonstrated by immunocytochemistry in agarose embedded synaptosomes, J. Cell Biol. 96:1355-1373 (1983).

de Jonge, H. R. and Rosen, O. M., Self-phosphorylation of cyclic guanosine 3':5' -monophosphate-dependent protein kinase from bovine lung, J. Biol. Chem. 252:2780-2783 (1977).

Demaille, J. G. and Pechere, J. -F., The control of contractility by protein phosphorylation, Adv. Cyclic Nucleotide Res. 15:337-371 (1983).

de Peyer, J. E., Cachelin, A. B., Levitan, I. B., and Reuter, H., Ca^{2+}-activated K^+ conductance in internally perfused snail neurons is enhanced by protein phosphorylation, Proc. Natl. Acad. Sci. USA 79: 4207-4211 (1982).

DeRiemer, S. A., Strong, J. A., Albert, K. A., Greengard, P., and Kaczmarek, L. K., Enhancement of calcium current in Aplysia neurons by phorbol ester and protein kinase C, Nature 313:313-316 (1985).

Dolphin, A. C. and Greengard, P., Serotonin stimulates phosphorylation of Protein I in the facial motor nucleus of rat brain, Nature 298:76-79 (1981a).

Dolphin, A. C. and Greengard, P., Neurotransmitter and neuromodulator-dependent alterations in phosphorylation of Protein I in slices of rat facial nucleus, J. Neurosci. 1:192-203 (1981b).

Douglas, W. W. and Ritchie, J. M., A technique for recording functional activity in specific groups of myelinated and non-myelinated nerve trunks, J. Physiol. Lond. 138:19-30 (1957).

Edelman, A. M., Raese, J. D., Lazar, M. A., and Barchase, J. D., Tyrosine hydroxylase: studies on the phosphorylation of a purified preparation of the brain enzyme by the cyclic AMP-dependent protein kinase, J. Pharmacol. Exp. Ther. 216:647-653 (1981).

Ehrlich, Y. H., Whittemore, S. R., Garfield, M. K., Graber, S. G., and Lenox, R. H., Protein phosphorylation in the regulation and adaptation of receptor function, Prog. Brain Res. 56:375-396 (1982).

Forn, J. and Greengard, P., Depolarizing agents and cyclic nucleotides regulate the phosphorylation of specific neuronal proteins in rat cerebral cortex slices, Proc. Natl. Acad. Sci. USA 75:5195-5199 (1978).

Fried, G., Nestler, E. J., DeCamilli, P., Stjärne, L., Olson, L., Lundberg, J. M., Hökfelt, T., Ouimet, C. C., and Greengard, P., Cellular and subcellular localization of Protein I in the peripheral nervous system, Proc. Natl. Acad. Sci. USA 79:2717-2721 (1982).

Gispen, W. H. and Routtenberg, A., eds., Brain Phosphoproteins, Prog. in Brain Res., Vol 56, Elsevier, Amsterdam (1982).

Goldenring, J. R., Gonzalez, B. and DeLorenzo, R. J., Isolation of brain Ca^{2+}-calmodulin tubulin kinase containing calmodulin binding proteins, Biochem. Biophys. Res. Commun. 108:421-428 (1982).

Gordon, A. S., Davis, C. G., Milfay, D., and Diamond, I., Phosphorylation of acetylcholine receptor by endogenous membrane protein kinase in receptor-enriched membranes of Torpedo californica, Nature 267:539-540 (1977).

Gurley, L. R., D'Anna, J. A., Halleck, M. S., Barham, S. S., Walters, R. A., Jett, J. J., and Tobey, R. A., Relationships between histone phosphorylation and cell proliferation, Cold Spring Harbor Conf. Cell Prolif. 8:1073-1093 (1981).

Hathaway, G. M. and Traugh, J. A., Casein kinases-multi-potential protein kinases, Curr. Top. Cell Reg. 21:101-127 (1982).

Hemmings Jr., H. C., Nairn, A. C., Aswad, D. W., and Greengard, P., DARPP-32, a dopamine and adenosine 3':5'-monophosphate-regulated phospho-protein enriched in dopamine-innervated brain regions. II. Purification and characterization of the phosphoprotein from bovine caudate nucleus, J. Neurosci. 4:99-110 (1984a).

Hemmings Jr., H. C., Greengard, P., Tung, H. Y. L., and Cohen, P., DARPP-32, a dopamine-regulated neuronal phosphoprotein, is a potent inhibitor of protein phosphatase-1, Nature 310:503-505 (1984b).

Hemmings Jr., H. C., Walaas, S. I., Ouimet, C. C., and Greengard, P., "Structure and Function of Dopamine Receptors" in Receptor Biochemistry and Methodology, (eds. Creese, I. & Fraser, C. M.), vol. 9, Alan R. Liss, Inc., New York (1985).

Hoch, D. B., Dingledine, R. J., and Wilson, J. E., Long-term potentiation in the hippocampal slice: possible involvement of pyruvate dehydro-genase, Brain Res. 302:125-134 (1984).

Hook, V. Y. H., Stokes, K. B., Lee, N. M., and Loh, H. H., Possible nuclear protein kinase regulation of homologous ribonucleic acid polymerases from small dense nuclei of mouse brain during morphine tolerance-dependence, Biochem. Pharmacol. 30:2313-2318 (1981).

Huang, C.-K., Browning, M. D., and Greengard, P., Purification and charac-terization of protein IIIb, a mammalian brain phosphoprotein, J. Biol. Chem. 257:6524-6528 (1982).

Huganir, R. L., Miles, K., and Greengard, P., Phosphorylation of the nico-tinic acetylcholine receptor by an endogenous tyrosine-specific pro-tein kinase, Proc. Natl. Acad. Sci. USA 81:6968-6972 (1984).

Huttner, W. B., DeGennaro, L. J., and Greengard, P., Differential phospho-rylation of multiple sites in purified Protein I by cyclic AMP-depen-dent and calcium dependent-protein kinases, J. Biol. Chem. 256:1482-1488 (1981).

Huttner, W. B., Schiebler, W., Greengard, P., and DeCamilli, P., Synapsin I (protein I), a nerve terminal-specific phosphoprotein. III. Its association with synaptic vesicles studied in a highly purified synap-tic vesicle preparation, J. Cell Biol. 96:1374-1388 (1983).

Joh, T. H., Park, D. H., and Reis, D. J., Direct phosphorylation of brain tyrosine hydroxylase by cyclic AMP-dependent protein kinase: mechanism of enzyme activation, Proc. Natl. Acad. Sci. USA 75:4744-4748 (1978).

Johnson, E. M., Ueda, T., Maeno, H. and Greengard, P., Adenosine 3':5'-monophosphate-dependent phosphorylation of a specific protein in synaptic membrane fractions from rat cerebrum, J. Biol. Chem. 247:5650-5652 (1972).

Johnson, E. M., Nuclear protein phosphorylation and the regulation of gene expression, Handb. Exp. Pharmacol. 58:(Part I) 507-533 (1982).

Kaczmarek, L. K., Jennings, K. R., Strumwasser, F., Nairn, A. C., Walter, U. L., Wilson, F. D., and Greengard, P., Microinjection of catalytic subunit of cyclic AMP-dependent protein kinase enhances calcium action potentials of bag cell neurons in cell culture, Proc. Natl. Acad. Sci. USA 77:7487-7491 (1980).

Kennedy, M. B. and Greengard, P., Two calcium/calmodulin-dependent protein kinases which are highly concentrated in brain, phosphorylate Protein I at distinct sites, Proc. Natl. Acad. Sci. 78:1293-1297 (1981).

Kennedy, M. B., McGuinness, T., and Greengard, P., A calcium/calmodulin-dependent protein kinase from mammalian brain that phosphorylates synapsin I: Partial purification and characterization, J. Neurosci. 3:818-831 (1983).

Kikkawa, U., Takai, Y., Minakuchi, R., Inohara, S., and Nishizuka, Y., Cal-cium-activated, phospholipid-dependent protein kinase from rat brain. Subcellular distribution, purification and properties, J. Biol. Chem. 257:13341-13348 (1982).

Krebs, E. G. and Beavo, J. A., Phosphorylation-dephosphorylation of enzymes, Ann. Rev. Biochem. 48:932-959 (1979).

Krueger, B. K., Forn, J., and Greengard, P., Depolarization-induced phosphorylation of specific proteins, mediated by calcium ion influx, in rat brain synaptosomes, J. Biol. Chem. 252:764-2773 (1977).

Kuhn, D. M. and Lovenberg, W., Role of calmodulin in the activation of tryptophan hydroxylase, Fed. Proc. 41:2258-2264 (1982).

Langan, T., Phosphorylation of liver histone following the administration of glucagon and insulin, Proc. Natl. Acad. Sci. 64:1276-1283 (1969).

Lee, R. H., Brown, B. M., and Lolley, R. N., Protein kinases of retinal rod outer segments: identification and partial characterization of cyclic nucleotide dependent protein kinase and rhodopsin kinase, Biochem. 20: 7532-7538 (1981).

Llinas, R. R., Calcium in synaptic transmission, Scientific American 247: 56-65 (1982).

Llinas, R. R., The squid giant synapse, Curr. Top. Mem. Trans. 22:519-546 (1984).

Llinas, R. R., Steinberg, I. Z., and Walton, K., Presynaptic calcium currents in squid giant synapse, Biophys. J. 33:289-322 (1981).

Llinas, R. R., McGuinness, T. L., Leonard, C. S., Sugimori, M., and Greengard, P., Intraterminal injection of synapsin I or calcium/calmodulin-dependent protein kinase II alters neurotransmitter release at the squid giant synapse, Proc. Natl. Acad. Sci. USA 82:3035-3039 (1985).

Marchmont, R. J. and Houslay, M. D., Insulin triggers cyclic AMP-dependent activation and phosphorylation of a plasma membrane cyclic AMP phosphodiesterase, Nature 286:904-906 (1980).

McGuinness, T. L., Lai, Y. L., Greengard, P., Woodgett, J. R., Cohen, P., A multifunctional calmodulin-dependent protein kinase: similarities between skeletal muscle glycogen synthase and brain synapsin I kinase, FEBS Lett. 163:329-334 (1983).

McGuinness, T. L., Lai, Y. L., and Greengard, P., Ca^{2+}/calmodulin-dependent protein kinase II: Isozymic forms from rat forebrain and cerebellum, J. Biol. Chem. 260:1696-1704 (1985).

Mobley, P. and Greengard, P., Evidence for widespread effects of noradrenaline on axon terminals in the rat frontal cortex, Proc. Natl. Acad. Sci. USA 82:945-947 (1985).

Nairn, A. C. and Greengard, P., Purification and characterization of brain Ca^{2+}/calmodulin-dependent protein kinase I that phosphorylates synapsin I, Neurosci. Abstr. 1029 (1983).

Navone, F., Greengard, P., and DeCamilli, P., Synapsin I in nerve terminals: Selective association with small synaptic vesicles, Science 226:1209-1211 (1984).

Nestler, E. J. and Greengard, P., Dopamine and depolarizing agents regulate the state of phosphorylation of Protein I in the mammalian superior cervical sympathetic ganglion, Proc. Natl. Acad. Sci. USA 77:7479-7483 (1980).

Nestler, E. J. and Greengard, P., Nerve impulses increase the state of phosphorylation of protein I in the rabbit superior cervical ganglion, Nature 296:452-454 (1982).

Nestler, E. and Greengard, P., Protein phosphorylation in the nervous system, Wiley, New York (1984).

Oestreicher, A. B.., van Duin, M., Zwiers, H., and Gispen, W. H., Cross-reaction of anti-rat B-50: Characterization and isolation of a "B-50 phosphoprotein" from bovine brain, J. Neurochem. 43:935-943 (1984).

Osterrieder, W., Brum, G., Hescheler, J.., Trautwein, W., Flockerzi, V., and Hofmann, F., Injection of subunits of cyclic AMP-dependent protein kinase into cardiac myocytes modulates Ca^{2+} current, Nature 298:576-578 (1982).

Ouimet, C. C., McGuinness, T. L., and Greengard, P., Immunocytochemical localization of calcium/calmodulin-dependent protein kinase II in rat, Proc. Natl. Acad. Sci. USA 81:5604-5608 (1984).

Rangel-Aldao, R. and Rosen, O. M., Dissociation and reassociation of the phosphorylated and non-phosphorylated forms of adenosine 3':5' -monophosphate-dependent protein kinase from bovine cardiac muscle, J. Biol. Chem. 251:3375-3380 (1976).

Roberts S., Ribosomal protein phosphorylation and protein synthesis in the brain, Prog. Brain Res. 56:195-211 (1982).

Schiebler, W. Jahn, R., Doucet, J. -P., Rothlein, J., and Greengard, P., Synapsin I (protein I) binds specifically and with high affinity to highly purified synaptic vesicles from rat brain, J. Biol. Chem. (in press) (1986).

Schlichter, D.. J., Casnellie, J. E., and Greengard, P., An endogenous substrate for cGMP-dependent protein kinase in mammalian cerebellum, Nature 273:61-62 (1978).

Schulman, H. and Greengard, P., Stimulation of brain membrane protein phosphorylation by calcium and an endogenous heat-stable protein, Nature 271:478-479 (1978a).

Schulman, H. and Greengard, P., Ca^{2+}-dependent protein phosphorylation system in membranes from various tissues, and its activation by "calcium-dependent regulator," Proc. Natl. Acad. Sci. USA 75:5432-5436 (1978b).

Sharma, R. K., Wang, T. H., Wirch, E., and Wang, J. H., Purification and properties of bovine brain calmodulin-dependent cyclic nucleotide phosphodiesterase, J. Biol. Chem. 255:5916-5923 (1980).

Shichi, H. and Somers, R. L., Light-dependent phosphorylation of rhodopsin. Purification and properties of rhodopsin kinase, J. Biol. Chem. 253: 7040-7046 (1978).

Sieghart, W., Forn, J., Schwarcz, R., Coyle, J. T., and Greengard, P., Neuronal localization of specific brain phosphoproteins, Brain Res. 156: 345-350 (1978).

Stadel, J. M., Nambi, P., Shorr, R. G. L., Sawyer, D. F., Caron, M. G., and Lefkowitz, R. J., Catecholamine-induced desensitization of turkey erythrocyte adenylate cyclase is associated with phosphorylation of the β-adrenergic receptor, Proc. Natl. Acad. Sci. USA 80:3173-3177 (1983).

Teichberg, V. I. and Changeux, J. -P., Evidence for protein phosphorylation and dephosphorylation in membrane fragments isolated from the electric organ of Electrophorus electricus, FEBS Lett. 74:71-76 (1977).

Thomas, G., Activation of multiple S6 phosphorylation and its possible role in the alteration of mRNA expression, Prog. Brain Res. 56:179-194 (1982).

Treiman, M. and Greengard, P., D-1 and D-2 dopaminergic receptors regulate protein phosphorylation in the rat neurohypophysis, Neurosci. 15:713-722 (1985).

Tsou, K. and Greengard, P., Regulation of phosphorylation of Proteins I, III_a and III_b in rat neurohypophysis in vitro by electrical stimulation and by neuroactive agents, Proc. Natl. Acad. Sci. USA 79:6075-6079 (1982).

Ueda, T., Attachment of the synapse-specific phosphoprotein, Protein I, to the synaptic membrane: A possible role of the collagenase-sensitive region of Protein I, J. Neurochem. 36:297-300 (1981).

Ueda, T., Maeno, H., and Greengard, P., Regulation of endogenous phosphorylation of specific proteins in synaptic membrane fractions from rat brain by adenosine 3':5' monophosphate, J. Biol. Chem. 248:8295-8305 (1973).

Ueda, T. and Greengard, P., Adenosine 3':5' monophosphate-regulated phosphoprotein system of neuronal membranes. I. Solubilization, purification, and some properties of an endogenous phosphoprotein, J. Biol. Chem. 252:5155-5163 (1977).

Walaas, S. I., Aswad, D. W., and Greengard, P., A dopamine- and cyclic AMP-regulated phosphoprotein enriched in dopamine-innervated brain regions, Nature 301:69-71 (1983).

Wise, B. C., Guidotti, A., and Costa, E., Phosphorylation induces a decrease in the biological activity of the protein inhibitor (GABA-modulin) of γ-aminobutyric acid binding sites, Proc. Natl. Acad. Sci. USA 80:886-890 (1983).

Wu, W. C. -S., Walaas, S. I., Nairn, A. C., and Greengard, P., Calcium-phospholipid regulates phosphorylation of M_r "87K" substrate protein in brain synaptosomes, Proc. Natl. Acad. Sci. USA 79:5249-5253 (1982).

Yamauchi, T. and Fujisawa, H., Tyrosine 3-monooxygenase is phosphorylated by Ca^{2+}-calmodulin-dependent protein kinase, followed by activation by activator protein, Biochem. Biophys. Res. Comm. 100:807-813 (1981).

Young, J. Z., The functioning of giant nerve fibers of the squid, J. Expl. Biol. 15:170-185 (1938).

Zwiers, H., Veldhuis, D., Schotman, P., and Gispen, W. H., ACTH, cyclic nucleotides and brain protein phosphorylation in vitro, Neurochem. Res. 1:669-677 (1976).

Zwiers, H., Jolles, J., Aloyo, V. J., Oestereicher, A. B., and Gispen, W. H., ACTH and synaptic membrane phosphorylation, Prog. Brain Res. 55:405-417 (1982).

Zwiller, J. M., Revel, O., and Basset, P., Evidence for phosphorylation of rat brain guanylate cyclase by cyclic AMP-dependent protein kinase, Biochem. Biophys. Res. Commun. 101:1381-1387 (1982).

REGULATION OF THE PHOSPHORYLATION AND DEPHOSPHORYLATION

OF A 96,000 DALTON PHOSPHOPROTEIN (P96) IN INTACT

SYNAPTOSOMES

Phillip J. Robinson

Merrell Dow Research Institute, 2110 E. Galbraith Road
Cincinnati, OH 45215

SUMMARY

When intact rat brain synaptosomes are depolarized there is a significant increase in the phosphorylation of many proteins, and a rapid dephosphorylation of a 96,000 dalton protein termed P96. The mechanisms governing
dephosphorylation are shown to be distinct from the mechanisms leading to
increased phosphorylation of proteins such as synapsin I. Depolarization-
dependent P96 dephosphorylation was found to be rapid (preceding the phosphorylation of synapsin I) and fully reversible, and required both depolarization and calcium entry. The phosphorylation of P96 was specifically
increased by fluphenazine and by the calcium channel agonist (BAY K 8644)
and antagonist (verapamil) by unknown mechanisms. Phosphorylation was also
increased in the presence of dibutyryl cAMP indicating some role for cAMP-
dependent protein kinase in P96 labeling. Preliminary evidence also raises
the possibility of a role for protein kinase C. The characteristics of this
unique synaptosomal protein suggest that it may play an important role in
nerve terminal function.

BACKGROUND

Five distinct classes of protein kinase (PK) enzymes have been recognized in brain, according to the agent which regulates their activity -
cAMP (PKA), cGMP (PKG), calcium/calmodulin (CM-PK), calcium-phospholipid/
diacylglycerol (PKC), or no known agent. PKA, CM-PK and PKC comprise the
dominant forms of protein kinases in synaptosomes. There are a number of
substrates of these enzymes in synaptosomes, many of which are unique to
nerve cells and many are substrates of more than one PK. Substrates of PKA
include MAP2, synapsin Ia and Ib, protein IIIa and IIIb, and DARPP-32 (Nestler and Greengard, 1984), substrates of CM-PK include MAP2, synastin Ia and
Ib, and tubulin, while substrates of PKC include MAP2, the "87k" protein,
B-50, and myelin basic proteins (Wu et al., 1982; Dosemeci and Rodnight,
1982; Aloyo et al., 1983). All of these proteins are identifiable in one-
dimensional polyacrylamide gels by their phosphorylation characteristics
and peptide maps from lysed synaptosomal subcellular fractions (Dunkley et
al., 1986). In many cases (such as synapsin I, protein III, "87k", and
B-50) these phosphoproteins are also identifiable in intact respiring
synaptosomes, where their phosphorylation may be regulated by various physiological stimuli (Dunkley et al., 1986).

Synaptosomes represent an excellent model in which to study these processes since these pinched off nerve terminals are functionally capable of stimulated neurotransmitter release. Protein phosphorylation in intact synaptosomes has been clearly shown to be regulated via the intracellular second messengers calcium and cAMP and has been proposed to play important roles in modulating neurotransmitter release (Llinas et al., 1985). The following sections will briefly review some of the mechanisms involved to show how the complex interactions of many nerve terminal enzymes and regulators may ultimately link a physiological stimulus to the phosphorylation of specific intracellular proteins.

Nerve terminals possess a variety of hormone and neurotransmitter receptors that, upon activation, may initiate alterations in calcium fluxes or activate synthesis of cAMP. In synaptosomes receptor-mediated events are exclusively presynaptic, and examples exist where receptor stimulation leads to the phosphorylation of specific intracellular proteins (Michaelson et al., 1979). Greengard's group have clearly demonstrated that some of the underlying mechanisms include activation of PKA. For example, in synaptosomes from the rat facial nucleus which have a defined serotonergic input, the phosphorylation of synapsin I is stimulated upon application of serotonin (Dolphin and Greengard, 1981). A fragment of synapsin I that is selectively phosphorylated by PKA, termed phosphopeptide 1, was selectively affected, suggesting activation of this PK. Similar results have been obtained for synapsin I and protein III in slices from many brain regions and with many other neurotransmitter systems (Nestler and Greengard, 1984).

The dynamics of calicum entry and removal from synaptosomes is also an important regulator of protein phosphorylation. Synaptosomes incubated with calcium under resting conditions will slowly take up calcium at a rate proportional to the concentrations (Nachshen and Blaustein, 1980) yet the intracellular free calcium concentration is strictly maintained at about 10^{-7}M. Calcium must be then extruded from the terminals by Na/Ca exchange or by a specific calcium pumping ATPase (Ca-ATPase) found in the plasma membrane or on intracellular vesicles (Chan et al., 1984). These resting equilibria are accompanied by large alterations in the basal phosphorylation of a number of synaptosomal phosphoproteins such as synapsin I (Robinson and Dunkley, 1985).

When synaptosomes are depolarized there is a rapid initial calcium influx that is complete within one second, followed by a slower influx for the subsequent 20 seconds (Nachshen and Blaustein, 1980). This calcium influx, apart from initiating neurotransmitter release, triggers the increased phosphorylation of many synaptosomal phosphoproteins (Krueger et al., 1977; DeLorenzo et al., 1979; Michaelson et al., 1979; Robinson and Dunkley, 1983). The first step in the underlying molecular mechanism is the rise in intrasynaptosomal calcium followed by activation of at least two calcium-dependent mechanisms. Firstly calcium activates calmodulin and CM-PK and thereby stimulates phosphorylation of synapsin I, tubulin and other proteins. This has been demonstrated since calmodulin inhibitors such as trifluoperazine and fluphenazine prevent the phosphorylation increases (DeLorenzo, 1981; Robinson et al., 1984). More direct evidence came from phosphopeptide mapping of synapsin I, which revealed depolarization-dependent increases in the phosphorylation of phosphopeptide 2 (and other fragments) which is known to be phosphorylated only by CM-PK (Huttner and Greengard, 1979). The second mechanism involves calcium activation of PKC, which phosphorylates a specific "87k" substrate in synaptosomes (Wu et al., 1982).

After depolarization-dependent increases, protein phosphorylation is returned to basal levels. When calcium entry is terminated by the addition of extracellular EGTA, protein phosphorylation is rapidly reversed (Krueger et al., 1977; Michaelson and Avissar, 1979), suggesting that calcium removal inactivates protein kinases and allows already active phosphatases to dephosphorylate the proteins. However, when calcium entry is not terminated these proteins may be extensively dephosphorylated to well below basal level and it has been proposed that this may involve activation by calcium of protein phosphatases, or calcium inactivation of protein kinases (Robinson and Dunkley, 1983, 1985).

As well as activating protein phosphorylation, depolarization-dependent calcium influx activates protein dephosphorylation. The alpha subunit of pyruvate dehydrogenase (PDH) is a mitochondrial phosphoprotein present inside intact synaptosomes (Dennig and Sieghart, 1984) which is slowly dephosphorylated upon depolarization-dependent calcium entry (Robinson and Dunkley, 1983). Dephosphorylation of PDH occurs via activation of a specific calcium-stimulated protein phosphatase and results in increased activity of PDH, thereby stimulating ATP production. Another dephosphorylating system is the rapid and extensive dephosphorylation of a 96,000 dalton synaptosomal protein termed P96 (Robinson and Dunkley, 1983, 1985). The identity of this protein and the mechanism of its dephosphorylation remain largely unknown, however those mechanisms are clearly distinct from any of those described above. The aim of this report is to review some recent studies on the phosphorylation/dephosphorylation of P96 that have begun to decipher some of the characteristics of the protein and its phosphorylation mechanism. The regulatory systems governing both the phosphorylation and dephosphorylation of P96 are quite distinct from those regulating proteins such as synapsin I or protein III.

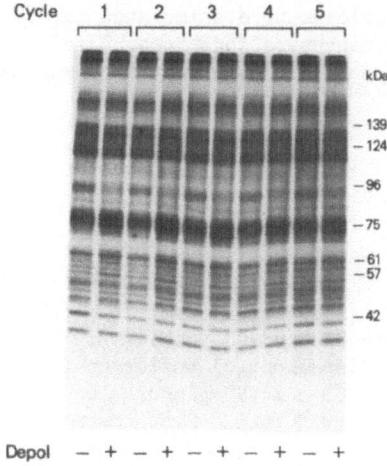

FIG. 1: Effect of several cycles of depolarization on protein phosphorylation. Striatal synaptosomes were depolarized for 30 seconds with 41 mM K$^+$ in the presence of 0.1 mM calcium. The synaptosomes were rapidly pelleted in a microfuge and resuspended in low K$^+$ buffer for 5 minutes to equilibrate prior to another depolarization. This was repeated for five cycles. Several phosphoproteins are indicated, P75 has been identified as synapsin Ib, P57 as protein IIIb and P42 as PDH (see text).

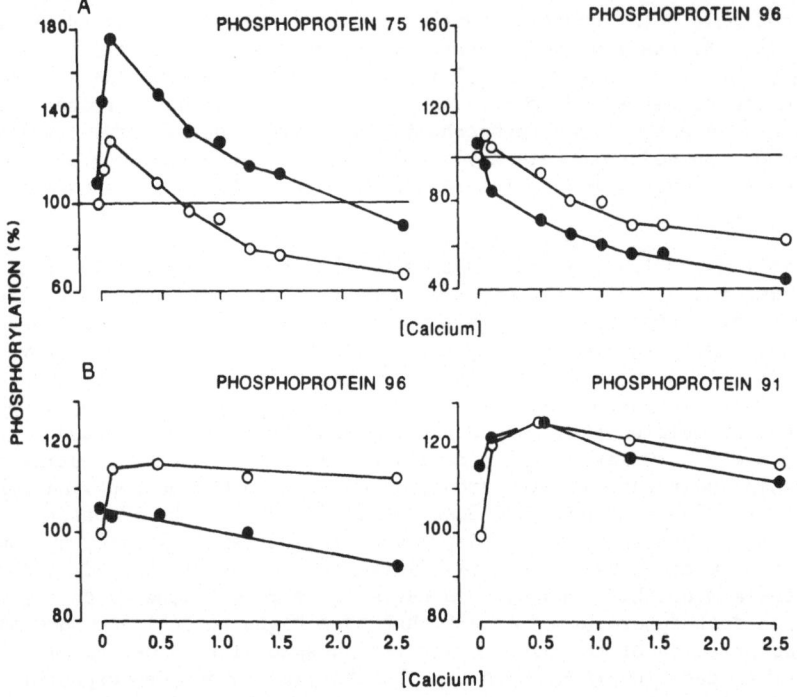

FIG. 2: Effects of calcium concentration during the prelabeling or during
the incubation. A: Cortical synaptosomes were prelabeled in the
presence of various calcium concentrations from 0 to 2.5 mM followed by 5 sec incubation in control (open circles) or 41 mM K[+]
buffers (closed circles). Phosphorylation in the absence of exogenous calcium was set at 100%. B: Synaptosomes prelabeled in
the absence of calcium were stimulated for 15 sec with pulses of
various calcium concentrations (open circles) or calcium plus 41
mM K[+] (closed symbols). Data in A from Robinson and Dunkley,
1985.

METHODS

An intact P2 pellet was isolated from rat cerebral cortex or striatum
as described previously (Robinson and Dunkley, 1983). The pellet was finally resuspended to 5 mg/ml in a control buffer containing (in mM) Na^+ 143,
K^+ 4.7, Mg^{2+} 1.2, Cl^- 126, HCO_3 24.9, glucose 10, and various calcium concentrations from 0 to 2.5, or EGTA 1. Synaptosomal samples were prelabeled
for 45 minutes at 37°C with 0.5 mCi/ml $^{32}P_i$ (Robinson and Dunkley, 1983, 1985;
Robinson et al., 1984). Aliquots were then incubated for various times in
41 mM K^+ buffers, or control buffers containing veratridine or other agents
from 5 seconds to 5 minutes. When drugs such as fluphenazine or calcium
channel agonists and antagonists were employed they were present for a 10
minute pre-incubation step prior to brief incubations in control or depolarizing buffers (Robinson et al., 1984). Reactions were terminated by addition of SDS stop reagent, solubilized proteins were separated on 7.5-15%
gradient polyacrylamide gels, and protein phosphorylation was determined by
quantitative densitometry of autoradiographs (Robinson and Dunkley, 1983).

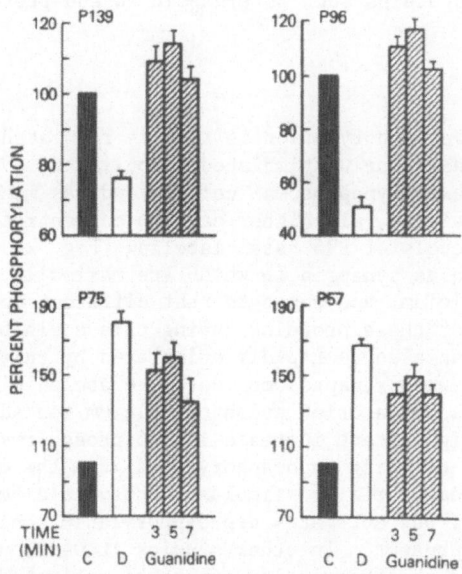

FIG. 3: Effect of guanidine on protein phosphorylation. Striatal synaptosomes prelabeled in the presence of 0.1 mM calcium were incubated with control (C) or 41 mM K^+ (D) buffers for 10 sec, or with 5 mM guanidine for the times shown.

RESULTS

Depolarization

Depolarization of synaptosomes with either K^+ or veratridine results in the rapid dephosphorylation of P96 (Fig. 1, lanes 1 and 2). P96 dephosphorylation appeared to be an "all-or-none" phenomenon, and intermediate phosphorylation states were rarely seen. Time course studies have shown that the effect occurs within the fastest time that was reproducibly measured (5 seconds) and there was no further decline, except in the background of the autoradiographs (Robinson and Dunkley, 1983). This dephosphorylation was unaffected by an extracellular calcium concentration above a threshold of 100 μM and usually preceded the maximal increased phosphorylation of other proteins such as synapsin Ib (Robinson and Dunkley, 1983; 1985). Depolarization-dependent P96 dephosphorylation was almost independent of the strength of the depolarizing stimulus of K^+ or veratridine. When veratridine was used to initiate depolarization, P96 was maximally dephosphorylated at all concentrations from 1 to 100 μM, however synapsin Ib responded to weak depolarizing strengths with very small phosphorylation increases and responded to 30 μM veratridine with much greater increases (Robinson and Dunkley, 1983).

In order to assume that there is physiological relevance to depolarization-dependent P96 dephosphorylation it was necessary to show that it was fully reversible. Thus, when synaptosomes that were depolarized with high K^+ were resuspended in low K^+ control buffers, P96 regained its original phosphorylation level (Fig. 1). When these "repolarized" synaptosomes were subjected to a further depolarizations P96 was rapidly dephosphorylated each time, indicating a dynamic process that is highly attuned to the synaptosomal membrane potential. The depolarization-dependent increase in phos-

159

phorylation of other proteins such as synapsin Ib and protein IIIb was also reversible (Fig. 1).

Calcium

The state of P96 phosphorylation is tightly regulated by calcium in severaal ways. The basal (or unstimulated) labeling of P96 was dependent on the concentration of extrasynaptosomal calcium and was maximal in the presence of only 10 μM calcium, all higher calcium concentrations resulted in progressively lower levels of P96 basal labeling (Fig. 2A). This in contrast to proteins such as synapsin Ib which are maximally labeled in the presence of 100 μM calcium, and suggests that different molecular mechanisms underly the labeling of these proteins during this equilibration period. P96 phosphorylation can also be rapidly stimulated by raising extrasynaptosomal calcium. Thus, when synaptosomes were prelabeled without exogenous calcium for 45 minutes and a brief pulse of calcium was added for only 15 seconds there was a significant increase in P96 phosphorylation (Fig. 2B). P96 was still capable of being dephosphorylated when the calcium stimulus was accompanied by a depolarizing stimulus. Thus, calcium regulates the basal labeling of P96, but activates dephosphorylation only when a depolarizing event is also present. In other studies it was found that depolarization-dependent P96 dephosphorylation cannot be activated by depolarization in the absence of calcium (Robinson and Dunkley, 1983; 1985), suggesting that depolarization is necessary, but not sufficient, to activate the P96 dephosphorylation mechanism.

To explore further the question of how P96 dephosphorylation is regulated, experiments were designed to alter intrasynaptosomal calcium levels without activating the calcium channel or depolarizing the synaptosomal membrane. In previous studies it has been shown that activation of protein

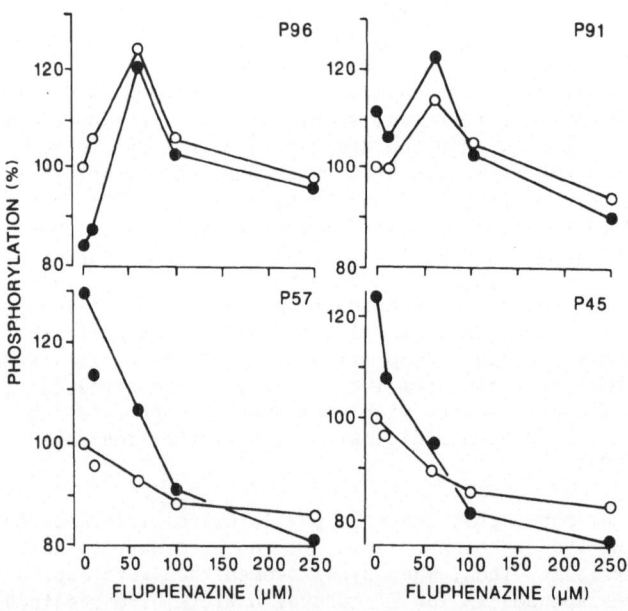

FIG. 4: Effect of fluphenazine on protein phosphorylation. Cortical synaptosomes prelabeled with 0.1 mM calcium were preincubated for 10 minutes with various fluphenazine concentrations then incubated for 5 seconds in control (open circles) or 41 mM K^+ buffer (closed circles). Note that P45 has tentatively been identified as B-50 (Dunkley et al., 1986).

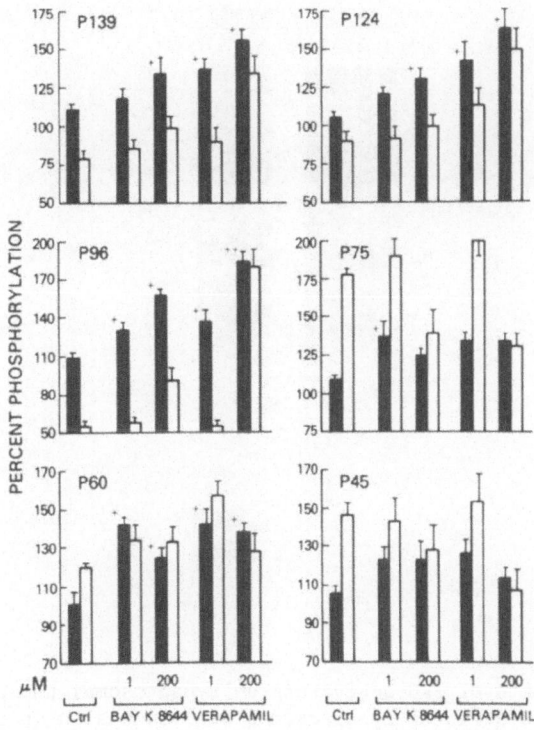

FIG. 5: Effect of BAY K 8644 and verapamil on protein phosphorylation. Striatal synaptosomes prelabeled with 0.1 mM calcium were preincubated for 10 minutes with BAY K 8644 or verapamil followed by 30 seconds incubation in control (shaded columns) or depolarizing buffers (open columns). Data from Robinson and Lovenberg, 1986.

phosphorylation on depolarization is dependent on the rise in intracellular calcium rather than on the mechanism of calcium's entry (Krueger et al., 1977; Michaelson and Avissar, 1979; DeLorenzo, 1981). In the current study intracellular calcium levels were raised by incubating synaptosomes in the presence of the metabolic poison guanidine. Guanidine raises intracellular calcium, even in calcium-free media, by release from intracellular stores such as mitochondria and endoplasmic reticulum (Boadle-Biber, 1982). Incubation of intact synaptosomes with 5 mM guanidine for 5 to 7 minutes resulted in increased phosphorylation of synapsin Ib and protein IIIb to almost the same level as a depolarizing stimulus, but did not activate P96 dephosphorylation, (Fig. 3). Another phosphoprotein whose phosphorylation is also decreased on depolarization, P139, was also unaffected by guanidine. The result clearly dissociates the mechanism of P96 dephosphorylation from the activation of CM-PK and PKC. P96 dephosphorylation therefore requires both a depolarizing event <u>and</u> the presence of raised intracellular calcium.

Drug Effects

The basal phosphorylation of P96 can be altered by at least two classes of drugs - phenothiazine and calcium channel agonists and antagonists. In both situations the phosphorylation of P96 and only few other proteins is selectively increased. When synaptosomes were preincubated for 10 minutes in the presence of the phenothiazine, fluphenazine (60 μM), there was a large and relatively specific increase in P96 phosphorylation (Fig. 4).

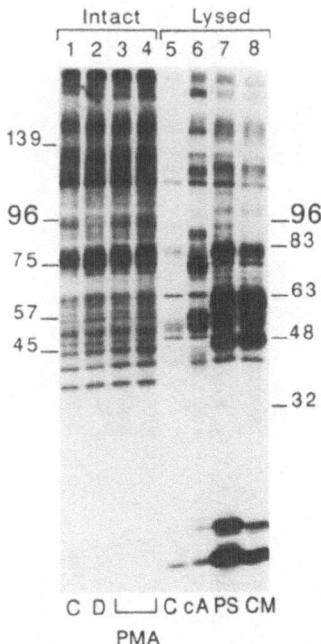

FIG. 6: Some effects of PKC activators on synaptosomal phosphoproteins.
A: Intact striatal synaptosomes were incubated in control (C) or
depolarizing buffer (D) for 30 seconds, or with phorbol-12-myri-
state-13-acetate (PMA), 500 ng/ml) for 2 (lane 3) or 3 minutes
(lane 4). B: Striatal synaptosomal cytoplasm was phosphorylated
in the presence of $[\gamma\ ^{32}P]$ ATP for 30 seconds with 1 mM EGTA (C),
50 µM cAMP (cA), 0.1 mM calcium plus 12.5 µg/ml calmodulin (CM) or
calcium plus 20 µg/ml phosphatidylserine (PS).

The effect was also seen for P91, but not for proteins that were normally
increased on depolarization such as protein IIIb, P45 (Fig. 4) and synapsin
Ib. This stimulation was selective for P96 only in the presence of 0.1 mM
calcium (Robinson et al., 1984). The underlying mechanism may be related
to alterations in intracellular messengers such as cAMP or IP3. P96 phosphorylation was also increased by the calcium channel agonist, BAY K 8644,
and antagonists, verapamil and nifedipine (Robinson and Lovenberg, 1986).
A 10 minute preincubation of synaptosomes with 1 or 200 µM BAY K 8644
resulted in significantly increased P96 phosphorylation, but did not alter
depolarization-dependent dephosphorylation (Fig. 5). Only 3 other phosphoproteins were increased, P139, P124 and P60, but phosphoproteins such as
synapsin Ib (Fig. 5) protein IIIb, and P45 were not significantly altered.
The calcium channel antagonists verapamil (Fig. 5) and nifedipine also produced dose-dependent increases in P96 phosphorylation. Verapamil also
blocked depolarization-dependent P96 phosphorylation, as expected of a drug
that prevents calcium influx (Nachshen and Blaustein, 1979). Since none of
the agents directly activated CM-PK, PKA or PKC, the effect on P96 is more
likely to be mediated by intracellular messengers such as calcium or cAMP.

Protein Kinase

The PK responsible for P96 phosphorylation remains to be identified,
but several lines of evidence suggest a role for calcium- and cAMP-stimula-
ted protein kinases. Dibutyryl cAMP is a membrane-penetrating form of cAMP

162

that is able to activate PKA in intact cells. When intact synaptosomes were stimulated for 10 min with 10 mM dibutyryl cAMP, P96 phosphorylation was significantly increased (Table 1). Phosphorylation of synapsin Ib (a known substrate for this kinase) was also increased, but P45 (which is not a substrate for this enzyme) was not. To control for the known effects of dibutyryl cAMP on calcium fluxes (Henquin and Meissner, 1983), synaptosomes were first preincubated with 250 µM fluphenazine, which inactivates CM-PK and PKC (Robinson et al., 1984). Under these conditions dibutyryl cAMP still increased the phosphorylation of P96 (Table 1), suggesting that PKA does play some role in basal P96 phosphorylation.

A role for PKC in P96 phosphorylation is currently being investigated. Preliminary experiments have provided indirect support for a role for this enzyme. When intact synaptosomes were incubated for 2 to 3 minutes with the phorbol ester, phorbol-12-myristate-13-acetate (PMA), a small increase in P96 phosphorylation was observed (Fig. 6). A range of other phosphoproteins were also affected, notably P45, P54, and P139. When synaptosomes are lysed and proteins are labeled with exogenous $[\gamma-^{32}P]ATP$ usually the same proteins observed in intact synaptosomes are labeled (Dunkley et al., 1986). In previous studies no phosphoprotein labeled with $[\gamma-^{32}P]ATP$ comigrated with P96 (Dunkley et al., 1986). However, recent results have suggested that this may have been due to proteolysis of several proteins and of PKC during the lysis step (Robinson and Lovenberg, unpublished). Figure 6 (right) shows that, when protease inhibitors are present, a phosphoprotein of molecular weight 96,000 is phosphorylated in synaptosomal cytosol by PKC. However, it will not be clear whether P96-lysed and P96-intact are the same phosphoprotein until phosphopeptide analysis has been undertaken.

DISCUSSION

P96 is a major synaptosomal phosphoprotein observed upon incubation of intact synaptosomes with $^{32}P_i$. The prominant characteristic of the protein is that upon depolarization P96 is dephosphorylated in contrast to the increased phosphorylation of many other phosphoproteins. Dephosphorylation

TABLE 1: Effect of Dibutyrl cAMP on the Phosphorylation of P96, Synapsin IB and P45.

	INCUBATION	PREINCUBATION	PROTEIN PHOSPHORYLATION (%)		
			P96	P75	P45
A.	Control	---	100	100	100
	Depolarized		45.9 ± 2.2	183 ± 5.7	151 ± 5.0
	Bt$_2$cAMP		128.9 ± 4.01	140 ± 3.3	106 ± 3.4
B.	Control	Flu	108 ± 6.3	100 ± 3.1	94 ± 3.3
C.	Control	Flu + Bt$_2$cAMP	128 ± 5.8	129 ± 1.8	94 ± 3.3
	Depol.		121 ± 6.2	130 ± 3.8	96 ± 4.9

Striatal synaptosomes were prelabeled with 0.1 mM calcium and incubated for 30 seconds in control or depolarizing buffer, or 5 minutes with dibutyryl cAMP (Bt$_2$cAMP). A, no preincubation; B, 10 minute preincubation with 250 µM fluphenazine; C, 10 minute preincubation with fluphenazine, with Bt$_2$cAMP present during the final 5 minutes. n=9 in A and 3 in B and C.

may occur by activation of a P96 phosphatase, by inactivation of P96 kinase or dissociation of P96 and its kinase, or as part of an enzymic reaction catalyzed by P96 (e.g. as in ATPases and turnover of acylphosphate). Since depolarization has been shown to activate CM-PK and PKC it seems less likely that protein kinase inactivation is involved in the mechanism.

Dephosphorylation of P96 is a fully reversible event that occurred even after 5 cycles of depolarization and repolarization and is not due to irreversible processes such as proteolysis, but to a cyclic phenomenon linked to the plasma membrane potential. P96 dephosphorylation was dissociated from the other effects of depolarization on increasing protein phosphorylation, because it was shown to be dependent on more than just a rise in intrasynaptosomal free calcium. Voltage-sensitive calcium channels, and extracellular calcium pools can be bypassed by the use of the mitochondrial poison guanidine, which causes release of calcium from intracellular stores (Boadle-Biber, 1982). Guanidine promoted increased phosphorylation of synapsin I and protein III, but did not stimulate P96 dephosphorylation. Thus, increased phosphorylation can be dissociated from activation of calcium channels, but P96 dephosphorylation cannot be dissociated. Therefore, dephosphorylation is dependent upon both depolarization and calcium entry.

P96 phosphorylation was found to be regulated by both cAMP and calcium in intact synaptosomes. Stimulation of P96 phosphorylation occurred by addition of calcium to synaptosomes prelabeled in its absence, as well as upon addition of dibutyryl cAMP. In both cases the stimulation was relatively small, suggesting either that P96 is near maximally labeled under resting conditions or that these are not the major regulators of P96 kinase. The preliminary finding that P96 from intact synaptosomes comigrates with a cytoplasmic P96 labeled with $[\gamma-^{32}P]$ATP by PKC raise the possibility that PKC is the major P96 kinase.

In these studies P96 phosphorylation was significantly increased by two classes of drugs. Firstly, the phenothiazine fluphenazine produced a very selective increase in P96 labeling at lower concentrations than were found to inhibit intracellular calcium-dependent PK. The effect is likely to be mediated via presynaptic receptors but is not completely understood. Secondly, the calcium channel agonist BAY K 8644 and antagonists verapamil and nifedipine produced a specific increase in P96 labeling (as well as 3 other phosphoproteins, Robinson and Lovenberg, 1986). It is interesting to note that BAY K 8644 can also enhance the turnover of inositol phospholipids in brain cortical slices (Kendall and Nahorski, 1985), and it is possible that such a mechanism could be related to the increases in P96 phosphorylation.

Although the identity and function of P96 remain unknown there are three areas of synaptic events where a role could be considered: i) in depolarization-dependent calcium entry, ii) in the modulation of neurotransmitter release and iii) in the recovery of nerve terminals from a depolarizing stimulus. In the first situation a role for P96 in some aspect of synaptosomal ion channel regulation is possible, but the role would more likely be a regulatory one rather than P96 being a molecular component of voltage-sensitive ion channels, since P96 is largely cytoplasmic (not shown). The second role of modulating neurotransmitter release is attractive, but cannot readily be tested until the protein has been purified. The third possible role, in the area of synaptosomal recovery after a stimulus, encompasses calcium extrusion. In depolarized nerve terminals Ca-ATPases are activated and are responsible for removing the calcium signal by pumping calcium into intrasynaptosomal vesicles, or out of the terminal. It is relevant to note that a neuronal form of Ca-ATPase has recently been purified and has a subunit molecular weight of 94kd (Chan et al., 1984).

In summary, the data presented in this report highlights many characteristics of the phosphorylation of P96 that distinguish it from other synaptosomal phosphoproteins. P96 is a unique phosphoprotein that responds to depolarization with a dephosphorylation event that is distinct because both calcium entry and depolarization are required. However, P96 dephosphorylation is fully reversible and can be stimulated by various drugs, by calcium and by cAMP. These characteristics stongly suggest a physiological role for P96 in some aspect of nerve terminal function. Future research should focus on identifying the protein kinase(s) and protein phosphatase(s) that delicately regulate the phosphorylation of this intriguing protein.

REFERENCES

Aloyo, V. J., Zwiers, H. and Gispen, W. H., Phosphorylation of B-50 protein by calcium-activated phospholipid-dependent protein kinase and B-50 protein kinase, J. Neurochem. 41:649 (1983).

Boadle-Biber, M. C., Further studies on the role of calcium in the depolarization-induced activation of tryptophan hydroxylase: effect of verapamil, tetracaine, haloperidol, and fluphenazine, Biochem. Pharmacol. 31: 2495 (1982).

Chan, S. Y., Hess, E. J., Rahamimoff, H. and Goldin, S. M., Purification and immunological characterization of a calcium pump from bovine brain synaptosomal vesicles, J. Neurosci. 4:1468 (1984).

DeLorenzo, R. J., The calmodulin hypothesis of neurotransmission, Cell Calcium 2:399 (1981).

DeLorenzo, R. J., Freedman, S. D., Yohe, W. B. and Maurer, S. C., Stimulation of Ca^{2+}-dependent neurotransmitter release and presynaptic nerve terminal protein phosphorylation by a calmodulin-like protein isolated from synaptic vesicles, Proc. Natl. Acad. Sci. 76:1838 (1979).

Dennig, G. and Sieghart, W., Apparent identity of the alpha-subunit of pyruvate dehydrogenase and the protein phosphorylated in the presence of glutamate in P2 fractions of rat cerebral cortex, J. Neural. Transm. 59:119 (1984).

Dolphin, A. C. and Greengard, P., Neurotransmitter- and neuromodulator-dependent alterations in phosphorylation of protein I in slices of rat facial nucleus, J. Neurosci. 1:192 (1981).

Dosemeci, A. and Rodnight, R., The effect of digestion with phospholipase C on intrinsic protein phosphorylation in synaptic plasma membrane fragments, FEBS Lett. 139:22 (1982).

Dunkley, P. R., Baker, C. M. and Robinson, P. J., Depolarization-dependent protein phosphorylation in rat cortical synaptosomes: characterization of active protein kinases by phosphopeptide analysis of substrates, J. Neurochem. 46, in press (1986).

Henquin, J. C. and Meissner, H. P., Dibutyryl cyclic AMP triggers Ca^{2+}-dependent electrical activity in pancreatic beta cells, Biochem. Biophys. Res. Commun. 112:614 (1983).

Huttner, W. B. and Greengard, P., Multiple phosphorylation sites in protein I and their differential regulation by cyclic AMP and calcium, Proc. Natl. Acad. Sci. 76:5402 (1979).

Kendall, D. A. and Nahorski, S. R., Dihydropyridine calcium channel activators and antagonists influence depolarization-evoked inositol phospholipid hydrolysis in brain, Eur. J. Pharmacol. 115:31 (1985).

Krueger, B. K., Forn, J. and Greengard, P., Depolarization-induced phosphorylation of specific proteins, mediated by calcium ion influx, in rat brain synaptosomes, J. Biol. Chem. 252:2764 (1977).

Llinas, R., McGuinness, T. L., Leonard, C. S., Sugimori, M. and Greengard, P., Intraterminal injection of synapsin I or calcium/calmodulin-dependent protein kinase II alters neurotransmitter release at the squid giant synapse, Proc. Natl. Acad. Sci. 82:3035 (1985).

Michaelson, D. M. and Avissar, S., Ca^{2+}-dependent protein phosphorylation of purely cholinergic Torpedo synaptosomes, J. Biol. Chem. 254:12542 (1979).

Michaelson, D. M., Avissar, S., Kloog, Y. and Sokolovsky, M., Mechanism of acetylcholine release: possible involvement of presynaptic muscarinic receptors in regulation of acetylcholine release and protein phosphory-lation, Proc. Natl. Acad. Sci. 76:6336 (1979).

Nachshen, D. A. and Blaustein, M. P., Some properties of potassium-stimu-lated calcium influx in presynaptic nerve endings, J. Gen. Physiol. 79:1065 (1980).

Nestler, E. J. and Greengard, P., "Protein phosphorylation in the nervous system", John Wiley and Sons, New York (1984).

Robinson, P. J. and Dunkley, P. R., Depolarization dependent protein phos-phorylation in rat cortical synaptosomes: factors determining the magnitude of the response, J. Neurochem. 41:909 (1983).

Robinson, P. J. and Dunkley, P. R., Depolarization dependent protein phos-phorylation and dephosphorylation in rat cortical synaptosomes is modu-lated by calcium, J. Neurochem. 44:338 (1985).

Robinson, P. J. and Lovenberg, W., Calcium channel agonists and antagonists regulate protein phosphorylation in intact synaptosomes, Neurosci. Lett. in press (1986).

Robinson, P. J., Jarvie, P. E. and Dunkley, P. R., Depolarization dependent protein phosphorylation in rat cortical synaptosomes is inhibited by fluphenazine at a step after calcium entry, J. Neurochem. 43:659 (1984).

Wu, W. C. S., Walaas, I. S., Nairn, A. C. and Greengard, P., Calcium/phos-pholipid regulates phosphorylation of a M_r "87k" substrate protein in brain synaptosomes", Proc. Natl. Acad. Sci. 79:5249 (1982).

PHOSPHORYLATION AND DEPHOSPHORYLATION OF NEUROFILAMENT

PROTEINS IN RETINAL GANGLION CELL NEURONS <u>IN VIVO</u>

R. A. Nixon* and Susan E. Lewis *

*Ralph Lowell Laboratories, Mailman Research Center, McLean Hospital, Belmont, MA 02178 and Department of Psychiatry and Program in Neuroscience, Harvard Medical School, Boston, MA 02115

INTRODUCTION

The cytoskeleton of most higher eukaryotic cells is composed of three filamentous systems, which serve varying dynamic and structural roles in cellular function. Two of these systems, the 80 Å microfilaments and the 230 Å microtubules, are composed of subunit proteins that are phylogenetically highly conserved. By contrast, the 100 Å intermediate filaments are encoded by a large multigene family, the members of which are differentially expressed in different tissues (Fuchs and Hanukoglu, 1983). On the basis of biochemical and immunological criteria, five major classes of intermediate filaments have been defined (Lazarides, 1980). These include keratin filaments, found in cells of epithelial origin; desmin filaments, predominantly found in smooth, skeletal and cardiac muscle cells; vimentin filaments, present in cells of mesenchymal origin; glial filaments, constituents of certain glial cell types; and neurofilaments, present in many differentiated neurons of vertebrates and invertebrates. This classification emphasizes the tissue specificity of intermediate filaments, although it is now known that subunits from more than one class may coexist in some tissues at certain developmental stages (Osborn et al., 1980; Drager, 1983).

Neurofilaments are composed of three subunits with apparent molecular weights by SDS-polyacrylamide gel electrophoresis (SDS-PAGE) of 70 kilodaltons (kD), 145 kD and 200 kD in a molar ratio of approximately 6:2:1 (Chiu et al., 1983; Shelanski and Liem, 1979). Since SDS-PAGE strongly overestimates the molecular size of NFPs due to the high phosphate content and other physicochemical properties of these polypeptides (Kaufman et al., 1984), they will be referred to by the more general descriptors L ("low"), M ("middle"), and H ("high") subunits in this paper. Biochemical studies of dissociated neurofilaments demonstrate that NFP-L can assemble by itself into neurofilaments. The higher molecular weight subunits, however, co-assemble only in the presence of NFP-L (Geisler and Weber, 1981; Liem and Hutchison, 1982). Assembly studies and electronmicroscopic observations on the attachment to neurofilaments of antibodies that react preferentially with H, M or L (Sharp et al., 1982; Hirokawa et al., 1984; Willard and Simon, 1981) have supported the concept that NFP-L forms the central core, and NFP-H composes a more peripheral structure which may represent the radially projecting "side arms" that interconnect neurofilaments and other cytoskeletal elements. Although NFP-M is also peripherally situated, its disposition within the filament and contribution to filament function are least clear.

Despite their different molecular weights, each of the NFP subunits shares important features of its structure with other intermediate filament subunits (Geisler et al., 1983; Weber et al., 1984). These include a highly conserved central domain, approximately 300 amino acid residues in length, which consists of four richly α-helical domains separated by non-helical spacer segments. The central domain is flanked by hypervariable non-α-helical end-domains of variable size and sequence (Steinert et al., 1985; Fuchs and Hanukoglu, 1983). At least a portion of these terminal regions is necessary for the end-to-end linkage of subunits to form a protofibril, and for the interprotofibrillar interactions required to form the 80 to 100 Å diameter of the neurofilament (Fuchs and Hanukoglu, 1983; Steinert et al., 1985). In addition, the end-domains, which are predominantly represented on the surface of the assembled filament, are believed to specify functions unique to a particular subunit or intermediate filament subclass (Weber et al., 1983; Steinert et al., 1985).

In addition to similarities of peptide structure, intermediate filaments of various subclasses are modified after synthesis by addition of phosphate groups to the polypeptide backbone. The L subunit of neurofilaments, like the "core" subunit of desmin (O'Connor et al., 1981), vimentin (Steinert et al., 1982) and glial (Wong et al., 1984) filaments, contains one or several phosphorylated sites. The M and H subunits, however, are extensively phosphorylated, containing as many as 9-24 and 22-100 phosphate groups per molecule, respectively (Julien and Mushynski, 1982; Jones and Williams, 1982; Wong et al., 1984; Carden et al., 1985). The phosphate groups are located predominantly on the end-domains that are exposed on the surface of the neurofilaments (Julien et al., 1983; Carden et al., 1985). In view of the presumed roles of the end-domains, these observations raise the possibility that phosphorylation may regulate not only aspects of neurofilament assembly but also certain interactions between neurofilaments and other neuronal constituents.

Electron microscopy has revealed neurofilaments to run parallel to the long axis of the axon for distances of many micrometers. Along their lengths, neurofilaments are extensively cross-linked to each other and to microtubules, mitochondria, endoplasmic reticulum and vesicles by 20-nM cross-bridges (Hirokawa et al., 1984). Although these images reinforce the notion of the neurofilament system as a complex lattice of linear and radial elements, the static representation provided by electron microscopy belies the dynamic nature of this lattice which is implied by microscopic and biochemical observations of axons in vivo. Neurofilaments in most neurons are continually synthesized and transported along axons (Hoffman and Lasek, 1975; Black and Lasek, 1980; Willard, 1983). In some neurons, transported neurofilaments leave the moving phase and become integrated into a stationary axonal cytoskeletal network (Nixon, 1983; Nixon and Logvinenko, 1986), which is regionally specialized along axons (Nixon et al., 1982; Brown et al., 1982). The translocation of neurofilaments and the extensive reorganization processes within the axon suggest, therefore, that interactions between neurofilaments and other cytoskeletal proteins, and membranous organelles may be highly dynamic and tightly regulated. If phosphorylation of NFPs were a regulatory step in mediating such interactions, it might be expected that phosphate groups would not have a static association with the filaments but would instead exhibit differential turnover, possibly with a characteristic timing and location-specificity within the neuron.

Two protein kinases have been identified to phosphorylate NFPs in vitro. A cAMP-dependent protein kinase co-purifying with brain microtubule proteins prepared by cycles of assembly and disassembly preferentially phosphorylates NFP-M in addition to exhibiting significant activity toward microtubule-associated proteins and tubulin (Leterrier et al., 1981). A kinase with similar properties also phosphorylates keratin (Gilmartin et al., 1984), vimentin, and desmin (O'Connor et al., 1981) subunits. A cAMP- and calcium-independent protein kinase, which copurifies with neurofilaments (Sheckel and Lasek, 1982; Julien et al., 1983), phosphorylates all three NFP subunits. Two-dimensional peptide map analyses have shown that the two protein kinases do not phosphorylate identical sites on NFP-M (Julien et al., 1983).

Recent findings suggest that phosphorylation of NFPs may be topographically segregated within some neurons. Pulse-labeling kinetics for NFPs in cultured chick spinal cord neurons, when interpreted in light of earlier neurofilament transport data, suggest that NFP-M appears in perikarya in an unphosphorylated form and is subsequently phosphorylated during transport along proximal neurites (Bennett and DiLullo, 1985). Similarly, in the squid, kinase activity toward NFPs in homogenates of stellate ganglion cell bodies is low compared to the level in axoplasm from giant axons. This regional difference may be due in part to the presence in ganglion cell bodies of an endogenous inhibitor of the calcium- and cAMP-independent protein kinase that phosphorylates neurofilaments (Pant et al., 1986). These biochemical findings extend earlier immunocytochemical studies involving monoclonal antibodies that distinguish phosphorylated forms of NFP-H from relatively non-phosphorylated or dephosphorylated ones. These studies demonstrated selective immunostaining of most neuronal cell bodies by antibodies recognizing dephosphorylated NFP-H and preferential axonal staining by antibodies against phosphorylated epitopes on this subunit (Sternberger and Sternberger, 1983). By the same experimental approach, the normal segregation of highly phosphorylated neurofilaments in axons has been found to be disrupted in some neurons in pathological states associated with perikaryal accumulation of neurofilaments. In brains from patients with Alzheimer's disease (Cork et al., 1986), some neuronal perikarya containing neurofibrillary tangles were immunostained by certain monoclonal antibodies in a series that recognize phosphorylated epitopes on NFP-H. Staining of neurofilamentous accumulations in the perikarya of motor neurons in aluminum-intoxicated rabbits has also been reported in studies with the same monoclonal series (Troncoso et al., 1986). Whether these abnormal neuronal distributions of phosphorylated NFPs reflect antecedent metabolic events in the development of neurofibrillary pathology or secondary responses to neuronal injury or degeneration (Drager and Hofbauer, 1984) remains an unanswered question of considerable clinical importance.

Although the addition of phosphate to NFPs is amply documented, the question of whether these phosphates are permanently associated with NFP subunits or undergo continual turnover has not previously been addressed. Phosphatases are particularly active in nervous tissue (Wallace et al., 1980), and the enzymatic properties of some forms are compatible with a role for these enzymes in regulating the phosphorylation state of intermediate filaments (Goto et al., 1985; Cooper et al., 1985; King et al., 1984; Tallant and Cheung, 1983; Yang and Fong, 1985).

In the following sections, we describe our observations on the temporal and topographical characteristics of NFP phosphorylation in mammalian CNS neurons in vivo (Nixon and Lewis, 1986; Nixon et al., 1986a, 1986b). In these studies, we examined the progressive modification of newly synthesized NFPs in mouse retinal ganglion cell (RGC) neurons after pulse-labeling RGC perikarya in vivo with [^{32}P]orthophosphate or radiolabeled amino acids. In addition to NFP phosphorylation, we investigated the possible turnover of phosphate groups on NFPs in RGC neurons in vivo by examining the metabolic fate of [^{32}P]-labeled phosphate groups associated with neurofilaments in relation to the fate of the neurofilament polypeptide backbone labeled with [^{3}H]-proline. Improved methods for intravitreally injecting the same dose of radioisotopes into different mice enabled us to quantitate the relative numbers of radiolabeled phosphate groups added and removed from NFPs during the period when the radiolabeled neurofilaments resided within axons.

Our results indicate that the steady-state phosphate content of each NFP subunit is regulated by a dynamic balance between the processes of phosphorylation and selective dephosphorylation. This balance is modulated differently in various sites within the neuron. We observed that phosphate addition to NFPs is extensive before the polypeptides have advanced beyond very proximal axonal levels. The modification of NFPs continues, however, as neurofilaments are transported along the entire length of the axon. These modifications included dephosphorylation at different rates for each NFP subunit as well as phosphorylation at the same or additional sites on the polypeptides. The possibility is raised that this complex pattern of NFP modifications may be involved in regulating the interactions between neurofilaments and other cytoskeletal proteins during axoplasmic transport and during the process of integrating neurofilaments into the axonal cytoskeleton.

MATERIALS AND METHODS

Isotope Injections

Radiolabeled compounds were injected intravitreally into anesthetized male and female C57BL/6J mice aged 10-14 weeks with a micropipette apparatus (Nixon, 1980). Mice received 0.20 µl of phosphate-buffered saline pH 7.4 containing 15-25 µCi of L[2,3-^{3}H]proline (spec. act. 30-50 Ci/mmol), 50-100 µCi of L[^{35}S]methionine (spec. act. 400 ci/mmol) or 50-100 µCi of [^{32}P]orthophosphate (spec. act. 1000 ci/mmol), purchased from New England Nuclear (Boston, MA). In double isotope studies of phosphate turnover, each mouse in groups of 40-60 animals received 0.20 µl of phosphate-buffered saline, pH 7.4, containing 15 µCi of L[2,3-^{3}H]-proline and 50 µCi of [^{32}P]orthophosphate using calibrated micropipettes. Additional precautions were taken in these experiments to reduce the usual variation in amounts of radioisotope incorporated by RGC neurons (Nixon and Logvinenko, 1986).

Tissue Preparation

Mice were killed by cervical dislocation followed by decapitation. After the brain was cooled, the meninges were removed and the optic tract on each side was severed at a point 2.5 mm from the superior colliculus. The dissected length of primary optic pathway was 9 mm long and consisted of the optic nerves severed at the scleral surface of the eye, the optic chiasm and lengths of the optic tract extending to but not including terminals in the lateral geniculate nucleus. In the phosphate turnover studies, this 9-mm optic pathway segment is referred to as the "axonal window." In additional experiments to establish the distributions of radiolabeled axonally transported NFPs, optic pathway samples were cut into consecutive 1.1-mm segments on a micrometer-calibrated slide. All manipulations were performed at 0°C or with the tissue in the frozen state.

Immune Precipitation and Immunoaffinity Chromatography of NFPs

Retinas or optic pathways from mice injected intravitreally with [^{35}S]-methionine were homogenized in cytoskeletal buffer (50 mM Tris at pH 7.4, 0.5% Triton, 5 mM EDTA, 2 mM PMSF, 50 µg/ml leupeptin) and centrifuged, yielding soluble and insoluble fractions. The soluble fractions were made 1% SDS. Insoluble fractions were resuspended in cytoskeletal buffer containing 0.9 M sucrose, and, after centrifugation, the floating myelin pad was removed. The final pellet was resuspended in 1% SDS. Soluble and insoluble fractions were immunoprecipitated with a 1:100 dilution of antibody followed by reaction with protein-A Sepharose (Nixon et al., 1986a; Fischer et al., 1986). Two polyclonal antisera from rabbits immunized to purified bovine NFP-H (R97) (Dahl, 1983) and to chicken brain NFPs (R39) (Dahl and Bignami, 1977) were provided by Dr. Doris Dahl. Under these immunoprecipitation conditions, both antisera cross-react selectively with phosphorylated and dephosphorylated forms of all three NFP subunits (Nixon et al., 1986a). Additional polyclonal antisera raised in rabbits immunized to the individual NFP subunits (Brown et al., 1983) were provided by Dr. Charles Marotta.

Polyacrylamide Gel Electrophoresis

One-dimensional SDS-PAGE was carried out on 320-mm slab gels using 3-7% or 5-15% linear polyacrylamide gradients (Brown et al., 1981). Two-dimensional polyacrylamide gels containing SDS were prepared as previously described using an 8-18% polyacrylamide gradient (Brown et al., 1981).

Identification of Radioactive Proteins

NFPs were identified on gels stained with Coomassie brilliant blue by comparing their migration relative to that of cytoskeletal protein standards. Complete separation of each NFP subunit from other proteins under the conditions of electrophoresis employed was demonstrated earlier (Nixon and Logvinenko, 1986). [^{3}H]proline-labeled proteins on Coomassie-stained gels were detected by fluorography (Nixon et al., 1982). Proteins labeled with [^{32}P]-orthophosphate or [^{35}S]methionine were detected by autoradiography.

Quantitation of Radioactivity Incorporated into NFPs

In preliminary experiments, radiolabeled proteins were first detected on gels by fluorography or autoradiography to confirm their positions. In most experiments, the protein bands were cut directly from gels stained with Coomassie brilliant blue but not fluorographed. Gel slices were counted in a Beckman LS-7000 counter equipped with external standard and automatic quench correction modes. Selected samples were recounted following the addition of an aliquot of [^{3}H]toluene or [^{32}P]orthophosphate to determine efficiencies and isotope crossover. Tritium efficiency was 55%, and [^{32}P] efficiency, 98%. The [^{32}P] to [^{3}H] crossover averaged 6.4%, and the [^{3}H] to [^{32}P] crossover, 1.5%. These values from different samples of the same gel did not vary significantly. [^{32}P] dpm were further adjusted for isotope decay. Finally, each radioactive band was corrected for background radioactivity in each gel lane since preliminary studies showed that for any area of gel containing no discernible radiolabeled bands by fluorography, a low level of radioactivity was measured that was proportional to the total radioactivity of proteins loaded onto the gel. The appropriate background radioactivity calculated from the total protein radioactivity loaded and the area of the gel slice containing the radiolabeled protein band was subtracted. Background ranged form 8% to 20% and 10% to 20% of the [^{32}P] or [^{3}H] radioactivity, respectively, in each protein.

Isolation of Neurofilament Proteins

A protein fraction enriched in neurofilaments was prepared from mouse CNS white matter by axonal flotation (Liem et al., 1978; Brown et al., 1981). In experiments involving alkaline phosphatase digestion, cytoskeletal protein fractions were prepared by the method of Chiu and Norton (1982) from mouse optic pathways dissected from mice 2 days after intravitreal injection of [^{32}P]orthophosphate (100 μCi/eye) or from mouse CNS white matter.

Alkaline Phosphatase Digestion

   In early experiments, bovine intestinal alkaline phosphatase was incubated at 37°C with cytoskeletal protein fractions (48 units/mg cytoskeletal
protein) in 0.1 M Tris buffer, pH 8.0, containing phenylmethylsulfonyl fluoride 0.1 mM, leupeptin 50 µg/ml, N-ethylmaleimide 1 mM and aprotinin 0.1%.
The reaction was carried out in dialysis tubing against the same buffer to
minimize inhibition of the enzyme by phosphate released during digestion.
Despite the use of inhibitors, protease contamination of the phosphatase preparation from any commercial source produced small amounts of proteolytic
fragments during the NFP digestion that were not observed when E. coli alkaline phosphatase was used. In recent experiments, cytoskeletal protein
fractions were incubatd with 1 unit of E. coli alkaline phosphatase (Sigma)
per mg protein at 37°C for various intervals up to 18 hr using the incubating
conditions of Carden et al. (1985). In experiments involving [^{32}P]-labeled
cytoskeletal proteins, the digested samples were subjected to electrophoresis
on 3-7% polyacrylamide gels. The bands corresponding to the neurofilament
proteins were cut directly from gels stained with Coomassie blue and radioactivity in each was measured as described above.

RESULTS

Neurofilament Proteins Are Extensively Phosphorylated in RGC Neurons In
Vivo

   Neurofilaments in axons of mouse RGC neurons are composed of proteins
with approximate molecular weights of 200 kD, 140 kD and 70 kD (Nixon and
Logvinenko, 1986) (Fig. 1). The H- and M-NFPs can each be resolved into
at least three distinct species with molecular masses of 190 kD to 200 kD
and 140 kD to 145 kD, respectively (Nixon et al., 1982; Lewis and Nixon,
1985), although the microheterogeneous forms of each subunit were considered
as one protein except when specified.

   When mice were injected intravitreally with [^{3}H]proline or [^{35}S]methio-
nine, radiolabeled NFPs in the Group V (SCa) wave of axoplasmic transport
advanced into RGC axons at the level of the optic nerve between 1 and 4 days
after injection, along with more than 50 other radiolabeled proteins composing the Group IV (SCb) transport phase (Fig 1A, 1C). After intravitreal
injection of [^{32}P]orthophosphate, the three NFP subunits were among a group
of only five proteins in RGC axons that were intensely radiolabeled (Fig.
1B, 1D). They were identified as the neurofilament triplet proteins by the
following criteria: their characteristic molecular weights (Fig. 1A, 1B),
their exclusive presence in Triton-insoluble preparations of cytoskeletal
proteins (Nixon and Lewis, 1986), their migration on two-dimensional polyacrylamide gels to the same position as unlabeled NFP subunits (Fig. 1C,
1D) and their cross-reactivity with specific antibodies to NFPs in immunoprecipitation studies (Nixon et al., 1986a; also see below).

Phosphorylation Influences The Electrophoretic Properties of NFPs

   The high content of phosphate associated with NFP-H and NFP-L alters
the migration of these polypeptides in SDS-polyacrylamide gels (Fig. 2).
When NFPs in cytoskeletal protein fractions prepared from mouse spinal cord
were incubated with alkaline phosphatase, the effect on the migration of
NFP-H on one-dimensional gel was noted within 2 min as a progressive broa-
dening of the stained band, which, by 10 min, spanned the molecular weight
range of 175-200 kD. Incubation for 80 min or more eliminated stained protein at 200 kD and generated a single band of 160 kD (Fig. 2A, 2B). During
the same interval, NFP-M displayed a gradual shift of band mobilities from
145 kD, 143 kD and 140 kD to 142 kD and 139 kD. NFP-L exhibited no change

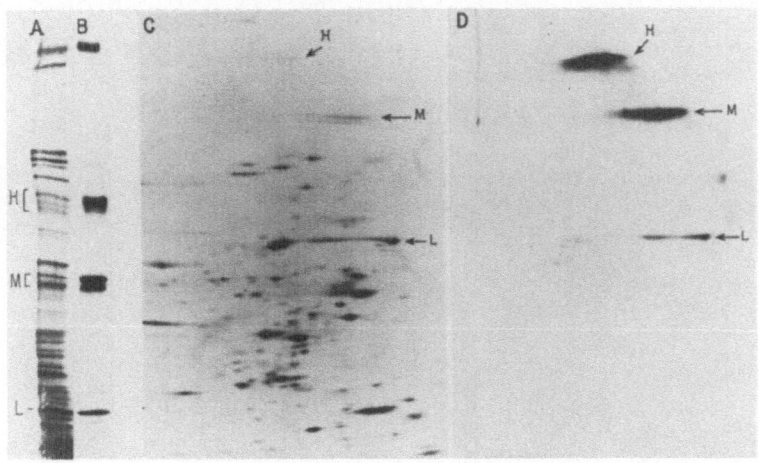

FIG. 1: [^{35}S]methionine and [^{32}P]orthophosphate labeled proteins comprising
the slow phases of axoplasmic transport in RGC axons composing the
mouse optic nerve and optic tract. Optic pathways were obtained 6
days after mice were injected intravitreally with [^{35}S]methionine
(panels A and C) or [^{32}P]orthophosphate (B and D). Proteins in
these samples were separated by one-dimensional SDS-PAGE on gels
containing a 3-7% acrylamide gradient (panels A and B) or by two-
dimensional SDS-PAGE on gels containing an 8-18% acrylamide gradient.
After electrophoresis, radiolabeled proteins were visualized by autoradiography. The positions of the NFP subunits, H, M, L, are indicated. Microheterogeneity of NFP-H and NFP-M can be seen (panel A).

in electrophoretic mobility on one-dimensional gels. These changes in mobility of NFP-H and NFP-M on SDS-gels do not reflect alterations of actual
molecular weight since the presence of phosphate groups on the molecule,
particularly in concert with the presence of carboxy-terminal regions with
unusual amino acid composition, causes anomalous migration on SDS-gels (Kaufman et al., 1984). Dephosphorylation, therefore, yields apparent molecular
weights closer to the values established by analytical gel filtration or
sedimentation equilibrium centrifugation methods (Kaufman et al., 1984).

    In addition to altering the mobilities of NFPs on one-dimensional SDS-
gels, phosphate groups markedly altered the charge of individual NFP subunits.
By two-dimensional PAGE, NFP-H appeared as a long streak spanning a pH range
of 6.1 to 5.6 (Fig. 2C). As alkaline phosphatase digestion proceeded, however, the charge of NFP-H became progressively more basic until a stable
isoelectric point of 6.6-6.5 was reached after 80 min of incubation (Fig.
2D). As these changes in isoelectric point progressed, the configuration of
the NFP-H spot became more oblique as the most basic of the heterogeneous
forms exhibited a more rapid mobility in the molecular weight dimension of
the gel (i.e., lower apparent molecular weight) than acidic forms of NFP-H.
After 80 min of incubation, the entire population of NFP-H molecules displayed a more rapid mobility as well as more basic pI compared with untreated
NFP-H (Fig. 2D). In comparison with the NFP-H subunit, NFP-M and NFP-L displayed smaller shifts toward a slightly more basic pI during alkaline phosphatase digestion (Nixon et al., 1986a). Separate experiments using [^{32}P]-
labeled NFPs incubated with alkaline phosphatase under the same conditions
demonstrated that 50-60% of the radiolabeled phosphate groups were released
from each subunit during an 80-min incubation. The isoelectric points
observed at shorter incubation times were associated with release of 20-30%

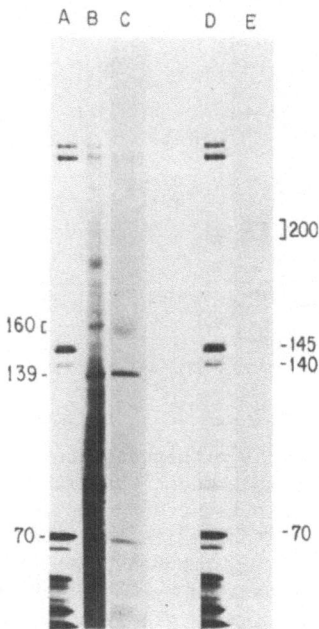

FIG. 2: Effects of alkaline phosphatase treatment on the electrophoretic
mobilities of NFP subunits. Neurofilament-enriched cytoskeletons
(Chiu and Norton, 1982) incubated for 80 min with alkaline phosphatase from bovine intestine (panel B) or E. coli (panel D) were
compared with unincubated cytoskeletons (panels A and C). The samples illustrated in panels A and B were analyzed by SDS-PAGE on
gels containing 3-7% polyacrylamide gradients. Proteins were visualized with Coomassie Brilliant Blue. Samples in panels C and D
were analyzed by two-dimensional SDS-PAGE on gels containing 8-18%
polyacrylamide gradient. Proteins were visualized by silver staining. The apparent molecular weights of NFP subunits before and
after alkaline phosphatase digestion are indicated (panels A and B,
respectively). On the two-dimensional gels, positions of the three
NFP subunits H, M, L, are indicated, and the prominent shift in
isoelectric point and apparent molecular weight of NFP-H is shown
by arrows. A smaller isoelectric point shift of NFP-M can also be
noted. NFP-L is not well visualized in this figure, although by
Coomassie-blue staining, a slight shift in isoelectric point toward
a more basic pH value is seen. The band corresponding to alkaline
phosphatase is evident on the gels in panels B-D.

of the [^{32}P]-phosphate groups from the NFP subunits (Nixon et al., 1986b).

Early Stages of Phosphorylation of NFPs

After radiolabeled amino acids are injected intravitreally into mice,
incorporation of radioactivity into retinal proteins reaches a level by 4 hr
(Nixon, 1980). In order to identify newly synthesized NFPs within the total
pool of radiolabeled retinal proteins, polyclonal antibodies raised against
NFPs were used to immunoprecipitate the triplet proteins selectively. The
proteins isolated by immunoprecipitation were then examined by SDS-PAGE (Fig.
3). Within 2 hr after [^{35}S]methionine injection, the retina contained highly
radiolabeled NFP-immunoreactive proteins with apparent molecular weights of
160 kD, 139 kD and 70 kD (Fig. 3C), corresponding to the values for exten-

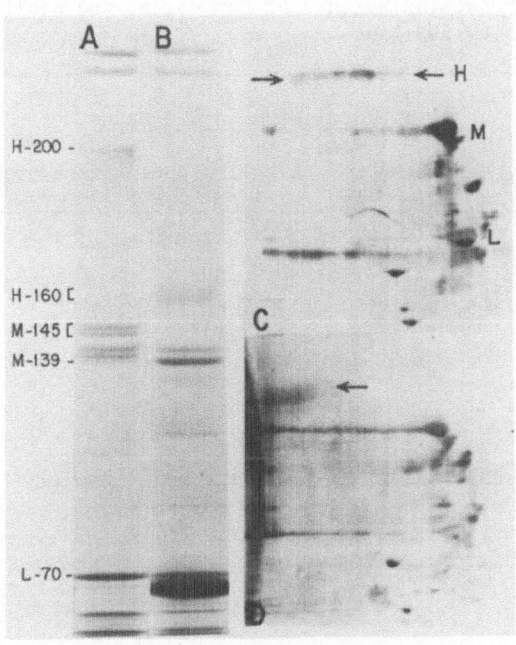

FIG. 3: Analysis of newly synthesized NFPs in retina and optic path way by
immunoprecipitation with NFP antibodies. Retinas dissected from
mice at 2 hr after intravitreal injection of [^{35}S]methionine and
optic pathways from mice at 6 days after a similar injection were
immunoprecipitated with a polyclonal antibody that cross-reacts
with the three NFP subunits in their phosphorylated and dephosphorylated form (R39) (Dahl and Bignami, 1977; Nixon et al., 1986).
The immunoreactive proteins were compared by SDS-PAGE (3-7% acrylamide gradient) with total radiolabeled proteins in retina or
cytoskeletal protein preparations from optic pathway. Lanes A and
B depict autoradiograms of proteins from neurofilament-enriched
cytoskeletons prepared from radiolabeled optic pathways, demonstrating the major axonal NFP forms at positions corresponding to molecular weights of 190-200 kD, 140-145 kD and 70 kD. Lane B illustrates the total radiolabeled protein pool in the retina at 2 hr
after intravitreal injection of [^{35}S]methionine. Radiolabeled species at 160 kD and 139 kD are prominent. The 70 kD NFP cannot be
seen on this overexposed autoradiogram. Lane C illustrates NFP-
immunoreactive proteins immunoprecipitated from the radiolabeled
retina shown in lane B. Prominent bands at 160 kD, 139 kD and 70
kD can be seen. Lane E depicts NFP-immunoreactive proteins immunoprecipitated from radiolabeled optic pathways 6 days after [^{35}S]-
methionine injection. Protein bands corresponding to the 200 kD,
140-145 kD and 70 kD bands in axonal cytoskeleton preparations
(lane D) are visible.

sively dephosphorylated axonal NFPs (Fig 2B, Table 1). No radioactive bands
corresponding to the 200 kD and 140-145 kD forms of the NFP-H or NFP-M subunits present in RGC axons were detected (Fig. 3C). Similar analyses after
intravitreal injection of [^{32}P]orthophosphate showed that the 70 kD, 139 kD
and 160 kD proteins did not contain [^{32}P] label (Nixon et al., 1986a), which
supported the notion that they represent relatively unphosphorylated forms
of the NFP triplet.

TABLE 1. Apparent Molecular Weights of Unmodified and Modified NFPs as
Determined by SDS-PAGE

NFP Subunit	Total NFPs in RGC Axons	
	Untreated[a]	Alkaline phosphatase treated[b]
H	195 - 200	160
M	145, 143, 140	142, 139
L	70	70

NFP Subunit	Newly Synthesized NFPs [^{35}S]Methionine[c]	
	Retina	RGC axons
H	160	190 - 200
M	139	145, 143, 140
L	70	70

NFP Subunit	Newly Synthesized NFPs [^{32}PO$_4$][c]	
	Retina	RGC axons
H	NRD	190 - 200
M	NRD	145, 143, 140
L	NRD	70

Apparent molecular weights were determined by one-dimensional SDS-PAGE on
linear 4-7% polyacrylamide gradients (Nixon et al., 1983) using the following
molecular weight standards (BioRad) electrophoresed in adjacent lanes: myosin (210 Kd), β-galactosidase (130 Kd), phosphorylase B (94 Kd), bovine
serum albumin (68 Kd), ovalbumin (43 Kd). [a]NFPs prepared from mouse optic
pathway by the method of axonal flotation (Brown et al., 1981; Liem et al.,
1978). [b]NFPs from optic pathway prepared by axonal flotation were incubated
for 120 min with E. coli alkaline phosphatase as previously described (Carden et al., 1985). [c]Mice were injected intravitreally with [^{35}S]methionine
or [^{32}P]orthophosphate (see Materials and Methods). Retinas were obtained
2 h after injection and RGC axons, after 2-6 days. NFPs were immunoprecipitated from homogenates of each tissue using polyclonal antibodies cross-reactive with all three NFP subunits (R39, R97) or with NFP-M and were then subjected to SDS-PAGE. [d]NRD = no radioactivity detected.

The three NFP-immunoreactive bands completely disappeared from the
retina by 2 days after injection, leaving behind in retina the majority of
radiolabeled proteins (Nixon et al., 1986a). During the same 2-day postin-

jection period, three radiolabeled NFP-immunoreactive proteins with the characteristic molecular weights of NFP-L, NFP-M and NFP-H in CNS white matter
(i.e., 70 kD, 140-145 kD and 200 kD) appeared in the optic nerve and persisted for several weeks as they were axonally transported (Fig. 3D, 3D).
When [^{32}P]orthophosphate was administered intravitreally, the same proteins
were heavily radiolabeled (Fig. 1B, 1D).

NFPs Continue To Be Modified During Axoplasmic Transport

Based on the observation that phosphate groups may influence the charge
and electrophoretic patterns of the proteins, we sought evidence for continued posttranslational modification of NFPs by kinases or phosphatases along
RGC axons by noting changes in the isoelectric points of NFPs as they were
axonally transported. Two-dimensional electrophoretic analyses of optic
nerve and optic tract proteins from mice injected intravitreally with [^{32}P]-
inorganic phosphate (Nixon et al., 1986b) indicated that two posttranslational processes, resulting in net charge alterations, affected the individual NFP subunits. In turn, the expression of these posttranslational
events depended upon two factors: the time interval during which NFPs
resided within an axonal segment, and the location of these proteins along
the length of the axon.

We first examined the effect of residence time within the same axonal
region on NFP modifications. Within optic nerve, [^{32}P]-radiolabeled NFP-L
shifted 0.2 pH units toward the basic end of gels between 1 and 6 days after
injection (Fig. 4A). During the same interval, the charge of NFP-H became
slightly more acidic (0.2 pH units) (Fig. 4A). This NFP-H modification, however, was more dependent upon the location of labeled NFP-H along the length
of the axon than upon the residence time at a particular site along the axon.
For example, comparisons of NFP-H at proximal axonal levels (optic nerve)
and at distal axonal levels (optic tract) in mice after the same 6-day post-
injection interval (Fig. 4B) demonstrated an acidic shift of 0.3 pH units

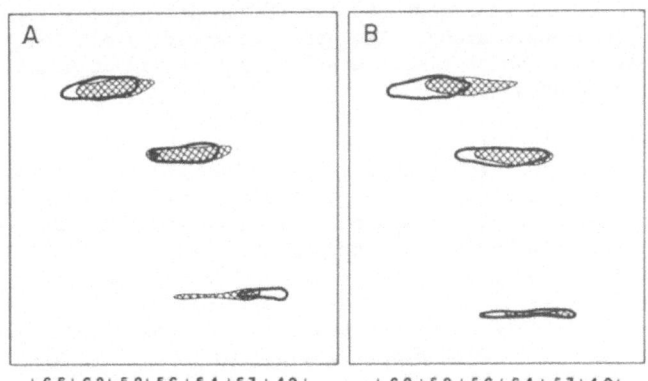

| 6.5 | 6.2 | 5.8 | 5.6 | 5.4 | 5.3 | 4.9 | pH | 6.2 | 5.8 | 5.6 | 5.4 | 5.3 | 4.9 |

FIG. 4:   Modifications in the charge of NFP subunits during axoplasmic transport. Mice were injected intravitreally with [^{32}P]orthophosphate
and optic nerves and optic tracts were analyzed by two-dimensional
electrophoresis after 1 day or 6 days. Panel A shows the relative
gel positions of the H, M and L NFP subunits in optic nerve at 1 day
(heavy line, unshaded area) and at 6 days (light line, shaded area).
Panel B compares the gel positions of H, M and L in the optic nerve
(heavy line, unshaded area) and in the optic tract (light line,
shaded area) at 6 days after intravitreal administration of [^{32}P]-
orthophosphate. Similar isoelectric point shifts have been observed
when NFPs were labeled with [^{3}H]proline (Nixon et al., 1986b).

for this subunit in distal axonal sites. A smaller but consistent acidic shift in the pI of NFP-M was also observed in optic tract compared with optic nerve at 6 days after injection. By contrast, little difference in the charge of NFP-L was observed at these two axonal levels.

These alterations in NFP charge, together with studies of limited alkaline phosphatase digestion, suggested that changes in net phosphorylation state continue to occur as NFPs are axonally transported and are deposited along axons. The differences in direction of the charge shifts suggest that while the 70 kD NFP undergoes a net dephosphorylation shortly after entering axons, NFP-H and NFP-M appear to accept additional phosphate groups during the same period.

Phosphate Groups On NFPs Are Selectively Turned Over During Axoplasmic Transport

Although the foregoing experiments revealed possible changes in net phosphorylation state of NFP subunits, they provided relatively little information about whether phosphate groups are permanently associated with NFP subunits or undergo continual turnover. We therefore sought direct evidence for turnover of phosphate groups on individual NFP subunits in vivo by quantitating the release of [^{32}P]phosphate groups from NFPs during axoplasmic transport. Groups of 50-60 mice were injected intravitreally with [^{32}P] inorganic phosphate and [^{3}H]proline in a fixed ratio under conditions designed to minimize variations in the amount of radioactivity delivered to RGC neurons of different mice. Standardization of the axonal window and the radioisotope dose were critical factors enabling levels of [^{32}P] NFPs or [^{3}H] NFPs to be meaningfully compared at different postinjection intervals.

Individual NFP subunits labeled with [^{3}H]proline behaved similarly and exhibited kinetics expected of axonally transported proteins entering the axonal window. The [^{3}H] radioactivity of each subunit within the axonal window increased 2.5-fold between day 1 and day 4 (Fig. 5) as labeled NFPs were observed in axonal transport studies (Nixon and Logvinenko, 1986) to gradually move into optic axons. The specific activity of [^{3}H] NFPs reached a plateau by day 4, at which time the NFP wavefront had advanced to the middle of the axonal window.

The kinetics of [^{32}P]phosphate groups on NFPs differed substantially from the behavior of the NF polypeptide backbone. The [^{32}P] radioactivity of NFPs within the 9-mm axonal window peaked by day 1 and then, depending on the subunit, decreased or remained constant during the next 4 days (Fig. 5). The [^{32}P] level in NFP-L decreased 42% (p<0.001) between 1 and 3 days and 52% by the fifth day after injection. During the same 5-day interval, the [^{32}P] radioactivity of NFP-M decreased 37% (p<0.001). Studies on the distribution of [^{32}P] NFPs along RGC axons at 1-6 days after injection indicated that loss of [^{32}P]phosphate groups was not due to axoplasmic transport of the NFP subunits beyond the axonal window. Unlike the lower molecular weight subunits, NFP-H contained the same level of [^{32}P] radioactivity between 1 and 5 days after injection, which indicated that the turnover of phosphate groups on this subunit during axonal transport is relatively slow or absent.

DISCUSSION

These studies demonstrate that newly synthesized NFP subunits in RGC neurons are modified in several stages by phosphorylation and selective dephosphorylation. Within 2 hr after the administration of [^{35}S]methionine intravitreally to mice, newly synthesized NFP subunits were identified in

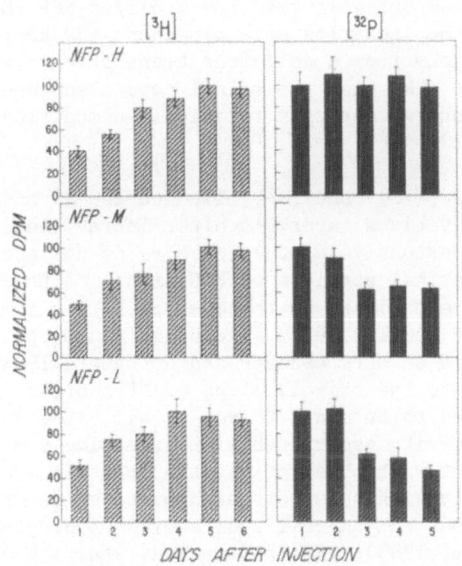

FIG. 5: Differential rates of phosphate turnover on NFP subunits during axoplasmic transport. Groups of mice injected with identical amounts
of [³H]proline and [³²P]orthophosphate in a fixed ratio were sacrificed at daily intervals from 1 to 6 days. Levels of [³H]proline
and [³²P]orthophosphate associated with neurofilaments present
within a 9-mm length of optic nerve and optic tract ("axonal window") were determined after SDS-PAGE. Data on the kinetics of [³H]-
labeled NFPs (left side) and [³²P]-labeled NFPs (right panels) are
presented separately. Each time point represents the mean from 8
to 12 separate determinations. The S. E. M. values are indicated
by vertical bars. Absolute radioactivity levels associated with
each NFP subunit range from 800 to 1800 dpm for tritium and 300 to
600 dpm for [³²P].

the retina as Triton-insoluble proteins with electrophoretic mobilities corresponding to molecular weights of 160 kD, 139 kD and 70 kD. Although these
proteins migrated to different positions on SDS-gels than NFP subunits isolated from white matter axons, several lines of evidence indicated that they
represent the relatively unmodified counterparts of the extensively phosphorylated NFPs in axons. First, the three subunits were selectively immunoprecipitated with a polyclonal anti-NFP antibody that cross-reacts with NFP
subunits isolated from white matter and with dephosphorylated forms of these
polypeptides generated by alkaline phosphatase digestion. Additional antibodies that crossreact with the phosphorylated and in vitro dephosphorylated
forms of individual NFP subunits also precipitated the same three proteins
from among hundreds of radiolabeled proteins in retinal homogenates (Nixon
et al., 1986a). Second, in vitro dephosphorylation of the 200 kD, 140-145
kD and 70 kD NFP forms isolated from axons yielded polypeptides with similar
mobilities on one-dimensional SDS-gels to the 160 kD, 139 kD and 70 kD proteins radiolabeled in the retina in vivo. Finally, the NFP-immunoreactive
polypeptides in the retina disappeared during the first 3 days after synthesis concomitantly with appearance of the radiolabeled 200 kD, 140-145 kD
and 70 kD NFP-immunoreactive proteins in the optic nerve.

These observations are corroborated by recent evidence that the appearance of a phosphorylated form of NFP-M in cultures of chick spinal cord
neurons may be preceded by synthesis of an unphosphorylated form which

179

migrates more rapidly on SDS-gels than the modified NFP (Bennett and DiLullo, 1985). In addition, the isolation of a prominent 139 kD form of NFP-M immunoaffinity chromatography from a cell-free translation mixture containing wheat germ factors and mRNA from rat spinal cord also reinforces the notion that the 139 kD species represents a relatively unmodified form of NFP-M (Strocchi et al., 1982).

The low levels of [^{35}S]- and [^{32}P]-labeled 200 kD and 145 kD NFPs observed in retina at various intervals after intravitreal injection of isotope indicate that extensively modified NFPs do not accumulate in RGC perikarya and intra-retinal portions of RGC axons. Although axonless horizontal cells and efferent fibers to the eye are highly immunoreactive with NFP antibodies (Drager et al., 1984), including monoclonal antibodies to phosphorylated epitopes on NFPs (Drager and Hofbauer, 1984), the contribution of these sources to the radiolabeling of NFPs after intravitreal injection is considered to be small (Nixon et al., 1986a). The finding of low levels of radioactivity associated with extensively modified NFPs in RGC perikarya is consistent with observations that certain monoclonal antibodies that react specifically with phosphorylated epitopes on NFP-H decorate neurofilaments in axons but not in cell bodies of many neuronal types (Sternberger and Sternberger, 1983). In this regard, staining of RGC perikarya and proximal levels of RGC axons in mouse retina is light and increases greatly at axonal levels near the optic disc and in the optic nerve when antibodies against phosphorylated epitopes on NFP-H are used (Drager et al., 1984; Drager and Hofbauer, 1984).

In contrast to the characteristics of NFPs immediately after synthesis, the NFPs radiolabeled with [^{32}P]orthophosphate that later appeared in axons at the optic nerve level exhibited the same electrophoretic mobilities as the 200 kD, 140-145 kD and 70 kD NFPs isolated from CNS white matter. Studies of [^{35}S]-methionine-labeled NFPs also suggest that extensively modified NFP subunits predominate in axons at the optic nerve and optic tract level. Therefore, if the slower mobility of axonal NFPs on SDS-gels is related directly or indirectly to an increased state of phosphorylation, these results suggest that NFPs are extensively phosphorylated before axonal tranport or as they are transported along the most proximal levels of RGC axons.

As NFPs advanced beyond intra-retinal portions of RGC axons, they continued to undergo extensive processing, including addition and/or removal of phosphate groups, in a characteristic pattern for each NFP subunit. The most striking modification involved a progressive shift in the isoelectric point of NFP-H toward a more acidic pH value which was greatest when the NFPs reached distal axonal sites. This charge shift and the smaller acidic shift in isoelectric point of NFP-M most likely reflect addition of one or more phosphate groups to the polypeptides since alkaline phosphatase treatment of [^{32}P]phosphate-labeled axonal NFPs induced comparable shifts in the charge of these subunits in the opposite direction (i.e., toward more basic pH values) and concomitant release of [^{32}P] radioactivity. Charge shifts equivalent in magnitude to those seen in vivo could be induced by removing about 20-30% of the [^{32}P]phosphates initially incorporated by intravitreal injection of [^{32}P]orthophosphate, suggesting that the number of phosphate groups added to these subunits along axons may be small compared to the number added before or as the subunits entered axons.

The shift in the isoelectric point of NFP-L toward a more negative charge during axoplasmic transport suggests that this subunit undergoes a net dephosphorylation after it enters axons. This conclusion is supported by alkaline phosphatase digestion studies and is consistent with the observation that more than half of the [^{32}P]-labeled phosphate groups incorporated into NFP-L from an intravitreal injection of [^{32}P]orthophosphate were lost during axoplasmic transport of neurofilaments along RGC axons.

Although NFP-H and NFP-M both appeared to gain phosphate groups during axoplasmic transport, they differed markedly with respect at which phosphate groups turned over. The loss of more than 50% of the initially incorporated [^{32}P]phosphate groups on NFP-M during the first 4 days of axonal transport implies a relatively high rate of exchange of [^{32}P]phosphate groups with unlabeled phosphates in axons and, possibly, a partial rearrangement of phosphate group topography on the polypeptides. In contrast to NFP-M, [^{32}P] levels associated with NFP-H in axons during the first 5 days of transport did not decrease; however, phosphate turnover may be underestimated in this study since [^{32}P] losses were calculated independently of changes in levels of [^{3}H] NFPs. The kinetics of [^{3}H] NFPs suggest that NFPs containing radio-labeled phosphates may continue to enter the axonal window gradually during the first 3 to 4 days after isotope injection. The on-going entry of [^{32}P]-labeled NFPs would, therefore, tend to mask a slow rate of phosphate turnover on NFP-H during this time interval.

These results demonstrate that the phosphate content on each NFP sub-unit depends on a dynamic balance between phosphorylation and dephosphory-lation, which is differentially modulated in various cellular sites (e.g., perikaryon, proximal, middle and distal axons). In addition to the early and intermediate stages of NFP processing observed in these studies, it is also possible that posttranslational modification of NFPs within axons con-tinues beyond the 6-day postinjection interval examined here. In this regard, neurofilaments that are incorporated into the stationary cytoskele-ton exhibit a half-life in axons of greater than 50 days (Nixon and Logvi-nenko, 1986). Since stationary NFPs represent a significant proportion of the total NFP pool in RGC axons, late modifications of this population could significantly influence the content of phosphate groups on NFP subunits mea-sured chemically. In the steady-state, the content and topography of phos-phate groups on NFPs at a given neuronal site are, therefore, expected to reflect the activities of early, intermediate and late posttranslational events.

Changes in protein phosphorylation state provide an attractive potential mechanism to specify and coordinate interactions among neuronal cytoskeletal elements. Increasing numbers of cytoskeleton-associated proteins have been identified in neural and non-neural cells that apparently serve to intercon-nect the three filamentous systems and the plasma membrane. Many of these proteins are phosphorylated, including microtubule-associated proteins (Theurkauf and Vallee, 1983; Selden and Pollard, 1983; Murthy and Flavin, 1983); spectrin (fodrin) (Anderson, 1979; Goodman et al., 1984; Nixon, 1986); ankyrin (Bennett and Stenbuck, 1979; Lu et al., 1985) and synapsin I (Huttner et al., 1983). Phosphorylated sites within the latter three proteins are located on peripheral domains of the molecule (Anderson, 1979; Weaver et al., 1984; Huttner et al., 1983) where binding sites for certain other proteins have been demonstrated (Correas et al., 1986; Huttner et al., 1983; Lu et al., 1985). Although phosphorylation state may increase or decrease the affinity of a protein for other physiologically interacting molecules (Nest-ler and Greengard, 1984), an impression is emerging in the case of cytoske-letal proteins that a higher state of phosphorylation most commonly inhibits the interactive capabilities of the protein (e.g. binding to other proteins, promotion of protein assembly), while dephosphorylation stimulates these capabilities (Murthy and Flavin, 1983; Jameson et al., 1980; Jameson and Caplow, 1981; Nishida et al., 1981; Selden and Pollard, 1983; Lu et al., 1985; Huttner et al., 1983).

In view of these recent observations, phosphorylation and dephosphory-lation events may be considered as a potential mechanism for regulating interactions between neurofilaments and other cytoskeletal proteins. NFPs contain binding sites for microtubule-associated proteins (Heimann et al., 1985), which may mediate certain microtubule-neurofilament interactions (Leterrier et al., 1982), and possibly for fodrin (Siman and Lynch, 1985), another highly interactive phosphorylated protein (Goodman and Zagon, 1984; Levine and Willard, 1981; Nixon, 1983). It has recently been shown that, as newly synthesized neurofilaments enter RGC axons, a substantial number of them are deposited along axons into a non-uniform stationary cytoskeletal network (Nixon and Logvinenko, 1986) which is composed of various cytoskeletal elements (Nixon, unpublished observations). The translocation of neurofilaments, and the transition of some neurofilaments from a moving phase to the stationary cytoskeleton, imply a considerable requirement for coordinated, specific and potentially reversible binding of cytoskeletal proteins to individual NFP subunits. The observed time- and location-dependent changes in the phosphorylation state of each NFP subunit, as well as the high concentration of phosphorylated sites on peripheral domains of the neurofilament polypeptides, are characteristics expected of such a mechanism for regulating cytoskeletal protein binding.

ACKNOWLEDGMENTS

Studies from our laboratory were supported by grants from the U. S. Public Health Service (AG 02126, AG 05604, and NS 17535) and the Anna and Seymour Gitenstein Foundation.

REFERENCES

Anderson, J. M., Structural studies on human spectrin: Comparison of subunits and fragmentation of native spectrin, J. Biol. Chem. 254:939-944 (1979).

Bennett, G. S. and DiLullo, C., Slow posttranslational modification of a neurofilament protein, J. Cell Biol. 100:1799-1804 (1985).

Bennett, V. and Stenbuck, P. J., Identification and partial purification of ankyrin, the high affinity membrane attachment site for human erythrocyte spectrin, J. Biol. Chem. 254:2533-2541 (1979).

Black, M. M. and Lasek, R. J., Slow components of axonal transport: Two cytoskeletal networks, J. Cell Biol. 86:616-623 (1980).

Brown, B. A., Majocha, R. E., Staton, D. M. and Marotta, C. A., Axonal polypeptides cross-reactive with antibodies to neurofilament protein, J. Neurochem. 40:299-308 (1983).

Brown, B. A., Nixon, R. A., Strocchi, P. and Marotta, C. A., Characterization and comparison of neurofilament proteins from rat and mouse CNS, J. Neurochem. 36:143-153 (1981).

Brown, B. A., Nixon, R. A. and Marotta, C. A., Posttranslational processing of a-tubulin during axoplasmic transport in CNS axons, J. Cell Biol. 94:159-164 (1982).

Carden, M. J., Schlaepfer, W. W. and Lee, V. M.-Y., The structure, biochemical properties, and immunogenicity of neurofilament peripheral regions are determined by phosphorylation state, J. Biol. Chem. 260:9805-9817 (1985).

Chiu, F.-C. and Norton, W. T., Bulk preparation of CNS cytoskeleton and the separation of individual neurofilament proteins by gel filtration: Dye-binding characteristics and amino acid compositions, J. Neurochem. 39:1252-1260 (1982).

Chiu, F.-C., Goldman, J. E. and Norton, W. T., Biochemistry of neurofilaments, in: "Neurofilaments," C. A. Marotta, ed., University of Minnesota Press, Minneapolis, pp. 27-56 (1983).

Cooper, N. G. F., McLaughlin, B. J., Tallant, E. A. and Cheung, W. Y., Calmodulin-dependent protein phosphatase: Immunocytochemical localization in chick retina, J. Cell Biol. 101:1212-1218 (1985).

Cork, L. C., Sternberger, N. H., Sternberger, L. A., Casanova, M. F., Struble, R. G. and Price, D. L., Phosphorylated neurofilament antigens in neurofibrillary tangles in Alzheimer's disease, J. Neuropath. Exptl. Neurol. 45:56-64 (1986).

Correas, I., Leto, T. L., Speicher, D. W. and Marchesi, V. T., Identification of the functional site of erythrocyte protein 4.1 involved in spectrin-actin associations, J. Biol. Chem. 261:3310-3315 (1986).

Dahl, D., Immunohistochemical differences between neurofilaments in perikarya, dendrites and axons. Immunofluorescence study with antisera raised to neurofilament polypeptides (200 K, 150K, 70K) isolated by anion exchange chromatography, Exp. Cell Res. 149:397-408 (1983).

Dahl, D. and Bignami, A., Preparation of antisera to neurofilament protein from chicken brain and human sciatic nerve, J. Comp. Neurol. 176:645-657 (1977).

Drager, U. C., Coexistence of neurofilaments and vimentin in a neurone of adult mouse retina, Nature 303:169-172 (1983).

Drager, U. C., Edwards, D. L. and Barnstable, C. J., Antibodies against filamentous components in discrete cell types of the mouse retina, J. Neurosci. 4:2025-2042 (1984).

Drager, U. C. and Hofbauer, A., Antibodies to heavy neurofilament subunit detect a subpopulation of damaged ganglion cells in retina, Nature 309:624-626 (1984).

Fischer, I., Shea, T. B., Sapirstein, V. S. and Kosik, K. S., Expression and distribution of microtubule-associated protein 2 (MAP2) in neuroblastoma and primary neuronal cells, Dev. Brain Res. 25:99-109 (1986).

Fuchs, E. and Hanukoglu, I., Unraveling the structure of the intermediate filaments, Cell 34:332-334 (1983).

Geisler, N. and Weber, K., Self-assembly in vitro of the 68,000 molecular weight component of the mammalian neurofilament triplet proteins into intermediate-sized filaments, J. Mol. Biol. 151:565-571 (1981).

Geisler, N., Kaufman, E., Fischer, S., Plessman, U. and Weber, K., Neurofilament architecture combines structural principles of intermediate filaments with carboxy-terminal extensions increasing in size between triplet proteins, EMBO J. 2:1295-1302 (1983).

Gilmartin, M. E., Mitchell, J., Vidrich, A. and Freedberg, I. M., Dual regulation of intermediate filament phosphorylation, J. Cell Biol. 98:1144-1149 (1984).

Goodman, S. R. and Zagon, I. S., Brain spectrin: A review, Brain Res. Bull. 13:813-832 (1984).

Goodman, S. R., Zagon, I. S., Whitfield, C. F., Casoria, L. A., Shohet, S. B., Bernstein, S. E., McLaughlin, P. J. and Laskiewicz, T. L., A spectrin-like protein from mouse brain membranes: Phosphorylation of the 235,000-dalton subunit, Am. J. Physiol. 247 (Cell Physiol. 16):C61-C73 (1984).

Goto, S., Yamamoto, H., Fukunaga, K., Iwasa, T., Matsukado, Y. and Miyamoto, E., Dephosphorylation of microtubule-associated protein, τ factor, and tubulin by calcineurin, J. Neurochem. 45:276-283 (1985).

Heimann, R., Shelanski, M. L. and Liem, R. K. H., Microtubule-associated proteins bind specifically to the 70-kDa neurofilament protein, J. Biol. Chem. 260:12160-12166 (1985).

Hirokawa, N., Glicksman, M. A. and Willard, M. B., Organization of mammalian neurofilament polypeptides within the neuronal cytoskeleton, J. Cell Biol. 98:1523-1536 (1984).

Hoffman, P. N. and Lasek, R. J., The slow component of axonal transport: Identification of major structural polypeptides of the axon and their generality among mammalian neurons, J. Cell Biol. 66:351-366 (1975).

Huttner, W. B., Schiebler, W., Greengard, P. and de Camilli, P., Synapsin (protein I), a nerve terminal-specific phosphoprotein. III. Its association with synaptic vesicles studied in a highly purified synaptic vesicle preparation, J. Cell Biol. 96:1374-1388 (1983).

Jameson, L. and Caplow, M., Modification of microtubule steady-state dynamics by phosphorylation of the microtubule-associated proteins, Proc. Natl. Acad. Sci. USA 78:3413-3417 (1981).

Jameson, L. Frey, T., Zeeberg, B., Dalldorf, F. and Caplow, M., Inhibition of microtubule assembly by phosphorylation of microtubule-associated proteins, Biochemistry 19:2472-2479 (1980).

Jones, S. M. and Williams, R. C., Jr., Phosphate content of mammalian neurofilaments, J. Biol. Chem. 257:9902-9905 (1982).

Julien, J.-P. and Mushynski, W. E., Multiple phosphorylation sites in mammalian neurofilament polypeptides, J. Biol. Chem. 257:10467-10470 (1982).

Julien, J.-P., Smoluk, G. D. and Mushynski, W. E., Characteristics of the protein kinase activity associated with rat neurofilament preparations, Biochim. Biophys. Acta 755:25-31 (1983).

Kaufmann, E., Geisler, N. and Weber, K., SDS-PAGE strongly overestimates the molecular masses of the neurofilament proteins, FEBS Lett. 170:81-84 (1984).

King, M. M., Huang, C. Y., Chock, P. F., Nairn, A. C., Hemmings, H. C., Jr., Chan, K.-F. J. and Greengard, P., Mammalian brain phosphoproteins as substrates for calcineurin, J. Biol. Chem. 259:8080-8083 (1984).

Lazarides, E., Intermediate filaments as mechanical integrators of cellular space, Nature 283:249-256 (1980).

Leterrier, J.-F., Liem, R. K. H. and Shelanski, M. L., Preferential phosphorylation of the 150,000 molecular weight component of neurofilaments by a cyclic AMP-dependent microtubule-associated protein kinase, J. Cell Biol. 90:755-760 (1981).

Levine, J. and Willard, M., Fodrin: Axonally transported polypeptides associated with the internal periphery of many cells, J. Cell Biol. 90:631-643 (1981).

Lewis, S. E. and Nixon, R. A., Microheterogeneity of the 200,000 dalton neurofilament protein (NFP), Trans. Amer. Soc. Neurochem. 16:245 (1985).

Liem, R. K. H. and Hutchison, S. B., Purification of individual components of the neurofilament triplet: Filament assembly from the 70,000-dalton subunit, Biochemistry 21:3221-3226 (1982).

Liem, R. K. H., Yen, S.-H., Salomon, G. D. and Shelanski, M. L., Intermediate filaments in nervous tissue, J. Cell Biol. 79:637-645 (1978).

Lu, P.-W., Soong, C.-J. and Tao, M., Phosphorylation of ankyrin decreases its affinity for spectrin tetramer, J. Biol. Chem. 260:14958-14964 (1985).

Murthy, A. S. N. and Flavin, M., Microtubule assembly using the microtubule-associated protein MAP-2 prepared in defined states of phosphorylation with protein kinase and phosphatase, Eur. J. Biochem. 137:37-46 (1983).

Nestler, E. J. and Greengard, P., "Protein Phosphorylation in the Nervous System," John Wiley and Sons, New York (1984).

Nishida, E., Kuwaki, T. and Sakai, H., Phosphorylation of microtubule-associated proteins (MAPs) and pH of the medium control interaction between MAPs and actin filaments, J. Biochem. 90:575-578 (1981).

Nixon, R. A., Proteolysis of neurofilaments, in: "Neurofilaments," C. A. Marotta, ed., University of Minnesota Press, Minneapolis, pp. 117-154 (1983).

Nixon, R. A., Fodrin degradation by calcium-activated neutral proteinase (CANP) in retinal ganglion cell neurons and optic glia: Preferential localization of CANP activities in neurons, J. Neurosci. 6:1264-1271 (1986).

Nixon, R. A. and Lewis, S. E., Differential rates of phosphate turnover on neurofilament subunits in retinal ganglion cell neurons in vivo (submitted for publication) (1986).

Nixon, R. A. and Logvinenko, K. B., Multiple fates of newly synthesized neurofilament proteins: Evidence for a stationary neurofilament network distributed nonuniformly along axons of retinal ganglion cell neurons, J. Cell Biol. 102:647-659 (1986).

Nixon, R. A., Brown, B. A. and Marotta, C. A., Posttranslational modification of a neurofilament protein during axoplasmic transport: Implications for regional specialization of CNS axons, J. Cell Biol. 94:150-158 1982).

Nixon, R. A., Lewis, S. E., Dahl, D. and Marotta, C. A., Early stages in the posttranslational modification of neurofilament proteins by phosphate in retinal ganglion cell neurons. Submitted for publication (1986).

Nixon, R. A., Lewis, S. E. and Marotta, C. A., Posttranslational modification of neurofilament proteins by phosphate during axoplasmic transport in retinal ganglion cell neurons in vivo (submitted) (1986).

O'Connor, C. M., Gard, D. L. and Lazarides, E., Phosphorylation of intermediate filament proteins by cAMP-dependent protein kinases, Cell 23:135-143 (1981).

Osborn, M., Franke, W. and Weber, K., Direct demonstration of the presence of two immunologically distinct intermediate-sized filament systems in the same cell by double immunofluorescence microscopy: Vimentin and cytokeratin fibers in cultured epithelial cells, Exptl. Cell Res. 125:37-46 (1980).

Pant, H. C., Gallant, P. E. and Gainer, H., Characterization of a cyclic nucleotide- and calcium-independent neurofilament protein kinase activity in axoplasm from the squid giant axon, J. Biol. Chem. 261:2968-2977 (1986).

Selden, S. C. and Pollard, T. D., Phosphorylation of microtubule-associated proteins regulates their interaction with actin filaments, J. Biol. Chem. 258:7064-7017, (1983).

Sharp, G. A., Shaw, G. and Weber, K., Immunoelectronmicroscopical localization of the three neurofilament triplet proteins along neurofilaments of cultured dorsal root ganglion neurones, Exp. Cell Res. 137:403-413 (1982).

Shecket, G. and Lasek, R. J., Neurofilament protein phosphorylation. Species generality and reaction characteristics, J. Biol. Chem. 257:4788-4795 (1982).

Shelanski, M. L. and Liem, R. K. H., Neurofilaments, J. Neurochem. 33:5-13 (1979).

Siman, R. and Lynch, G., Fodrin: Skeletal protein cross-linker in rat brain subcellular fractions, Neurosci. Abstr. 185: Vol. 11, p. 775 (1985).

Steinert, P. M., Wantz, M. L. and Idler, W. W., O-phosphoserine content of intermediate filament subunits, Biochemistry 21:177-183 (1982).

Steinert, P. M., Steven, A. C. and Roop, D. R., The molecular biology of intermedite filament, Cell 42:411-419 (1985).

Sternberger, L. A. and Sternberger, N. H., Monoclonal antibodies distinguish phosphorylated and nonphosphorylated forms in situ, Proc. Natl. Acad. Sci. USA 80:6126-6130 (1983).

Strocchi, P., Dahl, D. and Gilbert, J. M., Studies on the biosynthesis of intermediate filament proteins in the rat CNS, J. Neurochem. 39:1132-1141 (1982).

Tallant, E. A. and Cheung, W. Y., Calmodulin-dependent protein phosphatase: A developmental study, Biochemistry 22:3630-3635 (1983).

Theurkauf, W. E. and Vallee, R. B., Extensive cAMP-dependent and cAMP-independent phosphorylation of microtubule-associated protein 2, J. Biol. Chem. 258-7883-7886 (1983).

Troncoso, J. C., Sternberger, N. H., Sternberger, L. A., Hoffman, P. N. and Price, D. L., Immunocytochemical studies of neurofilament antigens in the neurofibrillary pathology induced by aluminum, Brain Res. (in press) (1986).

Wallace, R. W., Tallant, E. A. and Cheung, W. Y., High levels of a heat-labile calmodulin-binding protein (CaM-BP$_{80}$) in bovine neostriatum, Biochemistry 19:1831-1837 (1980).

Wang, E., Intermediate filament associated proteins, in: "Intermediate Filaments," E. Wang, D. Fischman, R. K. H. Liem, and T.-T. Sun, eds., Ann. N. Y. Acad. Sci. 455:32-56 (1985).

Weaver, D. C., Pasternack, G. R. and Marchesi, V. T., The structural basis of ankyrin function: II. Identification of two functional domains, J. Biol. Chem. 259:6170-6175 (1984).

Weber, K., Shaw, G., Osborn, M., Debus, E. and Geisler, W., Neurofilaments, a subclass of intermediate filaments: Structure and expression, <u>Cold Spring Harbor Symp. Quant. Biol</u>. 48:717-729 (1983).

Willard, M., Neurofilaments and axonal transport, in: "Neurofilaments," C. A. Marotta, ed., University of Minnesota Press, Minneapolis, pp. 86-116 (1983).

Willard, M. and Simon, C., Antibody decoration of neurofilaments, <u>J. Cell Biol</u>. 89:198-205 (1981).

Wong, J., Hutchison, S. F. and Liem, R. K. H., An isoelectric variant of the 150,000-dalton neurofilament polypeptide. Evidence that phosphorylation state affects its association with the filament, <u>J. Biol. Chem</u>. 259: 10867-10874 (1984).

Yang, S.-D. and Fong, Y.-L., Identification and characterization of an ATP MG-dependent protein phosphatase from pig brain, <u>J. Biol. Chem</u>. 260: 13464-13470 (1985).

EXTRACELLULAR PROTEIN PHOSPHORYLATION IN NEURONAL RESPONSIVENESS AND ADAPTATION

Yigal H. Ehrlich

The Neuroscience Research Unit - Department of Psychiatry, the Department of Biochemistry, and the Cell-Biology Program, University of Vermont College of Medicine, Burlington, VT 05405

INTRODUCTION

The significant role of protein phosphorylation systems in the regulation and modulation of multiple neuronal functions has been extensively documented in numerous studies over the last three decades. Beginning with the demonstration by Heald (1957) that brief depolarization of respiring brain slices can cause a significant increase of phosphate incorporation into cerebral proteins, progress in this line of investigation has led to the conclusion that the cyclic process of phosphorylation/dephosphorylation of proteins represents a ubiquitous target for diverse agents which produce rapid and transient changes in neuronal activity (reviewed by Greengard, 1978; Rodnight, 1983). More recent studies, carried out with identified neurons of invertebrates, have begun to provide direct evidence for the role of specific phosphoproteins in certain well defined neuronal functions (reviewed by Nestler and Greengard, 1983). Thus, in the chain of events that occurs intracellularly subsequent to the activation of second-messenger generating systems by neurotransmitters, hormones, growth factors and trophic agents, phosphoproteins constitute a crucial link essential for the process of stimulus-response coupling (for most recent reviews see Nishizuka, 1986 and chapter by Greengard et al., in this volume). In addition, phosphorylative activity has been recognized as a site of molecular adaptation in neurons, since it was shown that modifications in the process of protein phosphorylation are induced by inputs which cause long-lasting alterations in brain function (reviewed by Ehrlich, 1979, 1984, see also chapters by Lovinger and Routtenberg and by Alkon and Naito in this volume). The finding that the phosphorylation of receptors for certain neurotransmitters plays a role in the process of desensitization (see Sibley and Lefkowitz, this volume), has added a new dimension to our understanding of mechanisms whereby protein phosphorylation contributes to processes underlying neuronal adaptation. Taken together with the reports that several ion-channels (Nestler and Greengard, 1983) and neuronal cell adhesion molecules (N-CAM's; Edelman, 1983) are phosphorylated by intracellular protein kinases, these studies served to focus the attention on the potential involvement of functional phosphoproteins that traverse the plasma membrane in both short-lived and long-term regulation of neuronal function.

Detailed characterization of the regulation of the activity of various enzymes by phosphorylation (Krebs and Beavo, 1979) revealed that phosphate incorporation into several sites in the same protein can be catalyzed by different protein kinases, with differing and sometimes opposing functional consequences. Another line of investigation (see below) has provided evidence that ATP, the co-substrate of protein kinase, is secreted by neurons to the extracellular environment. These findings provide a basis for investigating the possibility that underlined extracellular protein kinases may play an important role in the nervous system by phosphorylating outfacing sites in, for example, receptors, ion-channels and N-CAM's. Several criteria for demonstrating the extracellular location of enzymes have been established in recent years. Furthermore, enzymatic and immunological probes necessary to demonstrate unequivocally that protein modifications occur at the outer surface of the plasma membrane have become available, and can now be used in experiments designed to determine whether surface and transmembarne neuronal proteins are phosphorylated by extracellular protein kinases. The functional significance of such regulatory mechanisms is highlighted by the realization that the extracellular environment is not inert, but exhibits intense metabolic activity involved in development, in cell communication and feedback regulation (Kreutzberg et al., 1986). Thus, the fine-tuning that protein phosphorylation activity provides in the modulation of intracellular events may be equally important when exerted extracellularly. This chapter reviews in-brief the evidence on the existence of ecto-protein kinase(s) at cell-surfaces, summarizes recent reports form our laboratories on the extracellular protein phosphorylation systems operating in neuronal cells, and discusses the potential of this line of investigation to provide new insights into molecular mechanisms that operate in neuronal development, regulate neuronal responsiveness, and may be involved in processes underlying synaptic plasticity.

Ecto-protein Kinase(s) at the Surface of Cultured Cells

An ecto-enzyme is operating at the cell surface. By definition, ecto-enzymes constitute part of the structure of the plasma membrane or are tightly attached to it, and their catalytic site faces the extracellular environment. A set of criteria that should be satisfied for conclusive demonstration of an ecto-enzyme has been formulated (Karnovsky, 1986). In the case of protein kinase, the phosphorylation of exogenous proteins (added to the medium) by extracellular ATP would substantiate the location of an ecto-kinase, provided that it can be shown that the activity is not exerted exclusively by enzymes originating from cells that break during cell preparation procedures. It should also be mentioned that since platelet activation can cause the release of a soluble protein kinase (Ehrlich and Kornecki, 1986; 1987), the possibility that the presence of this enzyme in serum had contributed to findings reported prior the advent of cell growth in chemically defined media should be considered.

Previous reports from several laboratories have shown that ecto-protein kinase activity can be detected on the outer surface in various types of cultured cells. Although in early studies not all of the considerations outlined above were followed, each has provided some clues concerning the regulation of certain cellular functions by extracellular protein phosphorylation. These reports described studies of 3T3 cells (Mastro and Rozengurt, 1976), of Ehrlich-ascites tumor cells (Ronquist et al., 1977), fibroblasts (Chiang et al., 1979), fat cells (Kang et al., 1978), hepatocytes (Sommarin et al., 1981), leukocytes (Emes and Crawford, 1982), macrophages (Amano et al., 1984) and epididymal spermatozoa (Haldar and Majunder, 1986). Kubler et al. (1982), using cultured HeLa cells, have provided the most convincing evidence available to date for the existence of an ecto-protein kinase and endogenous substrates for its activity at the surface of intact cells. Conclusive evidence for the role of ecto-protein kinase activity in a well defined cellular function is not available yet, but several investigators have associated the phosphorylation of surface proteins with selective changes in the permeability of plasma membranes that are induced under certain conditions by extracellular ATP (reviewed by Ehrlich et al., 1982). It should be emphasized here that a main question that has not been addressed in most of the reports cited above is the source of extracellular ATP that would be utilized by the studied ecto-protein kinases under physiological conditions. In contrast, studies of extracellular protein phosphorylation in cells that are well known to secrete ATP upon stimulation have not been reported until recently, and these are described below in greater detail.

Exocytosis of ATP and its Role in Neuronal Function

In a recent review, Gordon (1986) points out that although ATP has been detected in the extracellular environment of various tissues, only three cell types are known to store ATP within secretory vesicles, and release it by exocytosis upon cell stimulation. These are neurons, chromaffin cells of the adrenal medulla, and platelets. In these cells, depolarization or receptor activation produce a pulse of extracellular ATP which is synchronized with the cycles of cell activation. The secreted ATP can then provide feedback control over the activity of the releasing cell, as well as a signal that is transmitted to target cells. The duration and extent of this signal is determined by the frequency and strength of the stimulus which induces the release reaction, and by the activity of extracellular ATPases, ADPases and 5'nucleotidases, which hydrolyze adenine nucleotides (Zimmermann et al., 1986; Williams, 1987). In neurons, adrenal medullary cells and platelets the timing of ATP release is closely associated with cellular activation, and it may be expected, therefore, that in these cells extracellular ATP would play a role in the regulation of cellular responsiveness to environmental stimulation.

The initial observation implicating extracellular ATP in cell function was made in neuronal systems. Holton and Holton (1954) first reported that ATP may be secreted from nerve terminals, and suggested that it plays a role in neurotransmission. Subsequent studies have shown that ATP is stored in synaptic vesciles of cholinergic and adrenergic neurons (e.g. Silinsky, 1975; Winkler, 1976; Castel et al., 1984), and that ATP is secreted in association with classical neurotransmitters at certain synapses and neuromuscular junctions (Silinsky and Hubbard, 1973; Silinsky, 1975; Winkler, 1976; Phillis and Wu, 1981). Burnstock (1972, 1975) has reported findings indicating that in addition to the release of ATP as a cotransmitter, there may exist purinergic neurons that secrete ATP as the principal neurotransmitter. Storage of ATP in the dense granules of platelets and in secretory vesicles of adrenal cells is likewise well documented, as well as its exocytosis upon stimulation of these cells (see Gordon, 1986 for a comprehensive review).

Since extracellular ATP can be rapidly hydrolyzed to adenosine, which is a potent neuroregulator in its own right, it has been questioned whether secreted ATP molecules have a direct physiological function. The finding of transmission systems in which ATP itself is more potent than adenosine (Burnstock, 1981) has clarified this controversy. It is widely accepted today that adenine nucleotides exert their effects by interaction with two types of purinergic receptors. The type named by Burnstock P_1 is most sensitive to adenosine, and responds to ATP only after it has been hydrolyzed. Purinergic receptors of the P_2 type are stimulated more potently by ATP than by adenosine (Burnstock, 1981; Williams, 1987). Many of the responses mediated by P_2-purinoreceptors can be elicited by non-hydrolyzable analogs of ATP, and thus appear to involve ATP-binding proteins at the cell surface, which operate by activating transduction systems as all classical receptors. Such mediation was found in neurons and muscles (ibid), as well as in non-excitable cells, for example in hepatocytes (Charest et al., 1985). On the other hand, it has been shown that extracellular ATP can exert modulatory effects which require native, hydrolyzable ATP. This modulation may be mediated by the action of ATP-utilizing enzymes, such as extracellular protein kinase(s).

Modulation of neuronal function by extracelular ATP has been reported to occur both pre- and post-synaptically, where native ATP molecules were found to exert, respectively, inhibitory and excitatory actions. In studies of sympathetic ganglia, Silinsky and Ginsborg (1983) have found significant inhibition of the quantal secretion of acetylcholine by ATP, applied extracellularly to preganglionic nerves. In those cells where ATP was found to produce this inhibition, even higher concentrations of adenosine were without effect. Moreover, an unhydrolyzable analog of ATP also had no effect on acetylcholine release in this preparation. Silinsky and Ginsborg (1983) have concluded that the mechanism underlying these inhibitory effects of ATP does not involve adenosine receptors nor P_2-purinergic receptors. In a different study, utilizing chick myoblasts and myotubes, Hume and Honig (1986) have found that ATP applied at micromolar concentrations has a potent depolarizing action. This excitatory action of ATP was not mimicked by adenosine, nor by AMP or ADP. Moreover, several nonhydrolyzable ATP analogs tested by Hume and Honig (1986) had no depolarizing actions on myotubes. In contrast, the analog ATP-gamma-S had similar post-synaptic effects to those found with native ATP. While ATP-gamma-S resists hydrolysis by ATPases, it can be used as a potent co-substrate by synaptosomal protein kinases (Whittemore et al., 1984). Future studies should determine whether some of the post-synaptic actions of ATP, secreted at neuromuscular junctions, are mediated by extracellular protein phosphorylation systems. Similarly, studies of pure neuronal populations could determine the role of these regulatory systems in the control of pre-synaptic events. The initial phase of such investigations must be the identification and characterization of neuronal ecto- and/or exo-protein kinases.

Neuronal Ecto-protein Kinase Activity and its Endogenous Substrates

Conclusive evidence for the existence of an ecto-protein kinase and identification of the endogenous substrates for its activity require that the studies be carried out with intact, viable cells. Furthermore, investigation of a homogenous population of cloned cells would enable to carry out experiments designed to demonstrate causal relationships between the activity of a specific protein phosphorylation system and the modulation of defined neuronal functions by extracellular ATP. Therefore, we have initiated our studies of ecto-protein kinase with a cell line of neuronal origin, clone NG108-15. These hybrid cells can be differentiated in culture and exhibit numerous functions of mature neurons, including evoked acetylcholine release and formation of functional synapses (reviewed by Nirenberg et al., 1983; Hamprecht et al., 1985). We have first reported that these neural cells demonstrate extracellular protein phosphorylation activity (Ehrlich et al., 1982), and then provided evidence that intact NG108-15 neural cells, grown and differentiated in a chemically defined (serum-free) medium, have an ecto-protein kinase and identified its specific substrates at the cell surface (Ehrlich et al., 1986a).

The proteins phosphorylated at the surface of neural cells of the clone NG108-15, and some of the evidence that their phosphorylation is catalyzed by a membrane-bound ecto-protein kinase are shown in Fig. 1. Complete details of these studies have been reported recently (Ehrlich et al., 1986a,b). In-brief, cells were cultured in a chemically defined, serum-free medium in 96-well cluster plates. The reactions were carried out in a buffer mimicking the physiological extracellular environment, and initiated by adding radiolabeled ATP to the wells immediately after rinsing the attached cells. It was determined that these reaction conditions had no adverse effects on cell viability. The reaction medium was supplemented

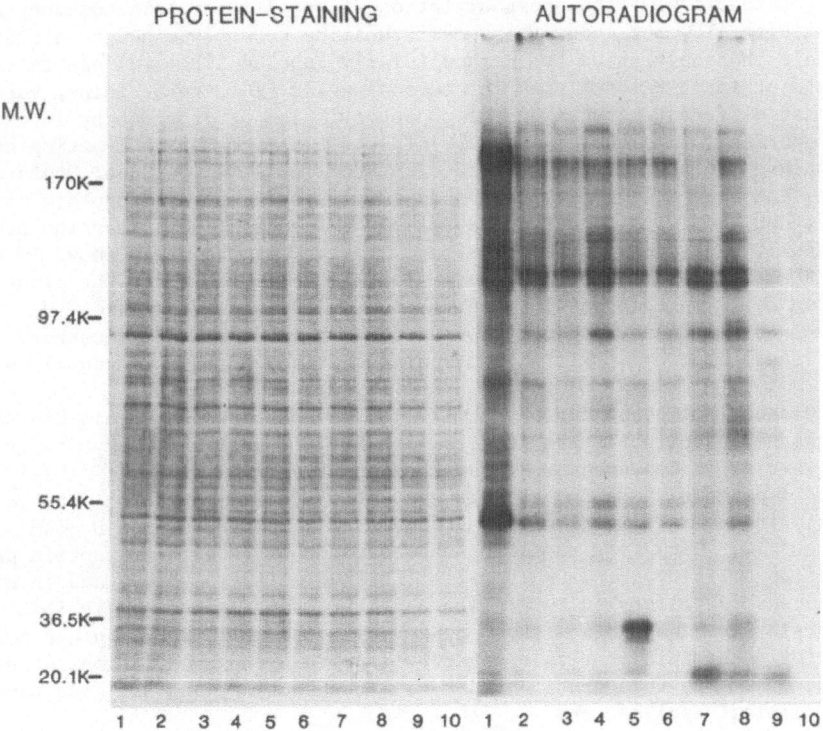

PROTEIN-STAINING   AUTORADIOGRAM

M.W.

170K—

97.4K—

55.4K—

36.5K—

20.1K—

1 2   3 4 5 6 7 8 9 10   1 2   3   4   5   6   7   8   9 10

FIG. 1: Phosphorylation of surface proteins by ecto-kinase in cloned neural cells. Cells of the line NG108-15 were grown in a chemically
defined medium and assayed while attached to the dish after 2
washes in a modified Krebs-Ringer buffer (KR). Under basal conditions (lanes 4 and 8) the reaction was initiated by adding γ-^{32}P-
ATP (final concentration 1μM) and stopped (by SDS solubilization)
after 10 min incubation at 37°C. Lane 1: KR supplemented with
4mM MnCl$_2$; lane 2: 10μM AppNHp added; lane 3: 1μM verapamil;
lane 4: KR only (basal); lane 5: alpha casein added; lane 6:
1μM veratridine; lane 7: 0.001% trypsin; lane 8: Basal conditions after 10 min preincubation in KR; lane 9: Reaction after
10 min preincubation with 0.01% trypsin; lane 10: 10 min reincu-
bation with trypsin after the reaction. The detailed procedures
used for cell growth, reaction protocols and gel electrophoresis
are as described in Ehrlich et al., (1986a,b).

with various additives (see legend to Fig. 1)), used to provide evidence
for the ecto-enzymatic nature of the observed phosphorylative activity and
to determine some of its characteristics. The main findings obtained in
assays such as shown in Fig. 1 and other experiments reported elsewhere
(Ehrlich et al., 1986a,b) were: (a) The spectrum of proteins phosphorylated in intact NG108-15 cells incubated with extracellular ATP and the time-
course of their phosphorylation was different from that obtained by metabolic labeling of intracellular ATP pools. (b) Exogenous proteins (e.g.
alpha-casein) added to the medium were phosphorylated by attached cells,
and competed with the cell surface phosphoproteins for the ecto-kinase
activity. (c) Mild trypsinization, that had no detectable effects on the
pattern of protein staining by Coomassie blue (lanes 9 and 10 in Fig. 1),
eliminated surface protein phosphorylation. (d) Several agents, for exam-

ple Mn^{++}-ions and AppNHp (Fig. 1) were found to selectively stimulate or inhibit, respectively, the phosphorylation of specific protein components by extracellular ATP, but under the same incubation conditions had no effects on protein phosphorylation by intracellularly labeled ATP. (e) Agents that stimulate or inhibit ion-fluxes in neuronal cells (KCl, veratridine, verapamil) had different effects on the phosphorylation of proteins by intra versus extracellular labeled ATP. (f) In the plasma-membrane fraction and in cytoplasm isolated from NG108-15 cells (Ehrlich et al., 1986b; Davis and Ehrlich, in-preparation) prominent phosphorylation of several protein components, not seen with intact cells, was in evidence. These reaction products should have appeared as major bands in the autoradiogram shown in Fig 1 if the activities measured in these assays were dependent on the presence of broken cells in the reaction medium, but such was not the case. We believe that all these findings provide the combined evidence necessary for proving that an ecto-protein kinase operates at the surface of neural cells.

Additional evidence for the activity of neuronal ecto-protein kinase was obtained in collaborative studies with Drs. E. Bock and O. Nybroe from the University of Copenhagen. These studies have also provided initial clues to a neuronal function of this enzymatic system. Specific antibodies were used in several experimental paradigms to demonstrate that D_2-CAM (Lyles et al., 1984) is a specific substrate for extracellular protein phosphorylation activity in NG108-15 cells (Ehrlich et al., 1986a), and in cerebral neurons in primary culture. The ability of an antibody to inhibit the phosphorylation of N-CAM's without penetrating the cells should prove most useful for determining the role of surface protein phosphorylation in neuronal development, regeneration and synaptic sprouting, and may also have future clinical applications.

Differentiation of NG108-15 in culture by a treatment known to induce synapse-competence was associated with increased ecto-protein kinase activity (Ehrlich et al., 1986a), suggesting that this enzymatic system may be involved in the regulation of neuronal specific functions. Indeed, using the procedures for culturing and differentiation in a defined chemical environment of primary neuronal cells from the central nervous system (CNS, Weiss et al., 1986), we have found recently that primary cortical and neostriatal neurons from embryonic mouse brain exhibit a pattern of extracellular protein phosphorylation which is similar to that demonstrated in differentiated NG108-15 cells. Moreover, surface phosphorylation of a protein with apparent molecular mass of 110-120Kd (average 116Kd) which is a major substrate of ecto-kinase in NG108-15 cells (Fig. 1) and in non-neural cells (Kubler et al., 1982), is not prominent in primary CNS neurons. Extracellular protein phosphorylation systems in the brain thus may have specialized functional roles. Such investigation, however, must be preceded by evidence that CNS neurons secrete ATP to the extracellular environment. Fig. 2 demonstrates the procedures implemented in our laboratory for measuring depolarization-induced vesicular exocytosis from CNS neurons in primary culture. The two main criteria employed were the dependence of K$^+$-induced neurotransmitter release on extracellular calcium, and the inhibition of veratridine-induced secretion by tetrodotoxin. Both criteria have been fulfilled for GABA release from cultured striatal neurons (Fig. 2), maintained in a chemically defined medium for 14-18 days after dissection from 14-day mouse embryos. Quantitative luciferin-luficerase reactions were carried out to determine ATP concentration in aliquotes of release reactions collected as described in the legend to Fig. 2. In these experiments we have found that over 75% of the ATP secreted by differentiated CNS neurons stimulated with KCl was dependent on extracellular calcium, and about 80% of the veratridine-induced ATP release was inhibited by tetrodotoxin (Zhang et al., 1987). Together with the demonstration of protein phosphorylation by extracellular ATP in these cells, the results provide the basis and an appropriate model system for investigating the

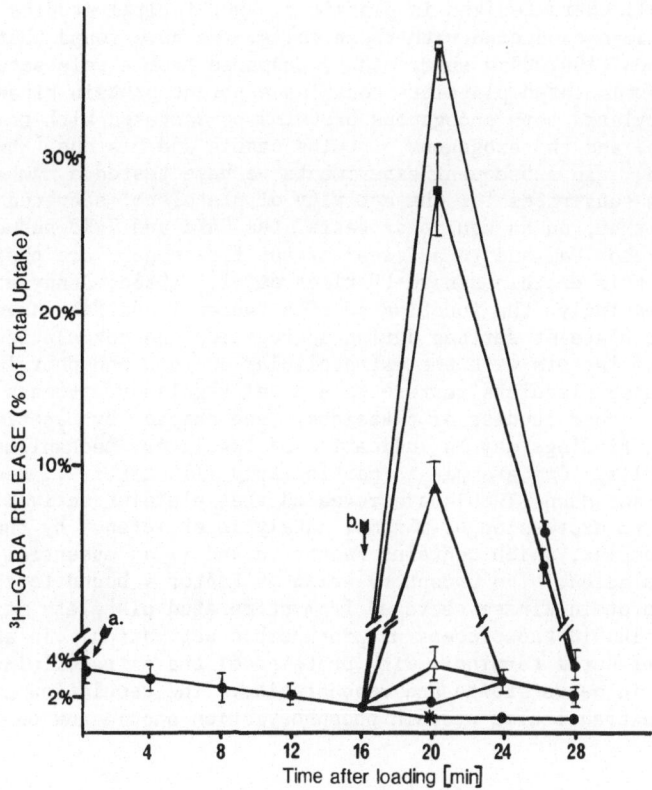

FIG. 2: Depolarization-induced GABA release from neostriatal neurons differentiated in primary culture. Striata dissected from 14 day
mice embryos were dissociated, plated (10^6/well) and the neurons
maintained in a chemically defined medium for 18-21 days. The
neurons were loaded with 0.5µCi/well of [^{3}H]GABA during 10 min
preincubation in Krebs-Ringer (KR) medium. After 2 gentle washes
with KR (at point a in the figure), basal release was monitored
over a 16 min period at 4 min intervals. Evoked release was initiated at time-point b by the additions listed below. A 4 min
releasate was collected followed by 2x4 min incubations with KR
medium. Release data are expressed as percentage of total [^{3}H]-
GABA uptake by the cells in each well. Key - Closed circles: only
KR medium added (n=6; ±SEM). Closed triangles: 50mM KCl in KR,
with NaCl adjusted (n=6). Open triangles: 50mM KCl in Ca^{++}-free
KR (n=3). Closed squares: 100µM veratridine in KR (n=8). Open
squares: 100µM veratridine in Ca^{++}: free KR (n=3). Asterisk:
100µM veratridine + 1µM tetrodotoxin (n=6).

involvement of ecto-protein kinase in brain function.

Release of a Soluble Exo-protein Kinase by Stimulated Cells

Exo-enzymes are present in the extracellular environment in a soluble
form, and this definition does not distinguish between dissociation of an
ecto-enzyme from the cell surface and secretion of enzyme molecules from
intracellular storage sites (Kreutzberg et al., 1986). In this context it
is important to note that the storage vesicles of secretory cells are known
to contain proteins which are secreted to the extracellular environment
during exocytosis. The release of granular proteins has been extensively

studied and well characterized in platelets. Our intital studies in this investigation were conducted with these cells. We have found that the soluble fraction (100,000xg supernatant) prepared from a releasate obtained from thrombin-stimulated platelets contains a potent protein kinase activity which phosphorylated both endogenous proteins co-secreted with this kinase from platelets, and the exogenous proteins casein and histone (Ehrlich and Kornecki, 1986). In subsequent experiments we have tested a number of plasma proteins as substrates for the activity of platelets' secreted kinase. We have found that, on an equimolar basis, the 94Kd and 74Kd subunits of coagulation Factor Va, and to a lesser extent fibrinogen, are preferred substrates of this protein kinase (Ehrlich et al., 1986c; Jenny et al., 1986). Interestingly, the function of both Factor V and fibrinogen involves binding to the platelet surface during aggregation and coagulation. Interaction of these factors with the extracellular protein phosphorylation systems of activated platelets represents a novel regulatory process in hemostatis. As in other studies of platelets, (see chapter by Lapetina in this volume), these findings may be indicative of regulatory mechanisms operating in neuronal cells. One example is particularly illustrative. Recent studies by Tracy and Mann (1986) have revealed that platelet activation is required for the expression of maximal catalytic efficiency by the prothrombinase complex, which contains Factor Va and is an essential component of blood coagulation. The phosphorylation of Factor V bound to the platelet surface by a protein kinase secreted from stimulated platelets may prove to be a missing link in the process of prothrombin activation. In analogy, interactions of nerve terminals with proteins of the extracellular matrix are important in nerve growth and regeneration. The regulation of these processes by extracellular protein phosphorylation should now be investigated.

In preliminary studies utilizing differentiated NG108-15 cells, we have detected release of casein-kinase and histone-kinase activity upon depolarization of these cells with KCl or veratridine (Ehrlich and Kornecki, 1986)). The presence of significant protease activity in the extracellular environment of stimulated nerves is a major problem that still needs to be overcome in this investigation. Nonetheless, we are encouraged by the finding that differentiated, primary CNS neurons co-secrete ATP with neurotransmitters (GABA) when depolarized by KCl or veratridine (Zhang et al., 1987). In continuing these studies, we plan to utilize CNS neurons in primary culture as target in the investigation of protein kinase secreted by stimulated neuronal cells. The effort required for overcoming some of the problems encountered in these studies appears to be highly warranted due to their potential to reveal new means of intercellular communication in the nervous system.

Extracellular Protein Phosphorylation in Synaptic Plasticity

The processes underlying synaptic plasticity are known to involve both long-term alterations in molecular mechanisms as well as changes in the ultrastructure of synaptic connections. Early studies on the involvement of mammalian protein phosphorylation systems in these processes implicated specific phosphoproteins endogenous to synaptic membranes in long-term changes associated with learning experience (Ehrlich et al., 1977), narcotic-dependence (Ehrlich et al., 1978), chronic administration of neuroleptic drugs (Ehrlich, 1979), and electroconvulsive treatment (Ehrlich et al., 1980). In recent years much progress has been made in the elucidation of mechanisms involved in the production of long-term changes in synapses by investigating the process of long-term potentiation (LTP; see chapters by Baudry and Lynch and by Lovinger and Routtenberg in this volume). The lasting changes associated with LTP are induced by brief episodes of high frequency, repetitive stimulation delivered at precise intervals (ibid). Studies by Lynch, Baudry and their colleagues have demonstrated that excessive increase

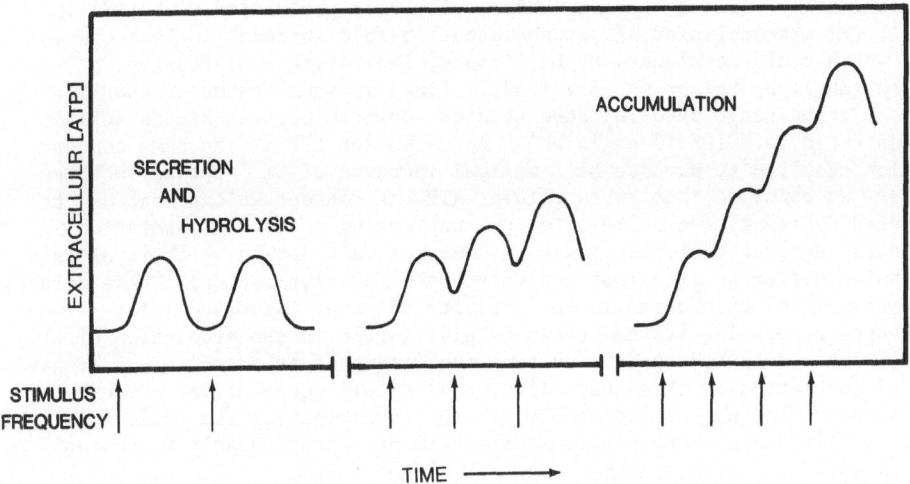

FIG. 3: Accumulation of secreted ATP in the synaptic cleft. This figure depicts the fate of extracellular ATP secreted by nerve-terminals after neuronal stimulation at various frequencies, drawn on the basis of experimental findings and theoretical considerations detailed in the text.

of intracellular Ca^{++}-levels is a crucial step in the induction of LTP. Obviously, the signal(s) which triggers the exaggerated increase in intracellular free Ca^{++}-ions in the chain of events leading to LTP must be different from those operating during routine transmission. A unique event that can be produced by a train of repetitive stimuli is the accumulation of extracellular ATP in the synaptic cleft to levels far greater than those obtained during routine transmission. This is presented schematically in Fig. 3, which illustrates that after a single stimulus the secreted ATP is rapidly hydrolyzed by ecto-ATPase and would, therefore, not accumulate. Upon repetitive stimulation, the change in balance between secretion and hydrolysis could cause accumulation of extracellular ATP, the extent of which would depend, at least in part, on the interval between stimuli. Electrophysiological studies of cholinergic transmission reported by Silinsky (1975; 1984) enabled to estimate the concentrations of extracellular ATP produced in these situations. Accordingly, a single stimulus which releases one quantum of neurotransmitter will produce in the synaptic cleft an ATP concentration of about 0.8 μM. After repetitive and especially tetanic-stimulation, ATP could accumulate in the cleft to concentrations of 100-400μM. We have tested, therefore, the ability of extracellular ATP in the concentration range of 100μM to 1mM to provide a signal which produces changes in second-messenger systems involved in the production of LTP.

Homogenous cell populations of clone NG108-15 and of the cholinergic clone N1E-115 were used to determine the effects of ATP on the uptake of Ca^{++}-ions, on the levels of intracellular free calcium, on intracellular cyclic AMP and cyclic GMP generation, and on phosphoinositide metabolism. Extracellular ATP in the concentration range of 0.1-1.0mM were found to cause up to six-fold increase in the uptake of $^{45}Ca^{++}$, and this effect was additive with the uptake induced by depolarizing the cells with KCl (Ehrlich et al., 1982; 1986b; 1987). Use of the probes aequorin and quin-2 verified that extracellular ATP induces a significant rise in intracellular free Ca^{++} and that this rise is associated with activation of guanylate cyclase (Enrlich et al., 1987). At the same concentration range, extracellular ATP produced a decrease in PGE_1-stimulation of intracellular cyclic

AMP formation (Ehrlich et al., 1986b). In cells prelabeled with $[^3H]$-ino-
sitol, 1mM extracellular ATP caused about 15-fold increase in IP accumulation which could be blocked by La^{+++}-ions, indicating mediation by Ca^{++}-
uptake (Ehrlich, Snider et al., 1987). Finally, when tested in the ATP
concentration range used in these studies, an ecto-protein kinase activity
was detected in NG108-15 cells which had a Km for ATP at the same concentration required to produce half-maximal increase of Ca^{++}-uptake in these
cells. We conclude that extracellular ATP, at concentrations that can be
produced by repetitive stimulation but not during routine transmission,
can cause an exaggerated increase in neuronal Ca^{++}-uptake which involves a
mechanism different from that activated by K^+-depolarization. These effects
are accompanied by changes in the activity of several intracellular second-
messenger generating systems shown to play a role in the production of LTP.
The evidence obtained to-date on the involvement of an ecto-protein kinase
in the initiation of this chain of events is only correlative. Nonetheless, these findings provide the basis for investigating the role of extracellular protein phosphorylation systems in processes underlying neuronal
plasticity.

CONCLUSIONS AND FUTURE DIRECTIONS

In cells which store ATP within secretory vesicles and release it by
exocytosis upon cell stimulation, extracellular protein phosphorylation
systems may play an important role in cellular activation and intercellular communication. Secretion of ATP has been well documented in three cell
types: neurons, chromaffin cells of the adrenal medulla and platelets. In
each of these cell types we have found extracellular protein phosphorylation
systems and surface proteins phosphorylated by extracellular ATP. Two types
of extracellular protein kinases have been detected: a membrane bound ecto-
protein kinase and a soluble, secreted exo-protein kinase. The ecto-protein
kinase has properties consistent with involvement in the feedback regulation
of cellular activation. The secreted protein kinase can phosphorylate proteins of the extracellular matrix and surface proteins of target cells, and
thus may serve as an enzyme with the role of first messenger in cell-cell
communication.

Studies of extracellular protein phosphorylation systems may shed new
light on molecular mechanisms which regulate short-lived responses as well
as long-term adaptive alterations in synapses of the peripheral and central
nervous system. Initial findings in this new line of investigation suggest
several areas in which significant progress may be made in the near future.
These include studies on the phosphorylation of N-CAM by ecto-protein kinase
and its role in neuronal adhesion, development and maturation (Ehrlich et
al., 1986a; Bock et al., in-preparation). The modulation of acetylcholine
release by extracellular ATP (Silinsky and Ginsborg, 1983) may involve surface phosphoproteins, and this possibility can be tested in homogenous populations of differentiated NG108-15 cells examined under depolarizing conditions (Ehrlich et al., 1986a). Studies implicating extracellular protein
phosphorylation in the regulation of norepinephrine uptake into nerve endings (Hendley et al., 1987) can be carried out on the level of a specific
ecto-protein kinase, when conducted with cultured cells originating from
the adrenal medulla (Chaffee et al., 1987). The role of neuronal exo-
protein kinase in the regulation of post-synaptic events could be best
studied by detailed examination of the excitatory actions of extracellular
ATP on myotubes (Hume and Honig, 1986). Investigation of the responsiveness
of cultured CNS neurons to extracellular ATP accumulated during tetanic-
stimulation may contribute to our understanding of the etiology of disorders such as epilepsy, and finally, the exciting possibility that extracellular protein phosphorylation systems may play a role in processes underlying synaptic plasticity provides a new direction in studies aimed to elucidate molecular mechanisms activated by learning experience and involved
in the formation of memory.

ACKNOWLEDGMENTS

Studies from the author's laboratory described here were supported by grant no. 84-0331 from the Air Force Office of Scientific Research and, in part, by NSF grants BNS82-09265 and BNS85-07238.

REFERENCES

Amano, F., Kitagawa, T. and Akamatsu, Y., Protein kinase activity on the cell surface of a macrophage-like cell line, Biochem. Biophys. Acta. 803:163-173 (1984).

Burnstock, G., Purinergic nerves, Pharmacol. Rev. 24:509-581 (1972).

Burnstock, G., Purinergic transmission. In Iverson, L. L., Iversen, S. D. and Snyder, S. H. (Eds.): Handbook of Psychopharmacology, Vol 5, New York: Plenum, pp. 131-194 (1975).

Burnstock, G., Neurotransmitters and trophic factors in the autonomic nervous system, J. Physiol. 313:1-35 (1981).

Castel, M., Gainer, H. and Pellman, H. D., Neuronal secretory systems, Int. Rev. Cytol. 88:303-359 (1984).

Chaffee, J. E., Hendley, E. and Ehrlich, Y. H., The regulation of ectokinase-mediated protein phosphorylation in PC12 cells, J. Neurochem., in press (1987).

Charest, R., Blackmore, P. F. and Exton, J. H., Characterization of responses of isolated rat hepatocytes to ATP and ADP, J. Biol. Chem. 260:15789-15794 (1985).

Chiang, T. M. and Kang, E. S. and Kang, A. H., Ecto-protein kinase activity of fibroblasts, Arch. Biochem. Biophys. 195:518-525 (1979).

Edelman, G. M., Cell adhesion molecules, Science 219:450-457 (1983).

Ehrlich, Y. H., Phosphoproteins as specifiers for mediators and modulators in neural function, Advances in Experimental Medicine & Biology 116: 75-102 (1979).

Ehrlich, Y. H., Protein phosphorylation: Role in the function, regulation and adaptation of neural receptors. In: A. Lajtha (Ed.), Handbook of Neurochemistry, Plenum, New York, Vol. 6, pp. 541-574 (1984).

Ehrlich, Y. H. and Kornecki, E., Secretion of protein kinase by stimulated nerve cells and activated platelets, Trans. Am. Soc. Neurochem. 17:134 (1986).

Ehrlich, Y. H. and Kornecki, E., Extracellular Protein Phosphorylation Systems in Cellular Responsiveness. In: G. Sato, W. L. McKeehan, and M. Cabot (Eds.), Mechanisms of Signal Transduction by Hormones and Growth Factors, Alan R. Liss, New York (in press) (1987).

Ehrlich, Y. H., Rabjohns, R. and Routtenberg, A., Experiential-input alters the phosphorylation of specific proteins in brain membranes, Pharmacology, Biochemistry and Behavior 6:169-174 (1977).

Ehrlich, Y. H., Bonnet, K. A., Davis, L. G. and Brunngraber, E. G., Decreased phosphorylation of specific proteins in neostriatal membranes from rats after long-term narcotic exposure, Life Sciences 23: 137-146 (1978).

Ehrlich, Y. H., Reddy, M. N., Keen, P., Davis, L. G. and Brunngraber, E. G., Transient changes in the phosphorylation of cortical membrane proteins after electro-convulsive shock, J. Neurochem. 34:1327-1330 (1980).

Ehrlich, Y. H., Whittemore, S. R., Garfield, M. K., Graber, S. G. and Lenox, R. H., Protein phosphorylation in the regulation and adaptation of receptor function. In: W. H. Gispen, and A. Routtenberg (Eds.), Progress in Brain Research, Vol. 56, Amsterdam, Elsevier/North Holland, pp. 375-396 (1982).

Ehrlich, Y. H., Davis, T., Bock, E., Kornecki, E. and Lenox, R. H., Ecto protein kinase activity on the external surface of intact neural cells, Nature (London) 320:67-70 (1986a).

Ehrlich, Y. H., Garfield, M. G., Davis, T. B., Kornecki, E., Chaffee, J. E. and Lenox, R. H., Extracellular Protein Phosphorylation Systems in the Regulation of Neuronal Function. In: W. H. Gispen and A. Routtenberg (Eds.), Phosphoproteins in Neuronal Function. Progress in Brain Research, Elsevier, Holland, Vol. 69, pp. 197-208 (1986b).

Ehrlich, Y. H., Kornecki, E., Jenny, R., Cierniewski, C. S. and Mann, K. G., Regulation of platelet activation and blood coagulation by extracellular protein phosphorylation systems, Blood 68:Suppl. 1, 315a (1986c).

Ehrlich, Y. H., Snider, R. M., Garfield, M. G., Kornecki, E. and Lenox, R. H., Modulation of neuronal signal transduction systems by extracellular ATP, J. Neurochem. in press (1987).

Emes, C. H. and Crawford, N., Ecto-protein kinase activity in rabbit peritoneal polymorphonuclear leucocytes, Biochem. Biophys. Acta 717:98-104 (1982).

Gordon, J. L., Extracellular ATP: effects, sources and fate, Biochem. J. 233:309-319 (1986).

Greengard, P., Cyclic Nucleotides, Phosphorylated Proteins and Neuronal Function, Raven Press, New York (1978).

Haldar, S. and Majumder, G. C., Phosphorylation of external cell-surface proteins by an endogenous ecto-protein kinase of goat epididymal intact spermatozoa, Biochem. Biophy. Acta 887:291-303 (1986).

Hamprecht, B., Glaser, T., Reiser, G., Bayer, E. and Propst, F., Culture and characteristics of hormone-responsive neuroblastoma x glioma hybrid cells, Meth. Enzymol. 109:316-341 (1985).

Heald, P. J., The incorporation of phosphate into cerebral phosphoproteins promoted by electrical impulses, Biochem. J. 66:659-663 (1957).

Hendley, E. D., Whittemore, S., Chaffee, J. and Ehrlich, Y. H., Regulation of norepinephrine uptake by adenine nucleotides and divalent cations: role for extracellular protein phosphorylation, submitted for publication (1987).

Holton, F. A. and Holton, P., The capillary dilator substances in dry powders of spinal roots: a possible role of ATP in chemical transmission from nerve endings, J. Physiol. London 126:124-140 (1954).

Hume, R. I. and Honig, M. G., Excitatory action of ATP on embryonic chick muscle, J. Neuroscience 6:681-690 (1986).

Jenny, R. J., Ehrlich, Y. H., Kornecki, E. and Mann, K. G., Kinase mediated phosphorylation of Factor V/Va, Circulation 74:1645 (1986).

Kang, E. S., Gates, R. E. and Farmer, D. M., Localization of the catalytic subunit of a cyclic-AMP-dependent protein kinase(s) and acceptor proteins on the external surface of the fat cell membrane, Biochem. Biophys. Res. Commun. 83:1561-1569 (1978).

Karnovsky, M. L., In: Kreutzberg et al., pp. 3-16, in this Reference List (1986).

Krebs, E. G. and Beavo, J. A., Phosphorylation-dephosphorylation of enzymes, Annu. Rev. Biochem. 48:923-959 (1979).

Kreutzberg, G. W. and Reddington, M. and Zimmermann, H. (Eds.), Cellular Biology of Ectoenzymes, Berlin: Springer-Verlag (1986).

Kubler, D., Pyerin, W. and Kinzel, V., Protein kinase activity and substrates at the surface of intact HeLa cells, J. Biol. Chem. 257:322-329 (1982).

Lyles, J. M., Linnemann, D. and Bock, E., Biosynthesis of the D2-cell adhesion molecule: Post-translational modifications, intracellular transport and developmental changes, J. of Cell Biol. 99:2082-2091 (1984).

Mastro, A. M. and Rozengurt, E., Endogenous protein kinase in outer plasma membrane of cultured 3T3 cells, J. Biol. Chem. 251:7899-7906 (1976).

Nestler, E. D. and Greengard, P., Protein phosphorylation in the brain, Nature 305:583-588 (1983).

Nirenberg, M., Wilson, S., Higashida, H., Rotter, A., Krueger, K., Busis, N., Ray, R., Kenimer, J. G. and Adler, M., Modulation of synapse formation by cyclic adenosine monophosphate, Science 221:331-338 (1983).

Nishizuka, Y., Studies and perspectives of protein kinase C, Science 233:
305-312 (1986).

Phillis, J. W. and Wu, P. H., Roles of adenosine and adenine nucleotides
in the central nervous system. In: Daly, J. W., Kuroda, Y., Phillis, J. W., Shimizu, H. and Ui, M., (Eds.), Physiology and Pharmacology of Adenosine Derivatives, Raven Press, New York, pp. 219-236
(1981).

Rodnight, R., Protein kinases and phosphatases. In: A. Lajtha (Ed.),
Handbook of Neurochemistry, Plenum, New York, pp. 195-217 (1983).

Ronquist, G., Agren, G., Eklund, S. and Wernstedt, C., Cyclic 3'-5-GMP independent protein kinase at the outer surface of intact Ehrlich cells,
Ups. J. Med. Sci. 82:1-5 (1977).

Silinsky, E. M., On the association between transmitter secretion and the
release of adenine nucleotides from mammalian motor nerve endings, J.
Physiol. 346:243-256 (1975).

Silinsky, E. M., On the mechanism by which adenosine receptor activation
inhibits the release of acetylcholine from motor nerve endings, J.
Physiol., 346:243-256 (1984).

Silinsky, E. M. and Hubbard, J. I., Release of ATP from rat motor nerve
terminals, Nature (London) 243:404-405 (1973).

Silinski, E. M. and Ginsborg, B. L., Inhibition of acetycholine release
from preganglionic frog nerves by ATP but not adenosine, Nature (London) 305:327-328 (1983).

Sommarin, M., Henriksson, T. and Jergil, B., Cyclic AMP-dependent protein
phosphorylation on the surface of rat hepatocytes, FEBS Lett. 127:285-
289 (1981).

Tracy, P. B. and Mann, K. G., A model for assembly of coagulation factor
complexes on cell surfaces: prothrombin activation on platelets. In:
D. R. Phillips and M. A. Schuman (Eds.), Biochemistry of Platelets,
Academic Press, Orlando, FL, pp. 296-318 (1986).

Weiss, S., Pin, J. P., Sebben, M., Kemp, D. E., Sladeczek, F., Gabrion, J.
and Bockaert, J. Synaptogenesis of cultured striatal neurons in serum-
free medium: a morphological and biochemical study, Proc. Natl. Acad.
Sci. USA 81:2238-2242 (1986).

Whittemore, S. R., Graber, S. G., Lenox, R. H., Hendley, E. D. and Ehrlich,
Y. H., Activation of adenylate cyclase by preincubation of rat cere-
bral-cortical membranes under phosphorylating conditions: Role of
ATP, GTP and divalent cations, J. Neurochem. 42:1685-1696 (1984).

Williams, M., Purine receptors in mammalian tissues, Ann. Rev. Pharmacol.
Toxicol., Vol. 27, in press (1987).

Winkler, H., The composition of adrenal chromaffin granules: an assessment
of controversial results, Neuroscience 1:65-80 (1976).

Zhang, J., Kornecki, E., Jackman, J. and Ehrlich, Y. H., Depolarization-
induced secretion of ATP from mature CNS neurons in primary culture,
J. Neurochem., in press (1987).

Zimmermann, H., Grondal, E. J. H. and Keller, F., In: Kreutzberg et al.,
pp. 35-48, see this Reference List (1986).

EXPRESSION OF RAT BRAIN EXCITATORY AMINO ACID RECEPTORS IN XENOPUS OOCYTES

Richard A. Lampe, Leonard G. Davis, and Michael J. Gutnick

Neurobiology Group, Medical Products Department, E. I.
du Pont de Nemours and Company, Experimental Station, E400
Wilmington, DE 19898

SUMMARY

Xenopus laevis oocytes when injected with rat brain mRNA synthesize neuronal receptors that can be analyzed electrophysiologically. After a post-injection incubation period of 24-72 hours, L-glutamic acid, kainic acid and quisqualic acid caused a dose dependent (10-100 μM) depolarization of the oocyte membrane. The voltage and conductance changes associated with kainate activation were distinguishable from those seen for L-glutamate or quisqualate. There was no response to L-aspartate application and an inconsistent response to N-methyl-D-aspartate.

Upon fractionation of the mRNA on sucrose gradients, transcripts greater than 2 Kb in length were obligatory for the synthesis of excitatory amino acid receptors. The electrophysiological response of injected oocytes exposed to L-glutamate was similar to that of native oocytes when exposed to muscarinic agents. This similarity may reflect the activation of the same ionophore and suggests that the active mRNA fraction for glutamate responsiveness either encodes for a binding protein that can be assembled along with native ion channels into the oocyte membrane or encodes for a glutamate binding site with a similar channel.

INTRODUCTION

Despite intensive effort, the isolation and purification of homogeneous receptor proteins from nervous tissue utilizing classical purification techniques has proven difficult. An alternative to these techniques is a molecular biological approach starting with the identification and isolation of an mRNA species coding for a particular neuroreceptor. Previous studies have demonstrated that Xenopus laevis oocytes are capable of translating exogenous mRNA with a high degree of fidelity to produce functionally active proteins (Gurdon et al., 1971; Labarca and Paigen, 1977; Sumikawa et al., 1981; Soreq et al., 1982; Simmen et al., 1984; Pure et al., 1984; Werner et al., 1985; Soreq, 1985). Based on these observations, we sought to use this cellular translation system as a bioassay to isolate and characterize brain mRNA species coding for receptor proteins. We have concentrated our efforts on the isolation of a glutamate receptor mRNA from rat brain since the glutamate receptor has been implicated in learning (Lynch and Baudry, 1984), cell death (Wieloch, 1985), and epilepsy (Baldino et al., 1986; Hablitz and

Langmoen, 1986) and because neuronal responses to glutamate are ubiquitous. Moreover, Xenopus oocytes do not respond intrinsically to glutamate receptor agonists (Gundersen et al., 1984; Houamed et al., 1984; Sumikawa et al., 1984; Parker et al., 1985). We report that excitatory amino acid receptors can be functionally incorporated into the oocyte membrane following the injection of oligo-dT enriched, size selected rat brain RNA.

METHODS

Isolation and Culture of Xenopus Oocytes: Ovarian sacs were surgically removed from ice-anesthetized Xenopus laevi through a small incision in the abdominal cavity and placed immediately into a petri dish containing modified Barth's solution (MBS). Since only a small percentage of total ovarian tissue is removed, the wound is sutured and the animal reused. The excised ovarian tissue was divided into small clumps (20-30 oocytes/clump) which were rinsed thoroughly in MBS. After teasing away all surrounding follicular cells, individual, healthy oocytes (stages 5-6, Dumont classification, (Dumont, 1972)) were collected and cultured in small petri dishes (10-20 oocytes/dish) maintained at 18-20°C in MBS that was changed daily.

Preparation of Poly A$^+$ RNA: RNA was isolated from total brain and from liver using the guanidine isothiocyanate and CsCl centrifugation method (Davis et al., 1986). Oligo-dT affinity chromatography (Aviv and Leder, 1972) was used to select for brain poly A$^+$ RNA (mRNA) which was then size separated (Robinson et al., 1983) into 30 different fractions on a linear sucrose density gradient (15-40%, 35000 rpm, 15 hr, 20°C). RNA samples were resuspended, after ethanol precipitation, in autoclaved H_2O at a concentration of 0.5-1.0 mg/ml for injection into oocytes.

Microinjection into oocytes of Poly A$^+$ RNA or H_2O: Oocytes were placed in small holding wells in a plexiglass plate and impaled with micropipettes (8-10 micron tip diameter) containing either poly A$^+$ RNA or H_2O (control). Delivery of material was accomplished with a nanoliter pump (World Precision Inst) set at a flow rate of 100 nl/min. One minute of delivery resulted in an injection of approximately 50-100 ng of RNA per oocyte. To allow for mRNA translation and protein transport to the plasma membrane, oocytes were cultured in MBS for 24-72 hours prior to electrophysiological measurements.

Electrophysiological Recordings: Oocytes were placed in a bath containing Frog Ringers Buffer (FRB; 118 mM NaCl, 1.9 mM KCl, 2 mM $CaCl_2$, 2.5 mM Tris-Cl, pH 7.4) with a flow rate of ca. 2 ml/min. Recordings were obtained under two electrode current clamp. Microelectrodes with an average resistance of 10-20 megaohms were used to measure membrane voltage or to deliver current. Impalements were signaled by an initial negative voltage deflection in excess of 20 mV which usually increased gradually until a stable level of -40 to -60 mV was reached. Only oocytes with a stable resting membrane potential in excess of -35 mV were included. In the current clamp mode, one second, 20 namp constant current pulses were delivered every four seconds to monitor changes in conductance. Glutamate receptor agonists were prepared in FRB and administered by direct application to the bath with a pipet. Amplified signals were recorded with a two channel strip chart recorder.

RESULTS

The native oocyte membrane is sensitive to certain neurotransmitters, but not all (Kusano et al., 1982). Under current clamp, control oocytes injected with H_2O responded to applications of acetylcholine (ACh) and adenosine (Aden) with depolarizations and hyperpolarizations, respectively.

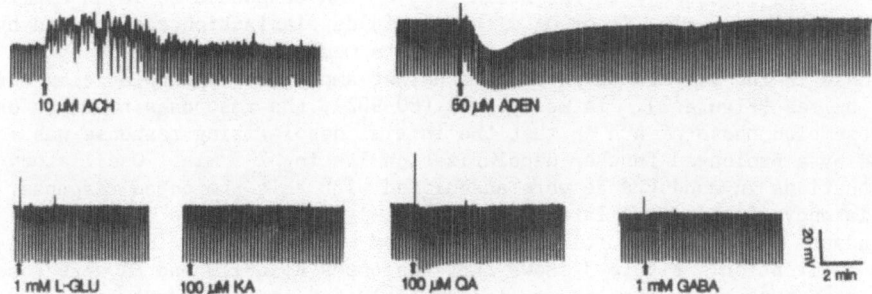

FIG. 1: Drug evoked changes in membrane potential and conductance for
         oocytes injected with water 48 hr prior to recording. 20 namp
         hyperpolarizing current pulses were administered in this, and sub-
         sequent experiments. Note that acetycholine (ACH) elicited a rapid
         depolarization followed by a prolonged depolarization associated
         with an oscillatory increase in conductance, whereas adenosine
         (ADEN) elicited a hyperpolarization and an increase in conductance.
         No responses were observed following application of excitatory
         amino acids or GABA. The resting membrane potential was -50 mV.

These intrinsic responses were used as a measure of oocyte viability in all
recordings. Control oocytes were unresponsive to the amino acid neurotransmitters L-glutamate (L-glu) and GABA and to the glutamate receptor agonists
kainate (KA) and quisqualate (QA) (Figure 1).

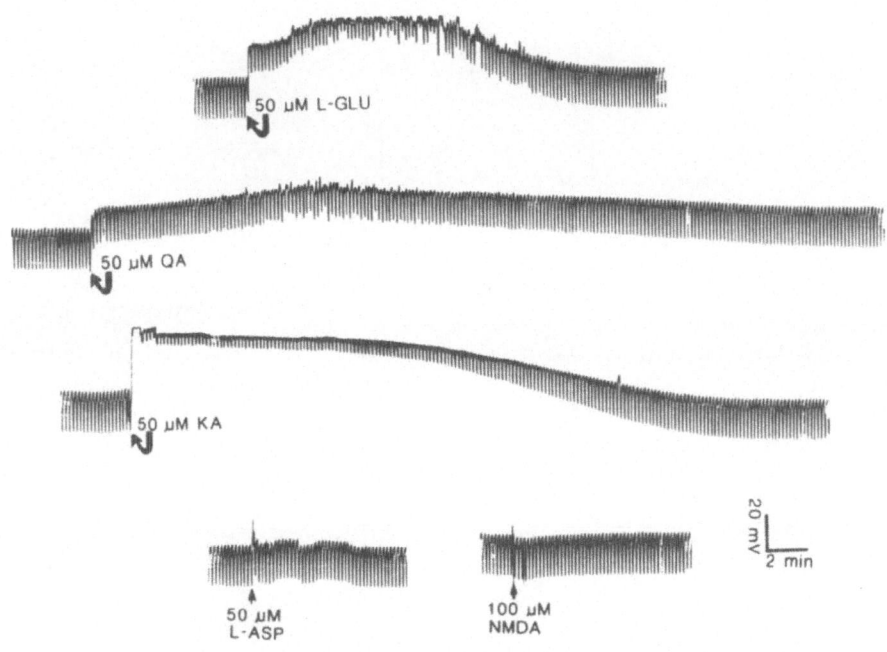

FIG. 2: Oocytes were injected with brain mRNA and their response to puta-
         tive excitatory amino acid agonists recorded. Note the delayed
         oscillation in conductance in response to glutamate (L-GLU) and
         quisqualate (QA) but not kainate (KA). L-aspartate (L-ASP) and
         N-methyl-d-aspartate (NMDA) were ineffective in this cell.

Oocytes injected with rat brain poly A$^+$ RNA responded to 50 μM applications of either L-glu, KA or QA with a rapid depolarization accompanied by an increase in the conductance of the oocyte membrane as indicated by a decrease in the voltage produced by constant amplitude hyperpolarizing current pulses (Figure 2). In most cases (80-90%), the responses to L-glu or QA resembled those of ACh in that the initial depolarizing response was followed by a prolonged further depolarization lasting 2-5 min. Oscillatory fluctuations in conductance were associated with this prolonged response. The latency of this oscillation, approximately 2-3 min., was not dosage dependent. The delayed oscillatory response was never seen following kainate application. Figure 3 shows that responses to L-glu and KA were dose dependent suggesting that these ligands activated specific, saturable receptors. Attempts to inhibit these agonist-induced responses with putative, excitatory amino acid antagonists such as glutamate diethyl ester, cis-2,3-piperidine dicarboxylic acid, or γ-D-glutamyl glycine proved unsuccessful.

In most experiments, the excitatory amino acids L-aspartate and N-methyl-d-aspartate (NMDA) were ineffective at concentrations up to 1mM. However, Figure 4 shows recording from one of two cells in which clear responses to 100 μM NMDA were seen following injection with fractions 23-30 rat brain mRNA (see below). The NMDA response was biphasic, consisting of an initial 5-10 mV depolarization which was followed within 30 seconds by a second, larger depolarization. Both phases of the response were accompanied by a marked increase in membrane conductance. The specificity of the NMDA response was ascertained by demonstrating that it was reversibly blocked by prior application of 50 μM 2-amino-5-phosphono-valeric acid (2-APV), while responses to L-glu and KA were unaffected. Moreover, the NMDA response was also selectively and reversibly blocked by adding 2 mM Mg^{++} to the bathing

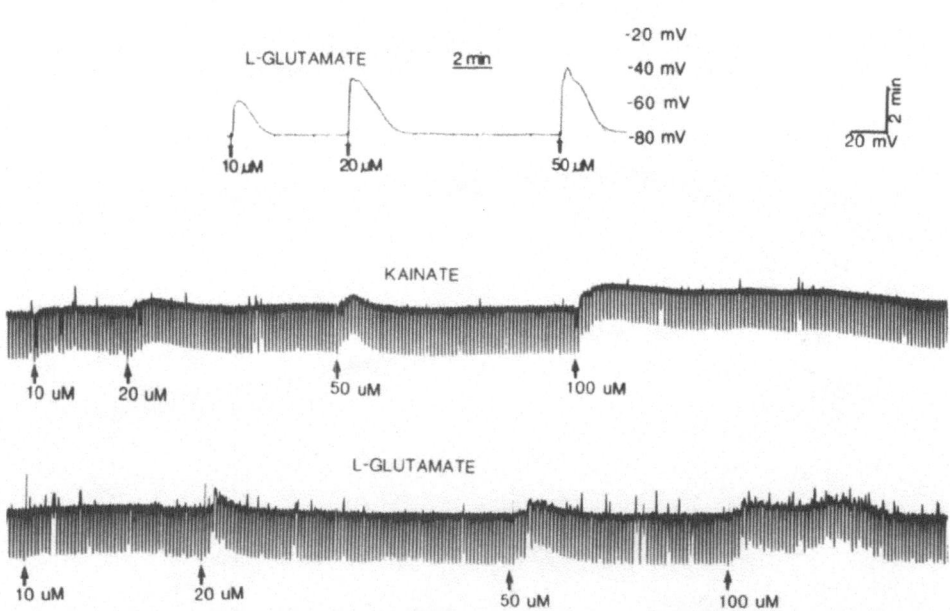

FIG. 3: Magnitude of depolarization and conductance change evoked by glutamate (L-GLU) and kainate (KA) was dose dependent. Single electrode recording of membrane potential following L-GLU application is shown in the top trace. The bottom recordings obtained from the same oocyte, illustrate the dose dependent responses to KA and L-GLU using current clamp analysis. Incubation time for translation of poly A$^+$ RNA (50 ng) was 48 hr.

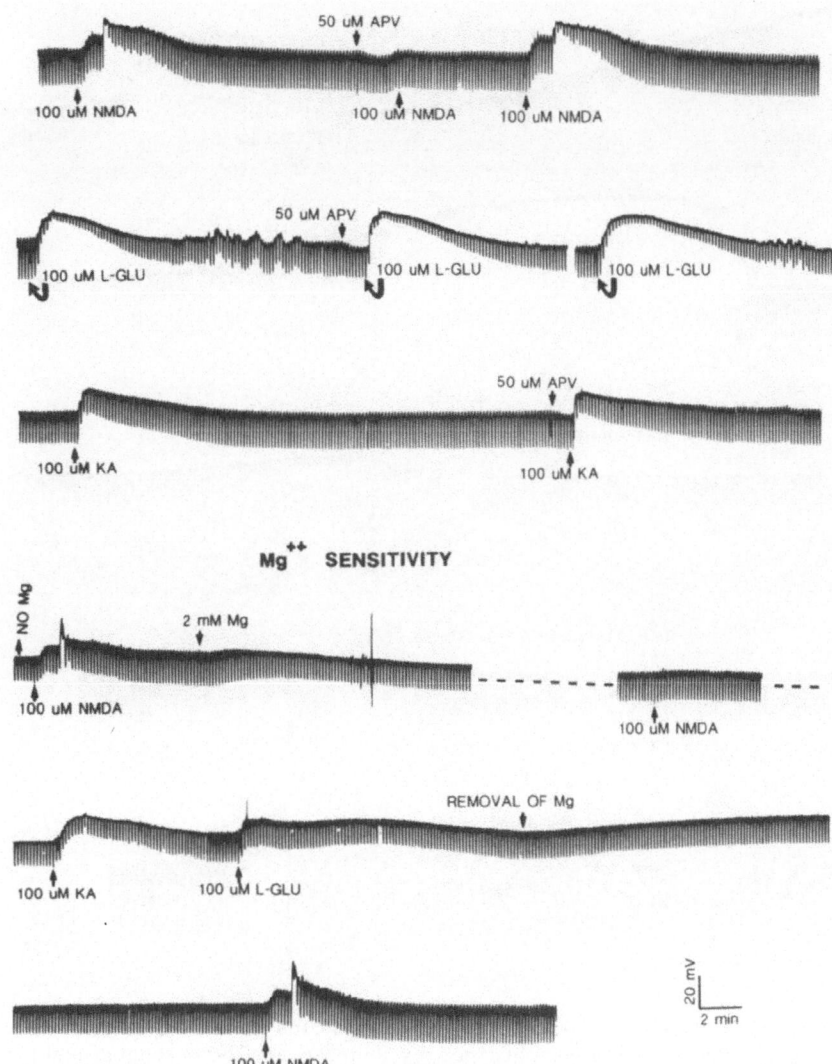

FIG. 4: Recordings from an oocyte, injected with fractions 23-30 of rat
brain mRNA, that is responsive to NMDA as well as glutamate (L-GLU)
and kainate (KA). The response characteristics for each agent are
different indicating that potentially three distinct receptor
sites or transduction systems are involved. Selective inhibition
of the NMDA response by either 2-amino-5-phosphono-valeric-acid
(2-APV) or Mg^{++} is the distinquishing feature of the NMDA subclass
of excitatory amino acid receptors. Incubation time for translation of poly A^+ RNA (40 ng) was 72 hr.

medium. Hereto, responses to other amino acids were unaffected.

Sucrose density fractionation of rat brain poly A^+ mRNA resulted in
specific mRNA species coding for L-glu and KA responses in fractions num-
bered 23-30 (>2 kb in length). Oocytes injected either with rat brain poly
A^+ RNA fractions 1-16 or 15-22 or with rat liver poly A^+ RNA were unresponsive to excitatory amino acid receptor agonists (Figure 5). The viability

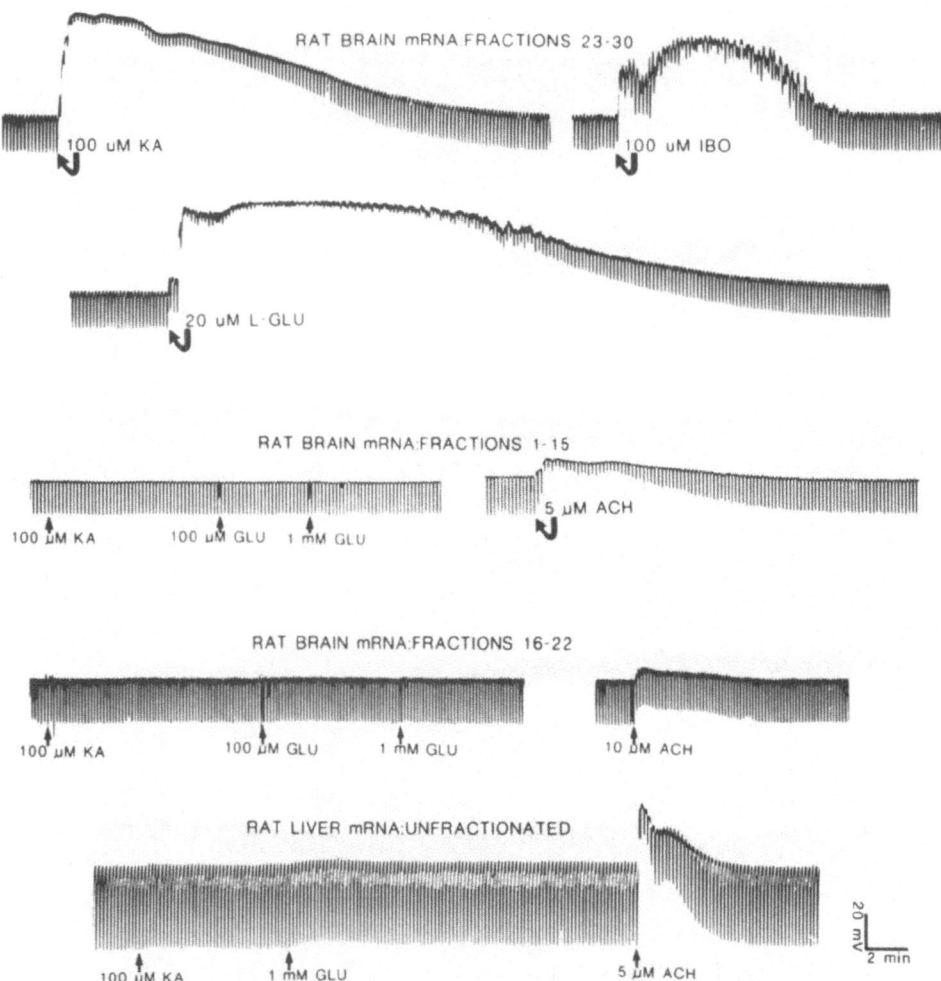

FIG. 5: Oocyte responses to excitatory amino acids following injection of
brain mRNA fractions obtained by sucrose gradient centrifugation.
Note that fractions 23-30 contained the appropriate mRNA for a
glutamate (L-GLU), kainate (KA) or ibotenate (IBO) response. Neither the other fractions from brain (1-15; 16-22) nor liver poly
A^+ RNA induced responsiveness to either L-GLU or KA (nor IBO; data
not included). The oocytes used in each experiment were viable
as demonstrated by their reponses to ACh.

of these nonresponding oocytes was confirmed by demonstrating an endogenous
response to ACh (Figure 5).

DISCUSSION

The present experiments confirm that Xenopus laevis oocytes, when
injected with specific fractions of rat brain poly A^+ RNA, can provide a
bioassay system that reproducibly translates this exogenous mRNA into proteins and incorporates the resulting proteins into neurotransmitter-sensi-
tive components of their membrane (Sumikawa et al., 1984). These new mem-

brane components can be monitored electrophysiologically by their dose dependent response to various excitatory amino acid agonists (L-glu, KA, QA, and IBO), indicating that they possess properties shared by functional glutamate receptors. A single, rapid depolarization was associated with KA application, whereas responses to L-glu or QA were composed of at least 2 distinct components, a short latency depolarization followed by a late onset oscillatory potential. Although it is not certain what accounts for the distinctive late onset oscillation seen following either L-glu or QA exposure, its latency suggests the involvement of a second messenger transduction system. The late-oscillation was also observed in native, H_2O-injected, oocytes in response to ACh application. According to Oron et al., 1985, the activation of muscarinic receptors results in an increase in inositol triphosphates causing a mobilization of intracellular Ca^{++} and activation of Cl^- channels. They attributed the oscillatory component to periodic, synchronous release of Ca^{++}. It is conceivable that the L-glu or QA receptors described here are coupled to a similar system. The fact that KA does not induce this oscillation may indicate that heterogenous mRNA species coding for distinct excitatory amino acid receptor subtypes exist. Alternatively, this may reflect differential processing of the same transcript, variations in post-translational processing of the receptor proteins, or variations in the manner in which the newly synthesized receptor proteins are inserted into the oocyte membrane and coupled to ion channel systems.

Discriminating between these possibilities was hampered by a lack of selective pharmacological tools for QA or KA binding sites. Although potent, selective antagonists do exist for NMDA sensitive binding sites, detectable expression of this subclass of receptors by the oocytes has proven difficult. In two cases, however, using poly A^+ RNA from fractions 23-30, definitive responses to NMDA were obtained. Selective, reversible blockade of NMDA elicited responses was achieved by the application of 2-APV or with the addition of Mg^{++} to the bath. These characteristic responses are similar to those seen with NMDA in the CNS where Mg^{++} is known to cause a voltage-dependent block of the ionophore, (Mayer et al., 1984; Nowak et al., 1984) and 2-APV is an inhibitor of ligand binding (Foster and Fagg, 1984). This raises the possibility that the NMDA binding site and its ionophore are part of the same complex similar to the nicotinic acetylcholine receptor (Noda et al., 1983). The reason NMDA responses are infrequently observed in the oocytes is unknown. It may relate to a low abundance of NMDA receptor mRNA and/or to a difficulty in posttranslational processing of this protein within the oocyte. NMDA-specific binding studies in injected oocytes may help to unravel this problem. In the two cases where an NMDA response was observed, it was stably maintained throughout the experiment and, it is therefore unlikely that desensitization is the reason it is rarely observed.

In conclusion, the oocyte bioassay can be used to elucidate the biophysical basis of receptor subtypes as well as to identify regions of these membrane proteins that are critical to receptor function. For example, the purification (Karlin, 1980; Conti-Tronconi et al., 1982) and cloning of the nicotinic acetylcholine receptor (Sumikawa et al., 1982; Noda et al., 1983; Claudio et al., 1983; Devillers-Thiery et al., 1983; Mishina et al., 1985) has demonstrated the essential sequences of this receptor protein that are required for ligand binding and ion transduction. Unfortunately, only a few of the many neurotransmitter receptors have been purified (Dixon et al., 1986; Kubo et al., 1986) and given the powerful molecular biological techniques available, we sought to evaluate whether these procedures might facilitate the isolation and characterization of the receptor(s) for the excitatory amino acid glutamate. Our results demonstrate that electrophysiological analyses of oocyte membrane properties after injection of exogenous rat brain poly A RNA is an appropriate bioassay for identifying a receptor clone(s) that encodes the mRNA for the glutamate receptor(s).

This approach may also prove feasible for other functional receptors.

ACKNOWLEDGMENTS

We would like to thank B. Wolfson, G. Christoph, and F. Baldino, Jr., for their discussions and comments; C. Molaison and A. Callahan for their technical assistance; and S. Vari for preparation of the manuscript. The present address of Dr. J. Gutnick is: Unit of Physiology, Faculty of Health Science, Ben-Gurion University, 84 105 Beersheva, Israel.

REFERENCES

Aviv, H. and Leder, P., Purification of biologically active globin messenger RNA by chromatography on oligothymidylic acid-cellulose, Proc. Natl. Acad. Sci. USA 69:1408-1412 (1972).

Baldino, F., Jr., Wolfson, B., Heinemann, U. and Gutnick, M. J., An n-methyl-d-aspartate (NMDA) receptor antagonist reduces bicuculline-induced depolarization shifts in neocortical explant cultures, Neuroscience Lett. 70:101-105 (1986).

Claudio, T., Ballivet, M., Patrick, J. and Heinemann, S., Nucleotide and deduced amino acid sequences of Torpedo californica acetylcholine receptor γ subunit, Proc. Natl. Acad. Sci. USA 80:1111-1115 (1983).

Conti-Tronconi, B. M., Hunkapiller, M. W., Lindstrom, J. M. and Raftery, M. A., Subunit structure of the acetylcholine receptor from Electrophorus electricus, Proc. Natl. Acad. Sci. USA 79:6489-6493 (1982).

Davis, L. G., Dibner, M. D. and Battey, J. F., Guanidine isothiocyanate preparation of total RNA. In: Basic Methods in Molecular Biology, Elsevier (New York), pp. 130-135 (1986).

Devillers-Thiery, Giraudat, A. J., Bentaboulet, M. and Changeux, J. P., Complete mRNA coding sequence of acetylcholine binding α-subunit of Torpedo marmorata acetylcholine receptor: a model for the transmembrane organization of the polypeptide chain, Proc. Natl. Acad. Sci. USA 80:2067-2071 (1983).

Dixon, R. A. F., Kobilka, B. K., Strader, D. J., Benovic, J. L., Dolhman, H. G., Frielle,1 T., Bolanowski, M. A., Bennett, C. D., Rands, E., Duhl, R. E., Mumford, R. A., Slater, E. E., Sigal, I. S., Caron, M. G., Lefkowitz, R. J. and Strader, C. D., Cloning of the gene and cDNA for mammalian β-adrenergic receptor and homology with rhodopsin, Nature 321:75-79 (1986).

Dumont, J. N., Oogenesis in Xenopus laevis (Daudin), J. Morphol. 136:153-180 (1972).

Foster, R. C. and Fagg, G. E., Acidic amino acid binding sites in mammalian neuronal membranes: Their characteristics and relationship to synaptic receptors, Brain Res. Rev. 7:103-164 (1984).

Gundersen, C. B., M ledi, R. and Parker, I., Messenger RNA from human brain induces drug - and voltage - operated channels in Xenopus oocytes, Nature 308:421-424 (1984).

Gurdon, J. B., Lane, C. D., Woodland, H. R. and Marbaix, G., Use of frog eggs and oocytes for the study of messenger mRNA and its translation in living cells, Nature 233:177-182 (1971).

Hablitz, J. J. and Langmoen, I. A., N-methyl-d-aspartate receptor antagonists reduce synaptic excitation in the hippocampus, J. Neurosci. 6: 102-106 (1986).

Houamed, K. M., Bilbe, G., Smart, T. G., Constante, A., Brown, D. A., Barnard, E. A. and Richards, B. M., Expression of functional GABA, glycine and glutamate receptors in Xenopus oocytes injected with rat brain mRNA, Nature 310:318-321 (1984).

Karlin, A., Molecular properties of nicotinie acetylcholine receptors. In: The Cell Surface and Neuronal Function, edited by C. W. Cotman, G.

Poste, and G. L. Nicolson, Amsterdam: Elsevier/North Holland Biomedical Press, pp. 191-260 (1980).

Kubo, T., Kuzuhiko F., Mikami, R., Maeda, A., Takahashi, H., Mishina, M., Haga, T., Haga, K., Ichiyama, A., Kangawa, K., Koyima, M., Matsuo, H., Hirose, T. and Numa, S., Cloning, sequencing and expression of complementary DNA encoding the muscarinic acetylcholine receptor, Nature 323: 411-416 (1986).

Kusano, K., Miledi, R. and Stinnakre, J., Cholinergic and catecholaminergic receptors in the Xenopus oocyte membrane, J. Physiol. 328:143-170 (1982).

Labarco, C. and Paigen, K., mRNA-directed synthesis of catalytically active mouse β-glucuronidase in Xenopus oocytes, Proc. Natl. Acad. Sci. USA 74:4462-4466 (1977).

Lynch, G. and Baudry, M., The biochemistry of memory: A new and specific hypothesis, Science 224:1057-1063 (1984).

Mayer, M. L., Westbrook, G. L. and Guthrie, P. B., Voltage-dependent block by Mg^{++} of NMDA responses in spinal cord neurones, Nature 309:261-263 (1984).

Mishina, M., Tobimatsu, T., Imoto, K., Tanaka, K., Fujita, Y., Fukuda, K., Kurasaki, M., Takahashi, H., Morimoto, Y., Hirose, T., Inayama, S., Takahashi, T., Kuno, M. and Numa, S., Location of functional regions of acetylcholine receptor α-subunit by site-directed mutagenesis, Nature 313:364-369 (1985).

Noda, M., Takahashi, H., Tanabe, T., Toyosato, M., Kibyotani, S., Furutani, Y., Hirose, T., Takashima, H., Inayama, S., Miyata, T. and Numa, S., Structural homology of torpedo californica acetylcholine receptor subunits, Nature 302:528-532 (1983).

Nowak, L., Bregestovski, P., Ascher, P., Herbet, A. and Prochiantz, A., Magnesium gates glutamate-activated channels in mouse central neurones, Nature 307:462-465 (1984).

Oron, Y., Dascal, N., Nadler, E. and Lupu, M., Inositol 1,4,5-trisphosphate mimics muscarinic response in Xenopus oocytes, Nature 313:141-143 (1985).

Parker, I., Sumikawa, K. and Miledi, R., Messenger RNA from bovine retina induces baumate and glycine receptors in Xenopus oocytes, Proc. R. Soc. Lond. B225:99-106 (1985).

Pure, E., Luster, A. D. and Unkeless, J. C., Cell surface expression of murine, rat, and human Fc receptors by Xenopus oocytes, J. Exp. Med. 160:606-611 (1984).

Robinson, R. R., Germain, R. N., McKean, D. J., Mescher, M. and Seidman, J. G., Extensive polymorphism surrounding the murine La αβ chain gene, J. Immun. 131:2025-2031 (1983).

Simmen, F. A., Schulz, T. A., Headon, D. R., Wright, D. A., Carpenter, G. and O'Malley, B. W., Translation in Xenopus oocytes of messenger RNA from A431 cells for human epidermal growth factor receptor proteins, DNA 3:393-399 (1984).

Soreq, H., Parvari, R. and Silman, I., Biosynthesis and secretion of catalytically active acetylcholinesterase in Xenopus oocytes microinjected with mRNA from rat brain and from Torpedo electric organ, Proc. Natl. Acad. Sci. USA 79:830 (1982).

Soreq, H., The biosynthesis of biologically active proteins in m-RNA microinjected Xenopus oocytes, CRC Critical Rev. in Biochem. 18:199-238 (1985).

Sumikawa, K., Houghton, M., Emtage, J. S., Richards, B. M. and Barnard, E. A., Active multi-subunit acetylcholine receptor assembly by translation of heterologous mRNA in Xenopus oocytes, Nature 292:862-864 (1981).

Simikawa, K., Parker, I. and Miledi, R., Partial purification and functional expression of brain mRNA coding for neurotransmitter receptors and voltage-operated channels, Proc. Natl. Acad. Sci. USA 81:7994-7998 (1984).

Sumikawa, K., Houghton, M., Smith, J. C., Belll, L., Richards, B. M. and
    Barnard, E. A., The molecular cloning and characterization of cDNA
    coding for the α subunit of the acetylcholine receptor, Nucl. Acid Res.
    10:5809-5822 (1982).
Wieloch, T., Hypoglycemia-induced neuronal damage prevented by an N-methyl-
    D-aspartate antagonist, Science 230:681-683 (1985).

TYROSINE AVAILABILITY: A PRESYNAPTIC FACTOR CONTROLLING CATECHOLAMINE RELEASE

J. D. Milner and R. J. Wurtman

Department of Brain and Cognitive Sciences
Massachusetts Institute of Technology
Cambridge, MA 02139

Considerable evidence has been available for many years that low doses of exogenous tryptophan, which elevate brain tryptophan but keep its levels within their normal range, can increase brain serotonin and 5-hydroxyindoleacetic acid concentrations (Wurtman and Fernstrom, 1976; Fernstrom and Wurtman, 1971). Moreover, food consumption had been shown some years ago to influence brain tryptophan, and thereby brain serotonin levels, with carbohydrate-rich meals increasing (Fernstrom and Wurtman, 1971) and protein meals decreasing (Fernstrom and Wurtman, 1972) serotonin synthesis, and it has been proposed that this coupling of food composition to serotonin synthesis allows serotonin-dependent behaviors to be affected by eating (Fernstrom and Wurtman, 1974; Young, 1983).

It was also found more than a decade ago, that consumption of supplemental tyrosine (Wurtman et al., 1974) or choline (Cohen and Wurtman, 1975; Haubrich et al., 1975) could affect the syntheses of their neurotransmitter products, the catecholamines and acetylcholine. However these responses were, until recently, poorly characterized, and were observed to be considerably less consistent than that of serotonin to supplemental tryptophan: Tyrosine administration only sometimes increased brain levels of catecholamine metabolites, and even then it failed to affect those of dopamine or norepinephrine; choline administration only sometimes affected brain acetylcholine levels (Blusztajn and Wurtman, 1983), and there was no acetylcholine metabolite to measure as evidence that more acetylcholine molecules were turning over. Moreover, no direct evidence was available that administration of tyrosine or choline could enhance either the release of their neurotransmitter, nor the postsynaptic responses to the transmitters.

This article summarizes evidence accumulated during the past decade that tyrosine availability can indeed affect catecholamine synthesis and release. It also comments on possible uses of supplemental tyrosine to amplify catecholamine-mediated neurotransmission, and thereby affect performance.

Early Studies Relating Tyrosine to Catecholamine Production

The failure of tyrosine administration to increase brain levels of its neurotransmitter products had led virtually all investigators to assume that tyrosine hydroxylase was fully saturated with its amino acid substrate in vivo, - this in spite of the fact that estimates of the enzyme's Km for

211

TABLE 1: Tyrosine Levels in Tissues and Fluids

Species	Tissue or Fluid	[Tyrosine]	Comments	Reference
Rat	Brain	47-65 nmol/g	Fasted	Fernstrom & Faller, 1978; Glaeser et al., 1983.
Rat	Brain	47-173 nmol/g	Non-fasted	Carlsson & Lindqvist, 1978; Fernstrom & Faller, 1978; Glaeser et al., 1983.
Rat	Plasma	70±80 µM	Adult	Brosnan et al., 1984; Chirigos et al., 1960.
Rat Fern-	Plasma	188±21 µM	Neonatal	Fernstrom & strom, 1981.
Rat Fern-	Plasma	233±2 µM	Fetal	Fernstrom & strom, 1981.
Dog Fern-	CSF	16±3 µM	-	Fernstrom & strom, 1981.
Primate	CSF	4-5 µM	Males, females	Young & Ervin, 1984.
Human	CSF	8-10 µM	Normal	Fernstrom & Fernstrom, 1981; Hagenfeldt et al., 1984; Perry et al., 1975.
Human	CSF	19-20 µM	Parkinsonian	Growdon et al., 1982.
Human	Plasma	50-60 µM	Normal, fasting	Hagenfeldt et al., 1984; Perry et al., 1975.

tyrosine in vitro [50 to 125 µM, depending upon whether tetrahydrobiopterin or a synthetic alternate was used as the cofactor (e.g., Morgenroth et al., 1975; Joh et al., 1978)] were not very much lower than whole-brain tyrosine levels [which vary between 100 and 200 μM, depending on the protein content of the meal most recently consumed (Table 1)]. Catecholamines were, however, known to be stored within multiple metabolic compartments, including some with slow turnover times; this raised the possibility that supplemental tyrosine might actually accelerate the synthesis of a particular "pool" of dopamine or norepinephrine, but that this pool constituted too small a fraction of the total catecholamine store to allow detection. Hence experiments were performed to determine whether changes in tyrosine levels might affect rates of catecholamine synthesis, as estimated from the accumulation of dihydroxphenylalanine (DOPA) in brains of animals pretreated with a decarboxylase inhibitor (Wurtman et al., 1974; Carlsson and Lindqvist, 1978). Under such conditions, DOPA accumulation was shown to be accelerated when brain tyrosine levels were increased (by giving rats tyrosine), and diminished when tyrosine was reduced [by giving rats other large neutral amino acids, like tryptophan, valine, or parachlorophenylalanine, which compete with tyrosine for passage across the blood-brain-barrier (Pardridge, 1971)]. Such observations provided estimates of the Km for tyrosine's hydroxylation in vivo and showed that this process could indeed be affected when tyrosine levels varied within their normal range. However, the use of a decarboxylase inhibitor - which also diminishes catecholamine synthesis in nerve terminals - rendered problematic the extrapolation of these data to physiologic states, since the drug probably also reduced both the end-product inhibition of tyrosine hydroxylase and the release of newly-formed catecholamine into synapses. What was needed was an experimental approach that allowed catecholamine synthesis to be estimated, in the presence of varying brain tyrosine concentrations, without concurrently disturbing that synthesis. When one such approach was tried - that of measuring dopamine metabolites in striata of animals given tyrosine - no effect of tyrosine was observed (Scally et al., 1977), [although tissue dopamine levels were subsequently shown to rise in striata and cortices of rats given large doses of tyrosine methyl-esther (Oishi and Szabo, 1984)]. However, if the animals were concurrently given haloperidol, a dopamine receptor antagonist that accelerates nigrostriatal firing (Bunney et al., 1973) and can lower tyrosine levels in striata (Westerink and Wirix, 1983), catecholamine production did exhibit precursor dependence: striatal levels of homovanillic acid (HVA) varied directly with those of brain tyrosine, while dopamine levels themselves remained constant (Scally et al., 1977). These observations were interpreted as indicating that a given catecholaminergic neuron might or might not be responsive to having more or less tyrosine, depending on its level of activity.

The ability of tyrosine supplementation to enhance the synthesis of catecholamines in, and their release from rapidly firing neurons, (but not from relatively quiescent cells) has since been affirmed using a variety of experimental manipulations (Table 2): Thus, tyrosine administration increases brain levels of the norepinephrine metabolite methoxy-hydroxy-phenylethylglycol sulfate (MHPG-SO4) in cold-stressed rats (Gibson and Wurtman, 1978) and in brains and brainstems of spontaneously hypertensive rats (SHR's) (Sved et al., 1979; Yamori et al., 1980) but not in those of control, normotensive animals, nor in SHR's given both tyrosine and another large neutral amino acid, valine (Sved et al., 1979), which competes with tyrosine for uptake into the brain (Pardridge, 1971). Increases in MHPG-SO4 after tyrosine treatment have also been observed in rats given yohimbine (Gibson, 1977), an alpha$_2$ antagonist, or in animals stressed by tail-shock (Reinstein et al., 1984). DOPA accumulation (after decarboxylase inhibition) is accelerated following tyrosine administration in straita of rats given gamma-butyrolactone (Sved and Fernstrom, 1981) (which blocks dopamine's release and the consequent activation of presynaptic inhibitory autoreceptors), and in the median eminence of animals that received exogenous prolactin (Sved, 1980) (which presumably activates the tuberoinfundibular dopaminergic neurons via a short feedback loop). Following a lesion

TABLE 2: Tyrosine Administration and Catecholamine Synthesis and Release

Tissue	Treatment	Biochemical Index	Tyrosine Effect	Reference
Striatum	Haloperidol	DOPAC, HVA	+60%	Scally et al., 1977.
Striatum, Limb. forb.	Haloperidol	DOPA	+15%	Carlsson and Lindqvist, 1978; Westerink and Wirix, 1983.
Striatum	NS tract lesions	DOPAC, HVA	+60%	Melamed et al., 1980.
Striatum	γ-butyrolactone	DOPA	+25%	Sved and Fernstrom, 1981.
Whole brain	Cold stress	MHPG-SO$_4$	+70%	Gibson and Wurtman, 1978.
Whole brain	SHRs	MHPG-SO$_4$	+40%	Sved et al., 1979.
Brainstem, forebrain	SHRs	MHPG-SO$_4$	+15%	Yamori et al., 1980.
Striatum, hypothalamus	Reserpine	DOPAC, HVA	+40%	Sved et al., 1979.
Whole brain	Yohimbine	MHPG-SO$_4$	+35%	Gibson, 1977.
Med. eminence	Prolactin	DOPA	+30%	Sved, 1980.
Hippocampus, hypothalamus	Tail shock	MHPG-SO$_4$	+40%	Reinstein et al., 1984.
Whole brain	Amfonelic acid, spiperone	DOPAC	+30%	Fuller and Snoddy, 1980.

that destroys about 80% of the nigrostriatal tract unilaterally [and thus accelerates the firing of the surviving neurons (Agid et al., 1973)] tyrosine administration increases dopamine release on the lesioned side, as estimated from the ratios of dihydorxyphenylacetic acid (DOPAC) or HVA to dopamine, or to tyrosine hydroxylase activity, but fails to affect either index on the intact side (Melamed et al., 1980). The tyrosine effect is, once again, blocked by valine, and is unassociated with changes in dopamine levels. Tyrosine administration also increases brain levels of dopamine metabolites in animals pretreated with reserpine (Sved et al., 1979), amfonelic acid, or spiperone (Fuller and Snoddy, 1980), all of which, like haloperidol, are thought to accelerate nigrostriatal firing; in prefrontal and cingulate cortex a low dose of tyrosine also increaes dopamine levels in animals not concurrently receiving drug treatments (Tam and Roth, 1984). Tyrosine increases levels of dopamine metabolites in light-activated rat retinas in vivo (Gibson et al., 1983), but not when animals are in darkness.

Direct evidence that physiologic variations in tyrosine availability can affect dopamine release has recently been obtained using an experimental system in which superfused slices of rat striatum are subjected to electrical pulses (20 Hz, 2ms) of varying train length; the amount of endogenous dopamine released into the medium is correlated with its tyrosine concentration.

Tissues are subjected to two trains of 600 or 1800 pulses, each lasting 30 or 90 seconds respectively, and separated by 30 minutes; dopamine release during the second (S2) period is expressed as a decimal fraction of the amount released during the initial (S1) period. When slices are superfused with Krebs bicarbonate buffer (which lacks tyrosine or any other amino acid), the S2/S1 ratio is 0.75-0.80, depending on the number of pulses (that is, dopamine release during S2 declined by 20-25%). A tyrosine concentration in the superfusate of at least 20 μM is needed to maintain an S2/S1 ratio of unity in slices stimulated for 30 seconds, while at least 40 μM is needed in tissues stimulated for 90 seconds. Tissues that have been stimulated while superfused with the tyrosine-free buffer display major reductions in tyrosine content (up to 50%), as well as in dopamine itself (25%) (Milner and Wurtman, 1984; Milner and Wurtman, in press).

Since dopaminergic terminals comprise only a small percentage of the total cellular mass of the striatum, this major depletion of striatal tyrosine suggests either that the amino acid becomes depleted within non-catecholaminergic cells, as well as within these terminals, or that most of the tyrosine normally present in the striatum is confined within dopaminergic terminals. In experiments designed to examine the latter possibility, these terminals were destroyed unilaterally by injecting the neurotoxin 6-hydroxydopamine into the substantia nigra. Even though the dopamine content of the ipsilateral striatum was depleted by more than 95%, its tyrosine levels were unchanged, indicating that striatal tyrosine is not preferentially localized within the dopaminergic terminals, and suggesting that the tyrosine depletion that occurred when the superfused slices were stimulated reflected mobilization of the amino acid from non-dopaminergic as well as from dopaminergic cells.

In vivo, dopaminergic nerve terminals are, of course, perfused not with a tyrosine-free solution but with tyrosine-containing blood; moreover, as discussed below, circulating tyrosine is able to enter the brain, its entry catalyzed by a facilitated diffusion system that it shares with other large, neutral amino acids (Pardridge, 1971). Hence it would not be expected that even the prolonged conversion of tyrosine to dopamine would cause its depletion, at least to the extent seen in vitro. However, the rate at which tyrosine diffuses from the plasma into dopaminergic nerve terminals is retarded both by the amino acid's limited water-solubility and by competition between it and other circulating amino acids for attachment to the blood-brain barrier transport site (Pardridge, 1971) and to neuronal membranes (Guroff et al., 1961). Hence the possibility remains that tyrosine in nerve terminals may fall, after prolonged neuronal firing, to levels sufficient to slow catecholamine synthesis. In that circumstance the ability of supplemental tyrosine to enhance catecholamine synthesis would be explained not so much by its ability to increase the substrate-saturation of tyrosine hydroxylase but by blocking the decrease that would otherwise occur. The inability of striatal dopaminergic terminals to sustain transmitter output without exogenous tyrosine contrasts with the ability of cholinergic neurons in the same tissue to continue making their neurotransmitter even when exogenous choline is lacking. Cholinergic terminals continue to release unchanged amounts of acetylcholine, even after 30 minutes of continuous stimulation, when superfused with the Krebs solution (which also, of course, lacks choline) (Maire and Wurtman, in press). The source of choline for this acetylcholine synthesis is probably a "reservoir" in the form of membrane phosphatidylcholine (PC) (Wurtman et al. 1985); the PC is hydrolyzed to free choline, which is then released into the extracellular space and then taken back up into the cholinergic terminal for acetylation. Apparently the protein in nerve terminals is unable to serve in an analogous manner as a reservoir for tyrosine.

Tyrosine and Sympatho-Adrenal Cells

Tyrosine availability has also been shown to affect catecholamine synthesis in peripheral tissues. Its acute or chronic (8 days) administration to cold-exposed rats caused dose-related increases in urinary norepinephrine and epinephrine (Alonso et al., 1980); valine or leucine failed to elicit similar responses, and, when administered with tyrosine, blocked the increases (Agharanya and Wurtman, 1982). The increase in urinary epinephrine was also blocked by bilateral adrenalectomy (Agharanya and Wurtman, 1982). In contrast, the rise in urinary norepinephrine was amplified in rats whose sympathetic terminals had been partially destroyed by prior administration of 6-hydroxydopamine (Agharanya and Wurtman, 1982), suggesting that an increase in their firing rates had occurred, rendering them more responsive to the amino acid. Tyrosine administration also caused increases in urinary dopamine; these were unaffected by adrenalectomy or 6-hydroxydopamine. Administration of oral tyrosine (33 mg/kg prior to each meal) to human subjects also increased urinary levels of the three catecholamines (Agharanya et al., 1981). After a single dose of the amino acid (100 or 150 mg/kg), urinary levels of the catecholamines and their principal metabolites all increased, with time-courses that paralleled the rise in blood tyrosine levels (Alonso et al., 1982). These observations were interpreted as suggesting that, in humans, both central and sympathoadrenal catecholamine synthesis are precursor-responsive.

Enhancement of sympathoadrenal catecholamine synthesis underlies tyrosine's ability to restore blood pressure in rats in hemorrhagic shock (Conlay et al., 1981). This response is blocked by bilateral adrenalectomy, performed immediately prior to testing (Conlay et al., 1981), or by pretreatment with carbidopa or phentolamine (Conlay et al., 1985), and is not simulated by other large neutral amino acids (like valine or leucine) (Conlay et al., 1985). [Tyrosine administration to hypotensive rats also elevates levels of epinephrine in the adrenal medulla, and of norepinephrine in the spleen (Conlay et al., 1985)]. That tyrosine's pressor effect is not mediated by its conversion to the sympathomimetic amine tyramine was demonstrated by its lack of effect (unlike tyramine) in rats receiving a ganglionic blocker (hexamethonium); its persistent pressor activity (again unlike tyramine) in reserpinized hypotensive rats; and its failure to elevate plasma tyramine levels when raising blood pressure (Conlay et al., 1984).

The fact that a given dose of tyrosine can raise blood pressure in hypotensive animals, lower it in spontaneously-hypertensive rats, and have little or no effect in normotensive animals (or people) has been interpreted as resulting from tyrosine's ability to enhance catecholamine synthesis only in neurons (or chromaffin cells) undergoing prolonged physiological activity: in hypotensive rats, the sympathoadrenal cells are active, and thus tyrosine-responsive; hence giving the amino acid potentiates their release of catecholamines, restoring blood pressure. In SHR, these peripheral neurons may be negatively quiescent, suppressed by the now-active brain-stem noradrenergic neurons that control sympathetic outflow (Philippu et al., 1980); hence tyrosine administration, by enhancing norepinephrine synthesis within the brain stem, further reduces sympathetic activity, causing blood pressure to fall. The possible utility of oral tyrosine in treating hypertension is currently under investigation (Bossy et al., 1983); its use in treating shock (which would require parenteral administration) is hampered by its very poor water-solubility.

Plasma and Tissue Tyrosine

Published data on the levels of tyrosine in various tissues and body fluids are summarized in Table 1 (Brosnan et al., 1984; Fernstrom and Fernstrom, 1981; Young and Ervin, 1984; Hagenfeldt et al., 1984; Perry et al., 1975; Growdon et al., 1982; Chirigos et al., 1960). Tyrosine recently taken up from the extracellular space may be used preferentially for dopamine synthesis (Kapatos and Zigmond, 1977) (i.e., in contrast to tyrosine already present in the cytoplasm). This would allow acute changes in the plasma amino acid pattern to have relatively more of an effect on the synthesis and release of this neurotransmitter. The constituents of the plasma that affect brain tyrosine are not only the amino acid itself, but also the other large, neutral amino acids (LNAA, primarily tryptophan, phenylalanine, valine, isoleucine, and leucine) that compete with tyrosine for entry into the brain (Pardridge, 1971; Chirigos et al., 1960). Brain tyrosine levels sometimes correlate only poorly with plasma tyrosine concentration (e.g., after a protein-rich meal) but apparently are always well-correlated with the "plasma tyrosine ratio" (to other LNAA) (Fernstrom and Faller, 1978). Hence, an increase in plasma levels of the branched-chain amino acids - such as would result from insulin deficiency or insensitivity - could decrease both the transport of tyrosine across the blood-brain-barrier, and its subsequent conversion to catecholamines. Both dietary (Fernstrom and Faller, 1978; Glaeser et al., 1983) and pharmacological (Ablett et al., 1984) manipulations of plasma LNAA levels have been shown to cause the predicted changes in brain tyrosine levels. Probably the most effective way to increase brain tyrosine is to administer the amino acid orally, along with sufficient carbohydrate to elicit insulin secretion and, thereby, to lower plasma levels of the other LNAA (Mauron and Wurtman, 1982).

Recent studies have obtained evidence that a variety of severe stresses [e.g., immobilization (Milakofsky et al., 1985); hemorrhage (Conlay et al., 1985)] can selectively increase the plasma tyrosine ratio. This may reflect a shunting of blood away from the liver, where tyrosine is metabolized, and could serve to increase the amino acid's availability.

Firing Frequency, Tyrosine-Dependence, and Allosteric Changes in Tyrosine Hydroxylase

An additional mechanism by which increased activity can cause catecholaminergic neurons to become tyrosine - sensitive involves kinetic changes in tyrosine hydroxylase that result from its phosphorylation. This process is accelerated when neuronal activity increases. Phosphorylation of tyrosine hydroxylase can be catalyzed by any of several protein kinases, each of which acts selectively on particular amino acids in the enzyme protein (Niggli et al., 1984; Ames et al., 1978).

The enzyme activation occurring when neuronal activity increases (i.e., after in vitro depolarization of rat striatum) is dependent on calcium and calmodulin (El Mestikaway et al., 1983). It increases the enzyme's activity, without changing its affinity for tyrosine or the tetrahydrobiopterin cofactor, nor its susceptibility to end-product inhibition. A different, cAMP-dependent protein kinase can also phosphorylate tyrosine hydroxylase, increasing its affinity for its cofactor (but not tyrosine), and also perhaps decreasing its susceptibility to end-product inhibition (Lovenberg et al., 1975; Harris et al., 1972). Recent evidence suggests that in striatal dopaminergic nerve terminals, the dopamine released following depolarization acts via presynaptic autoreceptors to decrease local cAMP levels; this slows

TABLE 3: Effects of Tyrosine on Catecholamine-Mediated Phenomena

Animal model	Physiological parameter	Tyrosine effect	Reference
Rat	Open field behavior	Reversal of stress-induced inhibition	Lehnert et al., 1984.
SHRs	Blood pressure	Hypotensive	Sved et al., 1979; Yamori et al., 1980; Bossy et al., 1983.
Rat (hemorrhaged)	Blood pressure	Hypertensive	Conlay et al., 1981.
Dog	Ventricular arrhythmia	Preventative	Scott et al., 1981.
Rat	Renal hypertension	Hypotensive	Breshnahan et al., 1980.
Aged, anestrous female rats	Estrous cycling	Restored	Linnoila and Cooper, 1976.
Aged mice	Motor activity	Increased	Thurmond and Brown, 1984.
Mice	Swim test immobility	Ameliorated	Gibson et al., 1982.
	Open field behavior	Increased activity	

the phosphorylation of the tyrosine hydroxylase and decreases its activity (El Mestikaway and Hamon, 1985). Hence the physiologic role of the cAMP-dependent protein kinase in striatum may be to suppress tyrosine's hydroxylation in response to prolonged dopamine release. In contrast, the calcium-calmodulin-dependent protein kinase is activated when voltage-gated calcium channels open during membrane depolarization; catecholamine formation thereupon depends on the extent to which the tyrosine hydroxylase happens to be saturated with tyrosine. This, in turn, will depend on the enzyme's Km for tyrosine (which apparently does not change with phosphorylation), and on tyrosine levels within the nerve terminal. The latter have not been measured, but presumably they bear a relationship to whole-brain tyrosine levels (which, in turn, vary with the plasma tyrosine ratio, as discussed above).

Physiological Consequence of Tyrosine Administration

If tyrosine availability does indeed affect brain catecholamine synthesis and release, it should also be expected that it will also influence various behaviors and physiological processes that involve catecholaminergic neurotransmission. Some publications demonstrating such effects are described in Table 3 (Lehnert et al., 1984; Scott et al., 1981; Breshnahan et al., 1980; Linnoila and Cooper, 1976; Thurmond and Brown, 1984; Gibson et al., 1982). One such catecholamine-dependent process is the control of blood pressure, which is _elevated_ by the release of norepinephrine or epinephrine from sympathoadrenal cells, and either decreased or increased by norepinephrine release within the central nervous system, depending on the locus of this release (Philippu et al., 1980). In SHR's, tyrosine injection causes a marked fall in blood pressure, which is blocked by co-administration of other LNAA's (Sved et al., 1979). Under these conditions, brainstem levels of the major noradrenaline metabolite MHPG-SO4 are elevated. Similar findings are obtained when tyrosine is injected into the lateral ventricles of these animals (Yamori et al., 1980). These observations indicate a central mechanism for tyrosine's antihypertensive effect, [consistent, for example, with enhanced noradrenaline release from locus coeruleus neurons). Tyrosine administration to stressed rats also causes behavioral effects: when given via the diet prior to stress or by injection thereafter, it blocks the stress-induced fall in regional brain NE levels and amplifies the increase in MHPG-SO4; it also blocks such post-stress behavioral changes as diminished locomotor activity and a diminished tendency of the animal to manifest its "curiosity" by standing on its hind legs or poking its nose into a hole (Reinstein et al., 1984; Lehnert et al., 1984). Attempts to determine whether tyrosine may also affect behavior and performance in healthy humans are in their infancy. However, given the extreme non-toxicity of the amino acid, any positive effects that tyrosine is found to exert will almost certainly be useful.

ACKNOWLEDGMENTS

The studies described in this report that were performed in the author's laboratories were supported in part by grants from the National Aeronautics and Space Administration, the U. S. Air Force, and the Center for Brain Sciences and Metabolism Charitable Trust. The manuscript has been modified, with permission, from an article published in Biochemical Pharmacology.

REFERENCES

Ablett, R. F., MacMillan, M., Sole, M. J., Toal, C. B. and Anderson, G. H., J. Nutr. 114:835 (1984).

Agharanya, J. C., Alonso, R. and Wurtman, R. J., Am. J. Clin. Nutr. 34:82 (1981).

Agharanya, J. C. and Wurtman, R. J., Biochem. Pharmacol. 31:3577 (1982).

Agharanya, J. C. and Wurtman, R. J., Life Sci. 30:739 (1982).

Agid, Y., Javoy, F. and Glowinski, J., Nature (New Biol.) 245:150 (1973).

Alonso, R., Agharnaya, J. C. and Wurtman, R. J., J. Neur. Trans. 49:31 (1980).

Alonso, R., Gibson, C. J., Wurtman, R. J., Agharanya, J. C. and Prieto, L., Biol. Psychiatr. 17:781 (1982).

Ames, M. M., Lerner, P. and Lovenberg, W., J. Biol. Chem. 253:27 (1978).

Blusztajn, J. K. and Wurtman, R. J., Science 221:614 (1983).

Bossy, J., Guidoux, R., Milon, H. and Wurzner, H. P., 2. Ernahrungswiss 22: 45 (1983).

Breshnahan, M. R., Hatzinikolaou, P., Brunner, H. R. and Gauras, H., Am. J. Physiol. 239:H201 (1980).

Brosnan, J. T., Forsey, R. G. and Brosnan, M. E., Am. J. Physiol. 247:c450 (1984).

Bunney, B. S., Walters, J. R., Roth, R. H. and Aghajanian, G. K., J. Pharmacol. Exp. Ther. 185:560 (1973).

Carlsson, A. and Lindqvist, M., Naunyn Schmiedeberg's Arch. Pharmacol. 303: 157 (1978).

Chirigos, M., Greengard, P. and Udenfriend, S., J. Biol. Chem. 235:2075 (1960).

Cohen, E. L. and Wurtman, R. J., Life Sci. 16:1095 (1975).

Conlay, L., Maher, T. J. and Wurtman, R. J., Fed. Proc. 44:1577 (1985).

Conlay, L. A., Maher, T. J. and Wurtman, R. J., Brain Res. 333:81 (1985).

Conlay, L. A., Maher, T. J. and Wurtman, R. J., Life Sci. 35:1207 (1984).

Conlay, L. A., Maher, T. J. and Wurtman, R. J., Science 212:559 (1981).

El Mestikaway, S., Glowinski, J. and Hamon, M., Nature 302:830 (1983).

El Mestikaway, S. and Hamon, M., J. Neurochem. 44 (Suppl.), SIIID (1985).

Fernstrom, J. D. and Faller, D. V., J. Neurochem. 30:1531 (1978).

Fernstrom, J. D. and Wurtman, R. J., Science 173:149 (1971).

Fernstrom, J. D. and Wurtman, R. J., Science 174:1023 (1971).

Fernstrom, J. D. and Wurtman, R. J., Science 178:414 (1972).

Fernstrom, J. D. and Wurtman, R. J., Scient. Am. 230:84 (1974).

Fernstrom, M. H. and Fernstrom, J. D., Life Sci. 29:2119 (1981).

Fuller, R. W. and Snoddy, H. D., Neuroendocrinology 31:96 (1980).

Gibson, C. J., PhD Thesis, Massachusetts Institute of Technology (1977).

Gibson, C. J., Deikel, S. M., Young, S. N. and Binik, Y. M., Psychopharmacology 76:118 (1982).

Gibson, C. J., Watkins, C. J. and Wurtman, R. J., J. Neur. Trans. 56:153 (1983).

Gibson, C. J. and Wurtman, R. J., Life Sci. 22:1399 (1978).

Glaeser, B. S., Maher, T. J. and Wurtman, R. J., J. Neurochem. 41:1016 (1983).

Growdon, J. H., Melamed, E., Logue, M., Hefti, F. and Wurtman, R. J., Life Sci. 30:827 (1982).

Guroff, G., King, W. and Udenfriend, S., J. Biol. Chem. 236:1773 (1961).

Hagenfeldt, L., Bjerkenstedt, L., Edman, G., Sedvall, G. and Weisel, F. A., J. Neurochem. 42:833 (1984).

Harris, J. E., Baldessarin, R. J., Morgenroth, V. H. and Roth, R. H., Proc. Natl. Acad. Sci. 72:789 (1972).

Haubrich, D. R., Wang, P. F., Herman, R. L. and Clody, D. E., Life Sci. 17:975 (1975).

Joh, H., Park, D. H. and Reis, D. J., Proc. Natl. Acad. Sci. 10:4744 (1978).

Kapatos, G. and Zigmond, M., J. Neurochem. 28:1109 (1977).

Lehnert, H. R., Reinstein, D. K., Strowbridge, B. and Wurtman, R. J., Brain Res. 303:215 (1984).

Linnoila, M. and Cooper, R. L., J. Pharmacol. Exp. Ther. 199:477 (1976).

Lovenberg, W., Bruckwick, E. A. and Hanbauer, I., Proc. Natl. Acad. Sci. 72:2955 (1975).

Maire, J.-C. and Wurtman, R. J., J. Physiol. (Paris) (in press).

Mauron, C. and Wurtman, R. J., J. Neur. Trans. 55:317 (1982).

Melamed, E., Hefti, F. and Wurtman, R. J., Proc. Natl. Acad. Sci. 77:4305 (1980).

Milakofsky, L., Hare, T. A., Miller, J. M. and Vogel, W. H., Life Sci. 36:753 (1985).

Milner, J. D. and Wurtman, R. J., Brain Res. 301:139 (1984).

Milner, J. D. and Wurtman, R. J., Neurosci. Lett. (in press).

Morgenroth, V. H., Boadle-Biber, M. C. and Roth, R. H., Mol. Pharmacol. 11:27 (1975).

Niggli, V., Knight, D. E., Baker, P. F., Vigny, A. and Henry, J.-P., J. Neurochem. 43:646 (1984).

Oishi, T. and Szabo, S., J. Neurochem. 42:894 (1984).

Pardridge, W. M., Am. J. Physiol. 221:1629 (1971).

Perry, T. L., Hansen, S. and Kennedy, J., J. Neurochem. 24:587 (1975).

Philippu, A., Dietl, H. and Sinha, J. N., Naunyn-Schmiedeberg's Arch. Pharmacol. 310:237 (1980).

Reinstein, D. K., Lehnert, H. and Wurtman, R. J., Life Sci. 34:2225 (1984).

Scally, M. C., Ulus, I. H. and Wurtman, R. J., J. Neur. Trans. 43:103 (1977).

Scott, N. A., DeSilva, R. A., Lown, B. and Wurtman, R. J., Science 211:727 (1981).

Sved, A., Fernstrom, J. D. and Wurtman, R. J., <u>Life Sci</u>. 25:1293 (1979).

Sved, A. F., PhD Thesis, Massachusetts Institute of Technology (1980).

Sved, A. F. and Fernstrom, J. D., <u>Life Sci</u>. 29:743 (1981).

Sved, A. F., Fernstrom, J. D. and Wurtman, R. J., <u>Proc. Natl. Acad. Sci</u>. 76: 3511 (1979).

Tam, S. Y. and Roth, R. H., <u>Neurosci. Abstr</u>. 10:881 (1984).

Thurmond, J. B. and Brown, J. W., <u>Brain Res</u>. 296:93 (1984).

Westerink, B. H. C. and Wirix, E. J., <u>Neurochem</u>. 40:758 (1983).

Wurtman, R. J., Blusztajn, J. K. and Maire, J.-C., <u>Neurochem. Int</u>. 7:369 (1985).

Wurtman, R. J. and Fernstrom, J. D., <u>Biochem. Pharmacol</u>. 25:1691 (1976).

Wurtman, R. J. and Larin, F., Mostafapour, S. and Fernstrom, J. D., <u>Science</u> 185:183 (1974).

Yamori, Y., Fujiwara, M., Horie, R. and Lovenberg, W., <u>Eur. J. Pharmacol</u>. 68:201 (1980).

Young, S., in <u>Handbook of Neurochemistry</u>, (Ed. A. Lajtha), Vol. 3, 559, Plenum Publishing Corporation (1983).

Young, S. N. and Ervin, F. R., <u>J. Neurochem</u>. 42:1570 (1984).

MOLECULAR MECHANISMS CONTROLLING NOREPINEPHRINE-MEDIATED

RELEASE OF SEROTONIN FROM RAT PINEAL GLANDS

Richard F. Walker[1,2] and Vincent J. Aloyo[2]

[1]Smith Kline and French Laboratories, Department of Reproductive and Developmental Toxicology, L-64, Philadelphia, PA and [2]Medical College of Pennsylvania, Philadelphia, PA

SUMMARY

This study describes various elements of the mechanism controlling norepinephrine (NE)-mediated release of serotonin (5HT) from rat pineal glands. After radiolabelling the endogenous pool of pineal 5HT with ^{3}H-5HT, individual pineal glands were exposed to depolarizing buffers or those containing NE. Although ^{3}H-5HT was not released by 50mM potassium, efflux of the indoleamine was increased by NE. Alpha-adrenergic receptors mediate the effects of NE as indicated by the fact that phenylephrine but not isoproterenol, a beta receptor agonist, also enhanced ^{3}H-5HT release. This hypothesis is supported further by the fact that prazosin and phentolamine (alpha-antagonists) but not sotolal (beta-antagonist), inhibited the stimulatory effects of NE on 5HT release. In order to determine the intracellular second messenger involved in the 5HT release process, pineals were incubated with 8-bromo cAMP or the phorbol ester, PMA. PMA simulated the effects of NE and phenylephrine on ^{3}H-5HT efflux, while cAMP had no effect. Furthermore, calcium-, phospholipid-dependent protein kinase activities in pineal homogenates were responsive to NE. These findings suggest that 5HT secretion from rat pinealocytes occurs rapidly in response to NE signals that act through alpha-adrenergic receptors in concert with phospholipid dependent protein kinase(s). These molecular processes are different from those involved in melatonin metabolism and may represent a general mechanism for regulating 5HT release in the brain.

INTRODUCTION

Synaptic release of chemical transmitters is an event common to most neurons and serves as the basis for their contribution to brain function. Transmitter release is preceded by its synthesis, and both functions are regulated by a variety of extrinsic factors that play upon the neuron. Generally, these extrinsic factors are neurochemicals that fall into two basic functional categories. One category provides the primary stimulus to initiate and/or sustain a sequence of molecular events that ultimately generate the neuronal signal. A generalized description of this process begins with the activation of receptors on target neurons by chemical substances that are often different from the one(s) to be released. Receptor activation is coupled with the intracellular production of regulatory molecules that then alter enzyme activity. Through this action, second messengers indirectly catalyse metabolic reactions that ultimately control

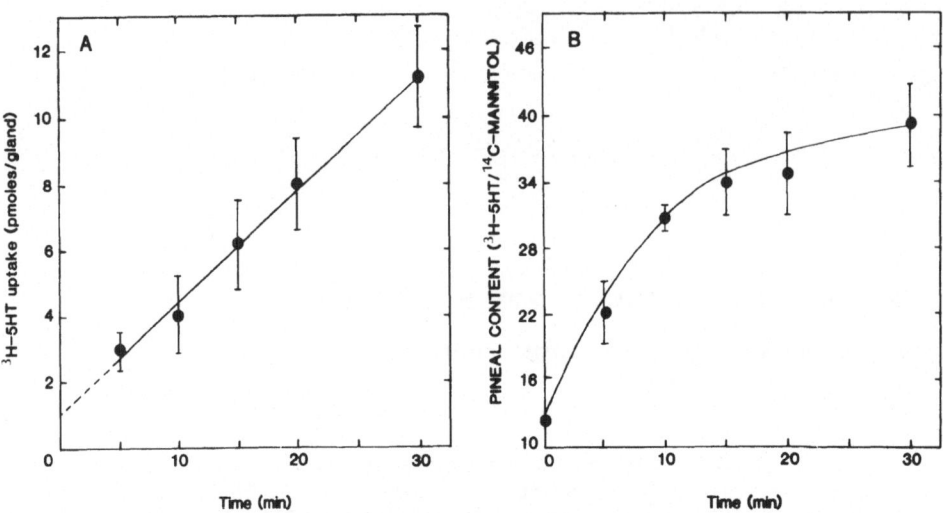

FIG. 1: Time Course for ^{3}H-5HT Accumulation by Rat Pineal Glands. Each
point represents the mean (± S.D.) for 3 pineal glands incubated in
10 μM ^{3}H-5HT and 424 μM ^{14}C-mannitol. (A) The net uptake of ^{3}H-5HT
per pineal gland was linear (r = 0.94) with significant increases
(p<0.001) occurring as a function of time. (B) The ratio of ^{3}H
to ^{14}C in media (time 0) is compared with those occurring in the
glands with increasing incubation time.

synthesis and/or release of the neurotransmitter.

The second category of extrinsic factors that influence neuronal function do so by modulating the responses initiated by the primary signal and so
attenuate or potentiate neurotransmitter output. This general plan is widely
recognized and elucidation of the precise mechanism(s) regulating different
types of neurons is a contemporary problem in neurobiology. The purpose of
this report is to present the results of a series of experiments that focused
upon events related to factors falling into the first category, i. e., the
primary stimulus for transmitter release. Specifically, the experiments were
designed to explore the cellular and molecular mechanisms controlling serotonin release in response to noradrenergic stimuli.

The Rat Pineal Gland as a Model for Studying Mechanisms of 5HT Secretion

Monamines are potent psychoneurogenic agents that participate in many
CNS activities. Of these molecules, 5HT is phylogenetically ancient, ubiquitous in distribution and it mediates a great number of diverse functions of
the brain (9). Thus, by understanding how serotoninergic neurons function,
it may be possible to gain insight into how a large part of the brain is
organized. Although 5HT-ergic projections are widely distributed throughout
the neuroaxis, their cell bodies are concentrated within the raphe nuclei
(1,31) which constitute a relatively small portion of the brain stem. Analysis of these nuclei is complicated by the fact that they are small, heterogeneous (2,5) and relatively difficult to isolate from single animals in
sufficient amounts for indepth study of their regulatory mechanisms. Thus,
an appropriate model for 5HT-ergic neurons would greatly facilitate such
studies.

Over the years, emerging data support the view that the pineal gland
may be useful as a model for studying regulatory mechanisms in serotoninergic

TABLE 1. Accumulation of ^{3}H-5HT in Pineal Glands from Intact and SGX Rats.

Media 5HT	Glandular 5HT (pmoles/gland)	
(μm)	Intact Rats	SGX Rats
10	11.20 ± 0.85	9.09 ± 2.0
	(3.13 ± 0.17)	(3.40 ± 0.12)
1	1.47 ± 0.09	1.64 ± 0.27
	(4.0 ± 1.0)	(3.6 ± 0.7)
0.1	0.163 ± 0.009	ND
	(5.3 ± 0.6)	

Pineal glands were incubated with ^{3}H-5HT for 30 minutes (10 μM) or 60 minutes (0.1 and 1 μM). Then glandular accumulation of ^{3}H-5HT as well as the ratios of ^{3}H to ^{14}C were calculated as described in Methods. The values represent the mean ± SEM of three determinations. The numbers in parenthesis represent the fold increase in the ratio of ^{3}H to ^{14}C in the glands as compared with that in the medium. The initial ratios were 12.5, 6.21 and 0.581 for the media containing 10 μM, 1 μM and 0.1 μM ^{3}H-5HT, respectively. Accumulation of ^{3}H-5HT by glands from intact and SGX rats were not significantly different.

neurons and/or for analyzing in general, molecular mechanisms of neuronal responsivity. Pinealocytes have been classified as paraneurons (13,42) and thus, may well serve as models for central 5HT neurons. By definition, paraneurons are derived from neuroectoderm, produce substances identical or related to real neurotransmitters and are capable of receptor-modulated responses. Paraneurons also contain synaptic vesicle-like or neurosecretory granules from which they release products in response to stimuli acting upon their membrane receptors.

Paraenchymal cells of the pineal gland fit all aspects of this definition. Pinealocytes metabolize and secrete 5HT (32), receive adrenergic innervation (24), and functionally resemble 5HT-ergic neurons of the midbrain raphe nuclei. Thus, the pineal gland consists of a homogeneous aggregation of 5HT-ergic cells whose neural innervation is uncomplicated, consisting basically of adrenergic fibers from the superior cervical ganglia (24,25). In addition, the pineal is easily removed from the brain without contamination by other tissues for study in vitro. These facts establish the suitability of the pineal for study of primary factors affecting synthetic and secretory functions in 5HT-ergic cell bodies. Finally, the pineal accumulates neuropeptides, which seem to alter the molecular regulatory mechanisms of the gland (39). This fact further supports use of the pineal to study modulatory mechanisms that influence 5HT synthesis and/or release. Thus, the general plan regulating pineal function is comparable to that for any neuronal population in the brain.

Pinealocytes produce and release 5HT in response to noradrenergic input (32). Since this type of regulatory interaction between NE-ergic and 5HT-ergic cells is known to exist within the brain (16,27), the use of the pineal

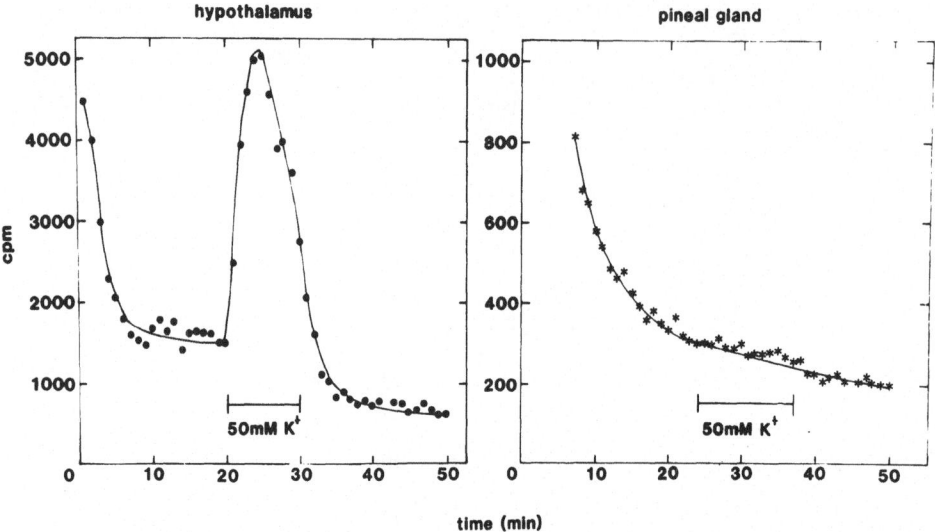

FIG. 2: Effect of K+-depolarization of [3]H-5HT Release From Rat Hypothalamus
On Pineal Glands. Rat hypothalamic slices (A) or intact pineal
glands (B) were incubated with 1 μM [3]H-5HT for 1 hr. Subsequently,
the tissues were perfused with buffer or buffer containing 50 mM K+
for the period indicated. The efflux of ^{3}H in each 1 min fraction
is shown.

as a model for analysis of neuronal regulation and adaptation becomes quite
apparent. Based upon such supporting evidence, the pineal gland was used in
the series of experiments summarized below to analyze the influence of adrenergic neurotransmitters on post-synaptic responses affecting 5HT dynamics
within pinealocytes. Emphasis was placed on the influence of adrenergic
transmitters on activation on specific receptors and second messengers controlling the secretion of 5HT.

5HT Secretion by the Rat Pineal Gland, in Vitro

Little is known of the efferent limb of the pineal neurosecretory axis
other than the fact that metabolites of tryptophan are released and these
compounds influence the activity of target sites within the brain. The
majority of studies focus on melatonin (MEL), probably because its unique
production in the pineal and retina allows certain reasonable assumptions
to be made about its source when measured in the blood. On the other hand
similar assumptions cannot be made about 5HT in the blood or cerebrospinal
fluid (CSF) since the amine is produced at various sites in the brain and
periphery. However, recent reports suggest that pineal 5HT enters the CSF
each day (11,37,38). Nonetheless, pineal mechanisms for 5HT and MEL synthesis have been studied almost entirely in preference to processes controlling their secretion. As a result of these efforts we know a great deal
about cellular and molecular events associated with the metabolism of pineal
tryptophan, especially for the synthesis of MEL.

First of all, the classical studies of Quay (28) that have since been
confirmed (15) defined the circadian rhythms of 5HT and MEL content in
pineals of rats. 5HT reaches peak levels at approximately midpoint in the
light phase, while the peak of pineal MEL occurs in darkness. Subsequent
studies (4) suggested that noradrenergic stimulation of the pineal increases

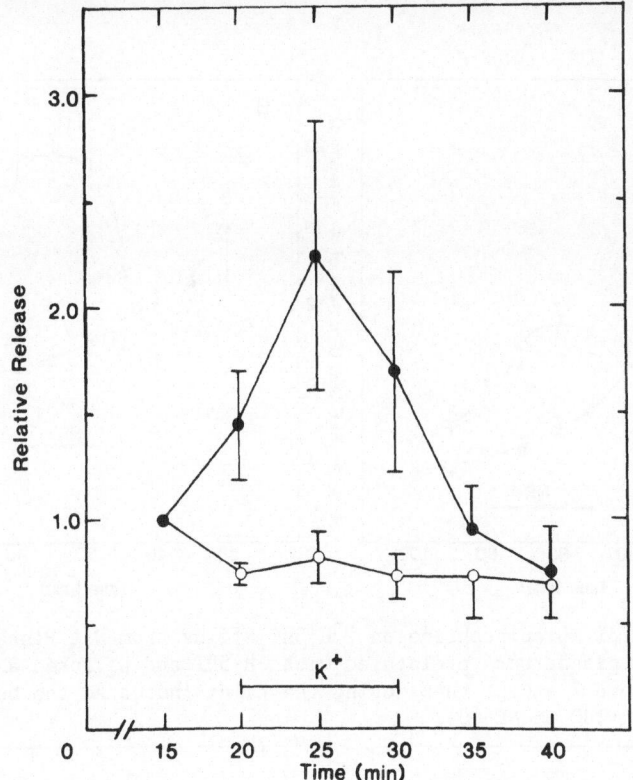

FIG. 3: Effect K$^+$ depolarization of ^{3}H-5HT on ^{3}H-NE Secretion From Rat
Pineal Glands. The internal pools of NE or 5HT were labeled by
incubating the pineal glands in 1 μM ^{3}H NE (o) or 1 μM ^{3}H 5HT (o)
for 1 hr. The efflux of ^{3}H was determined by the step-transfer
procedure at 5 min intervals. At the times indicated the medium
was changed to one containing 50 mM K$^+$. Relative release was calculated by setting the amount of ^{3}H release during the 5 min interval immediately proceeding incubation in 50 mM K$^+$ equal to 1 and
subsequent release was compared to this value. Each point represents the mean (±SEM) of four glands. K$^+$ significantly (p<.05)
stimulated NE release but not 5HT release.

in darkness and is necessary for the accelerated synthesis of MEL (18). It
was also discovered that the effects of norepinephrine (NE) on MEL synthesis
are initiated by pinealocyte β-receptors (20) and mediated by a cascade of
molecular events from the synthesis of cyclic adenosylmonophosphate (cAMP)
(8) to activation of a cAMP dependent protein kinase (10), through induction
of serotonin-N-acetyl-transferase (SNAT) synthesis (45). SNAT is the rate
limiting enzyme for MEL synthesis, thus regulating its production in the
pineal (19,23). In addition to this primary regulatory pathway, a molecular
process that is initiated by activation of α adrenoreceptors which in turn
potentiates the β receptor mediated induction of NAT has also been described
(20,22). In contrast to the cAMP related events initiated by β receptor
activation, the α receptor effects are linked to phosphatidylinositol (PI)
turnover (8a,14,35) and thus employ second messenger complexes that are
different from the primary pathway. Inhibitory effects of Ca^{++} and cyclic
guanosylmonophosphate (cGMP) on NAT induction have also been reported (43,
44), demonstrating the complex interaction of intracellular functions that
must be appropriately integrated for optimal synthesis of MEL by the pineal.
The balance of these multiple influences are expressed, presumably, as the
rhythmic changes in NAT induction and MEL content that occur in the pineal
each day (19,23).

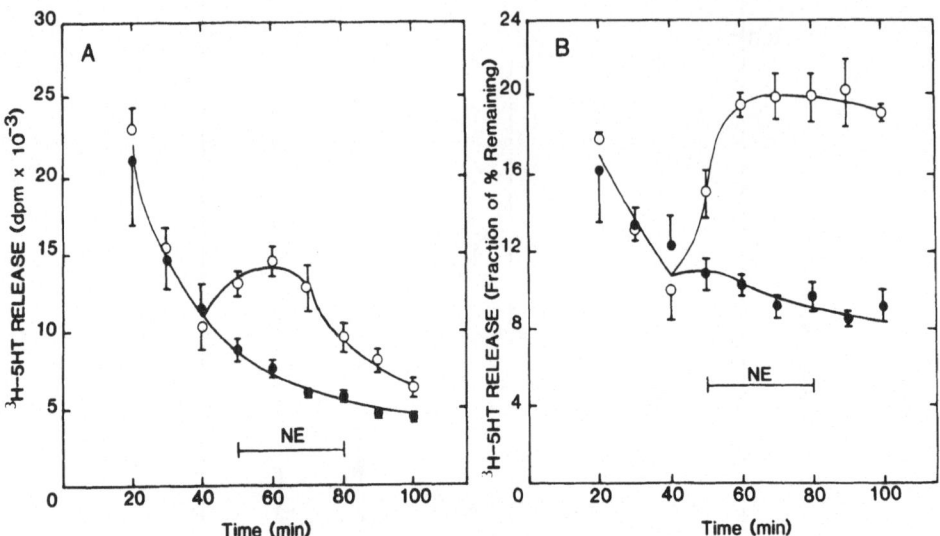

FIG. 4: Effect of Norepinephrine on [³H]5HT Efflux From Rat Pineal Glands.
Pineal glands were prelabeled with ³H-5HT and perfused as described
in figure 2 except that during the times indicated the buffer containing 100 μM NE (0).

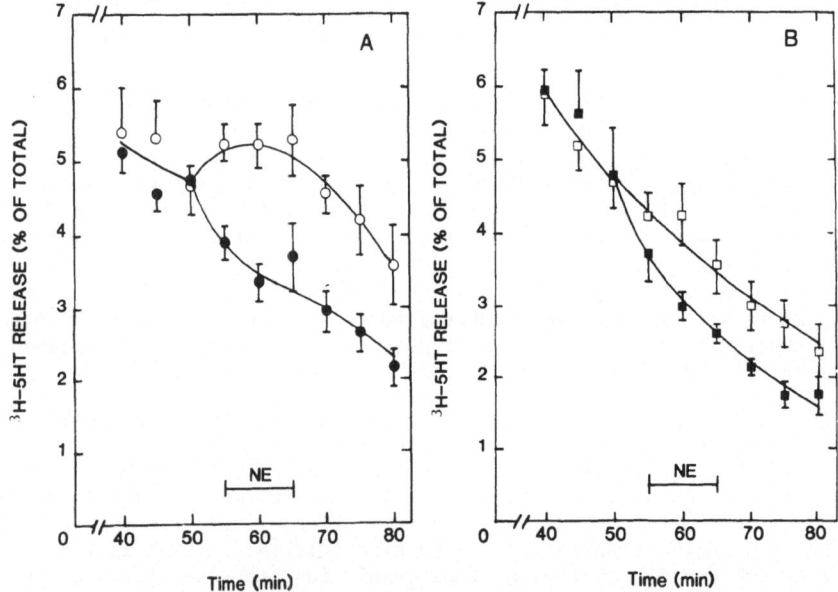

FIG. 5: NE-Stimulated Release of ³H-5HT From Intact And Denervated Pineal
Glands. Pineal glands were prelabeled with 1 μM ³H 5HT and efflux
was determined by the step-transfer procedure. During the times
indicated the buffer contained 100 μM NE (o,□). The efflux is
expessed as percent of total ³H accumulated by the gland during the
prelabelling period and the values represent the mean (±SEM) of 5 to
7 glands. (A) Intact glands presumably contained sympathetic nerve
terminals since they were collected from unoperated rats. (B) Denervated glands lack sympathetic terminals since they were collected
from rats in which the superior cervical ganglia were surgically
removed. Release from NE-stimulated (o□) glands was significantly
different (p<0.05) than release from non-stimulated (●,■) glands
during the same time interval.

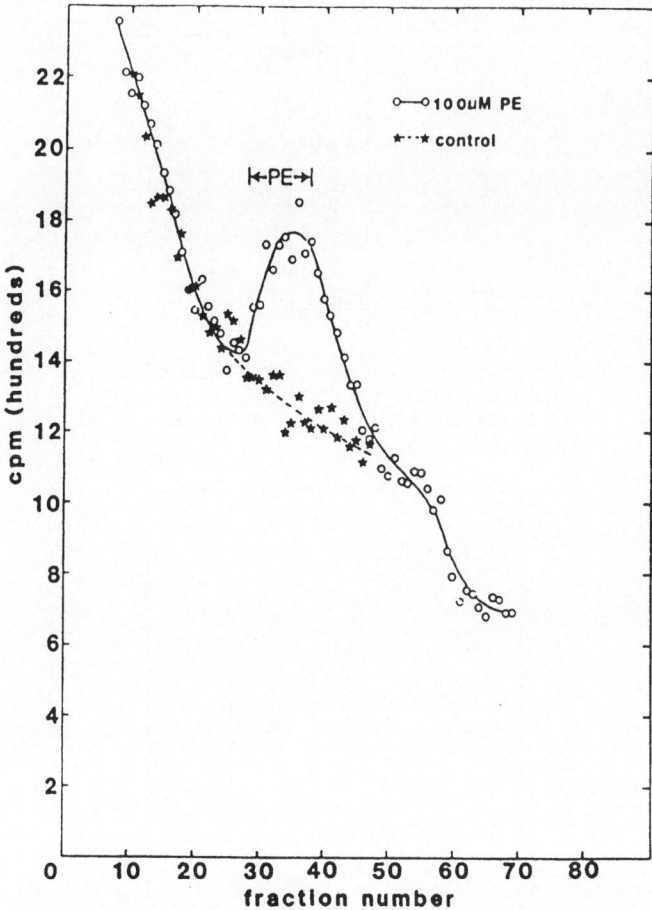

FIG. 6: Effect of Phenylephrine on ³H-5HT Efflux. Pineal glands were
labeled with ³H-5HT and perifused as described in figure 2 except
that during the interval indicated the buffer contained 100 μM
phenylephrine (o).

    Although an understanding of the metabolic and secretory pathways for
5HT and not MEL, is more relevant to how typical neurons function, this
aspect of pineal neurochemistry has been virtually unexplored when considering its potential as a model for serotoninergic neurons in the brain.
The work that has been done deals almost entirely with 5HT synthesis and
is briefly summarized below. The rate limiting enzyme for 5HT production
is tryptophan hydroxylase (TH) (12,17). Studies of this pineal enzyme
reveal that its activity is depressed by light (32), which also blocks NE
stimulation of the gland (4,26). This finding suggests that TH activity,
like that of NAT is increased by activation of β adrenergic receptors.
Support for this concept derives from the fact that propranolol, a β-antagonist, blocks activation of TH by NE (33). Furthermore, rhythmic increase
in TH activity occur each night (33,34). Since 5HT serves as a substrate
for MEL synthesis, coupled activation of TH and NAT appear logical. However, the literature fails to account for some rather disturbing facts that
may have significant implications in the pineal's role as a neurosecretory
organ. For example, peak accumulation of pineal 5HT precedes the rise in
TH activity and is associated with light not darkness (28,34). It has been
suggested that the afternoon decrease in pineal 5HT content results from its
use as a MEL precursor (19). However, the production of 5HT metabolites
during this interval are not stoichiometric so as to support the "precursor
hypothesis". During the interval from 1400-1900 h, pineal 5HT content drops
about 280 pmoles (28), while MEL does not rise at all. On the other hand,

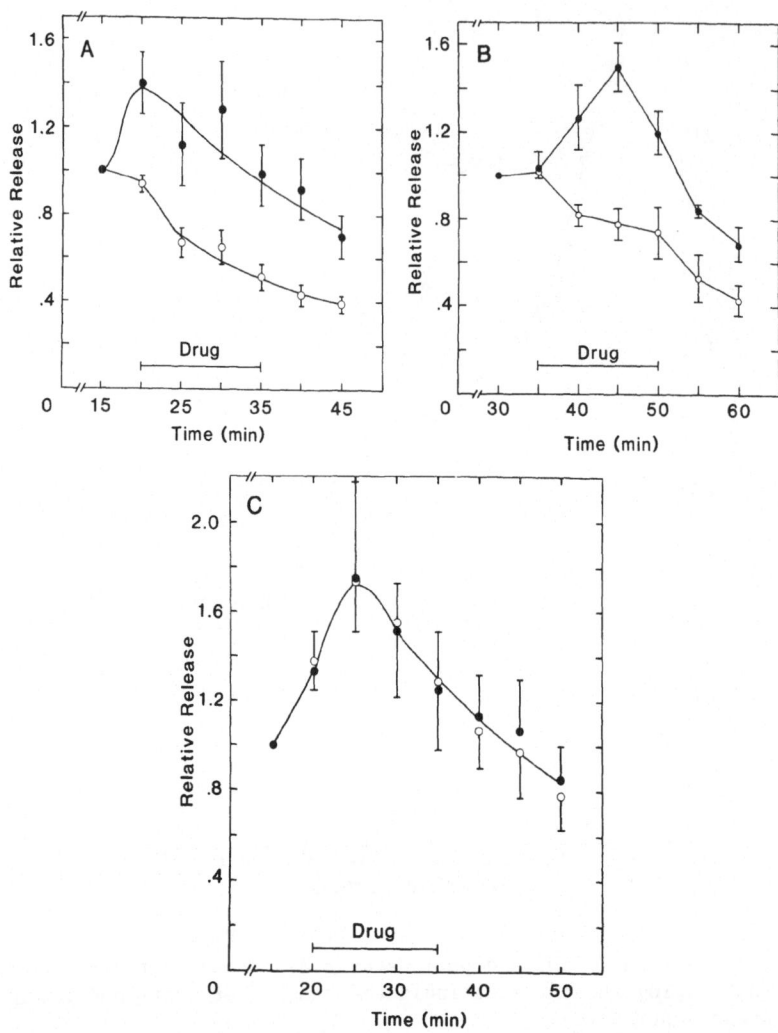

FIG. 7: Effects of α- and β-adrenoreceptor antagonists on NE-stimulated [3]H-
5HT release. Pineal glands were labeled with [3]H-5HT and relative
release was determined as described in Fig. 3. (A) During the
interval indicated the buffer contained 30 μM NE (o) or NE plus
30 μM prazonsin (o). (B) 30 μM phentolamine (o) or (C) 30 μM
sotolal (o). The values are the means (± SEM) of 3 to 4 glands.

MEL content of the pineal increases only about 10 pmoles after darkness (29).
Thus its production is neither temporally or quantitatively correlated with
5HT depletion. Furthermore, 5HIAA rises only about 25 pmoles and tryptophol
derivatives of 5HT are also insignificant during the post acrophase interval
of the 5HT circadian rhythm. These facts lead to the compelling conclusion
that a certain amount of 5HT is secreted from the pineal during the afternoon.

[3]H-5HT Uptake in Rat Pineal Glands

The first step in testing for 5HT release involved determination of
whether rat pineals possess an uptake mechanism for accumulating [3]H-5HT.
This question was of interest since complementary mechanisms for regulating
intracellular concentrations of 5HT should co-exist with those for secretion
of the indoleamine (6,7). Thus, rat pineal glands were incubated with seve-

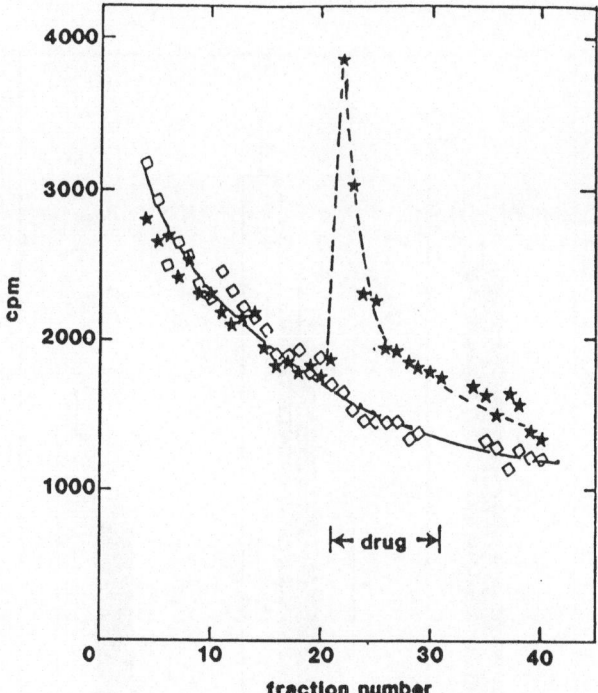

FIG. 8: ³H-5HT Release From Pineal Glands Perifused With Phorbol Ester or
cAMP. Pineal glands were pre-loaded with ³H-5HT and treated with
1 μM, PMA (*) or 1 mM 8-bromo-cAMP ( ) (both in 0.1% DMSO) during
the time interval indicated.

ral concentrations of ³H-5HT. In addition, the incubation media contained
¹⁴C-mannitol which was used to determine the volume of extracellular space
in each gland. Intracellular accumulation of ³H in pineal glands increased
linearly as a function of time for ³H-5HT at a concentration of 1-10uM (Fig
1A). In contrast, pineal content of ¹⁴C increased more slowly causing the
³H:¹⁴C ratio to increase for 30 minutes (Fig 1B). These changes indicated
that the pineals preferentially accumulate ³H-5HT and so demonstrate the
presence of a mechanism for 5HT uptake in the glands. The possibility that
 H-5HT accumulated in noradrenergic terminals rather than pinealocytes was
precluded by the finding that uptake was comparable in glands that were
denervated by superior cervical ganglionectomy one week prior to the study
(Table 1).

Having established that pinealocytes possess a mechanism for accumulating ³H-5HT, it was of interest to determine whether or not they also
possess a mechanism for its release. Pineals were preincubated with ³H-5HT
and then perifused with buffer containing 50mM potassium. Hypothalamic
tissue also prelabelled with the indoleamine was run in parallel to provide a basis for comparing ³H-5HT release in brain and pineal. As seen
in Figure 2, exposure of hypothalamic tissue to the hyperkalemic buffer
immediately stimulated ³H-efflux above basal levels. Tritium efflux reached
peak values within the first minute of exposure to potassium. In contrast,
insignificant amounts of ³H were released from comparably treated pineals.
In order to determine whether pineals were responsive to concentrations of
potassium that were depolarizing in brain, the experiment was repeated
using pineals prelabelled with 1uM ³H-NE or 1uM ³H-5HT. The data presented
in Figure 3, show that the rate of ³H-NE but not ³H-5HT efflux was increased
following perifusion of pineals with 50mM potassium. These findings demonstrate that the concentrations of ³H-5HT used in this study produced specific labelling of pinealocytes, and not adrenergic terminals. Furthermore,
the pinealocyte is relatively insensitive to potassium depolarization when

231

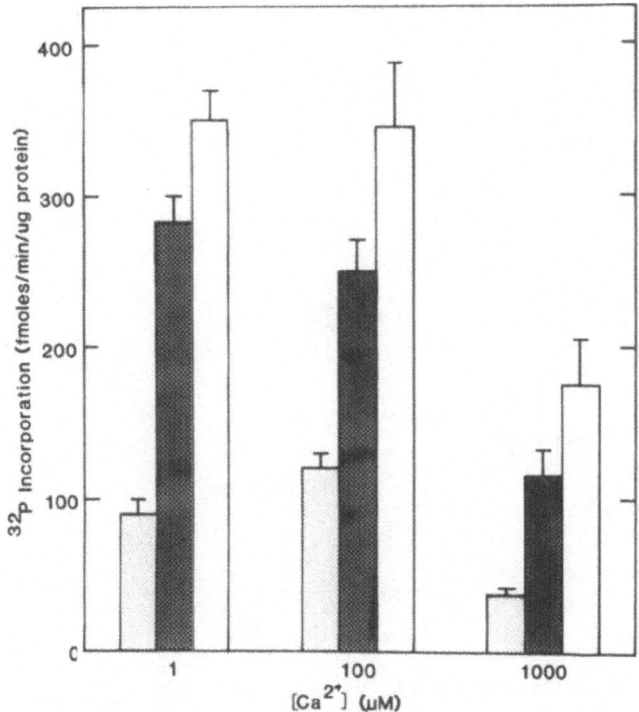

FIG. 9: Phospholipid-dependent Kinase Activity in the Soluble Fraction of Pineal Gland Homogenates. The kinase activity of the soluble pineal fraction was tested at 30°C in a reaction mixture containing 6 μl soluble fraction (.53 μg protein), 1mM EGTA, 10mM MgCl$_2$, 0.5 mg/ml Histone H1 10mM ATP (0.5 μCi ^{32}P-y-ATP) and 10mM Tris HCl pH 7.4 in a final volume of 60 μl. The free calcium concentration was varied by adding calcium chloride in excess of the EGTA present. Kinase activity was determined with no added lipid (▨), phosphatidylserine (20 μg/ml) (■) or phosphatidylserine (20 μg/ml) plus 1,2-diolein (0.6 μg/ml) (□). The results are expressed as fmole/min/μg soluble protein. The error bars represent the SD for 3 replicate determinations.

considering 5HT release in more typical neural tissue. Despite this poor response, ^{3}H-5HT efflux was significantly increased when pineals were perifused with buffer containing NE. As seen in Figure 4, ^{3}H efflux increased rapidly upon stimulation with NE, and release remained elevated throughout the period of exposure to the catecholamine. Proof that ^{3}H release occurred at least in part from pinealocytes derived from the fact that efflux of the isotope was also significantly increased by exposure of denervated glands to NE (Figure 5).

Alpha-adrenergic Receptors and 5HT Release

As previously mentioned, beta adrenoreceptors are recognized as mediating effects of NE on the cascade of events leading to MEL synthesis. Since we (3,40,41) established that NE also stimulates the release of 5HT from pineals, it was of interest to determine how the gland differentiates signals for 5HT release from those for MEL synthesis. In an attempt to understand which receptor subtype(s) contribute to the release process, pineals were perifused with phenylephrine (PHEN; an alpha agonist) or isoproterenol (ISO; a beta agonist) instead of NE. As seen in Figure 6, PHEN stimulated 5HT release in a manner that was similar to NE. In contrast,

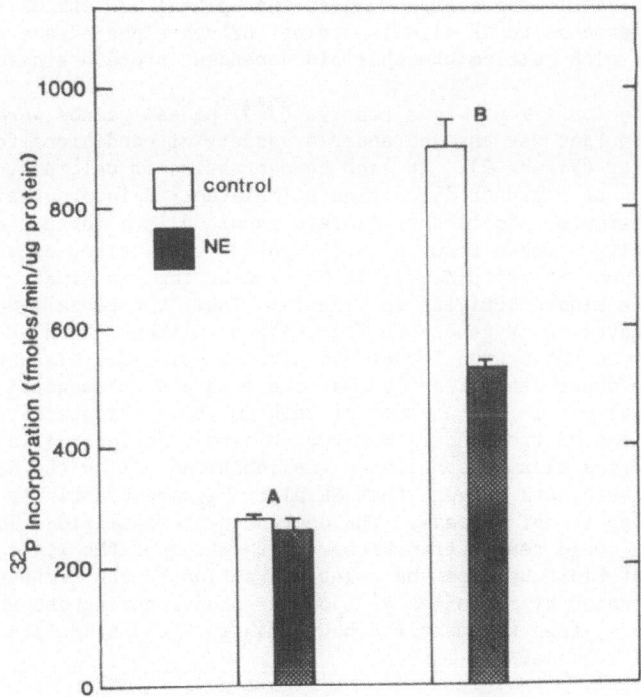

FIG. 10: Effect of Norepinephrine on Pineal Phospholipid-dependent Kinase
Activity. Pineal glands were treated at 37° with 10^{-4}M NE or
solvent for 10 min before the soluble fraction was prepared as
described in the text. The soluble fraction was assayed for
kinase activity in the absence of calcium (1mM EGTA) (A) or in
the presence of 1.1mM calcium chloride (i.e. 100 µM free calcium
plus phosphatidylserine (20 µg/ml and 1,2-diolein (0.6 µg/ml).
The results are presented as fmoles of phosphate incorporated/
min/µg soluble protein. The error bars represent the SD of 3
replicate determinations.

ISO did not stimulate ^{3}H-5HT release from the pineal glands. Thus, the
data suggest that noradrenergic signals rapidly stimulate 5HT release
from the pineal by an alpha-adrenergic mechanism in contrast to the beta-
adrenergic mechanism controlling melatonin synthesis. To test this hypothesis, glands (preloaded with ^{3}H-5HT) were exposed to NE or NE plus alpha
or beta receptor antagonists. The data presented in Figure 7 show that the
alpha-receptor antagonists, prazosin and phentolamine attenuate NE-stimulated efflux of ^{3}H-5HT while the β-receptor antagonist sotolal is without
effect. Taken together, these findings suggest that activation of alpha
receptors is a prerequisite to 5HT release from pineals exposed to NE.

Second Messengers Involved in NE-Mediated Release of 5HT

Although cyclic AMP controls intracellular processes governing MEL
metabolism in the pineal (21), different second messenger(s) may participate
in the mechanism for 5HT release. Support for this view derives from the
fact that activation of beta-adrenergic receptors initiates cAMP accumulation in the pineal (21), while activation of alpha receptors stimulates
phosphatidylinositol turnover (35). Based upon these observations pineals
were loaded with ^{3}H-5HT and perifused with buffer containing 8-bromo-cAMP
or phorbol 12-myristate 13-acetate (PMA) to determine if cAMP or phospholipid dependent kinases respectively, mediate the 5HT secretory response to
alpha-adrenoreceptor activation. As seen in Figure 8, PMA simulated the

effects of NE and phenylephrine, while ^{3}H-efflux was unchanged after exposure to 8-bromo-cAMP. These data suggest that pineal 5HT efflux is rapidly stimulated in response to NE signals acting through alpha-adrenergic receptors in concert with calcium/phospholipid dependent protein kinase(s).

In order to confirm previous reports (36), pineal glands were homogenized and supernatant was assayed under a variety of conditions for protein Kinase C activity (Figure 9). At each concentration of calcium that was tested, addition of phosphatidyl serine stimulated ^{32}P incorporation into histone. Furthermore, addition of diolein resulted in a further stimulation of kinase activity. These results, although not conclusive, establish that Kinase C is present in rat pineal. If NE were acting via Kinase C, then it should alter the kinase activity in pineals. Thus, rat pineals were incubated for 5 minutes in oxygenated buffer with or without 100uM NE under conditions similar to those used for efflux studies. Immediately after the incubation, the supernatant fraction was prepared and subsequently assayed for kinase activity. In the absence of calcium, basal kinase activity was not altered by the NE treatment. However, kinase C activity assayed in the presence of calcium plus phospholipids was inhibited 40% by the NE treatment (Figure 10). These data suggest that NE alters kinase C activity under conditions resulting in 5HT release. The decreased phospholipid-stimulated kinase activity could result from either degradation of the kinase or in the translocation of kinase C from the soluble fraction to the membrane bound pool as demonstrated by Sugden et al (36) for phenylephrine and PMA. If the latter occurs, then the membrane bound kinase C may stimulate protein(s) involved in 5HT release.

CONCLUSIONS

This report has focussed upon the potential utility of the pineal gland as a model for studying molecular mechanisms of neuronal responsivity. Specifically, we have shown that the pineal is capable of accumulating 5HT, and that its subsequent release is modulated by NE. The utility of the pineal as a model for studying monoamine interactions derives from the fact that this tissue is essentially a homogeneous aggregate of serotonin producing and secreting cells that are innervated almost exclusively by noradrenergic neurons. The pilot studies described in this report have been extremely productive, providing many clues as to how pineal receptors and second messengers discriminate a single noradrenergic signal into the separate functions of MEL synthesis and 5HT secretion. Although this specific type of discrimination subserving pinealocyte responsivity to a noradrenergic signal is undoubtedly unique to the pineal, the underlying principles are clearly relevant to many types of CNS interactions and particularly for monoamines. Thus, the pineal is particularly useful for studying certain types of brain mechanisms and its potential should be fully exploited.

ACKNOWLEDGMENTS

Support for this project was provided in part by a research grant from the Air Force Office of Scientific Research (AFOSR-85-0373).

REFERENCES

Aghajanian, G. K., Rosecrans, J. A. and Sheard, M.H., Serotonin: release in the forebrain by stimulation of mid brain raphe, Science 156:402-403 (1967).

Aghajanian, G. K., Wang, R. Y. and Baraban, J., Serotonergic and non-serotonergic neurons of the dorsal raphe: Reciprocal changes in firing induced by peripheral nerve stimulation, Brain Res. 153:169-175 (1978).

Aloyo, V. J. and Walker, R. F., Noradrenergic control of serotonin release from rat pineal glands, in vitro, Endocrinology (In Press) (1986).

Craft, C. M., Morgan, W. W. and Reiter, R. J., 24-hour changes in catecholamine synthesis in rat and hamster pineal glands, Neuroendocrinology 38:193-198 (1984).

Descarries, L., Beaudet, A., Watkins, K. C. and Garcia, S., The serotonin neurons in nucleus raphe dorsalis of adult rat, Anat. Rec. 193:520 (1979).

Ducis, I. and DiStefano, V., Evidence for a serotonin uptake system in isolated bovine pinealocyte suspension, Mol. Pharm. 18:438 (1980).

Ducis, I. and DiStefano, V., Characterization of serotonin uptake in isolated bovine pinealocyte suspension, Mol. Pharm. 18:447 (1980).

Ebadi, M. S., Weiss, B. and Costa, E., Adenosine 3',5'-monophosphate in rat pineal gland: increase induced by light, Science 170:188-190 (1970).

Eichberg, J., Shein, H. M., Schwartz, M. and Hauser, G., Stimulation of ^{32}P; incorporation into phosphatidylinositol and phosphatidylglycerol by catecholamines and β-adrenergic receptor blocking agents in rat pineal organ cultures, J. Biol. Chem. 248:3615-3622 (1973).

Essman, W. B., Serotonin in Health and Disease. Vol. II: Physiological Regulation and Pharmacological Action, Spectrum Publication, Inc. New York (1978).

Fontana, J. A. and Lovenberg, W., A cyclic AMP-dependent protein kinase of the bovine pineal gland, Proc. Natl. Acad. Sci. USA 68:2787-2790 (1971).

Garrick, N. A., Tamarkin, L., Taylor, P. L., Markey, S. P. and Murphy, D. L., Light and propranolol suppress the nocturnal elevation of serotonin in the cerebrospinal fluid of rhesus monkeys, Science 221:474 (1983).

Grahame-Smith, D. G., Tryptophan hydroxylation in brain, Biochem. Biophys. Res. Commun. 16:586-592 (1964).

Hansen, J. T. and Karasek, M., Neuron or Endocrine Cell? The Pinealocyte as a paraneuron, Prog. Clin. Biol. Res. 92:1-9 (1982).

Hauser, G., Shein, H. M. and Eichberg, J., Relationship of α-adrenergic receptors on rat pineal gland to drug-induced stimulation of phospholipid metabolism, Nature 252:482-483 (1974).

Hery, F., Rouer, E. and Glowinski, J., Daily variations of serotonin metabolism in the rat brain, Br. Res. 43:445-465 (1972).

Hillier, J. G., Martin, P. R. and Redfern, P. H., A possible interaction between the 24-hour rhythms in catecholamines and 5-hydroxytryptamine concentration in the rat brain, J. Pharm. Pharmacol, (Suppl. 2) 27:40P (1975).

Jequier, E., Robinson, D. S., Lovenberg, W. and Sjoerdsma, A., Further studies on tryptophan hydroxylase in rat brain stem and beef pineal, Biochem. Pharmacol. 18:1071-1081 (1969).

Klein, D. C., Circadian rhythms in indole metabolism in the rat pineal gland, In: C. S. Pittendrigh (Ed). Circadian Oscillations and Organization in Nervous System, MIT Press, Cambridge, pp. 509-515 (1975).

Klein, D. C. and Weller, J. L., Indole metabolism in the pineal gland: A circadian rhythm in N-acetyltransferase, Science 169:1093-1095 (1970).

Klein, D. C. and Weller, J. L., Adrenergic-adenosine 3', 5'-monophosphate regulation of serotonin N-acetyltransferase activity and the temporal relationship of serotonin N-acetyltransferase activity to synthesis of ^{3}H-N-acetyl-serotonin and ^{3}H-melatonin in cultured rat pineal gland, J. Pharmacol. Exp. Ther. 186:518-527 (1973).

Klein, D. C., Berg, G. R. and Weller, J. L., Melatonin synthesis: adenosine 3', 5'-monophosphate and norepinephrine stimulate N-acetyl-transferase, Science, 168:979-986 (1978).

Klein, D. C., Sugden, D. and Weller, J. L., Postsynaptic α-adrenergic receptors potentiate the β-adrenergic stimulation of pineal serotonin N-acetyltransferase, Proc. Natl. Acad. Sci. USA 80:599-603 (1983).

Klein, D. C., Weller, J. L. and Moore, R. Y., Melatonin metabolism: neural regulation of pineal serotonin: acetyl coenzyme A N-acetyltransferase activity, Proc. Nat. Acad. Sci. 68:3107-3110 (1971).

Moore, R. Y., The innervation of the mammalian pineal gland, Prog. Reprod. Biol. 4:1-29 (1978).

Moore, R. Y., The retinohypothalamic tract, suprachiasmatic hypothalamic nucleus and central neural mechanisms of circadian rhythm regulation, In: M. Suda, O. Hayaish and H. Nakagawa (Eds), Biological Rhythms and their Central Mechanisms, Elsevier, Amsterdam, pp. 343-354 (1979).

Morgan, W. W. and Reiter, R. J., Pineal noradrenaline levels in the Mongelian gerbil and in different strains of laboratory rats over a lighting regimen, Life Sci. 21:555-558 (1977).

Plaznik, A., Danysz W., Kostowki, W. Bidzinski, A. and Hauptmann, M., Interaction between noradrenergic and serotoninergic brain systems as evidenced by behavioral and biochemical effects of microinjections of adrenergic agonists and antagonists into the median raphe nucleus, Pharm. Biochem. Behav. 19:27-32 (1983).

Quay, W. B., Circadian rhythm in rat pineal serotonin and its modifications by estrous cycle and photoperiod, Gen Comp Endocrinol 3:473-479 (1963).

Quay, W. B., Circadian and estrous rhythms in pineal melatonin and 5-hydroxy-indole-3-acetic acid, Proc. Soc. Exp. Biol. Med. 115:710-713 (1964).

Reiter, R. J., The pineal and its hormones in the control of reproduction in mammals, Endocrine Reviews 1:109-131 (1980).

Saavedra, J. M., Distribution of serotonin and synthesizing enzymes in discrete areas of the brain, Fed. Proc. 36:2134-2141 (1977).

Shein, H. M. and Wurtman, R. J., Stimulation of ^{14}[C] tryptophan 5-hydroxylation of norepinephrine and dibutyryl adenosine 3',5'-monophosphate on rat pineal organ cultures, Life Sciences 10:935-940 (1971).

Shibuya, H., Toru, M. and Watanabe, S., A circadian rhythm of tryptophan hydroxylase in rat pineals, Brain Res. 138:364-368 (1978).

Sitaram, B. R. and Lees, G. J., Diurnal rhythm and turnover of tryptophan hydroxylase in the pineal glands of the rat, J. Neurochem. 31:1021-1026 (1978).

Smith, T. L., Eichberg, J. and Hauser, G., Postsynaptic localization of the alpha receptor-mediated stimulation of phosphotidylinositol turnover in pineal gland, Life Sci. 24:2179-2184 (1979).

Sugden, D., Vancecek, J., Klein, D. C., Thomas, T. P. and Anderson, W. B., Activation of protein Kinase C potentiates isoproterenol induced cyclic AMP accumulation in rat pinealocytes, Nature (1985).

Taylor, P. L., Garrick, N. A., Burns, R. S., Tamarkin, L., Murphy, D. L. and Markey, S. P., Diurnal rhythms of serotonin in monkey cerebrospinal fluid, Life Sci. 31:1993 (1982).

Taylor, P. A., Garrick, N. A., Tamarkin, L., Murphy, D. L. and Markey, S. P., Diurnal rhythms of N-acetylserotonin and serotonin in cerebrospinal fluid of monkeys, Science 228:900 (1985).

Tsang, D. and Martin, J. B., Effect of hypothalamic hormones on the concentration of adenosine 3',5'-monophosphate in the incubated rat pineal gland, Life Sci. 19:911-918 (1976).

Walker, R. F., Sparks, D. L., Slevin, J. and Rush, M. E., Temporal effects of norepinephrine on pineal serotonin in vitro, J. Pin. Res. 3:33-40 (1986).

Walker, R. F. and Aloyo, V. J., Norepinephrine stimulates serotonin secretion from rat pineal glands, in vitro, Br. Res. 343:188-190 (1985).

Weck, M. and Wake, K., The pinealocyte-a paraneuron. A Review, Arch. Histol. Jap. 40 (suppl)261-278 (1977).

Wilkinson, M., Inhibition of the noradrenergic induction of pineal N-acetyl-transferase by dibutyryl cyclic guanosine monophosphate and by ionphore X-537A, Neurosci. Lett. 2:29-33 (1976).

Wilkinson, M., The sensitivity of pineal gland β -receptors appears to be dependent upon calcium ions, Pflugers Arch. 373:209-210 (1978).

Winter, K. E., Morrissey, J. J., Loos, P. J. and Lovenberg, W., Pineal protein phosphorylation during serotonin N-acetyl-transferase induction, Proc. Natl. Acad. Sci. USA 74:1928-1931 (1977).

MOLECULAR MECHANISMS OF ACIDIC AMINO ACID RELEASE FROM

MOSSY FIBER TERMINALS OF RAT CEREBELLUM

David M. Terrian, Scott B. Bischoff,
Monica A. Schwartz* and Robert V. Dorman*

USAF School of Aerospace Medicine, Neurosciences
Function, Brooks AFB, Texas 78235-5301 and
*Department of Biological Sciences, Kent State
University, Kent, Ohio 44242

INTRODUCTION

The stimulus-induced release of neurotransmitters from nerve terminals requires rapid and reversible changes in the permeability properties of pre-synaptic membranes. Phospholipids, which provide the basis for the bilayer structure of biological membranes, have been implicated in chemical transmission. In particular, it has been suggested that an obligatory step in neurotransmitter release is the accumulation of unesterified, polyunsaturated fatty acids following the depolarization-induced degradation of phospholipids. Depolarization of cerebral synaptosomes results in the Ca^{2+}-dependent release of free arachidonic acid (Lazarewicz et al., 1983) from phosphatidylcholine and phosphatidylinositol (Majewska and Sun, 1982). The Ca^{2+}-dependent release of gamma-aminobutyric acid (GABA) from cerebral synaptosomes has also been shown to require the accumulation of free arachidonic acid and this release can be mimicked with exogenous arachidonic acid, but not other free fatty acids (Asakura and Matsuda, 1984). The release of neurotransmitters induced by exogenous fatty acids does not require Ca^{2+} (Rhoads et al., 1983), suggesting that the Ca^{2+} is required for phospholipid degradation and fatty acid accumulation.

The function of arachidonic acid accumulation in nerve terminal membranes may be to provide substrates for the formation of cyclooxygenase or lipoxygenase products. The production of such fatty acid derivatives has been shown to be involved in stimulus-secretion coupling in other cell types, such as platelets, but the role of prostaglandins and leukotrienes in neurotransmitter release remains unclear. Hedqvist (1973) suggested that prostaglandins produced in response to the activation of presynaptic autoreceptors provide a feedback mechanism to prevent further transmitter release. Alternatively, prostaglandins may be directly involved in the release process itself, since treatment of cerebral synaptosomes with acetylsalicyclic acid or indomethacin partially inhibits K^+-evoked catecholamine release (Bradford et al., 1983).

In an attempt to clarify some of the biochemical mechanisms involved in stimulus-induced neurotransmitter release we have employed a relatively homogeneous nerve terminal preparation, which provides an opportunity to correlate certain biochemical properties with specific neuronal elements (e.g.; the isolated cerebellar glomerulus). In previous studies, we demonstrated that isolated glomeruli possess specific high affinity uptake systems for [^{3}H]serotonin (Terrian et al., 1985) and [^{3}H]choline (Terrian et al., in press). The excitatory mossy fiber terminals provide the most likely site for serotonergic and cholinergic systems. The present results extend these findings to include the possibility that a relatively large proportion of mossy fibers utilize acidic amino acids as neurotransmitters. This relationship is supported by the evidence that cerebellar glomeruli contain sodium-dependent, high affinity uptake sites for D-[^{3}H]aspartate and that the depolarization-induced release of this amino acid is dependent on the presence of external Ca^{2+}.

Isolated cerebellar glomeruli also provide a useful _in vitro_ model for examining the presynaptic mechanisms which mediate neurotransmitter release. It is possible to eliminate diffusion barriers and the reuptake of released neurotransmitters, which are factors that can affect estimates of transmitter release, by exposing the glomeruli to superfusion conditions. We have employed the superfusion method to investigate the molecular mechanisms involved in the release of neurotransmitters from mossy fiber terminals. This report presents evidence to support a role for arachidonic acid as an intermediate in the depolarization-induced release of acidic amino acids from cerebellar mossy fiber terminals.

METHODS

Subcellular Fractionation

Glomerular particles were isolated from the cerebellar cortices of 175-250 g, male rats by differential and Ficoll-sucrose density gradient centrifugation as described by Terrian et al. (1985). The resulting pellets were obtained within 3.5 h and the morphological homogeneity of this preparation has been estimated to be 92.6 ± 3.9% (mean ± S.E.M.) (Terrian et al., 1985). An average of 5.5 ± 0.9 mg (mean ± S.E.M.) of glomerular protein were obtained per g wet wt. of starting tissue.

Uptake of D-[^{3}H]aspartate

D-[^{3}H]aspartic acid uptake was measured by a modification of the method described by Levi and Raiteri (1973). Glomerular particles were resuspended in 0.32 M glucose (7.64 ± 0.66 mg protein/ml; mean ± S.E.M.) and aliquots (50 ul) of this suspension were then equilibrated at 37°C for 5 min in 900 μl of oxygenated Krebs-Ringer medium (final concentrations: 128 mM NaCl, 5 mM KCl, 1.2 mM $MgSO_4$, 2.7 mM $CaCl_2$, 1 mM Na_2HPO_4, 20 mM HEPES; pH 7.35). A final volume of 1.0 ml was obtained by adding a known concentration of D-[^{3}H]-aspartate (New England Nuclear Corp.; spec. act., 4.0 Ci/mmol) and tissue samples were incubated an additional 5 min. Preliminary experiments confirmed that D-[^{3}H]aspartate uptake proceeded in a linear fashion under these conditions. Additional samples were kept at 0-5°C during the incubation period and radioactivity in these controls was subtracted from the total uptake observed at 37°C to correct for passive diffusion and nonspecific binding. Uptake was terminated by centrifugation in a microfuge for 45 sec, removing the supernatant and rinsing the pellet twice with ice-cold incubation medium. Trichloroacetic acid precipitates of the glomerular particles were assayed for protein (Lowry et al., 1951).

Release of D-[^{3}H]aspartate

Glomerular particles (3.2-4.8 mg protein/ml) were preloaded with D-[^{3}H]-aspartate at a final concentration of 0.2 μM (6.2 μCi) with the conditions described above, except that the volume of this suspension was increased four-fold. Aliquots (750 μl) were transferred to four thermostated (37°C) parallel superfusion chambers (Raiteri et al., 1974) and gently washed under moderate vacuum with 10 ml of the standard oxygenated medium (described above), containing 16 mM glucose. Glomeruli were then superfused at a rate of 0.5 ml/min and fractions were collected every 60 sec. Spontaneous efflux stabilized after 17 min of superfusion under these conditions and the standard medium was then replaced by a medium containing the test substance(s). The concentration of KCl in the medium was varied by equimolar substitution with NaCl; certain test compounds were added to the superfusion medium in dimethyl sulfoxide (DMSO) after DMSO was shown to have no effect on spontaneous efflux. All fractions were collected directly into scintillation counting vials and radioactivity was measured in an LKB Rackbeta liquid scintillation counter with a counting efficiency of approximately 36%. Glomerular particles retained on the Whatman GF/C filters, that were placed at the bottom of the superfusion chambers, were solubilized in 0.5 ml of 2% sodium dodecyl sulfate prior to counting. Greater than 75% of the radioactivity in the superfusates collected following K^+-evoked release was recovered as D-aspartate as shown by ion-exchange chromatography on a Waters^{TM} HPLC amino acid analysis system (data not shown).

The percentage of radioactivity released per 60 sec interval was calculated according to the method of Levi and Raiteri (1973). These data were plotted as a function of superfusion time to allow for graphic analysis. In addition, each curve was integrated and Student's t-test was used to detect significant differences in the levels of evoked release.

Metabolism of [^{3}H]arachidonic Acid

Glomerular membranes were isotopically labeled with [^{3}H]arachidonic acid (83.6 Ci/mmol; New England Nuclear) according to Kelleher and Sun (1985), with some modifications. Glomeruli were isolated from 10 to 15 rat cerebella prior to suspension in 5 ml of buffer (0.32 M sucrose, 50 mM Tris, 1 mM MgSO$_4$, 3 mM dithiothreitol, 0.1 mM CoA, 2.5 mM ATP, pH 7.3), which contained 2 μCi of [^{3}H]arachidonic acid. After radiolabeling for 15 min at 37°C, the glomeruli were split into four equal fractions prior to sedimentation in a microfuge for 15 sec. The pellets were resuspended in a buffer containing 0.32 M sucrose, 50 mM Tris, 1 mM MgSO$_4$ and 1 mg/ml fatty acid free bovine serum albumin. These suspensions were kept at 4°C for 15 min, to extract unincorporated isotope. Glomeruli were sedimented for 15 sec in a microfuge and the pellets were washed three times with the same buffer. These procedures were found to remove approximately 95% of the [^{3}H]arachidonic acid. The labeled glomeruli were then suspended in 1.0 ml of either a low K^+ (5 mM) or high K^+ (45 mM) buffer, with and without exogenous Ca^{2+} present. In the buffers which were devoid of divalent cations, Ca^{2+} and Mg^{2+} were replaced with equimolar concentrations of Na_2HPO_4. The suspensions were incubated at 37°C for the times indicated, when 100 μl aliquots were removed and the total lipids extracted according to Folch et al. (1957). Free fatty acids were separated by thin layer chromotography (Freeman and West, 1966), visualized with iodine vapors, scraped and counted for radioactivity. The results are expressed as percentage of matched 0 time controls, in order to correct for differences in the levels of isotope incorporated into the total glomerular fraction.

RESULTS

D-[^{3}H]asparate Uptake

Glomerular particles incubated in a modified Krebs-Ringer medium with 2×10^{-7} M D-[^{3}H]aspartate showed a rapid and linear rate of ^{3}H uptake over the

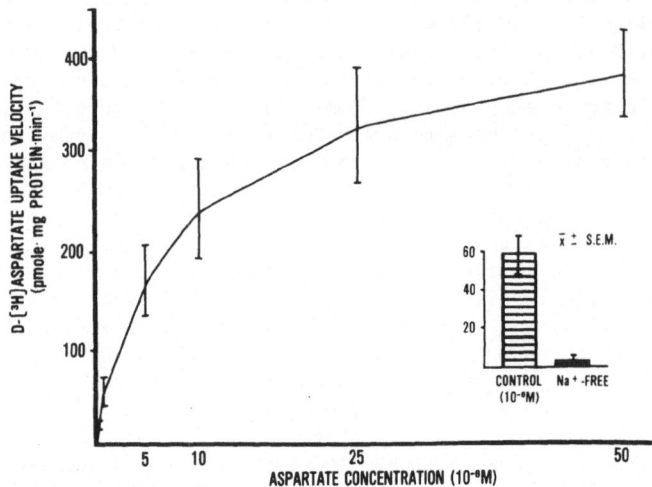

FIG. 1: Substrate concentration and sodium dependence of D-[^{3}H]-aspartate
uptake by cerebellar glomeruli. Glomerular particles were incubated
as described in the text and unlabeled D-aspartate was added to give
a range of concentrations from 0.2 to 50 uM. The sodium dependence
of D-[^{3}H]aspartate uptake was examined by equimolar substitution of
choline chloride for NaCl and measuring uptake in the presence of
1.0 uM radiolabeled D-aspartate (insert). Results are expressed as
means ± S. E. M. of triplicates from four experiments.

initial 10 min of incubation (data not shown). The uptake reached a peak at
553 pmole/mg glomerular protein within 20 min. The isotope was found to be
incorporated into membrane bounded structures, since hypotonic shock of the
glomeruli resulted in liberation of 89% of the D-[^{3}H]aspartate. D-[^{3}H]aspar-
tate concentrations between 2×10^{-7} and 5×10^{-5} M were used to show that the
uptake of the D-[^{3}H]-aspartate was saturable and the kinetics of this process
could be estimated according the Michaelis-Menton equations (Figure 1). The
kinetic parameters were determined by linear regression analysis of double-
reciprocal plots according to the format of Lineweaver-Burke. From this analysis it was determined that the high affinity uptake sites for D-[^{3}H]aspar-
tate have an estimated affinity (K_T) of 5.86 μM and a maximal uptake velocity
(V_{max}) of 384.6 pmole/mg protein/min. The sodium dependence of this uptake
process was tested at a substrate concentration of 2×10^{-7} M and the results
are also shown in Figure 1 (insert). Equimolar substitution of sodium ions
with sucrose in the incubation medium reduced D[^{3}H]aspartate uptake by 92.8
± 1.3% (mean ± S.E.M.).

D-[^{3}H]aspartate Release

The method used to quantify the effects of various compounds on D-[^{3}H]-
aspartate release is illustrated in Figure 2. The release response curves
were plotted as functions of time and the portions between 17 and 26 min of
superfusion were integrated (Figure 2a). The integrated value represents the
total release (spontaneous efflux + evoked release). The spontaneous efflux
was determined by continuous superfusion with a standard medium and the difference between that release and the release measured in buffers containing
various treatments represents the evoked release. This constant comparison
of spontaneous and evoked release allows for statistical, as well as graphi-
cal analysis of the data and provides a more objective basis for interpretation of the results.

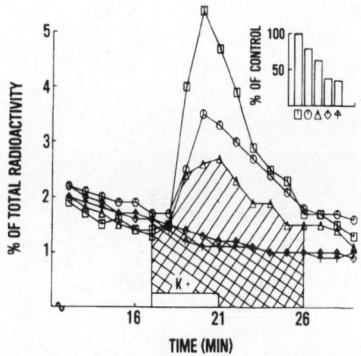

FIG. 2: Effects of external potassium on the release of D-[^{3}H]-aspartate.
A: Glomerular particles were preloaded with isotope as described
in the text, superfused (0.5 ml/min) for 17 min with standard
medium ($\Diamond$) and then exposed to various concentrations of KCl ($\square$ =
70 mM, $\oslash$ = 45 mM, $\triangle$ = 30 mM). The effects of DMSO ($\spadesuit$, final concentration = 0.5%) were also tested. Each curve represents the
average of four separate determinations. B: The relative contribution of spontaneous efflux to total release was quantitatively examined by integrating each response curve and expressing the results
as a percentage of the total response evoked by depolarization with
70 mM KCl.

The rate of spontaneous efflux was generally less than 1.6% per min of
the total radioactivity recovered immediately prior to stimulation, indicating that leakage from nerve terminals was minimal (Levi, 1984), and the level
of D-[^{3}H]aspartate release shows a graded reponse to increasing concentrations of exogenous K^+ (Figure 2a,b). Although the relative contribution of
spontaneous efflux to the total release diminishes as the K^+ concentration is
raised, about 50% of the total release measured at "physiological" K^+ concentrations (< 50 mM KCl) was attributable to the spontaneous efflux of the isotope (Figure 2b). Therefore, this baseline efflux was subtracted from the
total response and the difference was expressed as a percentage of the
release evoked under a specified control condition. DMSO alone, at the con-

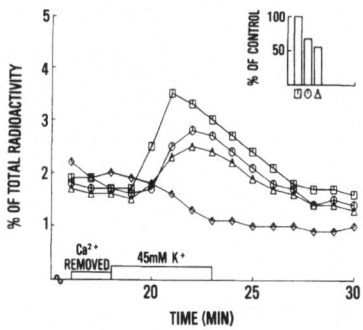

FIG. 3: Calcium dependence of K^+-evoked release of D-[^{3}H]aspartate. Superfusion with a standard medium was continued until spontaneous efflux
($\spadesuit$) was stable. The glomeruli were depolarized with 45 mM KCl, in
the presence ($\square$) and absence ($\oslash$) of Ca^{2+}. Verapamil ($\triangle$, 100 µM)
was added to the high K^+ medium at 17 min in the presence of Ca^{2+}
(2.7 µM). Each curve represents the average of duplicates from
three separate experiments.

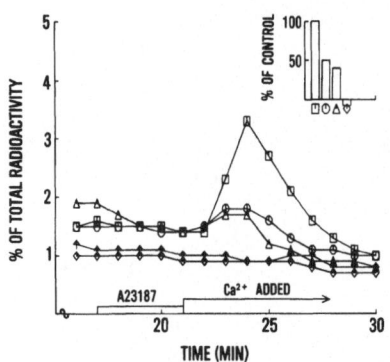

FIG. 4: Calcium-induced release of D-[^{3}H]aspartate in the presence of the
Ca^{2+} ionophore A23187. After superfusing glomeruli for 17 min
with standard medium, superfusion was continued with the indicated
changes. (◇= standard medium, ◆ = 0.5% DMSO; ▱= 50μM,◑= 30 μM
and △ = 10 μM A23187).

centration used in the treatment experiments (0.5%, v/v), was found to cause
no significant change in baseline release (Figure 2b).

Approximately 40% of the K$^+$-evoked release of D-[^{3}H]aspartate could be
eliminated by either the omission of Ca^{2+} or the addition of the Ca^{2+} antagonist verapamil (100 μM) to the superfusion medium (Figure 3). Both treatment
effects were significant (p<0.0005) and the presence of the Ca^{2+} ionophore
A23187 and 2.7 mM Ca^{2+} did stimulate D-[^{3}H]aspartate release in the absence
of membrane depolarization. In this case, increasing the concentration of
A23187 in the conditioning medium from 30 to 50 μM significantly (p<0.001)
enhanced the D-[^{3}H]-aspartate release induced by a fixed concentration of
Ca^{2+} (Figure 4). Also, the presence of veratrine (0.1 mg/ml) in the superfusion medium stimulated a significant amount of isotope release (p<0.0005)
and this form of depolarization-induced release was completely blocked by
the simultaneous addition of 1 μM tetrodotoxin (Figure 5).

Effects of Exogneous Fatty Acids on D-[^{3}H]aspartate Release

The effects of exogenous, unesterified fatty acids on D-[^{3}H]-aspartate

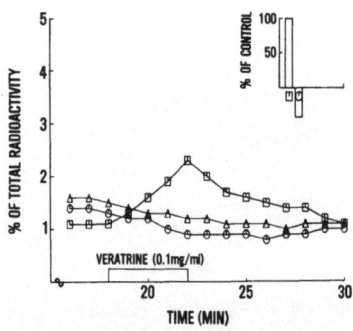

FIG. 5: Veratrine-induced release of D-[^{3}H]aspartate. Isotopically labeled
glomerular particles were superfused with a medium containing 0.1
mg/ml of veratrine, with (▱) or without (◑) 1.0 μM tetrodotoxin
(△ = spontaneous efflux). Curves represent the average of duplicates from three separate experiments.

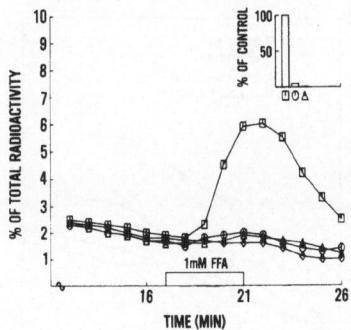

FIG. 6: Effects of exogenous fatty acids on D-[³H]aspartate release. Glomerular particles were superfused for 17 min with standard medium prior to addition of 1.0 mM fatty acids (⊡ = arachidonic acid, ⟳ = linoleic acid, △ = palmitic acid, ◇ = DMSO control; n = 4).

release from glomerular particles were tested. The fatty acids were solubilized in DMSO (final concentration of the DMSO was 0.5%) and were added the to control superfusion medium. All fatty acids were originally tested at 1 mM concentrations (Figure 6). The presence of palmitic acid had no effect on D-[³H]aspartate release (101% of unstimulated control value) and linoleic acid also did not significantly alter release (105% of control release). However, arachidonic acid stimulated a significant (p<0.0005) increase in D-[³H]aspartate release, which was 209% of the unstimulated controls. This release was dependent on the concentration of arachidonic acid up to 1 mM (Figure 7) and was not dependent on the presence of exogenous Ca^{2+} (Figure 9).

Effects of K^+-evoked Depolarization on The Accumulation of Unesterified [³H]arachidonic Acid

Depolarization of glomerular membranes with 45 mM KCl stimulated the accumulation of unesterified [³H]arachidonic acid and this effect was dependent on the presence of external Ca^{2+} (Figure 8). The level of free [³H]-arachidonic acid increased to 88% over the unincubated (0 min) control, in the absence of membrane depolarization, following 15 min of incubation. The level of free [³H]arachidonic acid increased more than two-fold over the unstimulated control, following 15 min of incubation, when glomeruli were

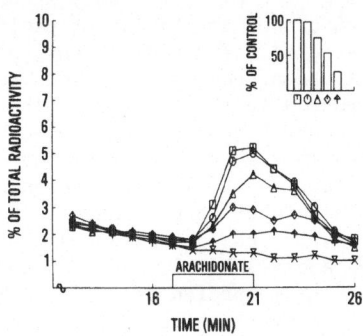

FIG. 7: Concentration dependence of the arachidonic acid-induced release of D-[³H]aspartate. Prelabeled glomeruli were exposed to various concentrations of arachidonic acid (⊡ = 2.5 mM, ⟳ = 1.0 mM, △ = 0.75 mM, ◇ = 0.50 mM, ⊀ = 0.025 mM, ✕ = DMSO control; n = 4).

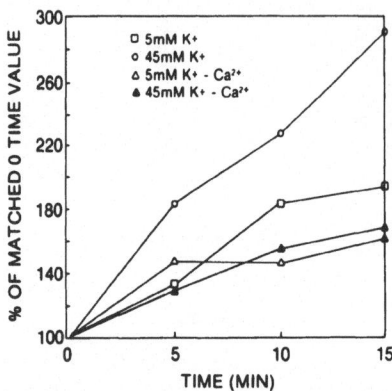

FIG. 8: K^+ -evoked accumulation of unesterified [^{3}H]arachidonic acid. Glomerular particles were prelabeled with [^{3}H]-arachidonic acid as
described in the text. They were then incubated for 0, 5, 10 and
15 min in various test buffers as indicated in the figure. Each
curve represents duplicate determinations from three separate experiments.

depolarized in the presence of Ca^{2+}. Omission of Ca^{2+} prevented the K^+-
evoked accumulation of [^{3}H]arachidonic acid, such that the observed labeling
was 33, 40 and 30% of the K^+-evoked values at 5, 10 and 15 min, respectively.

Effects of Cyclooxygenase Inhibitors on Exogenous Arachidonic Acid-induced Release of D-[^{3}H]aspartate

The presence of 750 µM arachidonic acid induced D-[^{3}H]aspartate release
that was 259% of the value obtained in the presence of the DMSO vehicle
alone. Evoked release was not significantly reduced when Ca^{2+} was omitted
from the perfusion medium. However, the release of D-[^{3}H]-aspartate was
reduced to 188% when 25 µM indomethacin was included with the arachidonic
acid and this value was further reduced to 118% in the presence of 25 µM
ibuprofen (Figure 9).

Effects of cyclooxygenase inhibitors on the K^+-evoked release of D-[^{3}H]aspartate

Cyclooxygenase inhibitors were added to the superfusion buffer containing a concentration of K^+ (45 mM) previously shown to evoke a submaximal
release (Figure 2), so that either inhibitory or excitatory effects of the
test compounds could be detected. Three different cyclooxygenase inhibitors
were tested, because it has been shown that cyclooxygenases from different
sources can respond differently to the various inhibitors. The inhibitors
were included at 25 µM concentrations and their effects on K^+-evoked D-[^{3}H]-
aspartate release are shown in Figure 10. Indomethacin and acetylsalicylic
acid had no significant effect on the K^+-evoked transmitter release at 25 µM
concentrations. However, at higher concentrations, both inhibitors did block
some of the K^+-evoked release. The inhibition was 39 and 31% in the presence
of 250 µM indomethacin and acetylsalicylic acid, respectively (data not
shown). The addition of 25 µM ibuprofen to the high K^+ medium caused a significant inhibition of D-[^{3}H]aspartate release (54% of the evoked release;
$p < 0.001$). This inhibition was dose dependent, since 125 µM ibuprofen caused
a 61% inhibition and 250 µM caused a 78% inhibition of evoked D-[^{3}H]aspartate
release (data not shown). The addition of a 5 µM 15-hydroxyeicosatetraenoic
acid (15-HETE; Seragen) to the depolarizing medium significantly potentiated
the K^+-evoked release (182% of stimulated controls; $p < 0.0005$; Figure 10).

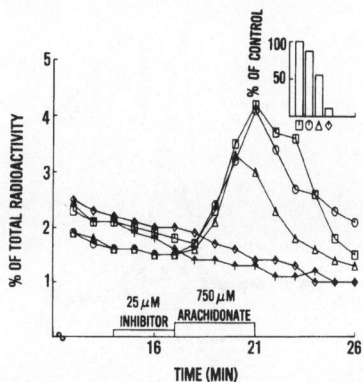

FIG. 9: Effects of cyclooxygenase inhibitors on arachidonic acid-induced release of D-[^{3}H]aspartate. Arachidonic acid (0.75 mM) was added in the presence (▯) or absence (◐) of Ca^{2+}. Indomethacin (△) or ibuprofen (◇) was added to the medium (25 μM) three min prior to the addition of arachidonic acid (◆ = DMSO control; n = 4).

Effects of Exogenous Prostaglandins and Leukotrienes on D-[^{3}H]aspartate Release

The effects of exogenous PGF_{2alpha}, PGE_2 (Sigma) and LTC_4 (Seragen) on D=[^{3}H]aspartate release were determined and compared to the release evoked with 45 mM K^+. PGF_{2alpha} induced a dose-dependent release of the D-[^{3}H]aspartate, since 0.5, 5.0 and 50 μm concentrations evoked release that was 161, 188 and 261% of that observed in the presence of 45 mM K^+, respectively (Figure 11). The presence of 50 and 100 μM PGE did stimulate D-[^{3}H]aspartate release, but that release was 78 and 94% of that observed in the presence of depolarizing K^+, respectively (Figure 11). Exogenous LTC_4 was also able to evoke release; the amount of release was 83, 85 and 95% of the K^+-evoked release, when leukotriene concentrations were 0.5, 1 and 5 μM, respectively (Figure 12).

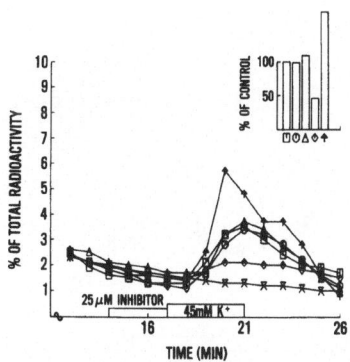

FIG. 10: Effects of cyclooxygenase inhibitors and 15-HETE on K^+-evoked release of D-[^{3}H]aspartate. Glomeruli were depolarized with 45 mM KCl at 17 min (▯). Idomethacin (◐), acetylsalicylic acid (▲) or ibuprofen (◈) was added to the medium (25 μM) three min prior to depolarization. 15-HETE (◆ , 5.0 μM) was added at the time of depolarization (✗ = 5 mM KCl control; n = 4).

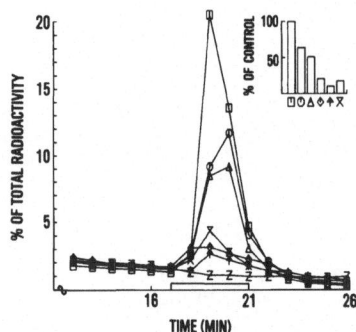

FIG. 11: Effects of exogenous prostaglandins on D-[^{3}H]aspartate release.
At 17 min, various concentrations of prostaglandins were added to
the standard medium in DMSO (Z) or the glomeruli were depolarized
with 45 mM KCl ($\Diamond$). PGF$_{2alpha}$ was tested at 50 μM ($\square$), 5.0 μM
(O) And 0.5 μM ($\triangle$) concentrations. PGE$_2$ was tested at 100 μM
(X) and 50 μM (4) concentrations (n = 4).

DISCUSSION

Mossy fibers comprise one of the two principal afferent systems projecting to the cerebellar cortex, where they terminate in a large glomerular synaptic complex. The distinctive and highly uniform architecture of
this structure has allowed investigators to isolate cerebellar glomeruli as
a reasonably homogeneous subcellular fraction (Hajos et al., 1975; Terrian
et al., 1985). The isolated cerebellar glomerulus appears to possess some
rather favorable characteristics for the study of neurotransmitter release,
when compared to tissue slice and synaptosomal preparations. Athough slices
contain intact neuronal circuits, which lend themselves to sophisticated
interdisciplinary investigations, their structural complexity and heterogeneity has frustrated attempts to correlate certain biochemical properties
with specific neuronal entities (Levi, 1984). Such correlations have also
been difficult to obtain in synaptosomal preparations, which are composed
of a biochemically heterogeneous population of nerve terminals with no
common site of origin within the intact neuronal circuitry. In contrast,
many synaptic articulations between the principal structural components of

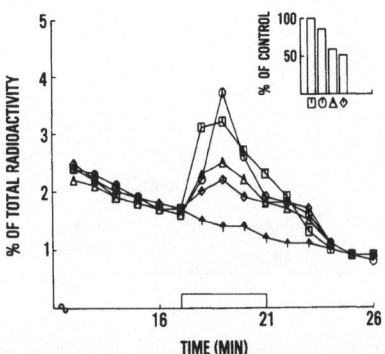

FIG. 12: Effects of exogenous leukotriene C$_4$ (LTC$_4$) on D-[^{3}H]-aspartate
release. At 17 min, various concentrations of LTC$_4$ (O = 5.0 μM,
$\triangle$ = 1.0 μM and $\Diamond$ = 0.5 μM) were added to the standard medium (4)
or the glomeruli were depolarized with 45 mM KCl in the absence
of LTC$_4$ ($\square$) (n = 4).

246

isolated glomeruli are retained. Large mossy fiber rosettes remain in contact with the numerous granule cell dendrites surrounding their periphery and Golgi axon terminals retain attachments to these same dendritic processes. The preservation of these neuronal interconnections, which collectively form a fundamental component of the cerebellar circuitry, liken the isolated glomerulus to a "microslice" composed of morphologically and physiologically well-defined structures. These characteristics enhance the ability to correlate the biochemical properties of glomeruli with their specific neuronal components. For example, with this preparation investigators have been able to demonstrate that the high affinity uptake sites for GABA are exclusively localized along the inhibitory Golgi axon terminals (Wilkin et al., 1974; Kelly et at., 1975).

The data reported in this paper indicate that glomerular particles also possess a high affinity uptake system capable of rapidly terminating the excitatory action of acidic amino acids. The estimated K^+ for D-$[^3H]$-aspartate uptake (5.86 µM) is in agreement with the value previously reported by Wilson et al., (1976) for $[^3H]$glutamate uptake by cerebellar glomeruli (4.9 µM). Estimates of the V_{max} for putative acidic amino acid transmitter uptake are also comparable and are 55-fold greater than those previously reported for either $[^3H]$serotonin (Terrian et al., 1985) or $[^3H]$choline (Terrian et al., in press) in this nerve ending preparation. Thus, while such uptake studies can not differentiate between the transmitter candidates glutamate and aspartate, they do suggest that uptake sites for these amino acids are either more abundant or more active than those for serotonin or choline. This finding is in agreement with immunohistochemical studies in which the serotonergic and cholinergic mossy fibers were described as having discrete patterns of innervation which only represent a small proportion of this nerve terminal population (Chan-Palay, 1977; Kan et al., 1980).

It should be noted that the rate of D-$[^3H]$aspartate accumulation observed in these experiments is likely to overestimate the activity of the mossy fiber system. Both glial contaminants and granule cell dendrosomes would be expected to contribute to determinations of D-$[^3H]$aspartate uptake (DeBarry et al., 1982; Levi et al., 1982). This limits the usefulness of D-aspartate as a false transmitter in uptake and release studies in crude synaptosomal preparations contaminated with glial fragments (Potashner and Gerard, 1983). However, glial cell uptake is thought to be significantly reduced by homogenization (Fonnum, 1984) and autoradiographic studies of cerebellar slices have been used to show that labeling of granule cell dendrites with D-$[^3H]$- asparate is minimal (Wilkin et al., 1982). It seems likely, therefore, that the considerable disparity between V_{max} values for the above uptake systems may be reflective of a relatively large subclass of mossy fibers which utilize acidic amino acids as transmitters.

To further assess the role of acidic amino acids as neurotransmitters in the cerebellar glomerulus, a series of studies were conducted to characterize D-$[^3H]$aspartate release. Preloaded glomerular particles exhibited a substantial non-Ca^{2+}-dependent release of the exogenous amino acid in response to membrane depolarization. Such an observation is common in studies of amino acid release (Levi, 1984) and it has been hypothesized that this component of the response could occur through mobilization of intracellular Ca^{2+} stores (Sandoval, 1980). If this were the case, it would not be possible to completely suppress evoked release by omitting exogenous Ca^{2+}. Alternatively, the non-Ca^{2+}-dependent release could originate from glial contaminants rather than neuronal structures (Levi et al., 1982). It should be added, however, that the depolarization-induced release of D-$[^3H]$aspartate was significantly potentiated by Ca^{2+} (Figure 3) and that this component of evoked release cannot be attributed to gliosomal contaminants, since (1) approximately 40% of the K^+-evoked release could be eliminated by either the omission of Ca^{2+} or the addition of verapamil to the superfusion medium,

(2) release was stimulated by Ca^{2+}, in the absence of membrane depolarization, when the glomeruli were pretreated with the ionophore A23187, and (3) veratrine induced a release which could be completely blocked by tetrodotoxin. Therefore, it appears as though the observed release is of neuronal origin and, considering the morphology of this nerve ending preparation, it is reasonable to assume that the mossy fiber terminals are the primary contributor to the Ca^{2+}-dependent release of radiolabeled D-aspartate. In view of these findings, isolated cerebellar glomeruli provide a useful model for the study of presynaptic mechanisms involved in the depolarization-induced release of neurotransmitters.

The effects of unesterified fatty acids on D-$[^3H]$aspartate release were examined, since it has been suggested that the Ca^{2+}-dependent accumulation of free arachidonic acid may be required for some forms of transmitter release (Majewska and Sun, 1982; Lazarewicz et al., 1983; Rhoads et al., 1983). Our results with cerebellar glomeruli are consistent with these reports, since we found that membrane depolarization caused a Ca^{2+}-dependent accumulation of unesterified arachidonic acid. The potential involvement of arachidonic acid in the release of D-$[^3H]$aspartate was further substantiated by the demonstration that exogenous arachidonic acid was able to mimic the effects of depolarizing KCl concentrations. This arachidonic acid-evoked release was dose dependent and did not depend on external Ca^{2+}. In contrast, exogenous palmitic and linoleic acids had no effect on transmitter release.

We interpret the data presented here to indicate that the accumulation of arachidonic acid is a required step for neurotransmitter release by the isolated mossy fiber terminals. It has also been suggested that conversion of unesterfied fatty acids to prostaglandins, or related compounds, is required for transmitter release. We employed cyclooxygenase inhibitors and products of the cyclooxygenase and lipoxygenase pathways to test this possibility.

It was clear that cyclooxygenase inhibitors affected the K^+-evoked release of D-$[^3H]$aspartate from glomerular particles. The effects of indomethacin, acetylsalicylic acid and ibuprofen were examined and it was found that all of these inhibitors could affect D-$[^3H]$aspartate release. However, ibuprofen was the most potent and showed a dose-dependent inhibition of the K^+-evoked release. It is not surprising that there was differential sensitivity of the glomerular cyclooxygenase(s) to the various inhibitors, since cylcooxygenases from other cell types have been shown to differ in their responses to inhibitory compounds (Tolman et al., 1983). In fact, there also appears to be differential sensitivity within the central nervous system. Bradford et al. (1983) have reported that indomethacin and acetylsalicylic acid inhibit the K^+-evoked release of dopamine from cerebral synaptosomes at inhibitor concentrations that had no effect on D-$[^3H]$aspartate release from this cerebellar preparation. However, similar concentrations of ibuprofen did significantly inhibit D-$[^3H]$aspartate release from the isolated glomeruli.

The involvement of prostaglandins in neurotransmitter release from the mossy fiber terminals was substantiated by the finding that exogenous prostaglandins were potent stimulators of D-$[^3H]$aspartate release in the absence of depolarizing K^+. We found that PGE_2 was able to partially mimic the effects of membrane depolarization, with 100 μM PGE_2 stimulating release that was 94% of that observed in the presence of 45 mM K^+, but the presence of PGF_{2alpha} stimulated D-$[^3H]$aspartate release that was greater than that observed in the presence of either a depolarizing concentration of KCl or PGE_2. Also, the PGF_{2alpha}-evoked release was dose dependent, with the range of concentrations tested (0.5 to 50 μM).

We also found that two of the products of the lipoxygenase pathway had an effect on D-[^{3}H]aspartate release. LTC_4 was able to partially mimic the effects of membrane depolarization, but again these effects were much less pronounced than those observed in the presence of PGF_{2alpha}. The presence of 15-HETE (5 µM) also stimulated release and in this case the evoked release was 184% of that observed with 45 mM KCl.

The question then remains as to which one(s) of the many arachidonic acid metabolites is most likely to be involved in the release of acidic amino acids from the cerebellar glomerulus. PGF_{2alpha} appears to be the best candidate for such involvement, since it elicited responses at the lowest concentrations and cyclooxygenase inhibitors were able to inhibit K^+-evoked D-[^{3}H]aspartate release. However, lipoxygenase products cannot be eliminated as potential candidates. That prostaglandins, and not lipoxygenase products, are required for evoked release is supported by the finding that ibuprofen, and to a lesser extent indomethacin were able to inhibit D-[^{3}H]aspartate release in the presence of 750 µM arachidonic acid. It has been shown that ibuprofen activates 5-lipoxygenase in neutrophils (Myers and Siegel, 1983) and also causes the accumulation of 15-HETE in human polymorphonuclear eukocytes (Vanderhoek and Bailey, 1985). If such effects are also occurring in the mossy fiber terminals, then 15-HETE is not directly involved in evoked release, but may be active by shunting free arachidonic acid through the cyclooxygenase pathway. In this context, it is important to note that free [^{3}H]arachidonic acid did accumulate in glomerular fractions prelabeled with the isotope, but not exposed to depolarizing conditions. Perhaps the presence of 15-HETE or LTC_4 allows for sufficient cyclooxygenase activity such that release can be evoked. This possibility will be tested by the inclusion of more specific inhibitors of the cyclooxygenase and lipoxygenase enzymes in the presence of depolarizing conditions and exogenous arachidonic acid.

ACKNOWLEDGMENTS

We would like to acknowledge the fine technical support of J. Grassel, M. Hofstetter, J. Stewart and N. Edgehouse. We would also like to thank Karen and Daniel D. for their consideration during the preparation of this manuscript. This research was supported by AFOSR grants #2312W3 (DMT) and #86-0045 (RVD).

REFERENCES

Asakura, T. and Matsuda, M., Efflux of 4-aminobutyric acid from and appearance of free arachidonic acid inside synaptosomes, Biochim Biophys Acta, 773:301-307 (1984).

Bradford, P. G., Marinetti, G. V. and Abood, L. G., Stimulation of phospholipase A_2 and secretion of catecholamines from brain synaptosomes by potassium and A23187, J Neurochem 41:1684-1693 (1983).

Chan-Palay, V., The indoleamine afferent axons to the cerebellum, in: "Cerebellar Dentate Nucleus: Organization, Cytology and Transmitters," Springer-Verlag, New York (1977).

deBarry, J., Langley, O. K., Vincendon, G. and Gombos, G., L-Glutamate and L-glutamine uptake in adult rat cerebellum: an autoradiographic study, Neuroscience, 7:1289-1297 (1982).

Folch, J., Lees, M. and Sloane-Stanley, G. H., A simple method for the isolation and purification of total lipids from animal tissues, J Biol Chem, 226:497-509 (1957).

Fonnum, F., Glutamate: a neurotransmitter in mammalian brain, J Neurochem, 42:1-11 (1984).

Freeman, C. P. and West, D., Complete separation of lipid classes on a single thin-layer plate, J Lipid Res, 7:324-327 (1966).

Hajos, F., Wilkin, G., Wilson, J. and Balazs, R., A rapid procedure for obtaining a preparation of large fragments of the cerebellar glomeruli in high purity, J Neurochem, 24:1277-1278 (1975).

Hedqvist, P., Autonomic neurotransmission, in: "The Prostaglandins," Plenum Press, New York (1973).

Kan, K., Chao, L. and Forno, L., Immunohistochemical localization of choline acetyltransferase in the human cerebellum, Brain Res, 193:165-171 (1980).

Kelleher, J. A. and Sun, G. Y., Enzymic hydrolysis of arachidonylphospholipids by rat brain synaptosomes, Neurochem Int, 7:825-831 (1985).

Kelly, J. S., Fabienne, D. and Schon, F., The autoradiographic localization of the GABA-releasing nerve terminals in cerebellar glomeruli, Brain Res, 85:225-259 (1975).

Lazarewicz, J. W., Leu, V., Sun, G. Y. and Sun, A. Y., Arachidonic acid release from K^+-evoked depolarization of brain synaptosomes, Neurochem Int, 5:471-478 (1983).

Levi, G., Release of putative transmitter amino acids, in: "Handbook of Neurochemistry," Plenum Press, New York (1984).

Levi, G., Gordon, R. D., Gallo, V., Wilkin, G. P. and Balazs, R., Putative acidic amino acid transmitters in the cerebellum I. depolarization-induced release, Brain Res, 239:425-445 (1982).

Levi, G. and Raiteri, M., GABA and glutamate uptake by subcellular fractions enriched in synaptosomes: critical evaluation of some methodological aspects, Brain Res, 57:165-185 (1973).

Lowry, O. H., Rosebrough, N. J., Farr, A. L. and Randall, R. J., Protein measurement with Folin phenol reagent, J Biol Chem, 193:265-275 (1951).

Majewska, M. D. and Sun, G. Y., Activation of arachidonylphosphatidylinositol and phosphatidylcholine turnover by K -evoked stimulation of brain synaptosomes, Neurochem Int, 4:427-433 (1982).

Myers, R. F. and Siegel, M. I., Differential effects of anti-inflammatory drugs on lipoxygenase and cyclooxygenase activities of neutrophils from a reverse passive Arthus reaction, Biochem Biophys Res Commun, 112:586-594 (1983).

Potashner, S. J. and Gerard, D., Kainate-enhanced release of D-[H]-aspartate from cerebral cortex and striatum: reversal by baclofen and pentobarbital, J Neurochem, 40:1548-1557 (1983).

Raiteri, M., Angelini, F. and Levi, G., A simple apparatus for studying the release of neurotransmitters from synaptosomes, Eur J Pharmacol, 25:411-414 (1974).

Rhoads, D. E., Osburn, L. D., Peterson, N. A. and Raghupathy, E., Release of neurotransmitter amino acids from synaptosomes: enhancement of calcium-independent efflux by oleic and arachidonic acids, J. Neurochem, 41:531-537 (1983).

Sandoval, M. E., Sodium-dependent efflux of [^{3}H]GABA from synaptosomes probably related to mitochondrial calcium mobilization, J Neurochem, 35:915-921 (1980).

Terrian, D. M., Butcher, W. I., Wu, P. H. and Armstrong, D. L., Isolation of glomeruli from areas of bovine cerebullum and comparison of [^{3}H]seroonin uptake, Brain Res, 14:469-475 (1985).

Terrian, D. M., Noison, E. L. and Thomas, W. E., Choline uptake by glomeruar synapses isolated from bovine cerebellar vermis, Brain Res, in press (1986).

Tolman, E. L., Fuller, B. L., Marinan, B. A., Capetola, R. J., Levinson, S. L. and Rosenthale, M. E., Tissue selectivity and variability of effects of acetaminophen on arachidonic acid metabolism, Prostanglandins, Leukotriends Med., 12:347-356 (1983).

Vanderhoek, J. Y. and Bailey, J. M., Postphospholipase activation of lipoxygenase/leukotriene systems, in: "Prostaglandins, Leukotrienes and Lipoxins," Plenum Press, New York (1985).

Wilkin, G. P., Garthwaite, J. and Balazs, R., Putative acidic amino acid transmitters in the cerebellum II. electron microscopic localization of transport sites, <u>Brain Res</u>, 244:69-80 (1982).

Wilkin, G., Wilson, J. E., Balazs, R., Schon, F. and Kelly, J. S., How selective is high affinity uptake of GABA into inhibitory nerve terminals?, <u>Nature</u>, 252:397-399 (1974).

Wilson, J. E., Wilkin, G. P. and Balazs, R., Metabolic properties of a purified preparation of large fragments of the cerebellar glomeruli: glucose metabolism and amino acid uptake, <u>J. Neurochem</u>, 26:957-965 (1976).

MOLECULAR MECHANISMS OF β-ADRENERGIC RECEPTOR

DESENSITIZATION

David R. Sibley, Jeffrey L. Benovic, Marc G. Caron
and Robert J. Lefkowitz

Howard Hughes Medical Institute, Departments of
Medicine, Biochemistry and Physiology, Duke
University Medical Center, Durham, North Carolina
27710

INTRODUCTION

Desensitization or adaptation is well known in biological regulation.
Also referred to as tachyphylaxis, tolerance or refractoriness, it is most
commonly observed as a loss of cellular responsiveness to a neurotransmitter
or drug after repeated or prolonged exposure to that agent. Examples of
systems in which desensitization is observed include chemotaxis of bacteria
or mammalian polymorphonuclear leukocytes, neurotransmission by various
neurotransmitters at synapses, stimulation of diverse physiological processes in eukaryotes by many drugs and hormones, and sensory perception. In
the context of clinical therapeutics, desensitization significantly limits
the efficacy of numerous pharmacological agents.

Common to most systems which display desensitization is the existence
of receptors which mediate the effects evoked by the specific stimuli.
Since such receptors constitute the first point of interaction of biologically active stimuli with cells, it is not unreasonable to suppose that
regulation of receptor function might constitute the basis for some forms
of desensitization.

Hormone-induced desensitization has been extensively investigated in
a variety of tissues and cells that contain β-adrenergic receptors coupled
to the stimulation of adenylate cyclase activity. Although the biochemical
mechanisms for producing β-adrenergic receptor coupled adenylate cyclase
desensitization appear to be diverse, two major categories of refractoriness have been identified. One type is referred to as "homologous" and is
characterized by the fact that only stimulation by the desensitizing hormone
is attenuated. In contrast, "heterologous" desensitization is characterized
by a diminished responsiveness to additional hormones and occasionally to
other activators of adenylate cyclase including guanine nucleotides and
fluoride.

Investigations of homologous desensitization in a variety of β-adrenergic receptor model systems have indicated that receptor-adenylate cyclase
uncoupling and sequestration of receptors from the cell surface is involved
in the refractoriness. Heterologous desensitization, by contrast, is not
associated with receptor sequestration or down regulation but instead

involves a functional uncoupling of the receptors and other components of the adenylate cyclase system.

Although the phenomenology of β-adrenergic receptor desensitization has been well described, the underlying molecular events have only recently begun to be elucidated. Our aim in this review is to focus on the amphibian and avian erythrocyte model systems employed in our laboratory as well as recent biochemical advances that have furthered our understanding of desensitization processes.

β -Adrenergic Receptor Desensitization In Intact Cell Systems

Homologous Desensitization

Homologous patterns of adenylate cyclase desensitization occur widely and have been extensively investigated. In this form of desensitization, only the subsequent response to the desensitizing hormone is attenuated while the efficacy of other hormones or nonhormonal activators such as guanine nucleotides and fluoride is unimpaired. Our laboratory has employed the frog erythrocyte as a model system with which to study homologous desensitization. When these cells are exposed to β-adrenergic catecholamines, there is a decrease in the number of assayable β-adrenergic receptors in the plasma membranes which parallels a decrease in hormone responsiveness.

Mukerjee et al., (1975; 1976) initially demonstrated that administration of isoproterenol in vivo led to a fall in [3H]dihydroalprenolol binding to frog erythrocyte membranes in addition to a decline in isoproterenol-stimulated adenylate cyclase activity. Mickey et al. (1975; 1976) also showed a similar decrement in [3H]dihydroalprenolol binding and isoproterenol-stimulated adenylate cyclase activity after in vitro desensitization of frog erythrocytes. In these studies, the loss in [3H]dihydroalprenolol binding was due to a decrease in receptor number rather than a change in affinity for the ligand. Moreover, there were no alterations in prostaglandin E_1- stimulated or fluoride-stimulated adenylate cyclase activities indicating that the catecholamine-induced desensitization was of the homologous type.

In addition to the decrease in β-adrenergic receptors in desensitized frog erythrocyte membranes, there is another receptor alteration which occurs upon desensitization. This involves the ability of the receptors to "couple" with the stimulatory guanine nucleotide binding regulatory protein (N_s). β-Adrenergic receptor interactions with this protein can be monitored by radioligand binding techniques. This is done by assessing the ability of an agonist to promote the formation of a high affinity ternary complex of hormone, receptor and N_s (HRN). This complex is manifested in high affinity, nucleotide-sensitive agonist competition curves of radiolabeled antagonist binding. Alternatively, the receptor-N_S complex can be detected directly with radiolabelled agonists such as [3H]hydroxybenzylisoproterenol. This high affinity complex of agonist, receptor and guanine nucleotide regulatory protein is an intermediate in the activation of adenylate cyclase and is rapidly dissociated upon interaction with guanine nucleotides. After desensitization of frog erythrocytes with catecholamines, agonist competition curves are shifted to the right, indicating a diminished ability of agonists to promote high affinity receptor-N_s interactions (Wessels et al., 1979; Kent et al., 1979). It is this inability to form the coupled state of the receptor that is responsible for the observation that direct agonist binding with [3H]hydroxybenzylisoproterenol is even more diminished after desensitization of frog erythrocytes than antagonist [3H]dihydroalprenolol binding (Wessels et al., 1978; 1979).

Thus, there appear to be at least two major defects in β-adrenergic receptor function in desensitized frog erythrocytes. The first is a diminution of overall receptor number in the plasma membrane as assayed with radiolabelled antagonists while the second is an "uncoupling" of the receptors, apparent as rightward shifts in agonist competition curves and a pronounced loss of [^{3}H]hydroxybenzylisoproterenol binding.

Studies of homologous desensitization in cultured mammalian cell lines demonstrated phenomenology similar to that observed in frog erythrocytes. Perkins and colleagues (Su et al., 1979; 1980; Harden et al., 1979; Perkins, 1983) have shown, using astrocytoma cells, that incubation with β-adrenergic agonists for short periods of time leads to a rapid desensitization of isoproterenol-stimulated adenylate cyclase activity. Concomitantly, there is a shift to the right in agonist competition curves of the binding of [^{125}I]-iodohydroxybenzylpindolol, an antagonist ligand. This data is consistent with the diminished ability of the receptor to couple with N_s that is observed in desensitized frog erythrocyte membranes.

Unlike frog erythrocytes, however, short term (e.g. minutes) agonist exposure of astrocytoma cells does not lead to a loss of [^{125}I]iodohydroxybenzylpindolol binding despite the pronounced desensitization and uncoupling (Su et al., 1979; 1980; Harden et al., 1979; Perkins, 1983). If agonist exposure of the cells is continued for longer periods (e.g. hours), however, then a progressive loss in [^{125}I]iodohydroxybenzylpindolol appears. Similar findings have been demonstrated in other cultured cells including S49 lymphoma (Shear et al., 1976; Green and Clark, 1981; Green et al., 1981) and glioma (Homburger et al., 1980; Fishman et al., 1981; Frederich et al., 1983) cells. Thus, in the mammalian cell lines, desensitization and receptor uncoupling appear to temporally precede the actual loss in antagonist binding sites. The data described thus far indicate that homologous desensitization is closely associated with a physical sequestration of the receptor away from the cell surface and the other adenylate cyclase components.

The fate and location of the sequestered or down regulated receptors remains, for the most part, unknown. Chuang and Costa, using frog erythrocytes, initially demonstrated that associated with desensitization and loss of β-adrenergic receptor sites from the plasma membranes was an appearance of receptors in what they defined as the soluble fraction of the cell (Chuang and Costa, 1979; Chuang et al., 1980). This soluble fraction represented a 30,000 x g supernatant which likely contained small membrane particles perhaps containing the receptors. The agonist-promoted appearance of the receptors in this supernatant fraction was blocked by antagonists as was the desensitization (Chuang and Costa, 1979; Chuang et al., 1980).

Harden and collaborators (Harden et al., 1980; Waldo et al., 1983; Toews et al., 1984) have presented further evidence for a redistribution of β-adrenergic receptors from the plasma membrane fraction into smaller membrane particles during the desensitization process. Using 1321N1 astrocytoma cells, it was shown that after the initial, rapid uncoupling and desensitization of the β-adrenergic receptor and adenylate cyclase, all the receptors were still accessible to ligand binding at the cell surface and could be pelleted at 40,000 x g after lysing the cells. However, in the presence of concanavalin A, a sequestered light membrane receptor fraction could be separated from the plasma membranes and the remainder of the receptors and adenylate cyclase on sucrose gradients, indicating a shift in the membrane localization of the receptors. Similar findings have been reported by these authors and others using C6 glioma cells (Frederich et al., 1983; Hertel et al., 1983a,b).

Using the frog erythrocyte model system, our laboratory has shown that a portion of the desensitized, sequestered receptors can be recovered in a light membrane fraction obtained by centrifuging the cell cytosol at 158,000 x g for 1 h (Stadel et al., 1983b). This membrane fraction shows markedly diminished activities of typical plasma membrane marker enzymes including adenylate cyclase. The receptors in these light membrane particles appear to be totally uncoupled from the nucleotide regulatory protein and show only low affinity agonist binding. Reconstitution of cholate solubilized N_s activity from the membrane particles into S49 cyc$^-$ cell membranes (which lack N_s) shows, in fact, that the sequestered membrane fraction is nearly devoid of this protein component. Further characterization of the sequestered receptors by photoaffinity labeling and SDS-PAGE has indicated that the sequestered receptors are not grossly altered or degraded in any fastion. These findings thus suggest that desensitization in frog erythrocytes is associated with a sequestration of the receptors into a membrane domain or compartment that is devoid of the other components of the adenylate cyclase system.

The process of resensitization and the return of normal functioning receptors to the plasma membrane is different in various systems. In most cases, the receptors likely recycle to the cell surface and are reinserted into their normal location in the plasma membrane. Evidence for recycling of the sequestered receptors is the observation that reappearance of receptors at the cell surface can occur even when protein synthesis is blocked (Mukherjee et al., 1976; Su et al., 1976; Strulovici and Lefkowitz, 1984; Doss et al., 1981). In some systems, however, recovery of normal receptor number is attenuated or blocked by protein synthesis inhibitors (Doss et al., 1981; Morishima et al., 1980).

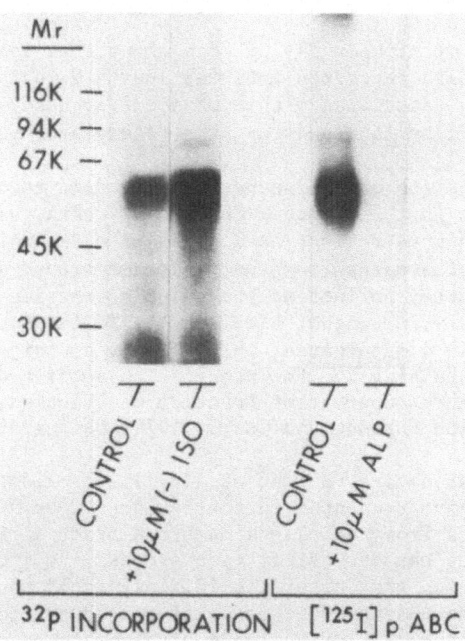

FIG. 1: SDS-PAGE of ^{32}P-labeled and [^{125}I]pABC-labeled β-adrenergic receptors from frog erythrocytes. For ^{32}P labeling, the cells were preincubated with [32]Pi prior to desensitization with 10 μM (-)isoproterenol (+ 10 μM (-)ISO). Subsequent to purification, the receptors in each sample were quantitated by [^{125}I]CYP binding and 4.3 pmol of receptor were loaded onto each lane. For [^{125}I]pABC labeling, frog erythrocyte membranes were incubated with [^{125}I]pABC in the absence (control) or presence of 10 μM alprenolol (+ 10 μM ALP).

Homologous desensitization thus seems to be composed of multiple related events serving to sequester or remove the receptor from its normal position in the plasma membrane. Although the receptor is eventually sequestered from N_S and C during the desensitization process, a central question relates to whether the receptor is functionally modified as well. Strulovici et al. (1983) have tested the functional activity of the sequestered receptors that can be recovered in light membrane particles in frog erythrocytes. By fusing these receptors to a cell (Xenopus laevis erythrocyte) that possesses the adenylate cyclase components N_S and C but lacks β-adrenergic receptors, the sequestered receptors were shown to be functionally active. Similar conclusions have been reached by Clark et al. (1985) in characterizing the sequestered receptors found in S49 cell light membrane fractions. Strasser and Lefkowitz (1985) have also found that treatment of desensitized S49 cells with the fusogen polyethylene gylcol is sufficient to restore the catecholamine-stimulated enzyme activity to control levels. Such studies suggest that homologous desensitization results primarily from the physical sequestration of the receptors. However, results of such studies must be interpreted in the light of the caveat that the reconstitution and/or fusion procedures might reverse a functional alteration.

In contrast, Kassis and Fishman (1984), using membrane-membrane and membrane-cell fusion techniques, have demonstrated that catecholamine-induced homologous desensitization in various cultured cells does indeed result in a functional alteration of the β-adrenergic receptor. Perkins and colleagues (Waldo et al., 1983; Toews et al., 1984) have also presented evidence that very short (1-2 min) exposure of astrocytoma cells to isoproterenol leads to desensitization of hormone-sensitive adenylate cyclase, yet the β-adrenergic receptors are not sequestered at this time. Moreover, pre-treatment of astrocytoma cells with concanavalin A blocks the sequestration process in these cells, yet does not block the desensitization and uncoupling events which occur maximally. These data suggest that a functional change in the receptor occurs early in the desensitization process but that this modification might be reversed by the time sequestration is complete. It is thus not clear if physical sequestration alone can completely account for the receptor alterations observed in homologous desensitization.

The biochemical mechanisms involved in homologous desensitization are not yet known with certainty. Studies of homologous desensitization in cell-free systems have indicated a requirement for phosphorylating conditions (Anderson and Jaworski, 1979; Salomon et al., 1981; Hunzicher-Dunn et al., 1979) and in at least one study (Hunzicher-Dunn et al., 1979) demonstrated that the desensitization could be reversed by phosphatase treatment. Recently, Sibley et al., (1985) have directly shown that homologous desensitization of adenylate cyclase in frog erythrocytes is associated with phosphorylation of the β-adrenergic receptor (Fig. 1). This phosphorylation is induced in a stereo-specific fashion by isoproterenol and is blocked by the β-adrenergic antagonist propranolol. The phosphorylation state of the receptor stoichiometrically increases three-fold to approximately 2 moles of phosphate per mole of receptor upon desensitization. Interestingly, prostaglandin E_1 (PGE_1) does not promote β-adrenergic receptor phosphorylation despite the fact that PGE activates adenylate cyclase in these cells (Sibley et al., 1985). This observation agrees well with the notion that homologous desensitization is not cAMP-mediated. We have recently demonstrated a similar phosphorylation of the β-adrenergic receptor using S49 lymphoma cells (Strasser et al., 1986). The nature of the protein kinase that phosphorylates the β-adrenergic receptor during homologous desensitization is currently under investigation (see below).

There are at least two potential mechanisms by which receptor phosphorylation could contribute to the homologous desensitization. One possibility is that the phosphorylation induces or triggers the sequestration of the receptor from the cell surface. Another possibility is that the phosphorylation results in a functional modification in the receptor protein such that it is less efficacious in activating adenylate cyclase. If the phosphorylation promotes an early occurring functional change in the receptor then, as discussed above, the receptor phosphorylation might be reversed upon sequestration. In fact, we have now been able to demonstrate that the phosphorylation of the sequestered β-adrenergic receptors in frog erythrocytes is indeed reversed toward basal levels (unpublished observations). Moreover, we have evidence that the relevant phosphatase responsible for dephosphorylating the β-adrenergic receptor is contained in the sequestered vesicle membrane compartment (unpublished observations). Thus, sequestration of the receptors may not only be a means of promoting desensitization but also a mechanism of resensitization through promoting receptor dephosphorylation and recycling back to the plasma membrane.

A major unanswered question about homologous desensitization is the precise cellular location of the sequestered or down regulated β-adrenergic receptors. One hypothesis is that the receptors are internalized into cytosolic vesicles via a coated pit mechanism as has been shown for peptide hormone receptors (Pastan and Willingham, 1981). The observation that upon lysis of desensitized cells the sequestered β-adrenergic receptors appear to have been translocated into light membranous particles is consistent with this possibility. However, Strader et al., (1984) have shown, using frog erythrocytes, that the production of these vesicular particles is very much dependent upon the method of cell lysis. When very "gentle" methods of cell lysis were used, the sequestered β-adrenergic receptors were shown to be localized to the plasma membrane fraction. This suggests that in the intact frog erythrocyte the receptors are never present in free floating cytoplasmic vesicles but rather are contained within structures which remain contiguous with the inner surface of the plasma membrane. Insel and coworkers have recently reached similar conclusions using the S49 lymphoma cell system (Mahan et al., 1985).

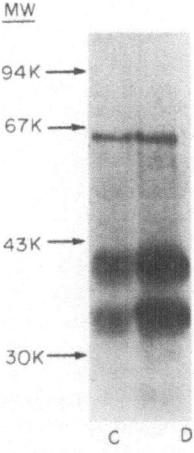

FIG. 2: SDS-polyacrylamide gel electrophoresis of ^{32}P-labeled β-adrenergic receptor peptides from control and isoproterenol-desensitized turkey erythrocytes. The two major bands seen at ≅ 39 kDa and 49 kDa represent the turkey erythrocyte β-adrenergic receptor peptides. Lane C shows the receptor peptides from control cells whereas Lane D represents those from isoproterenol desensitized cells.

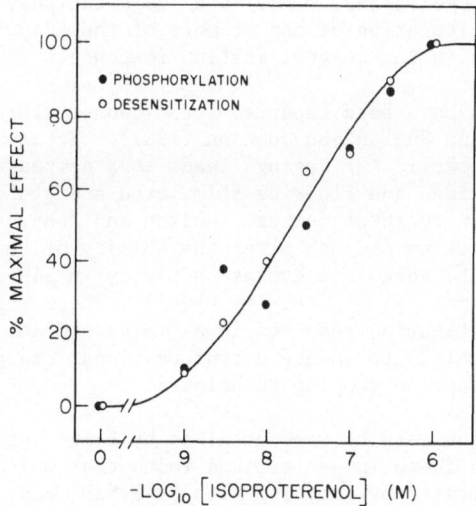

FIG. 3: Dose-response relationship for isoproterenol-induced receptor phosphorylation and adenylate cyclase desensitizatoin. Individual aliquots of ^{32}P-loaded cells were incubated with the indicated concentrations of isoproterenol prior to receptor purification and gel
electrophoresis. The phosphorylation data (•——•) is expressed as an
increase in the stoichiometric ratio of mol phosphate/mol receptor
over the control sample. The adenylate cyclase data (•——•) is
expressed as a percentage of the maximal desensitization of the
isoproterenol-sensitive enzyme activity.

Heterologous Desensitization

Heterologous forms of adenylate cyclase desensitization are known to
occur in a variety of tissues and cell types. This form of desensitization
is associated with a broad pattern of refractoriness in which the response
to multiple hormones and sometimes nonhormonal effectors is impaired. In
contrast to homologous desensitization which may be unimechanistic, heterologous desensitization likely occurs through more than a single mechanism.
Our laboratory has utilized the turkey erythrocyte as a model system in
which to examine heterologous desensitization mechanisms.

Hoffman et al., (1979) initially demonstrated that incubation of turkey
erythrocytes with isoproterenol produced an approximate 50% attenuation of
subsequent catecholamine-stimulated adenylate cyclase activity. Since only
β-adrenergic receptors are coupled to adenylate cyclase in these cells stimulation by other hormonal effectors could not be assessed. There were also
small but significant decrements in guanine nucleotide- and fluoride-stimulated enzyme activities observed as a result of desensitization. This suggests that the catecholamine induced desensitization in the turkey erythrocyte is of the heterologous type. There was no reduction in the binding
of the antagonist ligand [^{3}H]dihydroalprenolol in membranes derived from
treated cells (Hoffman et al., 1979).

Stadel et al. (1981) subsequently showed that catecholamine-promoted
desensitization in the turkey erythrocyte is associated with a functional
uncoupling of the β-adrenergic receptor. This was evidenced by an impaired
ability of the receptors to form a high affinity, guanine nucleotide-sensitive complex with agonists as detected in radioligand binding studies.
Importantly, the adenylate cyclase desensitization as well as the receptor
uncoupling were partially mimicked by incubating the erythrocytes with cell

permeable analogs of cyclic AMP. This data indicated that there may be a cyclic AMP-mediated alteration of one or more of the adenylate cyclase components which results in the desensitization response.

Similar findings have been reported for pigeon erythrocytes by Simpson and Pfeuffer (1980) and Hudson and Johnson (1981). Exposure of these cells to catecholamines or cyclic AMP analogs leads to a decrease in isoproterenol-, guanine nucleotide- and fluoride-stimulated adenylate cyclase activities with no change in receptor number. Hudson and Johnson additionally found that desensitization did not alter the ability of the solubilized N_s protein to reconstitute adenylate cyclase activity in S49 cyc$^-$ lymphoma membranes which lack functional N_s activity (1981). These workers did find, however, that desensitization resulted in a diminished ability of agonists and/or guanine nucleotides to induce a conformational change in the N_s protein as assessed by peptide mapping techniques.

Somewhat different results were obtained by Briggs et al. (1983) using turkey erythrocytes. These investigations found that when the solubilized N_s protein was quantitated by labeling with $[^{32}P]NAD^+$ and cholera toxin, desensitization resulted in a significant reduction of the ability of N to reconstitute enzyme activity in S49 cyc$^-$ membranes. These observations suggest that at a minimum, desensitization of avian erythrocytes is associated with a functional modification of the guanine nucleotide regulatory protein N_s.

Investigations of heterologous desensitization in mammalian cells have also indicated that functional modifications of N_s occur. Kassis and Fishman demonstrated that treatment with PGE_1 induced heterologous desensitiza-

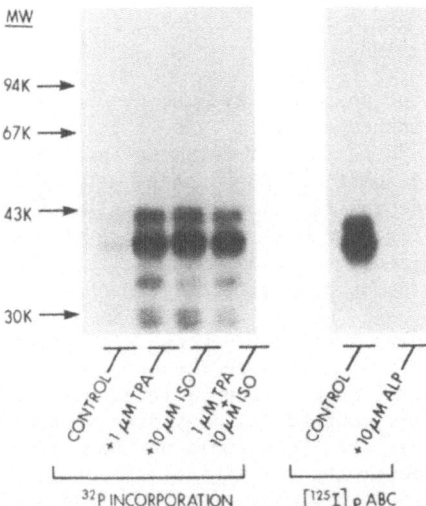

FIG. 4: SDS-polyacrylamide gel electrophoresis of ^{32}P-labeled and $[^{125}I]$-para-azidobenzylcarazolol ($[^{125}I]pABC$)-labeled β-adrenergic receptor peptides from duck erythrocytes. The two major peptides of molecular mass 40 kDa and 48 kDa represent the β-adrenergic receptor peptides in duck erythrocyte membranes as indicated in the right-hand panel through covalent incorporation of the photoaffinity probe, $[^{125}I]pABC$. For ^{32}P labeling, the cells were preincubated with ^{32}Pi prior to desensitization with either the phorbol diester, 12-0-tetradecanoyl phorbol-13-acetate (TPA), isoproterenol (ISO), or both TPA and ISO.

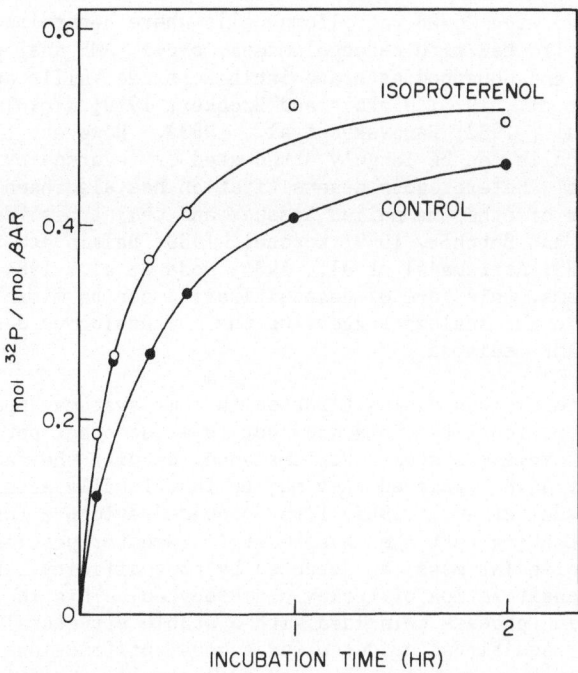

FIG. 5: Effect of isoproterenol on the time course of cAMP-dependent protein kinase catalyzed β-adrenergic receptor phosphorylation. Purified hamster lung β-adrenergic receptor (0.14 μM) was incubated
with the catalytic subunit of cAMP-dependent protein kinase (0.34 μM)
in the presence or absence of 20 μM (-)isoproterenol at 25°C for the
indicated period of time. Reactions were stopped by the addition
of SDS sample buffer followed by electrophoresis on a 10% SDS polyacrylamide gel. After drying and autoradiography, the gel was cut
and counted to determine the stoichiometry.

tion in human fibroblast cells and that N_1 solubilized from these cells was
less efficient in reconstituting adenylate cyclase activity in S49 cyc⁻ membranes when compared to controls (1982). Heterologous desensitization
induced by prostaglandin E_1 in liver (Garrity et al., 1983) and by human
chorionic gonadotropin in ovaries (Kirchik et al., 1983) also results in an
impaired functionality of N_s as determined with cyc⁻ reconstitution. In
contrast, Rich et al., (1984) have reported that glucagon-induced heterologous desensitization in MDCK cells was not associated with alterations in
the reconstitutive ability of N_s but instead involved increases in the apparent levels of the inhibitory guanine nucleotide regulatory protein, N_i.
This novel finding suggests that alterations in the N_i/N_s stoichiometry may
be one mechanism by which heterologous desensitization occurs.

Perkins and colleagues (Perkins, 1983; Su et al., 1976; Johnson et al.,
1978) have also investigated heterologous desensitization in clonal astrocytoma cells. Prolonged incubation with either catecholamines or prostaglandins diminished the subsequent capacity of both hormones to elevate
intracellular cyclic AMP levels. Incubation of the cells with dibutyryl
cyclic AMP produced refractoriness to both catecholamines and prostaglandins, suggesting that the heterologous desensitization might be cyclic AMP
mediated. When adenylate cyclase activity was examined in membranes from
desensitized cells, however, the heterologous form of desensitization was
no longer apparent. This phenomenon has precluded biochemical investigations of the heterologous desensitization in these cells. Similar findings

have been observed with C6-2B rat glioma cells where heterologous desensitization can be elicited with catecholamines, cyclic AMP analogs, cholera toxin, forskolin and phosphodiesterase inhibitors (de Vellis and Brooker, 1974; Terasaki et al., 1978; Nickols and Brooker, 1979; Nickols and Brooker, 1980; Moylan et al., 1982; Barovsky et al., 1983). However, the desensitization in these cells can be largely attenuated or reversed by protein synthesis inhibitors. Heterologous desensitization has also been shown to occur in a number of other mammalian tissues and cell types (Newcombe et al., 1975; Clark and Butcher, 1979; Koschel, 1980; Balkin and Sonenberg, 1981; Harden, 1983; Attramadal et al., 1984; Noda et al., 1984). In most, but not all systems, this form of desensitization can be mimicked by incubation with cyclic AMP analogs suggesting that heterologous desensitization is often cyclic AMP mediated.

Although heterologous desensitization in some systems appears to result in impaired N_S functionality, this does not rule out other potential lesions in the adenylate cyclase system. For instance, despite the fact that the receptors are not down regulated they may be functionally altered as in the case of N_S. Stadel et al., (1982) first provided evidence for this possibility by demonstrating that the β-adrenergic receptor peptides were of large apparent molecular mass, as detected by photoaffinity labeling after heterologous desensitization of turkey erythrocytes. This indicated that the desensitization process coincided with a stable structural modification of the receptor. Additional evidence for a covalent/functional modification of the receptor protein has been obtained by partially purifying the β-adrenergic receptors from desensitized turkey erythrocytes, reconstituting them into phospholipid vesicles and fusing them with Xenopus laevis erythrocytes (Koschel, 1980). It was shown that the β-adrenergic receptor from desensitized cells was less efficient than that from control cells in reconsti-

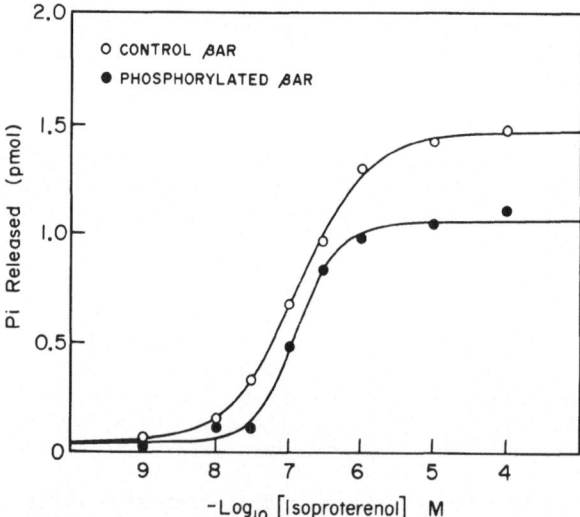

FIG. 6: Isoproterenol-prompted GTPase activity in phospholipid vesicles containing N_S and phosphorylated or control receptor. Phospholipid vesicles containing β-adrenergic receptor, N_S and the catalytic subunit of cAMP-dependent protein kinase were incubated with 30 mM Na phosphate, 10 mM Tris-HCl, pH 7.2, 100 mM NaCl, 5 mM $MgCl_2$, 5 mM p-nitrophenylphosphate, 50 μM AppNHp and ~250 μg of soybean phosphatidylcholine. Phosphorylation incubations additionally contained 50 μM ATP while control samples contained no ATP. After incubating for 2 hr at 25°C the samples were assayed for GTPase activity as a function of the isoproterenol concentration.

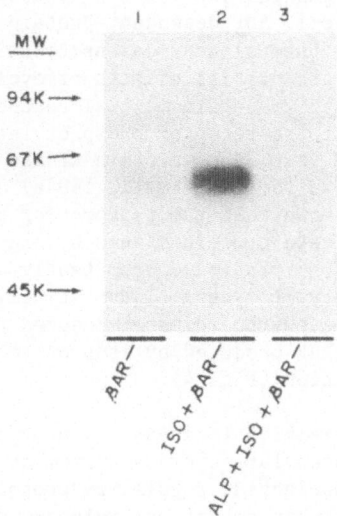

FIG. 7:   Phosphorylation of purified hamster lung β-adrenergic receptor by a
          supernatant fraction of lysed kin⁻ cells. Reconstituted receptor
          was incubated with lysed kin⁻ cell supernatant for 30 min at 30°C
          without (lane 1) or with additions of 10 μM (-)isoproterenol (lane
          2) and 10 μM (-)isoproterenol/20 μM (±)alprenolol (lane 3). The
          phosphorylated receptor was then purified before electrophoresis
          on a 10% SDS polyacrylamide gel. The molecular weight standards
          are shown x 1000 (K).

tuting catecholamine responsiveness of the Xenopus laevis adenylate cyclase
activity.

     Using [^{32}P]orthophosphate incorporation, it was subsequently shown that
during the desensitization process in turkey erythrocytes, the β-adrenergic
receptor undergoes phosphorylation (Stadel et al., 1983a). Sibley et al.,
(1984b) have investigated this phosphorylation process in detail. The
β-adrenergic receptor in turkey erythrocytes is stoichiometrically phosphorylated under basal conditions containing 0.7-1.0 mol phosphate/mol receptor
with this stoichiometry increasing to 2-3 mol/mol upon maximal desensitization (Sibley et al., 1984b). Fig. 2 shows an experiment comparing the
phosphorylated receptor peptides from control and isoproterenol desensitized cells. We have also shown that the phosphate/receptor stoichiometry
is tightly correlated with the degree of desensitization (Sibley et al.,
1984b). For instance, Fig. 3 shows the dose-response relationship for isoproterenol-induced receptor phosphorylation and adenylate cyclase desensitization. As can be seen, between 10^{-9}M and 10^{-6}M, isoproterenol promotes
desensitization of adenylate cyclase in parallel with receptor phosphorylation. In addition, the time courses for receptor phosphorylation and
adenylate cyclase desensitization are identical as are the rates of resensitization and the return of the phosphate/receptor stoichiometry to control
levels (Sibley et al., 1984b). Moreover, incubation of the cells with cyclic
AMP analogs causes submaximal phosphorylation of the β-adrenergic receptor
which is correlated with the partial desensitization of adenylate cyclase
which these analogs evoke (Sibley et al., 1984b). These data thus indicate
that in turkey erythrocytes, heterologous desensitization is tightly correlated with phosphorylation of the β-adrenergic receptor.

     The observation that cyclic AMP analogs partially reproduce the isopro-

terenol-promoted receptor phosphorylation and desensitization in turkey erythrocytes suggests that cyclic AMP-dependent protein kinase is, at least
partially, involved in the heterologous desensitization (see below). One
potential explanation for the partial effects of cyclic AMP analogs is that
agonist occupancy must occur to obtain maximal receptor phosphorylation.
Another possibility is that the receptor phosphorylation is not completely
mediated by cyclic AMP and that other protein kinase systems are involved
as well. In this regard, we (Sibley et al., 1984a) and others (Kelleher et
al., 1984) have recently shown that tumor-promoting phorbol diesters, compounds which potently activate protein kinase C, are capable of stimulating
β-adrenergic receptor phosphorylation concomitantly with adenylate cyclase
desensitization in avian erythrocytes. Interestingly, in duck erythrocytes
(Sibley et al., 1984a), the phorbol diester-induced receptor phosphorylation is nonadditive with that produced by isoproterenol, suggesting a common
mechanism or pathway of action (Fig. 4).

Heterologous desensitization in avian erythrocytes is thus associated
with modification in the adenylate cyclase system at both the level of the
receptor and the guanine nucleotide regulatory proteins. It seems reasonable to propose that the mechanisms of heterologous desensitization eluci-

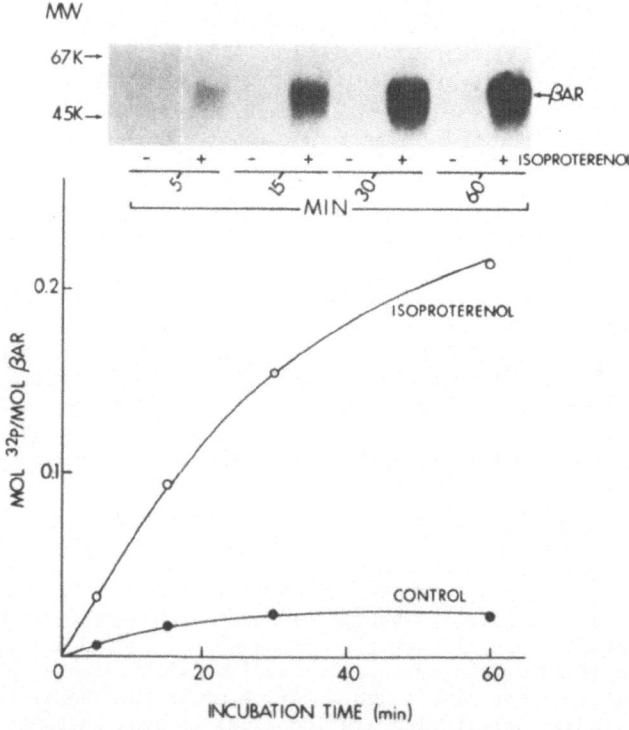

FIG. 8: Effect of isoproterenol on the time course of β-adrenergic receptor phosphorylation by βARK. Reconstituted β-adrenergic receptor
was incubated with DEAE-purified βARK in the presence (+) or
absence (-) or 2 μM (-)isoproterenol at 30°C for the indicated
period of time. Phosphorylated receptor was then purified before
electrophoresis on a 10% SDS polyacrylamide gel. The upper inset
shows the resulting autoradiogram after a 9 hr exposure of the
dried gel. Phosphorylation stoichiometires were determined by
counting the excised receptor bands. The molecular weight standards (MW) are shown x 1000 (K).

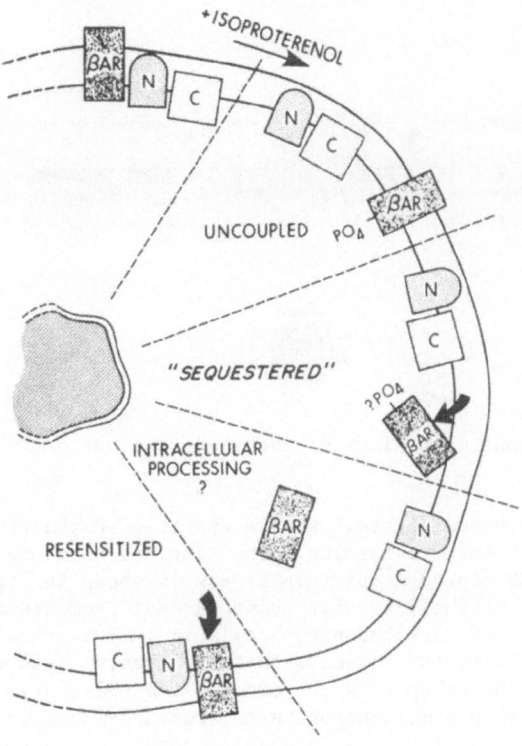

FIG. 9: Homologous desensitization of β-adrenergic receptor-coupled adenylate cyclase.

dated in avian erythrocytes are not unique and are operative in mammalian
cell types as well. Thus, one major means of achieving hetereologous
desensitization would be modification(s) in the N_s or N_i proteins as discussed earlier. Another major mechanism of heterologous desensitization
is modification of the receptor proteins. In this case, the effect is not
receptor sequestration or down regulation as in homologous desensitization
but rather a covalent modification producing a functional uncoupling of the
receptor from adenylate cyclase. Evidence from the avian erythrocytes indicates that phosphorylation is the likely candidate for the covalent modification. Moreover, due to the heterologous nature of the desensitization it
is probable that all of the different receptors coupled to adenylate cyclase
would be phosphorylated. It is likely that differing extents of receptor and
N_s/N_i modifications occur in different cell types and under different conditions. In some cells which exhibit heterologous desensitization, the modification in N_s might predominate. In other cells, such as erythrocytes,
modification of the receptor is the most relevant mechanism.

Protein Kinases Involved in β-Adrenergic Receptor Desensitization

In an attempt to directly assess what protein kinases are involved in
β-adrenergic receptor desensitization, we have studied the ability of several different protein kinases to phosphorylate the purified receptor (Benovic et al., 1985). In these studies, purified hamster lung β-adrenergic
receptor was incubated with [^{32}P]ATP, Mg^+ and several different protein
kinases. It was found thatcGMP protein kinase, myosin light chain kinase,
and casein kinases I and II were unable to phosphorylate the receptor.

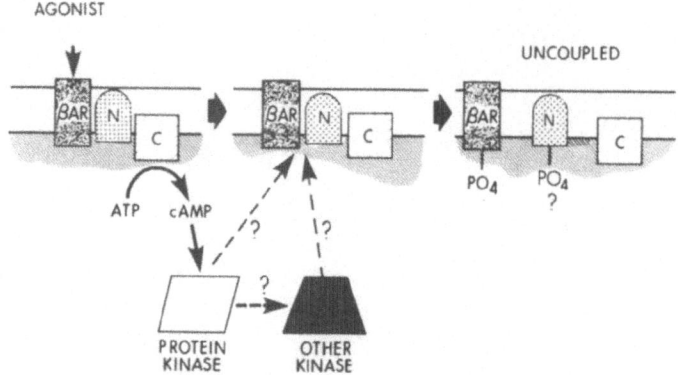

FIG. 10: Heterologous mechanisms of adenylate cyclase desensitization.

Conversely, cAMP-dependent protein kinase was able to phosphorylate the receptor in a stoichiometric fashion. The time course of receptor phosphorylation by the cAMP-dependent protein kinase is shown in Fig. 5. It is seen that inclusion of the β-agonist isoproterenol promotes a 2-3 fold increase in the rate of receptor phosphorylation. This effect can be blocked by β-antagonists and suggests that the agonist induces a conformational change in the receptor which exposes the phosphorylation site(s). The stoichiometry of ~0.5 mol phosphate/mol receptor seen in Fig. 5 was found to be very dependent on the incubation conditions. When the receptor used was initially reconstituted in phospholipid vesicles a stoichiometry of ~2 mol/mol could be attained. HPLC tryptic mapping of ^{32}P-labeled receptor revealed two major phosphorylation sites both at serine residues. In addition, the amino acid sequence deduced from cloning of the gene for the β-adrenergic receptor includes two consensus cAMP-dependent kinase phosphorylation sites (Dixon et al., 1986).

The functional significance of receptor phosphorylation was studied using ligand binding and reconstitution techniques. The β-adrenergic receptor, in addition to binding specific ligands, is able to interact with N_S in a manner which promotes N_S activation of adenylate cyclase. The interaction of purified receptor and N_S in phospholipid vesicles is promoted by agonists and may be monitored by measuring the GTPase activity of N_S. Fig. 6 demonstrates that phosphorylated β-adrenergic receptor has a diminished ability to interact with N_S with no apparent change in ligand binding. When the receptor is phosphorylated an ~25% decrease in isoproterenol-promoted GTPase activity is seen relative to control receptor. The concentrations of isoproterenol which promoted 50% of the maximum GTPase activities were 136 and 125 nM for control and phosphorylated receptor, respectively.

These studies provide a direct demonstration of regulation of the function of the isolated β-adrenergic receptor by cAMP-dependent protein kinase. Moreover, they suggest one possible general mechanism for heterologous regulation of adenylate cyclase coupled receptors. Cyclic AMP generated in response to agonist stimulation of the enzyme activates the cAMP dependent kinase. This kinase presumably phosphorylates multiple receptors, thus uncoupling them from productive interaction with N_S. While only the structure of the β-adrenergic receptor is currently known, it is reasonable to speculate that other adenylate cyclase coupled receptors have regions of strong homology since they all couple to the same N_S molecules. Thus, all these receptors might serve as substrates for the cAMP dependent kinase. Moreover, the ability of an agonist to change the conformation of the receptor in such a way as to make it a better substrate for the kinase provides

266

an additional mechanism for fine-tuning the regulation. Thus, in addition to phosphorylation and desensitization of a variety of receptors there might be even more profound alterations of the particular type of receptor which is occupied by its specific agonist.

While heterologous desensitization is cAMP-mediated and clearly involves the cAMP-dependent protein kinase, homologous desensitization, in contrast, is not cAMP-mediated. This has been best demonstrated in mutants (cyc⁻ and kin⁻) of S49 lymphoma cells that cannot mount an agonist promoted cAMP response. In these cells β-adrenergic receptor phosphorylation accompanies homologous desensitization in a manner identical to wild type cells (Strasser et al., 1986). This demonstrates that cAMP production and the cAMP-dependent protein kinase are not required for receptor phosphorylation and homologous desensitization in these cells. Thus, the relevant kinase(s) must either be stimulated by a second messenger other than cAMP or must preferentially phosphorylate the agonist-occupied form of the receptor. Accordingly, we have attempted to identify the kinase(s) involved in homologous desensitization using the S49 kin⁻ mutant as a model system (Benovic et al., 1986).

Initially we attempted to determine whether kin⁻ cell lysates, separated into particulate and soluble fractions, contained any kinase activity capable of phosphorylating the β-adrenergic receptor. Incubation of reconstituted hamster lung β-adrenergic receptor with the kin⁻ soluble fraction under phosphorylating conditions resulted in barely detectable phosphorylation of the receptor (Fig. 7, lane 1). However, when the receptor was occupied with the β-agonist isoproterenol, the phosphorylation was enhanced 5 fold (lane 2). The agonist effect could be completely blocked by incubation with the β-antagonist alprenolol (lane 3).

This kinase can be partially purified from the soluble fraction of lysed kin⁻ cells by successive chromatography on molecular sieve HPLC and DEAE-Sephacel. The partially purified β-adrenergic receptor kinase (βARK) is largely devoid of other protein kinase activities and does not phosphorylate standard kinase substrates such as casein and histones. Receptor phosphorylation by βARK is not affected by cAMP, cGMP, cAMP-dependent kinase inhibitor, Ca^{2+}/calmodulin or Ca^{2+}/phosphatidylserine. These results demonstrate that βARK is not a cAMP-dependent, cGMP-dependent, Ca^{2+}/calmodulin-dependent or Ca^{2+}/phospholipid-dependent protein kinase.

A time course of receptor phosphorylation by partially purified βARK in the presence or absence of isoproterenol is shown in Fig. 8. The striking increase in phosphorylation promoted by agonist clearly shows that agonist occupancy is virtually required for receptor phosphorylation. This is in contrast to receptor phosphorylation by the cAMP-dependent kinase where agonist occupancy results in only a 2-3 fold increase in the rate of phosphorylation, with no apparent change in the maximum stoichiometry obtained (Fig. 5).

These results provide evidence for a novel protein kinase which specifically phosphorylates the agonist-occupied β-adrenergic receptor. One of the most potentially interesting roles of the β-adrenergic receptor kinase might be its involvement in homologous desensitization. The mechanisms by which such receptor phosphorylation might lead to homologous desensitization are, however, currently unknown. One possibility is that the phosphorylated receptor has diminished ability to interact with the guanine nucleotide regulatory protein. Such a mechanism would be similar to that found for the cAMP-dependent protein kinase in heterologous desensitization precedes receptor sequestration would appear to support this hypothesis (see above). Alternatively, receptor phosphorylation might trigger sequestration of the functionally intact receptors away from the effector N_S-adenylate cyclase complex. These two mechanisms are, of course, not mutually exclusive. Purification of βARK may provide the key to ultimately elucidating the molecular events responsible for homologous desensitization.

SUMMARY

Multiple mechanisms seem to be involved in regulating the responsiveness of hormone receptor-coupled adenylate cyclase systems. These mechanisms at least involve the receptors and nucleotide regulatory proteins. With the recent development of methods for purifying the catalytic unit of the enzyme it will be possible to assess whether it is also a locus for such regulatory phenomena.

At least two major pathways of receptor regulation have been uncovered. Homologous desensitization (Fig. 9) involves the uncoupling and translocation of the receptors out of their normal plasma membrane environment. This process sequesters the receptors away from their effector, the regulatory and catalytic components of adenylate cyclase. The site of receptor sequestration is unclear and might lie within the plasma membrane or within the cell. The sequestered receptors can recycle to the cell surface or become down-regulated, perhaps being destroyed within the cell. Phosphorylation of the receptors through a β-adrenergic receptor kinase appears to be associated with homologous desensitization. This phosphorylation event may serve either to uncouple functionally the receptors or to trigger their sequestration from the cell surface or both.

In heterologous desensitization (Fig. 10), receptor function is regulated by phosphorylation in the absence of receptor sequestration or down-regulation. This covalent modification serves to functionally uncouple the receptors, that is, to impair their interactions with the guanine nucleotide regulatory proteins. Several protein kinases seem to be capable of promoting phosphorylation of the receptors including the cAMP-dependent kinase and protein kinase C. In addition to the receptor modification, heterologous desensitization seems to be associated with functional modifications (phosphorylation?) at the level of nucleotide regulatory proteins (N_s and N_i), (Fig. 10). Further studies of the mechanisms of desensitization of adenylate cyclase-coupled receptors are thus likely to help elucidate modes of regulation of a wide variety of receptor-coupled functions in diverse types of cells.

ACKNOWLEDGMENT

We thank Lynn Tilley for secretarial assistance. David R. Sibley is the recipient of NIH Postdoctoral Fellowship HL06631.

REFERENCES

Anderson, W. B. and Jaworski, C. J., Isoproterenol-induced desensitization of adenylate cyclase responsiveness in a cell-free system, J. Biol. Chem., 254:4596-4601 (1979).

Attramadal, H., Le Gac, F., Jahnsen, T. and Hansson, V., β-Adrenergic regulation of Sertoli cell adenylyl cyclase: Desensitization by homologous hormone, Mol. Cell. Endocrinol., 34:1-6 (1984).

Balkin, M. S. and Sonenberg, M., Hormone-induced homologous and heterologous desensitization in the rat adipocyte, Endocrinology, 109:1176-1183 (1981).

Barovsky, K., Pedone, C. and Brooker, G., Forskolin-stimulated cyclic AMP accumulation mediates protein synthesis-dependent refractoriness in

C6-2B rat glioma cells, J. Cyclic Nucleotide Prot. Phosphorylat. Res., 9:181-189 (1983).

Benovic, J. L., Pike, L. J., Cerione, R. A., Staniszewski, C., Yoshimasa, T., Codina, J., Caron, M. G. and Lefkowitz, R. J., Phosphorylation of the mammalian β-adrenergic receptor by cyclic AMP-dependent protein kinase: Regulation of the rate of receptor phosphorylation and dephosphorylation by agonist occupancy and effects on coupling of the receptor to the stimulatory guanine nucleotide regulatory protein, J. Biol. Chem., 260:7094-7101 (1985).

Benovic, J. L., Strasser, R. H., Caron, M. G. and Lefkowitz, R. J., -Adrenergic receptor kinase: Identification of a novel protein kinase that phosphorylates the agonist-occupied form of the receptor, Proc. Natl. Acad. Sci. USA, 83:2797-2801 (1986).

Briggs, M. M., Stadel, J. M., Iyengar, R. and Lefkowitz, R. J., Functional modification of the guanine nucleotide regulatory protein after desensitization of turkey erythrocytes by catecholamines, Arch. Biochem. Biophys., 224:142-151 (1983).

Chuang, D.-M. and Costa, E., Evidence for internalization of the recognition site of β-adrenergic receptors during receptor subsensitivity induced by (-)isoproterenol, Proc. Natl. Acad. Sci. USA, 76:3024-3028 (1979).

Chuang, D.-M., Kinnier, W. J., Farber, L. and Costa, E., A biochemical study of receptor internalization during β-adrenergic receptor desensitization in frog erythrocytes, Mol. Pharmacol. 18:348-355 (1980).

Clark, R. and Butcher, R. W., Desensitization of adenylate cyclase in cultured fibroblasts with prostaglandin E_1 and epinephrine, J. Biol. Chem., 254:9373-9378 (1979).

Clark, R. B., Friedman, J., Prashad, N. and Ruoho, A. E., Epinephrine-induced sequestration of the β-adrenergic receptor in cultured S49 WT and cyc lymphoma cells, J. Cyclic Nucleotide Prot. Phosphorylat. Res., 10:97-119 (1985).

de Vellis, J. and Brooker, G., Reversal of catecholamine refractoriness by inhibitors of RNA and protein synthesis, Science, 186:1221-1223 (1974).

Dixon, R. A. F., Kobilka, B. K., Strader, D. J., Benovic, J. L., Dohlman, H. G., Frielle, T., Bolanowski, M. A., Bennett, C. D., Rands, E., Diehl, R. E., Mumford, R. A., Slater, E. E., Sigal, I. S., Caron, M. G., Lefkowitz, R. J. and Strader, C. D., Cloning of the gene and cDNA for mammalian β-adrenergic receptor and homology with rhodopsin, Nature, (London), 321:75-79 (1986).

Doss, R. C., Perkins, J. P. and Harden, T. K., Recovery of β-adrenergic receptors following long-term exposure of astrocytoma cells to catecholamine: Role of protein synthesis, J. Biol. Chem., 256:12281-12286 (1981).

Fishman, P. H., Mallorga, P. and Talman, J. F., Catecholamine-induced desensitization of adenylate cyclase in rat glioma C6 cells: Evidence for specific uncoupling of beta-adrenergic receptors from a functional regulatory component of adenylate cyclase, Mol. Pharmacol., 20:310-318 (1981).

Frederich, R. C., Jr., Waldo, G. L., Harden, T. K. and Perkins, J. P., Characterization of agonist-induced β-adrenergic receptor-specific desensitization in C62B glioma cells, J. Cyclic Nucleotide Prot. Phosphorylat. Res., 9:103-118 (1983).

Garrity, M. J., Andreasen, T. J., Storm, D. R. and Robertson, R. P., Prostaglandin E-induced heterologous desensitization of hepatic adenylate cyclase: Consequences on the guanyl nucleotide regulatory complex, J. Biol. Chem., 258:8692-8697 (1983).

Green, D. A. and Clark, R. B., Adenylate cyclase coupling proteins are not essential for agonist-specific desensitization of lymphoma cells, J. Biol. Chem., 256:2105-2108 (1981).

Green, D. A., Friedman, J. and Richard, B. C., Epinephrine desensitization of adenylate cyclase from cyc⁻ and S49 cultured lymphoma cells, J. Cyclic Nucleotide Res., 7:161-172 (1981).

Harden, T. K., Agonist-induced desensitization of the β-adrenergic receptor-linked adenylate cyclase, Pharmacol. Rev., 35:5-32 (1983).

Harden, T. K., Cotton, C. U., Waldo, G. L., Lutton, J. K. and Perkins, J. P., Catecholamine-induced alteration in the sedimentation behavior of membrane-bound β-adrenergic receptors, Science, 210:441-443 (1980).

Harden, T. K., Su, Y.-F. and Perkins, J. P., Catecholamine-induced desensitization involves an uncoupling of beta-adrenergic receptors and adenylate cyclase, J. Cyclic Nucleotide Res., 5:99-106 (1979).

Hertel, C., Muller, P., Portenier, M. and Staehelin, M., Determination of the desensitization of β-adrenergic receptors by [^{3}H]CGP-12177, Biochem. J., 216:669-674 (1983a).

Hertel, C., Staehelin, M. and Perkins, J. P., Evidence for intravesicular β-adrenergic receptors in membrane fractions from desensitized cells: Binding of the hydrophilic ligand CGP-12177 only in the presence of alamethicin, J. Cyclic Nucleotide Prot. Phosphorylat. Res., 9:119-128 (1983b).

Hoffman, B. B., Mullikin-Kilpatrick, D. and Lefkowitz, R. J., Desensitization of beta-adrenergic stimulated adenylate cyclase in turkey erythrocytes, J. Cyclic Nucleotide Res., 5:355-366 (1979).

Homburger, V., Lucas, M., Cantau, B., Barabe, J., Penit, J. and Bockaert, J., Further evidence that desensitization of β-adrenergic sensitive adenylate cyclase proceeds in two steps: Modification of the coupling and loss of β-adrenergic receptors, J. Biol. Chem., 255:10436-10444 (1980).

Hudson, T. H. and Johnson, G. L., Functional alterations in components of pigeon erythrocyte adenylate cyclase following desensitization to isoproterenol, Mol. Pharmacol., 20:694-703 (1981).

Hunzicher-Dunn, M., Derda, D., Jungmann, R. A, and Birnbaumer, L., Resensitization of the desensitized follicular adenylyl cyclase system to luteinizing hormone, Endocrinology, 104:1785-1793 (1979).

Johnson, G. L., Wolfe, B. B., Harden, T. K., Molinoff, P. B. and Perkins, J. P., Role of β-adrenergic receptors in catecholamine-induced desensitization of adenylate cyclase in human astrocytoma cells, J. Biol. Chem., 253:1472-1480 (1978).

Kassis, S. and Fishman, P. H., Different mechanisms of desensitization of adenylate cyclase by isoproterenol and prostaglandin E$_1$ in human fibroblasts: Role of regulatory components in desensitization, J. Biol. Chem., 257:5312-5318 (1982).

Kassis, S. and Fishman, P. H., Functional alteration of the β-adrenergic receptor during desensitization of mammalian adenylate cyclase by β-agonists, Proc. Natl. Acad. Sci. USA, 81:6686-6690 (1984).

Kelleher, D. J., Pessin, J. E., Ruoho, A. E. and Johnson, G. L., Phorbol ester induces desensitization of adenylate cyclase and phosphorylation of the β-adrenergic receptor in turkey erythrocytes, Proc. Natl. Acad. Sci. USA, 81:4316-4320 (1984).

Kent, R. S., De Lean, A. and Lefkowitz, R. J., A quantitative analysis of beta-adrenergic receptor interaction: Resolution of high and low affinity states of the receptor by computer modeling of ligand binding data, Mol. Pharmacol., 17:14-23 (1979).

Kirchik, H. J., Iyengar, R. and Birnbaumer, L., Human chorionic gonadotropin-induced heterologous desensitization of adenylate cyclase from highly lutenized rat ovaries: Attenuation of regulatory N component activity, Endocrinology, 113:1638-1646 (1983).

Koschel, K., A hormone-independent rise of adenosine 3',5'-monophosphate desensitizes coupling of β-adrenergic receptors by adenylate cyclase in rat glioma C6-cells, Eur. J. Biochem., 108:163-169 (1980).

Mahan, L. C., Motolsky, H. J. and Insel, P. A., Do agonists promote rapid internalization of β-adrenergic receptors?, Proc. Natl. Acad. Sci. USA 82:6566-6570 (1985).

Mickey, J. C., Tate, R. and Lefkowitz, R. J., Subsensitivity of adenylate cyclase and decreased β-adrenergic receptor binding after chronic exposure to (-)isoproterenol in vitro, J. Biol. Chem., 250:5727-5729 (1975).

Mickey, J. V., Tate, T., Mullikin, D. and Lefkowitz, R. J., Regulation of adenylate cyclase-coupled beta-adrenergic receptor binding sites by beta-adrenergic catecholamines in vitro, Mol. Pharmacol., 12:409-419 (1976).

Morishima, I., Thompson, W. J., Robison, G. A. and Strada, S. J., Loss and restoration of sensitivity to epinephrine in cultured cells: Effect of inhibitors of RNA and protein synthesis, Mol. Pharmacol., 18:370-378 (1980).

Moylan, R. D., Barovsky, K. and Brooker, G., N^6, O^{2+}- dibutyryl cyclic AMP and cholera toxin-induced β-adrenergic receptor loss in cultured cells, J. Biol. Chem., 257:4947-4950 (1982).

Mukherjee, C., Caron, M. G. and Lefkowitz, R. J., Catecholamine-induced sub-sensitivity of adenylate cyclase associated with loss of beta-adrenergic receptor binding sites, Proc. Natl. Acad. Sci. USA, 72:1945-1949 (1975).

Mukherjee, C., Caron, M. G. and Lefkowitz, R. J., Regulation of adenylate cyclase coupled β-adrenergic receptors by β-adrenergic catecholamines, Endocrinology, 99:347-357 (1976).

Newcombe, D. S., Ciosek, C. P., Jr., Ishikawa, Y. and Fahey, J. V., Human synoviocytes: Activation and desensitization by prostaglandins and 1-epinephrine, Proc. Natl. Acad. Sci. USA, 72:3124-3128 (1975).

Nickols, G. A. and Brooker G., Induction of refractoriness to isoproterenol by prior treatment of C6-2B rat astrocytoma cells with cholera toxin, J. Cyclic Nucleotide Res., 5:435-447 (1979).

Nickols, G. A. and Brooker, G., Potentiation of cholera toxin-stimulated cyclic AMP production in cultured cells by inhibitors of RNA and protein synthesis, J. Biol. Chem., 255:23-26 (1980).

Noda, C., Shinjyo, F., Tomomura, A., Kato, S., Nakamura, T. and Ichihara, A., Mechanism of heterologous desensitization of the adenylate cyclase system by glucagon in primary cultures of adult rat hepatocytes, J. Biol. Chem., 259:7747-7754 (1984).

Pastan, I. H. and Willingham, M. C., Receptor-mediated endocytosis of hormones in cultured cells, Annu. Rev. Physiol., 43:239-250 (1981).

Perkins, J. P., Desensitization of the response of adenylate cyclase to catecholamines, in: "Current Topics in Membranes and Transport," Vol. 18, A. Kleinzeller, and M. B. Martin, eds., pp. 85-108, Academic Press, New York (1983).

Rich, K. A., Codina, J., Flloyd, G., Sekura, R., Hildebrandt, J. D. and Iyengar, R., Glucagon-induced heterologous desensitization of the MDCK cell adenylyl cyclase, J. Biol. Chem., 259:7893-7901 (1984).

Salomon, Y., Ezra, E. and Amir-Zaltsman, Y., The role of GTP in lutropin-induced desensitization of the GTP regulatory cycle and adenylate cyclase in the rat ovary, Adv. Cyclic Nucleotide Res., 14:101-109 (1981).

Shear, M., Insel, P. A., Melmon, K. L. and Coffino, P., Agonist-specific refractoriness induced by isoproterenol, J. Biol. Chem., 251:7572-7576 (1976).

Sibley, D. R., Nambi, P., Peters, J. R. and Lefkowitz, R. J., Phorbol die-sters promote β-adrenergic receptor phosphorylation and adenylate cyclase desensitization in duck erythrocytes, Biochem. Biophys. Res. Commun., 121:973-979 (1984a).

Sibley, D. R., Peters, J. R., Nambi, P., Caron, M. G. and Lefkowitz, R. J., Desensitization of turkey erythrocyte adenylate cyclase: β-Adrenergic receptor phosphorylation is correlated with attenuation of adenylate cyclase activity, J. Biol. Chem., 259:9742-9749 (1984b).

Sibley, D. R., Strasser, R. H., Caron, M. G. and Lefkowitz, R. J., Homologous desensitization of adenylate cyclase is associated with phosphorylation of the β-adrenergic receptor, J. Biol. Chem., 260:3883-3886 (1985).

Simpson, I. A. and Pfeuffer, T., Functional desensitization of β-adrenergic receptors of avian erythrocytes by catecholamines and adenosine 3', 5'-

phosphate, Eur. J. Biochem., 111:111-116 (1980).

Stadel, J. M., De Lean, A., Mullikin-Kilpatrick, D., Sawyer, D. D. and Lefkowitz, R. J., Catecholamine-induced desensitization in turkey erythrocytes: cAMP mediated impairment of high affinity agonist binding without alteration in receptor number, J. Cyclic Nucleotide Res., 7:37-47
(1981).

Stadel, J. M., Nambi, P. Lavin, T. N., Heald, S. L., Caron, M. G. and Lefkowitz, R. J., Catecholamine-induced desensitization of turkey erythrocyte adenylate cyclase: Structural alterations in the β-adrenergic
receptor revealed by photoaffinity labeling, J. Biol. Chem., 257:9242-
9245 (1982).

Stadel, J. M., Nambi, P., Shorr, R. G. L., Sawyer, D. F., Caron, M. G. and
Lefkowitz, R. J., Catecholamine-induced desensitization of turkey erythrocyte adenylate cyclase is associated with phosphorylation of the
β-adrenergic receptor, Proc. Natl. Acad. Sci. USA., 80:3173-3177
(1983a).

Stadel, J. M., Strulovici, B., Nambi, P., Lavin, T. N., Briggs, M. M., Caron,
M. G. and Lefkowitz, R. J., Desensitization of the β-adrenergic receptor
of frog erythrocytes: Recovery and characterization of the down-regulated receptors in sequestered vesicles, J. Biol. Chem., 258:3032-3038
(1983b).

Strader, C. D., Sibley, D. R. and Lefkowitz, R. J., Association of sequestered beta-adrenergic receptors with the plasma membrane: A novel
mechanism for receptor down regulation, Life Sci., 35:1601-1610 (1984).

Strasser, R. H. and Lefkowitz, R. J., Homologous desensitization of β-adrenergic receptor coupled adenylate cyclase: Resensitization by polyethylene glycol treatment, J. Biol. Chem., 260:4561-4564 (1985).

Strasser, R. H., Sibley, D. R. and Lefkowitz, R. J., A novel catecholamine-
activated adenosine cyclic 3', 5'-phosphate independent pathway for
β-adrenergic receptor phosphorylation in wild-type and mutant S_{49} lymphoma cells: Mechanism of homologous desensitization of adenylate
cyclase, Biochemistry, 25:1371-1377 (1986).

Strulovici, B., Cerione, R. A., Kilpatrick, B. F., Caron, M. G. and Lefkowitz, R. J., Direct demonstration of impaired functionality of a purified desensitized β-adrenergic receptor in a reconstituted system,
Science, 225:837-840 (1984).

Strulovici, B. and Lefkowitz, R. J., Activation, desensitization, and recycling of frog erythrocyte β-adrenergic receptors: Differential perturbation by in situ trypsinization, J. Biol. Chem., 259:4389-4395
(1984).

Strulovici, B., Stadel, J. M. and Lefkowitz, R. J., Functional integrity of
desensitizied β-adrenergic receptors: Internalized receptors reconstitute catecholamine-stimulated adenylate cyclase activity, J. Biol.
Chem., 258:6410-6414 (1983).

Su, Y.-F., Cubeddu, L. and Perkins, J. P., Regulation of adenosine 3':5'-
monophosphate content of human astrocytoma cells: Desensitization to
catecholamines and prostaglandins, J. Cyclic Nucleotide Res., 2:257-
270 (1976).

Su, Y.-F., Harden, T. K. and Perkins, J. P., Isoproterenol-induced desensitization of adenylate cyclase in human astrocytoma cells, J. Biol.
Chem., 254:38-41 (1979).

Su, Y.-F., Harden, T. K. and Perkins, J. P., Catecholamine-specific desensitization of adenylate cyclase: Evidence for a multistep process, J.
Biol. Chem., 255:7410-7419 (1980).

Terasaki, W. L., Brooker, G., de Vellis, J., Inglish, D., Husu, C.-Y. and
Moylan, R. D., Involvement of cyclic AMP and protein synthesis in catecholamine refractoriness, Adv. Cyclic Nucleotide Res., 9:33-52 (1978).

Toews, M. L., Waldo, G. L., Harden, T. K. and Perkins, J. P., Relationship
between an altered membrane form and a low affinity form of the β-adrenergic receptor occurring during catecholamine-induced desensitization,
J. Biol. Chem., 259:11844-11850 (1984).

Waldo, G. L., Northup, J. K., Perkins, J. P. and Harden, T. K., Characterization of an altered membrane form of the β-adrenergic receptor produced during agonist-induced desensitization, J. Biol. Chem., 258:13900-13908 (1983).

Wessels, M. R., Mullikin, D. and Lefkowitz, R. J., Differences between agonist and antagonist binding following beta-adrenergic receptor desensitization, J. Biol. Chem., 253:3371-3373 (1978).

Wessels, M. R., Mullikin, D. and Lefkowitz, R. J., Selective alteration in high affinity agonist binding: A mechanism of beta-adrenergic receptor desensitization, Mol. Pharmacol., 16:10-20 (1979).

LONG-TERM SYNERGISTIC REGULATION OF IONIC CHANNELS BY C-KINASE AND Ca^{2+}/CaM-TYPE II KINASE

Daniel L. Alkon and Shigetaka Naito

Section on Neural Systems, Laboratory of Biophysics
National Institute of Neurological and Communicative
Disorders and Stroke, National Institutes of Health
at the Marine Biological Laboratory, Woods Hole
Massachusetts

The distinct molecular identities of ionic channels within biological membranes are now being revealed. The acetylcholine receptor, a ligand-gated cation channel, was solubilized and purified from post-synaptic membranes of the ray Torpedo californica (Karlin, 1980; Changeux, 1981). Entire amino acid sequences for all of the acetylcholine receptor subunits were deduced from DNA sequence analysis of cDNA clones (Numa et al., 1983). Similarly, the voltage-sensitive Na$^+$ channel from rat brain has been solubilized and purified to homogeneity, and shown to consist of α (Mr 260,000), β_1 and β_2(Mr 39,000 and 37,000, respectively) subunits (Agnew et al., 1980; Weigele and Barchi, 1982; Hartshorne and Catterall, 1984). Consistent with the heterogeneity of tetrodotoxin binding sites, presence of at least three distinct Na$^+$ channels (I, II and III) in rat brain was suggested from a sequencing study of cDNA clones obtained from three distinct mRNAs for the α-subunit (Noda et al., 1986). The Ca^{2+} channel (a dihydropyridine-sensitive class) was also purified (Curtis and Catterall, 1985). None of the known K$^+$ channels have yet been purified, probably due to the unavailability of high affinity neurotoxins and a K$^+$ channel abundance which is small in comparison to that of the Na$^+$ channel and the acetylcholine receptor.

Channels, as integral membrane proteins, have also been shown to have subunit sites which, in vitro, undergo cyclic-AMP-dependent, Ca^{2+}-dependent, and/or lipid dependent phosphorylation. Within the acetylcholine receptor, for example, the γ and α subunits are phosphorylated by cyclic-AMP-dependent protein kinase (Huganir and Greengard, 1983), the α subunit by protein kinase C, and the β, γ, and α subunits by tyrosine-specific protein kinase (Huganir and Greengard, 1983). The α subunit of the Na$^+$ channel can serve as a substrate for both the cyclic AMP-dependent protein kinase (Costa et al., 1982) and protein kinase C (Costa and Catterall, 1984).

An in vivo regulatory role of phosphorylation on membrane channels has been implicated by experiments in which exposure of intact nerve cells or subcellular structures such as synaptosomes, to stimulating conditions such as depolarization, sensory input, neurotransmitters, etc., causes changes in phosphorylation of specific proteins. Perfusion of synaptosomal fractions with elevated external K$^+$ solutions, for example, produced changes of protein phosphorylation (Krueger et al., 1977). Photic stimulation of verte-

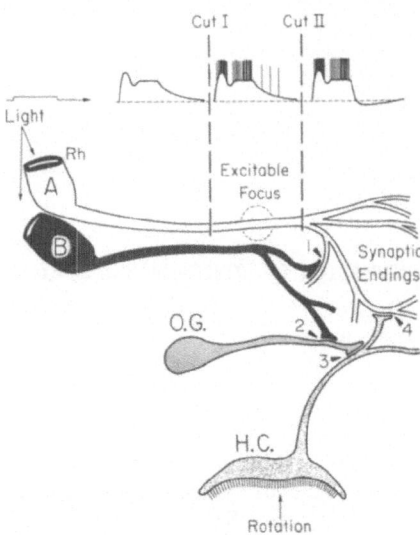

FIG. 1:   A) Excitation and inhibition of type B photoreceptor. Diagram of
          type A photoreceptor soma with rhabdome, excitable focus and termi-
          nal branches. Light depolarizes photoreceptors at rhabdomes (Rh).
          Type B photoreceptors (of which there are three) inhibit, at synap-
          tic endings, lateral type A, directly (1) and indirectly by inhibit-
          ing (2) optic ganglion (OG) cells (of which there are 13). Ipsila-
          teral hair cells (HC) receive less inhibition from optic ganglion
          cells (3) when the type B fires and thus increases its inhibition
          of the lateral type A (4). Hair cells also inhibit type A when
          they depolarize in response to rotation. In response to light, the
          type A cell depolarizes without impulses or afterhyperpolarization
          with cut I lesion. It depolarizes with impulses but without after-
          hyperpolarization with cut II lesion. The response of an intact
          type A cell is represented at upper right.

brate photoreceptors also clearly altered patterns of protein phosphorylation (Bownds et al., 1972; Szuts, 1985). Classical conditioning of the
nudibranch mollusc Hermissenda crassicornis changed phosphorylation of a
20,000 M. W. protein measured 2-3 hours after the conditioning experience,
but not control procedures (Neary et al., 1981). Finally, application of a
variety of neuropharmacologic agents produced phosphorylation differences
as well (c.f. Levitan, 1985).

     An in vivo regulatory role for phosphorylation of membrane channels
has received strong additional support from experiments which involve
application of protein kinases intracellularly to the cytoplasmic surface
of isolated cell membranes, or to artificial membranes containing reincor-
porated membrane channels. Injection of cyclic-AMP-dependent protein
kinase caused reduction (in some cases) of Aplysia K^+ currents (Castellucci et al., 1980; Kaczmarek et al., 1980), reduction of Hermissenda early
and late K^+ currents (Alkon et al., 1983), and an increase of an Aplysia K^+
current (DePeyer et al., 1982). Injection of Ca^{2+}/CaM-dependent kinase
also reduced Hermissenda I_A and $I_{Ca^{2+}-K^+}$ currents, but in a clearly Ca^{2+}-
dependent manner (Acosta-Urquidi et al., 1984; Sakakibara et al., 1985,
1986). Exposure to C-kinase activators (phorbol ester and OAG) caused
reduction of Ca^{2+} currents of chick dorsal root ganglion cells (Rane and
Dunlap, 1986) and C-kinase injection or bath application of phorbol ester

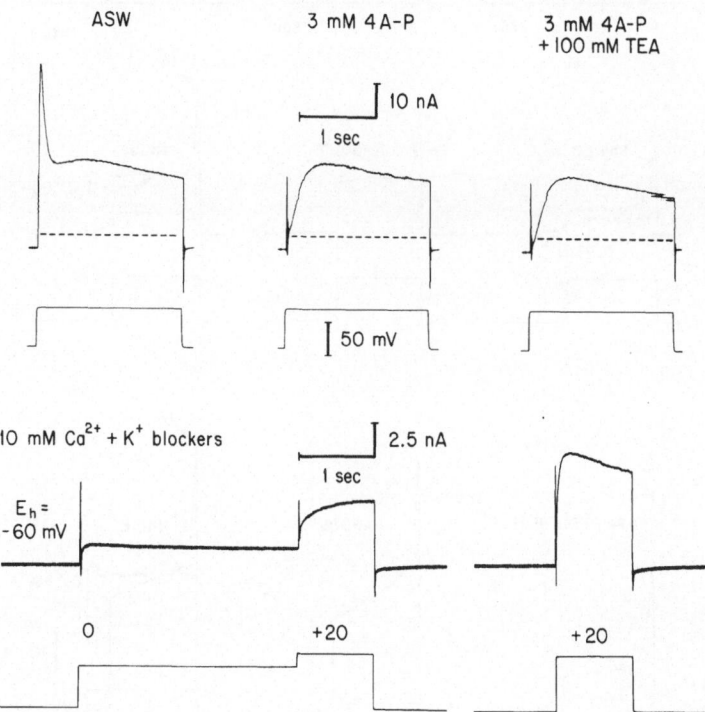

FIG. 2: A) Voltage-dependent outward currents across the membrane of the isolated type B cell soma. From left to right, ASW, 3 mM 4-amino-pyridine (4-AP) added to ASW, 4-AP and 100 mM tetraethylammonium (TEA) ion added to ASW. Note that addition of 4-AP and TEA remove only a small portion of the late outward current elicited by command to 0 mV from a holding potential of -60 mV. The dashed lines indicate the level of the non-voltage-dependent or 'leak' current. B) Voltage-dependent outward current in presence of 4-AP and TEA. Voltage-dependent activation of outward calcium-dependent K^+ current ($I_{Ca2^+-K^+}$) in type B photoreceptor, and reduction by pre-pulse depolarization. Current (top) and voltage (bottom) records from voltage-clamp experiment illustrate that a 3 sec depolarization to 0 mV elicits a small net outward current (top left), followed by a larger outward current when the membrane is stepped to +20 mV (top middle). The pre-pulse depolarization reduced by ~ 40% the outward current normally evoked by an 80 mV step from -60 to +20 mV (top right) and slowed the rise time as well. Bathing solutions included 10 mM 4-AP and 100 mM TEA to block the fast (I_A) and delayed (I_K) K^+ currents. ASW: artificial sea water.

caused enhancement of Ca^{2+} currents of _Aplysia_ bag cells (DeRiemer et al., 1985), and reduction of _Hermissenda_ K^+ currents (Farley and Auerbach, 1986; Alkon et al., 1986; Kubota et al., 1986). Isolated inside-out patches (i.e., with the cytoplasmic surfaces exposed to the external bathing medium) from _Helix_ neurons have been shown to contain calcium-dependent K^+ channels whose activity increased with bath application of cyclic-AMP-dependent protein kinase (Ewald, Williams, and Levitan, 1985). Similarly, cyclic-AMP-dependent protein kinase applied to the cytoplasmic surface of _Aplysia_ sensory cell inside-out patches regulates the activity of serotonin-sensitive or "S" channels (Shuster et al., 1985).

In still another type of experiment channel activity was recorded from crude membrane vesicle fractions which were reconstituted with exogenous phospholipid into bilayers (Wilmsen et al., 1983). Recording from single

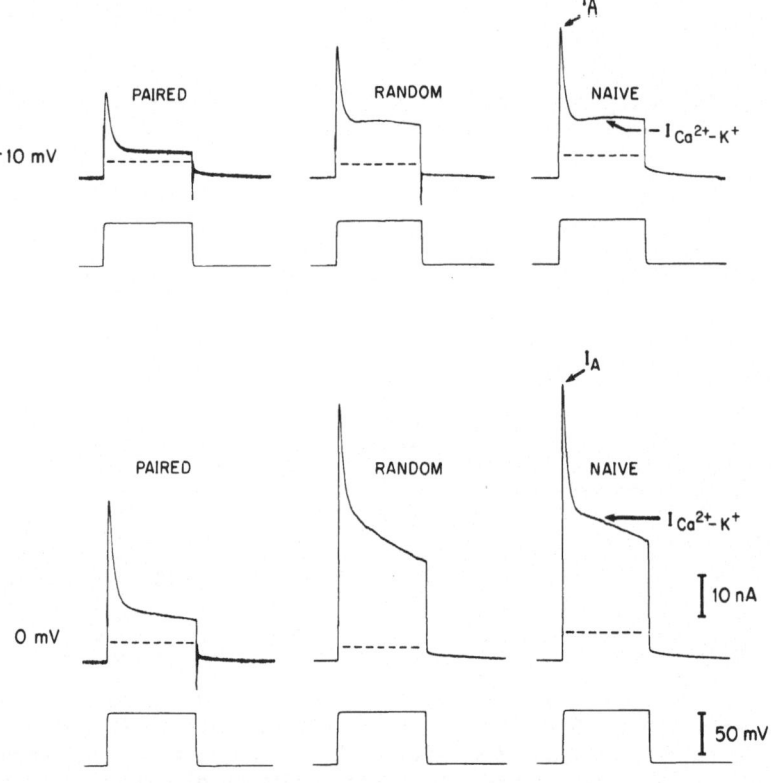

FIG. 3: Comparison of voltage-dependent K^+ currents measured in type B
somata isolated from paired, random, and naive animals. The
records were chosen to illustrate the reduction of I_A and $I_{Ca^{2+}-k^+}$
for paired as compared to random and naive animals (Alkon et al.,
1985).

channels has been achieved in this manner by a number of workers (Coronado
and Latorre, 1983; Suarez-Isla et al., 1983; Wilmsen et al., 1983). With
this technique, Ewald et al. (1985) have regulated activity of Ca^{2+}-depen-
dent K^+ channels by application of cyclic-AMP-dependent protein kinase.

Regulation of membrane channels, as mediated by phosphorylation,
appears to follow a much more prolonged time course than that of rapid molecular conformational changes which would underlie the constant opening and
closing of ionic channels within biological membranes. One time course of
channel regulation, that of minutes, was followed by increased bag cell
firing (Kaczmarek et al., 1978), by synaptic inhibition of neuronal bursting
activity (Levitan et al., 1979), and by sensitization-induced excitability
differences in sensory cells (Kandel and Schwartz, 1982). Transient hormonal and/or neurotransmitter elevation of cyclic-AMP have been implicated as
an important initial step in the generation of all three phenomena, namely
bag cell afterdischarge, inhibition of bursting, and sensitization-induced
excitability differences. The involvement of specific differences in protein phosphorylation in channel regulation, however, has not yet been determined, except that a study suggested that phosphorylation of two proteins
is related to the activation of a rectifier K^+-channel (Lemos et al., 1985).

Another time course, that of days, was followed by conditioning-specific reduction of K^+ currents across Hermissenda neural membranes (Alkon,
1979, 1980; Crow and Alkon, 1980; Alkon et al., 1982; West et al., 1982;
Farley and Alkon, 1982; Alkon et al., 1985). In the work with Hermissenda,

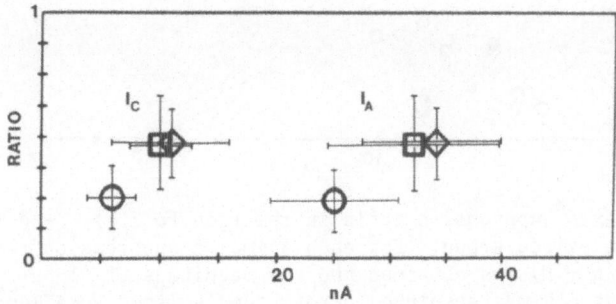

FIG. 4: Mean phototaxis suppression ratios in relation to ionic current magnitude. For individual animals of each group (Paired, O ; Random, □ ; and Naive, ◊) a suppression ratio (in the form B/A + B where A = post-treatment latency and B = pretreatment latency) was obtained and the magnitude of $I_{Ca^{2+}-k^+}$ (on the left and I_A was measured at -10 mV (absolute) across the isolated soma membrane of the medial type B cell. The values presented (± SD) are the mean ionic currents ($I_{Ca^{2+}-k^+}$ and I_A) measured in relation to the mean suppression ratio for each group. The Paired mean ratios and ionic currents are all clearly lower than for the Random and Naive groups (Alkon et al., 1985).

classically conditioned responses have been shown by a number of experimental criteria to be a causal consequence of persistent reduction of two specific voltage-dependent K^+ currents (Figs. 1-6). Reduction of these K^+ currents has a duration of not minutes but at least 2-3 days, a duration not previously observed for changes of ionic current flux across membranes of differentiated neurons. This conditioning-specific K^+ current reduction in Hermissenda represents biophysics within a new temporal domain - a domain of days or longer rather than msecs, seconds, or minutes. Such a biophysical domain would seem to require a particular cellular biochemistry as well - one involving reactions which are only slowly reversible and ultimately not reversible at all.

More recently, we have obtained evidence that a similar biophysical record of associative learning can be found in the mammalian brain (Fig. 7). The same long-lasting transformation of specific voltage-dependent K^+ channels, and perhaps the underlying biochemistry of such transformation, may provide a record of a learned stimulus relationship on days after conditioning the rabbit as it does after conditioning the mollusc Hermissenda.

Of particular interest in the Hermissenda conditioning was the possibility of reconstructing a sequence of biophysical and biochemical steps which ultimately lead to extremely long-lasting modulation of the ionic currents. Since Hermissenda classical conditioning in almost every important aspect closely resembles that of vertebrates (c.f. Alkon, 1984; Lederhendler et al., 1986), and since we have measured changes in vertebrate brain slices (Coulter et al., 1986; Disterhoft et al., 1986) which bear striking resemblance to changes measured in Hermissenda, perhaps a cellular sequence with some general relevance could be uncovered.

To reconstruct such a sequence, it was first necessary to identify a

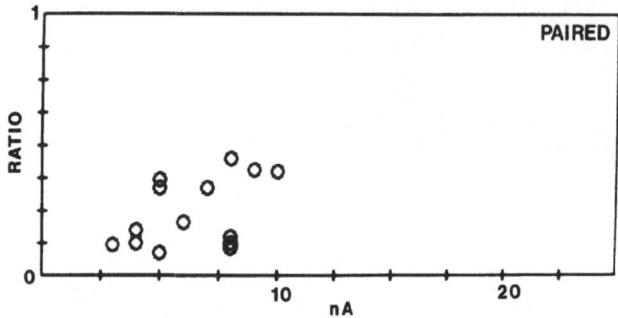

FIG. 5: Phototaxis suppression ratio in relation to $I_{Ca^{2+}-k^+}+50$ magnitude
for the paired group. For each animal a suppression ratio (in the
form A/A + B) was obtained and the magnitude of $I_{Ca^{2+}-k^+}$ was measured at -10 mV (absolute) across the isolated soma membrane of
the medial type B cell. The ratio is significantly correlated with
the current magnitude (Alkon et al., 1985).

locus where changes might bear a causal relationship to the acquisition and
retention of a classically conditioned response.

Changes within certain neurons, the type B photoreceptors, a locus of
convergence between the visual and vestibular pathways were, by a number of
criteria, primary or causal changes. These criteria can be described as
correlation, intrinsic localization, predictive value, and synthesis. Biophysical changes within type B cell somata were correlated with behavioral
modification measured in living animals, isolated nervous systems and isolated neurons. Demonstration that the biophysical changes were intrinsic to
identified membrane sites, here of the type B cell somata, was provided by
the finding that reduction of I_A and $I_{Ca^{2+}-k^+}$ could be found across the soma
membrane even after complete isolation of that membrane from intact conditioned (But not control) animals (Alkon et al., 1982; 1985). Furthermore,
the magnitude of I_A and $I_{Ca^{2+}-k^+}$ reduction predicted the degree of conditioning and the magnitude of changes of post-synaptic responses of inter-
neurons and motorneurons (Goh et al., 1985). Finally, by impaling type B
cells in living animals and producing (or synthesizing) the K^+ current
reduction, it was possible to produce the learning behavior measured days
later (Farley et al., 1983).

It was then possible to ask how these causally implicated type B cell
changes arise during conditioning, i.e., how are they stored and how are
they evoked on subsequent days by conditioned-stimulus presentations - i.e.,
how are they recalled?

Our data indicate that storage of associatively learned information
within the type B cell soma begins when:

(1) A single pairing of the CS and UCS, light and rotation, produces a
response unique to the pairing: enhanced and prolonged type B depolarization. This depolarization is the resultant of the integrated visual and
vestibular system's response to stimulus pairing.

(2) With repeated pairings depolarization of the type B cell becomes progressively greater and more prolonged - i.e., it accumulates. Accompanying
the depolarization is a marked and prolonged elevation of intracellular
Ca^{2+}.

280

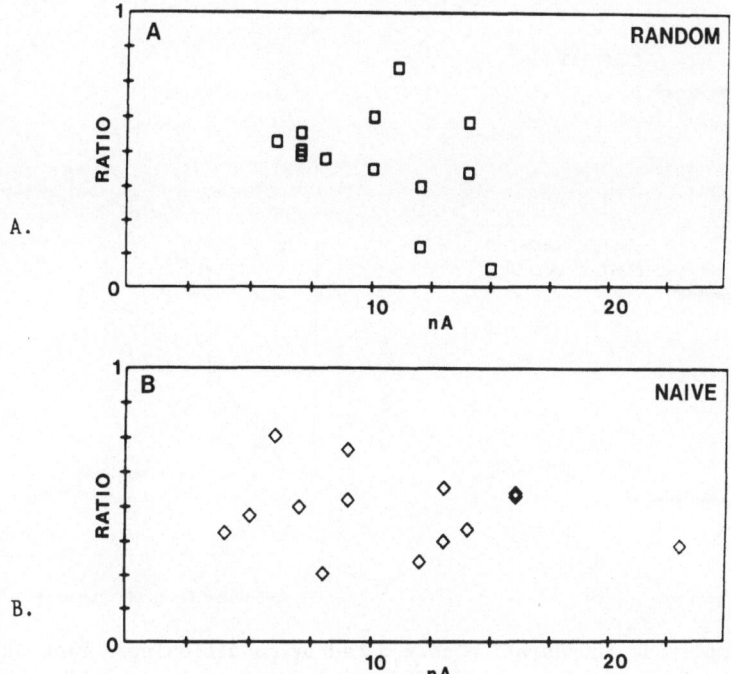

A.

B.

FIG. 6: A. Phototaxis suppression ratio in relation to $I_{Ca^{2+}-k^+}^{50}$ magnitude for the Random group. For each animal a suppression ratio
(in the form A/A + B) was obtained and the magnitude of $I_{Ca^{2+}-k^+}$
was measured at -10 mV (absolute) across the isolated soma membrane of the medial type B cell. The ratio is not significantly
correlated with the current magnitude (Alkon et al., 1985). B:
Phototaxis suppression ratio in relation to $I_{Ca^{2+}-k^+}^{50}$ magnitude
for the Naive group. For each animal a suppression ratio (in the
form A/A + B) was obtained and the magnitude of $I_{Ca^{2+}-k^+}$ was measured at -10 mV (absolute) across the isolated soma membrane of
the medial type B cell. The ratio is not significantly correlated
with the current magnitude (Alkon et al., 1985).

(3) Depolarization and elevated Ca^{2+} in turn cause inactivation of specific K^+ channels. This Ca^{2+}-mediated reduction of K^+ currents is a new biophysical phenomenon (Alkon et al., 1982, 1985) which became apparent when
we looked within the new biophysical domain relevant to acquisition and
retention of associative learning. Experimental manipulations which
increase Ca^{2+}_i (such as injection of Ca^{2+}_i under voltage-clamp through a
third microelectrode, injection of inositoltrisphosphate which releases
Ca^{2+}_i) causes the same K^+ channel inactivation. Manipulations which
decrease free Ca^{2+}_i (such as injection of a Ca^{2+}-chelator, EGTA) reduces
or prevents K^+ channel inactivation.

(4) Ca^{2+}_i elevation and depolarization do not persist but K^+ channel
inactivation continues at least for days after the training. <u>Persistent
reduction</u> of I_A and $I_{Ca^{2+}-k^+}$, then, is a cellular expression of the behaviorally manifest <u>retention</u> of the associative learning.

(5) Reduced <u>voltage-dependent activation</u> of I_A and $I_{Ca^{2+}-k^+}$ is the cellular expression of the behavioral phenomenon of <u>recall</u> of the learned association. I_A and $I_{Ca^{2+}-k^+}$ are only significantly activated when the type B

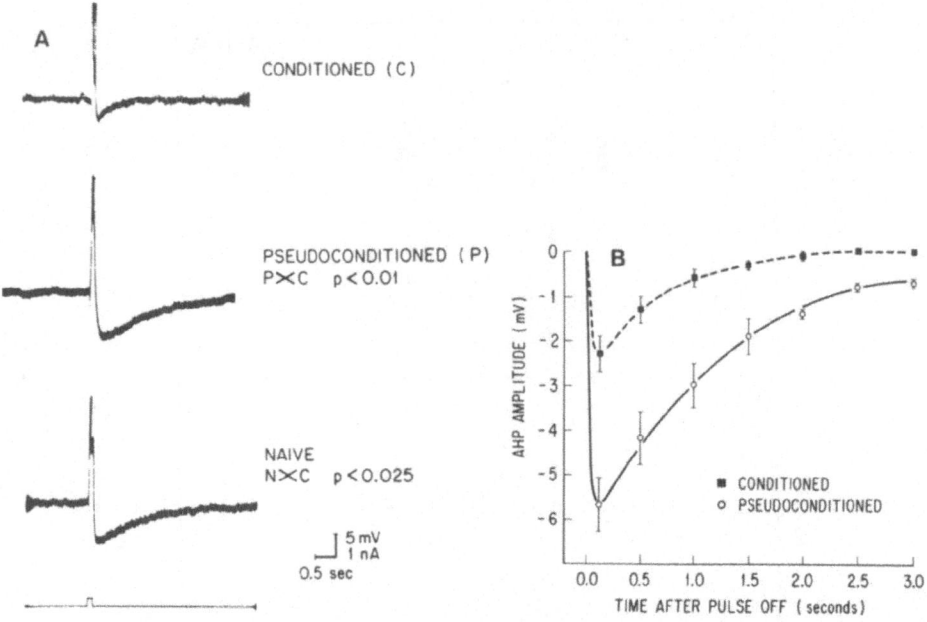

FIG. 7: AHP amplitude and duration is reduced by conditioning. Plot of
means ± 1 standard error of AHP amplitude at isochronal times after
stimulus offset for 18 pseudoconditioned cells (open circles) and
19 conditioned cells (filled squares) (Coulter et al., 1986).

cell depolarizes $\geq$ 30 mV in response to subsequent presentations of the
conditioned stimulus, light.

(6) Retained K^+ channel inactivation makes the type B cells more excitable
and through its inhibitory effects on a known neural pathway, can contribute to a new conditioned behavioral response.

Biochemical Substrates for Biophysical Records

The transition of K^+ channel inactivation lasting many minutes to
inactivation lasting several days appears to be mediated, at least in part,
by activation of Ca^{2+}-calmodulin-dependent type II protein kinase as well
as a Ca^{2+}- and lipid-activated C-kinase. The first evidence for such media-

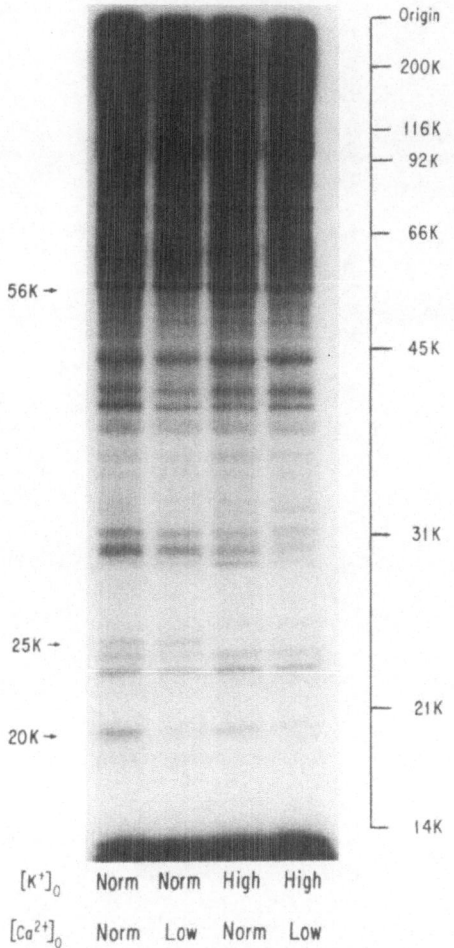

$[K^+]_0$ Norm   Norm   High   High

$[Ca^{2+}]_0$ Norm   Low   Norm   Low

FIG. 8: <u>Effect of high K^+ on in vivo protein phosphorylation of Hermissenda CNS in the presence and absence of $[Ca^{2+}]$</u>. <u>Hermissenda</u> central nervous systems (CNSs) were labeled with ^{32}P in ASW for 3 hrs and then exposed to 100 mM (high) K^+-ASW or 10 mM (normal) K^+-ASW containing 10 mM (normal) Ca^{2+} or 1 mM EGTA (low Ca^{2+}) in the presence of ^{32}P for 30 min. At the end of incubation, CNSs were homogenized in SDS-sample buffer and phosphoproteins were separated by SDS-polyacrylamide slab gel electrophoresis and visualized by autoradiography.

tion was provided by electrophoresis of eye samples (each containing only five photoreceptors) isolated from conditioned as well as randomized and unpaired control animals. Neary et al. (1981) found that there was a conditioning-specific increase in phosphorylation of a 20,000 M.W. protein. This change was measured 2-3 hours after the last training experience.

Subsequent research demonstrated that several phosphoproteins change their state of phosphorylation in response to exposure of the intact nervous systems to conditions which simulate the prolonged depolarization which occurs during acquisition of the learned behavior. Incubation in

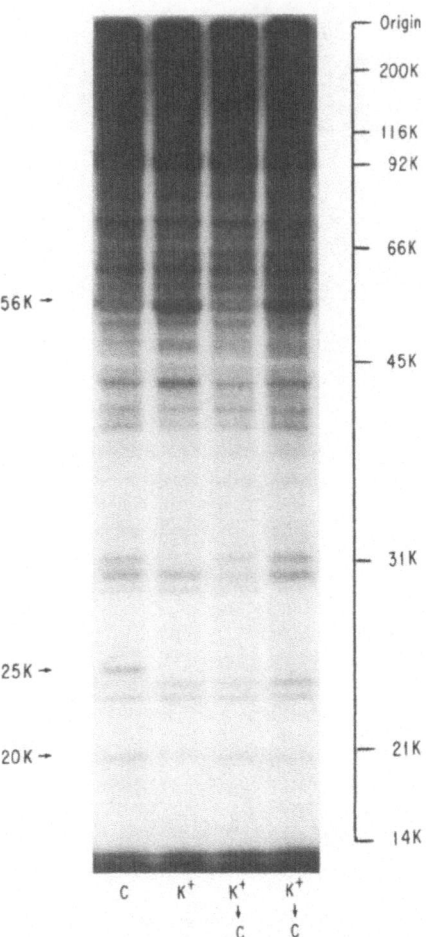

FIG. 9: <u>Long-lasting high K^+ effect on protein phosphorylation</u>. Prelabeled CNSs were treated with high K^+ (300 mM) for 30 minutes and
subsequently returned to ASW for 30 minutes. Radioactive phosphoproteins were analyzed as in Figure 8. C and K^+ stand for control
ASW and high K^+-ASW, respectively. Statistical analysis of additional similar experiments indicated that change (reduction) of
phosphorylation of 25,000 M.W. protein by high K^+ persisted for at
least 30 minutes after high K^+ depolarization was removed whereas
changes of that of 56,000 and 20,000 M.W. proteins did not.

elevated external K^+ (100-300 mM) for 10-30 minutes, for example, reduced
phosphorylation of the 25,000 and 20,000 M.W. proteins and increased phosphorylation of a 56,000 M.W. protein (Figs. 8, 9; Naito and Alkon, manuscript submitted). Trifluoperazine (TFP), an inhibitor of Ca^{2+}-dependent
phosphorylation (mediated both by Ca^{2+}/CaM-type II and C-kinases), caused
inhibition of phosphorylation of the 25,000 M.W. protein, and, to a lesser
extent, of the 20,000 M.W. protein as well as other proteins. In still
another <u>in vivo</u> study, activation of the C-kinase with phorbol ester caused
increased phosphorylation of a 22,000 M.W. protein and a more moderate
increase of 56,000 M.W. protein phosphorylation.

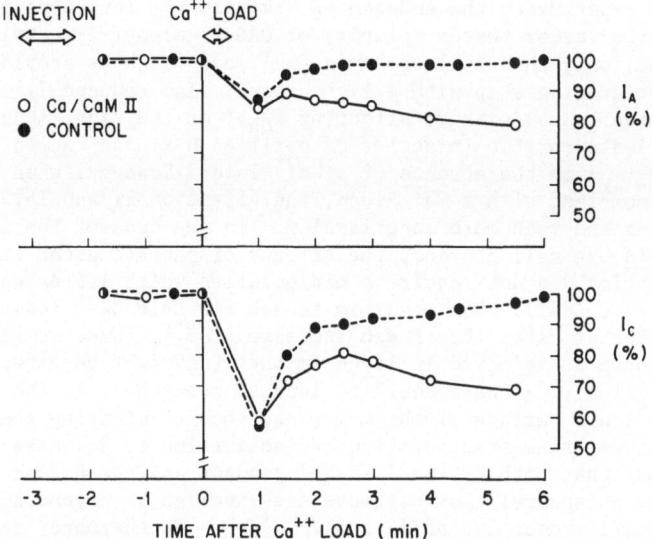

FIG. 10: Time course of I_A (upper panel) and I_C (lower panel) reduction by
CaM kinase II following a Ca^{2+} load. I_A and I_C were measured as
the peak outward currents ~ 20 msec (I_A) and 300-400 msec (I_C)
from the onset of a command depolarization to -5 mV absolute
(holding potential = -60 mV). Ca^{2+} loads were given before
(closed circles) and after iontophoretic injection of CaM kinase
II. I_A and I_C amplitudes before the Ca^{2+} load are normalized as
100%. Note that CaM kinase II injection prevents recovery of I_A
and I_C reduction after a Ca^{2+} load (Sakakibara et al., 1986).

In vitro analysis of Hermissenda nervous system homogenates revealed
25,000 M.W. protein, 56,000 M.W. protein and 20,000 M.W. protein as substrates for protein kinase C. The 20,000 M.W. protein, in addition, can
serve as a substrate for both cAMP-dependent and Ca^{2+}/CaM-dependent protein
kinases (Neary et al., in preparation). These results suggest that Ca^{2+}/
CaM-dependent and Ca^{2+}/lipid-dependent phosphorylation of low molecular
weight proteins may be involved at least in the early states leading to
persistent biochemical mechanisms for memory storage. To examine this possibility further, we asked whether it was possible to produce the same biophysical changes which occur during memory storage by manipulating these
implicated biochemical pathways. This proved to be the case with an unexpected degree of specificity. Iontophoretic injection of inositol trisphosphate (IP_3) for example (but not inositol monophosphate) caused reduction of I_A and $I_{Ca^{2+}-k^+}$ but not $I_{Ca^{2+}}$ or the light-induced I_{Na^+} across the
type B soma membrane. This reduction, accomplished under voltage clamp,
did not reverse for the recording period (up to three hours) and was particularly marked after a single depolarizing step (paired with light) sufficient to cause substantial Ca^{2+} flux across the soma membrane and a rise
of Ca^{2+}_i. Similarly, iontophoresis of Ca^{2+}/CaM-dependent kinase (phosphorylase kinase) as well as Ca^{2+}/CaM-dependent type II kinase, following a
single Ca^{2+} load, caused persistent reduction of the same two K^+ currents
reduced by conditioning, I_A and $I_{Ca^{2+}-k^+}$, without affecting the light-
induced Na^+ current (Fig. 10). Iontophoresis of heat inactivated enzyme
or enzyme inactivated by standing at room temperature was without effect.
Both the effects of IP_3 injection and Ca^{2+}/CaM-type II kinase injection were
prevented by the Ca^{2+}-dependent kinase inhibitor, TFP. TFP itself also
caused enhancement of I_A and $I_{Ca^{2+}-k^+}$, suggesting inhibition of endogenous
Ca^{2+}-dependent kinase(s) which normally contributes to I_A and $I_{Ca^{2+}-k^+}$ inactivation.

285

In other experiments the endogenous C-kinase was activated by perfusion with phorbol ester (water soluble) or OAG (1-oleoyl-2-acetylglycerol). This treatment, only when followed by a Ca^{2+} load (such as provided by pairing a depolarizing step with a light step), also reduced I_A and $I_{Ca^{2+}-k^+}$ (for up to three hours) without affecting $I_{Ca^{2+}}$ or the light-induced I_{Na^+}. Furthermore, iontophoretic injection of purified C-kinase caused reduction of I_A and $I_{Ca^{2+}-k^+}$ in the absence of a Ca^{2+} load. However, when C-kinase injection is combined with a Ca^{2+}-load, the effect on I_A and $I_{Ca^{2+}-k^+}$ would be much greater and much more long-lasting. In the case of the Ca^{2+}-current of Aplysia bag cell neurons, the effects of phorbol ester or protein kinase C injection did not require a manipulation which increases $[Ca^{2+}]_i$ (DeRiemer et al., 1985). One possible reason for this Ca^{2+} independence may be that phorbol ester itself can increase $[Ca^{2+}]_i$ (Wane et al., 1985). Also, if C-kinase is injected in large amounts (as might be expected for pressure injection), C-kinase could be located everywhere in the cell, including the inner surface of the membrane, thus eliminating the Ca^{2+}-dependence for membrane translocation and activation of C-kinase. These results suggest that both the Ca^{2+}/CaM-dependent and the phospholipid/Ca^{2+}-dependent phosphorylation pathways are involved in producing the conditioning-specific reduction of I_A and $I_{Ca^{2+}-k^+}$. Furthermore, since C-kinase activation by phorbol ester or OAG itself was without a significant effect, a synergistic interaction of the two phosphorylating pathways may be necessary for particularly long-lasting K^+ current reduction. In Hermissenda classical conditioning, activation of the C-kinase pathway may arise as a consequence of visual transduction since increased cleavage of PIP_2 and elevation of IP_3 and DG have been observed with stimulation of molluscan and Limulus photoreceptors (Szuts, in preparation).

Activation of the Ca^{2+}/CaM-dependent pathway arises as a result of the voltage-dependent Ca^{2+} flux due to paired light-induced and synaptic depolarization. However, the initiation of the interaction between two distinct sensory inputs occurs at post-synaptic loci of convergence within genetically constrained neural networks. This interaction consists first of electrical spread from one post-synaptic site (that of the UCS) to the other (of the CS). Consequent to this electrical interaction, post-synaptic second messengers are triggered and more long-lasting biochemical events follow. In other systems, e.g., hippocampus, synaptic activation itself (e.g., muscarinic effects) could be responsible for C-kinase activation (Nicoll et al., 1986). An analogous synergy between Ca^{2+}/CaM-dependent and C-kinase dependent phosphorylation in mediating sustained physiologic responses has already been demonstrated for platelet aggregation by Nishizuka and his colleagues (Kaibuchi et al., 1983) and for aldosterone secretion from adrenal cortical cells by Rasmussen and his associates (Kojima et al., 1984).

In other related studies of our laboratory, recent results suggest that membrane changes similar to those found for the type B cell are intrinsic to CA1 neurons, and occur only in hippocampal slices from classically conditioned rabbits. We reasoned that brain regions in which short latency neural correlates of acquisition and retention of the learning were obtained by Thompson and his colleagues in vivo with extracellular recording, might also be sampled with the brain slice technique.

The CA1 afterhyperpolarization, for example, measured within 300 msec of depolarization onset as shown by several workers, including Prince, Gustaffson, and others, is due to activation of a Ca^{2+}-dependent K^+ current. This CA1 afterhyperpolarization was significantly larger in amplitude and duration, and longer in time course of decay, for naive and pseudoconditioned animals as compared to that of conditioned animals. This condi-

tioning-specific difference occurred in the absence of between-group differences of M.P., input resistance of impulse amplitude. Thus, it is likely
that the same current, a Ca^{2+}-dependent K^+ current, in the same cellular
compartment, post-synaptic, is reduced for neurons of classically conditioned rabbits as was found for conditioned Hermissenda.

In conclusion, although we have a biophysical record of associative
memory, a record lasting at least several days now demonstrated in Hermissenda and consistent with our findings in rabbit hippocampus, we do not
have, (nor is such available for any system) a biochemical record. Molecular transformations which survive molecular turnover and permanently
encode associative memory have not yet been identified. One approach to
finding such transformations is illustrated by the work on Hermissenda in
which Ca^{2+}-mediated reduction of ionic currents extends into a new temporal
domain - one of days. By localizing such changes which are long-lasting
and can play a causal role in the storage of associative memory and then
tracing biochemical steps which lead to these relatively permanent biophysical records, ultimatly we may uncover more permanent biochemical storage
mechanisms.

It may be, for example, that Ca^{2+}/CaM-dependent and/or cyclic-AMP-
dependent phosphorylation predominate in channel regulation for many
minutes while a synergistic interaction of the Ca^{2+}/CaM-dependent and C-
kinase dependent phosphorylation extends channel regulation into the domain
of hours or longer. Channel changes lasting for days or longer may involve
still additional biochemical machinery such as that affecting protein synthesis and/or structural transformations, both of which we are presently
analyzing in the Hermissenda nervous system and rabbit brain hippocampal
slices.

REFERENCES

Abrams, T. W., Castellucci, V. F., Camardo, J. S., Kandel, E. R. and Lloyd,
P. E., Two endogenous neuropeptides modulate the gill and siphon withdrawal reflex in Aplysia by presynaptic facilitation involving cAMP-
dependent closure of a serotonin-sensitive potassium channel, Proc.
Nat. Acad. Sci. USA, 81:7956-7960 (1984).

Acosta-Urquidi, J., Alkon, D. L. and Neary, J. T., Ca^{2+}-dependent protein
kinase injection in a photoreceptor mimics biophysical effects of
associative learning, Science, 224:1254-1257 (1984).

Agnew, W. S., Moore, A. C., Levinson, S. R. and Raftery, M. A., Identification of a large molecular weight peptide associated with a tetrodotoxin
binding protein from the electroplax of Electrophorus electricus, Biochem. Biophys. Res. Commun. 92:860-866 (1980).

Alkon, D. L., Voltage-dependent calcium and potassium ion conductances: A
contingency mechanism for an associative learning model, Science, 205:
810-816 (1979).

Alkon, D. L., Membrane depolarization accumulates during acquisition of an
associative behavioral change, Science, 210:1375-1376 (1980).

Alkon, D. L., Calcium-mediated reduction of ionic currents: A biophysical
memory trace, Science, 226:1037-1045 (1984).

Alkon, D. L., Acosta-Urquidi, J., Olds, J., Kuzma, G. and Neary, J. T., Protein kinase injection reduces voltage-dependent potassium currents,
Science, 219:303-306 (1983).

Alkon, D. L., Kubota, M., Neary, J. T., Naito, S., Coulter, D. and Rasmussen, H., C-kinase activation prolongs Ca^{2+}-dependent inactivation of
K^+ currents, Biochem. Biophys. Res. Commun., 134:1215-1222 (1986).

Alkon, D. L., Lederhendler, I. and Shoukimas, J. J., Primary changes of
membrane currents during retention of associative learning, Science,
215:693-695 (1982).

Alkon, D. L., Sakakibara, M., Forman, R., Harrigan, J., Lederhendler, I. and Farley, J., Reduction of two voltage-dependent K^+ currents mediates retention of a learned association, Behav. Neural Biol., 44:278-300 (1985).

Bownds, M. D., Dawes, J., Miller, J. and Stahlman, M., Phosphorylation of frog photoreceptor membranes induced by light, Nature, 237:125-127 (1972).

Castellucci, V. F., Kandel, E. R., Schwartz, J. H., Wilson, F. D., Nairn, A. C. and Greengard, P., Intracellular injection of the catalytic subunit of cyclic AMP-dependent protein kinase simulates facilitation of transmitter release underlying behavioral sensitization in Aplysia, Proc. Nat. Acad. Sci. USA, 77:7492-7496 (1980).

Changeux, J.-P., The acetylcholine receptor: An "allosteric" membrane protein, Harvey Lect., 75:85-254 (1981).

Coronado, R. and LaTorre, R., Phospholipid bilayers made from monolayers on patch-clamp pipettes, Biophys. J., 43:231-236 (1983).

Costa, M. R. C., Casnellie, J. E. and Catterall, W. A., Selective phosphorylation of the α-subunit of the sodium channel by cAMP-dependent protein kinase, J. Biol. Chem., 257:7918-7921 (1982).

Costa, M. R. C. and Catterall, W. A., Phosphorylation of the αsubunit of the sodium channel by protein kinase C, Cell. Mol. Neurobiol., 4:291-297 (1984).

Coulter, D. A., Kubota, M., Disterhoft, J. F., Moore, J. W. and Alkon, D. L., Conditioning-specific reduction of CA1 afterhyperpolarization amplitude and duration in rabbit hippocampal slices, Soc. Neurosci. Abstr., 11:981 (1985).

Crow, T. J. and Alkon, D. L., Associative behavioral modification in Hermissenda: cellular correlates, Science, 209:412-414 (1980).

Curtis, B. M. and Catterall, W. A., Purification of the calcium antagonist receptor of the voltage-sensitive calcium channel from skeletal muscle transverse tubules, Biochemistry, 23:2113-2118 (1984).

DePeyer, J. E., Cachelin, A. B., Levitan, I. B. and Reuter, H., Ca^{2+}-activated K^+ conductance in internally perfused snail neurons is enhanced by protein phosphorylation, Proc. Nat. Acad. Sci. USA, 79:4207-4211 (1982).

DeRiemer, S. A., Strong, J. A., Albert, K. A., Greengard, P. and Kaczmarek, L. K., Enhancement of calcium current in Aplysia neurones by phorbol ester and protein kinase C, Nature, 313:313-316 (1985).

Disterhoft, J. F., Coulter, D. A. and Alkon, D. L., Conditioning-specific membrane changes of rabbit hippocampal neurons measured in vitro, Proc. Nat. Acad. Sci. USA, in press (1986).

Ewald, D. A., Williams, A. and Levitan, I. B., Modulation of single Ca^{2+}-dependent K^+ channel activity by protein phosphorylation, Nature, 315:503-506 (1985).

Farley, J. and Alkon, D. L., Associative neural and behavioral change in Hermissenda: consequences of nervous system orientation for light and pairing specificity, J. Neurophysiol. 48:785-807 (1982).

Farley, J. and Auerbach, S., Protein kinase C activation induces conductance changes in Hermissenda photoreceptors like those seen in associative learning, Nature, 319:220-223 (1986).

Goh, Y., Lederhendler, I. and Alkon, D. L., Input and output changes of an identified neural pathway are correlated with associative learning in Hermissenda, J. Neurosci., 5:536-543 (1985).

Hartshorne, R. P. and Catterall, W. A., The sodium channel from rat brain: purification and subunit composition, J. Biol. Chem., 259:1667-1675 (1984).

Huganir, R. L. and Greengard, P., cAMP-dependent protein kinase phosphorylates the nicotinic acetylcholine receptor, Proc. Nat. Acad. Sci. USA, 80:1130-1134 (1983).

Huganir, R. L., Miles, K. and Greengard, P., Phosphorylation of the nicotinic acetylcholine receptor by an endogenous tyrosine-specific protein kinase, Proc. Nat. Acad. Sci. USA, 81:6968-6972 (1984).

Kaczmarek, L. K., Jennings, K. and Strumwasser, F., Neurotransmitter modulation, phosphodiesterase inhibitor effects, and cyclic AMP correlates of afterdischarge in peptidergic neurons, Proc. Nat. Acad. Sci. USA, 75:5200-5204 (1978).

Kaczmarek, L. K., Jennings, K. R., Strumwasser, F., Nairn, A. C., Walter, U., Wilson, F. D. and Greengard, P., Microinjection of catalytic subunit of cyclic-AMP-dependent protein kinase enhances calcium action potentials of bag cell neurons in cell culture, Proc. Nat. Acad. Sci. USA, 77:7487-7491 (1980).

Kaibuchi, K., Takai, Y., Sawamura, M., Hoshijima, M., Fujikura, T. and Nishizuka, Y., Synergistic functions of protein phosphorylation and calcium mobilization, J. Biol. Chem., 258:6701-6704 (1983).

Karlin, A., Molecular properties of nicotinic acetylcholine receptors. In, The Cell Surface and Neuronal Function (Eds. Poste, G., Nicolson, G., and Cotman, C. W.) Elsevier/North Holland, Amsterdam, pp. 191-260 (1980).

Kojima, I., Kojima, K., Kreutter, D. and Rasmussen, H., The temporal integration of the aldosterone secretory response to angiotension occurs via two intracellular pathways, J. Biol. Chem., 259:14448-14457 (1984).

Krueger, B. K., Forn, J. and Greengard, P., Depolarization-induced phosphorylation of specific proteins, mediated by calcium ion influx, in rat brain synaptosomes, J. Biol. Chem., 252:2764-2773 (1977).

Kubota, M., Alkon, D. L., Naito, S. and Rasmussen, H., Regulation of membrane currents by injection of C-kinase, Soc. Neurosci. Abstr., in press (1986).

Lederhendler, I., Gart, S. and Alkon, D. L., Classical conditioning of Hermissenda: origin of a new response. J. Neurosci., in press (1986).

Lemos, J. R., Novak-Hofer, I. and Levitan, I. B., Phosphoproteins associated with the regulation of a specific potassium channel in the identified Aplysia neurons R15, J. Biol. Chem., 260:3207-3214 (1985).

Levitan, I. B., Phosphorylation of ion channels, J. Memb. Biol., 87:177-190 (1985).

Levitan, I. B., Harmar, A. J. and Adams, W. B., Synaptic and hormonal modulation of a neuronal oscillator: a search for molecular mechanisms, J. Exp. Biol., 81:131-151 (1979).

Malenka, R. C., Madison, D. V., Andrade, R. and Nicoll, R. A., Phorbol esters mimic some cholinergic actions in hippocampal pyramidal neurons, J. Neurosci., 6:475-480 (1986).

Neary, J. T. and Alkon, D. L., Protein phosphorylation/dephosphorylation and the transient, voltage-dependent potassium conductance in Hermissenda crassicornis, J. Biol. Chem., 258:8979-8983 (1983).

Neary, J. T., Crow, T. and Alkon, D. L., Change in a specific phosphoprotein band following associative learning in Hermissenda, Nature, 293:658-660 (1981).

Noda, M., Ikeda, T., Kayano, T., Suzuki, H., Takeshima, H. Kurasaki, M., Takahashi, H. and Numa, S., Existence of distinct sodium messenger channel RNAs in rat brain, Nature, 320:188-192 (1986).

Numa, S., Noda, M., Takahashi, H., Tanabe, T., Toyosato, M., Furutani, Y. and Kikyotani, S., Molecular structure of the nicotinic acetylcholine receptor, Cold Spring Harbor Symp. Quant. Biol., 48:57-69 (1983).

Rane, S. G. and Dunlap, K., Kinase C activator 1,2-oleoylacetylglycerol attenuates voltage-dependent Ca^{2+} current in sensory neurons, Proc. Nat. Acad. Sci. USA, 83:184-188 (1986).

Sakakibara, M., Alkon, D. L., Neary, J. T., DeLorenzo, R., Gould, R. and Heldman, E., Ca^{2+}-mediated reduction of K^+ currents is enhanced by injection of IP_3 or neuronal Ca^{2+}/calmodulin kinase type II, Soc. Neurosci. Abstr., 11:956 (1985).

Sakakibara, M., Alkon, D. L., DeLorenzo, R., Goldenring, J. R., Neary, J. T. and Heldman, E., Modulation of calcium-mediated inactivation of ionic currents by Ca^{2+}/calmodulin-dependent protein kinase II, Biophys. J., in press (1986).

Shuster, M. J., Camardo, J. S., Siegelbaum, S. A. and Kandel, E. R., Cyclic AMP-dependent protein kinase closes the serotonin-sensitive K^+ channels of Aplysia sensory neurons in cell free membrane patches, Nature, 313:392-395 (1985).

Suarez-Isla, B. A., Wan, K., Lindstrom, J. and Montal, M., Single channel recordings from purified acetylcholine receptors reconstituted in bilayers formed at the tip of patch pipets, Biochemistry, 22:2319-2323 (1983).

Szuts, E. Z., Light stimulates phosphorylation of two large membrane proteins in frog photoreceptors, Biochemistry, 24:4176-4984 (1985).

Wane, J. A., Johnson, P. C., Smith, M. and Salzman, E. W., Aequorin detects increased cytoplasmic calcium in platelets stimulated with phorbol ester or diacylglycerol, Biochem. Biophys. Res. Commun., 133:98-104 (1985).

Weigele, J. B. and Barchi, R. L., Functional reconstitution of the purified sodium channel protein from rat sarcolemma, Proc. Nat. Acad. Sci. USA, 79:3651-3655 (1982).

Wilmsen, U., Methfessel, C. Hanke, W. and Boheim, G., In, Physical Chemistry of Transmembrane Ion Motions, Elsevier/North Holland, Amsterdam. pp. 479-485 (1983).

A POSSIBLE SECOND MESSENGER SYSTEM FOR THE PRODUCTION OF

LONG-TERM CHANGES IN SYNAPSES

Michel Baudry, Peter Seubert, and Gary Lynch

Center for the Neurobiology of Learning and Memory
University of California, Irvine, CA 92717

INTRODUCTION

Interactions between neurons that leave physiological traces lasting more than a few milleseconds are typically explained by reference to a second messenger system. Candidates for second messengers in brain typically involve enzymes, most often protein kinases, and use activation sequences that vary considerably in complexity; for example, calcium/calmodulin activated kinases require only the presence of sufficient concentrations of calcium against an appropriate background while stimulation of the c-AMP dependent kinase is pictured as a series of steps that include protein translocation, activation of a cyclase, and so forth (Greengard, 1981). Recently a probable second messenger system of considerable complexity and involving a novel type of kinase has been identified (Berridge and Irvine, 1984). Activation of several types of transmitter and hormone receptors stimulates the turnover of membrane phosphatidylinositol (PI) with the formation of two breakdown products in the interior of the cell, one of which releases calcium from intracellular stores and a second that, together with calcium, causes the translocation and activation of protein kinase C. This system can be modulated at several stages, as indicated by the observation that certain membrane receptor classes suppress the activation by other receptors of PI turnover (Baudry et al., 1986). In this chapter we will review the hypothesis that a novel kind of second messenger system is found in brain that, when activated, causes irreversible changes in the fundamental structures and functions of the neuronal cytoskeleton.

Despite an extensive period of research, the functional consequences of the activation of the various protein kinases by the proposed second messengers are still poorly understood, but several possibilities are now under investigation. There is evidence that certain ion channels in neurons are altered by phosphorylation such that ionic currents are modified (Kandel and Schwartz, 1982; Alkon, 1984; Malenka et al., 1986). The acetylcholine receptor is also phosphorylated under some circumstances but the significance of this, while potentially great, is uncertain (Huganir and Greengard, 1983). Synaptic vesicles containing transmitters are thought to have extended phosphoproteins on their surface, the phosphorylated state of which may influence the likelihood of fusion with the inner surface of the axon terminal membrane, and therefore transmitter release (Nestler et al., 1984; Llinas et al., 1985).

Probably the most defined effect of a kinase based second messenger system in brain concerns mitochondrial metabolism. Pyruvate dehydrogenase (PDH) is the rate limiting step in the metabolism of glucose by mitochondria and in brain as elsewhere its activity is controlled by a phosphoprotein (aPDH); elevated calcium levels outside the mitochondria, among other conditions, activate a phosphatase which dephosphorylates aPDH and thereby increases the activity of the enzyme and the conversion of pyruvate into acetyl-CoA (Linn et al., 1969). The result of this is to increase the rate at which mitochondria sequester calcium (Browning et al., 1981a; Baudry et al., 1983).

These second messenger systems are rapidly activated (~second) and through activities like phosphorylation have effects that can last from fractions of seconds to perhaps several minutes (Greengard, 1981). They are thus well suited to have physiological consequences such as very long lasting after potentials and/or to produce transient changes in transmitter release (e.g., post-tetanic potentiation) and post-synaptic responses found after high frequency activity. The systems are inherently homeostatic in nature in that their end results are antagonized by other enzymes (e.g. phosphatases) (Ingebritsen and Cohen, 1983). At a maximum, covalent modifications of proteins can last no longer than the protein itself and most brain polypeptides have half-lives of hours or days (Dunlop et al., 1978). Yet it must be the case that neuronal interactions produce alterations that persist for extraordinarily long periods, perhaps as long as the life-span of the organism. Effects of this type are usually taken as evidence that second messenger systems can evoke "trophic" effects that involve changes in genomic expression and indeed there is evidence for this in the case of the neuropeptides. Recent work has shown that a period of seizures produced by focal lesions causes a profound change in the balance of neuropeptides in specific cell populations in hippocampus (Gall et al., 1981) that under some conditions lasts for at least several weeks (Gall et al., 1986). Subsequent studies revealed that the seizure selectively changes the production of messenger RNA for one of the peptides (White et al., 1986). Moreover, the genomic effects were found to be preceded by a massive, transient increase in ornithine decarboxylase activity (Baudry et al., 1986), an effect that is known to follow exposure of neurons to certain hormones and transmitters (Canellakis et al., 1979; Lewis et al., 1978).

However, it is almost certainly the case that extremely persistent changes in neuronal interactions occur that are localized to the synaptic region rather than appearing throughout the cell, as presumably happens with the type of trophic response just described. There are a variety of theoretical reasons for assuming that the encoding of memory requires modifications that are restricted to selected populations of synapses (Hebb, 1949; Feldman, 1981) and it seems unlikely that this type of specificity could be achieved by changes in the genome. Quite possibly related to this are studies showing that brief episodes of high frequency synaptic stimulation in hippocampus cause a very stable, synapse specific potentiation (LTP) of excitatory post-synaptic potentials (Swanson et al., 1982; Teyler and Discenna, 1984) and modifications in the number and ultrastructure of synaptic connections (Lynch and Baudry, 1983). Effects of this type must involve local but stable modifications of the dendritic cytoskeleton and this is not easily explained in terms of simple covalent modifications of structural proteins. We postulate that the above theoretical constraints can be accommodated by a chemical process involving a calcium activated protease that degrades specific elements of the synaptic cytoskeleton; since proteolysis is irreversible, at least in terms of the substrate proteins, this enzymatic reaction produces effects that should be very long lasting. We suggest that this system has a higher threshold than other second messenger systems, with which it very probably interacts, and produces stable structural changes that are superimposed upon the more transient effects of the

TABLE 1: Comparison of the properties of "classical" second messenger systems and the one proposed in this review.

	Classical 2nd Messenger	New 2nd Messenger
Activation Mechanism	Neurotransmitter-Hormones	Neurotransmitter-Hormones
Intermediate Mechanism	Enzymes (cyclase, kinase, phosphatase..)	Enzyme (calpain)
Targets	Enzymes, Structural Proteins	Enzymes, Structural Proteins
Time-course of Changes	Seconds----> hours	Seconds----> weeks
Threshold	Low	High
Reversibility	Mostly Reversible	Mostly Irreversible

other messengers (Table 1).

This review begins with a discussion of the probable organization of the neuronal cytoskeleton and then describes experiments showing that certain of its components are substrates for the protease. Evidence that proteolysis of structural proteins by the enzyme causes structural changes in simple cells and data suggesting that the same chemical processes occur in brain will then be considered. The next section of the paper takes up the question of whether the postulated second messenger system is activated by physiological stimuli and will provide a reasonably precise description of how unusual synaptic events might be linked to it. The review closes with a brief discussion of experiments concerned with the possible significance, behavioral and otherwise, of the postulated second messenger chemistry.

1) Organization of the Membrane Skeleton

A variety of cell functions are regulated by structural elements linking the cell surface and intracellular cytoplasmic components. In particular these structural elements participate in the regulation of the distribution of cell surface proteins and by their association with the cell cytoskeleton play an important role in the modification of cell motility and cell shape (see below) (Singer, 1974; Edelman, 1976; Nicolson, 1979). Since they form a protein meshwork, underlying the plasma membrane, which is insoluble in a variety of nonionic detergents, they have been defined as part of a submembraneous cytoskeleton or membrane skeleton (Yu et al., 1973; Cohen et al., 1977). Most of our knowledge of the structure, composition, and interactions between the proteins forming this membrane skeleton has been derived from studies with erythrocytes (Branton et al., 1981; Bennett, 1985), but in several years it has become evident that families of proteins with similar structures and functions are present in a variety of non-erythroid cells, including neurons (Bennett, 1985; Goodman and Zagon, 1985) (Fig. 1). As different cell types face different constraints and perform different functions, it is to be expected that various adaptations of a general design have evolved resulting in significant differences in the elements forming the membrane skeleton, and that caution should be

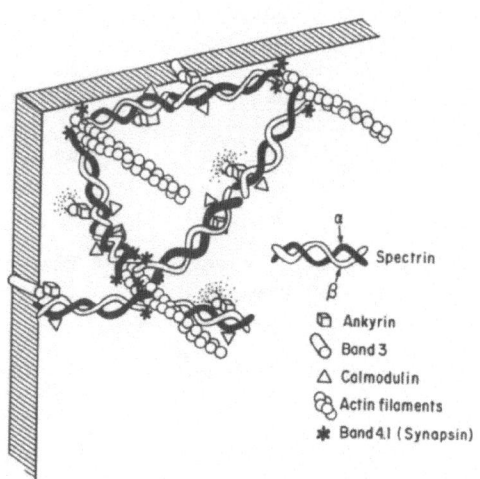

FIG. 1: Hypothetical model of the organization of the synaptic submembraneous cytoskeleton. In the model are represented several of the components forming the submembraneous cytoskeleton which are known to
be present or postulated to exist by analogy with the erythrocyte
model. Thus brain spectrin forms a filamentous meshwork which is
linked to the membranes via interaction with the brain equivalent
of 4.1 (possibly synapsin) and of ankyrin (itself bound to the brain
equivalent of band 3). Also shown is the association of calmodulin
with the α-subunit of spectrin. (Modified from Goodman and Zagon
(1985) and Bennett (1985).

taken before drawing general conclusions from the study of a particular
cell type. However, our understanding of the composition, structure, and
functions of the membrane skeleton of adult mammalian neurons is beginning
to approach that of its erythrocyte counterpart and comparisons between them
can be made with some confidence.

A major constituent of the neuronal membrane skeleton is a large
tetrameric protein which, as it was discovered independently by several
groups, has received a variety of names: fodrin (Levin and Willard, 1981),
calspectin (Sobue et al., 1982), brain calmodulin-binding protein (Davies
and Klee, 1981), brain spectrin (Goodman and Zagon, 1984). But as the
similarities of the brain protein with erythroid spectrin are becoming more
evident, we will adopt the suggestion of Goodman and Zagon (1984) and use
the term brain spectrin.

Brain spectrin consists of two equimolar subunits (α and β) with M_rs
of 240,000 and 230,000 respectively (see Goodman and Zagon, 1984, for a
review). The principle form of brain spectrin is thought to be an $\alpha_2\beta_2$
tetramer consisting of two heterodimers attached head to head forming a 200
nm flexible rod-like structure. Brain and erythrocyte spectrin share many
properties including the rod-like tetrameric form when visualized by rotary
shadowing electron microscopy (Glenney et al., 1982; Davis and Bennett,
1983), a remarkably high α-helix content (Burns et al., 1983), and the ability to bind to many of the same proteins (discussed below). In fact, the
individual subunits of brain and erythrocyte spectrin can be isolated and
mixed heterodimers formed which retain many of the functions of the native
molecules (Davis and Bennett, 1983). The primary structure of erythroid
spectrin contains multiple homologous sequences with a periodicity of 106

amino acids which may be folded in a triple helical structure (Speicher and Marchesi, 1984). It would not be surprising, given the above similarities, if brain spectrin shows a similar fundamental repeating unit despite the differences in peptide maps of brain and erythroid spectrin (Glenney et al., 1982; Glenney and Glenney, 1984). It was initially suggested that the subunits self-assemble to form the tetramer $(\alpha\beta)_2$ and possibly also higher-order oligomers (Calvert et al., 1980; Yoshino and Marchesi, 1984; Liu et al., 1984). However recent data have raised the possibility that spectrin assembly, at least in red blood cells, might be a more complex phenomenon and that both heterodimers and homo-oligomers are generated, the latter being targeted for proteolytic degradation (Woods and Lazarides, 1986).

In brain, spectrin represents 2-3% of the total proteins present in crude membrane preparations and 85% of brain spectrin is incorporated in the membrane skeleton (Davis and Bennett, 1983). Spectrin is present in all regions of mammalian brain, and neurons appear to contain more immuno-reactive spectrin than glial cells (Zagon et al., 1984). Levine and Willard (1981) have provided evidence that brain spectrin forms part of a dynamic lining prominent in the cortical cytoplasm of neurons. Spectrin is also prominent in highly purified postsynaptic densities (PSDs) (Carlin et al., 1983) indicating that it is present in dendritic spines since in cortical structures these microstructures provide the vast majority of PSDs.

Erythrocyte spectrin is thought to link actin filaments to the cell membrane by the following mechanism. The tails of the spectrin tetramer bind to F-actin and this interaction is stabilized by protein 4.1 (Ohanian et al., 1984). A site on the β subunit near the head junction binds to ankyrin (Tyler et al., 1979), which in turn is attached to the anion trans-porter band 3, an integral membrane protein (Bennett and Stenbuck, 1979).

Brain spectrin has been shown to interact in a similar fashion with the above proteins (Burns et al., 1983; Davis and Bennett, 1983). Interes-tingly, it has recently been shown that the brain equivalent of erythroid protein 4.1 is probably synapsin (Baines and Bennett, 1985), which may indicate that brain spectrin serves to link synaptic or other types of vesicles to the membrane skeleton. Brain ankyrin has been purified and shown to bind erythrocyte band 3 and the β subunit of brain spectrin (Davis and Bennett, 1984). Surprisingly, brain ankyrin also binds tubulin and thus has the potential to interconnect microtubules and the spectrin-actin network (Davis and Bennett, 1984). Brain spectrin also interacts with microtubules and induces the bundling of microtubules; the binding of spectrin to microtubules is decreased by the presence of microtubule asso-ciated proteins (MAPs), suggesting that spectrin and MAPs may interact in the regulation of microtubule structure (Ishikawa et al., 1983). Brain spectrin thus seems to occupy an ideal position to modulate the attachment of microtubules and microfilaments to the cytoplasmic membrane.

2) Regulation of the Organization of the Membrane Skeleton

Several factors are potentially involved in the regulation of the mul-tiple protein-protein interactions between the various elements of this complex membrane skeleton. Here we will note that several elements are targets for protein kinases and then discuss in detail recent work showing that many are also substrates for a calcium activated protease.

(a) phosphorylation reactions: Although most of the proteins consti-tuting the membrane skeleton can be phosphorylated by the cAMP-dependent or the Ca/calmodulin-dependent protein kinases, little is known concerning the functional significance of these phosphorylation reactions. In the case of spectrin, the β-subunit is phosphorylated by various protein kinases but so far no effect of phosphorylation on the association of spectrin to actin,

4.1, or ankyrin has been described (Bennett, 1985). For ankyrin, it was recently shown that phosphorylation reduces its affinity for the spectrin tetramer (Lu et al., 1985), suggesting that phosphorylation-dephosphorylation reactions might regulate the association of spectrin to membranes. Protein 4.1 or synapsin is also a major substrate for the cAMP-dependent protein kinase and it has been proposed that phosphorylation of synapsin regulates vesicular transmitter release (Nestler et al., 1984; Llinas et al., 1985). However, it is not known whether phosphorylation of 4.1 modifies its interactions with actin oligomers and with spectrin tetramers. MAP_2 is also a major substrate of cAMP-dependent protein kinase and MAP_2 phosphorylation has been shown to regulate microtubule assembly and also to inhibit MAP_2-actin association (Nishida et al., 1982; Sattilaro et al., 1982).

(b) Calcium-dependent reactions: Calcium plays a prominent role in the cycle of polymerization-depolymerization of microtubules and actin oligomers (Weisenberg, 1972). In most cases calcium mediates its effect through its association with calmodulin and allosteric interactions with calmodulin-binding proteins. Again, it is interesting to note that brain spectrin has a calcium-dependent calmodulin binding site on the α-subunit (Tsukita et al., 1983) and until recently no functional role had been reported for this calmodulin binding (but see below). While the above mentioned mechanism involves reversible alterations in protein structure, calcium may also play an important role in the organization of the membrane skeleton by its ability to stimulate calcium-dependent neutral proteases. Most of the proteins forming the membrane skeleton are substrates of calcium-dependent proteases (Sandoval and Weber, 1978; Malik et al., 1981; Zimmerman and Schlaepfer, 1982; Siman et al., 1984). For a long time, this characteristic of cytoskeletal proteins received little attention since it was generally believed that calcium-dependent protease required millimolar concentrations of calcium to be activated. The identification in a variety of tissues of a calcium-dependent protease activated by micromolar concentrations of calcium (Mellgren, 1980; Murachi et al., 1981) led us to reevaluate the potential roles of calcium-dependent proteolysis of cytoskeletal elements in the regulation of the organization of the membrane skeleton. Two distinct classes of calcium-dependent proteases (calpain) have now been purified from a variety of tissues (Yoshimura et al., 1983; Zimmerman and Schlaepfer, 1984); they require micromolar (calpain I) or millimolar (calpain II) concentrations of calcium. Both forms are composed of a heavy (Mr ~ 80,000) catalytic unit and a light (Mr ~ 30,000) peptide with unknown functions; the heavy catalytic subunit varies between the two calpains (Wheelock, 1982; Kitahara et al., 1984). The substrates of calpains include spectrin, MAP_2, neurofilament proteins, ankyrin and band 3; in other words most of the proteins forming the membrane skeleton are substrates of calpains, although the relative specificity and kinetic parameters for calpain I or II still remain to be studied.

The degradation of brain spectrin by calpain (I or II) results in the formation of a major proteolysis-resistant breakdown product (bdp) with a molecular weight of approximately 140,000 (Siman et al., 1984) (Fig. 2). It probably also results in the formation of small amounts of a variety of degradation products with different molecular weights not detected by Coomassie Blue staining because of their transient nature. Although calpain is not a calmodulin-dependent enzyme, calmodulin markedly increase the rate of brain spectrin degradation by calpain (Seubert et al., 1986). In addition, in the presence of calmodulin, additional breakdown products are generated. This effect is specific for calpain since calmodulin does not modify the rate of brain spectrin degradation by chymotrypsin. The effect of calmodulin is totally blocked by trifluoperazine, a calmodulin antagonist, which has been shown to prevent calmodulin binding to the α-subunit (Harris et al., 1986). Moreover, calmodulin has been shown to bind only

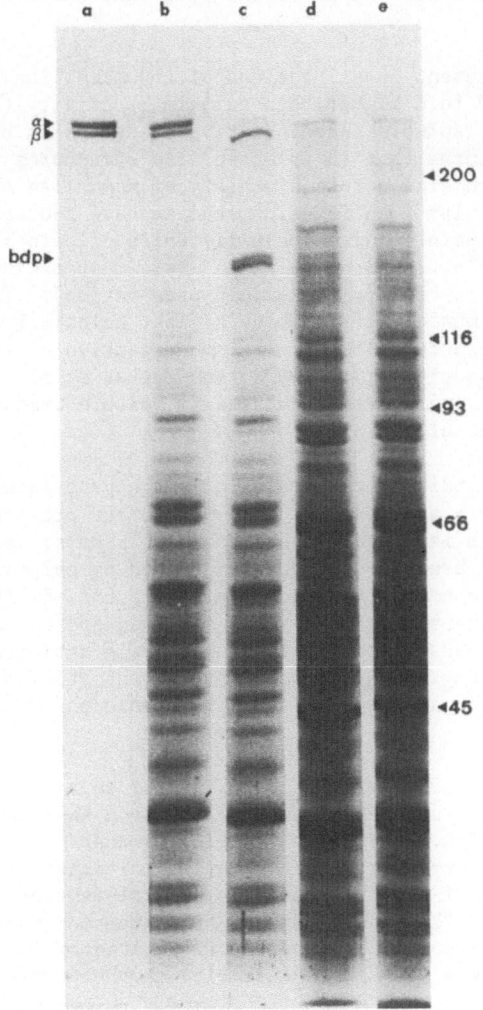

FIG. 2: Effect of calpain I on protein composition of Triton-insoluble
and Triton-soluble fractions of hippocampal synaptosomes. Hippocampal synaptosomes were prepared by differential and sucrose-
Ficoll gradient centrifugation. Triton-soluble and -insoluble
fractions were prepared according to Cohen et al., (1980), and
were incubated for 15 min at 30°C in the presence and absence of
purified calpain I (10 μg/ml). Proteins were solubilized in an
SDS solubilization solution (Baudry et al., 1981) and separated on
5-10% gradient polyacrylamide slab gels, stained with Coomassie
blue and destained with acetic acid a: spectrin obtained by immunoprecipitation with anti-spectrin antibodies; b, c: Triton-insoluble fractions; d, e: Triton-soluble fractions; c, e: incubations with calpain I (10 μg/ml) and 200 μM $CaCl_2$.

with low-affinity to erythrocyte spectrin (Bennett, 1985) and calmodulin
does not modify erythrocyte spectrin degradation by calpain. Calpain activity is also modulated by both an endogenous high molecular weight inhibitor, calpastatin (Murachi et al., 1981), and a low molecular weight acti-

vator protein (DeMartino and Blumenthal, 1982; Takeyama et al., 1986). The existence of these proteins supports the idea that calpain activity must be precisely modulated.

Calpains are present in all regions of the mammalian brain, calpain II activity being 10-20 fold higher than calpain I activity (Simonson et al., 1985). In addition, soluble calpain activity appears to be higher in non-telencephalic structures than in telencephalic structures and is inversely correlated with brain size across a variety of mammalian species (Baudry et al., 1986). Subcellular fractionation studies have indicated a clear distinction between calpain I and calpain II; while calpain II is predominantly present in the soluble fraction, calpain I is almost exclusively located in synaptosomal fractions (Baudry et al., in preparation). Immunohistochemical studies using a monoclonal antibody against calpain I have also provided evidence for the presence of calpain I-immunoreactive material in dendritic spines, sometimes associated with PSDs (Perlmutter et al., 1985). There is also evidence that calpain translocates from soluble fractions to the membranes (Pontremoli et al., 1985).

All these data indicate that calpain I is appropriately located to interact with spectrin and possibly also with other proteins of the membrane skeleton. This is especially the case in PSDs since it has been shown that the major breakdown product generated by calpain-induced degradation of spectrin is normally found in highly purified PSDs (Carlin et al., 1983). This suggests not only that spectrin and calpain are appropriately located in PSDs, but also that calpain is activated as a result of naturally occurring physiological activity; recent studies indicate that this process can be greatly accelerated by intense physiological activity (see below).

To summarize, it is clear that a variety of covalent modifications of the proteins constituting the membrane skeleton have the potential to modify its organization. It is also becoming evident that irreversible modifications of the membrane skeleton can be initiated by the stimulation by calcium of calpain I and the resulting partial degradation of several of structural proteins. The nature and extent of the modification of membrane cytoskeletal organization will probably be dictated by the way its elements re-assemble following a transient activation of the proteolytic enzyme.

3) Degradation of Spectrin by Calpain Mimics Effects Seen after Induction of Synaptic Potentiation

As discussed, the localization and substrate specificities of calpain make it likely that physiological events that elevate calcium levels above a threshold level (~ 5 uM) in spines will result in localized breakdown of the spectrin network. In support of this is evidence that activation of calpain in synaptic membrane fractions reproduces effects found after the induction of long-term potentiation by high frequency stimulation. As mentioned, LTP is correlated with changes in the numbers and dimensions of spine synapses (Lee et al., 1981; Chang and Greenough, 1984; Wenzel and Matthies, 1985) but there are no techniques for testing if calpain activation produces comparable effects on spines and dendrites. Accordingly, a search was conducted for a biochemical index that also correlates with LTP and which could be assayed in synaptic membranes, conditions under which attempts could be made to mimic the LTP correlate by stimulating the calpain-spectrin interaction. Glutamic acid is thought to be the transmitter in the synapses exhibiting the LTP phenomenon (Storm-Mathisen, 1977) and it seemed likely that alterations in spine synapses with attendant changes in cytoskeleton and surface chemistry would be reflected in altered processing of glutamate. The major hurdle to tests of this idea was developing a preparation in which a sufficient percentage of the synaptic population was potentiated for a biochemical effect to stand out above background. It has been estimated that not more than five percent of the synapses on a hippocampal neuron need be active to bring the cell to its firing threshold (Anderson et al., 1980). Therefore, even if a reasonably large population of neurons were potentiated, only a relatively small group of synapses needs be involved (this problem has not received the attention it deserves in experiments on biochemical correlates of LTP). Two modifications of standard in vitro slice methodology were employed in an effort to obtain the largest possible percentage of potentiated synapses (Lynch et al., 1982). First, a "minislice" technique was developed which used only the field CA1 of hippocampus minus the outermost dendritic segments; this served to restrict the total area of the synaptic field to a point that a reasonable percentage of it could be stimulated in a given preparation. Second, 24-36 contiguous stimulation points were used. In this way virtually the entire apical-basal extent of the Schaffer-commissural projections (by far the dominant input to the field CA1) could be stimulated. Multiple stimulation sites also permitted the investigator to insure that heterosynaptic depression was not elicited; that is, that the stimulation did not produce generalized effects in the target neurons. This is an important point since the depression effect may leave biochemical traces that could be mistaken for synapse-specific changes associated with LTP. Using these conditions, slices received potentiation or control stimulation after which crude synaptic membrane fractions were prepared using a miniaturized version of standard techniques and assays for binding of radiolabelled glutamate conducted. At the time these studies were conducted it was thought that the binding assay measured receptor number and affinities; it now appears that the assay samples sequestration of glutamate with unknown contributions from ligand receptor binding (Pin et al., 1984; Kessler et al., 1986). In any event, an increase in glutamate "binding" was obtained in membranes prepared from slices in which LTP was produced; control slices or slices that did not exhibit LTP were essentially unaffected (Lynch et al., 1982).

Comparable effects are found in fractions enriched in synaptic membrane fragments when they are exposed to low concentrations of calcium (Baudry and Lynch, 1979). Moreover, drugs and other conditions that block calpain completely inhibit this effect of calcium (Baudry and Lynch, 1980; Baudry et al., 1981). Thus calpain activation mimics a condition created by the induction of LTP. It remained then to determine if the effect of calpain was due to proteolysis of spectrin or some other substrate protein. Tests of this became possible when it was discovered that antibodies to spectrin protect it from calpain without inhibiting the activity of the enzyme. It was found that the antibodies prevented the calcium stimulation of glutamate binding and to a degree that was perfectly correlated with the extent to which they attached to cytoskeletal spectrin (Siman et al., 1985) (Fig. 3).

4) Possible Links between Physiological Events and Calpain Activation

The experiments described in the preceding section demonstrated that calpain mediated proteolysis of spectrin in vitro reproduced a biochemical effect seen after high frequency stimulation of synapses and the induction of LTP, but is the calpain-spectrin interaction triggered by physiological events in situ? This question is central to the idea of limited proteolysis as the effector of a second messenger system for producing stable changes and as reviewed earlier there is considerable indirect evidence pointing to a positive answer.

Experimental studies have now demonstrated that intense depolarization of hippocampus does result in the breakdown of synaptosomal cytoskeletal spectrin (Bodsch et al., 1986). In those experiments, radiolabelled methionine was infused into the ventricle for four or eight hours, after which

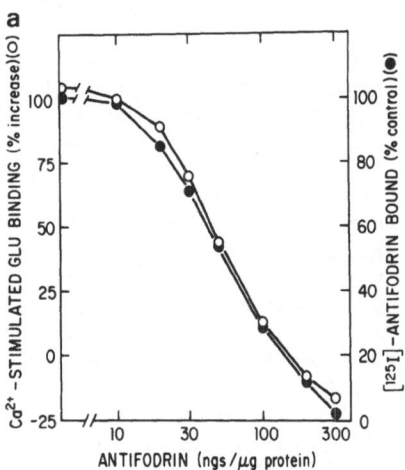 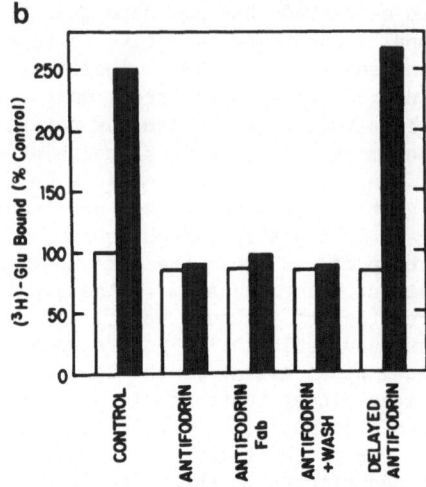

FIG. 3: Effect of antispectrin (antifodrin) antibodies on calcium-stimu-
lation of ³H-glutamate binding to cortical synaptic membranes. a)
Effects of various concentrations of anti-spectrin on ¹²⁵I-anti-
spectrin binding (closed circles) and Ca-stimulated glutamate binding (open circles). b) Effects of various incubation conditions
with antispectrin antibodies on Ca-stimulation of glutamate binding. Antispectrin antibodies (IgG), or monovalent antispectrin
Fab fragments blocked the increase in ³H-glutamate binding elicited by calcium (1 mM) (hatched bars) without affecting basal
binding (open bars). The blockade persisted in membranes pretreated with antispectrin and then washed free of unbound antibodies (antifodrin + wash), while antispectrin did not block the
increased binding when added after membrane incubation with calcium (delayed antifodrin).

synaptosomal cytoskeletal fractions were prepared and spectrin isolated by
immunoprecipitation. Autoradiograms of polyacrylamide gels were used to
assess the amount of newly synthesized spectrin present in samples from hippocampus and cortex. Nearly all the immunoprecipitated labelled peptides
were contained in the α and β spectrin bands after four hours of infusion
but a significant amount of the 140,000 dalton breakdown product was found
after eight hours. Depolarization of the hippocampus by a ten minute application of potassium greatly increased the amount of breakdown product found
at four hours (Bodsch et al., 1986a; Bodsch et al., 1986b). This breakdown
product has the same Mr as that resulting from calpain mediated proteolysis
of spectrin. These experiments provide the first experimental demonstration
that physiological stimuli accelerate the breakdown of the submembraneous
cytoskeletal network in brain and greatly increase the likelihood that the
postulated second messenger system is operative under certain conditions.

Studies are now in progress to determine if repetitive stimulation of
the type that induces LTP also causes the breakdown of spectrin. It is
interesting in this regard that intense depolarization, the stimulus used
to elicit proteolysis of spectrin, is very probably a necessary event for
long-term potentiation to be elicited. Collingridge and co-workers (Collingridge et al., 1983) as well as others (Harris et al., 1984; Morris et
al., 1986) made the very important observation that antagonists of the so-
called N-methyl-D-Aspartate (NMA) receptors potently and selectively block
LTP; these receptors are most unusual in that they (or the conductance
channels connected to them) are essentially inactive at resting membrane

potential and become functional only during depolarizing conditions (Nowak et al., 1984; Mayer et al., 1984). This suggests that induction of LTP typically involves a strong depolarization event which then transiently activates the NMA receptor and/or channel; additional stimulation would then produce effects that reflect physiological responses due to the action of both the normal receptor and the NMA sites, the summed values of which elicit the stable modification. There is an obvious link between these events and the calpain-spectrin system. Activation of the NMA receptor adds a chemically triggered calcium flux to the voltage-dependent calcium conductance that might occur at synapses. Thus we can imagine that the conditions needed to bring calcium levels in spines to the threshold for producing LTP within the time constraints imposed by the duration of high frequency bursts of stimulation involve both types of calcium channels. It might be noted here that intracellular injections of the calcium chelating compound EGTA into the target cell also prevent the occurrence of LTP without disturbing aspects of cell physiology other than those associated with calcium-mediated after potentials (Lynch et al., 1983). This provides fairly direct evidence that calcium accumulation does occur in the target cell during repetitive stimulation and is responsible for long-term changes. In all, the conditions needed for activation of cytoskeletal proteolysis by calpain correspond to those that occur during LTP-inducing stimulation and, as shown in the previous section, the effects produced by the calpain-spectrin interaction occur as correlates of LTP.

While it may prove possible to rigorously test the correspondence between the conditions that elicit LTP and those that stimulate calpain and cytoskeletal breakdown, it remains difficult to assay the role played by these biochemical events in structural changes and synaptic potentiation. A major problem lies in the fact that drugs that inhibit calpain are poorly characterized with regard to their entry and effects on intact cells or cell processes and data of this type are difficult to obtain. The discovery that depolarization greatly stimulates spectrin proteolysis resulting in the accumulation of the same breakdown product as seen after calpain activation may provide a test bed against which drugs could be evaluated.

5) Interactions with other Second Messenger Systems

Calpain as the effective enzyme of a second messenger system has the virtue of simplicity: it is activated by a single event (elevated calcium) and its effects are simply described (cleavage of proteins) and certain to be significant, at least at the level of protein chemistry. However, it seems inevitable that calpain activity and its effect on structural proteins will be modulated by more conventional second messenger systems. Note that calpain requires calcium levels that are considerably in excess of those needed for routine transmission-related physiology (e.g. transmitter release, stimulation of after-potentials) and the calcium-calmodulin kinase. Moreover, other second messengers are brought into action by events that occur as part of ongoing synaptic communication such as binding to receptors. Accordingly, we can assume that the exaggerated physiological stimuli needed to elicit a calpain response also produce effects on several second messenger systems. There are several routes through which these might regulate calpain and its interactions with cytoskeletal proteins, the most likely of which is calcium availability. As mentioned, ligand binding to certain types of receptors promotes phosphatidylinositsol turnover resulting in increased availability of inositol triphosphate (IP3) inside the cell and IP3 promotes calcium exchange with the endoplasmic reticulum. Recent studies have shown that this system is itself regulated by acidic amino acid binding sites including the NMA receptors and the suspected transmitter receptors (Baudry et al., 1986). Thus the combination of various receptor systems activated during intense synaptic activity might very well dictate the extent and duration of changes in intracellular calcium.

Kandel and his collaborators have made a strong case for the argument that kinase activation during and after synaptic transmission blocks potassium efflux channels and that this can serve to prolong the duration of calcium elevations (Kandel and Schwartz, 1982). Finally, the mitochondrial calcium buffer is also the target of a second messenger system involving a peptide released from the inner face of the membrane that affects the phosphorylated state, and thus the activity, of the rate limiting enzyme pyruvate dehydrogenase (Seals and Czech, 1980; Seals and Czech, 1981). It is not known if this system is present in brain but the phosphorylated state of PDH is affected by high frequency synaptic stimulation (Browning et al., 1981b).

Calpain's interactions with spectrin and other cytoskeletal proteins are also likely sites at which second messenger systems could act. As discussed, spectrin has binding sites for calmodulin and calmodulin greatly accelerates the breakdown of assembled spectrin without affecting the activity of the enzyme (Seubert et al., 1986). Thus availability of calmodulin could have an influence on the consequences of calpain activation and it is therefore of interest that soluble calmodulin is increased following occupancy of certain types of receptors (Hanbauer et al., 1979). Certain microtubule-associated proteins (MAPs) are phosphorylated by protein kinase C (Tsuyama et al., 1986), an event that might well be expected to produce conformational changes. Kinase C has been reported to be activated and translocated from the soluble to the membrane fraction following high frequency synaptic activity (Routtenberg et al., 1985; Akers et al., 1986).

The possibility that calpain activation might also influence other second messenger systems also deserves consideration. Protein kinase C is a substrate for calpain and is converted by it to an enzyme (kinase M) that does not require calcium or diacylgycerol for activation (Kishimoto et al., 1983); in a sense, calpain uses kinase C to create a novel type of phosphorylating enzyme that may be inactivated only through further proteolysis. The consequences of this could be profound since native kinase C has been postulated to be involved in several vital cell functions including regulation of pH and genomic expression (Berridge and Irvine, 1984). It is of interest in this regard that Alkon and collaborators have argued that some forms of learning in Hermissenda involve a chronic activation of kinases (Alkon, 1984).

The effect of calpain on kinase C emphasizes the point that the activity of the protease must be subject to a number of regulatory mechanisms since its irreversible actions offer any number of possibilities for cell dysfunction. The very feature that makes calpain an attractive candidate for a second messenger system directed at eliciting long-term changes also makes it a potential agent for causing cell pathology (Lynch et al., 1986).

6) Possible Consequences of the Postulated Second Messenger System to Neuronal Operation

The postulated second messenger system has been described in terms of the conditions that activate it and the structural and biochemical consequences it produces. This last section takes up the issues of possible functional significance of the system, questions which relate to both activation and consequences; that is, the functions of the second messenger can be deduced from knowing when it is brought into play and what lasting changes in neuronal and circuit physiology result from it.

The triggering event for calpain is a concentration of ionic calcium that is abnormal in the sense that it greatly exceeds that needed for other calcium sensitive functions. The most logical source for this calcium is the millimolar sized pool lying outside the cell. Since depolarization opens voltage-gated calcium channels in neurons, we have suggested that

TABLE II: Triggering Mechanisms for the Induction of LTP.

1)   Brief trains of high-frequency stimulation (4 pulses at 100 H)
"primes" a target pyramidal cell.   The effect develops in ~50 msec and
lasts < 2 sec.

2)   Stimulation of an input (similar or different) to a "primed" neuron
generates a prolonged EPSP.

3)   A burst given to a "primed" neuron allows the local build-up of depolarization and the activation of the voltage- and transmitter-dependent subclass of glutamate receptors: the NMDA receptors.

4)   Influx of calcium through NMDA receptor-associated channels and voltage-dependent channels provides a large and localized (to stimulated dendritic spines) increase in calcium concentration in dendritic spines.

an intensely depolarizing physiological event is the most obvious trigger
for calpain activation and cytoskeletal proteolysis (see above).   Also discussed earlier was the evidence that a certain class of receptor becomes
functional during depolarization which is probably linked to calcium influx,
thereby adding a second and parallel route for increasing intracellular
calcium.   It might be noted here that the nature of the dendritic spines
may facilitate the build-up of calcium.   Most spines possess a thin neck
with a peculiar membranous structure at its base (the spine apparatus) -
this arrangement may serve to restrict calcium diffusion out of the spine
head, the site at which synaptic contacts are made (Fifkova et al., 1983).
The idea that the restricted space of the spine could lead to calcium accumulation has been borne out in computer simulations; these suggest that
calcium levels in depolarized spines could rise to levels far in excess of
those needed for calpain activation (Koch and Poggio, 1983; Robinson and
Koch, 1984).   The idea that intense depolarization elicits breakdown of the
synaptosomal cytoskeleton has received experimental confirmation and indirect evidence suggests that calpain is involved (see above).   But what types
of physiological events would be expected to produce depolarization of this
type on a local level (that is, in the spine)?   Recent work testing the
efficacy of different patterns of stimulation in eliciting the long-term
potentiation effect has suggested one possibility.   Those studies found
that bursts of four pulses delivered to the Schaffer-commissural projections
of field CA1 of hippocampus will produce LTP if and only if they are preceded by a similar burst delivered 200 msec earlier (Larson et al., 1986).
Lesser amounts of LTP occurred if the bursts were given at shorter or longer
intervals.   The major difference in the responses to the two bursts is that
the EPSP's to the second were greatly prolonged and thus produced a much
more intense depolarization than is normally observed (Larson and Lynch,
1986).   Subsequent studies revealed that part of the added depolarization
is due to the activation of NMA receptors (Larson et al., in preparation)
(Table II).   Thus appropriately patterned stimulation results in a condition
in which very brief trains of stimulation produce an intense depolarization
response and thus the trigger for the calpain system.   It is indeed intriguing that the optimal interburst interval for this effect would be 200 msec
as this corresponds to the interwave interval of the hippocampal theta
rhythm, an EEG pattern that occurs when animals are actively exploring
their environments (Bland, 1986).   Moreover, the short bursts of stimulation used in these experiments correspond to the firing patterns of hippocampal pyramidal neurons in freely moving animals (Vanderwolf et al., 1975).
Possibly then the conditions needed for large calcium influxes and calpain
activation in hippocampus (at least) are set in place by the brain rhythms
that appear while an animal moves through and samples its environment.

Proteolysis of spectrin and other submembraneous structural proteins by calpain should cause a degree of disassembly and reassembly of the synaptosomal cytoskeleton. It is worth noting that calpain activation has been linked to changes in cytoskeletal structures and to the rapid proliferation of surface receptors in platelets (Jennings et al., 1981; Fox and Phillips, 1983; Baldassare et al., 1985) and cell shape changes in erythrocytes (Siman et al., 1986). The effects of calpain mediated degradation of synaptic cytoskeleton are unknown but, as discussed, it does reproduce a biochemical change that correlates with the stable synaptic potentiation (LTP) that appears after high frequency stimulation. It is not unreasonable to assume then that activation of the calpain-spectrin interaction in situ is responsible for both the biochemical and physiological changes that characterize long-term potentiation.

In any event, it seems unlikely that partial digestion of the synaptosomal cytoskeleton would leave the physiological operation of the synapse untouched. Thus physiological events that activate the postulated second messenger system would be expected to leave a lasting imprint on the circuitries in which they occur. This provides the conditions needed for a memory mechanism. It is rapid, spatially restricted and, because it involves structural changes, produces effects that could persist as long as the synapse itself. The absence of an adequately characterized pharmacology for calpain precludes satisfactory tests of the enzyme's role in memory. Nonetheless, it is of interest that chronic infusion of a calpain inhibitor into the cerebral ventricle causes a selective and profound inhibition of learning in spatial mazes (Staubli et al., 1984) and olfactory discrimination tests (Staubli et al., 1985) (Fig. 4).

It was noted earlier that any mechanism that involves degradation of the cytoskeleton carries with it a degree of risk. In accord with this, calpain has been implicated in nerve and muscle pathologies arising from several different conditions (Libby and Goldberg, 1978; Schlaepfer and Hasler, 1979). If for any reason activation of the enzyme were to occur outside spines in the primary shafts of dendrites, then disruption of transport and structural elements necessary to the survival of entire branches of the cell would become a real possibility. We have elsewhere made the argu-

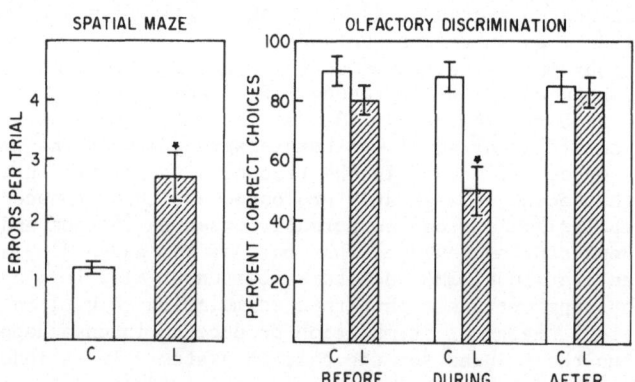

FIG. 4: Effect of leupeptin on spatial maze and smell discrimination learning. Leupeptin (4 mg) or saline (C) was constantly infused in the lateral ventricule of rats trained on an 8-arm radial maze or on a smell discrimination learning task. Left panel indicated the number of re-entries per trail in previously visited arms. Right panel indicates the scores of the rats before, during and after the implantation of leupeptin mimipumps. Data from Staubli et al., (Staubli et al., 1984; Staubli et al., 1985).

ment that several aspects of brain aging and age-related neuropathologies can be accounted for by perturbations in those systems responsible for regulating calpain (Lynch et al., 1986).

ACKNOWLEDGMENTS

Research supported by grants AFOSR 86-0099 to Gary Lynch and NSF 18427 to Michel Baudry. We wish to thank Jackie Porter for secretarial assistance.

REFERENCES

Akers, R. F., Lovinger, D. M., Colley, P. A., Linden, D. J. and Routtenberg, A., Translocation of protein kinase C activity may mediate hippocampal long-term potentiation, Science 231:587-589 (1986).

Alkon, D., Calcium-mediated reduction of ionic currents: a biophysical memory trace, Science 226:1037-1045 (1984).

Andersen, P., Silfvenius, H., Sundberg, S. H. and Sveen, O., A comparison of distal and proximal dendritic synapses on CA1 pyramids in guinea-pig hippocampal slices in vitro, J. Physiol. 307:273-299 (1980).

Baines, A. J. and Bennett, V., Synapsin I is a spectrin-binding protein immunologically related to erythrocyte protein 4.1. Nature 315:410-413 (1985).

Baldassare, J. J., Bakshian, S., Knipp, M. A. and Fisher, G. J., Inhibition of fibrinogen receptor expression and serotonin release by leupeptin and antipain, J. Biol. Chem. 260:10531-10535 (1985).

Baudry, M. and Lynch, G., Regulation of glutamate receptors by cations, Nature 282:748-750 (1979).

Baudry, M. and Lynch, G., Regulation of hippocampal glutamate receptors: evidence for the ivolvement of a calcium-activated protease, Proc. Nat. Acad. Sci. USA 77:2298-2302 (1980).

Baudry, M., Bundman, M., Smith, E. and Lynch, G., Micromolar calcium stimulates proteolysis and glutamate binding in rat brain synaptic membranes, Science 212:937-938 (1981a).

Baudry, M., Evans, J. and Lynch, G., Excitatory amino acids inhibit stimulation of phosphatidylinositol metabolism by aminergic agonists in hippocampus, Nature 319:329-331 (1986).

Baudry, M., Gall, C., Kessler, M., Alapour, H. and Lynch, G., Denervation-induced decrease in mitochondrial calcium transport in rat hippocampus, J. Neurosci. 3:252-259 (1983).

Baudry, M., Lynch, G. and Gall, C., Induction of ornithine decarboxylase as a possible mediator of seizure-elicited changes in genomic expression in rat hippocampus, J. Neurosci. (in press) (1986).

Baudry, M., Simonson, L., DuBrin, R. and Lynch, G., A comparative study of soluble calcium-dependent proteolytic activity in vertebrate brain, J. Neurobiol. 17:15-28 (1986).

Bennett, V., The membrane skeleton of human erythrocytes and its implication for more complex cells, Ann. Rev. Biochem. 54:272-304 (1985).

Bennett, V. and Davis, J., Erythrocyte ankyrin: immunoreactive analogues are associated with mitrotic structures in cultured cells and with microtubules in brain, Proc. Nat. Acad. Sci. USA 78:7550-7554 (1981).

Bennett, V. and Stenbuck, P. J., The membrane attachment protein for spectrin is associated with band 3 in human erythrocyte membranes, Nature 280:468-473 (1979).

Bennett, V., Davis, J. and Fowler, W. E., Brain spectrin, a membrane-associated protein related in structure and function to erythrocyte spectrin, Nature 299:126-131 (1982).

Berridge, M. J. and Irvine, R. F., Inositol Triphosphate, a novel second messenger in cellular signal transduction, Nature 312:315-321 (1984).

Bland, B. H., The physiology and pharmacology of hippocampal formation theta rhythms, Prog. in Neurobiol. 26:1-54 (1986).

Bodsch, W., Baudry, M. and Lynch, G., Activity-dependent breakdown of in vivo assembled spectrin in the neuronal cytoskeleton. (submitted for publication) (1986a).

Bodsch, W., Baudry, M. and Lynch, G., In vivo turnover of brain spectrin: stimulation of degradation by depolarization, Abst. Society for Neurosci. (in press) (1986b).

Branton, D., Cohen, C. M. and Tyler, J., Interaction of cytoskeleton proteins on the human erythrocyte membrane, Cell 24:24-32 (1981).

Browning, M., Baudry, M., Bennett, W. and Lynch, G., Phosphorylation-mediated changes in pyruvate dehydrogenase activity influence pyruvate-supported calcium accumulation by brain mitochondria, J. Neurochem. 36:1932-1940 (1981a).

Browning, M., Bennett, W., Kelly, P. and Lynch, G., The 40,000 Mr brain phosphoprotein influenced by high frequency synaptic stimulation is the alpha subunit of pyruvate dehydrogenase, Brain Res. 218:255-266 (1981b).

Burns, N. R., Ohanian, V. and Gratzer, W. B., Properties of brain spectrin (fodrin), FEBS Lett. 153:165-168 (1983).

Burridge, K., Kelly, T. and Mangeat, P., Nonerythrocyte spectrins actin-membrane attachment proteins occurring in many cell types, J. Cell Biol. 95:478-486 (1982).

Calvert, R., Bennett, P. and Gratzer, W. B., Properties and structural roles of the subunits of human spectrin, Eur. J. Biochem. 107:355-361 (1980).

Canellakis, E. S., Viceps-Madore, D., Kyriakidis, D. A. and Heller, J. S., The regulation and function of ornithine decarboxylase and of the polyamines, Curr. Top. Cell. Regul. 15:155-202 (1979).

Carlin, R. K., Bartelt, D. C. and Siekevitz, P., Identification of fodrin as a major calmodulin-binding protein in postsynaptic density preparations, J. Cell Biol. 96:443-448 (1983).

Chang, F. L. F. and Greenough, W. T., Transient and enduring morphological correlates of synaptic activity and efficacy change in the rat hippocampal slice, Brain Res. 309:35-46 (1984).

Cohen, C. M., Tyler, J. M. and Branton, D., Spectrin-actin associations studied by electron microscopy of shadowed preparations, Cell 21:875-883 (1980).

Cohen, R. S., Blomberg, F., Berins, K. and Siekevitz, P., The structure of postsynpatic densities isolated from dog cerebral cortex, J. Cell Biol. 74:181-191 (1977).

Collinridge, G. L., Kehl, S. J. and McLennan, H., Excitatory amino acids in synaptic transmission in the Schaffer-collateral-commissural pathway of the rat hippocampus, J. Physiol. 334:33-46 (1983).

Davies, P. J. A. and Klee, C. B., Calmodulin-binding proteins: a high molecular weight calmodulin-binding protein from bovine brain, Biochem. Int. 3:203-212 (1981).

Davis, J. and Bennett, V., Brain Spectrin. Isolation of subunits and formation of hybrids with erythrocyte spectrin subunits, J. Biol. Chem. 258:7757-7766 (1983).

Davis, J. Q. and Bennett, V., Brain Ankyrin: purification of a 72,000 Mr spectrin-binding domain, J. Biol. Chem. 259:1874-1881 (1984).

Davis, J. Q. and Bennett, V., Brain Ankyrin: a membrane-associated protein with binding sites for spectrin, tubulin and the cytoplasmic domain of the erythrocyte anion channel, J. Biol. Chem. 259:13550-13559 (1984).

DeMartino, G. N. and Blumenthal, D. K., Identification and partial purification of a factor that stimulates calcium-dependent proteases, Biochemistry 21:4297-4303 (1982).

Dunlop, D. S., Van Elden, W. and Lajtha, A., Protein degradation rates in regions of the nervous system, Biochem. J. 170:637-642 (1978).

Edelman, G. M., Surface modulation in cell recognition and cell growth, Science 192:218-226 (1976).

Feldman, J. A., A connectionist model of visual memory. In: "Parallel

Models of Associative Memory" (G. E. Hinton and J. A. Anderson, Eds.) Lawrence Erlbaum (Hillsdale, NJ) pp 49-82 (1981).

Fifkova, E., Markham, J. A. and Delay, R. J., Calcium in the spine apparatus of dendritic spines in the dentate molecular layer, Brain Res. 266: 163-168 (1983).

Fox, J. E. B. and Phillips, D. R., Stimulus-induced activation of the calcium-dependent protease within platelets, Cell Motility 3:579-588 (1983).

Gall, C., Brecha, N., Chang, T. and Karten, H., Localization of enkephalins in rat hippocampus, J. Comp. Neurol. 198:335-350 (1981).

Gall, G., Pico, R. and Lauterborn, J., Focal hippocampal lesions induce seizures and long-lasting changes in mossy fiver enkephalin and CCK immunoreactivity, Peptides (in press) (1986).

Glenney, J. R. Jr and Glenney, P., Comparison of spectrin isolated from erythrocyte and non-erythrocyte cells, Europ. J. Biochem. 144:529-539 (1984).

Glenney, J. R., Glenney, P. and Weber, K., F-actin binding and cross-linking properties of porcine brain fodrin, a spectrin-related molecule, J. Biol. Chem. 257:9781-9787 (1982).

Goodman, S. R. and Zagon, I. S., Brain spectrin: a review, Brain Res. Bull. 13, 813, 832 (1985).

Greengard, P., Intracellular signals in the brain, Harvey Lect. 75:277-331 (1981).

Hanbauer, I., Gimble, J. and Lovenberg, W., Changes in soluble calmodulin following activation of dopamine receptors in rat striatal slices, Neuropharm. 18:851-857 (1979).

Harris, A. S., Anderson, J. P., Yurchenco, P. D., Green, L. A. D., Ainger, V. J. and Morrow, J. S., Mechanisms of cytoskeletal regulation: functional and ontogenic diversity in human erythrocyte and brain beta spectrin, J. Cell Biochem. 30:51-69 (1986).

Harris, E. W., Ganong, A. H. and Cotman, C. W., Long-term potentiation in the hippocampus involves activation of N-Methyl-D-aspartate receptors, Brain Res. 323:132-137 (1984).

Hebb, D. O., The organization of behavior. Wiley (New York) (1949).

Huganir, R. L. and Greengard, P., CAMP-dependent protein kinase phosphorylates the nicotinic acetylcholine receptor, Proc. Nat. Acad. Sci. 80: 1130-1134 (1983).

Ingebritsen, T. S. and Cohen, P., Protein phosphatase: properties and role in cellular regulation, Science 221:331-338 (1983).

Ishikawa, M., Murofushi, H. and Sakai, H., Bundling of microtubules in vitro by fodrin, J. Biochem. 94:1209-1217 (1983).

Jennings, L. K., Fox, J. E. B., Edwards, H. H. and Phillips, D. R., Changes in the cytoskeletal structure of human platelets following thrombin activation, J. Biol. Chem. 256:6927-6932 (1981).

Kandel, E. R. and Schwartz, J. H., Molecular biology of learning: modulation of transmitter release, Science 218:433-443 (1982).

Kessler, M., Petersen, G., Vu, H. M., Baudry, M. and Lynch, G., Phe-Glu stimulated, chloride-dependent glutamate "binding" represents glutamate sequestration mediated by an exchange system (submitted for publication) (1986).

Kirino, T. and Sano, K., Changes in the contralateral dentate gyrus in mongolian gerbils subjected to unilateral cerebral ischema, Acta Neuropathol. 50:121-128 (1980).

Kishimoto, A., Kajikawa, N., Tabuchi, H., Shiota, M. and Nishizuka, Y., Calcium-dependent neutral proteases, widespread occurrence of a species of protease active at lower concentrations of calcium, J. Biochem. 90: 884-892 (1981).

Kitahara, A., Sasaki, T., Kikuchi, T., Yumoto, N., Hatanaka, M., Yoshimura, N. and Murachi, T., Large-scale purification of porcine calpain I and calpain II and comparison of proteolytic fragments of their subunits, J. Biochem. 95:1759-1766 (1984).

Koch, C. and Poggio, T., A theoretical analysis of electrical properties of spines, Proc. Roy. Soc. Lond. (B) 218:455-477 (1983).

Larson, J. and Lynch, G., Synaptic potentiation in hippocampus by patterned stimulation involves two events, Science 232:985-988 (1986).

Larson, J., Wong, D. and Lynch, G., Patterned stimulation at the theta frequency is optimal for induction of long-term potentiation, Brain Res. 368:347-350 (1986).

Lee, K., Schottler, F., Oliver, M. and Lynch, G., Brief bursts of high-frequency stimulation produce two types of structural change in rat hippocampus, J. Neurophysiol. 44:247-258 (1980).

Levine, J. and Willard, M., Fodrin: Axonally transported polypeptides associated with the internal periphery of many cells, J. Cell. Biol. 90:631-643 (1981).

Lewis, M. E., Laksmanan, J., Nagaiah, K., MacDonnell, P. C. and Guroff, G., Nerve growth factor increases activity of ornithine decarboxylase in rat brain, Proc. Natl. Acad. Sci. (U.S.A.) 75:1021-1023 (1978).

Libby, P. and Goldberg, A. L., Leupeptin, a protease inhibitor, decreases protein degradations in normal and diseased muscles, Science 199:534-536 (1978).

Lin, D. C., Flanagan, M. D. and Lin, S., Complexes containing actin and spectrin from erythrocyte and brain, Cell Motility 3:375-382 (1983).

Linn, T., Pettit, F. and Reed, L., Alpha-keto acid and dehydrogenase complexes. Regulation of the activity of the pyruvate dehydrogenase complex from bed kidney mitochondria by phosphorylation and dephosphorylation, Proc. Nat. Acad. Sci. U.S.A. 62:234-241 (1969).

Llinas, R., McGuinness, T. L., Leonard, C. S., Sugimori, M. and Greengard, P., Intraterminal injection of synapsin I or calcium/calmodulin-dependent protein kinase II alters neurotransmitter release at the squid giant synapse, Proc. Nat. Acad. Sci. 82:3025-3039 (1985).

Lu, P. W., Soong, C. J. and Tao, M., Phosphorylation of ankyrin decreases its affinity for spectrin tetramer, J. Biol. Chem. 260:14958-14964 (1985).

Lynch, G. and Baudry, M., Origins and manifestations of neuronal plasticity in the hippocampus. In Clinical Neurosciences (W. Willis, Ed.) Churchill-Livingstone Publishers, pp. 171-202 (1983).

Lynch, G., Halpain, S. and Baudry, M., Effects of high frequency synaptic stimulation on glutamate receptor binding studied with a modified in vitro hippocampus slice preparation, Brain Res. 244:101-111 (1982).

Lynch, G., Larson, J. and Baudry, M., Proteases, Neural Stability and Brain Aging: An Hypothesis. In Treatment Development Strategies for Alzheimer's Disease (T. Crook, R. Bartus, S. Ferris, and S. Gershan, Eds.) Mark Powley Assoc (Madison) (in press) (1986).

Lynch, G., Larson, J., Kelso, S., Barrionuevo, G. and Schottler, F., Intracellular injections of EGTA block the induction of hippocampal long-term potentiation, Nature 305:719-721 (1983).

Malenka, R. C., Madison, D. V., Andrade, R. and Nicoll, R. A., Phorbol esters mimic some cholinergic actions in hippocampal pyramidal neurons, J. Neurosci. 6:475-480 (1986).

Malik, M. N., Meyers, L. A., Iqbal, K., Sheikh, A. M., Scotto, L. and Wisniewski, H. M., Calcium activated proteolysis of fibrous proteins in central nervous system, Life Sci. 19:795-802 (1981).

Margolis, R. L. and Wilson, L., Opposite end assembly and disassembly of microtubules at steady state in vitro, Cell 13:1-8 (1978).

Mayer, M. L., Westbrook, G. L. and Guthrie, P. B., Voltage-dependent block by Mg^{2+} of NMDA responses in spinal and neurons, Nature 309:261-367 (1984).

Mellgren, R. L., Canine cardiac calcium-dependent proteases: resolution of two forms with different requirements for calcium, FEBS Lett. 109:129-133 (1980).

Murachi, T., Hatanaka, M., Yasumoto, Y., Nakayama, N. and Tanaka, K., A quantitative distribution study on calpain and calpastatin in rat tissues and cells, Biochem. Internat. 2:651-656 (1981).

Nestler, E. J., Walaas, S. I. and Greengard, P., Neuronal phosphoproteins, physiological and clinical implications, Science 225:1357-1364 (1984).

Nicolson, G. L., Topographic display of cell surface components and their role in transmembrane signaling, Curr. Top. Dev. Biol. 13:305-338 (1979).

Nishida, E., Kotani, S., Kuwaki, T. and Sakai, H., Phosphorylation of microtubule-associated proteins (MAPs) controls both microtubule assembly and MAPs-actin interaction. In Biological Functions of Microtubules and Related Structures (H. Sakai, H. Mohri, and G. Borisy, Eds.). Academic Press (Toyko, New York) pp 285-295 (1982).

Nowak, L., Bregestovski, P., Ascher, P., Herbet, A. and Prochiantz, A., Magnesium gates glutamate-activated channels in mouse central neurons, Nature 307:462-465 (1984).

Pin, J. P., Bockaert, J. and Recasens, M., The Ca^{2+}/Cl-dependent L-^{3}H-glutamate binding: a new receptor or a particular transport process? FEBS Lett. 175:31-36 (1984).

Perlmutter, L. S., Gall, C., Siman, R. and Lynch, G., Ultrastructural localization of a calcium-activated protease (calpain) in rat CNS: association with microtubules, mitochondria and synaptic elements, Abst. Soc. for Neurosci. 11:267 (1985).

Pontremoli, S., Salamino, F., Sparatore, B., Michetti, M., Sacco, O. and Melloni, E., Following association to the membrane, human erythrocyte pro-calpain is converted and released as fully activated calpain, Biochem. Biophys. Acta. 831:335-339 (1985).

Rasmussen, H. and Barrett, P. O., Calcium messenger system: an integrated view, Physiol. Rev. 64:938-984 (1984).

Robinson, H. and Koch, C., An information storage mechanism: calcium and spines, Artificial Intelligence Memo 779 (CBID Paper 004) MIT press (Cambridge) pp 1-14 (1984).

Routtenberg, A., Lovinger, D. and Steward, O., Selective increase in phosphorylation of a 47-kDa protein (F-1) directly related to long-term potentiation, Behav. and Neural Biol. 43:1-9 (1985).

Sandoval, I. V. and Weber, K., Calcium-induced inactivation of microtubule formation in brain extracts, Eur. J. Biochem. 92:463-470 (1978).

Sattilaro, R. F. and Dentler, W. L., The associaton of MAP-2 with microtubules, actin filaments, and coated vesicles. In Biological Functions of Microtubulin and Related Structures (H. Sakai, H. Mohri and G. Borisy, Eds.) Academic Press (Tokyo, New York) pp 297-309 (1982).

Seals, J. R. and Czech, M. P., Evidence that insulin activates an intrinsic plasma membrane protease in generating a secondary chemical mediator, J. Biol. Chem. 255:6529-6531 (1980).

Seals, J. R. and Czech, M. P., Characterization of a pyruvate dehydrogenase activator released by adipocyte plasma membranes in response to insulin, J. Biol. Chem. 256:2894-2899 (1981).

Seubert, P., Baudry, M. Dudek, S. and Lynch, G., Calmodulin stimulates the degradation of brain spectrin by calpain, Synapse (in press) (1986).

Siman, R., Baudry, M. and Lynch, G., Brain fodrin: substrate for the endogenous calcium-activated protease calpain I. Proc. Nat. Acad. Sci. (USA) 81:3276-3280 (1984).

Siman, R., Baudry, M. and Lynch, G., Glutamate receptor regulation by proteolysis of the cytoskeletal protein fodrin, Nature 315:225-227 (1985).

Siman, R., Baudry, M., and Lynch, G., Calcium-activated proteases as possible mediators of synaptic plasticity. In New Insights into Synaptic Function (G. Edelman, W. M. Cowan and W. Gall, Eds.) John Wiley, New York (in press) (1985).

Simonson, L., Baudry, M., Siman, R. and Lynch, G., Regional distribution of soluble calcium-activated proteinase activity in neonatal and adult rat brain, Brain Res. 327:153-159 (1985).

Singer, S. J., The molecular organization of membranes, Ann Rev. Biochem. 43:805-833 (1974).

Sobue, K., Kanda, K., Invi, M., Morimoto, k. and Kakiuchi, S., Actin polymerization induced by calspectin, a calmodulin-binding spectrin-like protein, FEBS Lett. 148:221-225 (1982).

Speicher, D. W. and Marchesi, V. T., Erythrocyte spectrin is comprised of many homologous triple helical segments, Nature 311:177-180 (1984).

Staubli, U., Baudry, M. and Lynch, G., Olfactory discrimination learning is blocked by leupeptin, a thiol-proteinase inhibitor, Brain Res. 337:333-336 (1985).

Staubli, U., Baudry, M. and Lynch, G., Leupeptin, a thiol-proteinase inhibitor, causes a selective impairment of spatial maze performance in rats, Behav. and Neural Biol. 40:58-69 (1984).

Storm-Mathisen, J., Localization of transmitter candidates in the brain: The hippocampal formation as a model, Prog. Neurobiol. 8:119-181 (1977).

Swanson, L. W., Teyler, T. J. and Thompson, R. F., Hippocampal long-term potentiation: mechanisms and implications for memory, Neurosci. Res. Prog. Bull. 20, No. 5 (1982).

Takeyama, Y., Nakanishi, H., Uratsuji, Y., Kishimoto, A. and Nishizuka, Y., A calcium-protease activator associated with brain microsomal-insoluble elements, FEBS Lett. 194:110-114 (1986).

Teyler, T. J. and Discenna, P., Long-term potentiation as a candidate mnemonic device, Brain Res. Rev. 7:15-28 (1984).

Tsukita, S., Ishikawa, H., Kurokana, M., Morimoto, K., Sobue, K. and Kakiuchi, S., Binding sites of calmodulin and actin on the brain spectrin, calspectrin, J. Cell. Biol. 97:574-578 (1983).

Tsuyama, S., Bramblett, G. T., Huang, K. P. and Flavin, M., Calcium/phospholipid-dependent kinase recognized sites in microtubule-associated protein 2 which are phosphoryalted in living brain and are not accessible to other kinases, J. Biol. Chem. 261:4110-4116 (1986).

Vanderwolf, C. H., Kramis, R., Gillespie, L. A. and Bland, B. H., Hippocampal rhythmical slow activity and neocortical low voltage fast activity: relations to behavior. In The Hippocampus, Vol 2., Neurophysiology and Behavior (R. L. Isaacson and K. H. Pribram, Eds.) Plenum, New York pp 101-128 (1975).

Weeds, A., Actin-binding proteins - regulators of cell architecture and motility, Nature 296:811-816 (1982).

Weisenberg, R. C., Microtubule formation in vitro in solutions containing low calcium concentrations, Science 177:1104-1105 (1972).

Wenzel, J. and Matthies, H., Morphological changes in the hippocampal formation accompanying memory formation and long-term potentiation. In Memory Systems of the Brain (N. Weinberger, J. McGaugh and G. Lynch, Eds.) The Guilford Press, New York, pp 150-170 (1985).

Wheelock, M. J., Evidence for two structurally different forms of skeletal muscle C^{2+}-activated protease, J. Biol. Chem. 257:12471-12474 (1982).

White, J. D., Gall, C. M. and McKelvy, J. F., Enkephalin biosynthesis and enkephalin gene expression are increased in hippocampal mossy fibers following a seizure-producing lesion (submitted for publication) (1986).

Woods, C. M. and Lazarides, E., Spectrin assembly in avian erythroid development is determined by competing reactions of subunit homo- and hetero-oligomerizatin, Nature 321:85-89 (1986).

Yoshimura, N., Kikuchi, T., Sasaki, T., Kitahara, A., Hatanaka, M. and Murachi, T., Two distinct Ca^{2+}-proteases (calpain I and calpain II) purified concurrently by the same method from rat kidney, J. Biol. Chem. 258:8883-8889 (1983).

Yoshino, H. and Marchesi, V. T., Isolations of spectrin subunits and reassociation in vitro: analysis by fluorescence polarization, J. Biol. Chem. 259:4496-4500 (1984).

Yu, J., Fischman, D. A. and Steck, T. L., Selective solubilization of proteins and phospholipids from red blood cell membranes by nonionic detergents, J. Supramol. Struct. 1:233-248 (1973).

Zagon, I. S., McLaughlin, P. J. and Goodman, S. R., Localization of spectrin in mammalian brain, <u>J. Neurosci</u>. 4:3089-3100 (1984).

Zimmerman, V. J. P. and Schlaepfer, W. W., Characterization of a brain calcium-activated protease that degrades neurofilament proteins, <u>Biochemis-
try</u> 21:3977-3983 (1982).

Zimmerman, V. J. P. and Schlaepfer, W. W., Calcium-activated neutral protease
(CANP) in brain and other tissues, <u>Progress in Neurobiol</u>. 23:63-78
(1984).

PROTEIN F1 AND PROTEIN KINASE C MAY REGULATE THE PERSISTENCE, NOT THE INITIATION, OF SYNAPTIC POTENTIATION IN THE HIPPOCAMPUS

David M. Lovinger and Aryeh Routtenberg

Cresap Neuroscience Laboratory, Northwestern
University, Evanston, IL

INTRODUCTION

Brain information storage likely involves enhanced neuronal responsiveness which persists for long periods of time following learning. Hebb
(1949) first postulated a mechanism for such enhanced responsiveness in
which repetitive activation of a synapse would produce persistent increases
in the efficacy of transmission at that synapse.

Long term potentiation (LTP) of the perforant pathdentate gyrus synapses was first observed by Bliss and Lomo (1973) who demonstrated that brief
periods of high frequency stimulation of perforant path axons led to long
lasting increases in the dentate gyrus granule cell responses to single perforant path test valleys. LTP can persist for days to weeks (Douglas and
Goddard, 1975) and is elicited by relatively short duration stimulus trains
presented at frequencies similar to the firing frequencies of bursting hippocampal neurons (Ranck, 1973). This paradigm thus fulfills three important
requirements for a model of increased synaptic efficacy such as that proposed by Hebb: 1) It involves a long lasting neuronal change which might
parallel the time course of some memories; 2) The increased synaptic efficacy is induced by repetitive activation of the synapse; and 3) Potentiation
similar to that induced using this paradigm could also be induced by neuronal activity observed in vivo.

Evidence from experiments examining LTP in animals trained on a spatial memory task suggests that potentiation is correlated with information
storage. Barnes (1979) demonstrated that the level of synaptic potentiation
observed in a given animal is related to the spatial memory exhibited by
that animal. It has also been shown that aged animals show both faster
memory decay and faster decay of potentiation (Barnes and McNaughton, 1980).
Thus the time course of spatial memory is associated with that of LTP, suggesting that increases in synaptic strength similar to those induced by high
frequency stimulation might play a role in information storage.

If synaptic potentiation observed using the LTP paradigm shares a
common mechanism of action with increases in synaptic efficacy which underlie memory then the induction of potentiation should alter performance on
a memory task. Berger (1984) has indeed demonstrated that rabbits given
potentiating stimulation in the hippocampus show more rapid conditioning
of the nictitating membrane reflex than rabbits given low-frequency stimulation which does not induce potentiation (Berger, 1984). This finding sug-

gests that some forms of information storage may be facilitated by the increased synaptic efficacy of LTP. High frequency stimulation leading to synaptic potentiation also affects the storage of newly acquired spatial information. McNaughton et al. (1986) have shown that animals which received potentiating high frequency stimulation have difficulty learning to search for reward in spatial locations other than those previously rewarded. However, spatial information acquired days before the onset of potentiation was not affected. These results are consistent with the idea that potentiation induced by high frequency stimulation involves synaptic mechanisms that are also involved in information storage lasting hours to days.

It is attractive to think that the molecular systems needed for the production of synaptic plasticity might be contained within the synapse itself since this would allow for alterations in transmission at individual synapses without the need for protein synthesis taking place in the nucleus and targeting of proteins for transport to specific synaptic sites (Routtenberg, 1982). Post-translational modifications of protein function might provide a mechanism for regulating increases in synaptic strength through alterations in molecules localized to the synapse.

Protein phosphorylation is a form of post-translational modification which has been suggested to play a role in synaptic plasticity underlying information storage (Routtenberg, 1979). The addition of inorganic phosphate to proteins by protein kinases is tightly coupled to the neurotransmitter-stimulated production of second messengers such as cAMP (see Nestler and Greengard 1986, for review). Therefore, phosphoproteins are likely involved in producing neuronal responses to synaptic transmission. Protein kinases and protein phosphatases (which remove phosphate groups from proteins), are stimulated by calcium suggesting a role for phosphoproteins in transmitter secretion (c.f. Delorenzo, 1981). Thus protein phosphorylation could regulate the efficacy of synaptic transmission by either pre- or post-synaptic mechanisms. In addition, the activity of protein kinases is especially high in brain synapses (Rodnight, 1982). This observation is consistent with the idea that increased protein phosphorylation is needed for the regulation of functions unique to the synapse, including the production of persistent increases in synaptic efficacy.

If protein phosphorylation plays a role in increased synaptic strength leading to information storage then phosphate incorporation into individual brain proteins might be altered following learning. Indeed earlier studies from this laboratory demonstrated that specific brain phosphoproteins exhibited increased phosphorylation in vitro in animals given behavioral training relative to untrained animals (Routtenberg et al., 1975, Morgan and Routtenberg 1981, Routtenberg, 1982b). One protein observed to be sensitive to behavioral training was originally termed band F (Routtenberg et al., 1975, Ehrlich et al., 1977) and is now called protein F1 (MW= 47kD, pI=4.5). This protein also exhibited altered phosphorylation in neonatal rats that were handled daily relative to rats that had no exposure to the handling experience (Cain and Routtenberg, 1983). Since this phosphoprotein was sensitive to training and altered by other behavioral experiences we postulated that it might play a role in synaptic changes relating to information storage. Therefore we investigated the relationship between protein F1 phosphorylation (as well as the phosphorylation of other proteins) and persistent changes in synaptic efficacy using the long term potentiation model.

II. Protein F1 Phosphorylation is Directly Related to the Persistence of LTP

We have studied the role of protein F1 phosphorylation in hippocampal

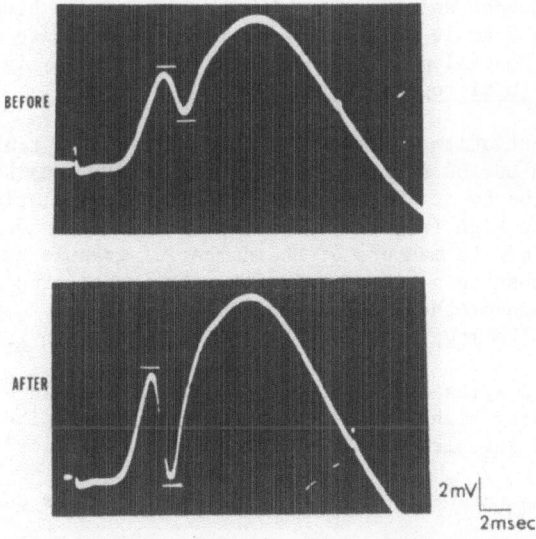

BEFORE

AFTER

2mV

2msec

FIG. 1: Long Term Potentiation (LTP) in the Intact Hippocampal Formation.
Field potentials recorded from the hilus of the dentate gyrus in
an animal subsequently used for in vitro phosphorylation assay.
Single 0.1 msec stimuli were delivered to the perforant path either
before (BEFORE) of 5 min after (AFTER) the delivery of 8 high frequency stimulus trains. The population spike amplitude was measured as the potential change delimited by the two horizontal bars.
The population EPSP slope was measured as the rate of rise of the
initial positive-going portion of the response (positive = up).

potentiation using both anesthetized and chronically implanted, freely
moving animals. Although potentiation has also been observed in the hippocampal slice preparation (Schwatzkroin and Wester, 1975), biochemical assays
performed on slices are compromised by necrosis in 25-50% of the slice tissue (Routtenberg 1984). Moreover, use of the in vivo preparations increases
the yield of potentiated tissue available for biochemical analysis. The
intact hippocampus contains multiple, rigidly laminated "lamellae" (Andersen
et al., 1971). By stimulating the perforant path fibers where they form a
"bottleneck" so as to maximize the activation of granule cell synapses
(Lomo, 1971) one can potentiate all of the perforant path-granule cell
synapses in a given lamella. It is not possible to stimulate this bottleneck in the hippocampal slice and thus the percentage of potentiated synapses in a given area of dentate gyrus will be smaller in the slice than in
vivo (Routtenberg, 1984). Furthermore, using the in vivo preparation one
can simultaneously potentiate synapses in a number of lamellae. Thus the
total volume of potentiated tissue available for biochemical analysis is
relatively large. Since the perforant path provides the major extrinsic
afferent input to the hippocampus (Hjorth-Simonsen, 1972) this procedure
also insures that each lamellae contains a large number of potentiated
synapse.

In our initial studies of phosphoproteins related to LTP we wished to
assay in vitro protein phosphorylation in the region of the hippocampus
containing potentiated synapses following high frequency stimulation of the
perforant path. Thus it was first necessary to determine the area of the
dentate gyrus innervated by the stimulated perforant path fibers. We
accomplished this by recording responses to single perforant path stimuli

at locations along the septo-temporal axis of the dentate gyrus prior to the induction of LTP in each animal. Only that region of the hippocampus in which dentate gyrus granule cell responses to perforant path stimulation were determined to be of local origin (as defined in Lomo, 1971) was used for the subsequent in vitro phosphorylation assay.

To induce potentiation we delivered 8 high frequency trains of stimulus pulses (1 train per 30 sec for 4 min). Each train consisted of 8 pulses delivered at a frequency of 400 Hz. We quantified potentiation (shown in Fig. 1) by measuring high frequency stimulation-induced increases in population spike amplitude (a measure of the number of granule cells firing synchronously in response to a given perforant path input), and population EPSP slope (a measure of the synaptic current generated by a given perforant path input). The magnitude of LTP was defined as the ratio of:

[The population spike amplitude or population EPSP slope observed at a given time following high frequency stimulation]/[The spike amplitude or EPSP slope observed just prior to high frequency stimulation].

The persistence (growth or decay) of potentiation was defined as the ratio of:

[The magnitude of potentiation at a given time after high frequency stimulation]/[The magnitude of potentiation observed immediately after such stimulation].

We first assayed in vitro protein phosphorylation 1 and 5 min after presentation of the last of the eight high frequency stimulus trains. We observed that protein F1 phosphorylation in dorsal hippocampal homogenate was increased in association with the response enhancement of LTP (Routtenberg, 1985a). Increased protein F1 phosphorylation was detected 5 min but not 1 min after LTP procedures (Routtenberg et al., 1985a). Since response enhancement can be observed as early as seconds after high frequency stimulation (McNaughton, 1982) the observed increase in protein F1 phosphorylation does not appear to be related to the potentiation at its onset, but develops minutes later. Increased protein F1 phosphorylation was not observed in thalamus, ventral hippocampus, or neostriatum. Thus the phosphorylation increase appears to be restricted to the brain area containing the potentiated synapses.

As Figure 2 demonstrates the increase in phosphate incorporation was selective for protein F1 among phosphoproteins in the 40-90 kD range observed in dorsal hippocampal tissue from animals exhibiting potentiation 5 min after high frequency stimulation. No other phosphoprotein assayed showed a significant phosphorylation increase associated with LTP (Routtenberg, 1985a).

We also observed that dorsal hippocampal protein F1 phosphorylation was directly related to the magnitude of spike amplitude potentiation 5 min after the end of high frequency stimulation (Figure 3). Thus, animals showing the greatest levels of potentiation also exhibited the greatest protein F1 phosphorylation while animals in which high frequency stimulation produced little potentiation exhibited protein F1 phosphorylation which did not differ from nonpotentiated controls. Protein F1 phosphorylation 1 min after LTP procedures was not significantly correlated with the magnitude of potentiation, providing further support for the idea that protein F1 phosphorylation is not related to potentiation at its onset. We observed no correlation between the phosphorylation of assayed proteins other than protein F1 and measures of response enhancement. Therefore, the potentiation-phosphorylation relationship appeared to be exclusive to protein F1.

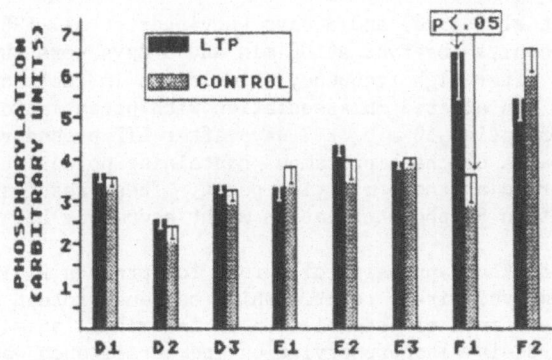

FIG. 2: LTP selectively increases protein F1 phosphorylation (2 min.).
A selective increase in the phosphorylation in vitro of protein F1
was observed in animals exhibiting potentiation 5 min after the
delivery of high frequency stimulation relative to animals given
low frequency, non-potentiating stimulation. No other protein
studied exhibited altered phosphorylation following LTP procedures
(n=10 in LTP group, n=8 in low frequency stimulated group). The
hippocampus was frozen by application of liquid nitrogen to a
retaining cup overlying the exposed cortical surface. The portion
of the dorsal hippocampus containing the perforant path-granule
cell synapses activated by stimulation in vivo was dissected in
the cold room (4°C) and homogenized in a 30mM potassium phosphate
buffer. Aliquots of this homogenate were then preincubated for 30
sec and incubated with a 5μM ATP solution containing (gamma-^{32}P)-
ATP at a temperature of 30°C. The reaction was terminated by the
addition of an SDS-containing solution. Final concentrations of
reaction constituents were: 2mM MG^{2+}, 1mM ethylenediaminetetra-
acetic acid (EDTA), 30 mM potassium phosphate, 1mg/ml hippocampal
homogenate protein, and 2mM dithiothreitol. Protein concentration
was determined by the Lowry (1954) method with bovein serum albumin
as standard. Aliquots containing 50μg of protein were layered on
10% polyacrylamide gels. The electrophoretically separated proteins
were stained, dried, and exposed to Kodak x-omat film. Autoradiographs were quantified densitometrically, and phosphorylation was
expressed as the percentage of the total densitometric area under
the peak, using a computer-based integration program. For presentation purposes the actual phosphorylation values for protein bands
D1, D2, and E3 for both the potentiated and control subjects have
been divided by 2, 5, and 10 respectively. Phosphorylation of
individual bands for LTP and control animals was compared using t
for uncorrelated means.

The observed increase in protein F1 phosphorylation might be important
for initiating or prolonging synaptic potentiation. Since phosphorylation
of this protein was not related to potentiation 1 min after end of the procedures used to induce potentiation it did not appear to play a role in
initiating increases in synaptic efficacy. Thus we investigated the possibility that protein F1 might exhibit a relationship with the persistence of
potentiation at longer times after high frequency stimulation. We studied
in vitro protein phosphorylation both 60 min after potentiating stimulation
in anesthetized rats, and 3 days after the induction of potentiation in
chronically implanted, freely-moving rats.

Dorsal hippocampal protein F1 phosphorylation was selectively increased

60 min (Lovinger et al., 1986) and 3 days (Lovinger et al. 1985) after LTP procedures. The results observed at 60 min and 3 days were quite similar to our findings 5 min after high frequency stimulation in that no other phosphoprotein assayed was altered in association with potentiation. Increased protein F1 phosphorylation 60 min or 3 days after LTP procedures was restricted to the area of the hippocampus containing potentiated synapses, as previously observed at the 5 min time point. These results indicated that increased protein F1 phosphorylation might accompany LTP lasting days.

Consistent with the hypothesis of a role for protein F1 in prolonging potentiation we observed direct relationships between protein F1 phosphorylation and the persistence of potentiation in individual animals. The relationship between protein F1 phosphorylation and persistence was found both 60 min (r=+0.606, df=14, p<.02) and 3 days (r=+0.66, df=8, p<.05) after high frequency stimulation. Figure 3 shows that this significant relationship at the three day time point can be observed using either the population spike amplitude or the population EPSP slope measure. Since the EPSP slope reflects current flow in the synaptic region, the significant relationship between this electrophysiological measure and protein F1 phosphorylation suggests that the function of this phosphoprotein may be related to the long lasting potentiation taking place at the synapse. These data indicate that protein F1 phosphorylation is greater in animals showing no decay of LTP or actual increases over initial LTP values than in animals showing loss of potentiation.

Protein F1 phosphorylation at the 60 min time point was not related to the magnitude of potentiation exhibited immediately after cessation of high frequency stimulation. This finding is consistent with the idea that protein F1 does not play a role in the initiation of response enhancement.

Phosphorylation of protein F1 in animals failing to exhibit potentiation following LTP procedures or in animals in which responses had returned to baseline amplitudes 60 min after high frequency stimulation did not differ from the phosphorylation observed in animals given low frequency, non-

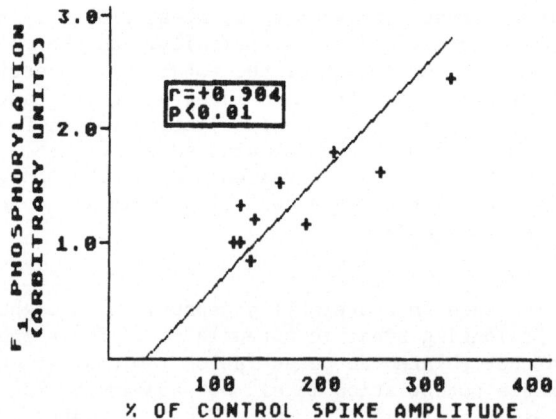

FIG. 3: Protein F1 phosphorylation predicts synaptic plasticity. Phosphorylation of protein F1 increases in direct proportion to the magnitude of the increase in the population spike observed during LTP (n=10). Procedures for dissection of hippocampus, phosphorylation of proteins in vitro, and separation of proteins by polyacrylamide gel electrophoresis were performed as described in Figure 2. Protein F1 phosphorylation values are expressed as the area under the densitometric peak (cm^2). Probability level for the regression coefficient was determined using a t-test.

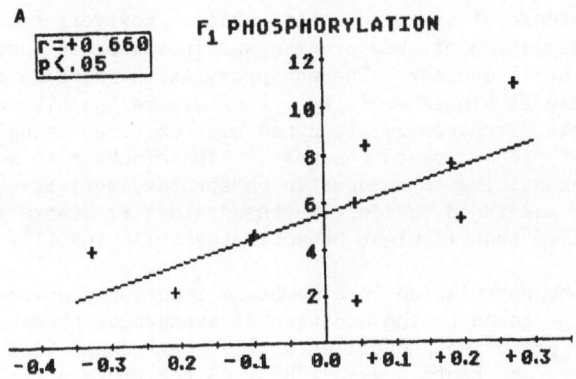

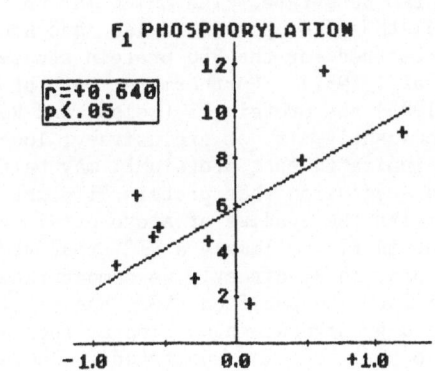

FIG. 4: Protein F1 phosphorylation is directly related to the growth or
decay of potentiation lasting 3 days in chronically implanted,
freely-moving animal. A. Fractional change in spike amplitude
from day 1 to day 3 after LTP. Dorsal hippocampal protein F1
phosphorylation varies in direct proportion of the growth or decay
of the change in the population spike amplitude (SA) over the 3
days following induction LTP procedures in perforant path synapses.
The fractional change in SA from day 1 to day 3 after high frequency stimulation was calculated using the formula: (Percent
change in SA day 1)-(Percent change in SA day 3)/Percent change
in SA day 1. B. Fractional change in EPSP amplitude from day 1
to day 3 after LTP. Dorsal hippocampal protein F1 phosphorylation
varies in direct proportion to the growth or decay of the change
in the population EPSP amplitude (EA) over the 3 days following
induction of potentiation in perforant path synapses. The fractional change in EA was calculated as in 4A, using the population
EPSP rather than the population spike. Note that the animals
demonstrating the highest protein F1 phosphorylation also demonstrated the greatest amount of growth of the population EPSP.

potentiating stimulation (Lovinger et al., 1986). Thus high frequency stimulation per se does not lead to increased protein F1 phosphorylation.
Rather it is the persistent increase in synaptic efficacy induced by such
stimulation which is associated with elevated phosphorylation of this protein.

III. Protein F1 is a Substrate for Protein Kinase C.

We had now identified a phosphoprotein which might function in regulating the persistence of synaptic potentiation. However, the mechanism by which the phosphorylation of this protein was increased in association with potentiation was still unclear. The phosphorylation increase might involve a change in protein F1 kinase activity. Furthermore the LTP-related increase in protein F1 phosphorylation had been observed using a brief (10 sec) in vitro assay sensitive to changes in kinase/substrate association and relatively insensitive to changes in phosphatase activity (Lovinger, et al., 1986). Thus we wished to identify the protein F1 kinase and investigate the possibility that it might be activated following LTP.

Protein F1 phosphorylation in a membrane fraction prepared from rat hippocampus was increased by the addition of exongenous phosphatidylserine (PS) or the tumor promoting phorbol ester 12-0-tetradecanoyl phorbol-13-acetate (TPA, Akers and Routtenberg, 1985). Since protein kinase C (PKC) requires PS for its activity (Takai, 1977) and is stimulated by phorbol ester (Castagna et al., 1982), these findings suggested that membrane-bound protein F1 is a PKC substrate. The molecular characteristics we observed for protein F1 (MW=47kD, pI=4.5, Nelson and Routtenberg, 1985) are similar to those observed for the B50 protein studied by Gispen and colleagues (Zwiers et al., 1982). Furthermore, both protein B50 (Oestreicher et al., 1985) and protein F1 (Nelson and Routtenberg, 1985) were found to be predominantly, if not exclusively, localized to plasma membrane. These data indicated that protein F1 may be identical to the B50 protein. Thus the suggestion that protein F1 might be phosphorylated by PKC was consistent with the studies of Aloyo et al. (1983) demonstrating that protein B50 (=protein F1) is likely a PKC substrate. To ensure that PKC can phosphorylate protein F1 directly we demonstrated that purified protein F1 is phosphorylated by purified PKC (Chan et al., 1985). Protein F1 is not phosphorylated by several known kinases such as type I or type II CaM-dependent kinase (Chan et al., submitted). Indeed, of the kinases tested thus far, only PKC has been shown to phosphorylate this protein.

IV. Protein Kinase C Redistribution from Cytosol to Membrane is Directly Related to the Persistence of Potentiation.

As discussed above the LTP-induced increase in protein F1 phosphorylation observed using the 10 sec in vitro phosphorylation suggests that an increase in PKC/F1 association in vivo takes place consequent to high frequency stimulation of the perforant path. Since protein F1 is localized to the plasma membrane such an association might involve an increase in membrane PKC activity. The discovery that PKC can be translocated from cytosol to membrane (Kraft and Andersen, 1983) provided us with a mechanism by which this kinase/substrate association at the membrane might take place.

Treatment with tumorogenic phorbol ester (Kraft and Andersen, 1983) or Ca^{2+} at μM concentrations (Wolf et al., 1985), increases PKC activity associated with the plasma membrane. This increase in membrane kinase C is accompanied by a decrease in cytosolic PKC activity, such that total PKC activity is not altered. Such a redistribution of PKC activity could reflect actual movement of cytosolic kinase to the membrane or an increase in the strength of association of kinase which was initially only very weakly bound to the membrane. This apparent translocation of PKC has been proposed to activate the kinase by moving it into association with membrane-bound phospholipids, (Takai et al., 1977), or unsaturated fatty acids (Murakami and Routtenberg, 1986).

To determine if PKC redistribution accompanies LTP we assayed kinase C activity in membrane and cytosolic fractions prepared from dorsal hippocampus of animals exhibiting potentiation following high frequency stimulation, animals given non-potentiating stimulation or animals given no sti-

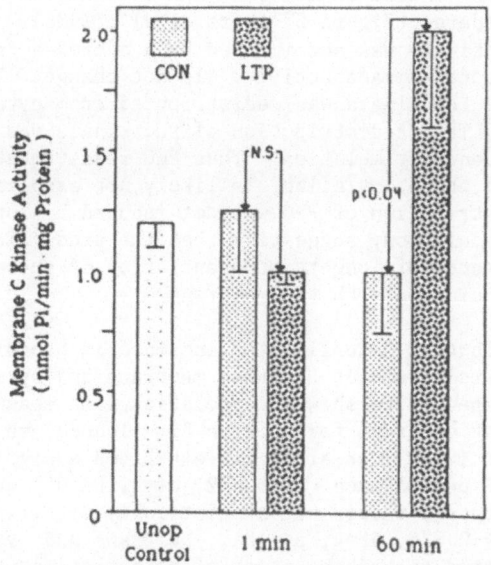

FIG. 5: Protein Kinase C activity is increased in dorsal hippocampal membranes one hour after LTP stimulation. The extent of perforant
path activation of the dentate gyrus was first determined as
described in the text. Eight high frequency stimulus trains (LTP),
or low frequency control stimulation (CON) was then delivered.
One minute or one hour after the cessation of LTP or CON stimulation the animals were rapidly frozen by whole body immersion in
liquid nitrogen. The innervated region of the dorsal hippocampus
was dissected, while frozen, and homogenized in 50mM Tris (pH 7.2),
0.1mM EDTA, and 10mg/ml leupeptin. To obtain particulate and
soluble fractions for each hippocampus, homogenates were spun at
100,000 x g for one hour, and the resulting supernatants were collected. The pellets were washed, and the supernatants were combined. The supernatants, containing PKC activity from the soluble
or cytosolic fraction, were then applied to a 0.4x1.0cm DEAE-cellulose minicolumn, equilibrated with 50mM Tris (pH 7.2), 2mM EGTA,
and 2 mM EDTA. The column was washed, and PKC activity was eluted
with 0.3M NaCl. To obtain particulate PKC activity, the 100,000 x
g pellet was resuspended in 50mM Tris (pH 7.2), 0.1% Triton x-100,
2mM EDTA, 2mM EGTA, and 10mg/ml leupeptin, and stirred for one hour
at 4°C. Following centrifugation at 100,000 x g for one hour to
remove debris, the supernatants, containing PKC activity extracted
from membranes, were applied to DEAE-cellulose minicolumns, and
PKC activity was eluted as before. PKC activity was determined
in the following assay mix: 50mM Tris (pH 7.2), 1.5mM $CaCl_2$,
0.5mM EGTA, 0.5mM EDTA, 100mg histone H1, 3-5mg enzyme preparation, and 0.5mM (gamma-^{32}P)-ATP (sp. act. = 50 c.p.m./pmol), +
100 mg/ml phosphatidylserine. The reaction was run for 10 min at
30°C, quenched by the addition of a saturated EDTA solution, and
reaction product was spotted onto phosphocellulose paper. The
papers were washed, and counted by liquid scintillation spectrometry. PKC activity was taken as the difference between activity
seen in the presence or absence of phosphatidylserine. All enzyme
assays were linear with respect to time and enzyme concentration.
All values are mean + s.e.m.; n=6 for all groups except naive
controls, where n=7.

mulation. We observed increased membrane PKC activity in hippocampus one hour after LTP procedures (Figure 5; Akers et al., 1986). This increase in membrane kinase C activity was accompanied by a decrease in cytosolic PKC activity such that total kinase activity did not change. These observations suggested that the kinase was redistributed from cytosol to membrane in association with LTP. Redistribution of the kinase was not observed 1 min after high frequency stimulation. Thus PKC redistribution, like increased protein F1 phosphorylation, is likely not associated with the onset of LTP. Redistribution of PKC was not induced by low frequency, non-potentiating stimulation, suggesting that the production of synaptic responses in the absence of long-term potentiation of these responses does not lead to PKC association with the membrane.

If PKC redistribution underlies the increase in protein F1 phosphorylation related to persistence of LTP then membrane and cytosolic kinase C activity might be expected to show a correlation with measures of persistence similar to that observed for protein F1. Indeed, we have observed that membrane PKC activity 1 hr after LTP exhibited a significant positive correlation with LTP persistence ($r=+0.852$, $df=4$, $p<.05$) while cytosolic PKC exhibited an inverse, though not statistically significant, correlation with persistence ($r=-0.726$, $df=4$, $p<.10$). Membrane and cytosolic PKC activities were not related to the magnitude of potentiation measured either immediately or 60 min after delivery of the last potentiating stimulus train.

Redistribution of PKC in association with the persistence of potentiation induced using the LTP paradigm should lead to a direct relationship between membrane-bound protein F1 and persistence measures. We tested this hypothesis by correlating protein F1 phosphorylation in hippocampal membranes prepared from potentiated animals with the persistence of LTP 5 min after high frequency stimulation. As predicted, we observed a significant relationship ($r=+0.74$, $df=6$, $p<.05$) between membrane protein F1 phosphorylation and the persistence of potentiation. This relationship was exclusive for protein F1 as no other phosphoprotein assayed exhibited a significant correlation.

The above data are consistent with the hypothesis that an increased association between membrane protein F1 and newly-relocated PKC might play a role in prolonging the increased synaptic efficacy of LTP (Routtenberg, 1985b). The molecular mechanisms which cause PKC redistribution following high frequency synaptic activation have recently been investigated in our laboratory (Akers and Routtenberg, submitted). It is attractive to think that such mechanisms might involve increased intracellular calcium levels resulting from high frequency stimulation (Morris et al., 1983; Morris et al., 1985) leading to activation of NMDA-type glutamate receptors coupled to Ca^{2+} channels (Collingridge, 1985) given that potentiation is blocked by intracellular EGTA (Lynch et al., 1986), and may be induced by brief increases in calcium concentration in hippocampal slices (Turner et al., 1982).

V. Activators of Membrane PKC Prolong Potentiation and Increase Protein F1 Phosphorylation.

The observations discussed to this point suggest a role for PKC and protein F1 in regulating LTP persistence, but we could not infer from such correlative evidence that increased membrane PKC activity or increased protein F1 phosphorylation would lead to prolonged potentiation. Thus we studied the effect of membrane PKC stimulators on potentiation. We hypothesized that substances which increase the amount of PKC associated with the membrane or which stimulate kinase already at the membrane should lead to persistent response potentiation when applied together with brief periods of high frequency stimulation.

322

We first tested this hypothesis by iontophoretically applying the PKC translocator/activator tetradecanoylphorbol 12,13-acetate (TPA) to the hippocampus prior to potentiation induced by the delivery of two trains of high frequency stimulation (Lovinger et al., 1985; Routtenberg, 1986). The spike amplitude and EPSP slope showed initial potentiation when the two high frequency stimulus trains were presented alone and consistently returned to their baseline levels 105 to 120 min after presentation of the second train. Following TPA application combined with high frequency stimulation the increases in spike amplitude did not decay back to baseline but rather increased above the levels observed immediately after high frequency stimulation (Fig. 6). TPA application also increased the persistence of EPSP slope potentiation, suggesting that the site of action of TPA might be located at perforant path-granule cell synapses. Potentiation measured using either the spike amplitude or EPSP slope values was not prolonged following application of a phorbol analogue (4-alpha phorbol) inactive with respect to PKC, or application of the Tris/DMSO vehicle solution.

To ensure that TPA applied to the hippocampus was active with respect to PKC we compared in vitro protein F1 phosphorylation in hippocampus from

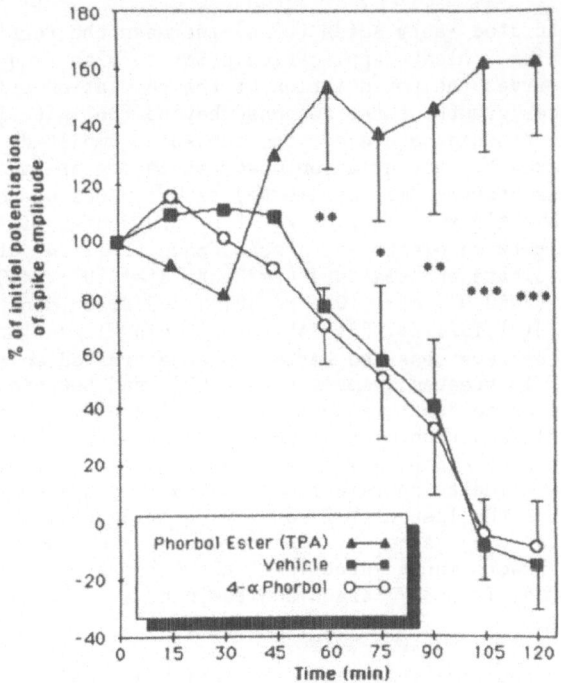

FIG. 6: Iontophoretic application of phorbol ester into dentate gyrus prevents decay and induces growth of potentiated response. Animals given TPA ejections prior to high frequency stimulation show a growth in potentiation beyond the initial increase in spike amplitude beginning at 45 min and persisting through the end of the 2 hour observation period. The percentage of initial potentiation seen in the TPA-ejected anials was significantly different (df= 1,10) with p<.025 (*), p<.005 (**) or p<.001 (***) from 4-alpha phorbol and vehicle (DMSO/Tris) controls at the last five time points.

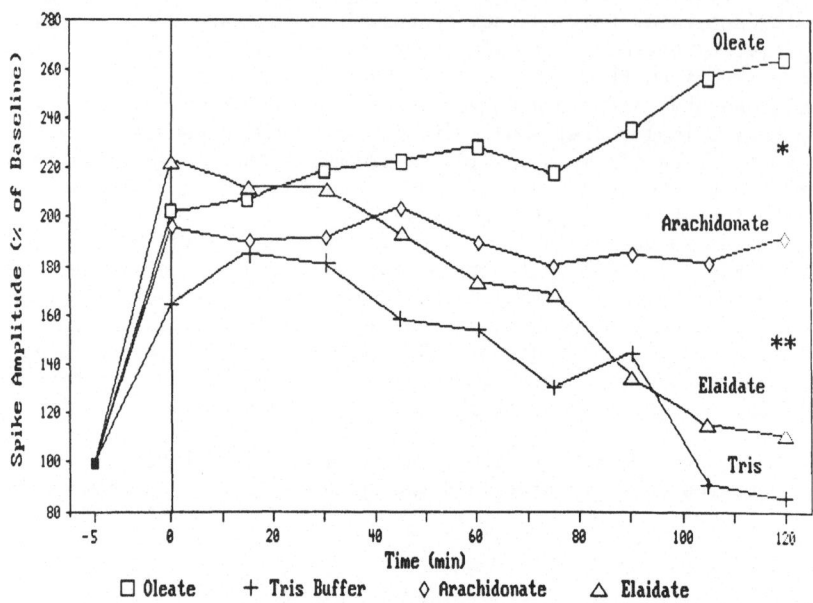

FIG. 7: Cis-Unsaturated Fatty Acids (UFAs) Increase the Persistence of
Potentiation. Oleate application prior to high frequency stimulation preserves the potentiation of the population spike response,
and induces growth of the response beyond the initial potentiation.
Following arachidonate ejection, the spike amplitude persisted
without growth. Potentiation observed in the presence of elaidate
(the trans stereoisomer of oleate) or Tris decayed to baseline.
Two-way ANOVA's run among pairs of groups showed significant differences between oleate and arachidonate (F=11.14, df=1,107,
p<.005), oleate and elaidate (F=23.08, df=1,107, p<.001), oleate
and Tris (F=60.97, df=1,107, p<.001), and arachidonate and elaidate
(F=4.36, df=1,107, p<.05). At the 120 min time point significant
differences were observed between oleate treated animals and elaidate or Tris treated animals (*, p<.005) and between animals
receiving arachidonate and those receiving elaidate or Tris (**,
p<.01). In addition oleate treated animals showed significant
increases in the population spike amplitude at 120 min after high
frequency stimulation relative to the amplitudes observed immediately after the last high frequency stimulus train (p=.036, Wil-
coxon matched pairs test). Standard errors of the mean at 120
min, which were representative of the other time points, were
oleate=37.5, Tris=8.7, arachidonate=16.6, and elaidate=9.2.

animals given combined high frequency stimulation and application of TPA,
4-alpha phorbol or vehicle. Protein F1 phosphorylation in vitro was
increased in TPA-treated, potentiated animals relative to animals given high
frequency stimulation and vehicle or 4-alpha phorbol application. The
ejected TPA presumably stimulated PKC phosphorylation of protein F1, perhaps by translocation of the kinase from cytosol to membrane.

To provide further support for the hypothesis that increased PKC activity prolongs existing potentiation it would be useful to assay the effects
of other PKC stimulators on the persistence of increased synaptic efficacy.
A recent discovery in our laboratory afforded us this opportunity. Murakami
and Routtenberg (1986) demonstrated that the cis-unsaturated fatty acids
(UFAs) oleic acid and arachidonic acid stimulate purified PKC in the absence

of Ca^{2+} and phospholipid. Oleic acid was more potent in this regard than arachidonic acid. Elaidic acid, the trans stereoisomer of oleic acid, had almost no PKC stimulatory effect. In a purified system containing PKC and protein F1, oleic acid stimulated phosphorylation of protein F1 (Chan et al., in preparation).

As a second test of the ability of PKC stimulators to prolong potentiation the cis-UFAs mentioned above were applied to the hippocampus prior to high frequency stimulation. The effects of cis-UFAs on persistence of potentiation (Linden, Muramami and Routtenberg, 1986a) were quite similar to those of TPA. Oleic acid and arachidonic acid increased the persistence of increases in spike amplitude, and oleic acid (the more potent PKC activator) caused potentiation of the spike amplitude and EPSP slope to increase above the level observed immediately after high frequency stimulation. Figure 7 illustrates the potency of UFAs in prolonging spike amplitude potentiation as oleic acid > arachidonic acid >> elaidic acid > vehicle, an order which closely paralleled (r=+0.991, df=2, p<.02) the potency of these substances for in vitro activation of PKC. The mechanism by which cis UFAs prolonged potentiation and even induced growth of the response is a current focus of our studies. The close correspondence between the UFA effects on PKC activity and their effects on potentiation suggests that these UFAs may increase the persistence of potentiation through their effects on PKC activity.

Oleic acid treatment in the absence of high frequency stimulation did not induce increases in response amplitude. Therefore this potent PKC activator does not appear to initiate potentiation.

More recent studies have demonstrated that oleic acid application to the region of the dentate molecular layer which contains the perforant path terminals also prolongs potentiation of the synaptic response (Linden, Murakami and Routtenberg, 1986b). Furthermore, the iontophoretic currents which eject enough oleic acid to prolong potentiation are considerably lower for ejections into the molecular layer than for hilar ejections. This observation suggests that lower concentrations of the compound are effective when the ejection is into the synaptic region than when the ejection site is distant from the synapses, consistent with the idea that oleic acid is acting at the perforant path synapses.

VI. Kinase C Translocation and Increased Protein F1 Phosphorylation: A Proposed Role in Persistent Synaptic Plasticity.

Our observations to date suggest that PKC and protein F1 act to increase the persistence or slow the decay of synaptic potentiation. Protein F1 phosphorylation increases accompany LTP from 5 min through 3 days after its onset and are correlated with the persistence of potentiation observed at 5 min, 60 min, or 3 days. Membrane and cytosolic PKC activity are altered in direct relation to the persistence of LTP observed 60 min after high frequency stimulation. Finally, PKC stimulators applied to the hippocampus prior to high frequency stimulation prolong potentiation.

The observations reviewed above further suggest that kinase C and protein F1 are not involved in the initiation of potentiation. PKC translocation (Akers et al., 1986) and increased protein F1 phosphorylation (Routtenberg et al., 1985) cannot be detected 1 min after high frequency stimulation and thus do not accompany potentiation at its onset. PKC activity is not related to the magnitude of LTP. Protein F1 phosphorylation, while correlated with the magnitude of potentiation at some time points, is most consistently related to persistence. Furthermore, PKC activity and protein F1 phosphorylation are not related to the initial magnitude of potentiation.

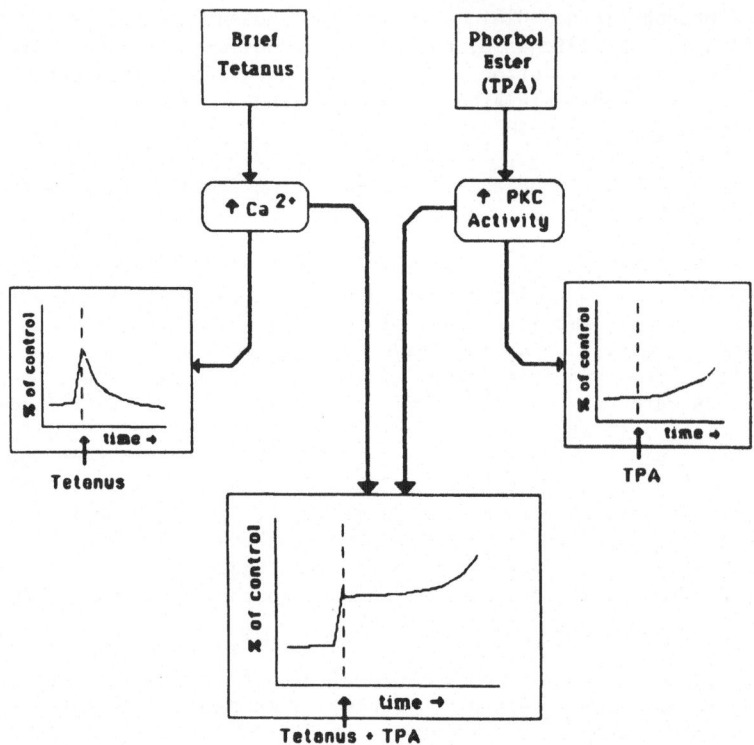

FIG. 8: Proposed Synergism of Calcium and Protein Kinase C May Lead to
Enduring Change in Synaptic Plasticity. Persistent potentiation
may require both increases in intracellular calcium and increased
protein kinase C activity. Increased Ca^{2+} influx resulting from
high frequency stimulation might induce increases in synaptic
efficacy which only persist for minutes, similar to the potentiation we observed following two trains of high frequency stimulation. Stimulators of PKC such as TPA do not potentiate the
response when applied to the hippocampus. However, when increased
intracellular Ca^{2+} is combined with increases in PKC activity a
prolonged response potentiation is observed.

Finally, stimulators of membrane PKC (TPA and cis-UFAs) do not induce potentiation in the absence of high frequency stimulation.

It has been demonstrated (Barnes, 1979) that senescent rats show less
persistent synaptic enhancement than adult rats following high frequency
activation of perforant path synapses. Given that PKC activity and protein
F1 phosphorylation are related to the persistence of enhancement we predicted that these senescent animals would exhibit a deficit in some aspect
of the PKC/protein F1 phosphorylation system. We have recently observed
(Lovinger et al., 1986) that senescent (24 month) rats exhibit decreased
protein F1 phosphorylation in dorsal hippocampus relative to adult (10
month) and adolescent (4 month) rats. We did not observe an age-related
decrease in the phosphorylation of the other protein bands examined. The
decrease in protein F1 phosphorylation in senescent animals was selective
for the dorsal hippocampus. We did not observe decreases in protein F1
phosphorylation in frontal and entorhinal cortex, or in cerebellar cortex.
We also observed that PKC activity was not altered in an age-related fashion
in dorsal hippocampus, and given the selectivity of the change for protein
F1, it is unlikely that phosphatase activity is altered. The decrease in

protein F1 phosphorylation in senescent animals may thus reflect a decrease in the amount of protein F1 in dorsal hippocampus. Such a loss of protein F1 with aging may contribute to the decrease in persistence of synaptic enhancement in old animals.

If synaptic enhancement induced using the LTP paradigm is a good model of information storage in the brain than PKC and protein F1 may function in increasing memory persistence. Indeed, evidence from studies of protein phosphorylation and behavioral training discussed in the Introduction to the present chapter indicate that protein F1 phosphorylation is increased in association with storage of memories. We have also observed increased protein F1 phosphorylation in animals trained on a spatial memory task (Akers and Routtenberg, in preparation). Based upon the finding that increased PKC activity is related to persistence of synaptic plasticity one might predict that PKC would be activated following training. From the findings presented above one might also predict that PKC activity and protein F1 phosphorylation could be directly related to the persistence of memory for correct performance on a behavioral task.

The mechanisms by which Protein F1 might function in regulating the persistence of LTP could involve growth of synaptic structures (Routtenberg, 1985b). Enlargement of synapses has been associated with LTP (Van Harreveld and Fifkova, 1975; Lee et al., 1980; Desmond and Levy, 1985). We have recently demonstrated that protein F1 is probably the same as a major axonal growth cone phosphoprotein pp46 studied by Pfenninger and coworkers (Nelson et al., 1985), suggesting a role for this protein in axon terminal growth (Nelson et al., in preparation). It now appears that the growth-related GAP 43 protein is also the same as protein F1 (Snipes, Freeman, Chan and Routtenberg, 1986). The coidentification of these growth related, presumably presynaptic proteins with protein F1 has led us to speculate that the protein may function in growth of presynaptic terminals related to LTP and memory storage (Routtenberg, 1985b). For example, PKC translocation leading to increased protein F1 phosphorylation might prolong increases in the size of perforant path-dentate gyrus synapses thus producing more persistent increases in synaptic efficacy.

REFERENCES

Akers, R. F. and Routtenberg, A., Kinase C phosphorylates a protein involved in synaptic plasticity, Br. Res. 334:147-151 (1985).

Akers, R. F., Lovinger, D., Colley, P., Linden, D. and Routtenberg, A., Translocation of protein kinase C activity may mediate hippocampal long term potentiation, Science 231:587-589 (1986).

Aloyo, V. J., Zwiers, H. and Gispen, W. H., Phosphorylation of B-50 protein by calcium-activated, phospholipid-dependent protein kinase and B-50 protein kinase, J. Neurochem. 41:649-653 (1983).

Andersen, P., Bliss, T. V. P. and Skrede, K. K., Lamellar organization of hippocampal excitatory pathways, Exp. Br. Res. 13:222-238 (1971).

Barnes, C. A., Memory deficits associated with senescence: A behavioral and electrophysiological study, J. Comp. Physiol. Psychol., 93:74-104 (1979).

Barnes, C. A. and McNaughton, B. L., Spatial memory and hippocampal synaptic plasticity in senescent and middle-aged rats. In D. G. Stein (ed.), The Psychobiology of Aging: Problems and Perspectives, Elsevier/Holland, Amsterdam 253-272 (1980).

Berger, T. W., Long-term potentiation of hippocampal synaptic transmission affects rate of behavioral learning, Science, 224:627-630 (1984).

Berridge, M. J., Rapid accumulation of inositol triphosphate reveals that agonists hydrolyse polyphosphoinositides instead of phosphatidylinositol, Biochem. J. 212:849-858 (1983).

Bliss, T. V. P. and Lomo, T., Long lasting potentiation of synaptic trans-
     mission in the dentate area of the anesthetized rabbit following sti-
     mulation of the perforant path, J. Physiol., 232:331-356 (1973).
Cain, S. and Routtenberg, A., Neonatal handling selectively alters the
     phosphorylation of a 47,000 mol. wt. protein in male rat hippocampus,
     Br. Res. 267:192-195 (1983).
Castagna, M., Takai, Y., Kaibuchi, K., Sano, K., Kikkawa, U. and Nishizuka,
     Y., Direct activation of calcium-activated, phospholipid-dependent
     protein kinase by tumor-promoting phorbol esters, J. Biol. Chem., 257:
     7847-7851 (1982).
Chan, S. Y., Murakami, K. and Routtenberg, A., Purification of a kinase C
     substrate: brain phosphoprotein F1 and the discovery of an endogenous
     kinase C inhibitory factor, Soc. Neurosci. Abstr., 11:926 (1985).
Collingridge, G. L., Long term potentiation in the hippocampus: mechanisms
     of initiation and modulation by neurotransmitters, Trends in Pharm.
     Sci., 6:407-411 (1985).
Delorenzo, R. J., Calcium, calmodulin, and synaptic function: Modulators
     of neurotransmitter release, nerve terminal protein phosphorylation
     and synaptic vesicle morphology by calcium and calmodulin, In: R.
     Tapia and C. W. Cotman (Eds.), Regulatory Mechanisms of Synaptic
     Transmission, Plenum Press, New York, 205-240 (1981).
Desmond, N. and Levy, W. B., Synaptic correlates of associative potentia-
     tion/depression: An ultrastructural study in the hippocampus, Br. Res.
     265 (1):21-30 (1983).
Douglas, R. M. and Goddard, G. V., Long-term potentiation of the perforant
     path-granule cell synapse in the rat hippocampus, Br. Res. 86:205-215
     (1975).
Ehrlich, Y. H., Rabjohns, R. R. and Routtenberg, A., Experiental-input
     alters the phosphorylation of specific proteins in brain membranes,
     Pharm. Biochem. Behav., 6:354-360 (1977).
Hebb, D. O., The Organization of Behavior, John Wiley and Sons, New York
     (1949).
Hjorth-Simonsen, A., Projection of the lateral part of the entorhinal area
     to the hippocampus and fascia dentata, J. Comp. Neurol., 146:219-232
     (1972).
Kraft, A. S. and Andersen, W. B., Phorbol esters increase the amount of
     Ca^{2+}, phospholipid-dependent protein kinase associated with the plasma
     membrane, Nature, 301:621 (1983).
Lee, K. S., Schottler, F., Oliver, M. and Lynch, G., Brief bursts of high-
     frequency stimulation produce two types of structural change in rat
     hippocampus, J. Neurophysiol., 44:247-258 (1980).
Linden, D. J., Murakami, K. and Routtenberg, A., A newly discovered protein
     kinase C activator (oleic acid) enhances long-term potentiation in the
     intact hippocampus, Br. Res., 379:358-363 (1986).
Linden, D. J., Murakami, K. and Routtenberg, A., Oleic acid, a protein
     kinase C activator, enhances hippocampal long-term potentiation, Soc.
     Neurosci. Abstr., 12:1169 (1986).
Lomo, T., Patterns of activation in a monosynaptic cortical pathway: The
     perforant path input to the dentate area of the hippocampal formation,
     Exp. Br. Res., 12:18-45 (1971).
Lovinger, D., Colley, P., Linden, D., Mukarami, K. and Routtenberg, A.,
     Phorbol ester, which induces protein kinase C (PKC) translocation to
     the membrane, prevents decay of long term potentiation, Soc. Neurosci.
     Abstr., 11:927 (1985).
Lovinger, D. M., Akers, R. F., Nelson, R. B., Barnes, C. A., McNaughton,
     B. L. and Routtenberg, A., A selective increase in the phosphorylation
     of protein F1, a protein kinase C substrate, directly related to three
     day growth of long term synaptic enhancement, Br. Res., 343:137-143
     (1985).
Lovinger, D. M., Colley, P., Akers, R. F., Nelson, R. B. and Routtenberg,
     A., Direct relation of long-duration synaptic potentiation to phospho-

rylation of membrane protein F1: A substrate for membrane protein kinase C, Br. Res., in press (1986).

Lovinger, D., Barnes, C. A., Mizumori, S. J. Y., Chan, S. Y., Linden, D., Murakami, K., Sheu, F-S. and Routtenberg, A., Protein F1, previously related to synaptic plasticity, exhibits decreased phosphorylation in senescent rat hippocampus, Soc. Neurosci., 12:1168 (1986).

Lowry, O. H., Rosebrough, N. J., Farr, A. L. and Randall, R. J., Protein measurement with the Folin phenol reagent, J. Biol. Chem., 193:265-275 (1951).

Lynch, G., Larson, J., Kelso, S., Barrioneuvo, G. and Schottler, F., Intra-cellular injections of EGTA block induction of hippocampal long-term potentiation, Nature, 305:719-721 (1984).

McNaughton, B. L., Long-term synaptic enhancement and short-term potentia-tion in rat fascia dentata act through different mechanisms, J. Phy-siol., 324:249-262 (1982).

McNaughton, B. L., Barnes, C. A., Rao, G., Baldwin, J. and Rasmussen, M., Long-term enhancement of hippocampal synaptic transmission and the acquisition of spatial information, J. Neurosci., 6 (2):563-571 (1986).

Morris, M. E., Krnjevic, K. and Ropert, N., Changes in free Ca^{++} recorded inside hippocampal neurons in response to fimbrial stimulation, Soc. Neurosci. Abstr., 9:395 (1983).

Morris, M. E., Krnjevic, K. and McDonald, J. F., Changes in intracellular free Ca ion concentration evoked by electrical activity in cat spinal neurons in situ, Neurosci., 14:563-580 (1985).

Murakami, K. and Routtenberg, A., Direct activation of purified protein kinase C by the unsaturated fatty acids (oleate and arachidonate) in the absence of phospholipids and Ca^{2+}, FEBS LETT., 192 (2):189-193 (1985).

Nelson, R. B. and Routtenberg, A., Characterization of the 47kD protein F1 (pI 4.5), a kinase C substrate directly related to neural plasticity, Exper. Neurol., 89:213-224 (1985).

Nelson, R. B., Routtenberg, A., Hyman, C. and Pfenninger, K. H., A phos-phoprotein, F1, directly related to neuronal plasticity in adult rat brain may be identical to a major growth cone membrane protein Soc. Neurosci. Abstr., 11:927 (1985).

Nestler, E. J. and Greengard, P., Protein Phosphorylation in the Nervous System, John Wiley and Sons, New York (1984).

Oestreicher, A. B., Zwiers, H., Schotman, P. and Gispen, W. H., Immunohis-tochemical localization of a phosphoprotein (B-50) isolated from rat brain synaptosomal plasma membranes, Brain Res. Bull., 6:145-153 (1981).

Ranck, J. B., Studies on single neurons in dorsal hippocampal formation and septum in unrestrained rats, Part I. Behavioral correlates and firing repertoires, Exp. Neurol. 461-555 (1973).

Rodnight, R., Aspects of protein phosphorylation in the nervous system with particular reference to synaptic transmission, In W. H. Gispen and A. Routtenberg (eds.), Prog. Br. Res. Vol. 56, Elsevier/Holland, Amster-dam, 1-25 (1982).

Routtenberg, A., Anatomical localization of phosphoprotein and glycoprotein substrates of memory, Progr. Neurobiol., 12:85-113 (1979).

Routtenberg, A., Memory formation as a post-translational modification of brain proteins. In: C.A. Marsan and H. Matthies (eds.), Mechanisms and Models of Neural Plasticity. Proc. VIth Intl. Neurobiol. IBRO Symposium on Learning and Memory, Raven Press, New York, pp. 17-24 (1982a).

Routtenberg, A., Identification and back-titration of brain pyruvate dehy-drogenase. In W. H. Gispen and A. Routtenberg (eds.), Prog. Brain Res. Vol. 56, Elsevier/Holland, Amsterdam, pp. 349-374 (1982b).

Routtenberg, A., Brain phosphoproteins, Kinase C and Protein F1 protagonists of plasticity in particular pathways. In G. Lynch, J. McGaugh, and N. Weinberger (eds.), Neurobiology of Learning and Memory, The Guilford Press, New York, 479-490 (1984).

Routtenberg, A., Synaptic plasticity and protein kinase C, In: W. H. Gispen and A. Routtenberg (Eds.), Phosphoproteins in the Nervous System, Elsevier/Holland, Amsterdam, 211-234 (1986).

Routtenberg, A., Lovinger, D. and Steward O., Selective increase in the phosphorylation of a 47kD protein (F1) directly related to long-term potentiation, Behav. Neural Biol. 43:3-11 (1985a).

Routtenberg, A., Protein kinase C activation leading to protein F1 phosphorylation may regulate synaptic plasticity by presynaptic terminal growth, Behav. Neural Biol., 44 (2):186-200 (1985b).

Routtenberg, A., Ehrlich, Y. H. and Rabjohns, R., Effect of a training experience on phosphorylation of a specific protein in neocortical and subcortical membrane preparations, Fed. Proc., 34:293 (1975).

Routtenberg, A., Colley, P., Linden, D., Lovinger, D., Murakami, K. and Sheu, F-S., Phorbol ester promotes growth of synaptic plasticity, Br. Res. 378:374-378 (1986).

Schwartzkroin, P. A. and Wester, K., Long-lasting facilitation of synaptic potential following tetanization in the hippocampal slice, Br. Res., 89:107-119 (1975).

Snipes, G. J., Freeman, J. A., Costello, B., Chan, S. and Routtenberg, A., Evidence that the growth-associated protein, GAP-43, and plasticity-associated protein, protein F1, are identical, Soc. Neurosci. Abstr., 12 (1986).

Takai, Y., Yamamoto, M., Inoue, M., Kishimoto, A. and Nishizuka, Y., A proenzyme of cyclic nucleotide independent protein kinase and its activation by calcium-dependent neutral protease from rat liver, Biochem. Biophys. Res. Comm., 77:542-550 (1977).

Teyler, T. J. and Discenna, P., Long-term potentiation as a candidate mnemonic device, Br. Res. Rev., 7:15-28 (1984).

Turner, R. W., Baimbridge, K. G. and Miller, J. J., Calcium-induced long-term potentiation in the hippocampus, Neurosci., 7:1411-1416 (1982).

Van Harreveld, A. and Fifkova, E., Swelling of dendritic spines in the fascia dentata after stimulation of the perforant fibers as a mechanism of post-tetanotic potentiation, Exp. Neurol., 49:736-739 (1975).

Zwiers, H., Jolles, J., Aloyo, V. J., Oestreicher, A. B. and Gispen, W. H., ACTH and synaptic membrane phosphorylation in rat brain, In: W. H. Gispen and A. Routtenberg (Eds.), Prog. Br. Res., Vol. 56, Elsevier/Holland, Amsterdam, 405-417 (1982).

We have recently observed (Lovinger et al. Soc. Neurosci Abstr., in press) that PKC inhibitors decrease the persistence of potentiation when applied to the dentate molecular layer before, 10 min after, or 1 hr after high frequency stimulation. Inhibitors had no effect on the initiation of potentiation. Inhibitors did not altersynaptic responses when applied either 4 hr after high frequency stimulation, or in the absence of such stimulation. These data strongly support the hypothesis of a role for PKC in the persistence, but not the initiation of potentiation, and suggest that the importance of PKC for persistence is maximal minutes after the onset of potentiation and decreases with time thereafter. We propose that the biochemical processes underlying synaptic enhancement consist of 3 separable stages: 1) Initiation by NMDA receptor activation and Ca2+ influx (Collingridge, 1985); 2) PKC activation and protein F1 phosphorylation necessary for maintenance over the first few hours after high frequncy stimulation; and 3) De novo protein synthesis required for persistence lasting hours to days.

ELECTROPHYSIOLOGIC RESPONSES AND ADENYLATE CYCLASE

ACTIVITIES OF MOUSE SPINAL CORD-DORSAL ROOT GANGLION

EXPLANTS RENDERED TOLERANT BY CHRONIC EXPOSURE TO

MORPHINE OR PERTUSSIS TOXIN

Stanley M. Crain[1] and Maynard H. Makman[2]

Departments of Neuroscience[1], Biochemistry[2] and
Molecular Pharmacology[2], Albert Einstein College of
Medicine, Yeshiva University, Bronx, New York

Explant cultures of fetal mouse spinal cord with attached dorsal-root ganglia (DRGs) (Fig. 1: Crain 1976) provide a valuable in vitro system for study of neurotransmitter modulation in the CNS. We have previously developed and utilized this system extensively for analysis of the actions of opioids on the electrophysiologic responses of these neurons and for study of the development of tolerance to opioid depressant effects (see review by Crain, 1984). Exposure of fetal mouse spinal cord-DRG explants to opioid alkaloid or peptide agonists resulted in stereospecific, naloxone-reversible, dose-dependent depression of sensory-evoked dorsal-horn synaptic-network responses within a few minutes (e.g. Fig. 2A; Crain et al., 1977, 1978). After chronic exposure to opioids, e.g. 2-3 days in 1 μM morphine (at 35°C), sensory-evoked dorsal-horn responses recovered, and they could then be elicited by DRG stimuli in the presence of opioids at concentrations 10- to 100-fold higher than required to depress a naive explant (Fig. 2B; (Crain et al., 1979). In addition, these opioid-tolerant explants developed significant cross-tolerance to serotonin (5HT) (Crain et al., 1982). The tolerant state did not develop if the explants were exposed to morphine at lower temperatures (e.g. 20°C for as long as 7 days; Fig. 2C). The data suggest that the sustained decrease in opioid sensitivity observed during chronic opioid exposure at 35°C is mediated by a temperature-dependent metabolic change in these neurons (Crain et al., 1979; Crain, 1984).

When cord-DRG explants were treated briefly with forskolin (10-50 μM)-- a selective activator of adenylate cyclase (AC) (Seamon & Daly, 1981)-- the usual depressant effects of opioids, as well as serotonin and norepinephrine, on sensory-evoked dorsal-horn responses were markedly attenuated (Fig. 3A; Crain et al., 1984, 1986a). Furthermore, exposure of cord-DRG explants to cyclic AMP (cAMP) analogs (0.1-10 mM) produced a decreased sensitivity to opioids and monoamines similar to that observed with forskolin (Fig. 3B).

The finding that "acute tolerance" to opioids is elicited by brief exposure of the explants to forskolin or cAMP analogs suggested the possibility that tolerance following chronic exposure to opioids may be mediated

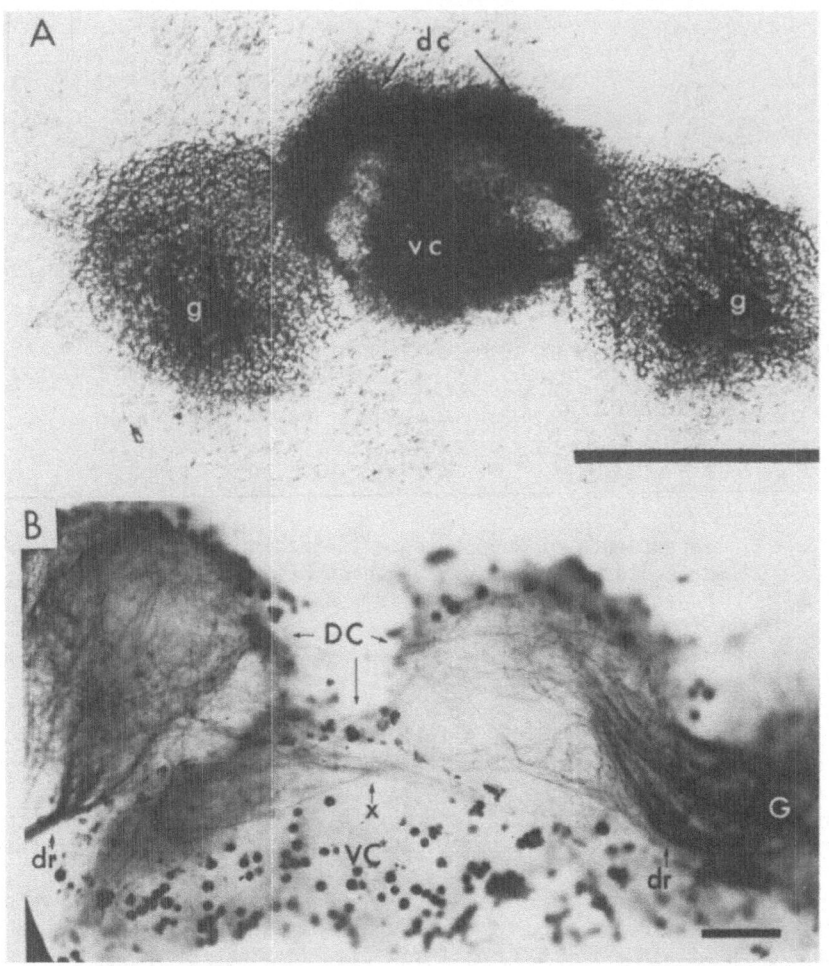

FIG. 1: Photomicrographs of 13-day fetal mouse spinal cord explants (transverse cross-sections, ca. 0.5mm thick) with attached DRGs. (A):
Living cord-DRG explant after 1 month in culture (grown on colla-
gen-coated coverglass in Maximow slide chamber in medium containing
serum and embryo extract; supplemented with NGF during first week
in vitro). Many hundreds of DRG neurons survive under these conditions and innervate cord neurons within multi-layered explant.
Scale: 1 mm. (B): Similar cord-DRG explant in which DRG neurites
were labeled by microinotophoretic injection of horseradish perox-
idase (HRP) into the DRGs (1 week in vitro). Note abundant DRG
neurites in dorsal roots (dr)--arising from DRGs (G; left DRG is
beyond field of view)--that ramify and arborize throughout the
dorsal cord regions, whereas almost none invade the equally available ventral region of the cord (VC). Note the decussation of two
orderly fascicles of DRG neurites (x), each projecting to the contralateral DC. Scale: 100 µm. (HRP-labeled perikarya in cord
are macrophages that have taken up HRP from the bathing fluid; no
back-filling of cord neurons occurred.) (A: from Crain & Peterson, 1974; B: from Crain, 1982).

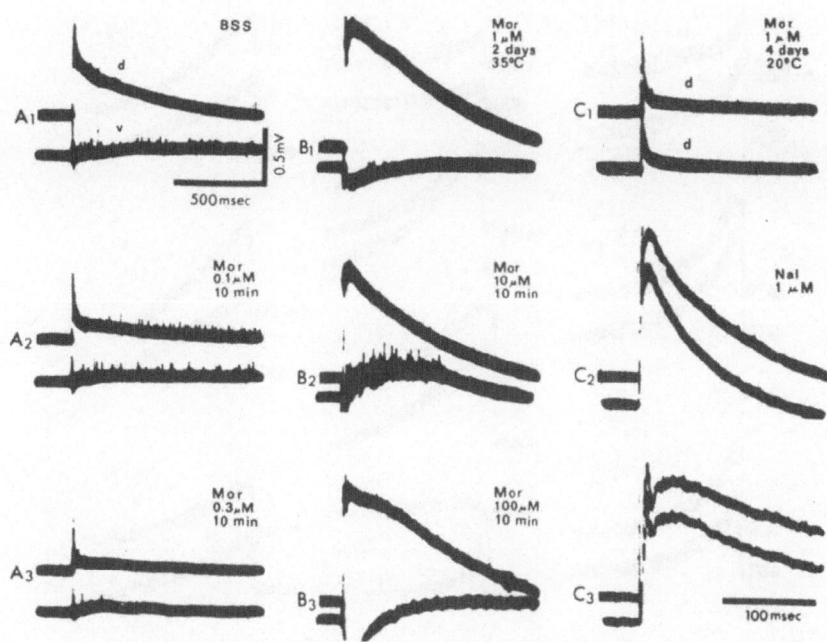

FIG. 2: Selective opioid depression of sensory-evoked dorsal-horn network
responses in cord-DRG explants and development of tolerance during
chronic opiate exposure (3 weeks in vitro). $\underline{A}$: Short-latency
negative slow-wave is evoked in dorsal cord (d̄) by DRG stimulus
(in Hanks' balance salt solution: BSS). $\underline{A}$: Within 10 min in 0.1
μM morphine, dorsal cord response is attenuated, whereas ventral
cord discharge (v) shows little change. $\underline{A}$: In 0.3 μM morphine,
dorsal cord slow-wave response is almost abolished. (No recovery
occurred during exposure to 0.3 - 1 μM morphine for several hrs.).
$\underline{B}$: Large dorsal-cord network response is evoked in another cord-
DRG explant after chronic exposure to 1 μM morphine for 2 days at
25°C (cf. $A_{2\ 3}$). $\underline{B}$: Dorsal cord response remains large even
after increasing morphine concentration to 10 μM. $\underline{B}$: Even after
increase to 100 μM morphine dorsal cord response shows no further
attenuation (10 min test; partial depression developed after longer
exposure at 100 μM level). $\underline{C}$: Only small dorsal-horn responses
can be evoked in another explant (2 sites) after exposure to 1 μM
morphine for 4 days at 20°C, resembling attentuation after acute
exposure to morphine (cf. A_2). $\underline{C}_2$: 1 μM naloxone restores characteristic dorsal cord responses within a few min; same recording
sites as in C_1. Sharp rising phases of these DRG-evoked dorsal
cord responses are shown at faster sweep (C_3). Note: upward
deflection indicates negativity at recording electrode; onset of
stimulus is indicated by first sharp pulse or break in baseline of
each sweep. All records show maximal responses evoked by large
stimuli with standard placements of recording microelectrodes
within the cord and stimulating electrodes within the DRG tissue.
All recordings were made after 10 min test periods unless otherwise
indicated. (From Crain et al., 1979.)

by enhancement of the AC/cAMP system. In order to explore this possibility,
we examined both the acute effects of opioids and the influence of chronic
morphine treatment on AC activity of the explants. Opioid inhibition of AC

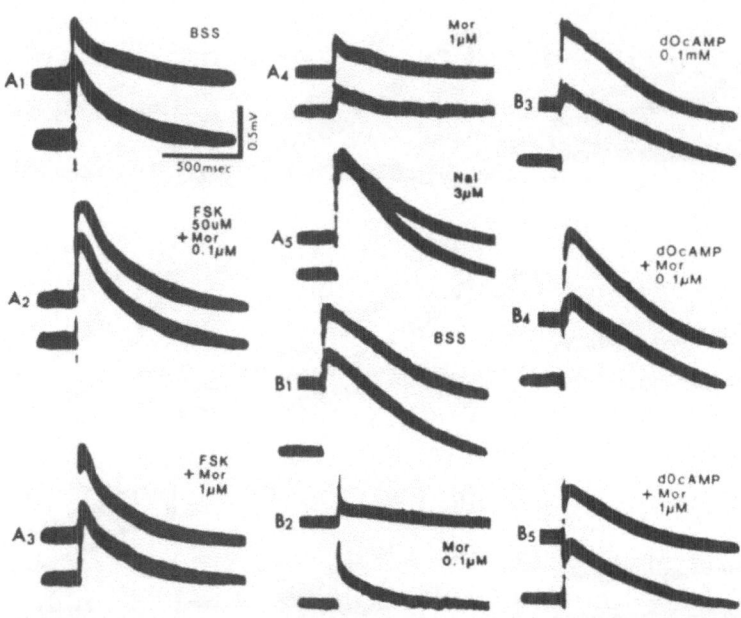

FIG. 3: Pretreatment of cord-DRG explants with forskolin or cAMP analogs
attenuates the depressant effects of morphine on sensory-evoked
dorsal-horn network responses. $\underline{A}_1$: dorsal cord responses are
evoked at two sites by single DRG stimulus (in BSS). $\underline{A}_2$: addition of 0.1 μM morphine, in the presence of forskolin, does not
produce the usual depression of the cord responses (cf. Figs. 2B$_2$,
3B$_2$, 5A$_2$). $\underline{A}_3$: even after raising morphine concentration to
1 μM, the cord responses are only slightly attenuated. $\underline{A}_4$: in
contrast, after withdrawal of forskolin, the cord responses become
progressively attenuated. $\underline{A}_5$: after introduction of naloxone,
cord responses are not only restored, but markedly enhanced in
amplitude and duration. $\underline{B}_1$: sensory-evoked dorsal-cord responses
at two sites in another explant (in BSS). $\underline{B}_2$: characteristic
depression of cord responses occurs within 10 min in 0.1 μM morphine. $\underline{B}_3$: After restoration of cord responses in BSS, exposure
to 0.1 mM dioctanoyl cAMP (dOcAMP) for 30 min produced no significant alterations in these responses. $\underline{B}_4$: addition of 0.1 μM morphine in the presence of dOcAMP, does not produce the usual depression of the cord responses (cf. Fig. 3B$_2$) even after a 20 min test
period. (Similar results were obtained with 10 mM dibutyryl cAMP.)
$\underline{B}_5$: after increasing the morphine concentration to 1 μM, prominent
cord responses are still generated in the presence of dOcAMP.
(From: Crain, Crain & Peterson, 1986a).

had previously been demonstrated to occur in certain cultured cell lines
(Sharma et al., 1975b; Blume et al., 1979; McLawhon, 1981a), as well as in
specific regions of the CNS (Walczak et al., 1979, 1981; Law et al., 1981;
Stefano et al., 1981; Cooper et al., 1982; Gentleman et al., 1983), but had
not been previously examined in cord or DRG tissues. AC activity of the
naive (untreated) explants was found to be stimulated by forskolin, and
forskolin-stimulated activity was significantly decreased in the presence of
levorphanol (Table 1) or morphine (data not shown), but not by dextrorphan
(Table 1) (Dvorkin et al., 1985). Naloxone blocked the inhibitory effect of
levorphanol. Also, in order for opioid inhibition to occur, the presence of

TABLE 1: Adenylate Cyclase Activity of Spinal Cord-Dorsal Root Ganglion Explants: Opioid Sensitivity and Influence of NaCl.

	Adenylate cyclase activity	% Stimulation over basal activity	% Inhibition of forskolin-stimulated activity due to opiod
I (with 40 mM NaCl)			
Basal (control)	0.77		
Levorphanol (2 μM)	0.95	23 ± 3	
Naloxone (2 μM)	1.10	32 ± 4	
Forskolin (10 μM)	2.41	211 ± 12	
Forskolin + Levorphanol (2 μM)	1.74	125 ± 18	28 ± 3
II (with 40 mM NaCl)			
Basal (control)	1.11		
Forskolin (10 μM)	2.58	132 ± 12	
Forskolin + Levorphanol (2 μM)	1.97	77 ± 14	42 ± 12
Forskolin + Levorphanol + Naloxone (2 μM)	2.73	146 ± 12	0 (+ 6 ± 8)
Forskolin + Dextrorphan (2 μM)	2.65	139 ± 21	0 (+ 5 ± 7)
III (without NaCl)			
Basal (control)	0.78		
Levorphanol (2 μM)	1.11	42 ± 10	
Naloxone (2 μM)	1.53	95 ± 24	
Forskolin (10 μM)	2.68	240 ± 45	
Forskolin + Levorphanol (2 μM)	2.61	236 ± 17	4 ± 7

Adenylate cyclase activity is expressed as nmoles cyclic AMP formed/mg protein/15 min. incubation. Values for percent stimulation or inhibition are means ± S.E.M.s. Group I represents 7-10 and II and III 3 separate experiments, each assayed in triplicate and for each of which 6-9 explant cultures were pooled. Prior to assay pooled explants were homogenized in 0.32 M Tris-maleate buffer (pH 7.4) containing 0.8 mM EGTA. Homogenates were centrifuged and the membrane (particulate) fraction resuspended in the same buffer for assay. Assays were carried out at 30°C for 15 min. in the presence of (final concentration) 80 mM Tris-maleate (pH 7.4), 10 mM theophylline, 2 mM MgSO4, 0.5 mM ATP, 50 μM GTP, 2 mM phosphoenopyruvate, 10 μg/ml pyruvate kinase, 20 μM ouabain and 0.2 mM EGTA with or without 40 mM NaCl and other additions as indicated. Cyclic AMP formed was measured as previously described. (Walczak et al. 1981).   (Makman et al., 1987).

Na^+ in the assay was required, as has been reported for inhibition of AC by various receptor agonists, including opioids, in other systems (Lichtenstein et al., 1979; Blume et al., 1979; Michael et at., 1980; Jacobs et al., 1981; Makman et al., 1982; Rodbell, 1984). Naloxone alone was stimulatory, suggesting the possibility that endogenous opioids (Chalazonitis et al., 1984) bound to some of the receptors (Hiller et al., 1978) in the membrane fraction of these explants exert a tonic inhibition of AC in the absence of naloxone [consonant with naloxone-enhancement of DRG-evoked dorsal-horn network responses in naive explants (Crain et al., 1977)]. In contrast to the inhibitory effect of levorphanol on forskolin-stimulated activity, in the absence of forskolin levorphanol alone exerted an unusual stimulatory effect. This stimulation was subsequently found to be more pronounced under certain conditions (see below).

Chronic morphine treatment of the explants resulted in an enhancement of AC activity that was evident under all conditions of assay (Fig. 4A). Also following chronic morphine treatment, the inhibitory effect of levorphanol on forskolin-stimulated AC was attenuated and the stimulatory effect of levorphanol on basal AC was enhanced (Fig. 4A; Table 2). Compensatory enhancement of the AC/cAMP system following chronic exposure to opioids has been reported for NG-108 neuroblastoma x glioma hybrid cells cultured in vitro (Sharma et al, 1975a, 1977). It is possible that the increase in AC of cord-DRG explants observed here contributes to or is responsible for the tolerance that develops to the depressant effects of opioids (Fig. 2B) since acute exposure of explants to forskolin or cAMP analogs rapidly attenuated the opioid-depressant effects (Fig. 3A, 3B).

Opioid, muscarinic,α_2-adrenergic and other receptor-mediated inhibitory effects on AC have been shown to require participation of a guanine nucleotide regulatory protein, G_i (or N_i) (Gilman '84). Receptor-mediated effects on AC or other signal transducing systems that require G_i are blocked by pertussis toxin (PTX, islet-activating protein), due to PTX-induced ADP-ribosylation of G_i (Gilman, 1984; Bogoch et al., 1984; Katada & Ui, 1982, Kurose et al., 1983; Hsia et al., 1984). Chronic exposure to cord-DRG explants to PTX led to enhancement of AC activity (Fig. 4B), comparable to that produced by chronic morphine (Fig. 4A). Furthermore, following PTX treatment, the inhibitory effect of levorphanol on forskolin-stimulated AC was attenuated and the stimulatory effect of levorphanol on basal activity was enhanced (Fig. 4B; Table 2), similar to the effects of chronic exposure to morphine.

FIG. 4: [a]Enhancement of adenylate cyclase activity of spinal cord-DRG explants following chronic exposure to morphine. In each of 4 separate experiments 6-9 control cultures (C) and 6-9 cultures treated with 1 μM morphine for 7 days (M) were pooled separately, homogenized and the particulate fractions assayed for adenylate cyclase activity as described in Table 1. Additions to the assay are indicated above the vertical bars. Activity values are means ± S.E.M. Values for % increase in activity due to exposure to morphine, based on paired control and treated cultures, are 79 ± 20 (basal) 65 ± 10 (2 μM naloxone), 126 ± 31 (2 μM levorphanol) 86 ± 16 (10 μM forskolin) and 142 ± 21 (10 μM forskolin + 2 μM levorphanol. [b]Enhancement of adenylate cyclase activity of spinal cord-DRG explants following chronic exposure to pertussis toxin. In each of 5 separate experiments 6-12 control cultures (C) and 6-12 cultures treated with pertussis toxin (10 μg/ml for 1 day or 1 μg/ml for 5-7 days) (P) were pooled separately, homogenized and the particulate fractions assayed for adenylate cyclase activity as described in Table 1. Additions to the assay are indicated above the vertical bars. Activity values are means ± S.E.M. Values for % increase in activity due to exposure to pertussis toxin, based on paired control and treated cultures, are 42 ± 12 (basal), 49 ± 15 (2 μM naloxone), 106 ± 20 (2 μM levorphanol) 47 ± 9 (10 μM forskolin) and 113 ± 17 (10 μM forskolin + 2 M levorphanol.

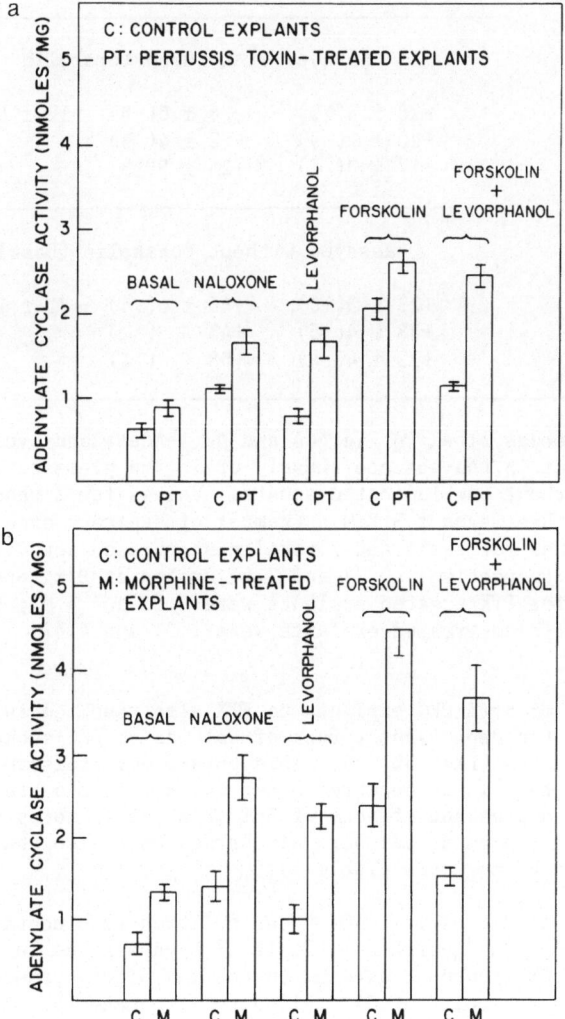

Figure 4.

TABLE 2: Attenuation of Inhibitory Effects of Levorphanol and Serotonin on Forskolin-Stimulated Adenylate Cyclase Activity Following Chronic Exposure of Explants to Pertussis Toxin (PTX) or Morphine

| | Percent change in adenylate cyclase activity due to agent | | |
| Agent | Control explants | Explants treated with: | |
		PTX	Morphine
assayed with 10 μM forskolin			
Levorphanol, 2 μM	$-28 \pm 3(10)$	$- 6 \pm 6(5)$	$-12 \pm 6(4)$
Serotonin, 10 μM	$-20 \pm 2(9)$	$- 2 \pm 4(3)$	
Carbachol, 10 μM	$-17 \pm 3(7)$	$-24 \pm 5(3)$	
assayed without forskolin (basal)			
Levorphanol, 2 μM	$+23 \pm 3(10)$	$+46 \pm 6(4)$	$+71 \pm 13(4)$
Serotonin, 10 μM	$+38 \pm 4(5)$	$+43$ (2)	
Carbachol, 10 μM	$+23 \pm 4(4)$	$+58$ (1)	

Explant treatments are described in Fig. 4A and 4B. Adenylate cyclase assays were carried out in the absence (basal) or in the presence of 10 μM forskolin, with or without the indicated agents. Values for % change in activity due to agent are means ± S.E.M.s (number of separate experiments in parentheses). Values for basal and forskolin-stimulated activity of control explants were respectively 0.77 and 2.41 nmoles cAMP/mg protein. Corresponding values for PTX-treated explants were 1.11 and 4.02; corresponding values for morphine-treated explants were 1.31 and 4.43.

Chronic exposure of cord-DRG explants to PTX also resulted in a marked attenuation of the depressant effect of opioids on DRG-evoked dorsal-horn network responses (Fig. $5B_{2-4}$). This physiologic tolerance was remarkably similar to that which occurred after chronic opioid treatment (Fig. 2B). The acute depressant effects of 5HT (Fig. 5B), norepinephrine, carbachol and oxotremorine on dorsal-horn discharges were also sharply attenuated in PTX-treated cultures (Table 3).

In addition to opioid-inhibited AC, these cultures also contained AC that was inhibited by 5HT and carbachol (Table 2) when assayed in the presence of forskolin. PTX treatment attenuated the inhibitory effect of 5HT as well as that of levorphanol, but PTX failed to attenuate the inhibitory effect of carbachol on AC (Table 2).

As was the case for opioids, 5HT and carbachol also stimulated basal AC activity and these stimulatory effects were appreciably enhanced following PTX treatment of the cultures (Table 2). These unusual stimulatory effects of opioids are not unique for the cord-DRG explants since we have also observed PTX-enhanced opioid stimulation of AC in cultured NCB-20 hybrid cells (Table 4), a cell line that contains delta-- as well as possibly kappa-- opioid receptors (McLawhon, 1981b). The receptors that may possibly be involved, as well as the functional significance of the stimulatory effects of these agonists on AC in cord-DRG explants, remains to be determined. These effects may correlate with preliminary electrophysiologic evidence of direct excitatory effects of opioids on DRG-evoked dorsal-horn

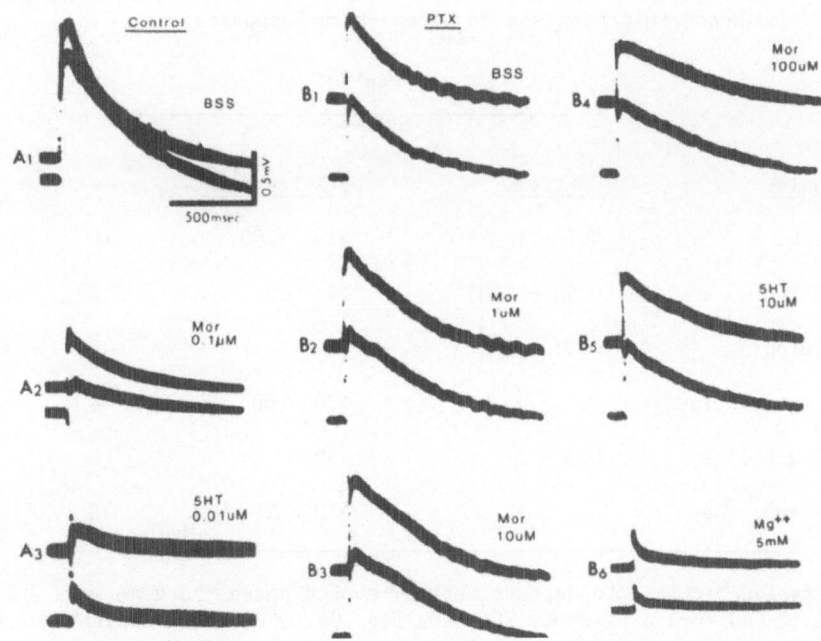

FIG. 5: Pretreatment of cord-DRG explants with PTX blocks the acute depressant effects of morphine and serotonin on sensory-evoked dorsal-
horn network responses. $\underline{A}$: Control explant. $\underline{A}_1$: Dorsal-cord
responses are evoked at two sites in control explant by a single
DRG stimulus (in BSS). $\underline{A}_2$: Responses are markedly attenuated
within a few min after exposure to 0.1 µM morphine. $\underline{A}_3$: Similar
depressant effects occurred in 0.01 µM serotonin (5HT) after ini-
tial restoration in BSS. $\underline{B}$: PTX-treated explant. $\underline{B}_1$: DRG-evoked
dorsal cord responses appear normal after 2-day treatment with
10 µg/ml PTX. $\underline{B}_2$: However, within a few min after introduction
of 1 µM morphine, the dorsal-horn responses are slightly <u>enhanced</u>
(in amplitude and duration) rather than depressed (cf. Fig. $1A_2$).
$\underline{B}_3$: Increasing the morphine concentration to 10 µM is ineffective
in depressing the cord responses, and only a slight depression of
cord responses occurs even in 100 µM morphine ($\underline{B}_4$). $\underline{B}_5$: Exposure
to 10 µM 5HT is also ineffective (whereas dorsal-horn responses in
control explants are severely attenuated at 0.01-0.1 µM levels,
e.g. $\underline{A}_3$). $\underline{B}$:$_6$ In contrast, increasing the Mg^{++} concentration of
the BSS from 1 to 5 mM results in characteristic blockade of the
cord responses, as occurs in control explants. (From Crain, Crain
& Makman, 1986b).

responses (e.g. Fig. $5B_2$) and on DRG action potentials in some of the PTX-
treated explants (Crain et al., 1986 a,b,c).

Based on the data presented here, PTX exerts a profound influence on
the depression by opioids of sensory-evoked dorsal-horn responses, as well
as on AC activity and opioid inhibition of AC in cord-DRG explants. These
data provide insights into some of the factors that may be involved in the
tolerance that develops in cord-DRG explants after chronic exposure to opioids. PTX also blocked the depressant effects of 5HT, α_2-adrenergic and
muscarinic agonists, as well as the inhibitory effect of 5HT on AC. These
results are in agreement with reports indicating that PTX-pretreatment of

TABLE 3: PTX-Blockade of Acute Depressant Effects of Opioid, Monaminergic, and Muscarinic Agonists on Dorsal-Horn Responses.

$ID_{50}(\mu M)^a$

Agonist	Control	PTX-Treated[b]	n*
morphine	0.1 - 1	>10 - 100	10
DADLE	0.03 - 0.3	>1 - 10	4
serotonin	0.01 - 0.1	>10	6
norepinephrine	0.1 - 1	>30 - 50	4
carbachol	0.3 - 3	>50	2
oxotremorine	0.1 - 1	>10 - 25	5

[a]Concentration required to depress sensory-evoked dorsal-horn network response in DRG-cord explant by 50% (see Fig. 2). [b]DRG-cord explants were pretreated with 10 µg/ml PTX for 1-2 days (see text). *Number of PTX-treated explants tested; n>8 for each agent in tests on control cultures. (From: Crain, Crain and Makman, 1986b.)

TABLE 4: Influence of Pertussis Toxin (PTX) Pre-treatment on Stimulatory and Inhibitory Components of D- ala-met-enkephalin (DALA)-Sensitive Adenylate Cyclase of NCB-20 Hybrid Cells

Treatment	Adenylate cyclase activity			
	Basal	DALA	Forskolin	Forskolin + DALA
Control	12	142 (+ 97%)	326	234 (-29%)
PTX, 0.1 µg/ml	123	433 (+242%)	1559	1283 (-18%)
PTX, 0.2 µg/ml	109	482 (+342%)	1434	1350 (- 6%)
PTX, 0.4 µg/ml	91	499 (+448%)	1179	1180 (0)

Cultures were treated with PTX for 16 hrs. Adenylate cyclase (AC) assays of particulate (membrane) fractions were carried out in the absence or presence of 10 µM DALA or 10 µM forskolin as described in Table 1. AC activity is expressed pmoles cAMP/mg protein. Values for % stimulation of basal and inhibition of forskolin-stimulated activity are given in parentheses.

guinea pigs selectively blocks the presynaptic inhibitory effects of morphine, norepinephrine and the 2-adrenergic agonist, clonidine, on peripheral motorneurons in the ileum, whereas responses to acetylcholine via nicotinic receptors are unaltered (Lujan et al. '84; Tucker '84). PTX was also recently shown to block the inhibitory effects of norepinephrine and gamma-aminobutyric acid on voltage-dependent Ca^{++}-channels of DRG neuron perikarya in dissociated cell cultures (Holz et al., 1986).

In conclusion, we propose that at least certain neuromodulatory effects of opioid, serotonergic, α_2-adrenergic and muscarinic agonists on primary afferent networks in the CNS may be mediated by binding to neuronal receptor subtypes that are negatively coupled via G_i to a common pool of AC. The resulting decrease in AC/cAMP levels may, in turn, regulate phosphorylation-dependent gates in K^+ and/or Ca^{++} channels of DRG neurons, so that Ca^{++} influx in presynaptic terminals as well as in the perikarya is attenuated (Crain et al., 1986a). Convergence of these diverse receptor subtypes on G_i-mediated second messenger systems in DRG synaptic terminals would thereby provide integrative modulation of transmitter release in nociceptive, and perhaps other, primary afferent networks in the CNS (Crain et al., 1986b). Our data do not, however, preclude G_i-mediated depressant actions via linkage to other second messengers (see revs. by Berridge and Irvine, 1984; Miller, 1985; Yajima et al., 1986) or directly to some ionic channels (e.g. Pfaffinger et al., 1985). In addition, PTX may influence G-proteins other than G_i (Sternweiss & Robishaw, 1984).

Our studies are also consonant with the hypothesis that neurons may develop tolerance and/or dependence during chronic opioid exposure by a compensatory enhancement of their AC/cAMP system following initial opioid depression of AC activity (Sharma et al., 1975a, 1977; Collier, 1980; Klee et al., 1984; Crain et al., 1986a). Further studies with this in vitro model system may help to determine the degree to which alterations in opiate receptor linkages to AC/cAMP and other second messenger systems are involved in the expression of physiological tolerance in the CNS.

ACKNOWLEDGMENTS

This work was supported by research grants, DA-02031 to S.M.C. and NS-09649 & AG-00374 to Maynard H. Makman. The cord-DRG cultures were prepared by Edith R. Peterson in facilities kindly provided by Dr. M. B. Bornstein (Department of Neurology). Assays of adenylate cyclase were carried out together with B. Dvorkin (Department of Biochemistry). Morphine sulfate was obtained from Dr. E. J. Simon (New York University).

REFERENCES

Berridge, M. J. and Irvine, R. F., Inositol trisphosphate, a novel second messenger in cellular signal transduction, Nature, 312:315-321 (1984).

Blume, A. J., Licktenstein, D. and Boone, G., Coupling of opiate receptors to adenylate cyclase: requirement for Na^+ and GTP, Proc. Natl. Acad. Sci. USA, 76:5626-5630 (1979).

Bogoch, G. M., Katada, T., Northup, J. K., Ui, M. and Gilman, A. G., Purification and properties of the inhibitory guanine nucleotide-binding regulatory component of adenylate cyclase, J. Biol. Chem., 259:3560-3567 (1984).

Chalazonitis, A., Groth, J., Simon, E. J. and Crain, S. M., Development of met-enkephalin immunoreactivity in organotypic explants of fetal mouse spinal cord and attached dorsal root ganglia, Devel. Brain Res. 12: 183-189 (1984).

Collier, H. O. J., Cellular site of opiate dependence, Nature (London), 283:625-629 (1980).

Cooper, D. F. M., Londos, C., Gill, D. L. and Rodbell, M., Opiate receptor-mediated inhibition of adenylate cyclase in rat striatal membranes, J. Neurochem., 38:1164-1167 (1982).

Crain, S. M., Neurophysologic Studies in Tissue Culture, Raven Press, New York (1976).

Crain, S. M., Role of CNS target cues in formation of specific afferent synaptic connections in organotypic cultures. In: Neuroscience Approached Through Cell Culture, Vol. II, S. E. Pfeiffer, ed., CRC Press, Florida, pp. 1-32 (1882).

Crain, S. M., Spinal cord tissue culture models for analyses of opioid analgesia, tolerance and plasticity, in Mechanisms of Tolerance and Dependence, C. Sharp, Ed., National Institute on Drug Abuse Research Monograph, U.S. Govt. Printing Office (ADM 84-1330), Washington, D.C., pp. 260-292 (1984).

Crain, S. M. and Peterson, E. R., Enhanced afferent synaptic functions in fetal mouse spinal cord-sensory ganglion explants following NGF-induced ganglion hypertrophy, Brain Research, 79:145-152 (1974).

Crain, S. M., Crain, B., Finnigan, T. and Simon, E. J., Development of tolerance to opiates and opioid peptides in organotypic cultures of mouse spinal cord, Life Sci., 25:1797-1802 (1979).

Crain, S. M., Crain, B. and Makman, M. H., Pertussis toxin blocks depressant effects of opioid, monoaminergic and muscarinic agonists on dorsal-horn network responses in spinal cord-ganglion cultures, (subm. for publ.) (1986b).

Crain, S. M., Crain, B., Peterson, E. R. and Simon, E. J., Selective depression by opioid peptides of sensory-evoked dorsal-horn network responses in organized spinal cord cultures, Brain Research, 157:191-201 (1978).

Crain, S. M., Crain, B. and Peterson, E. R., Development of cross-tolerance to 5-hydroxytryptamine in organotypic cultures of mouse spinal cord-ganglia during chronic exposure to morphine, Life Sci., 31:241-247 (1982).

Crain, S. M., Crain, B. and Peterson, E. R., Cyclic AMP or forskolin produces rapid "tolerance" to the depressant effects of opiates on sensory-evoked dorsal-horn responses in spinal cord-dorsal root ganglion (DRG) explants, Soc. Neurosci. Abstr., 10:111 (1984).

Crain, S. M., Crain, B. and Peterson, E. R., Cyclic AMP or forskolin rapidly attenuates the depressant effects of opioids on sensory-evoked dorsal-horn responses in mouse spinal cord-ganglion explants, Brain Res., 370:61-72 (1986a).

Crain, S. M., Peterson, E. R., Crain, B. and Simon, E. J., Selective opiate depression of sensory-evoked synaptic networks in dorsal-horn regions of spinal cord cultures, Brain Research, 133:162-166 (1977).

Crain, S. M., Shen, K. E. and Chalazonitis, A., Altered pharmacologic sensitivities of opioid-sensitive dorsal root ganglion (DRG) neurons rendered hyperexcitable by exposure of DRG-cord explants to forskolin or pertussis toxin. In N. Chalazonitis (Ed.), Inactivation of Hypersensitive Neurones, in press (1986c).

Dvorkin, B., Crain, S. M. and Makman, M. H., Increased adenylate cyclase activity in mouse spinal cord-dorsal root ganglion (DRG) explants rendered tolerant by chronic exposure to morphine, Soc. Neurosci. Abstr., 11:1198 (1985).

Gentleman, S., Parenti, M., Neff, N. H. and Pert, C. B., Inhibition of dopamine-activated adenylate cyclase and dopamine binding by opiate receptors in rat striatum, Cell. Mol. Neurobiol., 3:17-26 (1983).

Gilman, A. G., G proteins and dual control of adenylate cyclase, Cell, 36:577-579 (1984).

Hiller, J. M., Simon, E. J., Crain, S. M. and Peterson, E. R., Opiate receptors in cultures of fetal mouse dorsal root ganglia (DRG) and spinal cord: predominance in DRG neurites, Brain Research, 145:396-400 (1978).

Holz, G. G., Rane, S. G. and Dunlap, K., GTP-binding proteins mediate transmitter inhibition of voltage-dependent calcium channels, Nature, 319:
670-672 (1986).

Hsia, J., Moss, J., Hewlett, E. L. and Vaughan, M., ADP-ribosylation of adenylate cyclase by pertussis toxin, J. Biol. Chem., 259:1086-1090
(1984).

Jakobs, K. H., Aktories, K. and Schultz, G., Inhibition of adenylate
cyclase by hormones and neurotransmitters, Adv. Cyclic Nucleotide Res.,
14:173-187 (1981).

Katada, T. and Ui, M., Direct modification of the membrane adenylate cyclase
system by islet activating protein due to ADP-ribosylation of a membrane protein, Proc. Nat. Acad. Sci. USA, 79:3129-3133 (1982).

Klee, W. A., Milligan, G., Simonds, W. F. and Tocque, B., The role of adenyl
cyclase in opiate tolerance and dependence, in Mechanisms of Tolerance
and Dependence, C. Sharp, ed., U. S. Govt. Printing Office (ADM 84-
1330). Washington, D.C., pp. 109-118 (1984).

Kurose, H., Katada, T., Amano, T. and Ui, M., Specific uncoupling by islet-
activating protein, pertussis toxin, of negative signal transduction
via α-adrenergic, cholinergic, and opiate receptors in neuroblastoma x
glioma hybrid cells, J. Biol. Chem., 258:4870-4875 (1983).

Law, P. Y., Wu, J., Koehler, J. E. and Loh, H. H., Demonstration and characterization of opiate inhibition of the striatal adenylate cyclase,
J. Neurochem., 36:1834-1846 (1981).

Lichtenstein, D., Boone, G. and Blume, A., Muscarinic receptor regulation
of NG108-15 adenylate cyclase: requirement for Na^+ and GTP, J. Cyclic
Nucleotide Res., 5:367-375 (1979).

Lujan, M., Lopez, E. Ramirez, R., Aguilar, H., Martinez-Olmedo, M. A. and
Garcia-Sainz, J. A., Pertussis toxin blocks the action of morphine,
norepinephrine and clonidine on isolated guinea-pig ileum, Eur. J.
Pharmacol., 100:377-380 (1984).

McLawhon, R. W., Schoon, G. S. and Dawson, G., Possible role of cyclic AMP
in the receptor-mediated regulation of glycosyltransferase activated
in neurotumor cell lines, J. Neurochem., 37:132-139 (1981a).

McLawhon, R. W., West Jr., R. E., Miller, R. J. and Dawson, G., Distinct
high-affinity binding sites for benzomorphan drugs and enkephalin in
a neuroblastoma-brain hybrid cell line, Proc. Natl. Acad. Sci., USA,
78:4309-4313 (1981b).

Makman, M. H., Dvorkin, B. and Klein, P. N., Sodium ion modulates D_2
receptor characteristics of dopamine agonist and antagonist binding
sites in striatum and retina, Proc. Natl. Acad. Sci. USA, 79:4212-4216
(1982).

Michael, T., Hoffman, B. B. and Lefkowitz, R. J., Differential regulation
of the α_2-adrenergic receptor by Na^+ and guanine nucleotides, Nature
(Lond.), 288:709-711 (1983).

Miller, R. J., Second messengers, phosphorylation and neurotransmitter
release, Trends in Neurosci., 8:462-465 (1985).

Pfaffinger, P. J., Martin, J. M., Hunter, D. D., Nathanson, N. M. and
Hille, B., GTP-binding proteins couple cardiac muscarinic receptors
to a K channel, Nature, 317:536-538 (1985).

Rodbell, M., Structure-function problems with the adenylate cyclase system,
Adv. Cyclic Nucleotide Res., 17:207-214 (1984).

Seamon, K. B. and Daly, J. W., Forskolin: a unique diterpene activator of
cyclic AMP-generating systems, J. Cyclic Nucleot. Res., 7:201-224
(1981).

Sharma, S. K., Klee, W. A. and Nirenberg, M., Dual regulation of adenylate
cyclase accounts for narcotic dependence and tolerance, Proc. Natl.
Acad. Sci. USA, 72:3092-3096 (1975a).

Sharma, S. K., Nirenberg, M. and Klee, W. A., Morphine receptors as regulators of adenylate cyclase activity, Proc. Natl. Acad. Sci. USA, 72:
590-594 (1975b).

343

Sharma, S. K., Klee, W. A. and Nirenberg, M., Opiate-dependent modulation of adenylate cyclase, Proc. Natl. Acad. Sci. USA, 74:3365-3369 (1977).

Stefano, G. B., Catapano, E. J. and Kream, R. M., Characterization of the dopamine stimulated adenylate cyclase in the pedal ganglia of Mytilus edulis: interactions with etorphine, β-endorphins, D-ala-and methionine-enkephalin, Cell Mol. Neurobiol., 1:57-68 (1981).

Sternweis, P. C. and Robishaw, J. D., Isolation of two proteins with high affinity for guanine nucleotides from membranes of bovine brain, J. Biol. Chem., 259:13806-13813 (1984).

Tucker, J. F., Effect of pertussis toxin on normorphine-dependence and on acute inhibitory effects of normorphine and clonidine in guinea-pig isolated ileum, Br. J. Pharmacol., 83:326-328 (1984).

Walczak, S. A., Wilkening, D. and Makman, M. H., Interaction of morphine, etorphine and enkephalins with dopamine-stimulated adenylate cyclase of monkey amygdala, Brain Res., 160:105-116 (1979).

Walczak, S. A., Makman, M. H. and Gardner, E. L., Acetyl-methadol metabolites influence opiate receptors and adenylate cyclase in amygdala, Eur. J. Pharmacol., 72:343-349 (1981).

Yajima, Y., Akita, Y. and Saito, T., Pertussis toxin blocks the inhibitory effects of somatostatin on cAMP-dependent vasoactive intestinal peptide and cAMP-independent thyrotropin releasing hormone-stimulated prolactin secretion of GH_3 cells, J. Biol. Chem., 261:2684-2689 (1986).

SUPPLEMENTARY REFERENCES

Crain, S. M., Shen, K.-F. and A. Chalazonitis, Opioids excite rather than inhibit sensory neurons after chronic opioid exposure of mouse dorsal root ganglion-spinal cord explants. Soc. Neurosci. Abstr., 13 (1987) in press.

Makman, M. H., Dvorkin, B. and Crain, S. M., Modulation of adenylate cyclase activity of mouse spinal cord-ganglion explants by opioids, serotonin and pertussis toxin. Brain Res. (1987) in press.

Qiu, X.-C., Crain, S. M. and Makman, M. H., Serotonin receptor systems in spinal cord and sensory ganglia: relation to opioid action, tolerance and cross tolerance. Soc. Neurosci. Abstr. 13 (1987) in press.

BIOCHEMICAL AND FUNCTIONAL INTERACTIONS OF A SELECTIVE

KAPPA OPIOID AGONIST WITH CALCIUM

P. F. VonVoigtlander, M. Camacho Ochoa and
R. A. Lewis

CNS Research, The Upjohn Company, Kalamazoo
Michigan 49001

SUMMARY

The discovery of the selective kappa opioid receptor agonist, U-50488H, has provided a tool for the study of the mechanisms and function of the kappa receptor-effector. We have investigated the interactions of this compound with calcium in several biochemical and functional studies to assess the involvement of calcium mechanisms in the kappa receptor-linked effector. In rat brain synaptosomes, U-50488H attenuated the uptake of $^{45}Ca^{++}$ induced by K^+ (40 mM) depolarization. This effect was concentration-related (U-50488H 10^{-5} to 10^{-7}M), was apparent in short (8-second) but not longer (1-minute) term incubations, and did not occur in the presence of a non-polarizing concentration (5.6 mM) of K^+. Naloxone (10^{-7}M) did not block this effect of U-50488H (10^{-6}M), and higher concentrations (10^{-5}M) alone blocked calcium uptake. We have found that the binding of the depolarizing amino acid analog, kainic acid, is enhanced by $CaCl_2$. U-50488H (10^{-4} to 10^{-6}M) blocks this enhancement of ^{3}H-kainic acid binding in vitro and also blocks the in vivo effects of kainic acid. In mice, intravenous injection of kainic acid causes scratching, convulsions, and death, depending on the dose administered. U-50488H blocks all of these effects (ED50=4.5 mg/kg for antagonism of convulsions induced by 27.5 mg/kg kainic acid). The convulsions induced by intracerebroventricularly administered kainic acid are also blocked by U-50488H as are those induced by similarly administered Bay K 8644, a calcium channel activator. All of these anticonvulsant effects of U-50488H were antagonized by naltrexone. Together these data indicate that the kappa agonist U-50488H has functionally relevant interactions with depolarization-related Ca^{++} mechanisms in the central nervous system.

INTRODUCTION

The characterization of receptors and their effector mechanisms requires the use of highly selective agonists and antagonists for the receptors in question. This requirement has posed a serious problem to the study of the kappa opioid receptor; the prototype agonists originally used to define the receptor pharmacologically (Martin et al., 1976) lack such selectivity (VonVoigtlander and Lewis, 1982) as do the opioid antagonists. We have discovered (Szmuszkovicz and VonVoigtlander, 1982) and characterized (VonVoigtlander et al., 1983) a highly selective kappa opioid agonist, U-50488H, and used it and a close congener, U-69593, to study kappa mechanisms. At the level of receptor binding, U-50488H has been used to define the portion of the binding of less selective ligands that is to the kappa receptor (Lahti et al., 1982; James and Goldstein, 1984), and ^{3}H-U-69593 has been used to directly label kappa binding sites in the brain and spinal cord (Lahti et al., 1985). One of the major pharmacological results of kappa receptor activation is a loss of response to painful stimuli. We have used U-50488H to study the neuronal pathways involved in the expression of this activity in rodents, and have found that serotonergic pathways are required for this antinociceptive action of kappa agonists (VonVoigtlander et al., 1984). On an intermediate level of integration, the question of the effector or second messenger mechanisms activated by stimulation of kappa receptors in the brain remains unanswered. We and others have hypothesized an involvement of Ca^{++}-related mechanisms. For example, non-selective kappa agonists have been shown to block Ca^{++} efflux from red cell ghosts (Yamasaki and Way, 1983) and to cause a Ca^{++}-dependent inhibition of intact red cell deformation (Rhoads et al., 1985). In the guinea pig ileum (Cherubini and North, 1985), U-50488H decreases Ca^{++} conductance and consequently the release of acetylcholine. Similarly, the Ca^{++}-dependent action potential of dorsal root ganglia is attenuated by opioid peptides that are relatively selective for kappa receptors (Werz and MacDonald, 1985).

In the present studies we have taken a biochemical and functional approach to determine the involvement of Ca^{++} mechanisms in kappa receptor-mediated effects. Specifically, we have investigated the ability of U-50488H to block Ca^{++} uptake into synaptosomes, to alter the calcium stimulation of binding of the neuroexcitotoxin, kainic acid, to neuronal membranes and to block the Ca^{++}-dependent excitation (convulsions) induced by kainic acid and the Ca^{++} channel activator, Bay K 8644 (Schramm et al., 1983).

METHODS

Uptake of $^{45}Ca^{++}$

The methods used were modified from those of Gripenberg et al., (1980). Charles River male rats (150-250 g) were decapitated and the brains quickly removed and homogenized in ice cold sucrose solution (0.32M plus 0.05 mM EGTA, and 5.0 mM sodium Hepes, pH 7.4, 1 g wet tissue in 20 volumes). All subsequent steps were performed at 0°C. The tissues were homogenized at 900 rpm (12 strokes) in a glass-teflon homogenizer (0.15 mm clearance). The homogenate was diluted three times and centrifuged at 1000 xg for 10 minutes and the supernatant retained. The supernatants were centrifuged at 49,600 xg for 10 minutes and the supernatants discarded. The pellet was resuspended to obtain a crude synaptosome preparation which was layered on a Ficoll gradient (16 ml Ficoll 10%, and 24 ml Ficoll 4% in 0.32M sucrose). Ten milliliters of the crude preparation were layered on each tube. The gradients were centrifuged at 20,700 xg for 40 minutes in a Beckman SW 25.1 swinging bucket rotor.

The cellular material at the interface was collected and pooled (20 ml from each tube). The synaptosomes were centrifuged at 49,600 xg for 10 minutes. The supernatant was discarded and the pellets resuspended in incubation medium (5.6 mM KCl), homogenized at 900 rpm (12 strokes), and centrifuged at 49,600 xg for 10 minutes. The supernatant was discarded and the washing repeated twice. Protein contents were determined by Bio-Rad standard methods. Synaptosomes were equilibrated with incubation medium lacking $CaCl_2$ but containing 5.6 mM KCl to a protein concentration of 1.5-2.0 mg/ml.

Synaptosomes were preincubated at 37°C for 10 minutes in calcium-free medium, drug solutions added, and one minute later the synaptosomes were loaded with $^{45}Ca^{++}$ for 8 seconds. The uptake was initiated by rapidly mixing 0.1 ml of incubation medium supplemented with $^{45}Ca^{++}$ at a specific activity of 0.1 mCi mmol/1 whereafter the uptake was stopped by diluting the samples with 3.0 ml of ice cold incubation medium without added calcium but supplemented with 0.5 mM $LaCl_3$. Extracellularly bound calcium was then displaced by incubation at 0°C for 20 minutes in the incubation medium containing lanthanum.

Synaptosomes were collected by suction-filtration on Whatman GF/A filters 3.7 cm in diameter. The synaptosomes retained on the filters were washed with 20 ml fresh ice cold lanthanum medium (fractions of 5.0 ml:one to rinse the tube, three to rinse the filter) to eliminate unbound radioactive calcium. After drying by suction, the filters were placed in counting vials containing 15.0 ml of ACS scintillation medium.

The non-specific binding to filters was determined using incubation medium instead of synaptosome suspension. Control values were determined in 5.6 mM KCl incubation medium. Synaptosomes were depolarized by adding 74.4 mM KCl to the $^{45}Ca^{++}$ solution for a final 40 mM KCl concentration. Statistical significance was determined with t-test and $p < 0.05$ was considered indicative of significant changes. The filter discs were treated for at least two hours in incubation medium supplemented with 2.2 mM $CaCl_2$. This treatment decreased the non-specific binding of $^{45}Ca^{++}$. Incubation medium with lanthanum was prepared daily as well as the $^{45}Ca^{++}$ and the drug solutions. Incubation medium composition used was (mM): NaCl (136); KCl (5.6); $MgCl_2$ (1.3); Tris/HCl (20), pH=7.4); and glucose (11). In order to retain isoosmolarity of the medium, the concentration of sodium

TABLE 1.   Effect of U-50488H and verapamil·HCl upon Ca^{++} uptake by rat whole brain synaptosomes (Ficoll gradient purified)

A.   8 second $^{45}Ca^{++}$ incubation.

	$^{45}Ca^{++}$ Uptake (cpm) X ± S.D.	
	K^+ 5.6 mM	K^+ 40 mM
Control	874 ± 72	1949 ± 50*
Verapamil·HCl 10^{-6}M	790 ± 50	1856 ± 59**
U-50488H 10^{-5}M	815 ± 43	1766 ± 71**
Filter Blank	128 ± 24	162 ± 29

B.   1 minute $^{45}Ca^{++}$ incubation.

Control	2512 ± 22	4343 ± 208*
Verapamil·HCl 10^{-6}M	2509 ± 123	4400 ± 193
U-50488H 10^{-5}M	2423 ± 216	4310 ± 242
Filter Blank	138 ± 12	148 ± 7

* $p<0.01$ compared to 5.6 mM K^+ control.   ** $p<0.05$ compared to 40 mM K^+ control.

was decreased when that of potassium was increased (110 mM NaCl was used with 40 mM KCl). The equilibrated synaptosomes were regularly examined by an electron-microscopy technique. Extracellular mitochondria were rare and synaptosomes were morphologically intact.

^{3}H-Kainic Acid Binding Assay

Charles River CF-1 male mice were decapitated and the brain rapidly removed for preparation of crude synaptic membrane fraction as described by London and Coyle (1979). The tissue was homogenized in 200 ml of glass-distilled water at 700 rpm in a Potter-Elvehjem glass homogenizer fitted with a teflon pestle. The homogenate was dispersed with a Brinkman homogenizer at setting #6 for 30 seconds. Tissue suspensions were centrifuged at 16,000 rpm for 10 minutes. The membrane pellets were resuspended, recentrifuged, and washed sequentially in 100 ml of buffer (50 mM Tris/acetate, pH=7.1). The suspension in buffer was incubated for 30 minutes at 37°C to dissociate any glutamate bound to the membranes. After centrifugation for 10 minutes at 16,000 rpm, the pellet was suspended in buffer at 4°C at 0.1-0.5 mg/ml and used for the binding assay (modified from that of Simon et al., 1976). Aliquots of the crude synaptic membranes were incubated in triplicate at 4°C for 60 minutes in a 2 ml total volume. ^{3}H-Kainic acid (^{3}H-KA) in a volume of 100 ul, various concentrations of potential radioligands, displacers, and drugs were used. During preparation and incubation, the tubes were kept in ice-water baths maintained at 4°C. Incubation was prolonged for 60 minutes and terminated by filtration under vacumn with G/B Whatman filters (3.7 cm). The filters were rinsed twice with 5 ml of H_2O and then counted in 15 ml of ACS solution. Total specific binding of ^{3}H-KA was defined as the difference between total binding with radioligand

TABLE 2: Further evaluation of the effects of U-50488H upon $^{45}Ca^{++}$ uptake (8 second incubation) by Ficoll purified synaptosomes

	$^{45}Ca^{++}$ Uptake (cpm) X ± SD 40 mM K^+
Control	2832 ± 287*
U-50488H 10^{-5}M	2032 ± 232**
U-50488H 10^{-6}M	2410 ± 96**
U-50488H 10^{-7}M	2661 ± 96
Verapamil·HCl 10^{-5}M	1896 ± 106**
Nifedipine 10^{-5}M	2472 ± 76**
Naloxone HCl 10^{-5}M	2253 ± 88**
Control	2824 ± 99*
Naloxone·HCl 10^{-7}M	2912 ± 151
Naloxone·HCl + U-50488H 10^{-6}M	2621 ± 100**

* $p < 0.01$ compared to respective 5.6 mM K^+ controls (1215±123, 1452±82, and 2795±150 cpm). ** $p < 0.05$ compared to 40 mM K^+ control. N=4.

alone and non-specific binding measured in the presence of 0.1 mM unlabelled kainic acid.

In order to characterize separately the binding to high affinity and low affinity receptor binding sites, advantage was taken of the fact that the radioligand dissociates slowly ($t/\frac{1}{2}$=90 minutes) from high affinity sites, but nearly instantaneously from low affinity sites. After 55 minutes of incubation of the membranes with ^{3}H-KA, 200-nmoles of unlabelled kainic acid were added to some experimental tubes to bring the final concentration of unlabelled ligand to 0.1 mM. After 5 minutes of additional incubation, the tubes were filtered and the binding of ^{3}H-KA was measured. The addition of the 200 nmoles of unlabelled ligand 5 minutes prior to filtration resulted in total displacement of radioligand from low affinity binding sites, but did not affect non-specific binding; thus, the difference between binding in this preparation and those incubated in the presence of radioligand alone represents binding to low affinity receptor sites. The difference between binding occurring in the continuous presence of unlabelled ligand (non-specific binding) and that occurring with the 5-minute exposure to an excess of unlabelled ligand represents binding to high affinity receptor sites (London and Coyle, 1979). Protein concentration was measured by the Bio-Rad method.

Kainic Acid Antagonism

Male CF-1 Upjohn mice (20-25 gm) were dosed subcutaneously with the test compound in a volume dose of 10 ml/kg. Fifteen minutes later the mice received 27.5 mg/kg of kainic acid intravenously. This and the water-soluble test compounds were prepared in 0.9% saline; others were prepared in equimolar aqueous citric acid and, if insoluble in this medium, further suspended in 0.125% aqueous carboxymethylcellulose. Immediately after the second injection, the mice were housed individually in observation cubicles and scored over the next 30 minutes for clonic seizures of the forelimbs (pianoplaying convulsions). Animals not displaying this behavior were considered to be protected from kainic acid-induced seizures. Compounds were tested at two or more dose levels (at 0.3 log intervals), 10 mice/dose) and ED50's for the protection from convulsion was calculated by the method of Spearman and Karber.

Seizures induced by intraventricularly administerd kainic acid or Bay K 8644 were studied in a similar fashion. Male CF-1 mice (18-20 gm) were pretreated subcutaneously with the test compound and 15 minutes later received a free-hand injection of the convulsant directly into the left lateral cerebral ventricle. Kainic acid (0.08 ug) was so administered in 2 ul of aqueous 0.9% NaCl and Bay K 8644 (50 μg) in 10 μl of aqueous 0.45% NaCl and 50% dimethylsulfoxide. The mice were then observed over the next 30 minutes for the occurrence of clonic seizures. Six mice were tested at each dose level, and the quantal data used to calculate ED50 values was as indicated above.

RESULTS

Initially the uptake of $^{45}Ca^{++}$ into gradient-purified synaptosomes was assessed at different time intervals and concentrations of K^+ (Table 1). The uptake was time-related (greater at 1 minute than at 8 seconds) and enhanced by K^+-(40 mM) induced depolarization at both times. However, the Ca^{++} antagonist verapamil was only effective in attenuating the depolarization-induced Ca^{++} uptake in the shorter incubation. This suggests a different mechanism is involved in this than in the non-depolarized and longer term Ca^{++} uptake. U-50488H was similar to verapamil in attenuating the rapid Ca^{++} uptake by depolarized synaptosomes only. This effect is further

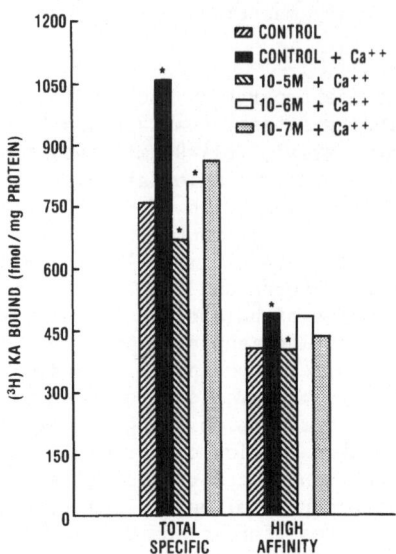

FIG. 1: Effect of U-50488H on (^{3}H)-kainic acid binding in the presence of
CaCl$_2$. Total specific (displaced by 55 minute incubation with
0.1 mM kainic acid) and high affinity (displaced by 5 minute
incubation with 0.1 mM kainic acid binding) were assessed in the
presence of various concentrations (10^{-5} to 10^{-7}M) of U-50488H
and/or 2.5 mM CaCl$_2$. Each assay was performed in triplicate.
* $p<0.05$ compared to the appropriate control, Student's t-test.

elucidated by the data presented in Table 2. U-50488H causes a concentration-related decrease in Ca^{++} uptake, and the magnitude of the effect compares favorably with that induced by similar concentrations of both verapamil and nifedipine. Interestingly, the narcotic antagonist naloxone
(10^{-5}M) causes a similar effect. This complicates efforts to demonstrate
that the U-50488H effect is related to an interaction at opioid receptors
(blocked by naloxone). A lower concentration (10^{-7}M) of naloxone that does
not directly alter Ca^{++} uptake does not block the U-50488H induced effect.

^{3}H-Kainic acid binding was studied in washed synaptic membranes. Preliminary kinetic determinations revealed a Kd of 21 nM and a Bmax of 1.3 pM/
mg protein for this ligand. Addition of 2.5 mM CaCl$_2$ led to a significant
enhancement of both total specific and high affinity binding (Figure 1).
U-50488H at concentrations that blocked Ca^{++} uptake (10^{-5} and 10^{-6}M) appears
to block the increment increase in binding induced by CaCl$_2$. Similar concentrations of U-50488H did not alter the binding of kainic acid in the
absence of CaCl$_2$ (data not shown). In the presence of CaCl$_2$, the ability
of U-50488H to block ^{3}H-kainic acid binding is more pronounced in the total
specific binding than in the high affinity binding.

The foregoing in vitro experiments suggest mechanisms whereby U-50488H
might display effects on depolarization related Ca^{++} metabolism. Some sub-
stantiation of the pharmacological significance of these effects is presented in Table 3. U-50488H is unique among the compounds tested in the
ability to block the seizures induced by the intravenous administration of
kainic acid. Standard anticonvulsants (phenytoin and phenobarbital), calcium antagonists (verapamil and bepridil), as well as less selective opioids (bremazocine and tifluadom) were not similarly active. Likewise, when
kainic acid was administered directly into the left lateral cerebral ventricle, U-50488H was unique in the ability to block the ensuing convulsion.

TABLE 3: Effect of U-50488H and various anticonvulsants, non-selective kappa agonists, and calcium antagonists on clonic seizures induced by the intravenous or intracerebroventricular administration of kainic acid to mice.

A. Intravenous Kainic Acid

Treatment	Anticonvulsant ED 50 mg/kg s.c. (95% CI)	
U-50488H	4.5	(3.1-6.4)
Phenobarbital Na	> 100	
Phenytoin Na	> 200	
Bremazocine	> 25	
Tifluadom HCl	> 50	
Bepridil HCl	> 10	
Verapamil HCl	> 10	
U-50488H + Saline	3.8	(2.7-5.4)
U-50488H + Naltrexone HCl (3 mg/kg)	14	(10-21)

B. Intracerebroventricular Kainic Acid

Treatment	Anticonvulsant ED 50 mg/kg s.c. (95% CI)	
U-50488H	28	(18-45)
Phenobarbital Na	> 100	
Phenytoin Na	> 100	
Bepridil HCl	> 10	
Verapamil HCl	> 10	
U-50488H + Saline	28	(19-41)
U-50488H + Naltrexone HCl (3 mg/kg)	56	(38-82)

In this case, the ED50 was somewhat higher than for the antagonism of intravenously administered kainic acid, but again, the standard anticonvulsants and calcium antagonist were ineffective in blocking these convulsions. In both of these models, the anticonvulsant property of U-50488H is significantly attenuated by treatment with naltrexone. The shift in the ED50 induced by the narcotic antagonist in the case of the intravenous kainic acid model is similar to the shift in the analgesic potency of U-50488H induced by the same dose of naltrexone (data not shown).

In a direct calcium-mediated seizure model (Table 4), U-50488H is also an effective anticonvulsant; convulsions induced by the calcium channel agonist Bay K 8644 are blocked by the kappa agonist. Again, as with the other convulsant assays, verapamil, phenytoin, and phenobarbital were ineffective. Likewise, the effects of U-50488H were also antagonized by naltrexone in this assay. Naltrexone was not proconvulsant at these doses in these assays; 3 and 10 mg/kg did not shift the convulsant dose 50 for either kainic acid or Bay K 8644 (data not shown).

DISCUSSION

The foregoing experiments indicate that U-50488H interacts with calcium. Most directly, it blocks the uptake of $^{45}Ca^{++}$ by depolarized synaptosomes and blocks the convulsions induced by the calcium channel agonist, Bay K 8644. U-50488H also blocks $CaCl_2$ activation of kainic acid binding and the convulsant action of this excitatory amino acid analog.

Whether these interactions are relevant to the kappa agonist properties of this compound is not entirely clear. For example, the ability of naloxone to block the effect of U-50488H on $^{45}Ca^{++}$ uptake could not be assessed due to the intrinsic activity of the antagonist on this parameter. On the other hand, the antagonism of kainic acid and Bay K 8644 induced convulsions by U-50488H is blocked by naltrexone. The shift in the anticonvulsant ED50 of U-50488H is similar to the shift in the analgesic ED50. This suggests that receptors involved in each of these effects have similar in vivo affinities for the opioid antagonist. Thus, both effects may be mediated by kappa receptors. However, other less selective kappa agonists do not share the anticonvulsant properties of U-50488H, despite exerting analgesic effects in part through the kappa receptor. Perhaps the anticonvulsant effects of U-50488H are mediated by a subtype of kappa receptor that is not insensitive to bremazocine and tifluadom.

The ability of U-50488H to block $^{45}Ca^{++}$ uptake into depolarized synaptosomes could be interpreted to indicate a similar mechanism as the calcium antagonists. Indeed, verapamil and nifedipine are similarly active. However, the calcium antagonists did not exert anticonvulsant activities at the non-toxic doses tested. Parenthetically, an earlier report (Shelton et al., 1985) indicated that Bay K 8644-induced seizures were blocked by verapamil, but the dose reported (50 mg/kg) was lethal in our test animals. More refined studies will be necessary to determine if the changes in Ca^{++} flux induced by these compounds differs from that by U-50488H, and if the latter is in fact mediated by an opioid receptor-controlled Ca^{++} channel. Such a mechanism can only be considered as hypothetical at this point.

U-50488H is an effective anticonvulsant against electroshock-induced seizures in the rat (Tortella et al., 1984). This anticonvulsant property

TABLE 4: Effect of U-50488H on seizures induced by intracerebroventricular administration of the calcium channel agonist Bay K 8644

Treatment	Anticonvulsant ED50 ED50 mg/kg s.c. (95% CI)	
U-50488H	45	(29-69)
Phenytoin Na	> 200	
Verapamil HCl	> 10	
U-50488H + Naltrexone HCl (3mg/kg)	71	(41-106)
U-50488H + Naltrexone HCl (10 mg/kg)	100	(64-156)

is also sensitive to opioid antagonists. The ability of U-50488H to alter kainic acid binding may be involved in this general anticonvulsant property, as well as in the ability of U-50488H to block kainic acid-induced seizures. Again, however, the underlaying mechanism may involve Ca^{++}; it is the $CaCl_2$-stimulated binding of ^3H-kainic acid that is blocked. More detailed kinetic studies of the interaction of U-50488H, Ca^{++}, and ^3H-kainic acid binding may help to clarify this relationship. The involvement of opioid receptors (as defined by naloxone antagonism) in this interaction also remains to be studied. A common Ca^{++}-related mechanism effecting both the transmembrane flux of Ca^{++} and the ability of Ca^{++} to alter membrane binding of excitatory transmitters may be involved.

U-50488H has recently been reported to have significant cerebroprotective properties in cerebral ischemia (Tang, 1985). Such ischemia results in Ca^{++} accumulation (Meldrum et al., 1985) and greatly elevated Ca^{++}-dependent release of glutamic acid (Drejer et al., 1985). This latter event may be causally related to the ischemia-induced neuronal degeneration as excitatory amino acid antagonists block ischemia-induced cellular damage (Meldrum et al., 1985; Simon et al., 1984). In light of the effects of U-50488H on Ca^{++} influx and excitatory amino acids reported here, it seems likely that the cerebroprotective effect of this compound is mediated by these mechanisms. However, U-50488H is also a potent water diuretic (Von-Voigtlander et al., 1983), and this mechanism appears to play a major role in the cerebroprotective actions of the compound (A. H. Tang, personal communication).

Although kainic acid has been used in both the binding and convulsant studies reported here, the interactions with U-50488H may not be specifically at this subtype of excitatory amino acid receptor or related effector. U-50488H also blocks quisqualic and N-methyl-asparatic acid-induced seizures (Lewis and VonVoigtlander, unpublished). Likewise, in the binding studies, the enhancement of ^3H-kainic acid binding may, in fact, represent binding at other excitatory amino acid sites. In this regard, $CaCl_2$ enhancement of binding has been associated primarily with the quisqulate receptor subtype (Mena et al., 1984; Greenmyre et al., 1983). Likewise, quisqualic acid is a more effective stimulant of polyphosphoinositide hydrolysis than is kainic acid (Sladeczek et al., 1985; Nicoletti et al., 1986). This reaction, which is catalyzed by phospholipase C and yields inositol phosphates and diacylglycerol, is a potential candidate for the mediation of the U-50488H interactions with Ca^{++} described here. This notion is strengthened by the observation that depolarization-induced Ca^{++} influx is controlled by the phospholipase C cascade (Harris et al., 1983). Further, it is possible that the ability of ionic (Ca^{++} and Cl^-) binding sites to modify kainate (or quisqulate) binding is in turn controlled by membrane phosphorylation via diacylglycerol stimulation of protein kinase C. This speculation is amenable to direct experimental investigation.

Thus, the studies we have done to date lead us to hypothesize that there is a subclass of kappa opioid receptor in brain that indirectly controls several Ca^{++}-mediated mechanisms through the phospholipase C cascade. This class of Ca^{++}-linked receptor may mediate the anticonvulsant, but not the analgesic, properties of the kappa agonist, U-50488H. We are continuing to direct experiments at the testing of this hypothesis.

ACKNOWLEDGMENTS

The authors express their appreciation to S. J. Crowder for secretarial assistance, and the respective pharmaceutical manufacturers for the donation of the following compounds: Bay K 8644 and nifedipine (Miles Laboratories), bepridil HCl (Norwich Laboratories), bremazocine and tifluadom (Sandoz), naloxone HCl and naltrexone HCl (Dupont).

REFERENCES

Cherubini, E. and North, R. A., Mu and kappa opioids inhibit transmitter release by different mechanisms, Proc. Nat. Acad. Sci. 82:1860-1863 (1985).

Drejer, J., Benveniste, H., Diemer, N. H. and Schousboe, A., Cellular origin of ischemia-induced glutamate release from brain tissue in vivo and in vitro, J. Neurochem. 45:145-151 (1985).

Greenmyre, J. T., Young, A. B. and Penny, J. B., Quantitative autoradiography of L-glutamate binding to rat brain, Neurosci. Lett. 37:155-160 (1983).

Gripenberg, J., Heinonen, E. and Jansson, S. E.,Uptake of radiocalcium by nerve endings isolated from rat brain: Pharmacological studies, Br. J. Pharmac. 71:273-278 (1980).

Harris, R. A., Fenner, D. and Leslie, S. W., Calcium uptake by isolated nerve endings: Evidence for a rapid component mediated by the breakdown of phosphatidylinositol, Life Sci. 32:2661-2666 (1983).

James, I. F. and Goldstein, A., Site-directed alkylation of multiple opioid receptors. I. Binding selectivity, Mol. Pharmacol. 25:337-342 (1984).

Lahti, R. A., VonVoigtlander, P. F. and Barsuhn, C., Properties of a selective kappa agonist, U-50488H, Life Sci. 31:2257-2260 (1982).

Lahti, R. A., Mickelson, M. M., McCall, J. M. and VonVoigtlander, P. F., (3H)-U-69593, A highly selective ligand for the K receptor, European J. Pharmacol. 109:281-284 (1985).

London, E. D. and Coyle, J. T., Specific binding of (3H)-kainic acid to receptor sites in rat brain, Mol. Pharmacol. 15:492-505 (1979).

Martin, W. R., Eades, C. G., Thompson, J. A., Huppler, R. E. and Gilbert, P. E., The effects of morphine- and nalorphine-like drugs in the non-dependent and morphine-dependent chronic spinal dog, J. Pharmacol. Exp. Ther. 197:517-532 (1976).

Meldrum, B., Evans, M., Griffiths, T. and Simon, R., Ischemic brain damage: The role of excitatory activity and of calcium entry, Br. J. Anaesth. 57:44-46 (1985).

Mena, E. E., Whittemore, S. R., Monaghan, D. T. and Cotman, C. W., Ionic Regulation of glutamate binding sites, Life Sciences 35:2427-2433 (1984).

Nicoletti, F., Meek, J. L., Iadarola, M. J., Chuang, D. M., Roth, B. L. and Costa, E., Coupling of inositol phospholipid metabolism with excitatory amino acid recognition sites in rat hippocampus, J. Neurochem. 46:40-46 (1986).

Rhoads, D. L., Yamasaki, Y. and Way, E. L., Opioids reduce human red blood cell deformability, Alcohol and Drug Res. 6:229 (1985).

Schramm, M., Thomas, G., Toward, R. and Franckowiak, G., Novel dihydropyridines with positive inotropic action through activation of Ca^{++} channels, Nature 303:535-537 (1983).

Shelton, R. C., Grebbe, J. A. and Freed, W. J., Calcium channel agonist-induced murine seizures, Soc. Neurosci. Abstr. 11:924 (1985).

Simon, J. R., Contrera, J. F. and Kuhar, M. J., Binding of (3H)-kainic acid, an analogue of L-glutamic acid, to brain membranes, J. Neurochem. 26: 141-147 (1976).

Simon, R., Swan, J. H., Griffiths, T. and Meldrum, B., Blockade of N-methyl-D-asparate receptors may protect against ischemic damage in the brain, Science 226:850-852 (1984).

Sladeczek, F., Pin, J.-P, Recasens, M., Bockaert, J. and Weiss, S., Glutamate stimulates inositol phosphate formation in striatal neurones, Nature 317:717-719 (1985).

Szmuszkovicz, J and VonVoigtlander, P. F., Benzeneacetamide amines: Structurally novel non-mu opioids, J. Med. Chem. 25:1125-1126 (1982).

Tang, A. H., Protection from cerebral ischemia by U-50488H, a specific kappa opioid analgesic agent, Life Sci. 16:1475-1482 (1985).

Tortella, F. C., Robles, L. and Holaday, J. W., Seizure-specific, dose- and time-dependent anticonvulsant profile for U-50488H, a novel kappa opioid agonist in rats, Soc. Neuro. Abstr. 10:408 (1984).

VonVoigtlander, P. F. and Lewis, R. A., U-50488H, a selective kappa opioid agonist: Comparison to other reputed kappa agonists, Prog. Neuro-Psychopharmacol. & Biol. Psychiat. 6:467-470 (1982).

VonVoigtlander, P. F., Lahti, R. A. and Ludens, J. H., U-50488H: A selective and structurally novel non-mu (kappa) opioid agonist, J. Pharmacol. Exp. Ther. 224:7-12 (1983).

VonVoigtlander, P. F., Lewis, R. A. and Neff, G. L., Kappa opioid analgesia is dependent on serotonergic mechanisms, J. Pharmacol. Exp. Ther. 231:270-274 (1984).

Werz, M. A. and MacDonald, R. L., Dynorphin and neoendorphin peptides decrease dorsal root ganglion neuron calcium-dependent action potential duration, J. Pharmacol. Exp. Ther. 234:49-56 (1985).

Yamasaki, Y. and Way, E. L., Possible inhibition of calcium pump of rat erythrocyte ghosts by K agonists, Life Sci. 33:723-726 (1983).

LONG-TERM INHIBITION OF KINDLED SEIZURES BY CHEMICAL AND

ELECTROPHYSIOLOGICAL TECHNIQUES: INSIGHTS INTO THE

KINDLING PROCESS?

H. A. Robertson and G. A. Cottrell

Department of Pharmacology, Faculty of Medicine
Dalhousie University, Halifax, Nova Scotia, Canada
B3H 4H7

The kindling phenomenon is frequently called a model of epilepsy but is more correctly a model of epileptogenesis. It is also a model for neuronal plasticity. For almost 20 years now, ever increasing numbers of neuroscientists have been fascinated by the fact that daily administration of a mild electrical stimulus which is initially without effect will lead to a progressive intensification of response culminating in a seizure. The importance of this phenomenon was first recognized by Goddard (1967) and the study of kindling by Goddard et al. (1969) remains the pivotal point in this area. Goddard (1967) used the term kindling as an analogy with the kindling or lighting of a fire. Strictly speaking, the kindling of a fire is not a good analogy for the process. However, the term now has the advantage of familiarity and certainly no other commonplace occurance suggests a closer analogy.

It is important to be precise in defining kindling. The term has sometimes been poorly defined. For example, Peterson and Albertson (1982) begin their review by describing kindling as "the emergence of progressively increasing epileptiform response to repeated convulsive stimuli." However, one important point about kindling is that the stimuli are not initially convulsive and when they are there is reason to believe that it is not kindling. It is also very important to note that in kindling the interstimulus interval plays an important role. Thus "repeated convulsive stimuli" may not in fact lead to kindling but may lead in some circumstances to inhibition of kindled seizures. The term "repeated" does not convey enough information. A more precise definition of kindling would be the phenomenon whereby repeated administration, to certain brain regions, at intervals greater than 1 - 2 hours, of an initially subconvulsive stimulus which could be electrical, chemical or physiological, resulting in progressive and permanent intensifictaion of seizure activity culminating in a generalized seizure. A precise definition is necessary to reflect the unique aspects of kindling. Thus the above definition includes the idea that kindling is produced by stimuli delivered at spaced intervals, that it is a permanent change, that the intensity of the stimuli is low and does not change while the response intensifies and that only certain brain regions support kindling.

Another important point to make as part of the definition of kindling is that kindling is not necessarily the same process as the triggering of a kindled seizure. Kindling refers to the process of development of supersensitivity to a stimulus of fixed intensity. It begins with a naive animal and ends with an animal which, when stimulated, has a generalized seizure and is therefore described as kindled. The seizure which the animal experiences is a kindled seziure but we have no reason for believing that this kindled seizure is qualitatively different from a seizure produced by an electric shock or an injection of pentylene tetrazole. Thus kindled seizures per se are probably of limited scientific interest while the process of kindling appears to be an important and unique window into the brain.

Classically, that is as first described by Goddard and his colleagues (1967, 1969), kindling is produced by delivering a daily electrical stimulus to a region of the brain, usually in the limbic system. The properties of the electrical stimulation are an important determinant of the development of the kindling. A typical paradigm consists of a daily administration of a 1-2 sec train of 1 msec biphasic square wave pulses at a frequency of 60 Hz via bipolar stimulating electrodes aimed at the amygdala. Stimulation with frequencies of 25, 60 and 150 Hz are equally effective in kindling but 60 Hz stimulation is most effective in eliciting a seizure from a kindled animal (Goddard, 1967). Generally, for kindling to occur the intensity of the stimulus must be sufficient to generate an afterdischarge at the site of stimulation. Usually a stimulator with a constant current controller is used, thus ensuring a constant level of stimulus. Current levels between 50 uA and 10 mA have been used to induce kindling and it is surprising that the rate of kindling is not dependent on the intensity of stimulation, at least when sine wave stimulation is used (Goddard et al., 1969). The important point appears to be the generation of an after-discharge, not the intensity of stimulation. In other words, an animal will kindle as fast with a stimulus of 100 uA as with a stimulus of 1 mA, so long as the 100 uA current induces an after-discharge. The interval between stimulations is a key variable in kindling; Goddard (1967) first observed kindling because he was stimulating the amygdala in rats at 24 hour intervals. In fact, if the interstimulus interval is less than 15-20 min, kindling will not develop (Goddard et al., 1969). A 1 sec stimulation of 1 msec pulses delivered at intervals ranging from 15 min to 7 days can produce kindled seizures (Goddard et al., 1969, Racine et al., 1973) but greater numbers of stimulations are required with inter-stimulus intervals of less than 1 hour (Goddard et al., 1969; Racine et al., 1973).

On the first stimulus, the animal will perhaps pause during the 1-2 sec stimulation and an EEG recorded from the amygdala will reveal a brief after-discharge lasting for perhaps 2-5 sec beyond the stimulus. At this stage there is generally no overt seizure activity. The stimulation must induce an after-discharge for kindling to occur. However, stimulation which is below the threshold for elicting an after-discharge also lowers the after-discharge threshold to the point where after-discharges and subsequently kindling occur (Racine, 1972a). With each daily stimulus the after-discharge duration increases in parallel with the appearance of a number of stages of seizure activity, depending on the area stimulated. With amygdala stimulation, rats generally pass through the following stages: 1) facial clonus; 2) head nodding; 3) forelimb clonus; 4) rearing and 5) rearing and falling (Racine, 1972b). These behavioral stages differ somewhat depending on the brain region stimulated. For example, in hippocampal kindling, stage 5 is better characterized by hindlimb clonus (GAC, unpublished observations).

Several recent reviews have discussed recent progress in kindling work in general (Peterson and Albertson, 1982; Goddard, 1983; McNamara, 1984). An older but comprehensive view of the phenomenon of kindling is given by Racine (1978). In this chapter, we are going to focus attention on long-term inhibition of kindled seizures, a little studied aspect of kindling. Of course "long term" inhibition of kindled seizures can be achieved by pharmacological means, using a benzodiazepine for example. Long-term inhibition as we will use the term refers to a long-lasting inhibition (1 to 10 or more days) resulting from a single intervention. There are reasons to believe that an understanding of the biological basis of long-term inhibition of kindled seizure will not only suggest mechanisms for controlling seizures but will cast some light on the process of kindling itself. Two types of post-stimulation seizure inhibition are seen; the first is short-term inhibition of seizures, usually lasting about 90 min. The second type, long-term inhibition, typically lasts for 1-10 days. One interesting feature of long-term inhibition is that it takes time (usually about 24 hours) to develop. The relationship between the two types of inhibition is not known but there are suggestions that they result from the same processes. Discussion of long-term inhibition must therefore include consideration of short-term inhibition of seizure activity. Both short- and long-term inhibition of seizures probably involve events occurring in the immediate (1-4 hr) post-seizure (IPP). Part of our interest in the subject arises from our view that events in the IPP are central to the development of kindling. This stems from one of the earliest observations in the kindling field, that to elicit kindling, stimuli must be spaced (Goddard et al., 1969). If stimuli must be spaced, it therefore follows that some event in the IPP is essential for the development of the kindling effect. There is a variety of experimental and even clinical evidence for this point. If we interfere in this IPP, we often alter the course of kindling development or alter the kindled seizure. It is therefore important to understand the processes occurring in the IPP.

Goddard et al. (1969) and Racine et al. (1972a,b) showed that amygdala kindling only occurred with interstimulus intervals of 20 min or greater with longer intervals (24 hr) being optimal. In other words, if stimulation is more frequent than every 15-20 minutes, the development of kindling is retarded or halted. Kindled seizures (and seizures in general) have anticonvulsant properties. Mucha and Pinel (1977) demonstrated that a single amygdala-kindled seizure produced an inhibition of seizures which dissipated over 90 min. Stimulation which did not produce an afterdischarge or a motor seizure resulted in no inhibition. Mucha and Pinel (1977) also made the important but still little-understood observation that this inhibitory effect of a single kindled seizure accumulates. One stimulation led to a 90 minute period of inhibition and 19 stimulations at 90 min intervals led to a 5 day period during which seizures were inhibited and after-discharge duration was attenuated. In fact, the attenuation in the afterdischarge duration could be detected for up to 10 days. With inhibitory effects of this duration, we must seriously consider whether we are seeing the return of the kindling effect or whether this is de novo kindling. Amygdala kindled seizures develop after between 5 and 15 daily stimulations but there is a pronounced effect of a primary kindling site on the rate of kindling from a secondary site (Racine, 1972b; Burnham, 1975). The second site usually requires fewer stimulations. Thus, it is possible that the effect of the 19 stimulations is to eliminate the primary site and the 5 days of stimulation is merely creating a "secondary" site, albeit in the primary location. This transfer effect of a primary kindling site on kindling from subsequent sites is possibly important in gaining an insight into the process of long-term inhibition of kindled seizures. Like long-term inhibition, this topic is also complicated. Transfer occurs between all limbic (amygdala, septal area and hippocampus) sites and Racine (1972b) and Burnham (1975) have suggested that the secondary sites are kindled by the propagated discharge from the primary sites. Thus, if its takes 10 stimulations to kindle the left amygdala and 5 of these stimulations are propagated to the right amygdala, it will take about 5 stimulations to the right amygdala to elicit a seizure. In addition, to the effect of the primary site on subsequent sites, there is an effect of secondary sites on the primary site. Goddard et al., (1969) and Burnham (1975) both found that the primary amygdala site did not rekindle on the first stimulation after secondary kindling of the contralateral amygdala. This negative transfer effect was found to work in both directions. First trial convulsions seldom occurred unless a 2 week rest period was interposed after the primary kindling suggesting that there is a slowly decaying inhibition of the contralateral amygdala. This slow-decaying inhibition probably does not originate from the most recently kindled amygdala as destruction of this amygdala does not eliminate the effect (McIntyre and Goddard, 1973). It was suggested that brain stem sites might be responsible for this effect. The reason for this suggestion was that the brain stem is clearly activated during convulsions and significantly, stimulation in the brain stem or cerebellum is

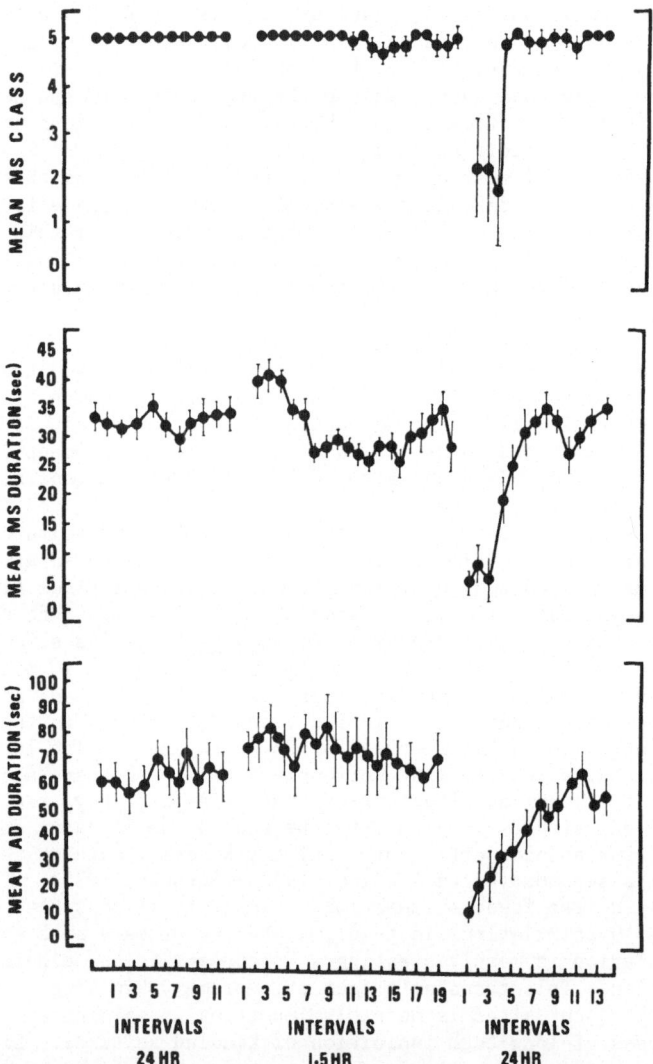

FIG. 1: Long-term inhibition of kindled seizures following multiple seizures. Mean after-discharge (AD) duration (bottom panel), mean motor seizure (MS) duration (middle panel), mean motor seizure class (top panel), elicited in kindled rats by the 11 daily baseline stimulations, the 19 1.5-hr interval stimulations, and the 13 daily test stimulations (S.E.M.). From Mucha and Pinel (1977).

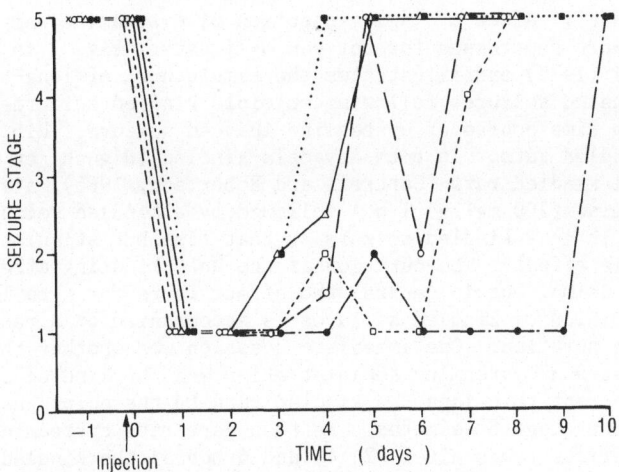

FIG. 2: The effect of cysteamine on kindled seizures. Data is presented
for individual hippocampal kindled rats over a 12 day period. All
animals were having consistent stage 5 seizures for at least 3 days
before the cysteamine injections. Day 0 is the day of cysteamine
injection. The ordinate represents the stage of the seizure modified after Racine (1972b).

inhibitory to kindled seizures (Goddard, pers. comm.; Robertson and Riives,
unpublished observations). However, it has been shown that sectioning the
forebrain commissures blocked most of this inhibitory effect of a second
site on the primary site (McIntyre, 1975; McCaughran et al., 1977). This
suggests inhibition from an ipsilateral site other than the amygdala. A
good candidate for this site might be that prepyriform area identified by
Piredda and Gale (1985) as an important site for inhibition of generalized
seizures. However, there are other possibilities. For example, it has
been shown that the substantia nigra is crucial to the development of kindled seizures (McNamara et al., 1983) and lesions of the substantia nigra
prevent experimentally induced seizures in general (Garant and Gale, 1983).
In addition to this anticonvulsant effect mediated via the substantia nigra,
there is also evidence that GABA terminals in the substantia innominata are
involved in an inhibitory action on seizure generalization of amygdala origin (Morita et al., 1985).

Long-term inhibition of kindled seizures following stimulation has
been observed after multiple kindled seizures (Mucha and Pinel, 1977;
Sainsbury et al., 1978) and after electroconvulsive shock (ECS) (Shao and
Valenstein, 1982; Handforth, 1982). Post-seizure inhibition has also been
noted after seizures induced by essence of absinthe (Elsberg and Stookey,
1923), by ECS in man (Kalinowsky and Kennedy, 1943) and cats (Essig et
al., 1963), and with seizure induced by hypothalamic stimulation (Herberg
and Watkins, 1966). Seizures induced by septal stimulation inhibit the
production of seizures by stimulation of the hypothalamus and vice versa
and seizures elicited by stimulation of either septum or hypothalamus are
blocked by a preceeding audiogenic seizure (Herberg et al., 1969). From
this work, it is evident that a seizure of one type can have an inhibitory
effect on another type of seizure, implying the existence of common pathways or inhibitory mechanisms.

Recently, Higuichi et al., (1983) demonstrated long-term inhibiton of
amygdala kindled seizures with injections of cysteamine (200 mg/kg, i.p.).
This long-term inhibition following cysteamine was similar in time course

to that seen with mutliple-kindled seizures (Mucha and Pinel, 1977; Sainsbury et al., 1978). After a single injection of cysteamine, amygdala-kindled seizures were suppressed for between 4-11 days. Fig. 1 is taken from
Mucha and Pinel (1977) and illustrates the time course of long-term suppression of kindled seizures following multiple-kindled seizures. Fig. 2
illustrates the time course of cysteamine-induced seizure inhibition in
hippocampal kindled rats. In both amygdala-kindled (Higuchi et al., 1983)
and hippocampal-kindled rats (Cottrell and Robertson, 1986), a single injection of cysteamine (200 mg/kg, i.p.) followed by a kindled seizure 4 hours
later will result in a kindled seizure at that time but stimulation 1 day
later has little effect. The duration of the anti-kindling effect varies
from animal to animal but in general the effect lasts for 4 to 11 days. The
long-term inhibition of kindled seizures is accompanied by a reduction of
after-discharge duration. One immediate question was whether this temporal
relationship between cysteamine administration and the kindled seizure was
obligatory. To test this idea, we kindled rats to the point that they were
having consistent stage 5 seizures. We then gave rats cysteamine (200 mg/
kg, i.p.) at various times after (2, 4, and 6 hours) the kindled seizure.
The animals were then tested at 24 hour intervals. The results are summarized in fig. 3. Cysteamine administered after the kindling stimulus still
produces the anti-kindling effect but it is correspondingly weaker and falls
off with as the kindling-to-cysteamine lengthens. This suggest that either
the cysteamine, a metabolite of the cysteamine or an effect of the cysteamine must be present at the time or close to the time of the kindling stimulus for long-term inhibition of kindled seizures to occur. One possibility is that this event associated with cysteamine which is responsible for
the long-term inhibition is the myoclonic seizure.

Higuchi et al. (1983) attributed this long-term inhibition of kindled
seizures to the ability of cysteamine to deplete brain somatostatin. Intra-
cerebral injections of antibodies to somatostatin had similar effects, presumably by preventing the actions of somatostatin. However, Cottrell and
Robertson (1986) have suggested that the long-term inhibition of kindled
seizures by cysteamine is not the result of somatostatin depletion. After
cysteamine treatment somatostatin levels decline monotonically while two

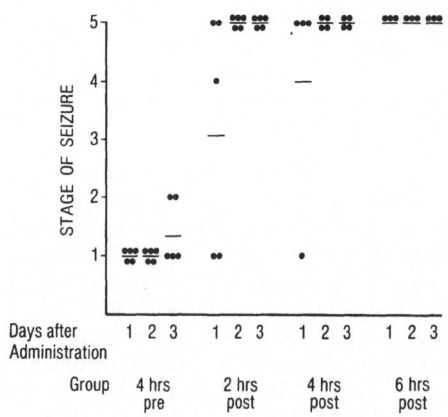

FIG. 3: The effect of injection-to-kindled seizure interval on long-term
inhibition of kindled seizures by cysteamine. The mean (-) and
individual (o) seizure stage is presented for each group 1, 2 and
3 days after injection of 200 mg/kg i.p. cysteamine. On the test
day, the four groups were injected with cysteamine 4 hr before, 2
hr after, 4 hr after or 6 hr after the kindling stimulation. All
rats were exhibiting reliable stage 5 seizures before administration of the drug.

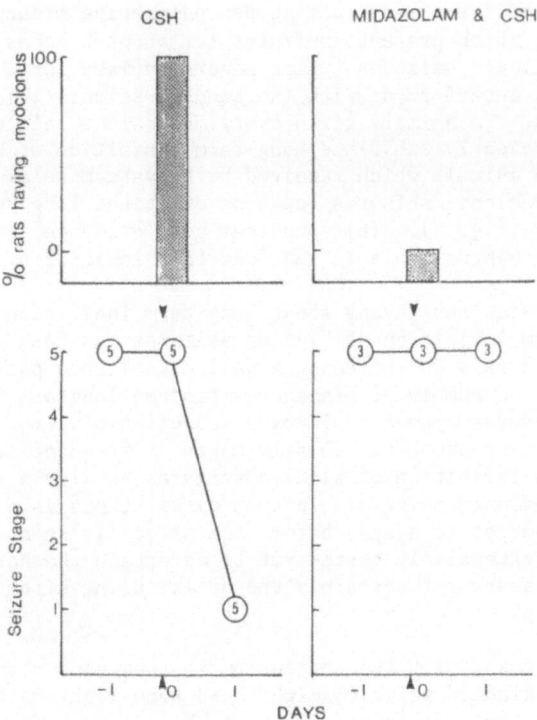

FIG. 4: Hippocampal Kindled Rats. Elimination of myoclonic seizure prevents the long-term inhibition after cysteamine. The upper panel
shows the percentage of rats having myoclonic seizures. Cysteamine (200 mg/kg, i.p.) was given to 2 groups of rats of hippocampal
rats, one of which was given saline and the other midazolam (5 mg/
kg, i.p.) 10 minutes before the cysteamine. The lower panel shows
the stage of motor seizure exhibited by these rats. The number of
rats per group is shown in the circles.

effects are seen on convulsive behavior. First, all kindled animals, both
amygdala-kindled and hippocampal-kindled, have severe myoclonic seizures
during the first 90 min after cysteamine (Cottrell and Robertson, 1986).
Naive animals seldom or never have such myoclonic seizures. Thus, kindled
animals respond differently to cysteamine injections. This may prove to be
a useful animal model for myoclonic epilepsy (Cottrell and Robertson, in
preparation). In addition to the time course of somatostatin differing from
the time course for the anti-kindling effect, there are a variety of other
reasons for rejecting the idea that cysteamine is acting via somatostatin
depletion. Cysteamine is not a very clean pharmacological tool; in addition to its actions on somatostatin it affects a number of other peptides
and also alters catecholamine and serotonin levels in brain. Cysteamine,
at the concentrations used to inhibit seizures, also affects calcium-dependent binding of [3H]-glutamate to brain membranes (Robertson et al., 1985).
Thus, it seems unlikely that somatostatin depletion alone is responsible for
the long-term inhibition.

One possibility raised earlier is that the myoclonic seizures produced
by the cysteamine uniquely in kindled animals leads to the long-term inhibition. Naive animals also have myoclonic seizures following cysteamine
but doses much higher than 200 mg/kg are necessary to produce them. To

test whether these myoclonic seizures produce the long term inhibition, we have taken advantage of the short-acting benzodiazepine midazolam. Use of this benzodiazepine which prevents seizures for about 2 hours permitted us to prevent the myoclonic seizures (which generally last for a maximum of 90 minutes) without interferring with the kindled seizure which we administered at 4 hours. In animals given cysteamine alone, all experienced myoclonus and subsequently exhibited long-term inhibition of kindled seizures. However, in animals which received both cysteamine and midazolam, none experienced myoclonic seizures and none exhibited long term inhibition of kindled seizures (fig. 4). This provides good evidence that the myoclonic seizure plays a central role in the long term inhibition.

In conclusion what can we say about long-term inhibition of kindled seizures? Long term inhibition of kindled seizures has been observed consistently and can be said to represent a well-established part of the kindling phenomena. A number of procedures produce longterm inhibition but all these procedures appear to involve induction of severe seizure activity of one sort or another. Thus 19 hours of frequent seizures will lead to a prolonged inhibition of kindled seizures as will a severe cysteamine-induced myoclonic seizure. In both cases, there is a requirement of a 24 hour rest period to elapse before the effect is seen. This latter point has not been extensively tested yet to ascertain whether a full 24 hour period is necessary but certainly the antikindling effect takes several hours to develop.

We began this essay with the contention that an understanding of long-term inhibition of kindled seizures might shed some light on the process of kindling. At this point, we cannot say that this is so. The mechanisms behind long-term inhibition of kindled seizures are currently as obscure as the kindling phenomenon itself. It appears reasonable that it is seizure activity itself that produces long-term inhibition of kindled seizures. It will be important next to identify the brain areas that are responsible for the long-term inhibitory effects. There are a number of candidates for such an area: the substantia nigra, the substantia innominata, the prepyriform area and even the amygdala itself is a suspect area.

ACKNOWLEDGMENTS

Supported by the Medical Research Council of Canada. G. A. Cottrell is a Fellow of the Dalhousie Medical Research Foundation.

REFERENCES

Burnham, W. M., Primary and "transfer" seizure development in the kindled rat, Can. J. Neurol. Sci., 2:417-428 (1975).

Cottrell, G. A. and Robertson, H. A., Induction and suppression of seizures by cysteamine in hippocampal kindled rats, Brain Res., 365:393-396 (1986).

Elsberg, C. A. and Stookey, B. P., Studies on epilepsy. I. Convulsions experimentally produced in animals compared with convulsive states in man, Arch. Neurol. Psychiat., 9:613-626 (1923).

Essig, C. F., Groce, M. E. and Williamson, E. L., Electroconvulsive threshold elevation from daily stimulation of adrenalectomized animals, Science, 140:828-829 (1963).

Garant, D. S. and Gale, K., Lesions of substantia nigra protect against experimentally induced seizures, Brain Res., 273:156-161 (1983).

Goddard, G. V., Development of epileptic seizures through brain stimulation at low intensity, Nature, 214:1020-1021 (1967).

Goddard, G. V., McIntyre, D. C. and Leech, C. K., A permanent change in brain function resulting from daily electrical stimulation, Exp. Neurol., 25:295-330 (1969).

Goddard, G. V., The kindling model of epilepsy, Trends in Neurosci., 6:275-279 (1983).

Handforth, A., Postseizure inhibition of kindled seizures by electroconvulsive shock, Exp. Neurol., 78:483-491 (1982).

Herberg, L. J. and Watkins, P. J., Epileptiform seizures induced by hypothalamic stimulation in the rat: resistance to fits following fits, Nature, 209:515-516 (1966).

Herberg, L. J., Tress, K. H. and Blundell, J. E., Raising the threshold in experimental epilepsy by hypothalamic and septal stimulation and by audiogenic seizures, Brain, 92:313-328 (1969).

Higuchi, T., Sikand, G. S., Kato, N., Wada, J. A. and Friesen, H. G., Profound suppression of kindled seizures by cysteamine: possible role of somatostatin in kindled seizures, Brain Res., 288:359-362 (1983).

Kalinowsky, L. B. and Kennedy, F., Observations in electric shock therapy applied to problems of epilepsy, J. Nerv. Ment. Dis., 98:56-67 (1943).

McCaughran, J. A., Jr., Corcoran, M. E. and Wada, J. A., A facilitation of secondary-site amygdaloid kindling following bissection of the corpus callosum and hippocampal commissure in rats, Exp. Neurol., 57:132-141 (1977).

McIntyre, D. C. and Goddard, G. V., Transfer, interference and spontaneous recovery of convulsions kindled from the rat amygdala, Electroencephalogr. Clin. Neurophysiol., 35:533-543 (1973).

McIntrye, D. C., Split brain rat: transfer and interference of kindled amygdala convulsions, Can. J. Neurol. Sci., 2:429-437 (1975).

McNamara, J. O., Rigsbee, L. C. and Galloway, M. T., Evidence that substantia nigra is crucial to neural network of kindled seizures, Europ. J. Pharmacol., 86:485-486 (1983).

McNamara, J. O., Kindling: an animal model of complex partial epilepsy, Ann. Neurol., 16 (suppl):S72-S76 (1984).

Morita, K., Okamoto, M., Seki, K. and Wada, J. A., Suppression of amygdala-kindled seizure in cats by enhanced GABAergic transmission in the substantia innominata, Exp. Neurol., 89:225-236 (1985).

Mucha, R. F. and Pinel, J. P. J., Postseizure inhibition of kindled seizures, Exp. Neurol., 54:266-282 (1977).

Peterson, S. L. and Albertson, T. E., Neurotransmitter and neuromodulator function in the kindled seizure and state, Prog. Neurobiol., 19:237-270 (1982).

Piredda, S. and Gale, K., A crucial epileptogenic site in the deep prepiriform cortex, Nature, 317:623-625 (1985).

Racine, R. J., Modification of seizure activity by electrical stimulation: I. after-discharge threshold, Electroencephalogr. Clin. Neurophysiol., 32:269-279 (1972a).

Racine, R. J., Modification of seizure activity by electrical stimulation: II. motor seizure, Electroencephalogr. Clin. Neurophysiol., 32:281-284 (1972b).

Racine, R. J., Burnham, W. M., Gartner, J. G. and Levitan, D., Rates of motor seizure development in rats subjected to electrical brain stimulation: strain and interstimulation interval effects, Electroencephalogr. Clin. Neurophysiol., 35:553-556 (1973).

Racine, R. J., Kindling: the first decade, Neurosurgery, 3:234-252 (1978).

Robertson, H. A., Peterson, M. R. and Cottrell, G. A., Inhibition of Ca^{2+}-dependent ^3H-glutamate binding in vivo and in vitro: a possible mechanism for the cysteamine-induced suppression of kindling, Soc. Neurosci. Abstr., 11:283 (1985).

Sainsbury, R. S., Bland, B. H. and Buchan, D. H., Electrically induced seizure activity in the hippocampus: time course for post-seizure inhibition of subsequent kindled seizures, Behav. Biol., 22:479-488 (1978).

Shao, J. and Valenstein, E. S., Long-term inhibition of kindled seizures by brain stimulation, Exp. Neurol., 78:376-392 (1982).

ALTERED REACTIVITY OF THE RAT ADRENAL MEDULLA FOLLOWING

PERIODS OF CHRONIC STRESS

J. P. Mitchell and P. R. Vulliet

Department of Physiology, Colorado State University
Fort Collins, Colorado 80523

INTRODUCTION

The chromaffin cells of the adrenal medulla are derived from embryonic neural tissues; specifically, the neural crest cells. These cells share many biochemical and functional properties with adrenergic neurons and are a useful model for studying many of the biochemical processes that occur in adrenergic neurons. These cells are known to be responsive to a variety of agonists and antagonists that also modulate sympathetic neuronal activity, contain the proteins necessary for the synthesis and release of the catecholamine neurotransmitters, and respond to growth factors that are known to regulate adrenergic growth. This tissue also contains several putative neurotransmitters and neurally active peptides. The primary physiological function of the chromaffin cells is to release epinephrine and norepinephrine in response to acetylcholine released from the preganglionic neuron.

A variety of physiological perturbations, including insulin induced hypoglycemia (Weiner and Mosimann, 1970), immobilization (Kvetnansky et al., 1970), exposure to cold (Chuang and Costa, 1974) and drug induced hypotension (Theonen et al., 1969a), will result in characteristic alterations in the adrenal medulla. Long term stress is known to produce certain biochemical changes in the adrenal medulla, such as increased adrenal weight (Cannon, 1939), increased levels of tyrosine hydroxylase (TH) (Thoenen et al., 1969b; Kvetnansky et al., 1970), and increased catecholamine stores (Kvetnansky et al., 1985). Activation of existing TH molecules, presumably through a mechanism involving protein phosphorylation, occurs following short term insulin treatment (Fluharty et al., 1985) and treatments involving the production of physiological pain (Masserano and Weiner, 1981). Although much is known about the biochemical changes that occur in the adrenal, little data has been presented on the effects of these treatments upon the functional properties (i.e. release of catecholamines) of this gland.

Kvetnansky et al. (1985) demonstrated that rats subjected to daily periods of immobilization exhibited a significantly greater rise in plasma epinephrine and norepinephrine in response to a novel stress than previously unstressed animals. This finding suggests that adrenal responsivity may be altered by this previous experience. However, the experimental approach employed in these studies did not allow the identification of whether this change occurs at the level of the central nervous system, the sympathetic nerves innervating the adrenal medulla, or in the adrenal chromaffin cells themselves.

We have used the isolated perfused rat adrenal preparation as described by Wakade (1981) to investigate whether the responsivity of the adrenal medulla to acetylcholine is altered in response to chronic stress. Experimental animals were treated with immobilization stress, dexamethasone administration, or insulin administration over several days and the adrenal glands evaluated for alteration in response to a fixed concentration of acetylcholine. The relative density of these neuronally derived cells makes this preparation an ideal method for the study of the biochemical factors that result in alteration of neuronal responsivity in adrenergic neurons.

MATERIALS AND METHODS

Animals

Male Sprague-Dawley rats (250-350 g) were housed 2 per cage on a 12 hour light/dark cycle with food and water available ad libitum. Each cage contained one control and one treated rat. All control animals received a volume of 0.9% saline equal to the volume of drug administered to the test animal. All surgical manipulations were performed in rats anesthetized with sodium pentobarbital (60 mg/kg, I. P.)

Perfusion of the Rat Adrenal Gland

The left adrenal gland of male Sprague-Dawley rats was removed and perfused by a modification of the method of Wakade (1981). After ligating the branches of the adrenal vein, a polyethylene catheter was inserted into the adrenal vein and advanced to the level of the adrenal gland. Perfusion was initiated with Krebs-bicarbonate (KB) buffer using a syringe pump at a flow rate of 0.36 ml/min. A small slit was cut in the adrenal cortex to allow the perfusate to exit the adrenal gland. The adrenal was then placed on a plexiglas chamber maintained at 38°C. Adrenals were perfused with KB buffer for 30 minutes prior to experimental manipulation. Agonist mediated release was evaluated by injecting 50 ug of acetylcholine in a volume of 0.1 ml of KB buffer into the perfusion stream and collecting the perfusate for 5 minutes. The perfusate was analysed for catecholamines as described below.

Epinephrine Determination

Epinephrine content was measured in adrenal glands homogenized in 20 volumes of 0.1 M $HClO_4$ containing 40 ng/ml 3,4-dihydroxybenzylamine (DHBA) as an internal standard. The homogenate was diluted 1000 fold and passed through a 0.45 um nitrocellulose filter. Epinephrine secretion was measured by diluting adrenal perfusate in an equal volume of 0.1 M $HClO_4$ and employing an alumina extraction procedure to partially purify and concentrate catecholamines. Samples were analysed on a Gilson HPLC system equipped with a Rainin C_{18} reversed phase column equilibrated in 0.1 M citric acid, 3% acetonitrile, 0.5mM EDTA, 25 mg/liter sodium octyl sulfate, pH 3.0, coupled to a BAS LC 4B electrochemical detector.

Data Analysis

Statistical significance was determined by Student's t-test (2-tailed). The t-statistic for r, the linear correlation coefficient, was calculated by the formula:

$$t = \sqrt{vr^2/(1-r^2)}$$

where v is the degrees of freedom, n-2.

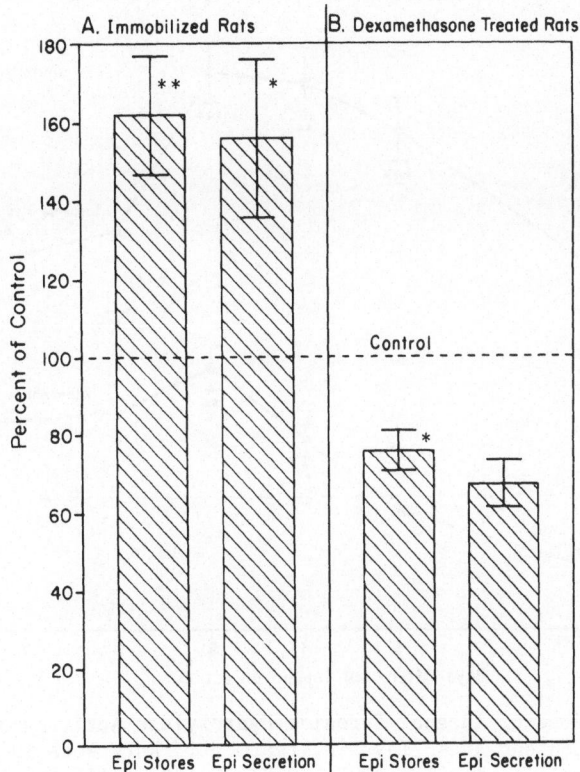

FIG. 1: Effect of Immobilization Stress and Dexamethasone Administration
on Adrenal Epinephrine Content and Acetylcholine Induced Epinephrine Secretion. A. Rats were immobilized daily for 2.5 hours/
day for 14 days in plexiglas cylinders stood on end (head end
down). Twenty-four hours after the last treatment, rats were anesthetized and the left adrenal perfused as described. Each bar
represents the mean of 9 animals ± standard error, expressed as
percent of control animals. Control Epi Stores = 15.9 ug/adre-
nal, control Epi Secretion = 29.9 ng/min (n = 11). B. Rats were
injected with dexamethasone (1 mg/kg, IP) daily for 14 days.
Twenty four hours after the last injection the animals were
treated as described in a. Each bar represents the mean of 4 animals ± S. E. Control Epi Stores = 17.3 ug/adrenal, control epinephrine secretion = 51.2 ng/min. * = p < 0.05; ** = p < 0.005.

RESULTS

The effect of chronic stress on the reactivity of the rat adrenal
gland was investigated in immobilized rats. One day after the last immobilization, the left adrenal gland was removed from the anesthetized animal
and evaluated for alterations in reactivity by perfusing with acetylcholine.
The amount of epinephrine released by 50 ug of acetylcholine was quantitated in the adrenal perfusate. Chronic immobilization stress resulted in
an increase of nearly 60% in the tissue level of epinephrine (figure 1a).
More importantly, the amount of epinephrine released in response to a fixed
amount of acetylcholine was increased by a similar magnitude. To determine
if the change in increased adrenal reactivity was mediated by long term
exposure to glucocorticoids, a group of rats was treated with dexamethasone,
a long acting potent glucocorticoid. In contrast to the results observed
with immobilization stress, the dexamethasone treated animals were found to

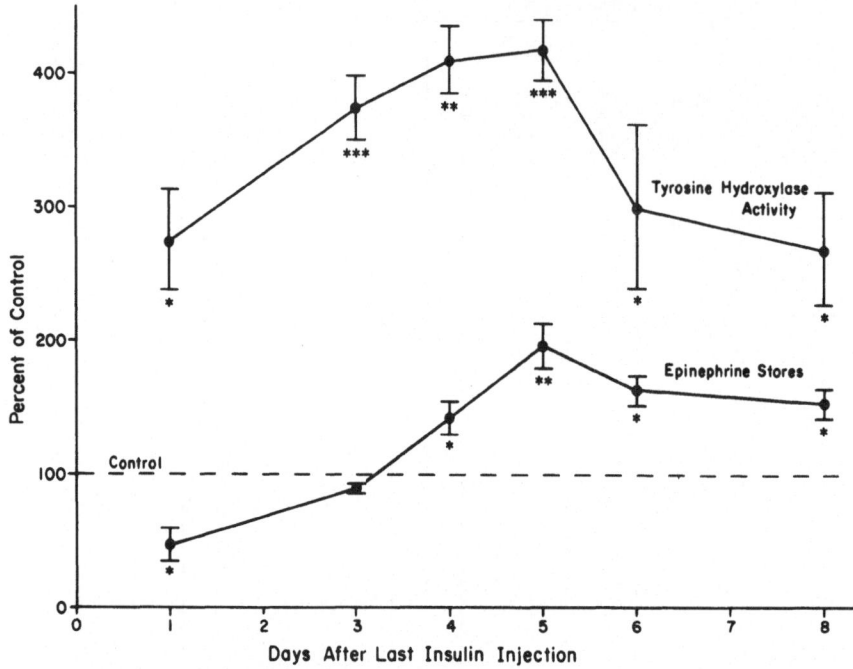

FIG. 2: Time Course of Adrenal Tyrosine Hydroxylase Activity and Epinephrine Content After Insulin Treatment. Rats were injected with
protamine zinc insulin (PZI, 10 U/rat, SC) daily for 4 days. At
various times after the last injection, animals were anesthetized
and both adrenal glands removed and stored at -70°C for 1-8 days.
Adrenals were homogenized in 20 volumes of 50 mM tris acetate, pH
8.0, 0.27 M sucrose, 40 mM NaF. A small aliquot (10 ul) of homogenate was diluted in 0.1 M HClO$_4$ for adrenal epinephrine determination. The remaining homogenate was centrifuged (35,000 x g,
30 min, 2°C) and small molecules were removed by Sephadex G-25
chromatography. The protein-containing fraction was assayed for
tyrosine hydroxylase (TH) activity using a coupled decarboxylase
procedure (Waymire et al., 1971). Assay conditions were: 0.1 mM
ferrous ammonium sulfate, 20 mM NaF, 1.0 mM 6-methyl-5, 6,7,8-
tetrahydropterine, 0.1 mM 1-^{14}C-L-tyrosine, 120 mM tris acetate,
pH 6.1, for 10 min at 37°C. Each point represents the mean of 3
animals ± standard error. Control TH activity = 0.54 nmoles DOPA/
adrenal/min, control epinephrine stores = 14.8 ug/adrenal (n = 4).
* = p < 0.05; ** = p < 0.005; *** = p < 0.0005.

have decreased epinephrine stores and decreased catecholamine release in
response to a standard amount of acetylcholine (figure 1b).

Chronic hypoglycemia, produced by treatment with a long acting insulin
preparation, was investigated for its effects on adrenal medullary catecholamine content. Employing the 4 day protocol described by Fluharty et al.
(1985), protamine zinc insulin (PZI) treated animals were evaluated for
functional changes in the adrenal medulla. In figure 2, the time course of
changes in adrenal catecholamine stores is presented. On the day following
the last PZI treatment, epinephrine was observed to be decreased to 48% of
control. Adrenal epinephrine stores were significantly reduced one day
after the last PZI injection, returned to approximately control levels by
day 3, and were significantly elevated on days 4-8. Tyrosine hydroxylase
was found to be increased 2.75 fold one day following the last PZI injec-

tion. A maximum induction of TH was observed at 5 days following the cessation of PZI treatment. This maximum increase in TH was coincident with
an "overshoot" of catecholamine stores.

Figure 3 illustrates the effects of the PZI treatment on plasma glucose, epinephrine stores and epinephrine secretion at 1,3 and 5 days following the last PZI treatment. This protocol significantly decreased the
plasma glucose on day 1. Epinephrine stores were reduced on days 1 and 3
and exceeded control values on day 5. Acetylcholine mediated catecholamine
secretion was significantly decreased on days 1 and 3.

The decrease in tissue epinephrine and acetylcholine mediated catecholamine secretion appeared to be correlated on day 1 and day 3. To examine
this possibility, the amount of epinephrine in the adrenal gland of each
rat in figure 3 was plotted versus the amount of epinephrine released by
depolarization with acetylcholine. This relationship is summarized in
figure 4. The mean value for control adrenal glands tested was 16.9 ug/
adrenal gland (as indicated by the arrow). Linear regression analysis was
used to determine the line of best fit. The correlation coefficient for
all the plotted values was 0.61. This value is statistically significant
($p < 0.0005$), indicating that a correlation exists between epinephrine
stores and acetylcholine induced epinephrine secretion.

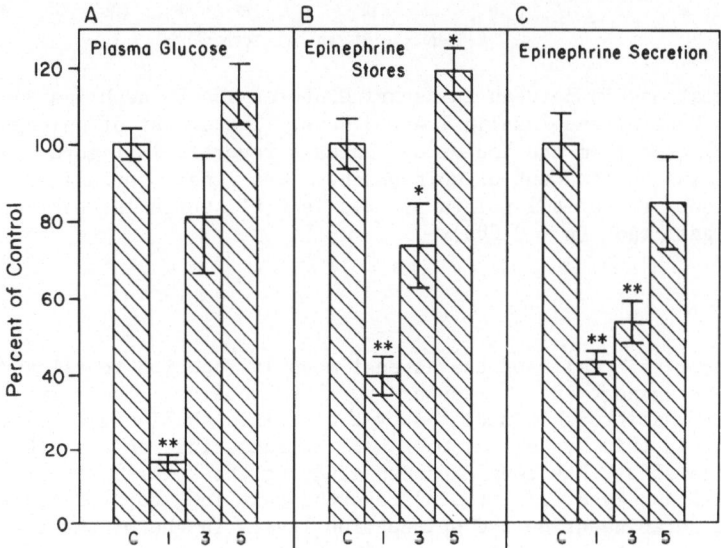

FIG. 3: Effect of Insulin Treatment on Plasma Glucose, Adrenal Epinephrine
Content, and Acetylcholine Induced Epinephrine Secretion. Rats
were treated with insulin or saline as described in Fig. 3. At 1,
3, and 5 days after the last injection animals were anesthetized
and the left adrenal gland perfused as described. Blood samples
were collected by cardiac puncture and plasma glucose determined
by a commercial assay kit (CentrifiChem, Baker Instruments).
Each bar represents the mean of 8 animals ± standard error. * =
$p < 0.06$; ** = $p < 0.005$. A. Plasma glucose levels were significantly reduced one day after the last insulin injection, but
not on days 3 and 5. Control glucose = 229 mg/100 ml. B. Epinephrine stores were significantly reduced on days 1 and 3, and significantly elevated on day 5. Control = 16.9 ug/adrenal. C.
Epinephrine secretion was significantly decreased on days 1 and
3. Control = 29.2 ng/min.

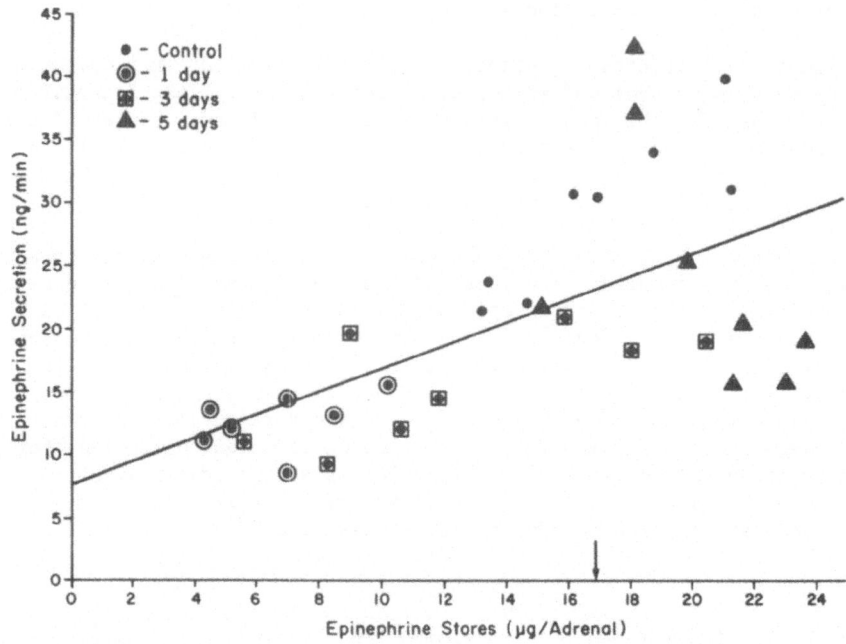

FIG. 4: Relationship Between Epinephrine Stores and Epinephrine Secretion
in Control and Insulin Treated Rats. The amount of epinephrine
released from the individual animals reported in figure 3 plotted
versus the content of epinephrine. The linear correlation coefficient, r, for all points plotted is 0.61, which is statistically
significant ($p < 0.005$).

DISCUSSION

The chromaffin cells of the adrenal medulla function to release epinephrine and norepinephrine in response to acetylcholine. The release of
these neurotransmitters is mediated by a nicotinic cholinergic receptor in
vivo. Although this system has been well characterized on a short term
basis, little effort has been expended studying the "plasticity of this
system." Kvetnansky et al., (1985) reported that long term stress will
enhance the norepinephrine and epinephrine levels in the plasma following
exposure to an acute novel stressor. Since these workers were using intact
animals, it was not possible to identify the specific site in the peripheral
or central nervous system that was reponsible for the increase in plasma
catecholamines.

Using a similar stress protocol involving long term immobilization of
rats and the perfused adrenal preparation, it is possible to determine if
part of this neuronal plasticity occurs at the level of the adrenal gland.
Significant increases in tissue catecholamine levels were observed following
14 days of immobilization. Quantitatively similar results were obtained
when the acetylcholine mediated epinephrine release from the adrenal gland
was measured.

Since this enhancement of adrenal reactivity may represent a long term
adaptation to stress possibly modulated by the glucocorticoid axis, animals
were treated with high levels of dexamethasone in an attempt to ascertain
whether the glucocorticoids affect adrenal reactivity. No evidence for

glucocorticoid mediated enhancement of adrenal medullary response was observed. A slight dimunition of catecholamine stores and acetylcholine releasable epinephrine was observed, presumably due to the dexamethasone mediated feedback inhibition of ACTH release, since ACTH is known to be involved in the maintenance of adrenal medullary function (Weinshilboum and Axelrod, 1970).

Insulin mediated hypoglycemia produced a biphasic effect upon tissue catecholamine content. One day following four days of PZI treatment, the stores were reduced 48% in response to the resulting hypoglycemia and TH activity was induced 275%. The catecholamine content increased with time to near control levels at three days following this treatment. At five days post insulin treatment, the catecholamine stores were elevated relative to control values, presumably in response to increased levels of TH activity. The acetylcholine mediated catecholamine release was highly correlated with content of the adrenal gland on days one and three.

It is clear from the data in figures 1a, 1b and 3c that the isolated perfused adrenal gland's response to a fixed dose of acetylcholine is modulated by the prior treatment that the animal has received. Treatments that resulted in a significant increase in tissue catecholamines resulted in a increase in the amount of catecholamines released. Insulin mediated hypoglycemia, which caused a reduction in tissue catecholamines, resulted in a decrease in the amount of acetylcholine mediated catecholamine release. The biochemical mechanisms responsible for the relationship between tissue catecholamine content and depolarization dependent catecholamine release are currently being investigated. It is anticipated that when this relationship has been fully elucidated, much knowledge will be gained about the molecular mechanisms regulating the synthesis and release of catecholamines.

CONCLUSIONS

1. The secretory response of the perfused rat adrenal medulla to a standard stimulus (50 ug acetylcholine) can be altered by pretreating the rat in various ways, including immobilization stress, dexamethasone administration, and insulin administration.

2. There is a positive correlation between the amount of epinephrine stored in the adrenal and the amount released in response to acetylcholine perfusion.

3. The altered responsivity of adrenal catecholamine secretion following periods of chronic stress can be at least partially explained by changes in the adrenal medulla itself, independent of processes occuring in the central nervous system or sympathetic nerves innervating the adrenal.

ACKNOWLEDGEMENTS

This study was supported by USAFOSR Grant #84-0122 and NSF Grant BNS 81-18957.

REFERENCES

Cannon, W. B., The Wisdom of the Body, Norton, New York, (1939).
Chuang, D. and Costa, E., Biosynthesis of tyrosine hydroxylase in rat adrenal medulla after exposure to cold, Proc. Natl. Acad. Sci. USA, 71: 4570-4574 (1974).

Fluharty, S., Snyder, G., Zigmond, M. and Stricker, E., Tyrosine hydroxylase activity and catecholamine biosynthesis in the adrenal medulla
   during stress, J. Pharmacol. Exp. Ther., 233:32-38 (1985).
Kvetnansky, R., Wise, V. K. and Kopin, I., Elevation of tyrosine hydroxylase and phenylethanolamine-N-methyl transferase by repeated immobilization, Endocrinology, 87:744-749 (1970).
Kvetnansky, R., Nemeth, S., Vigas, M., Oprsalova, Z. and Jurcovicova, J.,
   Plasma catecholamines in rats during adaptation to intermittent exposure to different stressors. In Stress: The Role of Catecholamines
   and Other Neurotransmitters., E. Usdin, R. Kvetnansky, and J. Axelrod,
   eds., Gordon and Breach Science Publishers, New York, pp. 537-562
   (1985).
Masserano, J. M. and Weiner, N., The rapid activation of tyrosine hydroxylase by the subcutaneous injection of formaldehyde, Life Sci., 29:
   2025-2029 (1981).
Thoenen, H., Mueller, R. A. and Axelrod, J., Trans-synaptic induction of
   adrenal tyrosine hydroxylase, J. Pharmacol. Exp. Ther., 169:249-254
   (1969a).
Thoenen, H., Mueller, R. A. and Axelrod, J., Increased tyrosine hydroxylase activity after drug induced alteration of sympathetic activity,
   Nature (Lond.), 221:1264-1266 (1969b).
Wakade, A. R., Studies on the secretion of catecholamines evoked by ace-
   tylcholine or transmural stimulation of the rat adrenal gland, J.
   Physiol., 313:463-486 (1981).
Waymire, J., Bjur, R. and Weiner, N., Assay of tyrosine hydroxylase by the
   coupled decarboxylation of DOPA formed from $1-^{14}C$-L-tyrosine, Anal.
   Biochem., 43:588 (1971).
Weiner, N. and Mosimann, W., The effect of insulin on the catecholamine
   content and tyrosine hydroxylase activity of cat adrenal glands,
   Biochem. Pharmacol., 19:1189-1199 (1970).
Weinshilboum, R. and Axelrod, J., Dopamine-B-hydroxylase activity in the
   rat after hypophysectomy, Endocrinology 87:894-897 (1970).

374

CHEMORECEPTION: PARAMECIUM AS A RECEPTOR CELL

Judith Van Houten and Robin R. Preston

Department of Zoology, University of Vermont
Burlington, Vermont 05405

In the sensory modalities of taste, smell and common chemical sense, there are receptor cells that make contact with the external environment and detect the presence of external chemical stimuli. The receptor cell is the site of stimulus recognition, which is thought to be mediated through binding of the stimulus to specific surface receptors and then transduction of this binding into "useful" electrical information. Information in this new form is passed on to higher order neurons and eventually is translated into a response. In order to study receptor cell function, it seems straightforward to isolate these receptor cells, identify the receptors among the membrane proteins and determine the ionic basis of receptor binding by conventional electrophysiology. However, there are limitations inherent in many of the chemosensory systems traditionally used to study chemoreception. Relatively small amounts of olfactory or taste epithelium limit the binding studies and biochemical studies necessary to identify receptor proteins; tissue is often of a mixed cell type, even when avaiable in quantity, making it difficult to be sure of the origins of putative receptor proteins (Price, 1981; Mooser, 1981; Cagan, 1981). Hence, indirect methods (e.g. treating the tissue with n-ethyl-maleimide to disrupt protein sulfhydryl bonds, and hence disrupting the chemoresponse, or demonstrating specificity and saturability of a response) are used to demonstrate that the receptor site is a protein. When a binding protein is identified, often it is not possible to draw the necessary behavioral correlations to demonstrate that the protein is the chemoreceptor. For example, a "green odorant" binding protein has been isolated from cow olfactory epithelium, but it will be difficult to demonstrate the involvement of this protein in cow chemo-esponse (Bignetti et al., 1985). Small cell size and tissues comprised of more than one cell type complicate electrophysiological studies as well, although the advent of patch clamping should circumvent the size limitations (S. Kleene, personal communication; Margolis et al., 1985).

This brings us to the use of single-cell organisms, such as Paramecium, to study chemoreception. Even though unicellular organisms have no stable cell contacts that characterize metazoan systems, there are compelling reasons to use unicells in the study of chemoreception, particularly in recepor cell function. Their hallmark is versatility and their most important attribute is the availability of mutants. In particular, cells can be grown in large, homogeneous populations to provide material for biochemical anasis; the cells are large for convenient electrophysiology; and mutant cell lines provide opportunities to apply a genetic dissection to the identification of chemoreceptors and other components of the chemosensory transduction pathway.

Among the eukaryotic unicells, the most work has been done on the slime mold Dictyostelium and the ciliates Tetrahymena and Paramecium. Dictyostelium is used for its chemoresponses to folic acid and cyclic AMP (cAMP), which change over developmental time (Gerisch, 1982). However, even in this highly convenient system, no receptor mutants or electrophysiological data are available. Tetrahymena rivals Paramecium in amenability to a variety of approaches for the study of membrane functions. However, studies of Tetrahymena chemoresponse still focus on assay methods and on mechanisms of the behavioral response (Levandowsky et al. 1984; Leick and Lellung-Larsen, 1985). We discuss here our progress with Paramecium as a chemoreceptor cell, and the development of techniques that will enable a genetic dissection of the chemosensory transduction pathway.

Paramecium has been called a "swimming neuron" (Machemer and dePeyer, 1977) and is a chemoreceptor cell. Paramecia respond to the external stimuli folic acid, acetate, lactate, cAMP and other compounds that signal the presence of their food, bacteria (Van Houten, 1978). We have developed a simple T-maze test to assay responses to these attractants and also to extremes of pH, high ionic strength and some organic membrane active compounds like quinidine, which are repellents (Van Houten, 1978). As an example, we will focus on responses to folic acid.

Folate Chemoresponse: Binding

It is the pterin portion of the folate molecule that is recognized by the cells in behavioral tests (Schulz et al., 1984). Attraction to folic acid is inhibited by cAMP, which shares only some structural features (Schulz et al., 1984). Folate is taken up by the cells and can be concentrated at least 50 fold from the surrounding media, as measured by radiobinding assay (Schulz and Van Houten, unpublished results). It is not possible to eliminate this uptake by low temperature or metabolic inhibitors without affecting cell integrity, therefore we measure binding of folate to cells as instantaneous binding, that is, by extrapolating the amount of ^{3}H-folate associated with the cells in centrifugation assays to time zero (Schulz et al., 1984). ^{3}H-folate binds specifically and saturably to whole cells with a K_d of approximately 29 μM (Schulz et al., 1984); binding shows the same specificity as the behavioral response and it is likely that a specific receptor exists for the binding of folate. While this may seem to be of low affinity to those who are used to dealing with neurotransmitters, it is in keeping with other external chemoreceptor systems. Estimates of half maximal response and K_ds for vertebrate and invertebrate olfactory and taste range from 1 nM up to 700 μM, with K_ds often in the μM range (Cagan, 1981; Hansen and Wieczorek, 1981; Price, 1981; Lancet, 1986). Similarly, paramecia respond only to relatively high levels of fermentation and other bacterial products when these are presented individually. We believe this to be a cell's means of coping with a chemically noisy environment (pond water) and to ensure a response only when bacteria are actually available. Therefore, the problems associated with biochemistry of low affinity receptors cannot necessarily be avoided but must be met head on if a variety of external chemoreceptors are to be identified and purified from Paramecium and other chemoreceptor cells.

Binding of folate to whole cells is primarily to the cell body and not to the cilia. Isolated cilia in filtration and centrifugation assays show <1% of the whole cell binding despite the fact that cilia are covered with approximately 50% of the surface membrane (Schulz et al., 1983; Schulz et al., 1984; Dunlap, 1977). Deciliated cells show normal folate-induced hyperpolarization (see below), which supports the notion that binding sites can be found on the cell body membrane.

A quantitative measure of folate binding is the amount of fluorescence of cells stained with FITC-folate, that is, folate conjugated to fluorescein through reaction with fluorescein isothiocyanate (Van Houten et al., 1985). The fluorescence of stained cells is easily distinguished from autofluorescence in blind tests when the cells are examined individually under oil immersion or in microtiter wells under a dissecting microscope outfitted with cut-off filters. The FITC-folate binding is specific for folate as evidenced by inhibition of fluorescence when cells are stained with FITC-folate in the presence of excess unconjugated folate. The binding is to the exterior of the cell and not to broken cells; cells permeabilized with Triton X-100 prior to staining have an increased fluorescence, probably attributable to access to the interior of the cell. Additionally, a mutant (d4-534) that has lost both attraction to folate and surface binding (see below) due to a single site mutation (DiNallo et al., 1982) does not show FITC-folate fluorescence much above autofluorescence (Van Houten et al., 1985).

At least three genes can be mutated to decrease or eliminate attraction to folate (DiNallo et al., 1982). These mutants have single site lesions that assort in Mendelian patterns and show specific losses of folate response, while responses to other attractants and repellents remain intact. Mutants in one particular complementation group (fol[1]) have lost the ability to bind ^{3}H-folate specifically and at normal levels (Schulz et al., 1984). These are the same mutants that can be distinguished from normal by FITC-folate staining above (Van Houten et al., 1985).

There are four revertants of one allele of the fol[1] complementation group, mutant d4-534. At least three of these revertants have second site mutations (Van Houten and White, unpublished results). Second site revertants open up the possibility of identifying chemosensory transduction pathway components that can be identified by no other means. Mutations in genes for these components may cause no or slight phenotypic changes in the chemoresponse of the cell. However, because they are mutations that suppress an abnormal chemoresponse phenotype and because they are not in the originally mutated gene, they must be mutants in genes that code for products that somehow interact with the original mutant gene product. D. J. L. Luck (Huang et al., 1982) has been especially productive using revertant analysis for a fine genetic dissection of the Chlamydomonas flagellum and we expect that revertants will be useful for analysis of the Paramecium chemoreception pathway as well.

Binding Proteins

The membrane protein(s) that are part of the chemoresponse pathway should include the receptor, which will bind folate, be exposed to external medium, and be defective in mutants missing the binding associated with chemoresponse. We have adapted a method to isolate cell body membranes (Schulz et al., 1986) and searched through the cell body membrane proteins for folate binding proteins by affinity chromatography, cross linking and FITC-folate binding. Affinity chromatography identified about 4 proteins that specifically elute with K_2-folate and not with buffers of similar ionic strength with KCl, K_2-glutamate (a component of the folate molecule but is neither an attractant and nor inhibitor of ^{3}H-folate binding), or galactose (the monomer of the Sepharose component of the affinity column). Methotrexate, a folate analog, is only a weak inhibitor of folate binding and response and selectively elutes only a subset of the folate binding proteins. Concanavalin A affinity columns were used to identify glycoproteins among the binding proteins and ^{125}I labeling of whole cells narrowed the field of putative receptor proteins to 3 (Schulz et al., 1986).

Clear identification of the chemoreceptor among these proteins awaits comparison with protein of null mutants that are allelic to fol[1] and verification of the protein's folate binding properties by crosslinking folate to receptor with immunodetection by anti-folate antibody on electroblots (Kershko, Sasner and Van Houten, unpublished results). Cells crosslinked with the N-hydroxysuccinimide ester of folate (Henderson and Zevely, 1984), are inhibited in their response to folate, but not to another stimulus, acetate (Table 1). Therefore, the chemoreceptors should be among the subset of surface proteins crosslinked and recognized on electroblots by antifolate antibodies (Towbin et al., 1979). We are currently producing antibodies against folate conjugated to keyhole limpit hemocyanin (KLH) (Langone, 1982; Hurn and Chantler, 1980). In ELISAs, the sera recognize folate conjugated to bovine serum albumin (BSA), but not BSA alone, which indicates that these sera will be useful in recognizing folate linked to proteins other than KLH.

Electrophysiological Correlates of Chemoreception

Binding of folate and other attractants is transduced into a hyperpolarization (Van Houten, 1979). Mutant d4-534 shows only a small hyperpolarization in folate, but normal hyperpolarization in other attractants. Therefore, membrane hyperpolarization is an integral part of the chemosensory motor response pathway.

There has been a long history of fine electrophysiological work on Paramecium's excitable membrane in order to describe the connection between membrane electrical events and changes in ciliary movement (Eckert, 1972; Naitoh, 1982; Kung and Saimi, 1982). A depolarization will move the membrane potential (V_m) toward threshold for an action potential, which brings with it a transient increase in intraciliary calcium. Calcium at $>10^{-6}$M causes the cilia to transiently reverse beating direction, causing a transient jerky turn in swimming path. A hyperpolarization moves V_m away from the threshold for action potentials, thereby decreasing action potential frequency. Hyperpolarization also increases ciliary beating frequency, which in turn increases swimming speed. Attractants generally hyperpolarize (Van Houten, 1979) and the changes in ciliary motility caused by hyperpolarization add up to a longer mean free path moving up the gradient of attractant and gradual accumulation up the gradient (Van Houten, 1978; Van Houten and Van Houten, 1982).

Chemosensitivity for folate is not uniform over the cell surface. Deciliated cells show normal size hyperpolarizations in folate and other attractants, confirming the binding data that indicate receptors are not primarily on the ciliary membrane (Preston and Van Houten, 1986a; Van Houten et al., 1983). Pressure perfusion of folate onto cells elicits maximum hyperpolarization in the anterior portion of the cell and particularly anterior ventral regions (Preston and Van Houten, 1986a). Gradients of receptors are not unprecedented in Paramecium: there is a very distinct mechanoreceptive gradient anterior to posterior (Ogura and Machemer, 1980). Highest chemosensitivity at the anterior of the cell may facilitate its movement into regions of attractant by eliciting fast smooth swimming as the cell is headed in the direction of the stimulus.

The ionic basis for the hyperpolarization remains elusive. The obvious candidates of Ca-dependent or voltage activated K and Na effluxes are clearly eliminated by ion substitution experiments: cells show normal hyperpolarizations to folate and other attractants in the absence of Na or K, or when external Na and K are simultaneously fixed to eliminate the driving force for net fluxes of both K and Na across the membrane. Membrane resistance in folate increases slightly and there is no obvious reversal potential, although we have not succeeded in voltage clamping the cells at the extreme V_m of +110 mV necessary to clamp at E_{Ca}. The folate-induced hyperpolarization is not likely to be due primarily to surface charge changes because of the specificity of the behavioral and hyperpola-

TABLE 1: Effect of Crosslinking Folate on Chemoresponse to the Stimuli
Folate and Acetate

Cell Treatment 1% DMSO +	I_{Che} Na$_2$-Folate	n	I_{Che} Na-OAc	n
5 µM Na$_2$folate	0.72 ± 0.06	12	0.67 ± 0.06	6
5 µM "activated" Na$_2$folate	0.54 ± 0.06	12	0.61 ± 0.06	6

Cells are treated with folate or "activated" folate in DMSO, washed in buffer, and tested in T-maze assays for response to folate (2.5 mM Na$_2$-folate vs 5 mM NaCl) and Na-OAc (5 mM Na-OAc vs 5 mM NaCl). Index of chemokinesis (I_{Che}) greater than 0.5 indicates attraction; less than 0.5 indicates repulsion; 0.5 indicates no response to stimulus.

rization responses. Polycations, which affect surface charge, and a surface charge mutant (Satow and Kung, 1981) do not perturb the folate-induced hyperpolarization. Studies of accumulation and hyperpolarization over a range of pH 5-8 indicate that the organic acid attractants can be fully charged and act as attractants (Schulz et al., 1985a,b). Therefore, while it is possible that the folate anion enters and directly hyperpolarizes by virtue of its net negative charge, it is not clear why pterine-6-carboxylic acid with no net charge should do likewise.

To explain the hyperpolarization, we are left with relatively few options that include: 1) receptor-mediated release of Ca from internal stores, which hyperpolarize by activating a Ca pump. This pump would have to be fast to cause the hyperpolarization as rapidly as we can perfuse the cell, but the enzymes of the rod outer segment have taught us that enzymes can account for electrical events that occur in milliseconds (Stryer, 1986). 2) Ca or folate activated H$^+$ efflux pump. We have not been able to change internal or external pH sufficiently to rule this out. It can be qualified that the putative H$^+$ efflux is not affected by amiloride in Na free solutions (Van Houten and Preston, 1985). In support of a role for Ca are Ca permeability changes of cells in acetate, another attractant. (Permeability to Ca cannot be tested in folate because of problems with precipitation.) However, the specificity of this permeability increase is difficult to sort out from the general effect of high ionic strength on hyperpolarization. A mutant, "Restless" (courtesy of E. Richard), cannot properly regulate V$_m$ in low K solutions and its V$_m$ plunges toward E$_k$ in low K solutions. These extremely hyperpolarized cells show an increased hyperpolarization in folate and acetate (Preston and Van Houten, 1986b). Plots of size of hyperpolarization vs V$_m$ of Restless extrapolate to near E$_{Ca}$ as reversal potential. Further clamping and study of folate mutants are necessary to solve the puzzle of the hyperpolarization.

Second Messengers

In order to investigate a possible role for Ca in the transduction pathway, we have turned to Quin-2, a calcium-sensitive fluorescent dye (Rink and Pozzan, 1985; Tsien et al., 1982). Paramecia take up and cleave the ester bond of the membrane permeable Quin-2/AM, trapping the Quin-2 inside the cell and making it available to act as a Ca indicator. Therefore, this dye and others will prove useful in examining internal free Ca levels by this very specific probe. It is difficult to manipulate external Ca for

TABLE 2. Effects of LiCl on Chemoresponse to Folate

Incubation	Duration of Incubation	I_{Che}	S. D.
2 mM LiCl	5 min	0.66	0.04
	15	0.60	0.05
	30	0.34	0.07
2 mM NaCl	30 min (Control)	0.77	0.06
4 mM LiCl	5 min	0.27	0.04
	15	0.16	0.06
	30	0.18	0.08
	60	0.24	0.04
4 mM NaCl	60 min (Control)	0.76	0.09
2 mM NaCl	30 min (Control)	0.67	0.04
1 mM Na$_2$folate		0.76	0.02
2 mM LiCl		0.55	
1 mM Ki$_2$folate		0.39	0.06

Cells were incubated in buffer with LiCl, NaCl (control), Li$_2$folate, or Na$_2$-folate (control). Cells were washed and tested for chemoresponse to 2.5 mM Na$_2$folate vs 5 mM NaCl in T-mazes. Data are averages of 3 experiments, except for the last incubation in 2 mM LiCl, which is the average of only two experiments.

conventional electrophysiological methods since external Ca cannot be com-pletely removed. Paramecia will not survive with Ca < 10^{-5} M, which still leaves a large driving force for Ca to enter the cell (E_{Ca}+110 mV).

With the possibility of increases in internal free Ca comes the possi-bility of involvement of a receptor-mediated inositol phospholipid turnover cycle. Inositol triphosphate (IP$_3$) is thought to liberate Ca from the endo-plasmic reticulum and thereby activate Protein Kinase C or other Ca depen-dent protein kinases (Nishizuka, 1984a,b). Lithium blocks this turnover cycle, inhibiting the renewal of phosphoinositol lipids for receptor-medi-ated degradation by phospholipase C (Berridge and Irvine, 1984; Nishizuka, 1984a,b). Lithium, but not Na, has a profound effect on Paramecium chemo-response (Table 2), giving a tantalizing hint of a role for phospholipids in chemoreception.

Adaptation

Paramecia, as other chemoreceptor cells, adapt to uniform concentra-tions of stimulus (Van Houten et al., 1982; Van Houten and Van Houten, 1982). The molecular mechanism of this adaptation is not evident. As in bacteria, it may involve methylation of receptor or transducer proteins (Hazelbauer and Harayama, 1983). S-adenosyl-L-methionine (SAM) is a major methyl donor in eukaryotic cells and SAM has an effect on the behavioral responses of Paramecium to folate and acetate. This effect is not mimicked by S-adenosyl-L-homocysteine, a very similar compound except for the trans-ferrable methyl group of SAM (Van Houten et al., 1984). Therefore, it is possible that methylation by SAM or even phosphorylation (as, for example, by protein kinase C) may mediate adaptation. There are mutants that apparently are abnormal in adaptation, that is, they are initially attracted to ammonium but fail to sustain this attraction (Van Houten et al., 1982). Such behavior is what an adaptation defect would cause in computer simulations of behavior (Van Houten and Van Houten, 1982). Mutants not affected by SAM also are available and between these sets of mutants the role of adaptation should be made more clear.

Other Chemoreception Systems in Paramacium

Folic acid is not the only organic stimulus for Paramecium. External cAMP, among other compounds, is an attractant. We study cAMP as external stimulus and have shown by HPLC methods that external cAMP is not detectably taken up into the cells and not detectably broken down when the phosphodiesterase inhibitor isobutylmethylxanthine (IBMX) is present (Smith et al., 1986). It is significant that cAMP acts externally because there have been recent reports of internal cAMP controlling ciliary beating frequency and perhaps hyperpolarization (Hennessey et al., 1985; Gustin et al., 1983; Schultz et al., 1984). These internal cAMP-induced changes look very similar to those induced by external attractants by virtue of the hyperpolarization they induce. However, we believe that internal cAMP is not involved in chemoreception because IBMX in combination with chemical stimuli does not potentiate the chemoresponses and IBMX, in both control and stimulus solutions, does not inhibit the chemoresponses (Table 3).

^{3}H-cAMP binds to cells in a saturable, specific fashion, but with low affinity ($K_d \sim 200$ μM; Smith et al., 1986). However, the K_d compares well with the half maximal concentration for behavioral responses to cAMP and for cAMP-induced hyperpolarization. We have available three mutants that are not normally attracted to cAMP (Gagnon and Van Houten, unpublished results). These mutants and the possibility of photoaffiity labeling the cAMP binding sites make the receptor approachable despite its low affinity. Cyclic AMP affinity columns consistently show one cAMP specific binding protein of approximately 48,000 dalton (Van Houten et al., 1986). This protein band specifically elutes with 5'-AMP, an inhibitor of cAMP chemoresponse behavior, but not with cGMP, which does not affect cAMP chemoresponse. In preliminary results, a protein of the same size is labeled with ^{32}P-8-azido-cAMP (Smith and Van Houten, unpublished results). However, we have not

TABLE 3. Effects of IBMX on Chemoresponse to Folate

Test Solution	Control Solution	I_{Che}	S. D.	n
IBMX does potentiate chemoresponse when added to stimulus solution:				
1 mM Na$_2$folate	2 mM NaCl	0.81	0.08	3
1 mM Na$_2$folate + 1 mM IBMX	2 mM NaCl	0.42	0.01	3
IBMX does not inhibit chemoresponse when added to both stimulus and control solutions:				
2.5 mM Na$_2$folate	5 mM NaCl	0.96	0.05	3
2.5 mM Na$_2$folate + 1 mM IBMX	5 mM NaCl + 1 mM IBMX	0.98	0.01	3

established the correlations necessary to establish a role for this protein in chemoreception.

CONCLUSIONS

The use of unicellular organisms circumvents problems accompanying the study of receptor cell function in some metazoan systems and, at the same time, allows the application of a variety of techniques and approaches to a single receptor cell type. Paramecium, in particular, presents opportunities to combine membrane protein biochemistry with electrophysiology and with the capability of generating and isolating chemoreceptor mutants to aid in the dissection of the chemosensory transduction pathway. The Paramecium system is still in the process of being developed as a receptor cell, but appears to have all the components necessary for a successful genetic dissection. The potential for the generation and analysis of mutants cannot be over emphasized because, in theory, each macromolecular component of the chemosensory transduction pathway can be mutated, whereas, it is not likely that there are sufficient numbers or specificities of pharmacological agents to achieve the same end. In addition, suppressor mutants often allow the only possible identification of components that interact with other pathway components, but that might not otherwise be identified. Lastly, the development of a genetic system paves the way for applying molecular techniques to the study of chemoreception, including molecular genetics (Margolis et al., 1985).

REFERENCES

Berridge, M. J. and Irvine, R. F., Inositol triphosphate, a novel second messenger in cellular signal transduction, Nature 312:315 (1984).

Bignetti, E., Cavaggioni, A., Pelosi, P., Persaud, K., Sorbi, R. and Tirindelli, R., Purification and characterization of an odorant-binding protein from cow nasal tissue, Eur. J. Biochem. 149:227 (1985).

Cagan, R., Recognition of taste stimuli at the initial binding interaction, in: "Biochemistry of Taste and Olfaction," R. Cagan and M. Kare, eds., Nutrition Foundation, Academic Press, NY (1981).

DiNallo, M., Wohlford, M. and Van Houten, J., Mutants of Paramecium defective in chemokinesis to folate, Genetics 102:149 (1982).

Dunlap, K., Localization of calcium channels in Paramecium caudatum, J. Physiol. 271:119 (1977).

Eckert, R., Bioelectric control of cilia, Science 176:473 (1972).

Gerisch, G., Chemotaxis in Dictyostelium, Ann. Rev. Physiol. 44:535 (1982).

Gustin, M., Bonini, N. and Nelson, D., Membrane potential regulation of cAMP: control mechanism for swimming behavior in the ciliate Paramecium, Soc. Neurosci. Abstr. 9:167 (1983).

Hansen, K. and Wieczorek, H., Biochemical aspects of sugar reception in insects, in: "Biochemistry of Taste and Olfaction," R. Cagan and M. Kare, eds., Nutrition Foundation, Academic Press, New York, (1981).

Hazelbauer, G. and Harayama, S., Sensory transduction in bacterial chemotaxis, Int. Rev. cytol. 81:33 (1983).

Henderson, G. and Zevely, E., Affinity labeling of the 5-methyl-tetrahydrofolate/methotrexate transport of protein by L1210 cells by treatment with an N-hydroxysuccinimide ester of methotrexate, J. Biol. Chem. 259:4558 (1984).

Huang, B., Ramanis, Z. and Luck, D. J. L., Suppressor mutations in Chlamydomonas reveal a regulatory mechanism for flagellar function, Cell 28:115 (1982).

Hennessey, T., Machemer, H. and Nelson, D., Injected cyclic AMP increases ciliary beat frequency in conjunction with membrane hyperpolarization, Eur. J. Cell. Biol. 36:153 (1985).

Hurn, B. and Chantler, S. M., Production of reagent antibodies, Meth. Enz. 70:104 (1980).

Kung, C. and Saimi, Y., The physiological basis of taxes in Paramecium, Ann. Rev. Physiol. 44:519 (1982).

Lancet, D., Vertebrate olfactory reception, Ann. Rev Neurosci. 9:329 (1986).

Langone, J., Radioimmunoassay of methotrexate, leucovorin, and 5-methyl-tetrahydrofolate, Meth. Enz. 84:409 (1982).

Leick, V. and Hellung-Larsen, P., Chemotaxis in Tetrahymena: the involvement of peptides and other signal substances, J. Protozool. 32:550 (1985).

Levandowsky, M., Chang, T., Kehr, A, Kim, J., Gardner, L., Tsang, L., Lai, G., Chung, C. and Prakash, E., Chemosensory responses to amino acids and certain amines by the ciliate Tetrahymena: a flat capillary assay, Biol. Bull. 167:322 (1984).

Machemer, H. and dePeyer, J., Swimming sensory cells: electrical membrane parameters, receptor properties and motor control in ciliated protozoa, Verh. Drsch. Zool. Ges. 1977:86 (1977).

Margolis, F. L., Sydor, W., Teitelbaum, Z., Blacher, R., Grillo, M., Rogers, K., Sun, R. and Gubler, U., Molecular biological approaches to the olfactory system: olfactory marker protein as a model, Chem. Senses 10:163 (1985).

Naitoh, Y., Protozoa, in: "Electrical Conduction and Behavior in 'Simple' Invertebrates," G. A. B. Shelton, ed., Clarendon Press, Oxford (1982).

Nishizuka, Y., The role of protein kinase C in cell surface signal transduction and tumour promotion, Nature 308:693 (1984a).

Nishizuka, Y., Turnover of inositol phospholipids and signal transduction, Science 225:1365 (1984b).

Ogura, A. and Machemer, H., Distribution of mechanoreceptor channels in the Paramecium surface membrane, J. Comp. Physiol. 135:233 (1980).

Price, S., Receptor proteins in vertebrate olfaction, in: "Biochemistry of Taste and Olfaction", R. Cagan and M. Kare, eds., Academic Press, NY, (1981).

Preston, R. R. and Van Houten, J. L., Localization of the chemoreceptive properties of the surface membrane of Paramecium tetraurelia, J. Comp. Physiol. in press (1986).

Preston, R. R. and Van Houten, J. L., Chemoreception in Paramecium tetraurelia: folate and acetate-induced membrane hyperpolarization, J. Comp. Physiol. submitted (1986).

Rink, T. and Pozzan, T., Using Quin2 in cell suspensions, Cell Calcium 6: 133 (1985).

Satow, Y. and Kung, C., Possible reduction of surface charge by a mutation in Paramecium tetraurelia, J. Membr. Biol. 59:179 (1981).

Schultz, J., Grünemund, R., von Hirschausen, R. and Schönfeld, U., Ionic regulation of cAMP levels in Paramecium tetraurelia, Febs. Lett. 167: 113 (1984).

Schulz, S., Denaro, M., Xypolyta-Bulloch, A. and Van Houten, J., Relationship of folate binding to chemoreception in Paramecium, J. Comp. Physiol. 155:113 (1984).

Schulz, S., Sasner, J. M. and Van Houten, J., In search of the folate chemoreceptor, J. Cell Biol. 101:302a (1985a).

Schulz, S., Preston, R. and Van Houten, J., Characterization of putative Paramecium chemoreceptors, Chem. Senses 10: in press (1985b).

Schulz, S., Sasner, J. M. and Van Houten, J., Folate binding proteins of the Paramecium surface membrane, Biochim. Biophys. Acta submitted (1986).

Smith, R., Gagnon, M. L., Preston, R. R., Schulz, S. and Van Houten, J., Correlation between cAMP binding and chemoreception in Paramecium, J. Comp. Physiol. submitted (1986).

Stryer, L., Cyclic GMP cascade of vision, Ann. Rev. Neurosci. 9:87 (1986).

Towbin, H., Staehelin, T. and Gordon, J., Electrophoretic transfer of proteins from polyacrylamide gels to nitrocellulose sheets: procedure and some applications, Proc. Natl. Acad. Sci. (USA) 76:4350 (1979).

Tsien, R., Pozzan, T. and Rink, T., Calcium homeostasis in intact lympho-
cytes: cytoplasmic free calcium monitored with a new, intracellularly
trapped fluorescent indicator, J. Cell Biol. 94:325 (1982).

Van Houten, J., A mutant of Paramecium defective in chemotaxis, Science 198:
746 (1977).

Van Houten, J., Two mechanisms of chemotaxis in Paramecium, J. Comp. Physiol.
127:167 (1978).

Van Houten, J., Membrane potential changes during chemokinesis in Paramecium,
Science 204:1100 (1979).

Van Houten, J. and Preston, R. R., Effects of amiloride on Paramecium chemoresponse, Chem. Senses 10: in press (1985).

Van Houten, J. and Van Houten, J., Computer analysis of Paramecium chemokinesis behavior, J. Theor. Biol. 98:453 (1982).

Van Houten, J., Martel, E. and Kasch, T., Kinetic analysis of chemokinesis of
Paramecium, J. Protozool. 29:226 (1982).

Van Houten, J., Schulz, S. and Denaro, M., Characterization and location of
folate binding sites involved in Paramecium chemoreception, J. Cell Biol.
97:469a (1983).

Van Houten, J., Wymer, J., Cushman, M. and Preston, R. R., Effects of S-
adenosyl-L-methionine on chemoreception in P. tetraurelia, J. Cell Biol.
99:242a (1984).

Van Houten, J., Smith, R., Wymer, J., Palmer, B. and Denaro, M., Fluorescein conjugated folate as an indicator of specific folate binding to
Paramecium, J. Protozool. 32:613 (1985).

Van Houten, J., Preston, R. R., Schulz, S., Sasner, J. M. and Smith, R.,
Chemoreceptors of Paramecium, Soc. Neurosci. Abstr. 12: in press (1986).

REDUCTION OF DOPAMINE RECEPTOR ACTIVITY DIFFERENTIALLY ALTERS

STRIATAL NEUROPEPTIDE mRNA LEVELS

Jesus A. Angulo, Greg R. Christoph, Robert W. Manning, Beth A. Burkhart, and Leonard G. Davis

Neurobiology Group, E. I. du Pont de Nemours and Co.
Medical Products Department, Experimental Station, E400
Wilmington, DE 19898

SUMMARY

We have investigated the effect of dopamine receptor blockade on striatal proenkephalin mRNA and protachykinin mRNA by Northern gel analysis and by in situ hybridization histochemistry. Chronic haloperidol treatment resulted in a 3.5 fold increase in striatal proenkephalin mRNA and a 30% decrease in protachykinin mRNA (no apparent change in α-tubulin mRNA was observed). The changes in mRNA levels for protachykinin and proenkephalin were uniform throughout the caudate-putamen of the rat as determined by in situ hybridization histochemistry. The results imply that altering receptor-mediated neurotransmitter functions can lead to profound, specific, and long-lasting alterations in neuronal gene expression.

INTRODUCTION

It has been long known that ligand-receptor interactions lead to concomitant electrophysiological and biochemical changes in the postsynaptic cell. For example, iontophoretic application of dopamine to striatal neurons results in receptor mediated facilitation or reduction of the electrical activity of those neurons (Siggins et al., 1976). Activation of D_1 or D_2 dopamine receptor subtypes also leads to stimulation or inhibition, respectively of the adenylate cyclase system (Kebabian and Calne, 1979). Presumably changes in cAMP levels after receptor stimulation lead to an altered activity of cAMP-dependent protein kinase (Ehrlich, 1979). Sequelae of ligand-receptor interactions can last from milliseconds (e.g., ion movement through channels) to hours (e.g., phosphorylation of proteins). We sought to determine whether changes in receptor activity could lead to longer term changes by altering mRNA levels (Fig. 1).

The well defined anatomical organization of the nigral-striatal pathway provides an in vivo model system to test whether blockade of dopamine receptors can alter striatal neuropeptide mRNA levels. The dopamine containing cells originating in the substantia nigra project to the striatum where they make direct synaptic contact with enkephalin- and tachykinin-containing neurons as well as other cell types (Kubofa et al., 1986; Jessel et al., 1978). The levels of both these neuropeptides have been shown by radioimmunoassays to change as a result of dopamine receptor blockade with

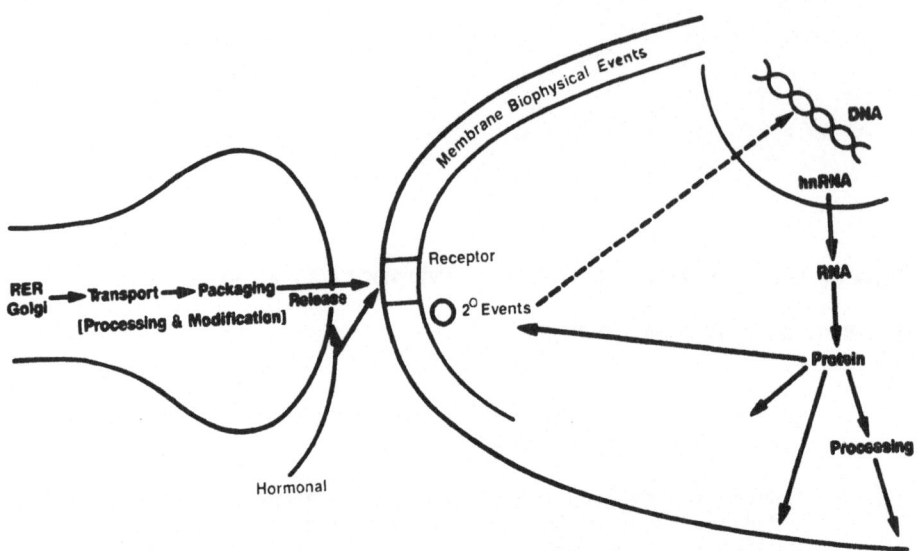

FIG. 1: Diagramatic representation of the sequence of events following receptor stimulation.

the dopamine antagonist haloperidol (Hong et al., 1978; Tang et al., 1983). Recent reports have demonstrated that chronic blockade of dopamine receptors affects the levels of proenkephalin and protachykinin mRNAs (Sabol et al., 1983; Bannon et al., 1986). The work presented here uses in situ hybridization and Northern blots to demonstrate that there is a differential effect on the mRNA levels for these two neuropeptides.

METHODS

Rats (n=8) were implanted with osmotic minipumps containing either haloperidol (37.5 μg/kg/hr, s.c.) or vehicle (3.5 M lactic acid). After 1,3, or 5 days they were sacrificed and the caudate nuclei were dissected over ice. Caudate tissues from two animals were pooled and total RNA was

FIG. 2: Tachykinin Nucleotide Sequences α- and β-Bovine vs. γ-Rat. A ^{32}P-labeled probe (39-mer) complementary to the bovine tachykinin sequence was used to screen a λgt11 rat brain cDNA library (Clontech, CA). Six clones were identified after an intial screening of approximately 5×10^{5} clones. After secondary screening one clone was found to give a positive hybridization signal. The λgt11 insert was subcloned into the EcoRI site of pBR322. Gel isolated fragments of decreasing size were generated by BAL 31 exonuclease digestion of the pBR322 subclone (Davis et al., 1986) and these were subcloned into the EcoRI-SmaI site of M13mp18. These M13 subclones were used for sequencing with the dideoxynucleotide-chain termination method of Sanger et al (1977). For comparison, the nucleotide sequences of the bovine α and β protachykinin mRNA forms are shown (Nawa et al., 1983). Amino acid translation based on the nucleotide sequence throughout the coding region for rat is also shown. The boxed regions indicate the tachykinin neuropeptides substance P and substance K and the regions where the complementary oligonucleotides were derived (see Figure 2 on next page).

```
   1
α              C    A GA      T A C
β              C    A GA      T A C
γ  ATGAAAATCCTCGTGGCGGTGGCGGTCTTTTTTCTCGTTTCCACT
   MetLysIleLeuValAlaValAlaValPhePheLeuValSerThr
                                |

              C      A        A          T C
              C      A        A          T C
   CAACTGTTTGCAGAGGAAATCGGTGCCAACGATGATCTAAATTAT
   GlnLeuPheAlaGluGluIleGlyAlaAsnAspAspLeuAsnTyr

                    C      G        A        C
                    C      G        A        C
   TGGTCCGACTGGTCCGACAGTGACCAAATCAAGGAGGCAATGCCG
   TrpSerAspTrpSerAspSerAspGlnIleLysGluAlaMetPro

   GA                G          T G
   GA                G          T G
   AGGCCCTTTGAGCATCTTCTTCAGAGAATCGCCCGAAGACCCAAG
   ArgProPheGluHisLeuLeuGlnArgIleAlaArgArgProLys

                    G                   TGATTCC
                    G                   TGATTCC
   CCTCAGCAGTTCTTTGGATTAATGGGCAAACGGGATGC.......
   ProGlnGlnPhePheGlyLeuMetGlyLysArgAspAl
   ─SP─

   TCAATTGAAAAGCAAGTGGCCCTGTTAAAGGCCCTTTA     A
   TCAATTGAAAAGCAAGTGGCCCTGTTAAAGGCCCTTTA     A
   ...................................TGGGCAT
                                      aGlyHis

        C  C T              ...................
        C  C T
   GGTCAGATCTCTCACAAAAGGCATAAAACAGATTCCTTTGTTGGA
   GlyGlnIleSerHisLysArgHisLysThrAspSerPheValGly
                                    ─SK─

                                              G T
                                              G T
   CTAATGGGCAAAAGAGCTTTAAATTCTGTGGCTTATGAAAGAAGC
   LeuMetGlyLysArgAlaLeuAsnSerValAlaTyrGluArgSer

   TG    G T T       A      G TACC     ..
   TG    G T T       A      G TACC     ..
   GCAATGCAGAACTACGAAAGAAGGCGTAAATAAACCCTGTAACGC
   AlaMetGlnAsnTyrGluArgArgArgLys

   .A  T    GG T T      A T GT AG  AA G
   .A  T    GG T T      A T GT AG  AA G
   ACTATCTATTCATCTCCATCTGTGTCCGCGAGCAGTGAGCGCTAA

     .G G CA  A         A              ...
     .G G CA  A         A              ...
   AATAAAAATGTGCGCTATGAGGAATGATTATTTATTTAATATCAA

   T      T   TT        G  ..GTA  TA
   T      T   TT        G  ..GTA  TA
   ATGTTGTTATGAGTGAAAAACTCAAAAAAGTGTTTATTTTTTCAT

          G   TATG      AAAA      T GTAATTCTAAT
          G   TATG      AAAA      T GTAATTCTAAT
   ATTGTGCCAATAAGCATTGTAATTCTAATGTGGT..........

   CTGA AA    T        C  G    T TC  CC
   CTGA AA    T        C  G    T TC  CC
   .GACCTCCTCAGACAGAAGTAGAAATTAGTTGTAACT..TCAGCA

            CA  A       A ACC AT GAA    AT T
            CA  A       A ACC AT GAA    AT T
   AAGCACAGTGTT.GATGGAGTTGTACAAGTTTGCCAGCGATGCAG

          C A  G  TCATTTCT T T  TG A C   CACAT
          C A  G  TCATTTCT T T  TG A C   CACAT
   TCTCCAAAGACAGAAA.......GGCTGCTGTGAGGGAGTGCAGG

   A  AC A  AAG AA     T  ACA A AGGC T     T
   A  AC A  AAG AA     T  ACA A AGGC T     T
   CGGCTCGTGCTGGAGGCAAGAAATCCTGTGTGTCTTGCGCTTCC

   .CAT   T   C   900
   .CAT   T   C   979
   CTTGGTTGCTTTTAT 841
```

prepared by the guanidine isothiocyanate method as described (Davis et al., 1986). Thirty micrograms of RNA were separated on 1% agarose-formaldehyde gels and blotted onto nitrocellulose (Davis et al., 1986).

These blots were probed with synthetic ^{32}P-labeled (Davis et al., 1986) oligonucleotide probes antinsense to rat proenkephalin mRNA (CTCCAC-GGGGTAAAGCTCATCCATCTTCTT; Howells et al., 1984) or rat protachykinin mRNA (Angulo et al., 1986; Fig. 2). The blots were also probed for α-tubulin (ATCTTTGGGGACCACATCACCACG; Lemischka et al., 1981) as a control. After washing the nonspecifically hybridized probe, the blots were autoradiographed and the intensity of the hybridization signals for proenkephalin mRNA, protachykinin mRNA or α-tubulin mRNA were quantified by densitometry. The mean and standard deviations were calculated and expressed as percent of control.

In other animals similarly treated with haloperidol or vehicle for 5 days, the animals were prepared for in situ hybridization to localize regional changes in these neuropeptide encoding transcripts. Adult male rats (200-220 g) were anesthetized with chloral hydsate and perfused through the aorta with 200 mL of ice-cold saline, then 150 mL of PLP (4% paraformaldehyde, 0.09 M lysine, and 1 mM sodium periodate in 0.1 M sodium phosphate buffer, pH 7.2), and subsequently with 150 mL of PLP/5% sucrose. The brains were postfixed in PLP/30% sucrose for 48 hr at 4^0C. Coronal brain sections (50 μm) were cut on a Lancer 1000 vibratome. The sections were digested at room temperature with proteinase K (1 μg/mL) for 10 min in 20 mM Tris-HCl (pH 7.5) containing 5 mM $CaCl_2$ and then delipidated sequen-

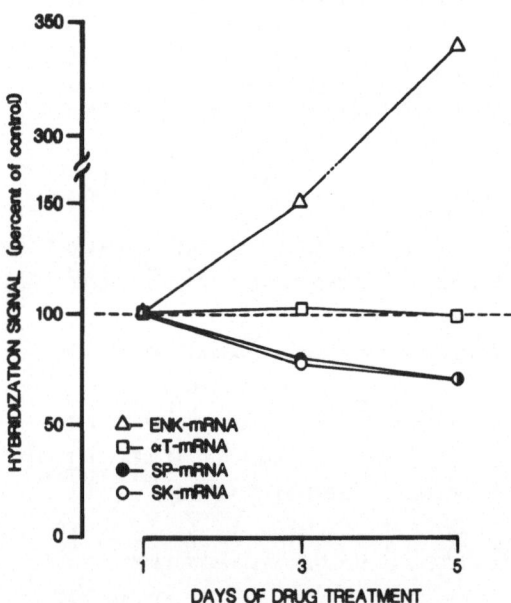

FIG. 3: Effect of haloperidol on the levels of striatal protachykinin mRNA and proenkephalin mRNA. Male rats (180 g) were anesthetized with halothane and osmotic minipumps were implanted (s.c.). The pumps delivered haloperidol (37.5 μg/kg/hr) or vehicle for up to five days. At the post-surgical times indicated, the rats were sacrificed and RNA was isolated from striatal tissue for Northern gel analysis. Optical densities of the hybridization signals for protachykinin (SP-probe, •; SK-probe, O), proenkephalin (Δ), and α-tubulin ($\square$) were normalized to time-matched vehicle-controls (100%). Each point represents the average of four experiments.

tially in the following ethanol solutions: 70%, 95%, 100%, 95%, 70% and then water for 20 sec each. The brain sections were transferred to a beaker containing 10 mL of hybridization solution (10X Denhardt's (Davis et al. 1986) and 100 μg/mL of denatured-sheared salmon sperm DNA) at a density of 5-8 sections/mL. Sufficient radiolabeled probe was added to yield 5.0 x 10^5 cpm/mL. The sections were hybridized overnight with mild shaking at 30°C and then washed 4 times for 15 min each (37°C) in 15 mL of 0.3 M sodium chloride and 30 mM sodium citrate buffer (pH 7.0). They are then placed in 0.15 M sodium chloride and 15 mM sodium citrate buffer (pH 7.0) for 30 min at 45°C. The sections were mounted on subbed slides and autoradiography was performed for 1-2 days using x-ray film (Kodak X-Omat RP) with intensifying screens (Du Pont Cronex Lightning Plus) at -70°C. Sections from treated and control animals were hybridized with either the ^{32}P-labeled oligonucleotide probes specific for rat proenkephalin mRNA or protachykinin mRNA.

RESULTS

Haloperidol treatment resulted in differential changes in neuropeptide-encoding mRNA levels (Fig. 3). Protachykinin mRNA levels progressively decreased to 70% of control levels concurrent with a 350% increase in proenkephalin mRNA levels. These progressive changes occurred while mRNA levels for α-tubulin remained unchanged. Total RNA recovered in each group was also unchanged during the treatment (data not shown).

In situ hybridization revealed that the change in mRNA levels in the caudate-putamen were uniform for protachykinin (Fig. 4c-4d), but for proenkephalin mRNA the increase was more prominent in the dorsal striatum (Fig. 4a-4b). In neither case were changes observed in other brain regions that contain these mRNA species.

DISCUSSION

The successful application of in situ hybridization histochemistry to detect neuropeptide mRNA represents a useful tool in determining the regional distribution of neuropeptide mRNA within structures such as the striatum. Chronic haloperidol treatment resulted in a uniform elevation of proenkephalin mRNA in the caudate-putamen of the rat, an observation consistent with the results of Shivers et al. (1986). A uniform decrease in protachykinin mRNA was also observed in the caudate-putamen following chronic haloperidol treatment. Furthermore, in situ hybridization provides a convenient method for analyzing how manipulation (e.g., drugs, lesions, stimulation) of specific afferents alters neuronal gene expression. For example, the effect of unilateral destruction of dopaminergic neurons on proenkephalin mRNA have been dramatically visualized by in situ hybridization (Angulo et al., 1986; Bloch et al., 1986; Young et al., 1986).

The haloperidol-induced reduction in the level of striatal protachykinin mRNA presumably causes the parallel decrease in nigral substance P detected by us and others (Hong et al., 1978), whereas the elevation of proenkephalin mRNA results in higher levels of enkephalin in striatum (Tang et al., 1983). The simultaneous reduction of protachykinin mRNA and the elevation of proenkephalin mRNA occurring within the striatum after dopamine receptor blockade indicate that receptor mediated events can lead to differential regulation of mRNA levels. Whether the mechanisms underlying these changes in mRNA are at the level of transcriptional regulation or RNA stability awaits investigation. Regardless, these events do not appear to be of a general nature because the changes were opposite in direction and α-tubulin remained unchanged. Although our data do not address whether

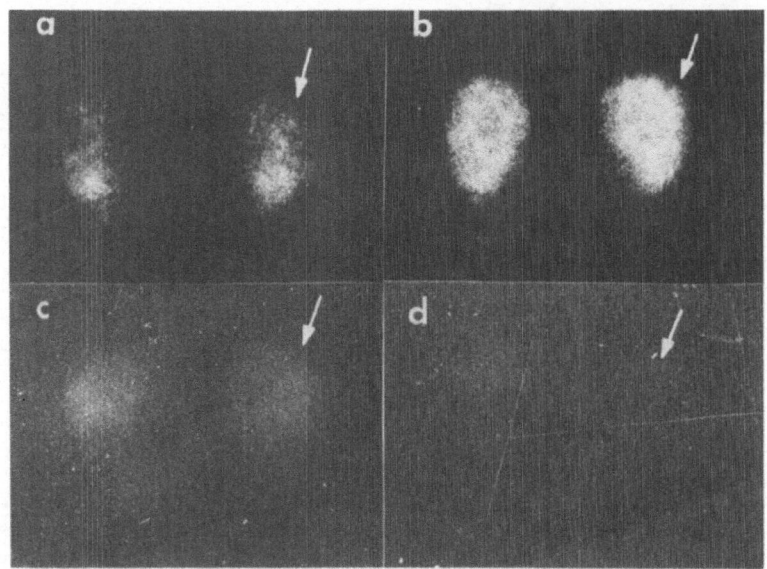

FIG. 4: <u>In situ</u> hybridization histochemistry: Effect of haloperidol treatment on striatal protachykinin mRNA and proenkephalin mRNA. Male
rats were anesthetized and implanted with osmotic minipumps as
described in Figure 3. After 5 days of haloperidol treatment, the
animals were sacrificed and brain sections were prepared for <u>in
situ</u> hybridization. Sections for vehicle-control and haloperidol
treatment were processed simultaneously and exposed to the same
x-ray film. a and c: Vehicle-control sections showing hybridization signal for proenkephalin mRNA and protachykinin mRNA, respectively. b and d: Coronal sections from haloperidol-treated rat
showing the increase in hybridization signal for proenkephalin
mRNA (b) and the decrease in signal for protachykinin mRNA (d).
Arrows point to the caudate-putamen.

tachykinin and enkephalinergic cell types in striatum are interconnected or
are independently affected by dopamine receptor blockade, albeit differentially, our results do imply that altering receptor-mediated neurotransmitter functions can lead to profound, specific, and long-lasting alterations
in neuronal gene expression. `

REFERENCES

Angulo, J. A., Davis, L. G., Burkhart, B. A. and Christoph, G. R., Reduction of striatal dopaminergic neurotransmission elevates striatal
proenkephalin mRNA, <u>Eur. J. Pharm.</u> 130:341-343 (1986).
Angulo, J. A., Manning, R. W., Davis, L. G., Burkhart, B. A. and Christoph,
G. R., <u>In situ</u> hybridization of substance P mRNA in rat brain using
synthetic oligonucleotide probes based on the nucleotide sequence of
the rat substance P mRNA, <u>Soc. Neurosci.</u> 12 (Abst. #379.6):1397
(1986).
Bannon, M. J., Lee, J.-M., Giraud, P., Young, A., Affolter, H.-U. and Bonner, T. I., Dopamine antagonist haloperidol decreases substance P,
substance K, and preprotachykinin mRNAs in rat striatonigral neurons,
<u>J. Biol. Chem.</u> 261:6640-6644 (1986).

Bloch, B., Popovici, T., Le Guellec, D., Normand, E., Chouham, S., Guitteny, A. F. and Bohlen, P., In situ hybridization histochemistry for the analysis of gene expression in the endocrine and central nervous system tissues: A 3-year experience, J. Neurosci. Res. 16:183-200 (1986).

Davis, L. G., Dibner, M. D. and Battey, J. F., Basic Methods in Molecular Biology, Elsevier (New York) (1986).

Ehrlich, Y. H., Phosphoproteins as specifiers for mediators and modulators of neural function. In: Modulators, Mediators, and Specifiers in Brain Function edited by Y. H. Ehrlich, J. Volavka, L. G. Davis, and E. G. Brunngraber (Plenum, New York), Adv. Exp. Med. and Biol. 116:75-101 (1979).

Hong, J. S., Yong, H. Y. and Costa, E., Substance P content of substantia nigra after chronic treatment with antischizophrenic drugs, Neuropharmacol. 17:83-85 (1978).

Howells, R. D., Kilpatrick, D. L., Bhatt, R., Monahan, J. J., Poonian, M. and Udenfriend, S., Molecular cloning and sequence determination of rat preproenkephalin cDNA: Sensitive probe for studying transcriptional changes in rat tissue, Proc. Natl. Acad. Sci. USA 81:7651-7655 (1984).

Jessell, T. M., Emson, P. C., Paxinos, G. and Cuello, A. C., Topographic projections of substance P and GABA pathways in the striato- and pallidonigral system, Br. Res. 152:487-498 (1978).

Kebabian, J. w. and Calne, D. B., Multiple receptors for dopamine, Nature 277:93-96 (1979).

Kubota, Y., Inagaki, S., Kito, S., Takagi, H. and Smith, A. D., Ultrastructural evidence of dopaminergic input to enkephalinergic neurons in rat neostriatum, Br. Res. 367:374-378 (1986).

Lemischka, I. P., Farmer, S., Racaniello, V. R. and Sharp, P. A., Nucleotide sequence and evolution of a mammalian alpha-tubulin messenger RNA, J. Mol. Biol. 151:101-120 (1981).

Nawa, H., Hirose, T., Takashima, H., Inayama, S. and Nakanishi, S., Nucleotide sequences of cloned cDNAs for two types of bovine brain substance P precursor, Nature 306:32-36 (1983).

Sabol, S. L., Yoshikawa, K. and Hong, J.-S., Regulation of methionine-enkephalin precursor messenger RNA in rat striatum by haloperidol and lithium, Biochem. Biophys. Res. Comm. 113:391-399 (1983).

Sanger, F., Nicklen, S. and Coulson, A. R., DNA sequencing with chain-terminating inhibitors, Proc. Natl. Acad. Sci. USA 75:5463-5467 (1977).

Shivers, B. D., Harlan, R. E., Romano, G. J., Howells, R. D. and Pfaff, D. W., Cellular localization of proenkephalin mRNA in rat brain: Gene expression in the caudate-putamen and cerebellar cortex, Proc. Natl. Acad. Sci. USA 83:6221-6225 (1986).

Siggins, G. R., Hoffer, B. J., Bloom, F. E. and Ungerstedt, U., Cytochemical and Electrophysiological studies of dopamine in the caudate nucleus, in: The Basal Ganglia edited by M. D. Yahr, Raven Press, NY 227-248 (1976).

Tang, F., Costa, E. and Schwartz, J. P., Increase of proenkephalin mRNA and enkephalin content of rat striatum after daily injection of haloperidol for 2 to 3 weeks, Proc. Natl. Acad. Sci. USA 80:3841-3844 (1983).

Young, W. S. III, Bonner, T. I., Brann, M. R., Mesencephalic dopamine neurons regulate the expression of neuropeptide mRNAs in the rat forebrain, Proc. Natl. Acad. Sci. USA 83:9827-9831 (1986).

B-50 PHOSPHORYLATION, PROTEIN KINASE C AND THE INDUCTION

OF EXCESSIVE GROOMING BEHAVIOR IN THE RAT

Louise H. Schrama, Pierre N. E. De Graan, A. Beate
Oestreicher and Willem Hendrik Gispen

Division of Molecular Neurobiology, Institute of Molecular
Biology and Medical Biotechnology and Rudolf Magnus
Institute for Pharmacology, University of Utrecht
Padualaan 8, 3584 CH Utrecht, The Netherlands

INTRODUCTION

Behaviorally active neuropeptides might affect synaptic plasticity by
changing the degree of phosphorylation of synaptic proteins. Neuronal
electrical activity and neurotransmission are accompanied by covalent modification of synaptic proteins through cyclic phosphorylation and dephosphorylation (c.f. Weller, 1979). Studies using behavioral paradigms similar
to those used to measure the behavioral effects of melanocortins (ACTH/MSH)
suggested that the acquisition of new information may be accompanied by
changes in the degree of phosphorylation of synaptic phosphoproteins
(Glassman et al., 1973). The original idea was to study the in vitro modulation of synaptic plasma membrane phosphorylation by $ACTH_{1-24}$ and its
behaviorally active fragments and to compare the structural requirements
of ACTH in this assay with those influencing the extinction of active
avoidance behavior (Greven and De Wied, 1973). In our first study along
this line, we noted that high concentrations of $ACTH_{1-24}$ indeed inhibited
the endogenous phosphorylation of several phosphoproteins in rat brain
synaptic membranes (Zwiers et al., 1976). These phosphoproteins were phosphorylated by a cyclic AMP-independent mechanism, at that time the most
important phosphorylation system studied (Zwiers et al., 1976). Our next
step was to investigate the nature of the affected substrate proteins and
their corresponding kinase(s).

The first characterized substrate protein and its kinase, B-50 and
B-50 kinase, were isolated in a complex from synaptosomal plasma membranes
(Zwiers et al., 1980). The B-50 kinase has been shown to be very similar,
if not identical to protein kinase C (Aloyo et al., 1982, 1983). In the
present paper we review our current knowledge on B-50 and discuss the evidence that the degree of B-50 phosphorylation may modulate receptor-mediated
polyphosphoinositide breakdown (Gispen et al., 1985a). The protein B-50
discussed in this paper is most likely identical to protein $\gamma5$ of Rodnight
(Rodnight, 1982, Gower and Rodnight, 1982), to p54 (Ca)p by Mahler (Mahler
et al., 1982), to protein F_1 by Routtenberg (Routtenberg et al., 1985,
Akers and Routtenberg, 1985, Gispen et al., 1986) and to pp46 described
by Pfenninger (Katz et al., 1985, De Graan et al., 1985). Furthermore we
will put emphasis on the possible role of kinase C and possibly B-50 phos-

393

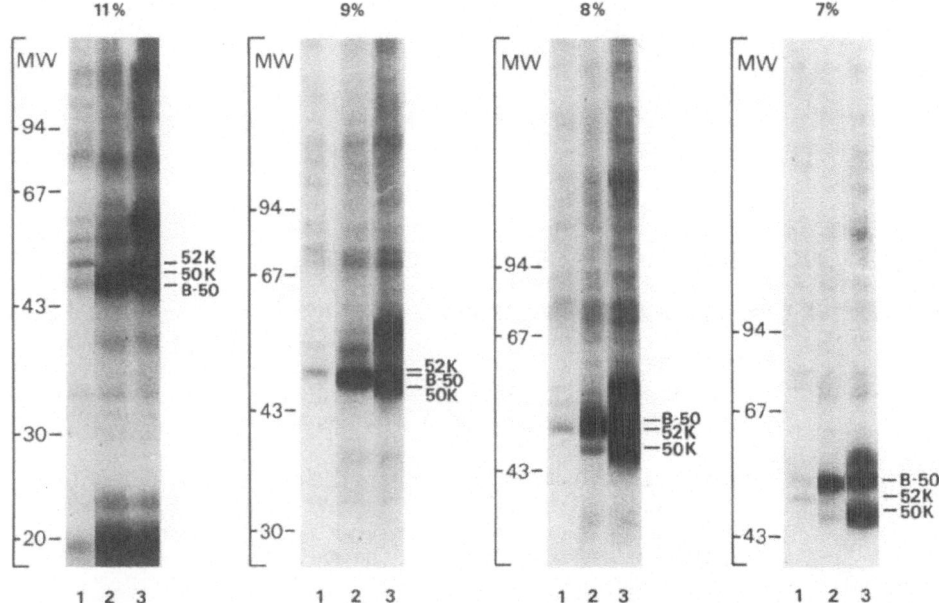

FIG. 1: SPM was phosphorylated in 10 mM Na-acetate, 10 mM Mg^{2+}-acetate, 0.5 mM EGTA (pH6.5) and varying Ca^{2+} concentrations for 15 sec at 30°C. The phosphorylation assay was carried out in the absence of Ca^{2+} (lanes 1), in the presence of 0.6 mM Ca^{2+} (lanes 2) and in the presence of 0.6 mM Ca^{2+} and 5U calmodulin in 25 µl (lanes 3). The proteins were separated on 4 different gels, containing 11% acrylamide and 0.2% bisacrylamide, and 9, 8 and 7% acrylamide containing the same ratio of acrylamide to bisacrylamide as the 11% gel. The numbers to the left indicate the position of the Pharmacia low molecular weight markers in kDa, the identity of the three phosphoproteins indicated to the right is explained in the text.

phorylation in some of the behavioral effects of ACTH.

Biochemical Characterization of B-50

    Initial phosphorylation studies on B-50 have been performed in synaptosomal plasma membranes, and these membranes have been used as the source for the isolation and characterization of both B-50 and B-50 kinase. B-50 is an acidic (pI 4.5) 48 kDa protein that is intimately associated with the synaptic plasma membrane, since it can only be solubilized in the presence of detergent (Zwiers et al., 1979, 1980). The B-50 protein displays heterogeneity both in its relative molecular weight on SDS-PAGE and in its isoelectric point. Depending on the percentage crosslinking of the gel and on the percentage of acrylamide in the gel used, B-50 can be found between 43 and 54 kDa. The lower the cross-linking of the gel and/or the percentage acrylamide used, the higher molecular weight B-50 displays (Fig. 1, Gower and Rodnight, 1982). The 11% gel contained 0.2% bisacrylamide, and the ratio of acrylamide to bisacrylamide was kept constant for the lower percentage gels (Fig. 1). We have distinguished between calcium- and calmodulin-dependent phosphorylation by incubating SPM either in the presence of EGTA (0.5 mM EGTA, lanes 1), in the presence of calcium (0.5 mM EGTA and 0.6 mM Ca^{2+}, lanes 2), or in the presence of 5 U calmodulin (lanes 3). The 50 kDa phosphoprotein is the lower molecular weight autophosphorylated subunit of the calcium/calmodulin-dependent protein kinase II (Schrama et al., 1986b) and the 52 kDa phosphoprotein is the major phosphorylated coated vesicle protein (Schrama et al., 1986b, De Graan et al., 1986b). These data indicate that the molecular weight of B-50 determined by SDS-PAGE will be higher, the lower the amount of acrylamide in the gel. This characteristic of the B-50 protein has added to the confusion on the nature of this protein in various laboratories.

Separation of purified B-50 on a narrow pH gradient (pH 5.0-3.5) in the first dimension and on SDS-PAGE in the second dimension, results in 4 distinctprotein spots (48 kDa on 11% acrylamide) which are partly interconvertible by extensive phosphorylation or dephosphorylation of B-50 (Zwiers et al., 1985). From these data the conclusion was drawn that B-50 contains at least 2 phosphorylatable sites, as confirmed by limited digestion of the protein with Staphylococcus aureus protease V-8 (Zwiers et al., 1985). The microheterogeneity displayed in isoelectric focussing is most probably not the result of the presence of a glycomoiety on B-50, since several methods failed to detect the presence of sugar residues (Zwiers et al., 1985).

Phosphorylation Characteristics of B-50

The phosphorylation of B-50 by its endogenous B-50 kinase in synaptosomal plasma membranes (SPM) is inhibited by ACTH and several of its congeners and this effect of ACTH was the result of inhibition of B-50 kinase rather than a stimulation of B-50 phosphatase (Zwiers et al., 1976, 1978). The activity of the kinase is not influenced by cyclic nucleotides (Zwiers et al., 1976), but is highly sensitive to calcium (Zwiers et al., 1980). Calmodulin in our hands, either purified from calf brain or obtained from a commerical source (Sigma, St. Louis, MO) does not stimulate nor inhibit the phosphorylation of B-50 tested at several calcium concentrations (Zwiers et al., 1980). This in contrast with the results of Gower et al. (1986), who showed that a small but significant stimulation of B-50 phosphorylation could be observed at 50-130 nM free calcium; a similar observation was made by the group of Pfenninger (Katz et al., 1985). Dunkley and co-workers (Dunkley et al., 1986) also showed that the phosphorylation of B-50 in synaptosomes with $^{32}P_i$ is calmodulin-sensitive. This discrepancy could be the result of another yet unidentified kinase phosphorylating B-50 present in crude brain extracts. It should also be beared in mind that inhibition of B-50 phosphorylation by calmodulin at low calcium concentrations could be the result of binding of calcium to calmodulin, thereby lowering the effective free calcium concentration (see also Gower et al., 1986).

The protein kinase phosphorylating B-50 is similar if not identical to protein kinase C (Aloyo et al. 1982, 1983). This conclusion was based on the following data: i) both kinases are cyclic nucleotide-independent, ii) out of a number of kinases, only B-50 kinase and protein kinase C were able to phosphorylate B-50, iii) exogenously added kinase C to native SPM preferably phosphorylated B-50, iv) both protein kinase C and B-50 kinase are sensitive to calcium, phospholipid, ACTH, chlorpromazine and can be activated by proteolytic breakdown by Ca^{2+}-dependent proteases, v) the peptide maps produced by Staphylococcus aureus protease V8 are identical (Aloyo et al., 1982, 1983).

Localization of B-50

Initial immunohistochemical localization of B-50 with antibodies raised against a rat brain membrane extract revealed its presence in neuropil areas of both cerebellum and hippocampus (Oestreicher et al., 1981), whereas it was absent from cell bodies in these regions. This pattern of distribution of B-50 immunoreactivity closely resembles that found by $[^3H]$-PDB (phorbol 12,13-dibyturate) binding in both cerebellum and hippocampus

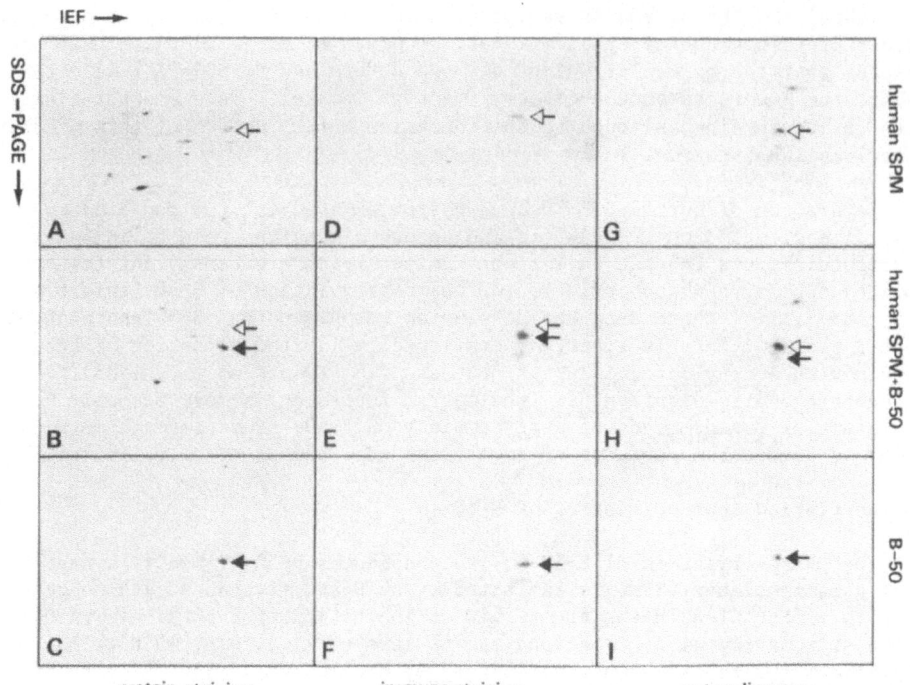

FIG. 2:   Human SPM was phosphorylated under standard conditions (Kristjansson et al., 1982) and phosphorylated tracer B-50 was prepared as
described by Oestreicher et al., (1986). Human SPM and/or rat
B-50 were separated by isoelectric focussing (IEF) in the first
dimension on a pH 9-4 gradient, followed by SDS-PAGE in the second
dimension. The position of rat B-50 is indicated by a filled arrow
head and the position of the human B-50-like protein by an open
arrow head.

(Woley et al., 1986). The binding of PDB is most probably an indicator for
the presence of protein kinase C, since this kinase is considered to be the
receptor for phorbol diesters (Castagna et al., 1982, Kikkawa et al., 1983,
Niedel et al., 1983). On the light microscopic level, this indicates that
B-50 and its protein kinase are co-localized in neuron-enriched areas.
Specific anti-B-50 immunoglobulines were purified from the crude antiserum
by affinity chromatography on a Sepharose-4B column conjugated with purified B-50 (Oestreicher et al., 1983). With the purified anti-B-50-IgGs,
B-50 could be demonstrated in brain homogenates of various mammals and
birds, whereas no cross-reactivity was found in fish and amphibians (Oestreicher et al., 1984). Adult human brain also contains a B-50-like protein, as analyzed by Western blot of a two-dimensional polyacrylamide gel.
Fig. 2 shows the migration of purified rat B-50 with human SPM after two-
dimensional separation on isoelectric point and molecular weight (panels
A-C, protein staining, panels G-1, autoradiogram) and the cross-reaction
of the human B-50-like protein with anti-B-50 antiserum on a Western blot
(panels D-F). The human B-50-like protein has a very similar isoelectric
point to rat B-50, but the molecular weight is slightly higher, about 52
kDa (Fig. 2, see also De Graan et al., 1986a). Such species differences
in M_r of B-50 have been reported earlier (Oestreicher et al., 1984).

    In the rat, the B-50 protein could only be detected in brain and spi-

TABLE 1: B-50 levels in total homogenates (RIA) and endogenous B-50 phosphorylation in SPM of various rat brain regions.

Brain region	B-50 pg/µg tissue*	Endogenous B-50 phosphory-lation in SPM**
Cortex cerebrum	43.2 ± 2.2	8.8 ± 0.4
Septum	84.4 ± 7.3	10.8 ± 0.2
Hippocampus	36.7 ± 2.4	7.2 ± 0.9
Periaquaductal gray	53.7 ± 5.1	n.d.
Cerebellum	17.6 ± 0.5	2.7 ± 0.1
Medulla spinalis	9.2 ± 1.1	0.4 ± 0.1

* n=9; Medulla spinalis n=6. ** Data obtained from Kristjansson et al.
(1982). n.d.: not determined.

nal cord with the crude B-50 antiserum and endogenous B-50 phosphorylation
(Kristjansson et al., 1982). Using the endogenous phosphorylation of B-50
in SPM prepared from several brain regions and two-dimensional analysis, a
clear regional distribution of B-50 phosphorylating activity was observed.
The order of decreasing endogenous phosphorylation activity was septum >
hippocampus/neocortex > cerebellum > medullla spinalis (Kristjansson et al.,
1982, Table 1). Recently, a radioimmunoassay (RIA) for B-50 was developed
(Oestreicher et al., 1986) and using this more accurate method to quantify
the regional distribution of B-50 we found that the distribution pattern was
the same as that observed by Kristjansson et al., (1982, Table 1). These
data support our earlier suggestion that the regional difference in endogenous B-50 phosphorylation reflects the content of B-50 rather than the
activity of B-50 kinase (Kristjansson et al., 1982).

Ultrastructural localization of B-50 was assessed in cryosections of
adult rat hippocampus with affinity-purified anti-B-50 lgGs and visualized
with protein A coupled to gold particles. These studies revealed that B-50
is predominantly associated with the presynaptic terminals and that the
postsynaptic protein A-gold staining was not different from that obtained
with preimmune lgGs (Gispen et al., 1985b). This indicates that the B-50
protein is a presynaptic protein presumably associated with the inner surface of the plasma membrane (Gispen et al., 1985b, see also Sörensen et
al., 1981).

B-50 Phosphorylation, Kinase C and Polyphosphoinositide Metabolism

Kinase C activation and B-50 phosphorylation

The nervous tissue specific protein B-50 is phosphorylated by protein
kinase C (see previous section, Aloyo et al., 1982, 1983). Since this protein kinase can be stimulated by phorbol esters and diacylglycerol (Takai
et al., 1979, Takai et al., 1985, Nishizuka, 1984), we studied the effects
of these modulators of protein kinase C on B-50 phosphorylation in SPM
either by endogenous B-50 kinase or by added purified protein kinase C.
The synthetic short chain diacylglycerol, 1,2-dioctanoylglycerol (DOG) stimulated the phosphorylation of B-50 at concentrations of 100-300 µM in isolated synaptic plasma membranes, either in the absence or in the presence
of exogenous protein kinase C. Comparable stimulation of protein kinase C
was achieved with 1 µM DOG when histone was used as the substrate (Eichberg
et al., 1986). Phorbol 12,13-dibutyrate (PDB) was a much more effective

stimulator of endogenous B-50 kinase activity than DOG. This phorbol diester stimulated the B-50 phosphorylation at 10 nM. Phorbol 12,13-diacetate
(PAA) and phorbol 12-myristate 13-acetate (PMA) were effective at concentrations above 1 μM. The 4α-phorbol did not affect B-50 kinase at concentrations as high as 100 μM (De Graan et al., manuscript in preparation). In
general both the effect of the short chain diacylglycerol DOG and of the
tumor promoting phorbol diesters could only be detected at low calcium concentrations (300 nM) and were optimal when ionic conditions approximating
intracellular concentrations in the brain were used (Eichberg et al., 1986).
These data further support the hypothesis that B-50 kinase is indeed identical to protein kinase C.

B-50 phosphorylation and polyphosphoinositide (PPI) metabolism

In response to receptor activation by a variety of hormones, neurotransmitters and other extracellular messages, hydrolysis of phosphatidylinositol 4,5-bisphosphate (PIP_2) by phosphodiesteratic cleavage leads to the
formation of two products, diacylglycerol (DAG) and D-myo-inositol 1,4,5-
trisphosphate (IP_3) (Berridge and Irvine, 1984). DAG is considered to be
the activator of protein kinase C (Nishizuka, 1984), while IP_3 is capable
of mobilizing Ca^{2+} from most probably the endoplasmic reticulum (Streb et
al., 1983). The simultaneous processes of protein phosphorylation and Ca^{2+}
mobilization are thought to constitute synergistic effects, which are integral to a large number of cellular responses (Nishizuka, 1984, Kikkawa et
al., 1986).

Some years ago we observed that ACTH, when added to isolated SPM, stimulated the phosphorylation of PIP_2 (Jolles et al., 1981, Table 2). In this
same membrane preparation it was already shown that ACTH inhibited B-50
phosphorylation (Zwiers et al., 1976, 1978, Table 2). The same effects of
ACTH were found in a membrane fraction prepared from human brain, where the
peptide inhibited the phosphorylation of a B-50-like protein of 52 kDa (see
Fig. 2) and stimulated PIP_2 phosphorylation (Table 1). Furthermore Jolles
et al. (1980) demonstrated that the degree of phosphorylation of B-50 modulated the activity of PIP-kinase in SP_{55-80}, a fraction obtained from the

TABLE 2: Modulation of polyphosphoinositide (PPI) metabolism and B-50 phosphorylation in a crude rat and human plasma membrane fraction by
$ACTH_{1-24}$

	Rat		Human	
	control	ACTH	control	ACTH
PIP_2	20.5±0.5	66.2±5.0*	15.4±1.8	47.0±1.0*
PIP	32.8±2.6	25.4±2.1	17.3±0.2	15.6±1.1
PI	9.8±1.3	7.2±1.3	5.6±1.1	8.3±1.6
PA	17.2±2.0	18.5±1.4	6.1±0.4	7.0±0.4
B-50	74.3±7.8	36.7±2.2*	27.3±1.0	10.1±0.4*

The incorporation of phosphate into the phospholipids is expressed as
pmol/μg protein; the incorporation of phosphate into B-50 is expressed as
peak height above background. The data are presented as mean ± S.E.M.
(n=3). *2p<0.02, determined with Students t-test.

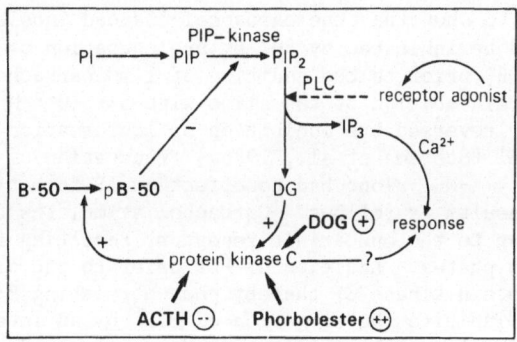

Fig. 3: Feedback control mechanism by B-50 phosphorylation in receptor-mediated hydrolysis of inositol phospholipids. For explanation of the
scheme see text.

isolation procedure of B-50 (Zwiers et al., 1980). This fraction contains
PIP-kinase activity as well as B-50 and its kinase, whereas phospholipase
C is absent. Therefore, these observations were interpreted as a direct
influence of the phosphoprotein B-50 on PIP-kinase activity. Again an
inverse relation existed, as in SPM incubated with ACTH, between the degree
of B-50 phosphorylation and PIP-kinase activity. Two further examples of
this inverse relationship were demonstrated with dopamine in the hippocampal slice (Jork et al., 1984) and with anti-B-50 IgGs in SPM (Oestreicher
et al., 1983).

PIP kinase (MW 45 kDa, IEP 5.8) has been purified from rat brain and
was identified by means of specific immunostaining and reduction of enzyme
activity due to interaction with affinity-purified anti-45 kDa protein
antibodies (Van Dongen et al., 1984, 1986). Purified B-50 was phosphorylated with protein kinase C or dephosphorylated with alkaline phosphatase
(Zwiers et al., 1985), after which both B-50 forms, the acidic phosphorylated form and the dephosphorylated basic form, were purified by flat-bed
isoelectric focussing on a narrow pH gradient. When the effect of phosphorylated or dephosphorylated B-50 was tested on the activity of purified PIP
kinase, the dephosphorylated form had no effect on PIP kinase activity,
wherease the identical amount of phosphorylated B-50 substantially dimi-
ished the formation of PIP_2. Non-specific protein-protein interactions
were prevented by the presence of bovine serum albumin (Van Dongen et al.,
1985).

These findings have led us to propose that B-50, B-50 kinase and PIP
kinase exist together in a multi-molecular complex in the presynaptic plasma
membrane and that the phosphorylation of B-50 exerts a regulatory effect on
PIP kinase. This implies that the nervous tissue specific phosphoprotein
B-50 may be part of a feedback control mechanism in the receptor-mediated
hydrolysis of inositol phospholipids, by regulating the amount of PIP_2 available for phosphidiesteratic cleavage after receptor activation (Fig. 3).
The hypothesis that phosphorylation of a kinase C substrate protein exerts
a negative feedback function is supported by several groups and in several
systems (see results from Vicentini et al., 1985, Labarca et al., 1984,
Watson and Lapetina, 1985, Orellana et al., 1985, Leeb-Lundberg et al.,
1985, Okano et al., 1985).

Recently we have been able to demonstrate that the inhibitory effect
of protein kinase C on receptor-mediated polyphosphoinoisitide hydrolysis
in hippocampal slices could be partially reversed by preincubation of the
slice with $ACTH_{1-16}$. In a similar system as employed by Labarca et al.,

(1984) we were able to show that the carbachol-induced inositol phosphate (IP) formation could be inhibited by 50% by preincubation of the slice for 10 min. with 1 μM PDB, prior to the addition of 1 mM carbachol (Schrama et al., 1986a). Pre-preincubation of the slice with 5×10^{-5} M $ACTH_{1-16}$-NH_2 for 20 min partially reversed the inhibition of IP formation by PDB in response to carbachol (Schrama et al., 1986a). Incubation of the hippocampal slice with $ACTH_{1-16}$-NH_2 alone had no effect on IP formation. We have interpreted these results as follows. Carbachol stimulates the hydrolysis of PIP_2 after binding to the muscarinic receptor, resulting in the formation of inositol phosphates. Addition of PDB prior to the carbachol stimulation activates protein kinase C, thereby phosphorylating B-50 (Fig. 3). Phosphorylated B-50 inhibits the formation of PIP_2 by an interaction with PIP kinase, which will lower the amount of PIP_2 available for receptor-mediated hydrolysis. The reversal of the inhibition by PDB by pre-preincubation of the slice with ACTH is most probably due to inhibition of B-50 phosphorylation by the peptide, resulting in a less inhibited PIP kinase thereby leaving more PIP_2 available for hydrolysis. In human platelets PMA also inhibits the degradation of PIP_2 in response to thrombin. The inhibitor of protein kinase C 1-(5-isoquinolinesulfonyl)-2-methylpiperazine (H7) could reverse the inhibitory effect of PMA on phosphoinositide turnover and Ca^{2+}-mobilization (Tohmatsu et al., 1986). Concomitant with this finding these

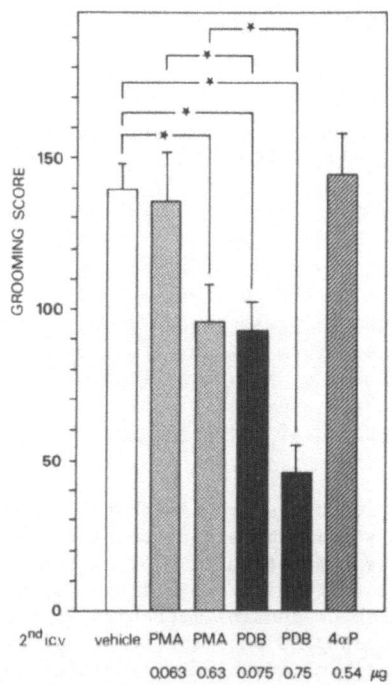

FIG. 4: Modulation of ACTH-induced excessive grooming by phorbol esters. All rats received an i.c.v. injection with 0.1 μg $ACTH_{1-24}$in 3 μl, followed at t = 10 min by a second injection with either vehicle (0.5% ethanol in saline), phorbol 12-myristate 13-acetate (PMA), phorbol 12,13-dibutyrate (PDB) or 4α-phorbol (4αP) at the amount indicated in a volume of 3 μl. The bars represent the mean grooming score ± SEM (number of animals was at least 5 in each group). The asterisk indicates a significant difference at $p < 0.05$ determined with one-factor analysis of variance followed by a supplemental t-test.

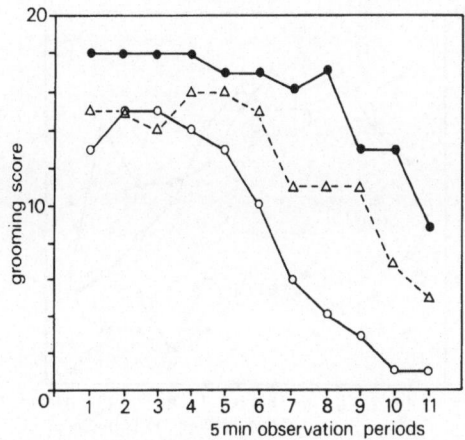

FIG. 5:   Effect of DOG on ACTH-induced excessive grooming. Animals received
a 1st i.c.v. injection with saline followed by vehicle (open circles), or a 1st injection with 0.1 μg ACTH$_{1-24}$ followed at 10 min by
either vehicle (closed circles) or 10 μg DOG (open triangles). The
mean grooming score over 5 min observation periods (maximal score
20, 5 rats per group) is given for a total grooming period of 55
min.

authors showed that the phosphorylation of the 20 kDa protein was reduced
to the basal level after preincubation of the platelets with H7 before addition of PMA, and that the phosphorylation of the 40 kDa protein was only
partially inhibited by pretreatment with H7. Whether one of these two
phosphoproteins in the platelet may serve a similar feedback function as
B-50 in nervous tissue remains to be proven.

Activators of Protein Kinase C and ACTH-Induced Excessive Grooming in the
Rat

In a variety of studies we have used the peptide ACTH$_{1-24}$ as a tool to
modulate the phosphorylation of B-50, irrespective of the question whether
such modulation represents a physiological mechanism by which the peptide
could affect the brain. The use of broken cell preparations and the relatively high peptide concentrations required to inhibit B-50 phosphorylation
(IC_{50} = 3 x 10^{-6} M) have cast some doubt as to the physiological importance
of the inhibition of protein kinase C by ACTH$_{1-24}$ in rat brain membrane preparations. Recent experiments on the sensitivity of endogenous B-50 phosphorylation to ACTH$_{1-24}$ in SPM indicate that under special conditions
ACTH$_{1-24}$ significantly inhibits B-50 phosphorylation at concentrations as
low as 10^{-7} M.

Evidence is accumulating to suggest that the proposed mechanism of
action of ACTH is of relevance to the induction of excessive grooming behavior (Gispen et al., 1975). First of all, it was observed that intracerebroventricular (i.c.v.) administration of ACTH$_{1-24}$ in rats followed by a
post-hoc endogenous phosphorylation assay of SPM prepared from the brains
of these rats, resulted in a dose- and time-dependent change in the phosphorylation of those phosphoproteins that were affected by ACTH when added
to the phosphorylation assay <u>in vitro</u> (Zwiers et al., 1977). Thus, the
effect of the peptide can be induced in the intact system and is not just
an artefact of broken cell preparations.

Secondly, we have shown that the structural requirements of ACTH for

401

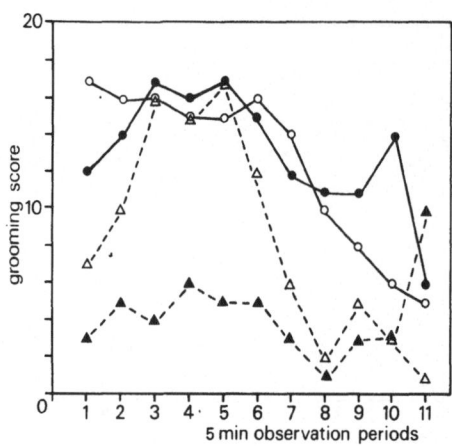

FIG. 6: Effect of phorbol esters on ACTH-induced excessive grooming. Animals received an i.c.v. injection with 0.1 μg ACTH followed at 10
min by a 2nd injection with vehicle (open circles), 0.63 μg PMA
(open triangles), 0.75 μg PDB (filled triangles) or 0.54 μg 4αP
(closed circles). For explanation of the grooming score see legend
to Fig. 5 and text.

the in vitro inhibition of B-50 kinase in SPM are very similar to those
required for the ACTH-induced excessive grooming behavior in the rat (Zwiers
et al., 1978, Gispen et al., 1979). Furthermore, one of the regions which
is also enriched in B-50 (see Table 1), the periaqueductal gray (PAG), is
known to receive peptidergic terminals containing peptides from the pro-
opiomelanocortin (POMC) family (Watson et al., 1978). The PAG is the primary target for the induction of excessive grooming behavior by ACTH-like
peptides (Spruijt et al., 1986). Lesions placed in the dorsal part of the
PAG, prevented the display of excessive grooming after i.c.v. injection of
$ACTH_{1-24}$.

Thirdly, we have recently reported that treatment of rats with DOG
suppresses ACTH-induced excessive grooming (Gispen et al., 1985c). Here,
we will describe more detailed information on the inhibition of ACTH-induced
grooming by both DOG and phorbol diesters.

One week prior to the grooming test, a polyethylene cannula was implanted in the third brain ventricle (Gispen and Isaacson, 1981). The rats
received an i.c.v. injection of either saline or $ACTH_{1-24}$(0.1 μg/3.0 μl) at
the beginning of the behavioral test. Ten min after the first injection the
rats received a second i.c.v. injection with either 0.5% ethanol in saline
(vehicle) or an injection with either DOG or phorbol diesters, both of which
were dissolved in vehicle (total injection volume 3 μl). After this second
injection the animals were placed individually in novel glass boxes and the
observation of the rats started 10 min after the second injection and lasted
for 55 min. Every 15th sec an observer scored whether the rats displayed
elements of the grooming repertoire, yielding a maximal score of 220. The
individual grooming scores were averaged per group and analyzed by one-factor of variance followed by an adjusted supplemental t-test. The results
of such an experiment for 3 different phorbol esters are given in Fig. 4.
In order to compare the relative potency of the 3 phorbol esters with DOG
(Gispen et al., 1985c), equimolar amounts to 1 μg DOG were used. As can
be seen from the data presented in Fig. 4, the non-active 4-α -phorbol did
not reduce the grooming score compared to the second injection with vehicle.
PDB was more effective in reducing the ACTH-induced excessive grooming than
PMA was.

In order to gain better insight on the time course of inhibition of excessive grooming by the phorbol diesters and by DOG, the grooming score over 5 min time periods is given in Figs. 5 and 6. The inhibition of the grooming by DOG is already visible during the first observation times and persists over the whole observation period (Fig. 5). The inhibition by 0.75 µg PDB resembled that of DOG, but again it is shown that PDB is a much more potent inhibitor than DOG (Fig. 6). PMA did inhibit grooming behavior at the onset of the observation period after which no effect was observed during 15 min (Fig. 6). From the 5 min observation periods 6-11, PMA again inhibited the excessive grooming behavior. It thus seems that the major effect of the active phorbol diesters and DOG is observed in the dopamine-sensitive portion of the peptide-induced excessive grooming behavior (Isaacson et al., 1983). From Fig. 5 and 6 it is also clear that the order of potency for inhibition of grooming by kinase C activators is PDB > PMA > DOG > 4-α-phorbol. This order of relative potency for inhibition of grooming behavior is very similar to that found for the stimulation of B-50 phosphorylation in SPM (see above, Eichberg et al., 1986, De Graan et al., in preparation). These data support the notion that modulation of protein kinase C activity may be part of the molecular mechanism underlying ACTH-induced excessive grooming.

CONCLUDING REMARKS

In nervous tissue the inositol cascade seems to be controlled by the phosphorylation of a membrane-bound protein, B-50, by the calcium phospholipid-dependent protein kinase C. This hypothesis is supported by several findings using both biochemical and behavioral paradigms. In the hippocampal slice system, preincubation of the slice with ACTH seems to prevent the phorbol diester-induced inhibition of IP formation in response to carbachol. In the whole animal, phorbol diesters have the capacity to reduce ACTH-induced excessive grooming. It remains to be proven, whether in both systems PIP_2 breakdown after receptor activation is modulated by the degree of B-50 phosphorylation and also whether the degradation of PIP_2 itself is involved in the induction of excessive grooming behavior in the rat by ACTH. Very recently, Farese and coworkers have been able to demonstrate in rat adrenal cells that ACTH at certain concentrations can activate both the cAMP and the IP_3-Ca^{2+} intracellular signalling system (Farese et al., 1986). Very low concentrations of $ACTH_{1-24}$, between 10^{-14}-10^{-11} M, significantly stimulated the formation of inositol phosphates, with a concomitant increase in the intracellular calcium concentration, whereas at concentrations above 10^{-10} M cAMP production was very pronounced.

As mentioned above several groups have independently shown that stimulation of protein kinase C by either phorbol diesters or synthetic diacylglycerols partially prevents the rise in cytosolic calcium in response to receptor stimulation. One of the explanations for this action of protein kinase C is that it phosphorylates a specific substrate protein thereby regulating the amount of PIP_2 available for phosphodiesteratic cleavage. For those cells containing B-50, phosphorylation of this protein might indeed fulfill this role. On the other hand, evidence has been gathered that stimulation of protein kinase C by either diacylglycerol analogues or phorbol diesters attenuates the calcium current through voltage-sensitive calcium channels (Di Virgilio et al., 1986, Rane and Dunlap, 1986). It would be of particular interest to measure the activation of protein kinase C and/or to add either protein kinase C itself or antibodies directed against the kinase to establish its role in modulating these voltage-sensitive calcium-channels. This is especially necessary since DeRiemer et al. (1985a) showed exactly the opposite effect, namely an increase in calcium current due to activated protein kinase C in _Aplysia_ bag cells. Moreover these authors could also demonstrate an endogenous enzyme in these cells, that was activated by PMA and had properties very similar to mammalian protein kinase C (DeRiemer et al., 1985b). The endogenous substrate proteins phosphorylated by this protein kinase C-like enzyme are not yet identified and it is therefore not clear whether the modulation of calcium channels by protein kinase C involves the phosphorylation of the channel itself or whether the inhibition is due to a secondary effect of protein kinase C via the polyphosphoinositide metabolism.

In order to understand more about the role of protein kinase C in the modulation of excessive grooming behavior induced by ACTH in the rat, we will have to establish whether our biochemical observations in the hippocampal slice also apply to the whole animal. In view of the fact that the PAG is involved in the expression of grooming behavior and the fact that ACTH-terminals as well as B-50 protein are present in this structure, we believe that studies in the PAG will help us to understand how ACTH induces excessive grooming behavior and learn more about the role of phosphoprotein B-50 in the nervous system.

REFERENCES

Akers, R. F. and Routtenberg, A., Protein kinase C phosphorylates a 47 M_r protein (F_1) directly related to synaptic plasticity, _Brain Res._ 334: 147-151 (1985).

Aloyo, V. J., Zwiers, H. and Gispen, W. H., B-50 protein kinase and kinase C in rat brain, _Prog. Brain Res._ 56:303-315 (1982).

Aloyo, V. J., Zwiers, H. and Gispen, W. H., Phosphorylation of B-50 by calcium-activated phospholipid-dependent protein kinase and B-50 protein kinase, _J. Neurochem._ 41:649-653 (1983).

Berridge, M. J. and Irvine, R. F., Inositol trisphosphate, a novel second messenger in cellular signal transduction, _Nature (London)_ 312:649-653 (1984).

Castagna, M., Takai, Y., Kaibuchi, K., Sano, U., Kikkawa, U. and Nishizuka, Y., Direct activation of calcium-activated, phospholipid-dependent protein kinase by tumor-promoting phorbol diesters, _J. Biol. Chem._ 257:7847-7851 (1982).

De Graan, P. N. E., Van Hooff, C. O. M., Tilly, B. C., Oestreicher, A. B., Schotman, P. and Gispen, W. H., Phosphoprotein B-50 in nerve growth cones from fetal rat brain, _Neurosci. Lett._ 61:235-241 (1985).

De Graan, P. N. E., Oestreicher, A. B., Schrama, L. H. and Gispen, W. H., Phosphoprotein B-50: localization and function, _Prog. Brain. Res._ 69: 37-50 (1986a).

De Graan, P. N. E., Schrama, L. H. and Gispen, W. H., Characterization of a 52 kDa phosphoprotein possibly related to long-term potentiation, in: "Proceedings VIIth International Neurobiological Symposium on Learning and Memory, Oct 28 - Nov 2 (1985), Magdeburg (GDR)", Pergamon Press, Oxford, in press (1986b).

DeRiemer, S. A., Strong, J. A., Albert, K., Greengard, P. and Kaczmarek, L. K., Phorbol ester and protein kinase C enhance calcium current in _Aplysia_ neurones, _Nature (London)_ 313:313-316 (1985a).

DeRiemer, S. A., Greengard, P. and Kaczmarek, L. K., Calcium/phosphatidylserine/diacylglycerol-dependent protein phosphorylation in the _Alplysia_ nervous system, _J. Neurosci._ 5:2672-2676 (1985b).

Di Virgilio, F., Pozzan, T., Wollheim, C. B., Vicentini, L. M. and Meldolesi, J., Tumor promoter phorbol myristate acetate inhibits Ca^{2+} influx through voltage-gated Ca^{2+} channels in two secretory cell lines, PC12 and RINm5F, _J. Biol. Chem._ 261:32-35 (1986).

Dunkley, P. R. and Robinson, P. J., Depolarization-dependent protein phosphorylation in synaptosomes: mechanisms and significance, _Prog. Brain Res._ 69:273-294 (1986).

Eichberg, J., De Graan, P. N. E., Schrama, L. H. and Gispen, W. H., Diocta-
noylglycerol and phorbol esters enhance phosphorylation of phosphoprotein B-50 in native synaptic plasma membranes, Biochem. Biophys. Res.
Commun. 136:1007-1012 (1986).

Farese, R. V., Rosic, N., Babischkin, J., Farese, M. G., Foster, R. and
Davis, J. S., Dual activation of the inositol-trisphosphate-calcium and
cyclic nucleotide intracellular signaling systems by adrenocorticotropin in rat adrenal cells, Biochem. Biophys. Res. Commun. 135:742-748
(1986).

Gispen, W. H., Wiegant, V. M., Greven, H. M. and De Wied, D., The induction
of excessive grooming behavior in the rat by intracerebroventricular
application of peptides derived from ACTH: structure-activity studies,
Life Sci. 17:645-652 (1975).

Gispen, W. H., Zwiers, H., Wiegant, V. M., Schotman, P. and Wilson, J. E.,
The behaviorally active neuropeptide ACTH as neurohormone and neuromo-
dulator: the role of cyclic nucleotides and membrane phosphoproteins,
Adv. Exp. Med. Biol. 116:199-224 (1979).

Gispen, W. H. and Isaacson, R. L., ACTH-induced excessive grooming in the
rat, Pharmacol. Ther. 12:209-246 (1981).

Gispen, W. H., Van Dongen, C. J., De Graan, P. N. E., Oestreicher, A. B. and
Zwiers, H., The role of phosphoprotein B-50 in phosphoinositide metabolism in brain synaptic plasma membranes, in: "Inositol and Phosphoinositides", J. E. Bleasdale, G. Hauser, J. Eichberg, eds, Humana Press,
Dallas (1985a).

Gispen, W. H., Leunissen, J. L. M., Oestreicher, A. B., Verkleij, A. J. and
Zwiers, H., Presynaptic localization of B-50 phosphoprotein: the ACTH-
sensitive protein kinase substrate involved in rat brain polyphosphoinositide metabolism, Brain Res. 328:381-385 (1985b).

Gispen, W. H., Schrama, L. H. and Eichberg, J., Stimulation of protein
kinase C reduces ACTH-induced excessive grooming, Eur. J. Pharmacol.
114:399-400 (1985c).

Gispen, W. H., De Graan, P. N. E., Chan, S. Y. and Routtenberg, A., Comparison between the neural acidic proteins B-50 and F_1, Prog. Brain Res.
69: in press (1986).

Glassman, E. Gispen, W. H., Perumal, R., Machlus, B. and Wilson, J. E., The
effect of short experiences on the incorporation of radioactive phosphate into synaptosomal and non-histone acid-extractable nuclear proteins from rat and mouse brain, in: "Proceedings 5th International
Congress Pharmacology", San Francisco, Vol. 4 (1973).

Gower, H. and Rodnight, R., Intrinsic protein phosphorylation in synaptic
plasma membrane fragments from the rat. General characteristics and
migration behavior on polyacrylamide gels of the main phosphate receptors, Biochim. Biophys. Acta 716:45-52 (1982).

Gower, H., Rodnight, R. and Branimer, M. J., Ca^{2+}-sensitivity of Ca^{2+}-depen-
dent protein kinase activities toward intrinsic proteins in synaptosomal membrane fragments from rat cerebral tissue, J. Neurochem. 46:440-
447 (1986).

Greven, H. M. and De Wied, D., The influence of peptides derived from cor-
ticotropin (ACTH) on performance. Structure-activity studies. Prog.
Brain Res. 39:429-442 (1973).

Isaacson, R. L., Hannigan, J. H., Brakkee, J. H. and Gispen, W. H., The
time course of excessive grooming after neuropeptide administration,
Brain. Res. Bull. 11:289-293 (1983).

Jolles, J., Zwiers, H., Van Dongen, C. J., Schotman, P., Wirtz, K. W. A.
and Gispen, W. H., Modulation of brain polyphosphoinositide metabolism
by ACTH-sensitive protein phosphorylation, Nature 286:623-625 (1980).

Jolles, J., Zwiers, H., Dekker, A., Wirtz, K. W. A. and Gispen, W. H., Cor-
ticotropin-(1-24)-tetracosapeptide affects protein phosphorylation and
polyphosphoinositide metabolism in rat brain, Biochem. J. 194:283-291
(1981).

Jork, R., De Graan, P. N. E., Van Dongen, C. J., Zwiers, H., Matthies, H. and Gispen, W. H., Dopamine-induced changes in protein phosphorylation and polyphosphoinositide metabolism in rat hippocampus, Brain Res. 291: 73-81 (1984).

Katz, F., Ellis, L. and Pfenninger, K. H., Nerve growth cones isolated from fetal rat brain, J. Neurosci. 5:1402-1411 (1985).

Kikkawa, U., Takai, Y., Tanaka, R., Miyake, R. and Nishizuka, Y., Protein kinase C as a possible receptor protein of tumor-promoting phorbol esters, J. Biol. Chem. 258:11442-11445 (1983).

Kikkawa, U., Kitano, T., Saito, N., Fujiwara, H., Nakanishi, H., Kishimoto, A., Taniyama, K., Tanaka, C. and Nishizuka, Y., Possible roles of protein kinase C in signal transduction in nervous tissues, Prog. Brain Res. 69:29-38 (1986).

Kristjansson, G. I., Zwiers, H., Oestreicher, A. B. and Gispen, W. H., Evidence that the synaptic phosphoprotein B-50 is localized exclusively in nerve tissue, J. Neurochem. 39:371-378 (1982).

Labarca, R., Janowsky, A., Patel, J. and Paul, S. M., Phorbol esters inhibit agonist induced ^{3}H-inositol-1-phosphate accumulation in rat hippocampal slices, Biochem. Biophys. Res. Commun. 123:703-709 (1984).

Leeb-Lundberg, L. M. F., Cotecchia, S., Lomasney, J. W., DeBernardis, J. F., Lefkowitz, R. J., and Caron, M. G., 1985, Phorbol esters promote α - adrenergic receptor phosphorylation and receptor uncoupling from inositol phospholipid metabolism, Proc. Natl. Acad. Sci. USA 82:5651-5655 (1985).

Mahler, H. R., Kleine, L. P., Ratner, N. and Sorensen, R. G., Identification and topography of synaptic phosphoproteins, Progr. Brain Res. 56: 27-48 (1982).

Niedel, J. E., Kuhn, L. J. and Vanderbark, G. R., Phorbol diester receptor copurifies with protein kinase C, Proc. Natl. Acad. Sci. USA 80:36-40 (1983).

Nishizuka, Y., The role of protein kinase C in cell surface signal transduction and tumor promotion, Nature (London) 308:693-697 (1984).

Oestreicher, A. B., Zwiers, H., Schotman, P.. and Gispen, W. H., Immunohistochemical localization of a phosphoprotein (B-50) isolated from rat brain synaptosomal plasma membranes, Brain Res. Bull. 6:145-153 (1981).

Oestreicher, A. B., Van Dongen, A. B., Zwiers, H. and Gispen, W. H., Affinity-purified anti-B-50 protein antibody: interference with the function of the phosphoprotein B-50 in synaptic plasma membranes, J. Neurochem. 41:331-340 (1983).

Oestreicher, A. B., Van Duin, M., Zwiers, H. and Gispen, W. H., Cross-reaction of anti-rat B-50: characterization and isolation of a 'B-50-phosphoprotein' from bovine brain, J. Neurochem. 43:935-943 (1984).

Oestreicher, A. B., Dekker, L. V. and Gispen, W. H., A radioimmunoassay for the phosphoprotein B-50: distribution in rat brain, J. Neurochem. 46: 1366-1369 (1986).

Okano, Y., Takagi, H., Nakashima, S., Tohmatsu, T. and Nozawa, Y., Inhibitory action of phorbol myristate acetate on histamine secretion and polyphosphoinositide turnover induced by compound 48/80 in mast cells, Biochem. Biophys. Res. Commun. 132:110-117 (1985).

Orella, S. A., Solski, P. A. and Brown, J. H., Phorbol ester inhibits phosphoinositide hydrolysis and calcium mobilization in cultured astrocytoma cells, J. Biol. Chem. 260:5236-5239 (1985).

Rane, S. G. and Dunlap, K., Kinase C activator 1,2-oleoylacetylglycerol attenuates voltage-dependent calcium current in sensory neurons, Proc. Natl. Acad. Sci. USA 83:184-188 (1986).

Rodnight, R., Aspects of protein phosphorylation in the nervous system with particular reference to synaptic transmission, Progr. Brain Res. 56:1-25 (1982).

Routtenberg, A., Lovinger, D. M. and Steward, P., Selective increase in phosphorylation state of a 47 kDa protein (F_1) directly related to long-term potentiation, <u>Behav. Neural Biol</u>. 43:3-11 (1985).

Schrama, L. H., De Graan, P. N. E., Eichberg, J. and Gispen, W. H., Feedback control of the inositol phospholipid response in rat brain is sensitive to ACTH, <u>Eur. J. Pharmacol</u>. 121:403-404 (1986a).

Schrama, L. H., De Graan, P. N. E., Wadman, W. J., Lopes da Silva, F. H. and Gispen, W. H., Long-term potentiation and 4-aminopyridine-induced changes in protein- and lipid-phosphorylation in the hippocampal slice, <u>Prog. Brain Res</u>. 69: in press (1986b).

Sörensen, R. G., Kleine, L. P. and Mahler, H. R., Presynaptic localization of phosphoprotein B-50, <u>Brain Res. Bull</u>. 7:57-61 (1981).

Spruijt, B. M., Cools, A. R. and Gispen, W. H., The periaqueductal gray: a prerequisite for ACTH-induced excessive grooming, <u>Behav. Brain Res</u>. in press (1986).

Streb, H., Irvine, R. F., Berridge, M. J. and Schulz, I., Release of Ca^{2+} from a nonmitrochondrial intracellular store in pancreatic acinar cells by inositol-1,4,5-trisphosphate, <u>Nature (London)</u> 306:67-69 (1983).

Takai, Y., Kishimoto, A., Kikkawa, U., Mori, T. and Nishizuka, Y., Unsaturated diacylglycerol as a possible messenger for the activation of calcium-activated, phospholipid-dependent protein kinase system, <u>Biochem. Biophys. Res. Commun</u>. 91:1218-1224 (1979).

Takai, Y., Kaibuchi, K., Tsuda, T. and Hoshijima, M., Role of protein kinase C in transmembrane signalling, <u>J. Cell. Biochem</u>. 29:143-155 (1985).

Tohmatsu, T., Hattori, H., Nagao, S., Ohki, K. and Nozawa, Y., Reversal by protein kinase C inhibitor of suppressive actions of phorbol-12-myristate-13-acetate on polyphosphoinositide metabolism and cytosolic Ca^{2+} mobilization in trombin-stimulated human platelets, <u>Biochem. Biophys. Res. Commun</u>. 134:868-875 (1986).

Van Dongen, C. J., Zwiers, H. and Gispen, W. H., Purification and partial characterization of the phosphatidylinositol 4-phosphate kinase from rat brain, <u>Biochem. J</u>. 223:197-203 (1984).

Van Dongen, C. J., Zwiers, H., De Graan, P. N. E. and Gispen, W. H., Modulation of the activity of purified phosphatidylinositol 4-phosphate kinase by phosphorylated and dephosphorylated B-50, <u>Biochem. Biophys. Res. Commun</u>. 128:1219-1227 (1985).

Van Dongen, C. J., Kok, J. W., Schrama, L. H., Oestreicher, A. B. and Gispen, W. H., Immunochemical characterization of phosphatidylinositol 4-phosphate kinase from rat brain, <u>Biochem. J</u>. 233:859-864 (1986).

Vicentini, L. M., Di Virgilio, F., Ambrosini, A., Pozzan, T. and Meldolesi, J., Tumor promoter phorbol 12-myristate, 13-acetate inhibits phosphoinositide hydrolysis and cytosolic Ca^{2+} rise induced by the activation of muscarinic receptors in PC12 cells, <u>Biochem. Biophys. Res. Commun</u>. 127:310-317 (1985).

Watson, S. J., Richard, C. W. III and Barchas, J. D., Adrenocorticotropin in rat brain: immunocytochemical localization in cells and axons, <u>Science</u> 275:226-228 (1978).

Watson, S. P. and Lapetina, E. G., 1,2-Diacylglycerol and phorbol ester inhibit agonist-induced formation of inositol phosphates in human platelets: possible implications for negative feedback regulation of inositol phospholipid hydrolysis, <u>Proc. Natl. Acad. Sci. USA</u> 82:2623-2626 (1985).

Weller, M., "Protein phosphorylation. The Nature, Function and Metabolism of Proteins, which Contain Covalently Bound Phosphorus", PION Ltd, London (1979).

Worley, P. F., Barbaban, J. M. and Snyder, S. H., Heterogeneous localization of protein kinase C in rat brain: autoradiographic analysis of phorbol ester receptor binding, <u>J. Neurosci</u>. 6:199-207 (1986).

Zwiers, H., Veldhuis, H. D., Schotman, P. and Gispen, W. H., ACTH, cyclic nucleotides, and brain protein phosphorylation <u>in vitro</u>, <u>Neurochem. Res</u>. 1:669-677 (1976).

Zwiers, H., Wiegant, V. M., Schotman, P. and Gispen, W. H., Intraventricular administered ACTH and changes in rat brain phosphorylation: a preliminary report, <u>in</u>: "Mechanism, Regulation and Special Functions of Protein Synthesis in the Brain", S. Roberts, A. Lajhta, and W. H. Gispen, eds. Elsevier/North-Holland Biomedical Press, Amsterdam (1977).

Zwiers, H., Wiegant, V. M., Schotman, P. and Gispen, W. H., ACTH-induced inhibition of endogenous rat brain protein phosphorylation <u>in vitro</u>: structure-activity, <u>Neurochem. Res</u>. 3:247-256 (1978).

Zwiers, H., Tonnaer, J., Wiegant, V. M., Schotman, P. and Gispen, W. H., ACTH-sensitive protein kinase from rat brain membranes, <u>J. Neurochem</u>. 33:247-256 (1979).

Zwiers, H., Schotman, P. and Gispen, W. H., Purification and some characteristics of an ACTH-sensitive protein kinase and its substrate protein in rat brain membrane, <u>J. Neurochem</u>. 34:1689-1699 (1980).

Zwiers, H., Jolles, J., Aloyo, V. J., Oestreicher, A. B. and Gispen, W. H., ACTH and synaptic membrane phosphorylation in rat brain, <u>Prog. Brain Res</u>. 56:405-417 (1982).

Zwiers, H., Verhaagen, J., Van Dongen, C. J., De Graan, P. N. E. and Gispen, W. H., Resolution of rat brain synaptic phosphoprotein B-50 into multiple forms by two-dimensional electrophoresis: evidence for multisite phosphorylation, <u>J. Neurochem</u>. 44:1083-1090 (1985).

MOLECULAR MECHANISMS OF NEURONAL EXCITABILITY: POSSIBLE INVOLVEMENT OF CaM KINASE II IN SEIZURE ACTIVITY

William C. Taft, James R. Goldenring* and
Robert J. DeLorenzo

Departments of Neurology and Pharmacology, Medical
College of Virginia - VCU, Richmond, Virginia 23298
and *Department of Surgery, Yale University School of
Medicine, New Haven, Connecticut 06510

INTRODUCTION

An understanding of the molecular mechanisms that underlie neuronal responsiveness is an important goal of contemporary neuroscience. The specific biochemical events that modulate excitability of neurons and neuronal systems will provide important insights into the complex regulatory mechanisms of the nervous system. In the clinical neurosciences, an understanding of these molecular events may provide significant avenues for the development of treatment protocols for diseases of the nervous system.

In this chapter we will examine the role of Ca^{2+} in regulating neuronal excitability and seizure discharge. Although the precise mechanisms for mediating the actions of Ca^{2+} are not completely understood, several recent advances have identified important biochemical systems that regulate the effects of Ca^{2+} on synaptic function. The identification of the Ca^{2+} binding protein, calmodulin, and the characterization of several Ca^{2+}-calmodulin dependent enzyme systems has provided a molecular approach to studying the effects of Ca^{2+} on cellular function (Cheung, 1980; Klee et al., 1980). Evidence will be presented that Ca^{2+}-calmodulin dependent protein phosphorylation is an important mediator of some of the neuromodulatory effects of Ca^{2+} on neuronal excitability and synaptic function. The identification and characterization of Ca^{2+}-calmodulin dependent protein kinase (CaM kinase II) has provided important insights into the molecular mechanisms that may mediate some of the effects of Ca^{2+} on neuronal function (Fukanaga et al., 1982; Goldenring et al., 1983; Bennett et al., 1983; McGuinness et al., 1983; Yamauchi and Fujisawa, 1983; Schulman, 1984). The role of CaM kinase II in modulating neuronal excitability will be discussed. Recent studies have demonstrated that direct injection of CaM kinase II into identified invertebrate neurons modulates synaptic activity (Llinas et al., 1985; Sakakibara et al., 1986). In addition, evidence will be presented that alterations in CaM kinase II activity are associated with the development of septal kindling, an experimental model of epilepsy and a paradigm for the study of long-term memory (Goldenring et al., 1986a). These observations suggest that changes in CaM kinase II activity are associated with an experimental model of epilepsy and may represent an important endogenous mechanism for the regulation of neuronal excitability.

Epilepsy and Neuronal Excitability

Epilepsy refers to the many varieties of recurrent seizures produced by excessive synchronous discharges of populations of central neurons in different regions of the brain (Gastaut et al., 1969; Jasper et al., 1969; Glaser, 1975). There are many types of epilepsy, and the possible causes of seizure disorders are numerous (Glaser, 1975). No single approach will allow us to fully understand the nature of these complex clinical phenomena or the mechanism of action of anticonvulsant drugs (Glaser et al., 1980; DeLorenzo, 1980). However, all the causes of epilepsy appear to involve alterations in the normal mechanism that regulate neuronal excitability. Thus, to obtain a molecular insight into epilepsy and develop optimal therapies for seizure disorders, it will be important to understand the basic aspects of regulating neuronal excitability and the interaction of pathological processes with normal modulatory mechanisms of neuronal excitability.

The regulation of neuronal excitability and modulation of synaptic transmission in the nervous system is complex and poorly understood. Present knowledge of the molecular processes and structural organization of synaptic connections and neuronal circuitry is not sufficient to provide clear explanations of the modulation of neuronal excitability. However, the expansion of research in the neurosciences in recent years has made considerable progress and defined several areas that are likely to yield new and meaningful information. One major advance has been in the understanding of calcium regulation of neuronal excitability (Katz and Miledi, 1969, 1970; Rubin, 1972). Recent investigations have provided insights into the molecular mechanisms of Ca^{2+} regulation of neuronal excitability and seizure discharge.

Calcium and Neuronal Excitability

Calcium plays a major role in the function of the nervous system and has been implicated as an important second messenger in brain (Katz and Miledi, 1969, 1970; Rubin, 1972). Initial studies focused on the role of Ca^{2+} in the regulation of neurotransmitter release and synaptic transmission, as well as in various forms of stimulus-secretion coupling in other tissues (Douglas, 1968). These processes are directly dependent on the entry of Ca^{2+} into the nerve terminal or secretory cell and the associated rise in cytoplasmic Ca^{2+} concentrations. Although the importance of Ca^{2+} in regulating synaptic function has been well documented, the precise molecular mechanisms which mediate the effects of Ca^{2+} on synaptic activity have not been elucidated.

The calcium binding protein, calmodulin, has been purified and characterized, and has been shown to mediate several of the second messenger effects of Ca^{2+} on specific cellular systems and neuronal excitability (Cheung, 1980; Klee et al., 1980). Calmodulin binds Ca^{2+} in physiological concentration ranges and can be released from brain membrane fractions with the Ca^{2+} chelator, ethylene glycol-bis tetra-acetic acid (EGTA). Calmodulin represents a large percentage of total brain protein (approximately 1%) and has been identified in various presynaptic and postsynaptic fractions, including synaptoplasm (DeLorenzo, 1980a, 1980b), synaptic vesicles (DeLorenzo, 1981a) and the postsynaptic density (Grab et al., 1979, 1980). The presence of calmodulin in such high concentrations in the synaptic region suggests an important role for this protein in mediating the effects of Ca^{2+} on both presynaptic and postsynaptic functions.

Calmodulin regulates numerous Ca^{2+}-mediated processes in neuronal tissue (Cheung, 1980; Klee, et al., 1980). Initial studies focused on the observation that Ca^{2+}-calmodulin regulates the phosphorylation of membrane proteins in brain and other tissues (Schulman and Greengard, 1978a, 1978b).

410

Since protein phosphorylation has been implicated as a major posttranslational mechanism that has important effects on cellular function and neuronal excitability (Schulman and Greengard, 1978b; Kennedy, 1983; Nestler et al., 1984), it was important to examine these phosphoproteins and the associated protein kinase systems. Our laboratory and others have shown that endogenous calmodulin plays an important role in synaptic protein phosphorylation and in the activity of depolarization-dependent synaptic processes (DeLorenzo, 1980a; DeLorenzo et al., 1982). These observations have led to the calmodulin hypothesis of neurotransmission, which states that Ca^{2+} enters the presynaptic terminal, binds to calmodulin and activates several Ca^{2+}-calmodulin regulated processes that mediate the effects of Ca^{2+} on neurotransmission and synaptic excitability (DeLorenzo, 1981b).

In developing a molecular approach to understanding the regulation of neuronal excitability, it is necessary to demonstrate that these Ca^{2+}-calmodulin dependent processes take place at the synaptic region. Initial investigations demonstrated Ca^{2+}-calmodulin regulated protein kinase systems in preparations of synaptic vesicles (DeLorenzo et al., 1979; DeLorenzo, 1982), synaptosomes (DeLorenzo, 1976), synaptoplasm (Burke and DeLorenzo, 1982) and in postsynaptic density fractions (Grab et al., 1981). Thus, the Ca^{2+}-calmodulin dependent kinase systems are present at the synapse and could endogenously mediate the effects of Ca^{2+} on synaptic activity. Evidence has accumulated that correlates this endogenous Ca^{2+}-calmodulin dependent protein kinase activity with neurotransmitter release and neuronal function (Burke and DeLorenzo, 1982; DeLorenzo et al., 1979). Therefore, it is important to focus in detail on Ca^{2+}-calmodulin dependent protein kinases to precisely determine their role in the regulation of neuronal excitability.

Isolation and Characterization of CaM Kinase II

Employing different substrates to isolate specific calmodulin-dependent kinases from brain, our laboratory (Goldenring et al.,1982; Goldenring et al., 1983) and others (Bennett et al., 1983; Fukanaga et al., 1982; McGuinness et al., 1983; Schulman, 1984; and Yamauchi and Fujisawa, 1983) purified to apparent homogeneity specific Ca^{2+}-calmodulin-activated protein kinases. Following the eventual characterization of these enzymes, it became apparent that each laboratory had isolated the same calmodulin kinase, with a wide range of substrate specificities. These enzymes are now collectively designated as type II calmodulin-dependent protein kinase (CaM kinase II) and have been suggested to mediate the effects of Ca^{2+} in a variety of neuronal preparations. The detailed characterization of this enzyme activity makes it possible to examine its role in intact neuronal systems and in various models of altered excitability, such as in kindling.

CaM kinase II has been the object of extensive investigation by numerous laboratories. They have demonstrated that CaM kinase II is highly enriched in nerve cells and have characterized it in soluble and particulate fractions from both pre- and postsynaptic elements, including the postsynaptic density (Goldenring et al., 1984; Kelley et at., 1984; Kennedy et al., 1983), dendrites (Ouimet et al., 1984), neuronal stomata (Ouimet et al., 1984), synaptic membranes (Schulman and Greengard, 1978b), microtubules (Larson et al., 1985; Vallano et al., 1985b) and synaptic vesicles (DeLorenzo et al., 1978; DeLorenzo, 1982). Several major neuronal proteins have been identified as substrates for CaM kinase II, including synapsin I (DeCamilli et al., 1983), microtubule associated protein 2 (MAP 2) (Goldenring et al., 1983), tubulin (Goldenring et al., 1983) and the neurofilament triplet polypeptides (Vallano et al., 1985a). CaM kinase II exists as a holoenzyme complex which chromatographs at a molecular weight of approximately 600,000 daltons on Sephacryl S-300 (Bennett et al., 1983; Goldenring et al., 1983). The enzyme has two auto-phosphorylating, calmodulin-binding subunits with molecular weights of 50,000 and 60,000 daltons. The isoelectric points of these two subunits are between 6.7 and 7.2. The enzyme subunits themselves are substrates for the activated enzyme (Bennett et al., 1983; Goldenring et al., 1983).

Several specific properties of CaM kinase II can be utilized to distinguish this enzyme from other kinase systems (Goldenring et al., 1983). The two-dimensional peptide map of CaM kinase II phosphorylated substrate protein is characteristic, and it is significantly different from the peptide map of the same substrate phosphorylated by other kinase systems. Thus, the tryptic peptide map of MAP-2 phosphorylated by cAMP-dependent protein kinase is significantly different from that of MAP-2 phosphorylated by CaM kinase II (Vallano et al., 1985b; Goldenring et al., 1985). Employing two-dimensional ^{125}I tryptic peptide mapping, the two principle autophosphorylating enzyme subunits exhibit significant homology. In addition, phosphoamino acid analysis of CaM kinase II phosphorylated beta tubulin demonstrates that the enzyme phosphorylates equally on threonine and serine residues, whereas other kinase systems phosphorylate principally on serine residues. Recently, specific monoclonal antibodies have been raised against highly purified preparation of CaM kinase II which recognize the 50,000 dalton subunit of the enzyme on immunoblots and which precipitate enzyme activity from soluble fractions (Goldenring et al., 1986b). These monoclonals can be used to confirm the presence of CaM kinase II in a variety of neuronal preparations.

CaM Kinase II and Neuronal Excitability

Protein phosphorylation has been implicated as a major posttranslational mechanism that has important effects on cellular function and neuronal excitability (Schulman and Greengard, 1978b; Kennedy, 1983; Nestler et al., 1984). Several protein phosphorylation systems have been described, principally those regulated by cyclic nucleotides and calcium (Greengard, 1978b; Nestler et al., 1984; Nestler and Greengard, 1983). Electrophysiological techniques have been employed to determine whether protein phosphorylation regulates neuronal excitability by examining the role of protein phosphorylation in the alteration of specific ion conductances. Initial efforts investigated the role of cAMP-dependent protein kinase. Studies in the mollusc, Aplysia californica (Castelluci et al., 1980; Seigelbaum et al., 1982; Kandel and Schwartz, 1982; Hawkins et al., 1983), demonstrated that cAMP-dependent protein phosphorylation regulates specific K^+ currents in several identified neurons. Injection of cAMP-dependent protein kinase or kinase inhibitors was also shown to regulate K^+ currents in Helix neurons (DePeyer et al., 1982) and Hermissenda photoreceptors (Alkon et al., 1983), as well as Ca^{2+} current in myocytes (Osterrieder et al., 1982). These results established the concept that the activity of specific ion channels in the cell membrane can be modulated by protein phosphorylation and demonstrated that the effects of cAMP on neuronal excitability (for review see Levitan et al., 1983) may be mediated by the activity of cAMP-dependent protein kinase.

Recent studies have investigated the role of Ca^{2+}-regulated protein kinases in the modulation of ion conductances. Principle interest has focused on the involvement of Ca^{2+}-dependent phospholipid-sensitive protein kinase (kinase C) (DeRiemer et al., 1985; Baraban et al., 1985; Farley and Averbach, 1986) and Ca^{2+}-calmodulin-dependent protein kinase (CaM kinase II) (Acosta-Urquidi et al., 1984a; Llinas et al., 1985; Sakakibara, et al., 1986), although other Ca^{2+}-dependent kinases are being investigated (Acosta-Urquidi et al., 1984b). The recent purification and characterization of CaM kinase II makes it possible to directly determine the effects of purified kinase on identified ion conductances in intact neurons (Acosta-Urquidi et al., 1984a; Llinas et al., 1985; Sakakibara et al., 1986).

Several independent observations have contributed to the hypothesis that CaM kinase II mediates some of the effects of Ca^{2+} on neuronal excitability and have provided important insight into the molecular regulation of neuronal responsiveness and seizure mechanisms. The 50,000 subunit of CaM kinase II has been shown to be homologous with the major postsynaptic density protein, suggesting an important role for this protein in postsynaptic function (Kennedy et al., 1983; Kelly et al., 1984; Goldenring et al., 1984). The activity of purified CaM kinase II is regulated by anticonvulsant compounds. The enzyme may mediate some of the effects of these compounds on neuronal excitability (DeLorenzo et al., 1981; Taft et al., 1985). Further, the direct injection of CaM kinase II into identified invertebrate neurons has shown that this enzyme modulates synaptic function in squid giant synapse (Llinas et al., 1985) and specific membrane ion conductances in Hermissenda neurons (Acosta-Urquidi et al., 1984a; Sakakibara et al., 1986). In addition, recent studies have demonstrated that alterations in CaM kinase II activity are associated with the development of septal kindling, an important experimental model of epilepsy and neuronal plasticity (Goldenring et al., 1986a). Thus, recent research has suggested that Ca^{2+}-calmodulin-regulated protein phosphorylation may play a major role in regulating neuronal excitability. The recent purification and characterization of CaM kinase II makes it possible to directly identify the involvement of this enzyme in various models of neuronal responsiveness.

Homology of CaM Kinase II and the Major Postsynaptic Density Protein

The postsynaptic density (PSD) is an electron dense structure attached to the cytoplasmic surface of the postsynaptic membrane at the synaptic junction (for review see Matus, 1981). Because of its unique location, it has been suggested that the postsynaptic density may be an important mediator of synaptic processing. A single 50,000 dalton protein comprises greater than 50% of postsynaptic density protein and has been designated the major postsynaptic density protein (mPSDp) (Kelly and Cotman, 1978). The mPSDp is an endogenous substrate for Ca^{2+}-calmodulin dependent protein phosphorylation and is one of several calmodulin binding proteins in the PSD (Grab et al., 1981). Since Ca^{2+}-calmodulin dependent protein phosphorylation is endogenous to the PSD, we investigated the possibility that the mPSDp may be the 50,000 dalton subunit of CaM kinase II (Goldenring et al., 1984).

The results of this investigation, and others (Kennedy et al., 1983; Kelly et al., 1984) indicated that mPSDp is homologous with the 50,000 dalton subunit of CaM kinase II and that PSD fractions also contain a protein corresponding with the 60,000 dalton subunit of CaM kinase II (Fig. 1). Homologies between these proteins were established based on several biochemical parameters: a) identical apparent molecular weights based on migration in one-dimensional SDS-PAGE gels; b) identical ^{125}I-calmodulin binding properties; c) Ca^{2+}-calmodulin dependent autophosphorylation; d) identical isoelectric points near neutrality; e) homologous two-dimensional tryptic peptide maps and f) similar phosphoamino acid-specific phosphorylation of tubulin. These observations indicate that CaM kinase II is the major protein component of the PSD and suggest that CaM kinase II activity plays an important role in synaptic function.

Inhibition of CaM Kinase II Activity by Anticonvulsant Compounds

The investigation and identification of the site of action of neuroactive drugs have yielded important insights into the mechanisms by which neurons respond to external stimuli. These pharmacological studies have led to the identification of important regulatory pathways associated with neuronal excitability. In order to identify the mechanism of action of anticonvulsant compounds, we examined the effects of various types of anti-

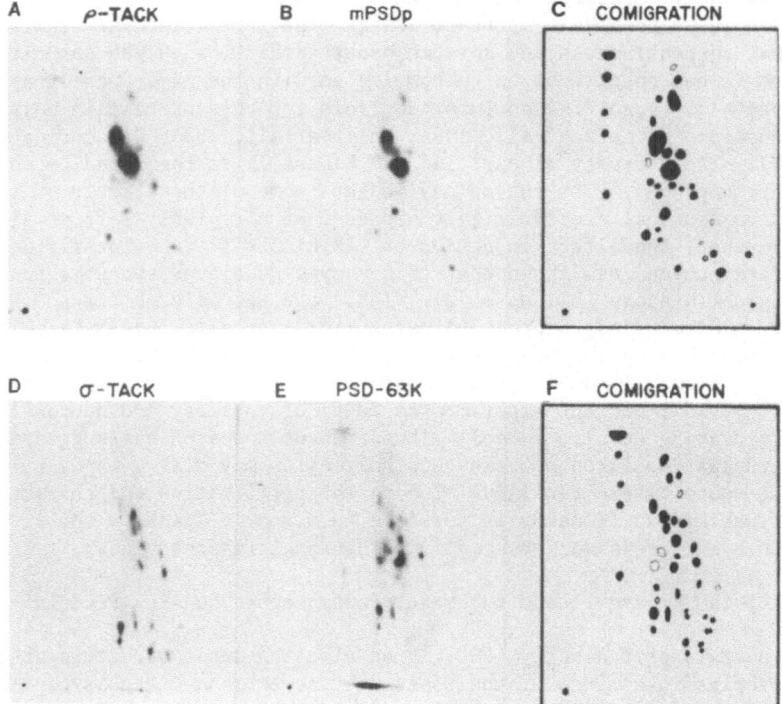

FIG. 1: Autoradiographic exposures of two-dimensional ^{125}I-tryptic peptide
maps of purified CaM kinase II and kinase endogenous to PSDs. (A)
50,000 dalton subunit of purified CaM kinase II; (B) 50,000 dalton
mPSDp; (C) composite showing areas of peptide homology between A
and B; (D) 60,000 dalton subunit of purified CaM kinase II; (E)
the 60,000 dalton PSD protein; (F) composite showing areas of peptide homology between D and E. The comigrating peptides in C and
F are represented by black spots. (Reproduced from Goldenring et
al., 1984).

convulsants on CaM kinase II activity. Anticonvulsants are widely utilized
clinically for their potent neuronal stabilizing properties which effectively depress the excessive neuronal excitability associated with seizure
disorders.

    Initial studies from our laboratory examined the effect of various
anticonvulsants, including benzodiazepines, phenytoin and carbamazepine, on
Ca^{2+}-calmodulin dependent kinase activity (DeLorenzo et al., 1981). These
studies demonstrated that a general property of anticonvulsants that regulate Ca^{2+} systems is the inhibition of Ca^{2+}-calmodulin dependent phosphorylation of neuronal proteins in brain membrane, cytosol and synaptic vesicle
fractions. The phosphorylation of numerous proteins was affected, but a
dramatic inhibition of phosphorylation in the 50,000 - 60,000 dalton range
was noted. Subsequent studies have shown that the phosphorylation in this
molecular weight range corresponds to the phosphorylation of the substrate,
tubulin, as well as the autophosphorylation of the subunits of CaM kinase
II. These observations suggested that the inhibition of CaM kinase II
activity in neuronal membranes may be a molecular mechanism by which anticonvulsants produce some of their neuronal stabilizing effects (DeLorenzo
et al., 1981).

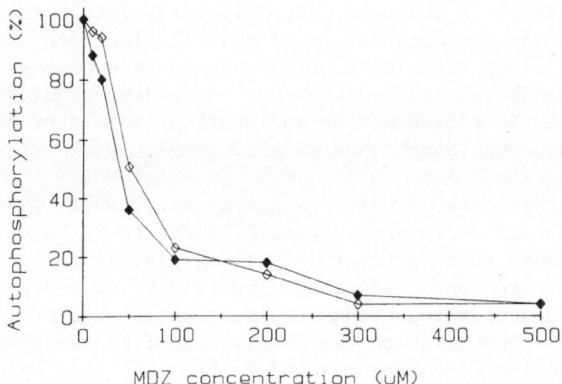

FIG. 2: Inhibition of autophosphorylation of purified CaM kinase II by medazepam. In vitro phosphorylation and separation by SDS-PAGE were as described (Goldenring et al., 1983). Data were obtained by densitometric scan of the autoradiographs. Closed and open symbols represent autophosphorylation of 50,000 and 60,000 dalton subunits of CaM kinase II, respectively.

To determine whether anticonvulsants inhibit kinase activity by directly affecting CaM kinase II itself or by interfering with other regulatory processes in the membrane, studies were conducted using purified preparations of CaM kinase II (Taft et al., 1985). Several variables were eliminated in this study, one of which was the presence of phosphatase activity in the membrane fractions. The purified preparations did not contain endogenous phosphatase activity that possibly could be affected by the presence of anticonvulsants. In these studies, we observed a similar pattern of anticonvulsant inhibition of CaM kinase II activity. Micromolar concentrations of benzodiazepines inhibited the autophosphorylation of purified cytosolic CaM kinase II (Fig. 2). The order of potency of the benzodiazepines tested (with IC_{50} values) was medazepam (32 μM) > diazepam (57 μM) > clonazepam (129 μM). The benzodiazepines also inhibited the phosphorylation of several major substrates by CaM kinase II, including tubulin, MAP 2 and neurofilament protein. The effects of benzodiazepines were concentration-dependent and stereospecific. Benzodiazepine inhibition of CaM kinase II autophosphorylation appears to be a direct effect on the kinase itself, and not an effect on Ca^{2+}, Mg^{2+} or calmodulin. These data suggest that the inhibition of CaM kinase II is a specific property of anticonvulsants and may be the molecular mechanism by which they produce some of their anticonvulsant and sedative effects. Since these compounds are known to diminish neuronal excitability, the inhibition of CaM kinase II may be the mechanism by which they exhibit their neuronal stabilizing properties.

Regulation of Neuronal Activity by Direct Intracellular Injection of CaM Kinase II

Results from our laboratory, in collaboration with Dr. Daniel Alkon, have shown that injection of highly purified CaM kinase II regulates ion conductances in Hermissenda photoreceptor cells (Acosta-Urquidi et al., 1984a; Sakakibara et al., 1986). The nudibranch mollusc, Hermissenda crassicornis, has been utilized as a model system for studying the electrophysiological and biochemical correlates of conditioned responses (Alkon, 1975, 1979, 1984; Farley and Alkon, 1980, 1982; Alkon et al., 1985). The effects of Ca^{2+} on several isolated membrane currents have been well characterized in Hermissenda photoreceptor cells (Alkon et al., 1982a; Alkon and Sakakibara, 1986), making this model particularly well-suited for the study of the effects of CaM kinase II on membrane conductances. In these studies, CaM kinase II is injected intracellularly into Hermissenda neurons, and its effect on ion conductances is documented electrophysiologically employing voltage clamp techniques. The results have demonstrated that kinase injection enhances and prolongs Ca^{2+}-dependent regulation of I_A, I_C and I_{Ca}, and that the kinase may be an important regulator of excitability in invertebrate neurons (Sakakibara et al., 1986).

Previous voltage clamp studies (Alkon et al., 1982b, 1984) have characterized two voltage-dependent outward K^+ currents in Hermissenda type B photoreceptor neurons which activate at voltage levels more negative than 0 mV: I_A and I_C. These conductances are reversibly reduced when the cell is subjected to a protocol which produces a substantial and prolonged rise in intracellular Ca^{2+} concentration (depolarization followed by illumination). CaM kinse II injection, when accompanied by the Ca^{2+} loading protocol, was followed by significantly greater and longer lasting reductions in I_A and I_C (Fig. 3). Repeated enzyme injections, each followed by a Ca^{2+} load, progressively reduced I_A and I_C values. Neither heat-inactivated enzyme nor vehicle alone produced these effects. Injection of CaM kinase II not accompanied by Ca^{2+} load also showed minimal or no effect, demonstrating clearly the Ca^+ dependence of kinase effects. A previous voltage clamp study (Alkon et al., 1984) demonstrated the presence of a single voltage-dependent inward current in the photoreceptor which is carried by Ca^{2+} (I_{Ca}). Injection of CaM kinase accompanied by Ca^{2+} load also resulted in reduced peak I_{Ca} current. Injection of control solutions produced no effect.

Illumination of the type B photoreceptor soma elicits a voltage-independent inward Na^+ current, I_{Na} (Alkon, 1979; Alkon et al., 1982a, 1982b, 1984). This current was measured after at least 10 min dark adaptation, and thereafter, at 3 min intervals in the dark. CaM kinase II injection was not followed by any change in I_{Na} other than a slight but not significant increase which typically occurs during dark adaptation. This slight increase was also observed following injection of inactivated enzyme.

Results of other studies are consistent with the interpretation that illumination of the type B photoreceptor releases Ca^{2+} from internal stores. This light-induced elevation of $[Ca^{2+}]_i$ was thought to cause prolonged inactivation of steady-state I_C elicited by a step depolarization (Alkon and Sakakibara, 1984; 1986). Enzyme injection (but not injection of inactivated kinase) caused a significant prolongation of the light-induced reduction of I_C. It was previously shown that light does not affect the sustained I_{Ca} (Alkon and Sakakibara, 1986). Thus, the enzyme-induced prolongation of light-elicited reduction of I_C is most likely due to a direct enzyme effect on I_C (and not due to an indirect enzyme effect on I_{Ca}).

Trifluoperizine (TFP) is known to inhibit CaM kinase II, protein kinase C and other enzymatic reactions. Although TFP is not a specific inhibitor of CaM kinase II, its actions on the Ca^{2+}-dependent kinase endogenous to the type B soma might be expected to produce effects opposite to those of enzyme injection. In fact, perfusion of the type B cell soma with 5 μM TFP consistently caused a marked increase of I_A and I_C and a shortening of the light-induced reduction of I_C. Perfusion with TFP also consistently eliminated or greatly reduced the effects of CaM kinase II injection with Ca loads on the outward K^+ currents. All of these results are consistent with the other observations reported here indicating that the CaM kinase II catalyzes the reaction(s) mediating or regulating inactivation of I_A and I_C.

These studies demonstrate that the injection of CaM kinase II into Hermissenda type B photoreceptor neurons alters ion conductances in these cells. Specifically, enzyme injection enhances Ca^{2+}-dependent regulation of I_A, I_C and I_{Ca}. These voltage clamp studies directly demonstrate a

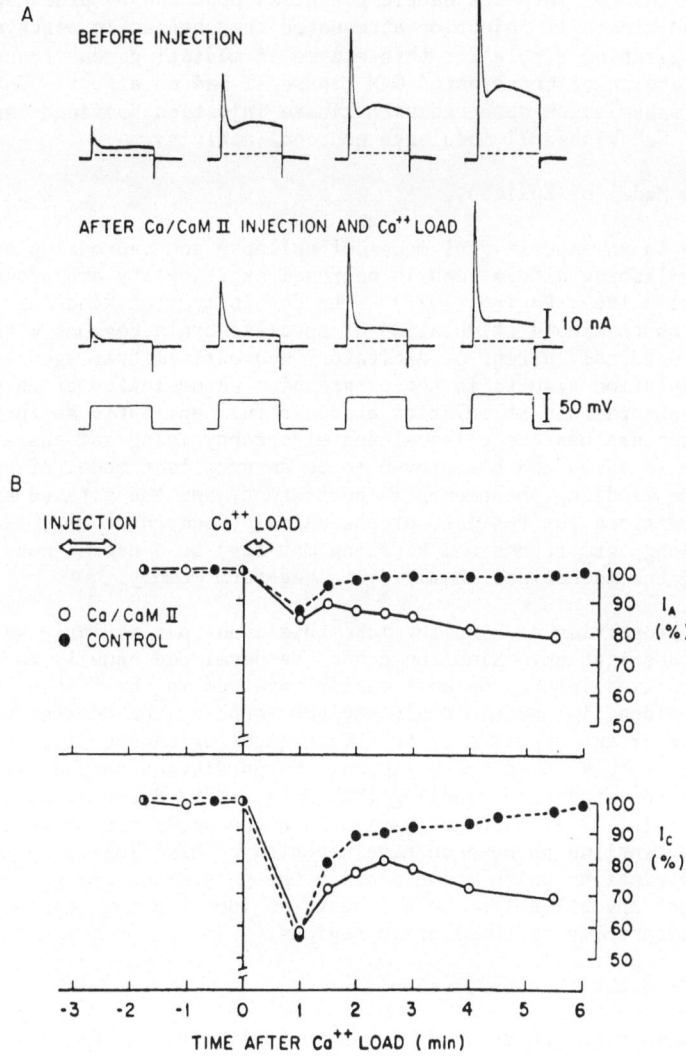

FIG. 3: Effect of CaM kinase II injection on I_A and I_C of the isolated
<u>Hermissenda</u> type B photoreceptor soma. (A) Outward K^+ currents
before and after CaM kinase II injection. Currents are markedly
reduced following CaM kinase II injection accompanied by a Ca^{2+}
load. (B) Time course of I_A (upper panel) and I_C (lower trace)
reduction following a Ca^{2+}load. I_A and $I_C$amplitudes before the
Ca^{2+} load are normalized as 100%. CaM kinase II injection prevents recovery of I_A and I_C reduction after a Ca^{2+} load. (Reproduced from Sakakibara et al., 1986).

molecular effect of CaM kinase II on membrane currents which are major
components of membrane potential. Alteration of these conductances can
produce profound changes in neuronal activity and may underlie CaM kinase
II effects on neuronal excitability.

Recent studies by Llinas et al. (1985) demonstrated that CaM kinase II
injection into presynaptic terminals at the squid giant synapse may mediate
synaptic facilitation. In this study, purified CaM kinase II was injected
into the presynaptic terminal of the squid (<u>Loligo pealii</u>) giant synapse
under voltage clamp conditions. The rate of neurotransmitter release was

determined by changes in postsynaptic potential produced by presynaptic stimulation. CaM kinase II injection attenuated the changes in postsynaptic potential, suggesting a role for this enzyme in mediating neurotransmitter release. Injection of inactivated CaM kinase II had no effect. The increase in synaptic transmission observed with kinase injection provided important evidence that CaM kinase II modulates neuronal activity.

Kindling as a Model of Epilepsy

Kindling is an experimental model of epilepsy and neuronal plasticity in which long-lasting alterations in neuronal excitability are produced (Goddard et al., 1969; Racine, 1978). The development of kindling involves the repeated subthreshold stimulation of specific brain regions with small amounts of electrical current or excitatory neurotransmitter agonists. This repeated stimulation results in the progressive accumulation of response until every subthreshold stimulation produces full epileptic seizure activity. The response has the clinical and electrophysiological characteristics of human seizures and has proven to be an excellent model of epileptic activity. The kindling phenomenon is persistent, and the altered excitability may be retained for the life of the animal (Goddard et al., 1969). Due to this long-term retention, kindling may also be a useful model for understanding the mechanisms involved in long-term memory.

Several properties of kindling make this model particularly well-suited for investigation. Kindling cannot be developed equally in all areas of the brain. Kindling can be most easily obtained in the limbic brain, but cannot be developed in certain brain regions, such as the cerebellum. This provides an important opportunity to investigate region-specific changes associated with the kindled brain region. No persistent morphological changes are associated with kindling (McNamara, 1978; Byrne et al., 1980; Wasterlain et al., 1982). Investigation into the molecular or anatomical basis for the kindling phenomenon have revealed no histological or ultrastructural alterations which could explain the persistent change in excitability. Recent investigations have focused on long-lasting changes in the biochemical properties of these brain regions.

Kindling and Protein Phosphorylation

Since protein phosphorylation has been implicated in the regulation of neuronal excitability and could prove to mediate the altered excitability associated with kindling, changes in the endogenous protein phosphorylation pattern in hippocampal membranes were investigated in the kindling model (Wasterlain and Farber, 1984). In these initial studies by Wasterlain's group, hippocampal membrane fractions obtained from control and kindled animals were examined for endogenous kinase activity using an in vitro phosphorylation assay employing [^{32}P]ATP. No differences in the protein phosphorylation pattern between control and kindled rats were observed with the in vitro addition of cAMP, cGMP or physiological concentrations of Ca^{2+}. However, Ca^{2+} plus calmodulin stimulated phosphorylation of several membrane proteins was significantly reduced in the kindled animals (Fig. 4). These results marked the first observation of a biochemical change associated with kindling.

The Ca^{2+}-calmodulin dependent phosphoproteins altered in the kindled membranes were in the 50,000 - 60,000 dalton range on SDS-PAGE gels. The amount of Ca^{2+}-calmodulin dependent phosphorylation of these proteins in the kindled membranes was significantly reduced compared to the control membranes (Fig. 4). The phosphorylation of several other proteins was also stimulated by Ca^{2+}-calmodulin, but these phosphoproteins were less affected by kindling, suggesting that the effects of kindling on Ca^{2+}-calmodulin dependent phosphorylation are more prominent in 50,000 - 60,000 dalton proteins. The differences between control and kindled animals were most pronounced in hippocampus and amygdala, but were less striking in cortex, basal ganglia and brain stem. No significant changes were observed in the cerebellum, where kindling cannot be achieved. This regional study suggested that the biochemical changes are specific to the kindling phenomenon and are not artifacts of membrane preparation. Thus, septal kindling produces a long-lasting decrease in the Ca^{2+}-calmodulin dependent phosphorylation of 50,000 - 60,000 dalton proteins in hippocampal membranes (Wasterlain and Farber, 1984). The description and characterization of these proteins and the enzyme system that phosphorylates them would provide an important molecular insight into the regulation of seizure discharge.

Alteration of CaM Kinase II Associated with Kindling

As indicated above, CaM kinase II has been implicated in the regulation of neuronal excitability and in the pathogenesis of seizure disorders. Several anticonvulsants have been shown to inhibit the Ca^{2+}-calmodulin dependent phosphorylation of brain membrane proteins in the 50,000 - 60,000 dalton range, corresponding to the autophosphorylating subunits of CaM kinase II (DeLorenzo et al., 1981; Taft et al., 1985). In addition, the direct injection of purified CaM kinase II modulates synaptic activities and regulates specific ion conductances (Acosta-Urquidi et al., 1984a; Llinas et al., 1984; Sakakibara et al., 1986). These observations suggested the possibility that the Ca^{2+}-calmodulin dependent phosphoproteins in the 50,000 - 60,000 dalton range sensitive to the development of kindling may represent the subunits of CaM kinase II.

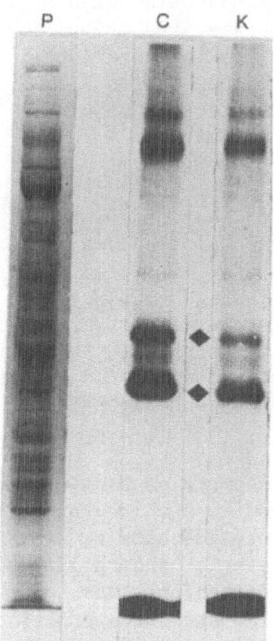

FIG. 4: Decrease of Ca^{2+}-calmodulin-dependent protein phosphorylation in kindled animals. Hippocampal synaptic plasma membrane (HSPM) protein from control and kindled rats were phosphorylated in vitro and separated by SDS-PAGE. (P) Coomassie blue stain of HSPM; (C) Autoradiography of HSPM phosphorylation from control rats; (K) Autoradiography of HSPM phosphorylation from kindled rats. (Reproduced from Goldenring et al., 1986a).

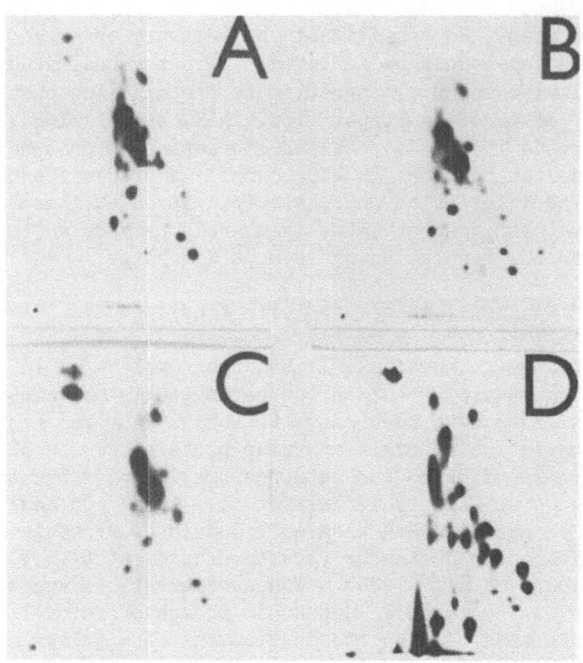

FIG. 5: Autoradiographic exposures of two-dimensional ^{125}I tryptic peptide
maps. (A) 50,000 dalton calmodulin binding protein in hippocampal
synaptic plasma membrane; (B) mPSDp; (C) 50,000 dalton subunit of
purified CaM kinase II; (D) beta-tubulin. Significant homology
exists between the 50,000 dalton HSPM protein (A), mPSDp (B) and
the 50,000 dalton subunit of CaM kinase II (C). (Reproduced from
Goldenring et al., 1986a).

We have recently reported that the 50,000 - 60,000 dalton hippocampal
phosphoproteins affected by septal kindling are homologous with the subunits
of CaM kinase II (Goldenring et al., 1986a). This conclusion was based on:
a) similar migration on SDS-PAGE gels; b) similar isoelectric points; c)
corresponding autophosphorylation properties; d) identical calmodulin binding characteristics and e) comparable tryptic peptide maps. These results
indicate that kindling produces long-lasting alterations in the activity of
CaM kinase II and support the hypothesis that CaM kinase II regulates neuronal excitability.

Several prominent phosphoproteins have migrations on one-dimensional
SDS-PAGE gels which correspond to a molecular weight in the range of 50,000
- 60,000 daltons, including the subunits of CaM kinase II, tubulin and the
B-50 protein, a major substrate for phospholipid dependent protein kinase C
(Aloyo et al., 1982). An important preface to determining whether the kin-
dling-sensitive phosphoproteins are the subunits of CaM kinase II is determining that they are not tubulin or B-50 protein. The 50,000 and 60,000
dalton phosphoproteins were compared to tubulin and B-50 protein by several
biochemical techniques. First, high resolution one-dimensional SDS-PAGE
gels showed that the 50,000 dalton protein affected by kindling had a different mobility than either beta-tubulin or B-50 protein. In addition, specific antibodies to beta-tubulin and B-50 protein did not cross react with
the 50,000 dalton kindling protein on immunostaining gels. Further, the
isoelectric points determined for beta-tubulin, B-50 protein and the 50,000

kindling protein did not correspond. Thus, the 50,000 dalton protein from hippocampal membrane was not homologous with the B-50 protein or beta-tubulin.

The resolution on one and two-dimensional gels of the 50,000 dalton phosphoprotein affected by kindling was, however, markedly homologous with CaM kinase II. Thus, calmodulin binding studies were performed on two dimensional gels to compare the calmodulin binding protein of CaM kinase II and the 50,000 dalton hippocampal membrane protein. Similar patterns of calmodulin binding were observed, each with isoelectric points near neutrality. In order to confirm the hypothesis that the 50,000 dalton hippocampal protein was the 50,000 dalton subunit of CaM kinase II, two dimensional ^{125}I tryptic peptide mapping was performed on these proteins (Fig. 5). The proteins were excised from two dimensional gels, iodinated and digested to completion with trypsin, as previously performed in this laboratory (Vallano et al., 1985b). The two dimensional tryptic peptide maps were essentially homologous for the 50,000 dalton phosphoprotein affected by kindling, the 50,000 dalton subunit of purified CaM kinase II and the 50,000 dalton major postsynaptic density protein (previously shown to be identical to the 50,000 dalton subunit of CaM kinase II). On the basis of these data, we concluded that the 50,000 dalton hippocampal protein affected by septal kindling is identical to the 50,000 dalton subunit of purified CaM kinase II. Thus, long-lasting alterations in the activity of CaM kinase II in hippocampal membranes were associated with septal kindling.

Recent reports have demonstrated that diazepam, a benzodiazepine anticonvulsant, inhibits the development of septal kindling in rats (Farber and Wasterlain, 1986). Diazepam also inhibits the autophosphorylation of purified and membrane-bound CaM kinase II, blocking phosphorylation in the 50,000 - 60,000 dalton range. These observations provide further evidence that the acquisition of septal kindling is mediated by alterations in the activity of CaM kinase II.

SUMMARY

A type II calmodulin-dependent protein kinase (CaM kinase II) has been characterized in the synaptic region and may mediate some of the effects of Ca^{2+} on neuronal excitability. The activity of CaM kinase II is inhibited by anticonvulsant compounds and may be the molecular basis of their neuromodulatory effects. The direct injection of purified CaM kinase II into invertebrate neurons has demonstrated that this kinase can directly alter specific ion conductances and neuronal activity. A long-lasting decrease in CaM kinase II activity is associated with septal kindling, an experimental model of epilepsy and long-term memory. In summary, CaM kinase II appears to be a central mediator of the effects of Ca^{2+} on neuronal function. Further investigation of this enzyme and its effects on neuronal activity may provide a molecular insight into an endogenous mechanism for modulating some of the effects of Ca^{2+} on neuronal excitability and may increase our understanding of the complex regulatory mechanisms that underlie the pathogenesis of seizure discharge and its regulation by anticonvulsant compounds.

REFERENCES

Acosta-Urquidi, J., Neary, J. T., Goldenring, J. R., Alkon, D. L. and DeLorenzo, R. J., Modulation of I_{Ca} and late K^+ currents by intrasomatic injection of Ca-calmodulin dependent kinase in Hermissenda giant neurons, Soc. Neurosci. Abstrs., 10:1129 (1984a).

Acosta-Urquidi, J., Alkon, D. L. and Neary, J. T., Ca^{2+}-dependent protein kinase injection in a photoreceptor mimics biophysical effects of associative learning, Science 224:1254-1257 (1984b).

Alkon, D. L., Neural correlates of associative training in Hermissenda, J. Gen. Physiol., 65:46-56 (1975).

Alkon, D. L., Voltage-dependent calcium and potassium conductances: a contingency mechanism for an associative learning model, Science 205:810-816 (1979).

Alkon, D. L., Shoukimas, J. J. and Heldman, E., Calcium mediated decrease of a voltage-dependent potassium current, Biophys. J., 40:245-250 (1982a).

Alkon, D. L., Lederhendler, I. and Shoukimas, J. J., Primary changes of membrane currents during retention of associative learning, Science 215: 693-695 (1982b).

Alkon, D. L., Acosta-Urquidi, J., Olds, J., Kuzma, G. and Neary, J., Protein kinase injection reduces voltage-dependent potassium currents, Science 219:303-306, (1983).

Alkon, D. L., Calcium-mediated reduction of ionic currents: a biophysical memory trace, Science 226:1037-1045, (1984).

Alkon, D. L., Farley, J., Sakakibara, M. and Hay, B., Voltage-dependent calcium and calcium-activated potassium currents of a molluscan photoreceptor, Biophys. J. 46:605-614 (1984).

Alkon, D. L. and Sakakibara, M., Prolonged inactivation of a Ca^{2+} dependent K^+ current but not Ca^{2+} current by light induced elevation of intracellular calcium, Soc. Neurosci. Abstr. 10:10 (1984).

Alkon, D. L., Sakakibara, M., Forman, R. R., Harrigan, J., Lederhendler, I. and Farley, J., Reduction of two voltage-dependent K^+ currents mediates retention of a learned association, Behav. Neural Biol. 44:278-300 (1985).

Alkon, D. L. and Sakakibara, M., Calcium activates and inactivates a photoreceptor soma K^+ current, Biophys. J., in press (1986).

Aloyo, V. J., Zweirs, H. and Gispen, W. H., B-50 protein kinase and kinase C in rat brain, Prog. Brain Res. 56:303-315 (1982).

Baraban, J. M., Snyder, S. H. and Alger, B. E., Protein kinase C regulates ionic conductance in hippocampal pyramidal neurons: electrophysiological effects of phorbol esters, Proc. Natl. Acad. Sci. USA 82:2538-2542 (1985).

Bennett, M. K., Erondu, N. E. and Kennedy, M. B., Purification and characterization of a calmodulin-dependent protein kinase that is highly concentrated in brain, J. Biol. Chem. 258:12735-12744 (1983).

Burke, B. and DeLorenzo, R. J., Calcium and calmodulin regulated endogenous tubulin kinase activity in synaptic nerve terminal preparations, Brain Res., 236:393-415 (1982).

Byrne, M. C., Gottlieb, R., and McNamara, J. O., Amygdala kindling induces muscarinic cholinergic receptor declines in a highly specific distribution within the limbic system, Exp. Neurol. 69:85-98 (1980).

Castelluci, V. F., Kandel, E. R., Schwartz, J. H., Wilson, F. D., Nairn, A. C. and Greengard, P., Intracellular injection of the catalytic subunit of cyclic AMP-dependent protein kinase simulates facilitation of transmitter release underlying behavioral sensitization in Aplysia, Proc. Natl. Acad. Sci. USA 77:7492-7496 (1980).

Cheung, W. Y., Calmodulin role in cellular regulation, Science 207:19-27 (1980).

DeCamilli, P., Camerson, R. and Greengard, P., Synapsin I (Protein I), a nerve terminal specific phosphoprotein I: Its general distribution in synapses of the central and peripheral nervous system demonstrated by immuno-fluorescence in frozen and plastic sections, J. Cell Biol. 96: 1337-1354 (1983).

DeLorenzo, R. J., Calcium-dependent phosphorylation of specific synaptosomal fraction proteins: possible role of phosphoproteins in mediating neurotransmitter release, Biochem. Biophys. Res. Commun. 71:590-597 (1976).

DeLorenzo, R. J., Freedman, S. D., Yohe, W. B. and Maurer, S. C., Stimulation of calcium-dependent neurotransmitter release and presynaptic nerve terminal protein phosphorylation by calmodulin and a calmodulin-like protein isolated from synaptic vesicles, Proc. Natl. Acad. Sci. USA 76:1838-1842 (1979).

DeLorenzo, R. J., Role of calmodulin in neurotransmitter release and synaptic function, Ann. N. Y. Acad. Sci. 356:92-109 (1980a).

DeLorenzo, R. J., Phenytoin: calcium-calmodulin-dependent protein phosphorylation and neurotransmitter release, in: "Antiepileptic Drugs: Mechanism of Action," G. H. Glaser, J. K. Penry, and D. W. Woodbury, eds., Raven, New York, pp 399-414 (1980b).

DeLorenzo, R. J., Calcium, calmodulin and synaptic function: modulation of neurotransmitter release, nerve terminal protein phosphorylations, and synaptic vesicle morphology by calcium and calmodulin, in: "Regulatory Mechanism of Synaptic Transmission," R. Tapie and C. W. Cotman, eds., Plenum, New York and London, pp 205-240 (1981a).

DeLorenzo, R. J., The calmodulin hypothesis of neurotransmission, Cell Calcium 2:365-385 (1981b).

DeLorenzo, R. J., Burdette, S. and Holderness, J., Benzodiazepine inhibition of the calcium-calmodulin protein kinase system in brain membrane, Science 213:546-549 (1981).

DeLorenzo, R. J., Gonzales, B., Goldenring, J. R., Bowling, A. C. and Jacobson, R., Ca^{2+}-calmodulin tubulin kinase system and its role in mediating the Ca^{2+} signal in brain, in: "Progress in Brain Research," Volume 56, W. H. Gispen and A. Routtenberg, eds., Elsevier Biomedical Press, Amsterdam, pp. 255-286 (1982).

DeLorenzo, R. J., Calmodulin in neurotransmitter release and synaptic function, Fed. Proc. 41:2265-2272 (1982).

DePeyer, J. E., Cachelin, A. B., Levitan, I. B. and Reuter, H., Ca^{2+}-activated K^+ conductance in internally perfused snail neurons is enhanced by protein phosphorylation, Proc. Natl. Acad. Sci. USA 79:4207-4211 (1982).

DeRiemer, S., Strong, J., Albert, K., Greengard, P. and Kaczmarek, L., Enhancement of calcium current in Aplysia neurones by phorbol ester and protein kinase C, Nature 313:313-316 (1985).

Douglas, W. W., Stimulus-secretion coupling: the concept and clues from chromaffin and other cells, Br. J. Pharmacol. 34:451-474 (1968).

Farber, D. B. and Wasterlain, C. G., Inhibition of kindled seizure by diazepam: mediation by phosphoproteins?, Proc. Natl. Acad. Sci. USA, in press (1986).

Farley, J. and Alkon, D. L., Membrane depolarization accumulates during acquisition of an associative behavioral change, Science 210:1375-1376 (1980).

Farley, J. and Alkon, D. L., Associative neural and behavioral change in Hermissenda: consequences of nervous system orientation for light and pairing specificity, J. Neurophysiol. 48:785-807 (1982).

Farley, J. and Auerbach, S., Protein kinase C activation induces conductance changes in Hermissenda photoreceptors like those seen in associative learning, Nature 319:220-223 (1986).

Fukunaga, K., Yamamoto, H., Matsui, K., Higashu, K. and Miyamoto, E., Purification and characterization of a Ca^{2+}-calmodulin-dependent protein kinase from rat brain, J. Neurochem. 39:1607-1617 (1982).

Gastaut, H., Jasper, H., Bancaud, J. and Waltregny, A., eds., "The Physiopathogenesis of the Epilepsies," Charles C. Thomas, Springfield, Illinois (1969).

Glaser, G. H., Epilepsy, in: "Recent Advances in Clinical Neurology," W. P. Matthews, ed., Churchill-Livingston, London (1975).

Glaser, G. H., Penry, J. K. and Woodbury, D. M., eds., "Antiepileptic Drugs: Mechanisms of Action," Raven Press, New York (1980).

Goddard, G. V., McIntyre, D. C. and Leech, C. K., A permanent change in brain function resulting from daily electrical stimulation, Exp. Neurol. 25:243-330 (1969).

Goldenring, J. R., Gonzalez, B. and DeLorenzo, R. J., Isolation of brain Ca^{2+}-calmodulin tubulin kinase containing calmodulin binding proteins, Biochem. Biophys. Res. Commun. 108:421-428 (1982).

Goldenring, J. R., Gonzalez, B., McGuire, J. S., Jr. and DeLorenzo, R. J., Purification and characterization of a calmodulin-dependent protein kinase from rat brain cytosol able to phosphorylate tubulin and microtubule-associated proteins, J. Biol. Chem. 258:12632-12640 (1983).

Goldenring, J. R., McGuire, J. S., Jr., and DeLorenzo, R. J., Identification of the major post-synaptic density protein as a homologous with the major calmodulin-binding subunit of a calmodulin-dependent protein kinase, J. Neurochem. 42:1077-1084 (1984).

Goldenring, J. R., Vallano, M. L. and DeLorenzo, R. J., Phosphorylation of microtubule-associated protein 2 at distinct sites by calmodulin-dependent and cyclic-AMP-dependent kinases, J. Neurochem. 45:900-905 (1985).

Goldenring, J. R., Wasterlain, C. G., Destreicher, A. B., deGraan, P. N. E., Farber, D. B., Glaser, G. and DeLorenzo, R. J., Kindling induces a long lasting change in the activity of a hippocampal membrane calmodulin-dependent protein kinase, Brain Res., in press (1986a).

Grab, D. J., Berzins, K., Cohen, R. S. and Siekevitz, P., Presence of calmodulin in postsynaptic densities isolated from canine cerebral cortex, J. Biol. Chem. 254:8690-8696 (1979).

Grab, D. J., Carlin, R. K. and Siekevitz, P., The presence and functions of calmodulin in the postsynaptic density, Ann. N. Y. Acad. Sci. 356:55-72 (1980).

Grab, D. J., Carlin, R. K. and Siekevitz, P., Function of calmodulin in postsynaptic density II. Presence of calmodulin-activable protein kinase activity, J. Cell Biol. 89:440-448 (1981).

Hawkins, R. D., Abrams, T. W., Carew, T. J. and Kandel, E. R., Differential classical conditioning of a defensive withdrawal reflex in Aplysia californica, Science 219:397-404 (1983).

Jasper, H. H., Ward, A. A., Jr. and Pope, A., eds., "Basic Mechanisms of the Epilepsies," Little, Brown, Boston (1969).

Kandel, E. R. and Schwartz, J. H., Molecular biology of learning: modulation of transmitter release, Science 218:433-443 (1982).

Katz, B. and Miledi, R., Spontaneous and evoked activity of motor nerve endings in calcium Ringer, J. Physiol (Lond) 203:689-706 (1969).

Katz, B. and Miledi, R., Further study of the role of calcium in synaptic transmission, J. Physiol. (Lond) 207:789-801 (1970).

Kelly, P. T. and Cotman, C. W., Synaptic protein: characterization of tubulin and actin and identification of a distinct postsynaptic density protein, J. Cell Biol. 79:173-183 (1978).

Kelly, P. T., McGuinness, T. L. and Greengard, P., Evidence that the major postsynaptic density protein is a component of a Ca^{2+}/calmodulin-dependent protein kinase, Proc. Natl. Acad. Sci. USA, 81:945-949 (1984).

Kennedy, M., Experimental approaches to understanding the role of protein phosphorylation in the regulation of neuronal function, Ann. Rev. Neurosci. 6:493-525 (1983).

Kennedy, M. B., Bennett, M. K. and Erondu, N. E., Biochemical and immunochemical evidence that the "major postsynaptic density protein" is a subunit of a calmodulin-dependent protein kinase, Proc. Natl. Acad. Sci. USA 80:7357-7361 (1983).

Klee, C. B., Crouch, T. H. and Richman, P. G., Calmodulin, Ann. Rev. Biochem. 49:489-515 (1980).

Larson, R. E., Goldenring, J. R., Vallano, M. L. and DeLorenzo, R. J., Identification of endogenous calmodulin-dependent kinase and calmodulin binding proteins in cold-stable microtubule preparations from rat brain, J. Neurochem. 44:1566-1574 (1985).

Levitan, I. B., Lemos, J. R. and Novak-Hofer, I., Protein phosphorylation and the regulation of ion channels, Trends Neurosci. 6:496-499 (1983).

Llinas, R., McGuinness, T. L., Leonard, C. S., Sugimori, M. and Greengard, P., Intraterminal injection of synapsin I or calcium/calmodulin-dependent protein kinase II alters neurotransmitter release at the squid giant synapse, Proc. Natl. Acad. Sci. USA 82:3035-3039 (1985).

Matus, A., The postsynaptic density, Trends Neurosci. 4:51-53 (1981).

McGuinness, T. L., Lai, Y., Greengard, P., Woodgett, J. R. and Cohen, P., A multifunctional calmodulin-dependent protein kinase, FEBS Lett., 163: 329-334 (1983).

McNamara, J. O., Selective alterations of regional beta-adrenergic receptor binding in the kindling model of epilepsy, Exp. Neurol. 61:582-591 (1978).

Nestler, E. J. and Greengard, P., Protein phosphorylation in the brain, Nature 305:583-588 (1983).

Nestler, E. J., Walaas, S. I. and Greengard, P., Neuronal phosphoproteins: Physiological and clinical implications, Science 225:1357-1364 (1984).

Osterrieder, W., Brum, G., Hescheler, J., Trautwein, W., Flockerzi, V. and Hofmann, F., Injection of subunits of cyclic AMP-dependent protein kinase into cardiac myocytes modulates Ca^{2+} current, Nature 298:576-578 (1982).

Ouimet, C. C., McGuinness, T. L. and Greengard, P., Immunocytochemical localization of calcium/calmodulin dependent protein kinase II in brain, Proc. Natl. Acad. Sci. USA 81:5604-5608 (1984).

Racine, R. J., Kindling: The first decade, J. Neurosurg. 3:234-252 (1978).

Rubin, R. P., The role of calcium in the release of neurotransmitter substances and hormones, Pharmacol. Rev. 22:389-428 (1972).

Sakakibara, M., Alkon, D. L., DeLorenzo, R. J., Goldenring, J. R., Neary, J. T. and Heldman, E., Modulating of calcium-mediated inactivation of ionic currents by Ca^{2+}/calmodulin-dependent protein kinase II, Biophys. J., in press (1986).

Schulman, H. and Greengard, P., Ca^{2+}-dependent protein phosphorylation system in membranes from various tissues, and its activation by "calcium-dependent regulator", Proc. Natl. Acad. Sci. USA 75:5432-5436 (1978a).

Schulman, H. and Greengard, P., Stimulation of brain membrane protein phosphorylation by calcium and an endogenous heat-stable protein, Nature 271:478-479 (1978b).

Schulman, H., Phosphorylation of microtubule-associated proteins by a calcium/calmodulin-dependent protein kinase, J Cell Biol 99:15-21 (1984).

Siegelbaum, S. A., Camardo, J. S. and Kandel, E. R., Serotonin and cyclic AMP close single K^+ channels in Aplysia sensory neurones, Nature 299: 413-417 (1982).

Siegelbaum, S. A. and Tsien, R. W., Modulation of gated ion channels as a mode of transmitter action, Trends Neurosci. 6:307-312 (1983).

Taft, W. C., Goldenring, J. R., Buckholz, T. M. and DeLorenzo, R. J., Benzodiazepine inhibition of purified CaM-dependent kinase, Pharmacologist 27:185 (1985).

Vallano, M. L., Buckholz, T. M. and DeLorenzo, R. J., Phosphorylation of neurofilament proteins by endogenous calcium/calmodulin dependent protein kinase, Biochem. Biophys. Res. Commun. 130:957-963 (1985a).

Vallano, M. L., Goldenring, J. R., Buckholz, T. M., Larson, R. E. and DeLorenzo, R. J., Separation of endogenous calmodulin- and cAMP-dependent kinases from microtubule preparations, Proc. Natl. Acad. Sci. USA 82: 3203-3206 (1985b).

Wasterlain, C. G., Morin, A. M. and Jonec, V., Kindling: a pharmacological approach, Electroencephalog. Clin. Neurophysiol. 36:264-273 (1982).

Wasterlain, C. G. and Farber, D. B., Kindling alters the calcium/calmodulin-dependent phosphorylation of synaptic plasma membrane proteins in rat hippocampus, Proc. Natl. Acad. Sci. USA 81:1225-1257 (1984).

Yamauchi, T. and Fujisawa, J., Purification and characterization of the brain calmodulin-dependent protein kinase (Kinase II), which is involved in the activation of tryptophan-5-mono-oxygenase, Eur. J. Biochem. 132:15-21 (1983).

CYTOSKELETAL PATHOLOGY IN NEURODEGENERATIVE DISEASES

William W. Pendlebury, David Munoz-Garcia
and Daniel P. Perl

Department of Pathology (Neuropathology), University
of Vermont College of Medicine, Medical Alumni Building
Burlington, Vermont 05405

INTRODUCTION

The neurodegenerative diseases are a large group of disorders of the central nervous system (CNS) which are characterized by cell death in specific populations of neurons. Affected neuronal populations differ in each disease, and causal factors are not well understood. Excluded from this definition are diseases due to known infectious agents, hypoxia, toxins, nutritional deficiencies, and those in which a metabolic defect has been identified. The three most common neurodegenerative diseases are Alzheimer's disease (presenile and senile forms), Parkinson's disease, and amyotrophic lateral sclerosis (known to the lay public as Lou Gehrig's disease). In all three diseases, the cytoplasm of certain neurons accumulate abnormal structures which are derived, at least in part, from cytoskeletal proteins. In addition, neuronal cytoskeletal pathology is seen in several less common neurodegenerative diseases, and the study of these diseases can provide valuable insights into the pathogenesis of the more common disorders. Since the cytoskeleton is critical for the maintenance of neuronal shape (Lazarides, 1980) and cytoplasmic transport mechanisms, including axonal transport (Bray & Gilbert, 1981), neurons exhibiting these abnormalities may not function normally. This review will address recent advances in the understanding of cytoskeletal abnormalities found in degenerative diseases of the CNS, with emphasis on immunocytochemical characterization of the ultrastructural morphology.

Alzheimer's Disease

In 1907, Alois Alzheimer described the pathological findings in the brain of a 51 year old woman who for the last four years of life suffered a progressive disease characterized by global deterioration of intellectual function (Alzheimer, 1907). Although the term Alzheimer's disease historically has been restricted to cases of dementia with onset prior to age 65, most patients with the clinical syndrome of senile dementia (Diagnostic and Statistic Manual of Mental Disorders, ed. III) show histopathological changes in the brain qualitatively identical to those described by Alzheimer. In the following discussion, the term dementia of the Alzheimer type (DAT) will refer to all patients with dementia whose brains show these characteristic pathologic features, irrespective of the age of onset.

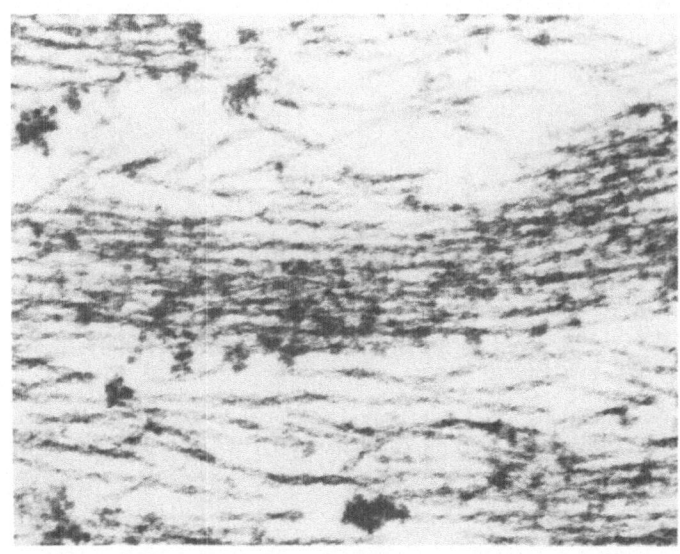

FIG. 1: Electron micrograph demonstrating the paired helical appearance of
fibrils forming a neurofibrillary tangle in a hippocampal neuron
from the brain of a DAT patient. Individual fibril thickness
varies from 10nm to 20nm with a periodicity of 80nm. X30,000.

The Neuropathology of DAT: An Overview

On histological examination, the brain of DAT patients shows neuronal
loss in the hippocampus (Ball, 1977), the neocortex (Terry et al., 1981),
and certain subcortical nuclei. Among the latter, the neuronal depopulation
in the nucleus basalis of Meynert of the basal forebrain has been studied
most extensively (Whitehouse et al., 1982). This nucleus is the major
source of cholinergic innervation to the cerebral cortex and hippocampus
(Johnston et al., 1979), and loss of neurons in this area can be correlated
with presynaptic cholinergic deficits, widely regarded as the most significant and consistent neurotransmitter abnormality in DAT (Davies et al.,
1976).

In addition to cell loss, three types of abnormal cytoplasmic forma-
tions appear in certain neuronal subsets. These include neurofibrillary
tangles (NFTs), granulovacuolar degeneration (GVD), and Hirano bodies.
Finally, amyloid deposits are found in the cortical neuropil (senile
plaques), and in cortical and meningeal vessels (congophilic angiopathy).
All of these abnormalities (to be described in the following sections), can
be found in small numbers and in restricted locations in the brains of
normal elderly individuals. Therefore, at the present time, the histopathologic criteria for the diagnosis of DAT is quantitative (Khachaturian,
1985).

Neurofibrillary Tangles (NFTs)

In tissue stained with silver impregnation methods, NFTs appear as masses of thick, twisted fibrils within the cytoplasm of hippocampal, neocortical, and subcortical neurons. Normal looking neurons are always present in
the vicinity of NFT-bearing cells. A few NFTs can be found in the hippocampal formation of normal elderly individuals (Dayan, 1970), but the presence
of large numbers of NFTs, particularly in the neocortex, is highly associ-

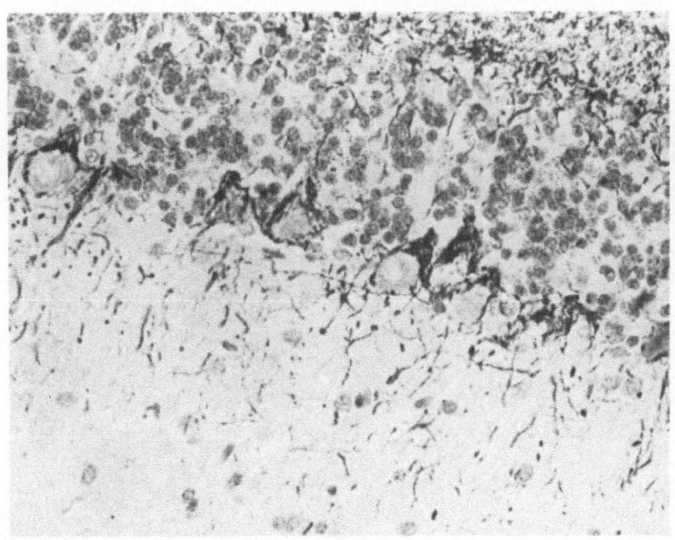

FIG. 2: Immunoperoxidase staining of normal rat cerebellar cortex with a monoclonal antibody directed against phosphorylated epitopes of normal neurofilaments. Distal axons, such as those forming the basket fibers, are stained.

ated with dementia (Blessed et al., 1968). It has been proposed that the formation of NFTs is the forerunner of cell death in DAT (Saper et al., 1985), that memory loss in DAT is the consequence of the isolation of the hippocampal formation by the development of NFTs in adjacent areas of the cortex (Hyman et al., 1984).

Ultrastructurally, the fibrils of NFTs show periodic reductions in diameter from 20 nm to 10 nm every 80 nm. The name given to these fibrils, paired helical filaments (PHFs) (Kidd, 1963; Wisniewski et al., 1976), reflects the appearance of PHFs as two twisted filaments (Figure 1). Although high resolution electron microscopy has shown that these fibrils are not neurofilaments, the term PHF has gained wide acceptance and will be used in this review. PHFs are highly resistant to proteolytic degradation, and insoluble in virtually every solvent, including sodium dodecyl sulfate (SDS) (Selkoe et al., 1982a). The lack of solubility of PHFs in solvents used for electrophoresis explains the surprising absence of differences between normal and DAT brains found in earlier studies; PHFs failed to enter the gel. One group has reported to have achieved partial solubilization of PHFs (Iqbal et al., 1984).

Regardless whether these results are confirmed by other laboratories, the high degree of insolubility of PHFs is not in doubt. This feature has been exploited for the preparation of polyclonal and monoclonal antibodies directed against PHFs. On immunocytochemical staining, a number of studies have shown that these antibodies decorate PHF-bearing structures, i.e. NFTs and the neurites of senile plaques (Ihara et al., 1983; Grundke-Iqbal et al., 1984; Wang et al., 1984). Although the anti-PHF antibodies do not decorate normal neurons (Ihara et al., 1983), it has been possible in some cases to eliminate anti-PHF activity by absorption with concentrated extracts of normal human brain (Grundke-Iqbal et al., 1985a). This suggests that PHFs contain an antigen present in low concentration in normal brain.

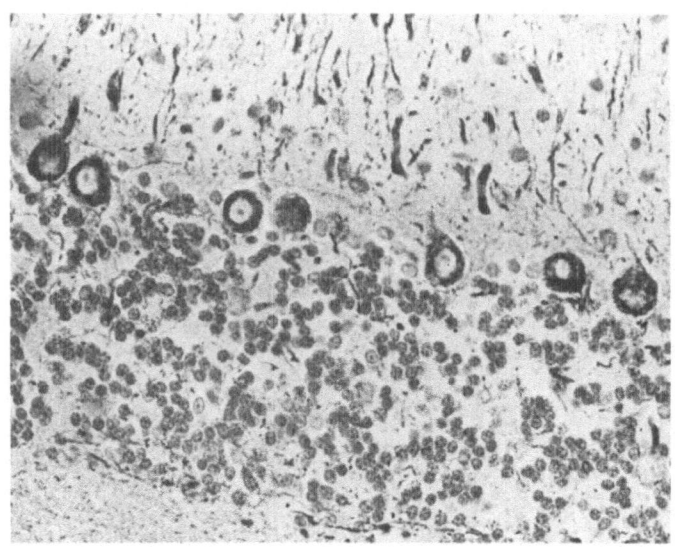

FIG. 3:   Immunoperoxidase staining of normal rat cerebellar cortex with a
          monoclonal antibody directed against non-phosphorylated epitopes
          of normal neurofilaments.   Purkinje cell perikarya are stained.

In addition to the PHF-specific component, NFTs are decorated by antibodies against microtubules (Grundke-Igbal et al., 1979; Yen et al., 1981).
The antigen in PHFs with which antimicrotubule antibodies react is not tubulin, but one or more of the microtubule associated proteins, including MAP-2
(Kosik et al., 1984) and tau.   PHFs have been conclusively shown by immunocytochemical staining to react with certain antibodies against the 200 kD
and 160 kD proteins of the neurofilament triplet (Anderton et al., 1982;
Autilio-Gambetti et al., 1983).   Furthermore, Perry et al. (1985) have
demonstrated by immunogold staining at the ultrastructural level that this
reactivity is associated with PHFs themselves, and persists after procedures that extract normal neurofilaments.   It is clear, however, that not
all antibodies that decorate normal neurofilaments recognize NFTs, particularly after SDS extraction of the latter (Ihara et al., 1983; Anderton et
al., 1982).

An exciting opportunity was recently provided by the preparation by
Sternberger and Sternberger (1983) of monoclonal antibodies that recognize
either phosphorylated or non-phosphorylated epitopes in the 200 kD and 160
kD proteins of neurofilaments.   They discovered by immunocytochemical staining, using these antibodies, that phosphorylated neurofilament proteins are
normally located in terminal axons, whereas non-phosphorylated neurofilament
proteins are located in the perikaryon, dendrites, and proximal axonal segments of neurons.   These workers subsequently showed that NFT-bearing neurons in DAT accumulate phosphorylated neurofilament proteins in the perikaryon (Sternberger et al., 1985), where they are normally absent. Our group
has confirmed these results utilizing a different set of monoclonal antibodies (Figures 2-4).

In summary, NFTs contain identified epitopes shared with cytoskeletal
components, as well as unidentified epitopes, presumably also polypeptides.
These unidentified components share antigenic determinants with the amyloid
core of senile plaques (see discussion of senile plaques below), demonstra-
ble only after isolation of the latter (Selkoe and Abraham, 1985).   How are

these diverse components bound together? Following the demonstration of
the insolubility of PHFs, Selkoe and collaborators proposed that such insolubility is due to intermolecular covalent cross-linking of proteins by
gamma-glutamyl-epsilon-lysine side chain residues (Selkoe et al., 1982b).
They were able to isolate from human brain a transglutaminase with activity
directed toward neurofilaments. Through the action of this enzyme, a network of insoluble fibrils was formed, but PHFs were not produced. This
covalent-link hypothesis of PHF insolubility has been challenged by an
alternative hypothesis (Kidd et al., 1985; Grundke-Iqbal et al., 1985b).
This latter hypothesis suggests that PHFs may form by non-covalent associations, like those found in amyloid (Glenner, 1980a, 1980b). PHFs stain
with Thioflavin S and show birefringence when stained with Congo red and
observed with cross-polarized light, a property shared by amyloid. Furthermore, it has recently been shown that the polypeptide chains in PHFs
have a Beta-pleated conformation (Kirscher et al., 1986).

Although NFTs composed of PHFs are found only in humans, they are not
restricted to the brains of DAT patients (Wisniewski et al., 1979). In
addition to a number of degenerative and genetically determined diseases
(including Down's syndrome) (Burger and Vogel, 1973), PHFs grouped into
NFTs are consistently found in the brains of patients suffering from the
chronic stages of infection with the virus of subacute sclerosing panencephalitis (Corsellis, 1951; Mandybur et al., 1977), and in the brains of old
boxers (dementia pugilistica) (Corsellis et al., 1973). Additionally, an
extraordinarily high degree of NFT formation is seen in the population of
the island of Guam and Kii peninsula of Japan (Chen, 1981). These two foci
were identified because of the remarkable incidence of amyotrophic lateral
sclerosis (ALS) and parkinsonism-dementia (PD) (Hirano et al., 1966). The
brains of patients with these diseases show the highest density of NFTs
seen in any condition; a lesser degree of NFT formation is found in 15% of
the normal population in these geographical sites (Anderson et al., 1979).

The nature of the molecular trigger that initiates the cascade of

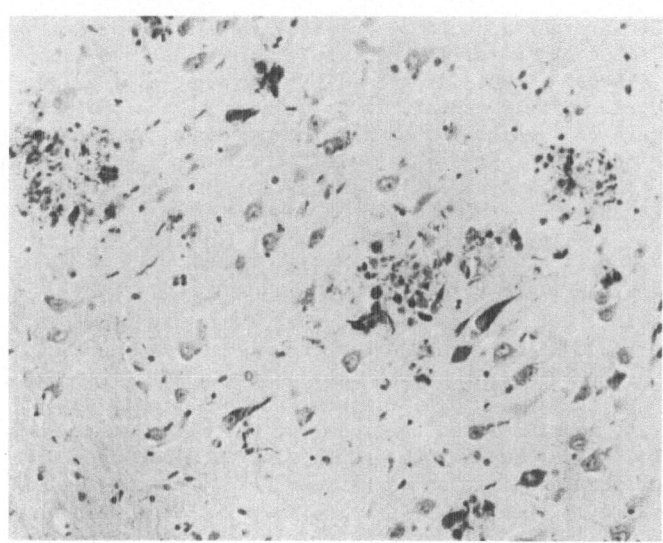

FIG. 4: Immunoperoxidase staining of a hippocampal section from the brain
         of a patient with DAT. The monoclonal antibody used in this expe-
         riment recognizes phosphorylated epitopes of normal neurofilaments.
         NFTs and neurites of senile plaques are stained.

events leading to the formation of PHFs and NFTs is unknown. Any hypothesis must take into account the fact that only some neurons develop NFTs, while adjacent ones do not. Utilizing a recently developed analytic probe (laser microprobe mass analyzer (LAMMA)), our group has demonstrated that NFT-bearing neurons in Guamanian ALS and PD contain high concentrations of aluminum and calcium (Perl et al., 1985). This confirms previous data demonstrating the cytoplasmic accumulation of calcium and aluminum in Guamanian ALS and PD, obtained through the use of X-ray spectrometry combined with scanning electron microscopy (SEM-XES) (Perl et al., 1982). LAMMA has allowed us to localize aluminum accumulations to the NFT itself. Aluminum also appears to accumulate in NFT-bearing neurons in DAT, but to a far lesser degree than in patients from Guam. Although Perl and Brody (1980) originally localized aluminum to the "nuclear region" of NFT-bearing neurons in DAT by SEM-XES, it is now clear that this more primitive method cannot provide precise subcellular localization. Studies in DAT utilizing LAMMA are in progress. Although aluminum might be one of the factors capable of initiating the formation of NFTs, it is also possible that, once formed, NFTs bind aluminum present in the medium.

The main obstacle to progress in the understanding of the origin and nature of NFTs has been the lack of an animal model. A few twisted fibrils, with much shorter periodicity than human PHFs, can be found in cerebral neurites in aged dogs, monkeys and rats (Knox et al., 1980). Similar twisted fibrils have been described in the dorsal root ganglia of rats chronically intoxicated with ethanol (Volk, 1980). The scarcity of these abnormal fibrils has not allowed for experimental exploitation. Injection of aluminum salts into the CSF of rabbits results in an acute encephalomyelopathy accompanied by the formation of neuronal cytoplasmic structures resembling NFTs at the light microscopic level (Klatzo et al., 1965; Wisniewski et al., 1967, 1980, 1982; Bugiani and Ghetti, 1982; Troncoso et al., 1982). However, both electron microscopy (Terry & Pena, 1965) and biochemical analysis (Selkoe et al., 1979) have conclusively demonstrated that aluminum-induced NFTs in the rabbit are made up of normal neurofilaments. Aluminum salts also induce neurofilament accumulation in human embryonic neurons in culture (Deboni et al., 1980).

Using a Sternberger monoclonal antibody (SMI31), directed against phosphorylated epitopes of neurofilament triplet proteins, our group (Munoz-Garcia et al., 1986), as well as others (Troncoso et al., 1985), have shown that aluminum-induced NFTs contain large numbers of phosphorylated neurofilaments (Figure 5A), which in normal circumstances are found only in terminal axons (Figure 5B). This parallels the similar finding in DAT (Sternberger et al., 1985), as previously discussed. In addition, an antigenic change takes place in non-phosphorylated neurofilaments that participate in the formation of aluminum-induced NFTs. Another of the Sternberger monoclonal antibodies (SMI 32) used by us recognizes an epitope of non-phosphorylated neurofilaments located in the perikaryon of the spinal anterior horn neurons of the rat, but not the normal rabbit. However, SMI-32 intensely stains aluminum-induced NFTs in the rabbit (Figure 6). Whether the antigenic change in the epitope recognized by SMI-32, and accompanying NFT formation, is due to chemical modification or conformational variation is not known. Recently, we have shown that, unlike DAT NFTs, aluminum-induced NFTs do not contain microtubule-associated proteins (Munoz-Garcia et al., 1986). Despite the substantial differences between DAT-type and aluminum-induced NFTs, some aspects of the latter, particularly the redistribution of phosphorylated neurofilaments, may provide a model system for the study of the cellular pathophysiology of human NFT formation.

The role of neurofilament phosphorylation in the pathogenesis of NFTs is unknown. Sayre et al (1985) have recently proposed that a modification of the ionic balance of neurofilament triplet proteins (induced by phospho-

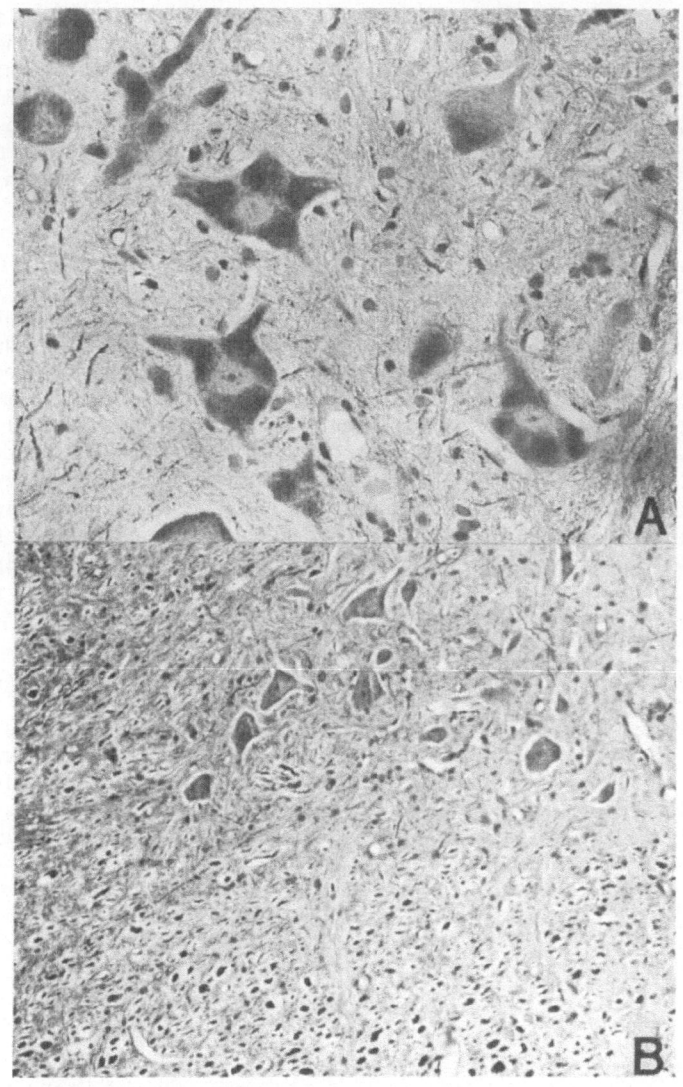

FIG. 5: Immunoperoxidase staining of rabbit spinal cord with a monoclonal
antibody directed against phosphorylated epitopes of normal neurofilaments. A. Aluminum-induced NFTs in anterior horn cells are
stained. B. Normal control rabbit shows no perikaryal staining
of anterior horn cells.

rylation of the hydroxyl groups in the tail piece of the protein), changes
the pattern of association with other neurofilaments and microtubules. This
is based on their studies of giant axonal neuropathies induced by neurotoxins. A hypothesis stating that altered phosphorylation can lead to cytoskeletal change is supported by the fact that vanadate, an agent known to
alter endogenous protein phosphorylating activity, induces the dissociation
of neurofilaments from microtubules as well as the aggregation of intermediate filaments in tissue culture (Sayre et al., 1985).

Granulovacuolar Degeneration

This histopathologic feature consists of one or several vacuoles, each 3-5 um in diameter and containing a basophilic granule, lying in the cytoplasm of pyramidal neurons in the hippocampus. The frequency of this lesion is correlated with the degree of dementia present in life (Ball and Lo, 1977). Electron microscopy reveals the granule to consist of osmophilic, finely granular material, and vacuole to be bounded by a unit membrane. Recently, Price et al. (1985) have shown that the granule is recognized by polyclonal and monoclonal antibodies directed against tubulin.

Hirano Bodies

Hirano bodies (Hirano, 1965) are brightly eosinophilic, elongated or rounded structures which appear adjacent to, and often indenting, neuronal perikarya. Although they can be found in brains in a variety of locations and diseases, as well as in the normal elderly, Hirano bodies are most commonly seen in the hippocampus of patients with DAT (Gibson and Tomlinson, 1977) and Guamanian PD and ALS (Hirano et al., 1968). These structures, located within neurites, appear ultrastructurally as alternating rows of cross-sectioned and longitudinally-sectioned fibrils, each 5-6 nm in diameter. Goldman et al., (1983a) first described the actin content of Hirano bodies, and this has now been confirmed by several laboratories (Figure 7). In addition, Peterson et al. (1985) have presented an animal model of Hirano body formation in the brindled mouse mutant.

Amyloid Deposits: Senile Plaques and Congophilic Angiopathy

Classical senile plaques, also referred to as neuritic plaques, consist of a rounded core of amyloid surrounded by a ring of distended neurites intermingled with astrocytes, microglial cells, and wisps of amyloid. They are found scattered in large numbers throughout the cerebral cortex and certain subcortical regions in brains from DAT, and in small numbers in the

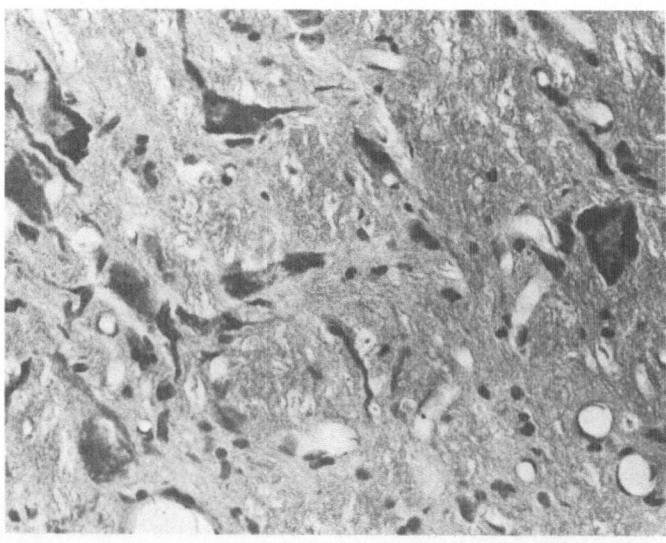

FIG. 6: Immunoperoxidase staining of rabbit spinal cord with a monoclonal antibody directed against non-phosphorylated epitopes of normal neurofilaments. The aluminum-induced NFTs in anterior horn cells are stained.

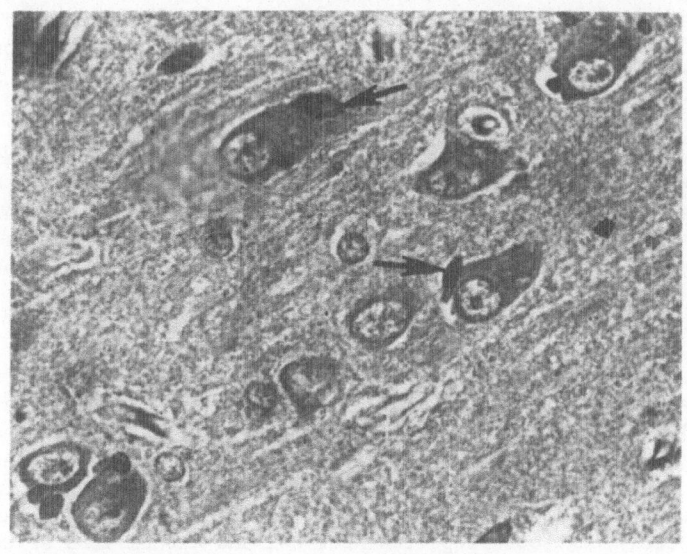

FIG. 7:   Section of hippocampus from the brain of a DAT patient.   Hirano
          bodies (arrows) are stained following immunocytochemical proces-
          sing with antiactin rabbit antiserum.

cortex of brains from some normal elderly individuals. The neurites, some
are cholinergic (Struble et al., 1982), are packed with secondary lysosomes
and abnormal fibrils (PHFs) identical to those in NFTs. Amyloid deposits
are also located in the walls of small arteries in the meninges and the
neocortex, resulting in the change called congophilic angiopathy (Sholtz,
1938). Although often associated with DAT, this lesion can be seen in isolation in normal elderly patients. Senile plaque amyloid has recently been
sequenced (Allsop et al., 1983; Masters et al., 1985). The amino acid
sequence is different from that of any other known protein, including other
forms of amyloid. In spite of earlier confusing results, immunocytochemical
studies have shown conclusively that plaque amyloid is different from amyloid of immunoglobulin origin, amyloid A, and several types of localized and
familiar amyloid. Preliminary results of sequencing indicate that plaque
amyloid may be identical to the amyloid of congophilic angiopathy, and surprisingly to PHF proteins (Kidd et al., 1985). In this respect, it is worth
mentioning that although PHF-specific antibodies do not stain amyloid core
of senile plaques in situ, they decorate the isolated plaque cores (Selkoe
and Abraham, 1985).

     Using energy dispersive x-ray microanalysis, Candy et al. (1986) have
reported high levels of aluminum and silicon in the cores of senile plaques.
However, our group has been unable to confirm these results using the previously mentioned technique, LAMMA (Stern et al., 1986).

Parkinson's Disease

     Parkinson's disease (PD) is characterized by tremor, rigidity, and
bradykinesia. Although a parkinsonian syndrome can result from viral infections and exposure to a large number of toxins and drugs (Schwab and England, 1968), most cases are idiopathic in nature. The brain of patients
dying with idiopathic PD shows severe neuronal cell loss in the substantia
nigra and other brain stem pigmented nuclei. The remaining neurons in
these nuclei often contain intracytoplasmic rounded eosinophilic concentric

bodies with a halo, and are called Lewy bodies. Ultrastructurally, the core of a Lewy body is made up of filaments, 7 to 8 nm in diameter, associated with granular material. The halo contains a looser arrangement of filaments (Duffy and Tennyson, 1965).

Several immunocytochemical studies have shown that the periphery of Lewy bodies stains with antibodies against neurofilaments (Goldman et al., 1983b), whereas the core may or may not appear stained in any given section (Figure 8). Unlike the selectivity of DAT NFTs, Lewy bodies seem to be recognized by all antibodies that decorate neurofilaments in adjacent axons.

Amyotrophic Lateral Sclerosis

Amyotrophic lateral sclerosis (ALS) is a degenerative disease in which there is selective death of the somatic motorneurons of the anterior horn of the spinal cord and medulla, and of the upper cortical motorneurons that give rise to the pyramidal tract. Patients show progressive muscle wasting and weakness in the face of normal sensation and preserved mental faculties. The cytoskeletal pathology is less conspicuous in ALS than in DAT and PD, and until recently textbooks reported neuronal and axonal loss as the only histopathology of ALS. It is now clear that the anterior horn motorneurons in this disease accumulate large amounts of structurally normal neurofilaments, both in the perikaryon and in the proximal axons (Hirano et al., 1984a). The latter accumulations take the form of spheroids (Delisle and Carpenter, 1984). These are probably early changes, since they are particularly prominent in cases with short duration of the disease. Our group (Greene et al., 1986), has shown by immunocytochemistry that the perikaryal neurofilaments are abnormally phosphorylated in a subpopulation of anterior horn motorneurons in ALS (Figure 9), thus linking the cytoskeletal abnormalities of ALS and DAT.

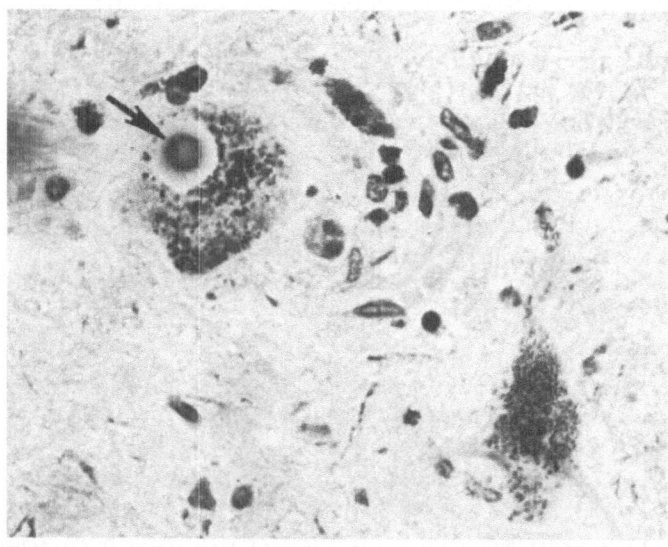

FIG. 8:   Section of substantia nigra from the brain of a patient with Parkinson's disease. The periphery of a Lewy body (arrow) is stained following immunocytochemical processing with human antineurofilament antiserum. Note that the core of the Lewy body does not appear stained.

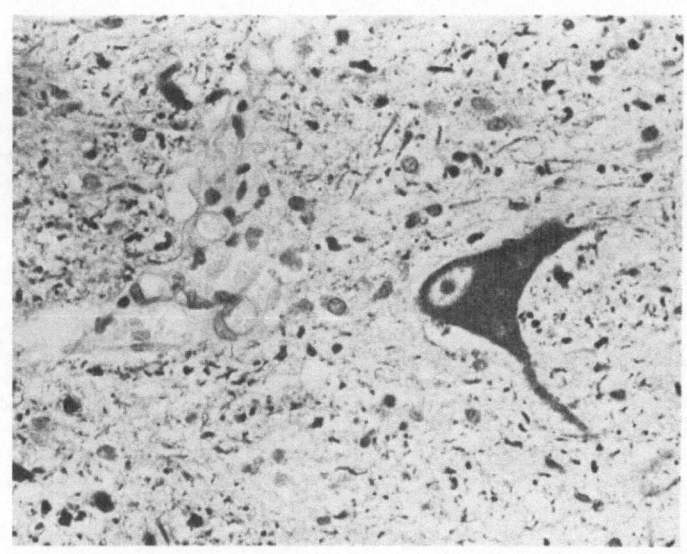

FIG. 9: Section of cervical spinal cord from a patient with amyotrophic
lateral sclerosis. An anterior horn cell demonstrates staining
following immunocytochemical processing with human antineurofilament antiserum.

Five per cent of cases of ALS are hereditary, and some of these
patients exhibit additional cytoskeletal abnormalities in motorneurons,
including Lewy-like bodies at the light microscopic level, and linear densities associated with robosomes at the untrastructural level (Hirano et
al., 1984b). The chemical nature of these abnormalities is unknown.

Other Neurodegenerative Diseases with Cytoskeletal Abnormalities

Progressive supranuclear palsy (PSP) is a degenerative disorder com-
bining clinical features of parkinsonism and a supranuclear opthalmoplegia.
At autopsy, instead of Lewy bodies, the remaining nerve cells in the substantia nigra contain NFT. On ultrastructural examination, PSP tangles
are made up of 15 nm straight fibrils (Teller-Nagel and Wisniewski, 1973).
The immunological reactivity of these tangles is identical to that of DAT
tangles with all of the antibodies (Yen et al., 1983) utilized to the present. Therefore, the biochemical basis for the ultrastructural difference
between the 15 nm fibrils of PSP and PHFs of DAT remains unclear.

Pick's disease is a dementing disorder, clinically similar to DAT, but
often affecting younger patients. The brain shows gross atrophy of the
frontal and temporal lobes, whereas the posterior regions of the brain are
spared. Many cortical and hippocampal neurons contain cytoplasmic ball-
shaped masses of argentophilic material. Ultrastructurally, these masses
consist of criss-crossing 15 nm fibrils, similar or identical to those of
PSP. Moreover, immunocytochemistry has shown no differences with DAT NFTs
(Munoz-Garcia and Ludwin, 1984), even at the level of antibodies against
purified NFTs. However, the existence of undisclosed differences between
the fibrils of Pick bodies and PHFs is suggested by the fact that Pick
bodies do not remain in the tissue after the death of the neurons containing
them, whereas undigested NFTs in the same circumstance form so called "ghost
tangles".

CONCLUSION

Only during the last five years have we started to obtain a viewpoint, incomplete as it is, of the chemical nature of the cytoskeletal abnormalities that characterize the most important neurodegenerative diseases. We could expect that a more complete characterization of these cytoskeletal changes will allow for an exploration of the mechanisms leading to their formation.

REFERENCES

Allsop, D., Landon, M. and Kidd, M., The isolation and amino acid composition of senile plaque core protein, Brain Res., 259:348 (1983).

Alzheimer, A., Uber eine eigenartige Erkrankung der Hirnrinde, Algemeine Zeitschrift Psychiatric, 64:146 (1907).

Diagnostic and Statistical Manual of Mental Disorders, ed 3, American Psychiatric Association, Task force on nomenclature and statistics, p. 111 (1980).

Anderson, F. H., Richardson, E. P., Okazaki, H. and Brody, J. A., Neurofibrillary degeneration on Guam: Frequency in chamorros and non chamorros with no known neurological disease, Brain, 102:65 (1979).

Anderton, B. H., Breinburg, D., Downes, M. J., Green, P. J., Tomlinson, B. E., Ulrich, J., Wood, J. N. and Kahn, J., Monoclonal antibodies show that neurofibrillary tangles and neurofilaments share antigenic determinants, Nature, 298:84 (1982).

Autilio-Gambetti, L., Gambetti, P. and Crane R. C., Paired helical filaments: relatedness to neurofilaments shown by silver staining and reactivity with monoclonal antibodies, in: "Biological Aspects of Alzheimer's Disease. Vol 15 Banbury Report", R. Katzman, ed., Cold Spring Harbor Laboratory, Cold Spring Harbor, NY (1983).

Ball, M. J., Neuron loss, neurofibrillary tangles and granulovacuolar degeneration in the hippocampus with aging and dementia. A quantitative study, Acta Neuropathol (Berl), 37:111 (1977).

Ball, M. J. and Lo, P., Granulovacuolar degeneration in the aging brain and in dementia, J. Neuropathol Exp Neurol., 36:474 (1977).

Blessed, G., Tomlinson, B. E. and Roth, M., The association between quantitative measures of dementia and of senile change in the cerebral gray matter of elderly subjects, Br J Psychiatry, 114:797 (1968).

Bray, D. and Gilbert, D., Cytoskeletal elements in neurons, Ann Rev Neurosci., 4:505 (1981).

Buigiani, O. and Ghetti, B., Progressing encephalomyelopathy with muscular atrophy, induced by aluminum powder, Neurobiol Aging, 3:209 (1982).

Burger, P. C. and Vogel, F. S., The development of the pathologic changes of Alzheimer's disease and senile dementia in patients with Down's syndrome, Am J Pathol., 73:457 (1973).

Candy, J. M., Klinowski, J., Perry, R. H., Perry, E. K., Fairbairn, A., Oakley, A. E., Carpenter, T. A., Atack, J. R., Blessed, G. and Edwardson, J. A., Aluminosilicates and senile plaque formation in Alzheimer's disease, Lancet 1:354 (1986).

Chen, L., Neurofibrillary change on Guam, Arch Neurol., 38:16 (1981).

Corsellis, J. A. N., Sub-acute sclerosing leucoencephalitis: clinical and pathological report of two cases, J Ment Sci., 97:570 (1951).

Corsellis, J. A. N., Bruton, C. J. and Freeman-Browne, D., The aftermath of boxing, Psychol Med., 3:270 (1973).

Davies, P. and Maloney, A. J. F., Selective loss of central cholinergic neurons in Alzheimer's disease, Lancet, 2:1403 (1976).

Dayan, A. D., Quantitative human studies in the aged human brain, Acta Neuropathol (Berl), 16:95 (1970).

DeBoni, U., Seger, M. and McLachlan, D. R. C., Functional consequences of chromatin bound aluminum in cultured human cells, Neurotoxicol., 1:65 (1980).

Delisle, M. B. and Carpenter, S., Neurofibrillary axonal swellings and amyotrophic lateral sclerosis, J. Neurol Sci., 63:241 (1984).

Duffy, P. E. and Tennyson, V. M., Phase and electron microscopic observations of Lewy bodies and melanin granules in the substantia nigra and locus ceruleus in Parkinson's disease, J. Neuropathol Exp Neurol., 24:398 (1965).

Gambetti, P., Shecket, G., Ghetti, B., Hirano, A. and Dahl, D., Neurofibrillary change in human brain. An immunocytochemical study with a neurofilament antiserum, J. Neuropathol Exp Neurol., 42:69 (1983).

Gibson, P. H. and Tomlinson, B. E., Number of Hirano bodies in the hippocampus of normal and demented people with Alzheimer's disease, J. Neurol Sci., 33:199 (1977).

Glenner, G. G., Amyloid deposits and amyloidosis: The B-Fibrilloses (1st of two parts), N Eng J Med., 302:1283 (1980a).

Glenner, G. G., Amyloid deposits and amyloidosis: The B-fibrilloses (2nd of two parts), N Eng J. Med., 302:1333 (1980b).

Goldman, J. E., The association of actin with Hirano bodies, J Neuropathol Exp Neurol., 42:146 (1983a).

Goldman, J. E., Yen, S., Chiu, F. and Peress, N., Lewy bodies of Parkinson's disease contain neurofilament antigens, Science, 221:1082 (1983b).

Greene, C., Munoz-Garcia, D., Perl, D. P. and Pendlebury, W. W., Accumulation of phosphorylated neurofilaments in the anterior horn motor neurons of ALS patients (abst), J Neuropathol Exp Neurol., in press (1986).

Grundke-Iqbal, I., Johnson, A. B., Wisniewski, H. M., Terry, R. D. and Iqbal, K., Evidence that Alzheimer neurofibrillary tangles originate from neurotubules, Lancet, 1:578 (1979).

Grundke-Iqbal, I., Iqbal, K., Tung, Y. C. and Wisniewski, H. M., Alzheimer paired helical filaments: Immunochemical identification of polypeptides, Acta Neuropathol (Berl), 62:259 (1984).

Grundke-Iqbal, I., Iqbal, K., Tung, Y. C. and Wisniewski, H. M., Alzheimer paired helical filaments: Cross-reacting polypeptide/s normally present in brain, Acta Neuropathol (Berl), 66:52 (1985a).

Grundke-Iqbal, I., Wang, G. P., Iqbal, K., Tung, Y. C. and Wisniewski, H. M., Alzheimer paired helical filaments: Identification of polypeptides with monoclonal antibodies, Acta Neuropathol (Berl), 68:279 (1985b).

Hirano, A., Pathology of amyotrophic lateral sclerosis, in: "Slow, Latent, and Temperate Infections. NINDB Monograph No. 2." D. C. Gajdusek, C. J. Gibbs eds., National Institutes of Health, Washington, D.C. (1965).

Hirano, A., Malamud, N., Elizan, T. S. and Kurland, L. T., myotrophic lateral sclerosis and parkinsonism-dementia complex on Guam, Arch Neurol., 15:35 (1966).

Hirano, A., Dembitzer, H. M., Kurland, L. T. and Zimmerman, H. M., The fine structure of some intraganglionic alterations. Neurofibrillary tangles, granulovacuolar bodies, and "rod-like" structures as seen in Guam amyotrophic lateral sclerosis and parkinsonism-dementia complex, J Neuropathol Exp Neurol., 27:167 (1968).

Hirano, A., Donnenfeld, H., Sasaki, S. and Nakano, I., Fine structural observations of neurofilamentous changes in amyotrophic lateral sclerosis, J Neuropathol Exp Neurol., 43:461 (1984a).

Hirano, A., Nakano, I. and Kurland, L. T., Fine structural study of neurofibrillary changes in family with amyotrophic lateral sclerosis, J Neuropathol Exp Neurol., 43:471 (1984b).

Hyman, B. T., Van Hoesen, G. W., Damasio, A. R. and Barnes, C. L., Alzheimer's disease: Cell-specific pathology isolates the hippocampal formation, Science, 225:1168 (1984).

Ihara, Y., Abraham, C. and Selkoe, D. J., Antibodies to paired helical filaments in Alzheimer's disease do not recognize normal brain proteins, Nature, 304:727 (1983).

Iqbal, K., Zaidi, T., Thompson, C. H., Merz, P. A. and Wisniewski, H. M.,

Alzheimer paired helical filaments: Bulk isolation, solubility, and protein composition, Acta Neuropathol (Berl), 62:167 (1984).

Johnston, M. V., McKinney, M. and Coyle, J. T., Evidence for a cholinergic projection to neocortex from neurons in basal forebrain, Proc Natl Acad Sci, USA, 76:5392 (1979).

Khachaturian, Z. S., Diagnosis of Alzheimer's disease, Arch Neurol, 42:1097 (1985).

Kidd, M., Paired helical filaments in electron microscopy of Alzheimer's disease, Nature, 197:192 (1963).

Kidd, M., Allsop, D. and Landon, M., Senile plaque amyloid, Interdiscipl Topics Geront., 19:114 (1985).

Kidd, M., Allsop, D. and Landon, M., Senile plaque amyloid, paired helical filaments, and cerebrovascular amyloid in Alzheimer's disease are all deposits of the same protein, Lancet, 1:278 (1985).

Kirschner, D. A., Abraham, C. and Selkoe, D. J., X-ray diffraction from intraneuronal paired helical filaments and extraneuronal amyloid fibers in Alzheimer disease indicates cross-B conformation, Proc Natl Acad Sci, USA, 83:503 (1986).

Klatzo, I., Wisniewski, H. and Streicher, E., Experimental production of neurofibrillary degeneration, J. Neuropathol Exp Neurol., 24:187 (1965).

Knox, C. A., Yates, R. D. and Chen, I-li, Brain aging in normotensive and hypertensive strains of rats, Acta Neuropathol (Berl), 52:7 (1980).

Kosik, K. S., Duffy, L. K., Dowling, M. M., Abraham, C., McCluskey, A. and Sekloe, D. J., Microtubule-associated protein 2: Monoclonal antibodies demonstrate the selective incorporation of certain epitopes into Alzheimer neurofibrillary tangles, Proc Natl Acad Sci, USA, 81:7941 (1984).

Lazarides, E., Intermedite filaments as mechanical integrators of cellular space, Nature, 283:249 (1980).

Mandybur, T. I., Nagpaul, A. S., Pappas, Z. and Niklowitz, W. J., Alzheimer neurofibrillary change in subacute sclerosing panencephalitis, Ann Neurol., 1:103 (1977).

Masters, C. L., Simms, G., Weinman, N. A., Multhaup, G., McDonald, B. L. and Beyreuther, K., Amyloid plaque core protein in Alzheimer disease and Down syndrome, Proc Natl Acad Sci, USA, 82:4245 (1985).

Munoz-Garcia, D. and Ludwin, S. K., Classic and generalized variants of Pick's disease; A clinicopathological, ultrastructural, and immunocytochemical comparative study, Ann Neurol., 16:467 (1984).

Munoz-Garcia, D., Pendlebury, W. W., Kessler, J. B. and Perl, D. P., An immunocytochemical comparison of cytoskeleton proteins in aluminum-induced and Alzheimer-type neurofibrillary tangles, Acta Neuropathol (Berl), in press (1986).

Perl, D. P., Gajdusek, C., Garruto, R. M., Yanagihara, R. T. and Gibbs Jr., C. J., Intraneuronal aluminum accumulation in amyotrophic lateral sclerosis and parkinsonism-dementia of Guam, Science, 217:1053 (1982).

Perl, D. P. and Brody, A. R., Alzheimer's disease: x-ray spectrometric evidence of aluminum accumulation in neurofibrillary tangle-bearing neurons, Science, 208:207 (1980).

Perl, D. P., Munoz-Garcia, D., Good, P. F. and Pendlebury, W. W., Intracytoplasmic aluminum accumulation in neurofibrillary tangle-bearing neurons: detection by laser probe mass analyzer (abst), Ann Neurol., 18:143 (1985).

Perry, G., Rizzuto, N., Autilio-Gambetti, L. and Gambetti, P., Paired helical filaments from Alzheimer disease patients contain cytoskeletal components, Proc Natl Acad Sci, USA, 82:3916 (1985).

Peterson, C., Suzuki, K., Kress, Y. and Goldman, J. E., Microfilament lattices (Hirano bodies) in brindled mice (abst). J Neuropathol Exp Neurol., 44:326 (1985).

Price, D., Struble, R. G., Altschuler, R. J., Casnova, M. F., Cork, L. C. and Murphy, D. B., Aggregation of tubulin in neurons in Alzheimer's disease (abst), J. Neuropathol Exp Neurol., 44:366 (1985).

Saper, C. B., German, D. C. and White, C. L., Neuronal pathology in the

nucleus basalis and associated cell groups in senile dementia of the
Alzheimer's type: possible role in cell loss, Neurol., 35:1089 (1985).

Sayre, L. M., Autilio-Gambetti, L. and Gambetti, P., Pathogenesis of experimental giant neurofilamentous axonopathies: A unified hypothesis based
on chemical modification of neurofilaments, Brain Res Rev., 10:69
(1985).

Scholz, W., Studien zur Pathologie der Hirngefasse: Die drusige Entartung
der Hirnarterien und capillaren, Z des Neurol Psychiatr., 162:694
(1938).

Schwab, R. S. and England Jr., A. C., Parkinson syndromes due to various
specific causes, in: "Diseases of the Basal Ganglia. Handbook of
Clinical Neurology", P. J. Vinken, G. W. Brun, (eds), Amsterdam, North-
Holland (1968).

Selkoe, D. J., Liem, R. K. H., Yen, S-H. and Shelanski, M. L., Biochemical
and immunological characterization of neurofilaments in experimental
neurofibrillary degeneration induced by aluminum, Brain Res., 163:235
(1979).

Selkoe, D. J., Ihara, Y. and Salazar, F. J., Alzheimer's disease: Insolubility of partially purified paired helical filaments in sodium dodecyl
sulfate and urea, Science, 215:1243 (1982a).

Selkoe, D. J., Abraham, C. and Ihara, Y., Brain transglutaminase: In vitro
crosslinking of human neurofilament proteins into insoluble polymers,
Proc Natl Acad Sci, USA, 79:6070 (1982b).

Selkoe, D. and Abraham, C., Biochemical analyses of senile plaque amyloid
cores purified by fluorescence activated cell sorting (abst), J Neuropathol Exp Neurol., 44:365 (1985).

Stern, A. J., Perl, D. P., Munoz-Garcia, D., Good, P. F., Selkoe, D. J. and
Abraham, C., Investigation of silicon and aluminum content in isolated
senile plaque cores by laser microprobe mass analysis (LAMMA) (abst),
J Neuropathol Exp Neurol., in press (1986).

Sternberger, L. A. and Sternberger, N. H., Monoclonal antibodies distinguish
phosphorylated and non-phosphorylated forms of neurofilaments in situ,
Proc Natl Acad Sci, USA, 80:6126 (1983).

Sternberger, N. H., Sternberger, L. A. and Ulrich, J., Aberrant neurofilament phosphorylation in Alzheimer disease, Proc Natl Acad Sci, USA, 82:
4274 (1985).

Struble, R. G., Cork, L. C., Whitehouse, P. J. and Price, D. L., Cholinergic
innervation in neuritic plaques, Science. 216:413 (1982).

Tellez-Nagel, I. and Wisniewski, H. M., Ultrastructure of neurofibrillary
tangles in Steele-Richardson-Olszewski syndrome, Arch Neurol., 29:324
(1973).

Terry, R. D. and Pena, C., Experimental production of neurofibrillary degeneration, J Neuropathol Exp Neurol., 24:200 (1965).

Terry, R. D., Peck, A., Deteresa, R., Schechter, R. and Horoupian, D. S.,
Some morphometric aspects of the brain in senile dementia of the Alzheimer type, Ann Neurol., 10:184 (1981).

Troncoso, J. C., Price, D. L., Griffin, J. W. and Parhad, I. M., Neurofibrillary axonal pathology in aluminum intoxication, Ann Neurol., 12:278
(1982).

Troncoso, J. C., Sternberger, L. A. and Sternberger, N. H., Immunocytochemical studies of neurofilament antigens in the neurofibrillary pathology induced by aluminum (abst), J Neuropathol Exp Neurol., 44:376
(1985).

Volk, B., Paired helical filaments in rat spinal ganglia following chronic
alcohol administration: An electron microscopic investigation, Neuropathol Appl Neurobiol., 6:143 (1980).

Wang, G. P., Grundke-Iqbal, I., Kascsak, R. J., Iqbal, K. and Wisniewski,
H. M., Alzheimer neurofibrillary tangles: Monoclonal antibodies to
inherent antigen(s), Acta Neuropathol (Berl), 62:268 (1984).

Whitehouse, P. J., Alzheimer's disease and senile dementia. Loss of neurons
in the basal forebrain, Science, 215:1237 (1982).

Wisniewski, H. M., Narkiewicz, O. and Wisniewski, K., Topography and dynamics of neurofibrillary degeneration in aluminum encephalopathy, <u>Acta Neuro-</u>pathol (Berl), 9:127 (1967).

Wisniewski, H. M., Narang, H. K. and Terry, R. D., Neurofibrillary tangles of paired helical filaments, <u>J. Neurol Sci</u>., 27:173 (1976).

Wisniewski, K., Jervis, G. A., Moretz, R. C. and Wisniewski, H. M., Alzheimer neurofibrillary tangles in diseases other than senile and presenile dementia, <u>Ann Neurol</u>., 5:288 (1979).

Wisniewski, H. M., Sturman, J. A. and Shek, J. W., Aluminum chloride-induced neurofibrillary changes in the developing rabbit: a chronic animal model, <u>Ann Neurol</u>., 8:479 (1980).

Wisniewski, H. N., Sturman, J. A. and Shek, J. W., Chronic model of neuro-fibrillary changes induced in mature rabbits by metallic aluminum, <u>Neurobiol Aging</u>, 3:11 (1982).

Yen, S-H. C., Guskin, R. and Terry, R. D., Immunocytochemical studies of neurofibrillary tangles, <u>Am J Pathol</u>., 104:77 (1981).

Yen, S. H., Horoupian, D. S. and Terry, R. D., Immunocytochemical comparison of neurofibrillary tangles in senile dementia of Alzheimer type, pro-gressive supranuclear palsy, and postencephalitic parkinsonism, <u>Ann Neurol</u>., 13:172 (1983).

MODULATION OF SCHWANN CELL ANTIGENS DURING WALLERIAN DEGENERATION AND REGENERATION IN THE ADULT, MAMMALIAN PERIPHERAL NERVE

Carson J. Cornbrooks and Timothy J. Neuberger

Department of Anatomy and Neurobiology, University of Vermont, College of Medicine, Burlington, Vermont 05405

INTRODUCTION

In the peripheral nervous system (PNS), the ability of injured neurons to recover and reform functional synapses depends primarily on cues from the environment. These cues may be in the form of neuronotrophic agents, such as nerve growth factor (NGF), which aid neuronal survival and/or neuronotropic factors, such as laminin, which assist in the elongation and orientation of regenerating neurites. The source of these molecules in the PNS may be gleaned from several experimental paradigms. Trauma which results in a complete discontinuity between the proximal and distal nerve stumps, often results in diminished neural regeneration. In contract, crush injury, in which the continuity of the connective tissue elements of the nerve are maintained, is associated with a greater opportunity for regeneration. Correspondingly, clinicians presently anastomose separated nerve stumps by suturing the peri- and epineuriums and may bridge larger gaps with autografts from another peripheral nerve (Michon, 1975). Neurobiologists now recognize that a solid matrix and soluble factors, both of which are produced by non-neuronal cells (fibroblasts and/or Schwann cells) are required constitutents within any terrain traversed by neurites regenerating in vivo (Politis et al., 1982; Longo et al., 1983; Williams et al., 1983; Davis et al., 1985; Schwab and Thoenen, 1985). Once viable neurons have extended neurites into a suitable terrain, functional regeneration can occur if: a) synaptogenesis proceeds with respect to the correct target and b) the re-establishment of a mature Schwann cell-neuron relationship is complete. With the exception of synapse formation, which is beyond the scope of the present chapter, the presence and proper metabolic function of Schwann cells appears to be a requirement for the presentation of many of the essential factors necessary for successful neuronal regeneration in the PNS.

Most likely derived from the neural crest, Schwann cells differentiate at the behest of neuronal contact by undergoing incredible alterations in their morphology and biochemistry as they ensheath or myelinate axons and form basal laminae (Noden, 1978; Bunge and Bunge, 1981). In the early stages of metamorphosis, proliferating, rounded glia modulate their shape by extending long processes in order to segregate bundles of axons (Webster et al., 1973; Billings-Gagliardi et al., 1974). In addition, the events leading to a permanent relationship with an axon(s) require alterations in morphology (Bunge et al., 1986). Previous studies have demonstrated that Schwann cells contain the cytoskeleton proteins actin, vimentin (Vim) and glial fibrillary acidic protein (GFAP) which are most likely involved in morphological events (Yen and Fields, 1981; Autillio-Gambetti et al., 1982; Jessen et al., 1984; de Nehaud et al., 1986). However, it remains unclear how these specific molecules participate in the various stages of glial differentiation. During these cytoarchitectural events, Schwann cells also expose a number of molecules to the peripheral nerve milieu in a sequential, ordered series of events. Of these molecules, some are secreted and organized into robust basal laminae which circumscribe all neuron-glial units, i. e. laminin (LAM), which a neuronotropic factor, type IV collagen, entactin and heparin sulfate proteoglycan (Cornbrooks, 1983; Cornbrooks et al., 1983; Carey et al., 1983; Eldridge et al., 1986). Yet other Schwann cell secreted molecules, possibly NGF, may function as neuronotrophic agents (Longo et al., 1983: Assouline et al., 1985). Schwann cells also produce a repertoire of membrane-associated antigens which are oriented toward the extracellular space. Evidence exists that at least one of these antigens, recognized by the C4 monoclonal antibody, is expressed on the surface of Schwann cells when they contact axons which do not induce myelination (Cornbrooks and Bunge, 1982). Therefore, during the developmental stages of differentiation, Schwann cells synthesize a number of molecules which are uniquely capable of functioning in both a trophic and tropic nature. These molecules may also support and nurture the neuron during regeneration.

How do Schwann cells, encased within the rigid constraints of a mature relationship with an axon(s) respond to the degeneration of their neuronal hosts? Orphaned peripheral glial cells undergo a number of well studied cellular changes. Some Schwann cells, primarily those which have formed myelination or have been directly injured, enter a short-lived proliferative stage. These cells fill the existing endoneurial channels and contribute to the cellular columns called the bands of Bungner (Weinberg and Spencer, 1978; Salzer and Bunger, 1980). Concomitantly, these glia phagocytose degenerating myelin and the synthesis of myelin-specific proteins is greatly diminished (Weinberg and Spencer, 1978; Poduslo et al., 1984). Preliminary evidence also suggests that Schwann cells separated from their neuronal hosts may synthesize NGF, a neuronotrophic factor (Assouline et al., 1985). After these initial response to neuronal injury, peripheral glia are demonstrably capable of reachieving a completely differentiated form during neural regeneration (Jenq and Coggeshall, 1986). Once contacted by regenerating axons, the induced Schwann cells probably utilize mechanisms similar to those employed during development to form a mature neuron-glial relationship.

At present, there is a relative paucity in our comprehension of the biochemical events experienced by orphaned Schwann cells after neural injury and the subsequent renewed contact with regenerating axons. Accordingly, we began preliminary experiments to characterize changes in known Schwann cell antigens when the axons within the nerve were: a) completely transected and b) subsequently allowed to regenerate into the distal nerve stump. In this chapter, we present evidence that there are distinct alterations in the distal nerve stump with respect to the distribution of Schwann cell-synthesized cytoskeletal proteins (GFAP), an extracellular matrix protein (LAM) and a membrane-bound antigen (C4). A preliminary report of these studies has been previously presented (Neuberger and Cornbrooks, 1986).

METHODS

Tissue procurement:

Adult, Holtzman rats were anesthetized by an intramuscular injection of ketamine:xylazine (87mg/kg:120mg/kg) and the right sciatic nerves were exposed close to the spinal column. The nerves were tightly ligated with 4.0 suture, completely transected just proximal to the thread and sutured into the muscle mass at least one centimeter away from the site of neural injury. The skin was closed with wound clips and the animals were maintained until the appropriate time point. Nerves were collected at 1, 3, 7, 21 and 42 days after transection. Experimental animals were anesthetized with an overdose of sodium pentobarbital (65mg/200g) and approximately 1 cm of sciatic nerve was harvested from both the left (control, nontransected) and right (experimental, transected) leg. Only the portion of the nerve distal to the site of transection was harvested from the experimental side. Fresh nerves were rapidly sandwiched between portions of the gastrocnemius muscle and immersed in liquid nitrogen. Alternatively, some animals were perfused by intracardiac puncture with 4% paraformaldehyde/0.1% glutaraldehyde in 0.1M phosphate buffer, pH 7.4 prior to the harvest of the nerves. Nerve/muscle samples were stored at -80°C until 10 micron frozen sections (cross and longitudinal) were cut using a Bright cryostat and dry mounted onto gelatin coated slides. Tissue on slides was stored in a -80°C freezer until they were stained by indirect immunohistochemical methods.

Treatment for Immunohistochemistry

Slides containing tissue sections were initially sprayed with teflon (Fluoroglide, VWR) over the non-tissue areas to minimize the area which was incubated with antibody solutions. The tissue was pretreated with cold acetone for 5 minutes, allowed to dry and incubated in 1% triton X-100 in Dulbecco's phosphate buffered saline for 15 minutes at room temperature. The samples were subsequently washed 3 times for 5 minutes each in cold Dulbecco's phosphate buffered saline containing 10% heat-inactivated horse serum (D-PSBH). Primary antibodies were diluted in D-PBSH which contained 4% DMSO and 0.02% sodium azide, bubbled on each sample and incubated at 4°C overnight in a sealed, moisture box. The primary antibodies consisted of: 1) anti-neurofilament (phosphorylated) diluted 1:1000; SMI-31 from Sternberger/Meyer, Inc., Jarettsville, Maryland; 2) anti-laminin diluted 1:100; from Gibco, Inc., Grand Island, New York; 3) anti-glial fibrillary acidic protein diluted 1:300; a generous gift from Dr. Lawrence Eng, Palo Alto, California; and 4) anti-C4 diluted 1:1000; a monoclonal antibody developed in our laboratory.

The following day, the tissue sections were washed 3 times for 5 minutes each in D-PBSH and the secondary antibody(s) diluted in D-PBSH was bubbled onto the teflon-free areas of slides and incubated at 37°C for 30 minutes in a sealed, moisture box. Secondary antibodies, conjugated either to rhodamine or fluorescein, were obtained from Cooper Biomedical, Inc. and diluted 1:100 (goat anti-rabbit, IgG) or 1:200 (goat anti-mouse, IgG). Finally, the tissue was washed 3 times for 5 minutes each with D-PBSH, fixed with 4% paraformaldehyde in 0.1M phosphate buffer pH 7.4 (if the animal was not previously perfused) and coverslipped with 5% n-propyl gallate/glycerol/ PBS to prevent photobleaching by the light source.

The processed tissue was observed with a Zeiss photomicroscope III equipped with epifluorescence and narrow band filters to view rhodamine and fluorescein. Photographs were recorded on Ilford HP film which was developed with Rodinal.

RESULTS

In control, uninjured sciatic nerves anti-neurofilament antibodies clearly demarcated the axoplasm of both non-myelinated and myelinated axons. In longitudinal sections of nerves, all axons contained the phosphorylated

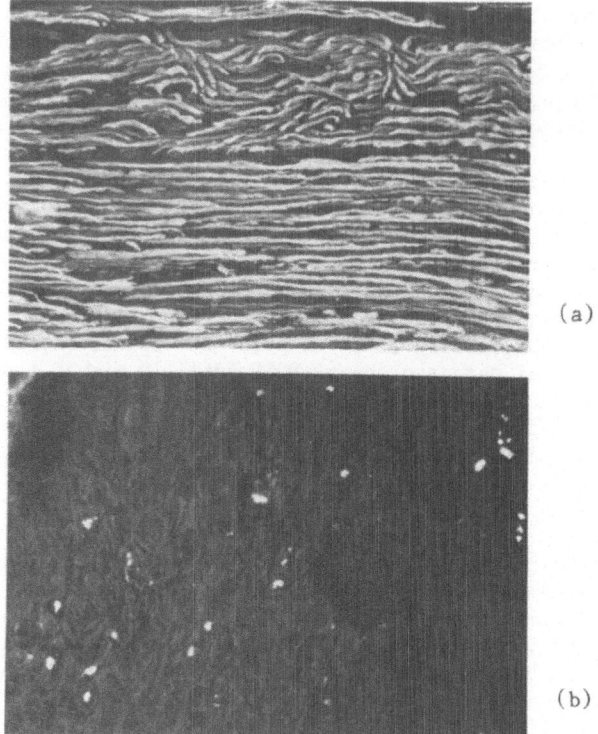

(a)

(b)

FIG. 1: Longitudinal sections of a) control sciatic nerve and b) nerve
transected 21 days prior to immunohistochemical staining. Both
sections were treated with a monoclonal antibody to delineate the
phosphorylated 200 kd and 150 kd neurofilament proteins. Magnification 160X.

neurofilament (NF) proteins which were distributed in a repetitive linear
series extending the entire length of the tissue section (Figure 1a). By 3
days post-transection, the NF+ material was distributed in randomly spaced
aggregations throughout the nerve (figure not shown). As time after transection increased, the amount of the antigens (the 200 kd and 150 kd proteins recognized by the SMI #31 monoclonal antibody) diminished such that
at the 21 day time point relatively few areas remained positively stained
(Figure 1b).

The laminin (LAM) glycoprotein within control nerves was localized to
discrete areas in the endoneurium and perineurium containing basal laminae
(see also Cornbrooks et al., 1983). In addition, when examined by phase
microscopy, brightly stained areas in the endoneurium were determined to
coincide with cytoplasm-rich areas of Schwann cells (Figure 2a). Cross
sections of nerve at 11 days post-transection revealed short blocks of LAM+
material (Figure 2b). Throughout all the time points examined, LAM+ tissue
was arranged in continuous tubes which extended parallel to the length of
the nerve (Figure 2c).

Antibodies directed against glial fibrillary acidic protein (GFAP)
recognized a distinct set of Schwann cells in the mature, control sciatic
nerves which were associated with non-myelinated axons (Figure 3a). This
class of Schwann cell was restricted to specific regions within the cross

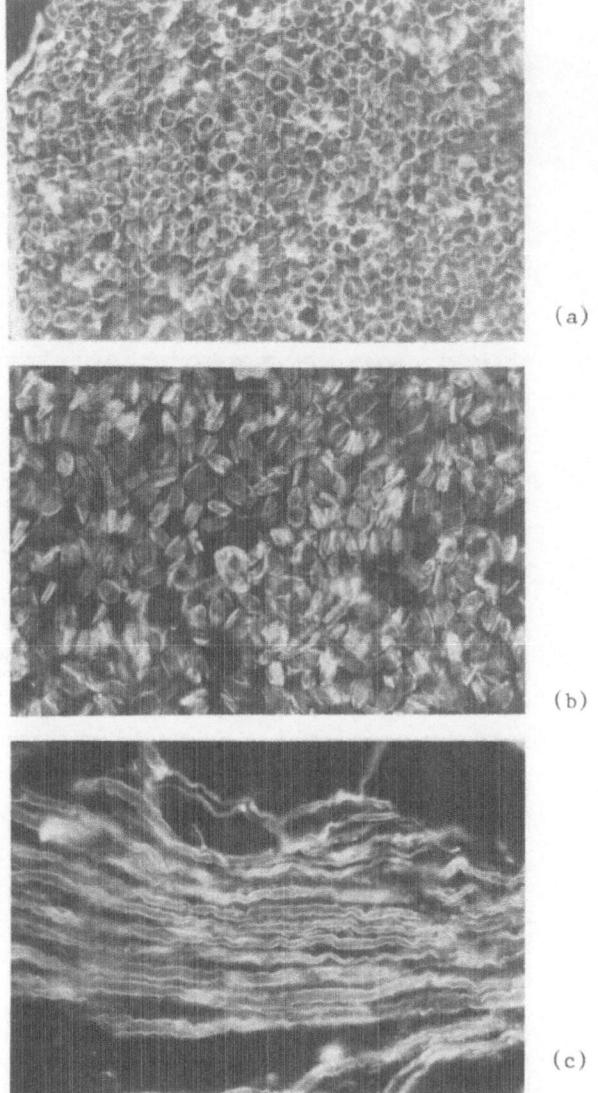

FIG. 2: Laminin immunoreactivity within cross sections (a and b) and a
longitudinal section of rat sciatic nerves (c). Tissue harvested
was from a) a control, nontransected nerve, b) the distal nerve
stump of an 11 day post-transected animal and c) a portion of a
transected nerve stained 42 days after injury. Magnification 250X.

sectional area of the nerve. In longitudinal sections, the GFAP+ cytoskeletal filaments extended long distances, at least 40 microns in length, were
characteristically continuous throughout the Schwann cell processes and surrounded the nucleus (Figure 3b). As the time after nerve transection
increased, there was a slight but progressive dimunition in the number of
Schwann cells which were stained positively with anti-GFAP. Moreover, the
obvious segregation of GFAP+ and GFAP- areas into domains seen in control
nerves disappeared after the nerve was severed. A representative sample

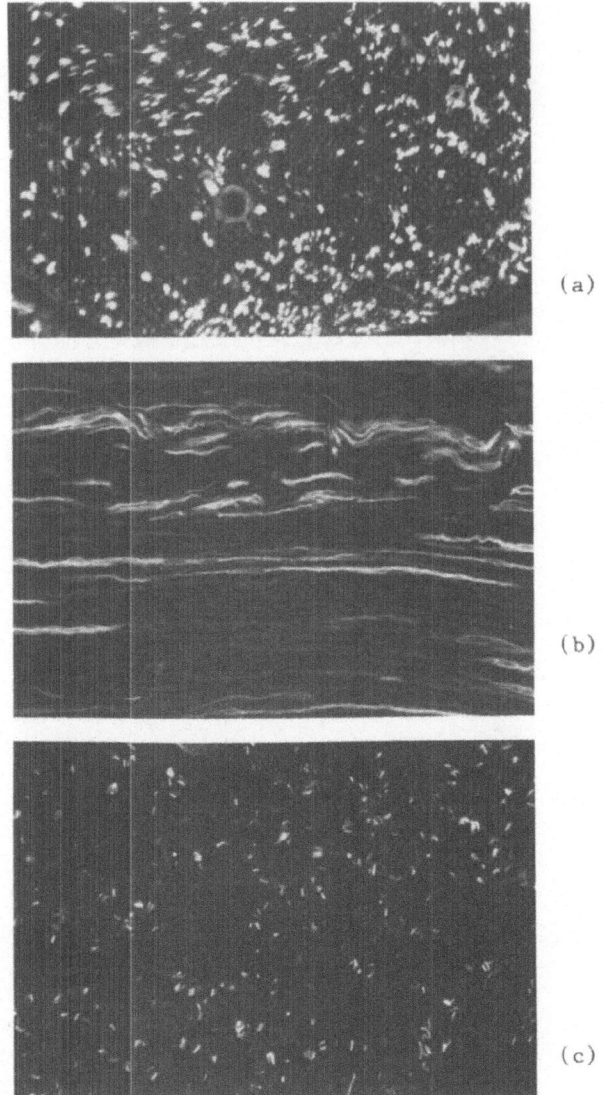

FIG. 3: Immunohistochemical staining of glial fibrillary acidic protein
within sciatic nerves removed from a control, nontransected nerve
sectioned a) transversly or b) longitudinally. In both cases,
staining is clearly segregated into GFAP+ and GFAP- domains within
the nerve. Plate c demonstrates the appearance of GFAP+ material
within the distal portion of the nerve transected 21 days prior to
staining. Compare the evenly distributed, shorted GFAP+ filaments
in plate c versus those in plate b. Magnification 160X.

from a nerve transected 21 days previously demonstrates the relative
decrease in GFAP+ figures (Figure 3c). In longitudinal sections, there was
also an obvious shortening of the overall length of the GFAP+ fibers. Examination of nerves 42 days post-transection revealed a virtual absence of
GFAP+ fibers. Examination of nerves 42 days post-transection revealed a

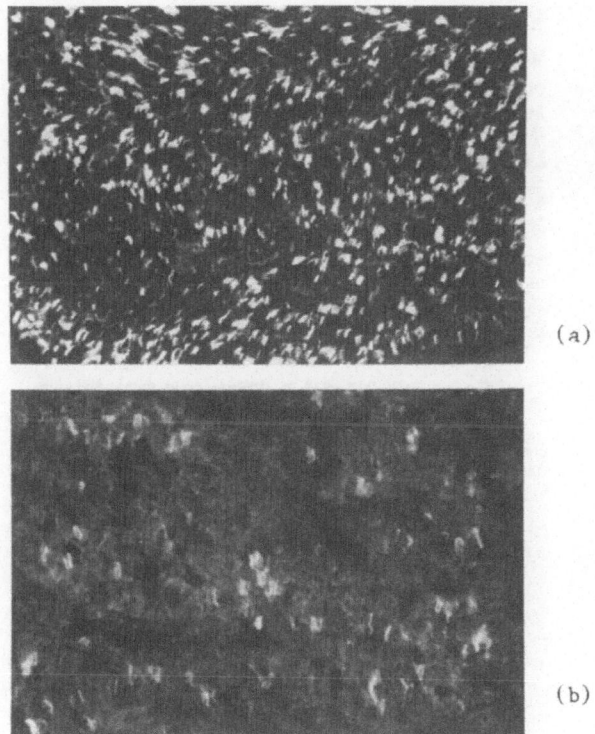

(a)

(b)

FIG. 4:   Fluorescent staining associated with antigens recognized by the
          monoclonal antibody C4 directed against a component synthesized by
          Schwann cells in contact with an axon(s). Pretreatment of the
          fresh, frozen transverse sections of sciatic nerve from a) the
          control animals revealed prominent staining in discrete domains of
          the nerve; whereas, b) similar sections obtained from the distal
          nerve stump 21 days post-transection revealed a dimunition of C4+
          tissue. Magnification 160X.

virtual absence of GFAP+ material with the occasional exception of small
aggregates of GFAP+ material surrounding the Schwann cell nucleus (figure
not shown).

    In uninjured nerves harvested from the left legs, the monoclonal anti-
body C4 recognized a specific antigen which was localized to segregated
domains within the endoneurium (Figure 4a). Using cryostat sections of
control nerves which were pretreated with 4% paraformaldehyde in 0.1M
phosphate buffer pH 7.4 prior to acetone and triton extraction, both C4 and
GFAP were colocalized to identical areas within the endoneurium (figure not
shown). In longitudinal sections, the C4 antibody delineated antigens which
were aligned in a longitudinal pattern parallel to the length of the nerve.
Sections taken from transected nerves revealed a progressive dimunition of
C4+ areas throughout the nerve (Figure 4b).

    Nerves which were actively undergoing reinnervation were also examined
for the presence of NF, GFAP and C4. Positive proof of the ingrowth of
new axons was directly obtained by using the monoclonal antibody directed
against phosphorylated NF proteins. This antibody revealed two types of
linear staining patterns, i.e. one with a very fine filament-like arrange-

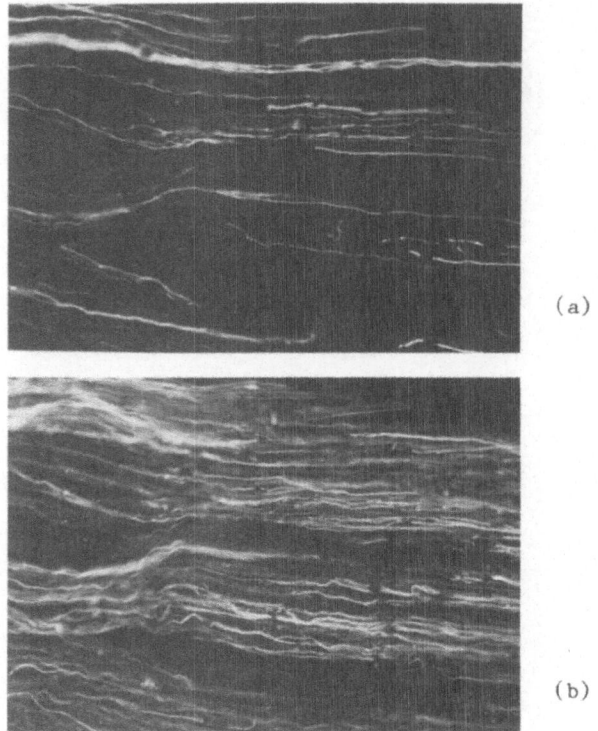

(a)

(b)

FIG. 5: Indirect immunohistochemical staining of the distal nerve stumps
containing regenerating axons. Plate a demonstrates the location
of several NF+ proteins and plate b shows the close proximity of
GFAP+ filaments on the same section of tissue. Arrows delineate
Schwann cell nuclei. Note that GFAP is associated with areas
which are NF+ as well as some NF- areas. Magnification 160X.

ment and a second, more prominent filament-like arrangement. Both patterns
were distributed in random areas throughout the sections of nerve examined
(Figure 5a). When the same section was examined with antibodies directed
against GFAP, a pattern of linear staining closely associated with the
prominently stained NF+ areas (but not the fine NF+ areas), circumscribed
Schwann cell nuclei and was also located in juxtaposition to areas beyond
the NF+ areas (figure 5b). Similar sections from transected nerves containing regenerating axons also demonstrated a striking coincidence of C4+
and GFAP+ fiber-like patterns (cf. Figures 6a & 6b). Curiously when pretreatment with paraformaldehyde was omitted prior to the acetone/triton
incubations, the C4+ staining was unaffected. This is in sharp contrast
to nerves without regenerating axons (control and transected) where such a
protocol i.e. pretreatment with acetone and triton, faithfully removed all
traces of the antigenic epitope recognized by the C4 monoclonal antibody.
In cases where reinnervation was documented by prominent NF+ staining, the
GFAP+ filaments were displayed as long, continuous fibers which closely
resembled those seen in the control, uninjured nerves (cf. Figures 2a and
6a).

450

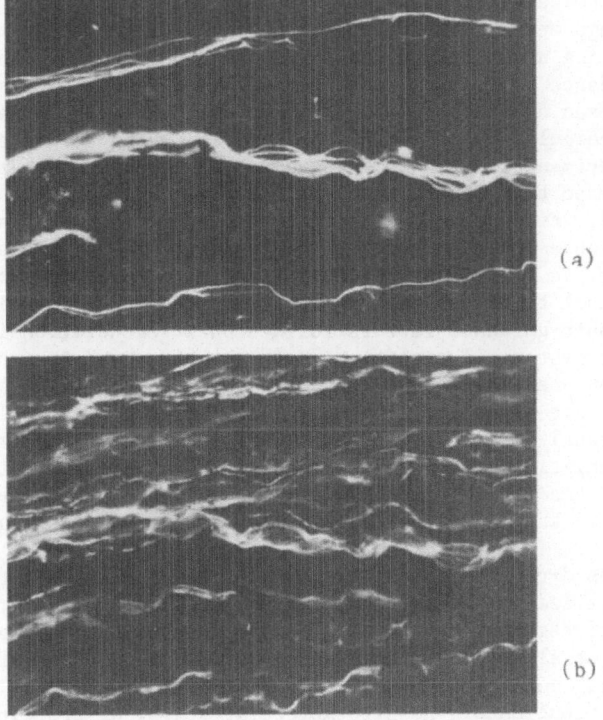

(a)

(b)

FIG. 6:   Localization of a) prominent GFAP+ filaments in nerves containing
NF+ regenerating axons and b) a number of C4+ stained areas in the
same tissue section. Note that some but not all of the GFAP+ areas
are closely associated with the C4+ filament-like structure. Magnification 400 X.

DISCUSSION

We have begun preliminary experiments to investigate the modulation of
Schwann cell-synthesized antigens in response to neuronal injury and regeneration. Our approach is to monitor characterized molecules using the
paradigm of complete nerve transection in combination with indirect immunohistochemical methods. By examining the portion of the nerve distal to the
site of transection at various times after the trauma, we can demonstrate
distinct alterations in the distribution of molecules localized in the
Schwann cell cytoskeleton and the plasma membrane. Furthermore, the normal
distribution pattern of at least one extracellular matrix component, laminin,
is also altered at various time points after nerve transection. Thus, it is
evident that a number of changes in the distribution of Schwann cell-synthesized cellular and extracellular molecules occur in response to neuronal
degeneration and subsequent regeneration.

It has been known since the early work of Ramón y Cajal that axonal
sprouts emanating from the proximal portion of transected nerves can grow
short distances and regenerate within the distal portion of the nerve
(Ramón y Cajal, 1928). Moreover, Cajal first proposed the release of a
neuronotrophic factor from Schwann cells which reside in the distal nerve
stump. Recent research has shown that regenerating axons apparently respond

to the presence of soluble factors released from the distal nerve stump and when given a choice will preferentially grow toward this source (Politis et al., 1982; Longo et al., 1983). Cajal (1928) also concluded that sprouts from regenerating axons were mechanically guided along the old (nerve) sheaths. Guidance of regenerating neurites by channels of basal laminae has recently been demonstrated to occur in culture using Schwann cell-derived extracellular matrix (ECM) (Cornbrooks, 1985). Furthermore, two different experimental approaches suggest that growth cones grow in contact with the inner portion of the basal lamina (Ide et al., 1983; Schwab and Thoenen, 1985). Evidence has also been presented demonstrating that the surfaces of viable Schwann cells provide a favorable substrate for the growth of neuritic processes (Noble et al., 1984; Fallon, 1985a;b). It thus appears that there are requirements for mechanical as well as soluble or neuronotrophic guidance factors for neuronal regeneration in the peripheral nervous system. The continued presentation of both types of factors to growing axons necessitates that viable nonneuronal cells, most likely Schwann cells, are metabolically functional and in close proximity to the exploring neuronal processes. After axonal pathfinding has proceeded, the successful completion of secondary events, such as the ensheathment and myelination of the regenerating axons, also requires that viable Schwann cells respond to the neuronal induction and by differentiating.

Antibodies directed against the 200k dalton and 150k dalton NF proteins were utilized to investigate both the breakdown of severed axons and the presence of regenerating axons. In agreement with the previous studies of Schlaepfer et al. (1984), there was a rapid degeneration of the high molecular weight neurofilament proteins within the axoplasm. Within three days, the distribution of the high molecular weight NF proteins within axoplasm became random and localized to progressively smaller and less numerous areas throughout the nerve. By 21 days post-transection there were few, if any, NF+ figure in the nerve sections examined. Since we used a monoclonal antibody which was directed toward the higher molecular weight, phosphorylated NF proteins, the possibility remains that the binding of this specific antibody did not accurately reflect the presence of the total complement of components of the degenerating axoplasm or the regenerating axons (Shaw and Weber, 1982). To date, current data indicates that the majority of the proteins within axons separated from neuronal somata degenerate (Schlaepfer et al., 1984). In some samples, we have also been able to identify numerous fine diameter NF+ fibers in the distal portions of the transected nerve which we believe are regenerating fibers. The mere presence of these figures illustrates: a) that the monoclonal antibody used can visualize at least a portion of the small diameter axons, b) the necessity of monitoring for the presence of axons in purported denervated nerves and c) the ability of the peripheral nerve stump to attract axonal sprouts from, as yet, unknown sources. At the moment, we cannot rule out the possibility that axons not recognized by the SMI-31 monoclonal antibody are also present in the nerve distal to the site of transection. However, in our studies, NF- areas coincide with areas of shorter GFAP+ filaments or the complete absence of GFAP. In contrast, most but not all prominent NF+ areas co-localize with longer GFAP+ filaments.

One of the hallmarks of PNS nerve development is the gross morphological and biochemical alterations of Schwann cells after contact with growing axons. The changes from an undifferentiated, rounded neural crest cell into a cell which produces the complex, orderly myelin lamellae, are profound. These changes, including proliferation, ensheathment, myelination and formation of basal laminae, occur after cues from the environment, i.e. contact with axons and ECM (Bunge and Bunge, 1981). These findings imply that chemical signals that are located extrinsically to the cell are transcribed via the glial plasma membrane into numerous metabolic events, some of which involve the cytoskeletal apparatus.

Schwann cells have been shown to contain vimentin (Vim), actin and glial fibrillary acidic protein (GFAP) but expression of these cytoplasmic filaments has not been correlated with specific developmental stages (Yen and Fields, 1981; Autillo-Gambetti et al., 1982; Jessen et al., 1984; de Nechaud et al., 1986). Filament proteins in peripheral glia may follow an analogous expression of that reported in astrocytes where Vim expression characterizes an immature stage and the subsequent and/or concomitant expression of GFAP signals a more mature stage (Bovolenta et al., 1984). However, Jessen and Mirsky (1984) have localized GFAP solely to non-myelinating Schwann cells in the mature nerve. This implies that upon maturation myelinating Schwann cells may utilize an altered schema of cytoskeletal arrangement. We have shown that the normal GFAP pattern is altered when Schwann cells are orphaned from neuronal contact. After nerve transection, the GFAP+ figures, normally clustered with patches of non-myelinated axons, redistribute to an evenly dispersed pattern. This suggests that GFAP is expressed by Schwann cells which are distributed throughout the injured nerve and is no longer segregated among a certain class (non-myelinating) of Schwann cells. It is conceivable that GFAP is expressed by Schwann cells that had previously produced myelin but are now associated with degenerating myelin.

In addition, to changes in the GFAP distribution, there are also changes in the shape of the GFAP+ filaments. The progressive decrease in the length of the GFAP+ filaments, which correlates with the time after nerve transection, may coincide with a gradual withdrawal of glial processes or a degradation of the filaments within the cytoplasm. In addition, the significance of GFAP+ caps on the nucleus at long time points post-transection is yet to be determined. It is conceivable that after neural injuries orphaned Schwann cells express less GFAP and revert to the Vim+ state which may characterize an immature, de-differentiated state. This hypothesis is supported by the report of an abundance of Vim+ material five weeks after nerve transection (Autillio-Gambetti et al., 1982). Correspondingly, when differentiation begins after glia are contacted by regenerating axons, the greater majority of Schwann cells in that area of the nerve appear to express GFAP (personal observation). Whether these GFAP+ filaments originate from a nuclear organizing center (the nuclear cap?) is not yet known. Later when axonal segregation progresses to a 1 to 1 relationship between the Schwann cell and an axon, the myelinating Schwann cells may lose or modify the GFAP within their cytoplasm. At present these theories must be tested by additional experiments.

The molecules which comprise basal laminae in peripheral nerves are synthesized, secreted and organized by differentiating Schwann cells as they segregate, ensheath and myelinate axons. The major constitutents, laminin, type IV collagen, heparin sulfate proteoglycan and entactin are produced by Schwann cells in a sequential, ordered fashion during development (for a review see Bunge and Bunge, 1981). In the mature nerve, the basal laminae form numerous linear channels which are aligned from the neuronal somata to the axonal terminals. We, like others (see Bignami et al., 1984), have demonstrated that ECM channels, i.e. LAM+ material, remain intact after nerve transection for at least 6 weeks. The LAM+ staining which occludes the basal lamina channels is most likely associated with the plasma membranes of the post-proliferative Schwann cells. By analogy, Schwann cells previously in contact with axons and a mature, LAM+ basal lamina, can be trypsinized and within 24 hours will re-express LAM on their membrane even in the absence of axonal contact (Cornbrooks et al., 1983). Thus regenerating axons encounter LAM displayed both on the Schwann cell membrane and in ECM channels when they venture into the peripheral nerve stump. LAM, a known neuronotropic factor probably assists in the elongation and orientation of these neuronal processes toward their targets (Baron-van Evercooren et al., 1982; Manthrope et al., 1983; Rogers et al., 1983; Lander et al., 1985; Davis et al., 1985). Preliminary evidence from tissue culture in which mature Schwann cell-neuron cultures containing a robust ECM were axotomized also indicates that type IV collagen, heparin sulfate proteoglycan and entactin remain distributed in channels for at least 6 weeks after axonal degeneration (unpublished observations).

The observation that the basal laminae channels are maintained for 6 weeks after nerve transection fails to discern whether or not the constituent molecules are actively maintained by Schwann cells. Whether or not viable Schwann cells must be present to metabolically maintain or even modify molecules in basal laminae after mammalian nerve transection and prior to neural regeneration has not been investigated. Preliminary biochemical data supports the idea that basal laminae are dynamic and not inert. When mature Schwann cell-neuron cultures are axotomized, Schwann cells which remain in contact with a stable ECM synthesize and secrete a number of molecules including LAM and type IV collagen (unpublished observations). To date, there is little evidence that basal laminae alone can support axonal regeneration if Schwann cells are not in close proximity and/or if a neuronotrophic factor(s) is not added.

It is clear from the work of Jessen and Mirsky (1984) that the non-myelinating Schwann cells express molecules, such as GFAP, that are distinct from those of the myelinating Schwann cells. The precise one to one colocalization of GFAP and the antigen recognized by the C4 monoclonal antibody identifies the C4 epitope as another molecule that characterizes the non-myelinating peripheral glia. In contrast to GFAP, the C4 antigen is distributed on the plasma membrane of mature Schwann cells and it is susceptible to treatment with trypsin, acetone or triton. However, culture experiments demonstrate that the antigen is not always on the Schwann cell membrane. This antigen is not expressed on the surface when C4+ Schwann cells are trypsinized and replated as a pure population of cells when the PNS glia produce myelin or when they are orphaned after axotomy (Cornbrooks et al., 1981; personal observations). Curiously, both C4 and GFAP are prominently redisplayed in the peripheral nerve stump as linear figures after regenerating axons become contiguous with quiescent Schwann cells. Additionally, the C4 antigen expressed during regeneration is distributed in such a way that is is not susceptible to acetone/triton. This suggests that the C4 antigen is bound to a molecule that is not extractable with triton such as a cytoskeletal filament. It is intriguing to speculate that a direct relationship exists between a cytoskeleton protein and a protein located in the membrane such as C4. If this were so, it would provide a method to transmit an extracellular signal across the membrane thereby facilitating an exchange of information which results in the alteration of the glial cell morphology. A similar transmembrane interaction was previously proposed by Sugrue and Hay (1982). Whether the co-localization of an antigen exposed on the Schwann cell membrane surface and a cytoskeletal element is a coincidence remains, to be determined.

Our initial studies on the response of the fully differentiated Schwann cell to the loss of axonal contact and subsequent renewal of axonal contact have demonstrated distinct changes in specific antigens in the glial cytoskeleton, plasma membrane and possibly the ECM. By using a combination of in vivo and in vitro paradigms where glial-neuronal relationships can be perturbed, it should be possible to characterize modulations in Schwann cells which facilitate the trophic and tropic aspects of neuronal regeneration and the reformation of the mature glial-neuronal relationships. Therefore we expect that this approach will promote our understanding of the molecular events which occur during regeneration as well as during development in the PNS. Furthermore, these studies should also contribute to improved techniques aimed at fostering regeneration in the PNS as well as the CNS.

REFERENCES

Asssouline, J., Bosch, E. P., Lim, R., Jensen, R. and Pantazis, N. J., Detection of a neurite promoting factor with some similarity to nerve growth factor in conditioned media from primary Schwann cells and astroglia, Abst. Soc. for Neurosci. 11:933 (1985).

Autillio-Gambetti, L., Sipple, J., Sudilousky, O. and Gambetti, P., Intermediate filaments of Schwann cells, J. Neurochem. 38:774-780 (1982).

Baron-van Evercooren, A., Kleinman, H. K., Ohno, S., Marangos, D., Schwartz, J. P. and Dubois-Dalcq, M. E., Nerve growth factor, laminin, and fibronectin promote neurite growth in human fetal sensory ganglia cultures, J. Neurosci. Res. 8:179-193 ((1982).

Bignami, A., Chi, N. H. and Dahl, D., Laminin in rat sciatic nerve undergoing Wallerian degeneration, J. Neuropathol. Exp. Neurol. 43:94-103 (1984).

Billings-Gagliardi, S., Webster, H. deF. and O'Connell, M. F., In vivo and electron microscopic observations on Schwann cells in developing tadpole nerve fibers, Am. J. Anat. 141:375-392 (1974).

Bovelenta, P., Liem, R. K. H. and Mason, C. A., Development of cerebellar astroglia: transitions in form and cytoskeletal content, Dev. Biol. 102:248-259 (1984).

Bunge, R. P. and Bunge, M. B., Cues and constraints in Schwann cell development. In Studies in Developmental Neurobiology, ed. W. Maxwell Cowan. New York: Oxford University Press pp. 322-353 (1981).

Bunge, R. P., Bunge, M. B. and Eldridge, C. F., Linkage between axonal ensheathment and basal lamina production by Schwann cells. In Annual Reviews for Neuroscience, eds. W. M. Cowan, E. M. Shooter, C. F. Stevens and R. F. Thompson. Palo Alto: Annual Reviews Inc. pp. 305-328 (1986).

Carey, D. J., Eldridge, C. F., Cornbrooks, C. J., Timpl, R. and Bunge, R. P., Biosynthesis of type IV collagen by cultured rat Schwann cells, J. Cell Biol. 97:473-479 (1983).

Cornbrooks, C. J., Expression and distribution of entactin in extracellular matrix of the adult rat peripheral nerve and peripheral nerve cells in culture, Soc. for Neurosci. Abstr. 9:346 (1983).

Cornbrooks, C. J., Pattern and growth rate of neurites in peripheral neuronal cultures are most pronounced on extracellular matrix preparations, Soc. for Neurosci. Abstr. 11:175 (1985).

Cornbrooks, C. J. and Bunge, R. P., A cell-surface specific monoclonal antibody to differentiating Schwann cells, Am. Soc. for Neurochem. Abstr. 13:171 (1982).

Cornbrooks, C. J., Carey, D. S., McDonald, J. A., Timpl, R. and Bunge, R. P., In vivo and in vitro observations on laminin production by Schwann cells, Proc. Natl. Acad. Sci. 80:3850-3854 (1983).

Davis, G. E., Varon, S., Engvall, E. and Manthorpe, M., Substratum-binding neurite promoting factors: relationships to laminin, Trends in Neurosci. 8:528-532 (1985).

de Nechaud, B., Gumpel, M. and Bourre, J. M., Changes in some myelin protein markers and in cytoskeletal components during Wallerian degeneration of mouse sciatic nerves, J. Neurochem. 46:708-716 (1986).

Eldridge, C. F., Sanes, J. R., Chiu, A. Y., Bunge, R. P. and Cornbrooks, C. J., Basal lamina-associated heparin sulfate proteoglycan in the rat peripheral nervous system: Characterization and localization using monoclonal antibodies, J. Neurocytol. (1986).

Fallon, J. R., Preferential outgrowth of central nervous system neurites on astrocytes and Schwann cells as compared with nonglial cells in vitro, J. Cell Biol. 100:198-208 (1985a).

Fallon, J. R., Neurite guidance by non-neuronal cells in culture: Preferential outgrowth of peripheral neurites on glial as compared to nonglial cell surfaces, J. Neurosci. 5:3169-3177 (1985b).

Ide, C., Tohyama, K., Yokota, R., Nitatori, T. and Onodera, S., Schwann cell basal lamina and nerve regeneration, Brain Res. 288:61-75 (1983).

Jenq, C. and Coggeshall, R. E., The effects of an autologous transplant on patterns of regeneration in rat sciatic nerve, Brain Res. 364:45-56 (1986).

Jessen, K. R. and Mirsky, R., Nonmyelin-forming Schwann cells coexpress surface proteins and intermediate filaments not found in myelin-forming cells: a study of Ran-2, A5E3 antigen and glial fribrillary acidic protein, J. Neurocytol. 13:923-934 (1984).

Jessen, K. R., Thorpe, R. and Mirsky, R., Molecular identity, distribution and heterogeneity of glial fibrillary acidic protein: an immunoblotting and immunohistochemical study of Schwann cells, satellite cells, enteric glia and astrocytes, J. Neurocytol. 13:187-200 (1984).

Lander, A. D., Fujii, D. K. and Reichardt, L. F., Purification of a factor that promotes neurite outgrowth: Isolation of laminin and associated molecules, J. Cell Biol. 101:898-913 (1985).

Longo, F. M., Manthorpe, M., Skaper, S. D., Lundborg, G. and Varon, S., Neuronotrophic activities accumulate in vivo within silicone nerve regeneration chambers, Brain Res. 261:109-117 (1983).

Manthorpe, M., Engvall, E., Ruoslahti, E., Longo, F. M., Davis, G. E. and Varon, S., Laminin promotes neuritic regeneration from cultured peripheral and central neurons, J. Cell Biol. 97:1882-1890 (1983).

Michon, J., Nerve suture today. In Traumatic Nerve Lesions, eds. J. Michon and J. Moberg. Edinburgh: Churchill-Livingston pp. 79-89 (1975).

Neuberger, T. J. and Cornbrooks, C. J., Modification of Schwann cell antigens after peripheral nerve transection. Am. Soc. for Neurochem. Abst. 17:159 (1986).

Noble, M., Fok-Seang, J. and Cohen, T., Glia are a unique substrate for the in vitro growth of central nervous system neurons, J. Neurosci. 4:1892-1903 (1984).

Noden, D., Interactions directing the migration and cytodifferentiation of avian neural crest cells. In The Specificity of Embryological Interactions, ed. D. Garrod. London: Chapman and Hall pp. 3-49 (1978).

Poduslo, J. F., Regulation of myelination: Biosynthesis of the major myelin glycoprotein by Schwann cells in the presence and absence of myelin assembly, J. Neurochem. 42:493-503 (1984).

Politis, M. J., Ederle, K. and Spencer, P. S., Tropism in nerve regeneration in vivo. Attractions of regenerating axons by diffusible factors derived from cells in distal nerve stumps of transected peripheral nerves, Brain Res. 253:1-12 (1982).

Ramón y Cajal, S., Degeneration and Regeneration of the Nervous System, English translation and reprint, 1959, Hafner Press, New York (1928).

Rogers, S. L., Letourneau, P. C., Palm, S. L., McCarthy, J. and Furcht, L. T., Neurite extension by peripheral and central nervous system neurons in response to substratum-bound fibronectin and laminin, Dev. Biol. 98:212-220 (1983).

Salzer, J. L. and Bunge, R. P., Studies of Schwann cell proliferations: I. An analysis in tissue culture of proliferation during development, Wallerian degeneration, and direct injury, J. Cell Biol. 84:739-752 (1980).

Schlaepfer, W. W., Lee, C., Trojanowski, J. Q. and Lee V. M.-Y, Persistence of immunoreactive neurofilament protein breakdown products in transected rat sciatic nerve, J. Neurochem. 43:857-864 (1984).

Schwab, M. E. and Thoenen, H., Dissociated neurons regenerate into sciatic but not optic nerve explants in culture irrespective of neurotrophic factors, J. Neurosci. 5:2415-2423 (1985).

Shaw, G. and Weber, K., Differential expression of neurofilament triplet proteins in brain development, Nature 298:277-279 (1982).

Sugrue, S. P. and Hay, E. D., Interaction of embryonic corneal epithelium with exogenous collagen, laminin, and fibronectin: role of endogenous protein synthesis, Dev. Biol. 92:97-106 (1982).

Webster, H. deF., Martin, J. R. and O'Connell, M. F., The relationships between interphase Schwann cells and axons before myelination: A quantitative electron microscopic study, <u>Dev. Biol</u>. 32:401-416 (1973).

Weinberg, H. J. and Spencer, P. S., The fate of Schwann cells isolated from axonal contact, <u>J. Neurocyt</u>. 7:555-569 (1978).

Williams, L. R., Longo, F. M., Powell, H. C., Lundborg, G. and Varon, S., Spatial-temporal progress of peripheral nerve regeneration within a silicone chamber: parameters for a bioassay, <u>J. Comp. Neurol</u>. 218:460-470 (1983).

Yen, S. and Fields, K. L., Antibodies to neurofilament, glial filament, and fibroblast intermediate filament proteins bind to different cell types of the nervous system, <u>J. Cell Biol</u>. 88:115-126 (1981).

A PHYSIOLOGICAL ROLE OF THE BENZODIAZEPINE/GABA RECEPTOR-CHLORIDE IONOPHORE COMPLEX IN STRESS

Hratchia Havoundjian[1,2], Ramon Trullas[1], Steven Paul[3], and Phil Skolnick[1]

[1]Laboratory of Bioorganic Chemistry, NIADDK, [2]Howard Hughes Medical Institute, and [3]Clinical Neuroscience Branch, NIMH, National Institutes of Health, Bethesda MD 20892

INTRODUCTION

Both direct and correlative evidence suggests that the principal pharmacologic actions of the benzodiazepines are mediated by high affinity, stereospecific recognition sites (receptors) found exclusively in tissue derived from the neural crest. The rapid advances in our understanding of the relationship of benzodiazepine receptors to (a subpopulation of) γ-aminobutyric acid (GABA, the principal inhibitory neurotransmitter in mammalian brain) receptors and an associated chloride channel have resulted in new insights about the molecular pharmacology of the benzodiazepines and other psychoactive drugs which share pharmacologic actions with the benzodiazepines (e.g. barbiturates, ethanol). In contrast, the physiological roles of this "benzodiazepine receptor complex" are not well understood, and the physiological relevance of benzodiazepine receptors has been questioned (cf. Guidotti, et al., 1983). This chapter will briefly summarize the evidence implicating the benzodiazepine receptor complex in the response to stress and anxiety, and present some recent findings from our laboratory which supports this hypothesis.

Evidence For A Physiological Role Of The Benzodiazepine Receptor Complex In Stress

A) Studies with benzodiazepine receptor "inverse agonists"

Benzodiazepine receptor "inverse agonists" (sometimes referred to as "active antagonists") such as 3-carboethoxy-B-carboline (B-CCE) and N-methyl-B-carboline-N-carboxamide (FG 7142) have been shown to produce a syndrome reminiscent of stress or "anxiety" in rodents and primates, including man. For example, in chair-adapted rhesus monkeys, intravenous administration of B-CCE produces a rapid, dramatic increase in heartrate and blood pressure which is accompanied by an activation of the pituitary-adrenal axis (evidenced by increases in plasma ACTH, cortisol, and epinephrine) and behaviors reminiscent of "anxiety" (including agitation, distress vocalization, piloerection, defecation, and a refusal of food or drink) (Ninan, et al., 1982). Pretreatment with the specific benzodiazepine receptor antagonist Ro 15-1788 (Hunkeler, et al., 1981) has been shown to block the somatic, endocrine, and behavioral effects of B-CCE, which strongly suggests the "anxiogenic" effects of this compound can be attributed to occupation of CNS benzodiazepine receptors rather than through a nonspecific, "arousing" effect. The behavioral effects of lower doses of B-CCE (≤ 1 mg/kg) are more subtle, manifest as highly individualized responses which appear correlated with the subsequent responses of an animal to "threatening" environmental stimuli (Insel, et al., 1984). Thus, if an animal were highly aroused and agitated when approached by a stranger, its' response to B-CCE would be more robust than an animal which was not highly aroused when confronted with the same stimulus.

Although it is more difficult to evaluate behavioral changes in rodents which could be construed as "anxiety", both behavioral and biochemical findings suggest B-carbolines elicit analogous responses to those seen in primates. For example, Corda, et al. (1983) have reported that B-carbolines reduce punished responding in a "thirsty rat conflict" test (Vogel, et al., 1971). This "proconflict" action is opposite to the effect of benzodiazepines in this paradigm. However, in order to see this effect, the shock intensity must be reduced to prevent a "floor" effect (a low rate of responding which would obscure a further reduction by the B-carbolines). A reduction in punished responding by B-carbolines has also been observed by deCarvalho, et al. (1983) using an analogous test in mice. File and co-workers (File, 1982; File and Lister, 1983; File and Pellow, 1984) have demonstrated B-carbolines such as B-CCE and FG 7142 reduce social interaction in rats, an action opposite to that observed with benzodiazepines and other anxiolytics. Furthermore, administration of FG 7142 has been reported to increase plasma corticosterone in rats (Pellow and File, 1985) which is antagonized by pretreatment with Ro 15-1788. Despite the similarities between the behavioral, somatic, and endocrine effects of B-carbolines in animals and some of the signs and symptoms of anxiety in man, these experiments do not provide conclusive evidence that administration of B-carbolines produces "anxiety". However, administration of FG 7142 to human volunteers (Dorow, et al., 1983) did increase muscle tension, autonomic activity, and produce extreme apprehension in two subjects that had measureable blood levels of this compound. The effects of FG 7142 were characterized by the subjects as feelings of "inner tension, excitation, and sensations of physical disturbance." The intravenous administration of a benzodiazepine almost immediately reversed both the autonomic hyperactivity (increased heart rate, blood pressure) and subjective feelings of anxiety.

Thus, it appears that the administration of a benzodiazepine receptor inverse agonist elicits a syndrome that resembles in many respects a generalized anxiety disorder. Although it may be argued that this mimicry of an anxiety-like state through occupation of benzodiazepine receptors supports a role for the benzodiazepine receptor complex in the physiological manifestation of anxiety, an equally persuasive argument can be made for the benzodiazepine receptor modulating GABA-gated chloride channels only under pharmacologic conditions (cf. Guidotti, et al., 1983) - i.e. when these sites are occupied by an exogenously administered ligand. The latter argument can readily be supported by the relatively late evolutionary appearance of benzodiazepine receptors [these sites first appear in chordates] (Nielsen, et al., 1978) which is clearly preceded by the appearance of GABA-gated chloride channels in invertebrates (e.g. Casida, et al., 1985; Takeuchi and Takeuchi, 1969). Since biochemical studies have shown that benzodiazepine and GABA receptors reside on the same protein with a molecular weight of 50-60 Kdalton (Gavish and Snyder, 1981; Sigel, et al., 1983; Schoch, et al., 1984; Schwartz, et al., 1985), it may be argued that "specific" recognition sites for benzodiazepines may be an evolutionary quirk that has proven fortuitous for the pharmaceutical industry. Clearly, the demonstration of an endogenous substance which subserves benzodiazepine receptors must be viewed as a critical element for assigning a physiological role to the benzodiazepine receptor and will ultimately define the physiological roles of the entire benzodiazepine receptor complex.

B) Isolation of putative endogenous ligands of the benzodiazepine receptor

Since the initial demonstration of benzodiazepine receptors in the central nervous system (Squires and Braestrup, 1977; Mohler and Okada, 1977), more than half-dozen endogenous substances have been described which inhibit radioligand binding to these sites (cf. Skolnick, et al., 1983). While most of these substances have neither been chemically identified nor extensively studied in vivo, several of these substances have been biochemically or pharmacologically characterized as "benzodiazepine-like", while others have properties that closely resemble inverse agonists. While the literature on putative endogenous ligands of the benzodiazepine receptor will not be reviewed, the evidence which supports a physiological role for two of these substances will be critically evaluated.

Guidotti, et al., (1983b) have described the presence of an ~11 kdalton protein termed DBI ("diazepam binding inhibitor") that competitively inhibits radioligand binding to benzodiazepine receptors. The rationale for isolation of this substance has been discussed in detail (Guidotti, et al., 1983a). This compound, which is found in relatively high concentrations in the brain compared to peripheral tissues (Guidotti et al., 1983a) has been shown to have a "proconflict" action in rats similar to that observed with B-carbolines (Corda, et al., 1983). The potency of DBI is not increased by GABA, and in this respect DBI resembles benzodiazepine receptor antagonists and some inverse agonists. DBI displaces $[^3H]$B-carboline binding with a K_i of ~5 µM; it is far less potent as an inhibitor of $[^3H]$benzodiazepine binding, and only inhibits binding of $[^3H]$benzodiazepines by ~50% (Ferrero, et al., 1986). The relatively large size of this compound and low affinity for benzodiazepine receptors (versus nM affinities for many physiologically important peptide neurohormones) remains a major obstacle in assigning a definitive physiological role to this substance. Tryptic digestion of DBI has resulted in the isolation of an octadecapeptide fragment, known as ODN. This compound has also been shown to have proconflict actions in rats, and the affinity of this compound for benzodiazepine receptors is somewhat higher (K_i ~1.5 µM versus $[^3H]$3-carbo-methoxy-B-carboline) (Ferrero, et al., 1986). ODN was initially not found in tissue extracts, but a recent report (Ferrero, et al., 1986) demonstrated ODN-like material in rat brain homogenates using rabbit antibodies that were directed against the amino terminal of ODN. The physiological significance of both DBI and ODN merits further investigation. Another class of identified endogenous substances that inhibit radioligand binding to benzodiazepine receptors are the purines inosine and hypoxanthine. The affinity of these compounds for benzodiazepine receptors is low (K_i ~800-900 µM) (Marangos, et al., 1979), and it is unlikely that concentrations approaching the K_i of either compound could be attained in vivo. Nonetheless, electrophysiological (MacDonald, et al., 1979) studies have demonstrated a dual action of inosine, with both benzodiazepine-like and antagonist actions. Parenteral or intraventricular administration of relatively large doses of this compound have been shown to possess anticonvulsant actions in rodents (Skolnick, et al., 1979; Marangos, et al., 1981). Antipunishment actions of inosine (which are blocked by Ro 15-1788) have been demonstrated in pigeons (Witkin and Barrett, 1985). Other studies (Slater and Longman, 1979; Crawley, et al., 1981) have shown that inosine can antagonize other of the pharmacologic actions of the benzodiazepines. Thus, although purines like inosine and hypoxanthine are endogenous and ligands of the benzodiazepine receptor, their role as endogenous ligands remains questionable.

There are also a number of intriguing behavioral observations which suggest the presence of an endogenous benzodiazepine receptor ligand. Perhaps the most consistent and convincing evidence for the release of an endogenous ligand stems from studies in patients suffering from hepatic encephalopathy (HE), where it was demonstrated that Ro 15-1788 ameliorates the neurologic symptoms of this disease (Scollo-Lavizzari and Steinman, 1985; Bansky, et al., 1985). In an experimental model of HE in rabbits, Ro 15-1788 can reverse the changes in visual evoked potential (VEP). Interestingly, the VEP in rabbits with experimentally induced HE resembles the VEP in rabbits administered a benzodiazepine (Bassett, et al. 1986). These studies suggest the presence of an endogenous benzodiazepine-like compound. Recently, Higgitt, et al., (1986) have demonstrated that in normal volunteers, Ro 15-1788 had a mixture of agonist (benzodiazepine-like) and other non-benzodiazepine-like effects on the variables studied. Although no clearcut dose-response relationship was obtained, the results suggested a predominance of benzodiazepine-like effects on physiological measures at the higher dose (100 mg/kg, p.o.), while the lower dose (30 mg/kg) produced a greater effect on behavioral and subjective dimensions. Interestingly, these latter effects were opposite of those normally found for benzodiazepines. Consistent with these findings in humans are the observations of File and co-workers (File, 1982; File, et al., 1982) who demonstrated that the benzodiazepine receptor antagonist Ro 15-1788 reduces social interaction (as is seen with B-carbolines) in rats. However, this effect is observed only over a narrow dose range of Ro 15-1788, and is not as robust an effect as is observed with B-carbolines. Ikonomidou, et al. (1985) have also reported an inverse agonist action of Ro 15-1788 (again, over a narrow dose range) to increase muscle tone in genetically spastic rats.

In marked contrast to these findings, most studies have failed to demonstrate pharmacological effects of Ro 15-1788 through a wide dose range (Hunkeler, et al., 1981; Haefely, 1983; Pieri and Biry, 1985). At relatively high doses of Ro 15-1788 (Dantzer and Perio, 1982; Greksch, et al., 1983; Kajima, et al., 1983), a slight "partial agonist" action (i.e. benzodiazepine-like) has been reported in a number of behavioral and pharmacological paradigms. Biochemical and electrophysiological studies have confirmed this partial agonist nature of Ro 15-1788 (Karobath, et al., 1983; Skerritt and MacDonald, 1983), with biochemical studies conducted in well-washed tissue preparations designed to remove any putative endogenous ligand. The many reports of the lack of "intrinsic" actions of Ro 15-1788 have generally been taken as evidence for the absence of an endogenous ligand. However, by analogy with the opiate receptor system, antagonists have also been found to be generally devoid of intrinsic actions, but can (for example) exacerbate pain sensation associated with nociceptive stimuli (Buchsbaum, et al., 1977; Levine, et al., 1978)-that is, during a period when an endogenous opioid is presumed to be released. Thus, the demonstration of "intrinsic" actions of Ro 15-1788 under certain circumstances could reflect an increased release of an endogenous benzodiazepine-like compound. Certainly, the presence of an endogenous ligand for benzodiazepine receptors remains an unresolved issue which merits further investigation.

C) Behaviorally-induced modification of the benzodiazepine receptor complex

It has been postulated that the physiological roles of the benzodiazepine receptor complex would be analogous to the pharmacological actions of the drugs which act at these sites. Hence, physiological roles for the benzodiazepine receptor complex in anxiety, sleep, seizure disorders, and maintenance of muscle tone have been proposed. Studies demonstrating that occupation of components of the benzodiazepine receptor complex by "antagonists" (e.g. picrotoxin at sites at or near the chloride ionophore, bicuculline at the GABA receptor, and B-carbolines such as B-CCE at benzodiazepine receptors) can result in "anxiety" (see previous section), convulsions, increased muscle tone, and increased sleep latency have been used to support this view. Therefore, many investigations attempting to define physiological roles for the benzodiazepine receptor complex have tried to demonstrate changes in the number and/or affinity of these sites in response to behavioral, phar-

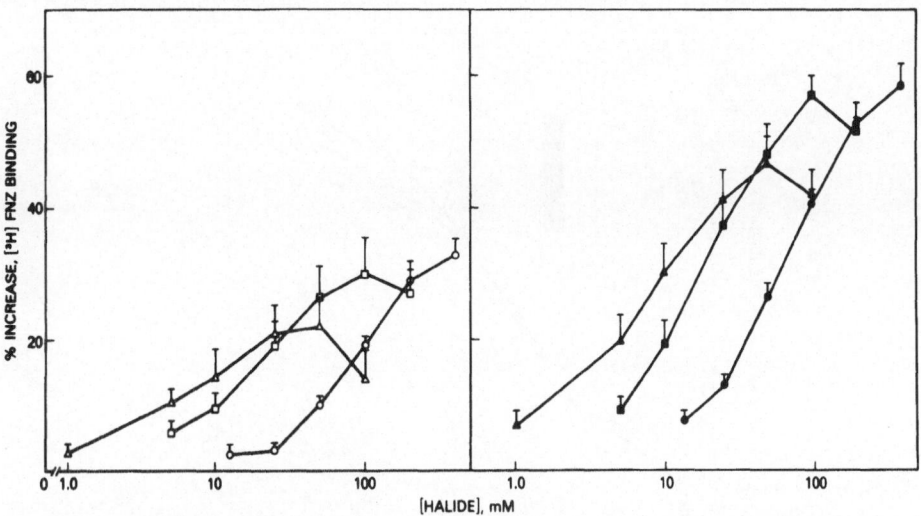

FIG. 1: Effects of swim-stress on halide-enhanced [^{3}H]flunitrazepam binding:
[^{3}H]flunitrazepam binding to cortical membranes prepared from non-
stressed rats (left panel) or swim-stressed (right panel) and
assayed in the presence of NaCl (circles), NaBr (squares), or NaI
(triangles). Values represent X±S.E.M. of 6-12 animals. The maximum enhancement in [^{3}H]flunitrazepam binding obtained using NaCl,
NaBr, and NaI was: 33±2%, 30±6%, and 22±5%, respectively in non-
stressed rats and 58±3%, 56±3%, and 46±4%, respectively in swim-
stressed rats. These differences were significantly different
($p<0.005$, Student's t-test). The EC_{50} values for NaCl, NaBr, and
NaI was 92, 17, and 5.4 mM, respectively in non-stressed animals
and 59, 16, and 6.2 mM, respectively in swim-stressed animals.
Animals were group housed in a colony room for at least 7 days
prior to experimentation. The concentration of [^{3}H]flunitrazepam
was ~0.7 nM. From Havroundjian, et al., (1986b).

macological, or physiological manipulations. Such studies attempting to
demonstrate that stress (i.e. exposure to an "anxiety" producing situation)
can alter either the number or apparent affinity of these sites have resulted
in small, bidirectional changes in either the number or affinity of benzodiazepine receptors (Lippa, et al., 1978; Braestrup, et al., 1979; Soubrie,
et al., 1980; Grimm and Hershkowitz, 1981; Lane, et al., 1982; Medina, et
al., 1983). However, if an analogy is made between the benzodiazepine
receptor complex and other signal transducing processes (for example, the
B-adrenoceptor coupled adenylate cyclase system), it is the effector rather
than the recognition component where activation or inhibition is observed.
Since the GABA-gated chloride channel can be viewed as the effector component of the benzodiazepine receptor complex, we initiated studies to determine whether stress could alter the benzodiazepine receptor coupled, GABA-
gated chloride channel.

In our initial studies, a brief (10 min.) ambient temperature swim was
used as the stressor (Havoundjian, et al., 1986a,b), and halide enhanced
[^{3}H]flunitrazepam binding [which reflects the coupling between chloride
channels and benzodiazepine receptors] (Costa, et al., 1979) was measured
in several brain areas. As previously reported (Costa, et al., 1979),
permeant anions such as CI$^-$, Br$^-$, and I$^-$ elicit a concentration dependent
increase in [^{3}H]flunitrazepam binding in well-washed membrane preparations.
However, the response to these ions in cortical, hippocampal, and cerebellar
membranes from stressed animals is dramatically increased in stressed ani-

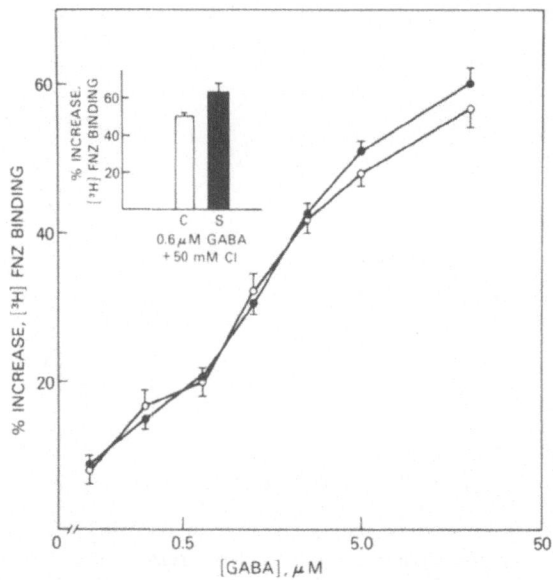

FIG. 2: Effects of swim-stress on GABA enhanced [3H]flunitrazepam binding:
Cortical tissue from non-stressed or swim-stressed rats was extensively washed in 50 mM Tris-Citrate buffer containing 100 mM NaCl.
The tissue was then resuspended in 50 mM Tris-Citrate buffer and
[3H]flunitrazepam binding (~0.7 nM) assayed in the presence and
absence of GABA. The maximum enhancement obtained in the presence
of 20 μM GABA was 57±3% and 60±2% in non-stressed and stressed rats,
respectively (n=6-7/group). The EC_{50} values for GABA were (estimated graphically) ~1.0 and ~1.1 μM in non-stressed and stressed rats,
respectively. Inset: GABA enhanced [3H]flunitrazepam binding
assayed in the presence of 50 mM NaCl. The GABA concentration in
these experiments was 0.6 μM. Open bars: non-stressed rats; closed
bars: stressed rats (n=4/group). The maximum enhancement of [3H]-
flunitrazepam binding under these conditions was 51±3% and 63±5%
(p<0.1) in non-stressed and stressed rats. From Havoundjian, et
al., (1986b).

mals compared to identically prepared tissues from non-stressed animals
(Havoundjian, et al., 1986a,b and unpublished observations) (Fig. 1).
These differences were manifest as an increase in both the efficacy [E_{max}
58±3% and 33±2% in stressed and control rats respectively (p<0.001) (~76%
increase)] and potency of Cl⁻ ions [EC_{50} 59±4 mM in stressed and 92±5 mM
in control rats, respectively (p<0.001)]. A similar increase in the E_{max}
of Br⁻ and I⁻ enhanced [3H]-flunitrazepam binding was observed in cortical
tissue from stressed rats (Fig. 1), but was not seen with ions such as
$C_2H_3O_2$ or HPO_4^{-2}, which are relatively impermeable through chloride channels. In contrast, no significant differences in GABA-enhanced [3H]fluni-
trazepam binding were observed between the stressed and control groups in
the absence of halide ions (Fig. 2). However, if GABA enhanced [3H]fluni-
trazepam binding was assayed in the presence of 50 mM Cl , a small increase
in binding was observed in the stressed group (Fig. 2, inset), which probably reflects the potentiation of GABA-enhanced [3H]benzodiazepine binding
by Cl⁻ (Supavilai, et al., 1982).

Costa, et al., (1979) first demonstrated that halide ions increase the
apparent affinity of radioligands for benzodiazepine receptors, but do not

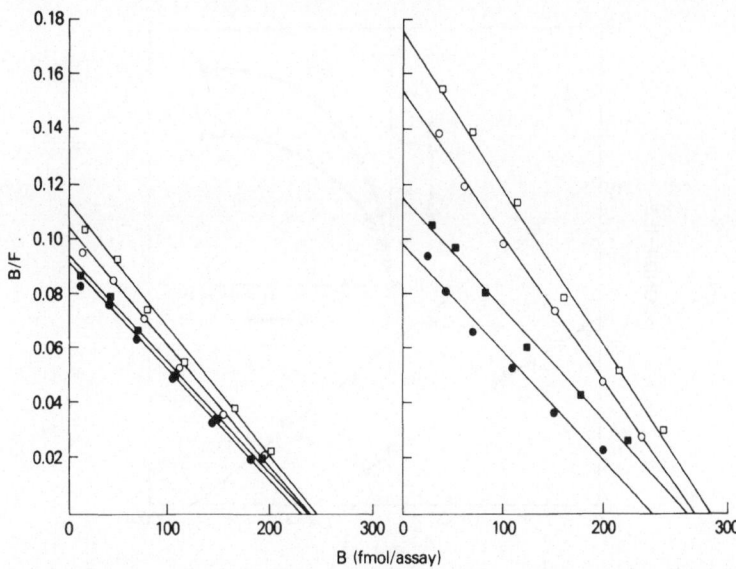

FIG. 3: Scatchard analyses of [3H]flunitrazepam binding in stressed and non-stressed rats. [3H]Flunitrazepam binding in the absence (closed circles) and presence of 25 mM (closed squares), 100 mM (open circles), and 400 mM (open squares) NaCl. Left panel: non-stressed rat; right panel: swim-stressed (10 min.) rat. This is a representative experiment repeated three times with similar results. The K_d values (nM) for control rats were: 2.64±0.20 (no NaCl), 2.47±0.18 (25 mM NaCl), 2.39±0.10 (100 mM NaCl), and 2.11±0.12 (400 mM NaCl). The increase in apparent affinity of [3H]flunitrazepam was statistically significant (p<0.05, paired t-test) at 400 mM NaCl. The B_{max} values in these animals were 3.15±0.22, 3.29±0.29, 3.56±0.44, and 3.56±0.30 pmol/mg protein, respectively. In stressed rats, the corresponding values for K_d (nM) were: 2.47±0.11 (no NaCl), 2.19±0.16 (25 mM NaCl), 1.84±0.04 (100 mM NaCl), and 1.65±0.09 mM (400 mM NaCl). The respective B_{max} values were: 3.05±0.12, 3.22±0.05, 3.49±0.12, and 3.60±0.07 pmol/mg protein. The presence of 100 and 400 mM NaCl elicited a significant reduction in the K_d of [3H]flunitrazepam in stressed rats (compared to no NaCl) (p<0.01 and <0.005, respectively). The B_{max} was significantly increased in stressed rats (p<0.01, paired t-test) by ≥100 mM NaCl compared to no NaCl. The K_d of [3H]flunitrazepam was significantly lower in stressed than control rats in the presence of 100 and 400 mM NaCl (p<0.01 and 0.02, respectively). Values represent X±SEM of at least three experiments. [3H]-Flunitrazepam concentrations in these experiments ranged from 0.3-9.2 nM. The animals were group housed in a colony room for at least seven days. Data from Havoundjian, et al., 1986a.

significantly alter the maximum number of binding sites. It is apparent (Fig. 3) that Cl⁻ ions have a much more robust effect in increasing the apparent affinity of [3H]flunitrazepam for benzodiazepine receptors in stressed animals (see Fig. 3 legend). Neither the B_{max} nor K_d of [3H]flunitrazepam was different between these groups in the absence of halide ions. However, a small but statistically significant increase in the B_{max} of [3H]-flunitrazepam was observed in the stressed, but not control rats (Fig. 3 legend) in the presence of ≥100mM Cl⁻.

These observations strongly suggest that either the chloride ionophore

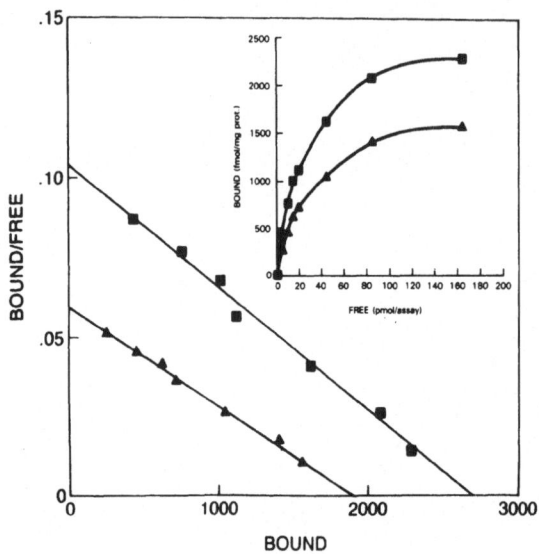

FIG. 4: Effects of stress on [^{35}S]TBPS binding: The K_d of [^{35}S]TBPS in cortical membranes of non-stressed (triangles) and swim-stressed (squares) rats was: 37.2±3.0 and 28.6±1.6 nM, respectively (p<0.05). The B_{max} values were 2030±56 and 2722±63 fmol/mg, respectively (p<0.001). These values are the X̄±SEM of four experiments. [^{35}S]TBPS (5 nM) was diluted with TBPS to achieve concentration of 5-165 nM. Animals were group housed in a colony room for at least seven days. Data from Havoundjian, et al., 1986a.

or the coupling between the chloride ionophore and benzodiazepine receptor is altered as a result of stress. We then examined the effects of stress on the binding of [^{35}S] tert-butylbicylophosphorothionate (TBPS), a picrotoxinin-like cage convulsant which labels a population of sites at or near the GABA/benzodiazepine receptor linked chloride channel (Squires, et al., 1983). A significant increase in both the number of binding sites and apparent affinity of this radioligand was observed in cortical membranes of stressed rats (Fig. 4). These observations were the first direct evidence that stress may affect the chloride ionophore component of the benzodiazepine receptor complex.

The effect of stress on halide-enhanced [^{3}H]benzodiazepine binding or [^{35}S]TBPS binding was observed after as little as four minutes of ambient temperature swim (the earliest time examined in this study) and was still present after thirty minutes of swim (Havoundjian, et al., 1986b).

These initial studies raised a number of important issues about the potential physiological relevance of these phenomena. It could be argued that these changes may be directly related to physical exertion rather than to the "anxiety" produced by subjecting the animal to a novel (and most certainly stressful) environment. However, during the course of our initial investigations, we sometimes observed an apparent effect of the order in which "naive", but not stressed animals were sacrificed. That is, the E_{max} of chloride to increase [^{3}H]flunitrazepam binding (in cortical membranes) increased progressively as the animals were sacrificed, despite a randomization of the order in which tissues were processed and assayed to eliminate the introduction of experimental bias. Thus, we subsequently restricted the number of animals sacrificed to two "naive" and two "stressed" animals per

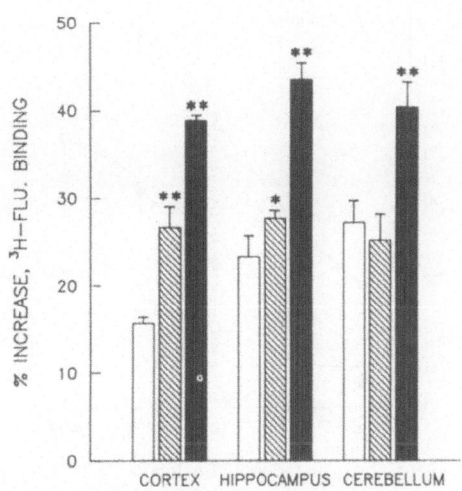

FIG. 5: Effects of "kill order" and swim-stress on chloride-enhanced [^{3}H]-
flunitrazepam binding: regional specificity. Rats were group
housed in an environmental chamber for at least seven days. The
"first kill" and "second kill" rats were removed from the same cage.
The stressed rats (10 min. ambient temperature swim) were removed
from a different cage. The tissue was prepared and [^{3}H]flunitra-
zepam binding assayed in the presence of 400 mM NaCl as described
(Havoundjian, et al., 1986a,b). Symbols: Open bars, first animal
killed; hatched bars, second animal killed; solid bars, swim-
stressed (10 min., ambient temperature) animals. Cerebral cortex:
"Second kill" and "stress" significantly different from first kill
(p<0.01 and p<0.01, respectively, paired t-test); Hippocampus:
"Second kill" and "stress" significantly different from first kill
(p<0.05 and p<0.01, respectively paired t-test). Cerebellum:
"Stress" significantly different from non-stressed (p<0.01); no
difference between "first" and "second" killed. A consistent "kill
order" effect was not observed if the "second" animal killed was
removed from a different cage in the same environmental chamber.
The interval between the "first kill" and "second kill" was ≤15
seconds. The concentrations of radioligand in these experiments
were ~0.7 nM. From Trullas, et al., 1986.

experiment (Havoundjian, et al., 1986a,b).

Nonetheless, these observations suggested that the benzodiazepine
receptor complex could be even more sensitive to "stress" than our initial
experiments with ambient temperature swim indicated. In our initial studies,
animals were housed in the laboratory's animal colony room (Havoundjian, et
al., 1986a,b). We examined this "kill order" phenomenon in more detail using
environmental chambers where the animals were housed similarly (i.e. group
housed in metal racks on a 12 hour light cycle with food and water freely
available) to our colony room but in an environment which eliminated much of
the visual, olfactory, and auditory stimuli usually encountered in a "common"
colony room. In this more carefully controlled environment, a significant
increase in chloride enhanced [^{3}H]flunitrazepam binding was observed in cortical and hippocampal (but not cerebellar) membranes from the second animal
removed from a common cage compared to the first. The level of chloride
enhanced [^{3}H]flunitrazepam binding in hippocampal and cortical membranes of
the second animal sacrificed was intermediate to that of the first animal
removed from that cage, and animals subjected to a 10 minute ambient tempe-

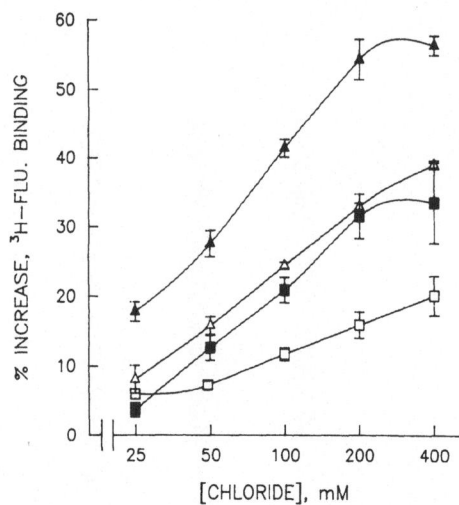

FIG. 6: Effects of housing on chloride-enhanced [^{3}H]flunitrazepam binding:
Cortical tissue was prepared and assayed as described (Havoundjian,
et al., 1986a,b). Values represent X±SEM from at least 4 animals.
Symbols: Squares, "non-stressed" animals housed in an environmental chamber (open symbols) or colony room (closed symbols); Triangles, animals housed in an environmental chamber (open symbols)
or colony room (closed symbols) subjected to a 10 minute ambient
temperature swim. The concentrations of radioligand in these experiments was ~0.7 nM. From Trullas, et al., 1986.

rature swim (Fig. 5). A similar effect of "kill order" was not consistently
observed if the second animal sacrificed was taken from a different cage in
the environmental chamber. A parallel "kill order" effect has also been
observed in the number and apparent affinity of [^{35}S]TBPS binding sites
(Trullas, et al., 1986). Thus, it appears that changes in benzodiazepine/
GABA receptor coupled chloride channels can occur in 8-12 seconds (the time
needed to remove one animal from the chamber, close the chamber, sacrifice
that animal, reopen the chamber, then remove and sacrifice the second animal), and can be provoked by a perturbation of the environment as "subtle"
as that of an investigator removing a cohort from a common cage. Similar
"minor" perturbation of the environment can also produce significant elevations in circulating levels of ACTH and corticosterone that are first
observed 2.5 and 5 minutes, respectively after application of the stressor
(Cook, et al., 1977; Guillemin, et al., 1977). These findings demonstrate
that "stress"-induced alterations in the benzodiazepine receptor complex
precede activation of both the pituitary-adrenal axis and other brain
transmitter/receptor systems that have been associated with stress (cf.
Ikeda and Nagatsu, 1985). Further, the alterations in both [^{35}S]TBPS binding and halide-enhanced [^{3}H]flunitrazepam binding cannot be solely attributed to increases in motor activity since this "kill order" effect was not
observed in cerebellar membranes, despite the robust increase in halide-
enhanced [^{3}H]flunitrazepam binding seen in this tissue after an ambient
temperature swim.

There also appears to be a tonic effect of environment on the benzodiazepine receptor complex, as illustrated in Figure 6. In these experiments,
chloride enhanced [^{3}H]flunitrazepam binding in cortical membranes prepared
from "control" and swim-stressed animals housed in our common colony room
were compared with tissue prepared from animals housed in the environmental
chamber. While significant, stressed-induced increases in chloride-enhanced

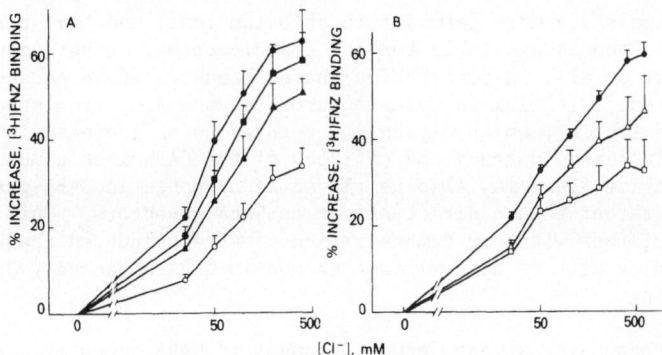

FIG. 7: Chloride enhanced [^{3}H]flunitrazepam binding in cerebral cortex: effects of pentobarbital and TBPS. Left panel: Potentiation of chloride enhanced [^{3}H]flunitrazepam binding by pentobarbital in non-stressed rats. Animals were housed in a colony room for at least one week prior to use. In this series of experiments, the maximum enhancement of [^{3}H]flunitrazepam binding (0.7 nM) by 400 mM NaCl was 34±4%. The maximum enhancement was increased by 100 μM (closed triangles), 200 μM (closed squares), and 400 μM (closed circles) pentobarbital to 51±3% (p<0.05 compared to no pentobarbital), 58±6% (p<0.02), and 63±7% (p<0.02), respectively. Right panel: Chloride enhanced [^{3}H]flunitrazepam binding in stressed rats: inhibition by TBPS. Animals were exposed to a 10 min. ambient temperature swim prior to sacrifice. In these experiments, the maximum enhancement of [^{3}H]flunitrazepam binding by 400 mM NaCl was 60±3%. Addition of 5 μM TBPS (open triangles) and 10 μM TBPS (open squares) reduced this enhancement to 47±8% (p<0.005) and 33±7% (p<0.005 compared to no addition of TBPS). Values represent the X̄±SEM of at least three animals per group. From Havoundjian, et al., 1986b.

[^{3}H]flunitrazepam binding were observed in both groups, the maximum enhancement (in both "control" and swim-stressed animals) was significantly higher in tissue prepared from animals housed in the colony room (Fig. 6) (Trullas, et al., 1986). Thus, the chloride ionophore component of the benzodiazepine receptor complex appears to be under both acute and tonic control.

There is evidence to suggest that stress-induced modification of the benzodiazepine receptor complex is a compensatory response to stressful or "anxiety"-provoking changes in the environment. For example, the maximum enhancement of [^{3}H]flunitrazepam binding by chloride ions in tissue prepared from swim-stressed animals can be mimicked in control animals by addition of 200-400 μM pentobarbital (Fig. 7). These concentrations of pentobarbital have been shown to directly increase chloride conductance (Nicoll and Wojtwoicz, 1980; Schulz and MacDonald, 1981). The hypothesis that stress induces alterations in GABA-gated chloride channels that resembles the addition of a barbiturate is consistent with the observations of Soubrie et al. (1980), who demonstrated that a brief exposure to ice-water increases the seizure threshold to picrotoxin (which, like TBPS), is thought to act at or near the chloride channel), but not to the glycine antagonist, strychnine. The effects of stress to increase the number and apparent affinity of [^{35}S] TBPS binding are also consistent with this view (Fig. 4), since studies from our laboratory have provided evidence that the apparent affinity of [^{35}S]TBPS is directly related to the permeability characteristics of GABA-gated chloride channels, while the increase in number of [^{35}S]TBPS is related to the number of channels in an "open" conformation (Havoundjian, et al., 1986c). Thus, the consistent stress-induced increase in the apparent affinity and number of [^{35}S]TBPS binding sites suggests that chloride channels from these animals are more permeant to chloride ions, and that there are more of these channels available for ion translocation. Other recent studies (Schwartz, et al., in preparation) have demonstrated an enhanced muscimol stimulated $^{36}Cl^-$ flux in synaptoneurosomes prepared from stressed animals. This finding, which may be interpreted as an increase in GABAergic "tone" (i.e. increased potency and efficacy of the GABAmimetic muscimol to stimulate chloride flux) may also be viewed as a change in GABA-gated chloride channels rather than a direct effect on GABA receptors. This interpretation is consistent with our demonstration of stress-induced changes in [^{35}S]TBPS binding with no concomittant change in GABA-enhanced [^{3}H]flunitrazepam binding.

Reports concerning stress induced changes in GABA receptors are, in general, as conflicting as those concerned with stress-induced changes in benzodiazepine receptors. For example, Soubrie, et al. (1980) reported no difference in [^{3}H]muscimol or GABA-enhanced [^{3}H]flunitrazepam binding following a brief, cold water swim stress. In contrast, Biggio and colleagues (Biggio, et al., 1981; Biggio, 1983; Concas, et al., 1985) have reported a consistent reduction in the B_{max} of low affinity GABA receptors after a brief exposure to footshock that is accompanied by an increase in the apparent affinity of [^{3}H]GABA for the remaining population of low affinity receptors. The effects of stress were reported to be more robust in "handling-habituated" than "naive" animals (Biggio, et al., 1981). Subsequently, these changes were attributed to an action at benzodiazepine receptors, since the in vitro addition of diazepam (5 μM reversed this effect, and the benzodiazepine antagonist Ro 15-1788 (100 nM) blocked the effect of diazepam. However, Biggio and colleagues assayed [^{3}H]GABA binding in the presence of 50 mM Cl$^-$, which, by analogy to our findings, might reflect an alteration at the chloride ionophore rather than at GABA receptors. Further, assessment of GABA receptors using [^{3}H]GABA or muscimol binding may not accurately reflect that subpopulation of low affinity (EC$_{50}$of GABA ~1 μM [Fig. 2] in the absence of halide ions) GABA receptors linked to the benzodiazepine receptor complex. Electrophysiological studies (Gallager, et al., (1984); Chan and Farb (1985) support the hypothesis that the subpopulation of GABA receptors which are functionally linked to benzodiazepine receptors is most accurately assessed by measuring GABA enhanced [^{3}H]benzodiazepine binding. Thus, it is possible that stress-induced changes in GABA receptors reported by some laboratories may reflect changes in the GABA-gated chloride channel rather than the GABA receptor. This hypothesis is currently under investigation in our laboratory.

Other evidence supporting a compensatory stress-induced change in GABA-gated, benzodiazepine receptor coupled chloride channels is derived from the similar effects of stress and in vitro application of benzodiazepines on [^{35}S]TBPS binding (Trullas, et al., in preparation). One possible explanation for this mimickry could be the stress-induced release of an endogenous modulator of the benzodiazepine receptor complex. GABA can mimic the effects of pentobarbital on halide-enhanced [^{3}H]flunitrazepam binding (i.e. produce an increase in the potency and efficacy of chloride ions) (Supavilai, et al., 1982) (Fig. 7), so it might be argued that this phenomenon is due to stress-induced increases in GABA turnover. However, GABA will reduce, rather than increase the number (and does not markedly affect the affinity) of [^{35}S]TBPS binding sites in our well-washed tissue preparation (Wong, et al., 1984). Thus, it is unlikely that an increase in GABA release or turnover could account for the effects of stress on [^{35}S]TBPS binding. The apparent in vitro mimickry of stress by benzodiazepines is currently under investigation using in vivo techniques.

In toto, these results suggest that the benzodiazepine receptor complex can be altered by stress. The very rapid change in the number and apparent affinity of [^{35}S]TBPS binding sites may seem somewhat surprising. However, our studies (Havoundjian, et al., 1986c) suggest that cage convulsants such as TBPS and picrotoxin bind directly to GABA-gated chloride channels. Thus, these rapid changes in the characteristics of [^{35}S]TBPS binding provoked by stress may reflect changes in the functional state (i.e. permeability and number) of GABA-gated channels. The speed at which these changes occur - preceeding the earliest measureable rises in ACTH, plasma corticosterone (Cook, et al., 1977; Guillemin, et al., 1977), or other "classical" neuro-transmitters (cf. Ikeda and Nagatsu, 1985) that have been implicated in stress suggest the benzodiazepine receptor complex may play a key role in modulating the involvement of both neurotransmitters and neurohormones in stress and anxiety.

ACKNOWLEDGMENTS

Hratchia Havoundjian was a Howard Hughes Medical Research Scholar in the Laboratory of Bioorganic Chemistry, NIADDK during these studies. Ramon Trullas received financial support from the Comite Conjunto Hispano-Norte-americano. Mr. M. Jackson provided excellent technical assistance.

REFERENCES

Bansky, G., Meier, P., Ziegler, W., Walser, H., Schmid, M. and Huber, M., Reversal of hepatic coma by benzodiazepine antagonist (Ro 15-1788), Lancet June 8, 1985, 1324-1325.

Bassett, M., Mullen, K., Skolnick, P. and Jones, E., Amelioration of hepatic encephalopathy by pharmacological antagonism of the GABA-benzodiazepine receptor complex in a rabbit model of fulminant hepatic failure, Hepatology, submitted (1986).

Biggio, G., The action of stress, B-carbolines, diazepam, and Ro 15-1788 on GABA receptors in the rat brain. In: Benzodiazepine recognition site ligands: Biochemistry and Pharmacology. G. Biggio and E. Costa (eds.), Raven Press, New York, p. 105-119 (1983).

Biggio, G., Corda, M., Concas, A., Demontis, G., Rosetti, Z. and Gessa, G., Rapid changes in GABA binding induced by stress in different areas of rat brain, Brain Res. 229:363-369 (1981).

Braestrup, C., Nielsen, M., Nielsen, E. and Lyon, M., Benzodiazepine receptors in brain as affected by different experimental stresses: the changes are small and not unidirectional, Psychopharmacol. 65:273-277 (1979).

Buchsbaum, M., Davis, C. and Bunney, W., Naloxone alters pain perception and somatosensory evoked potentials in human subjects, Nature 270:620-622 (1977).

Chan, C. and Farb, D., Modulation of neurotransmitter action: control of the γ-aminobutyric acid response through the benzodiazepine receptor, J. Neurosci. 5:2365-2372 (1985).

Concas, A., Corda, M. and Biggio, G., Involvement of benzodiazepine recognition sites in the footshock-induced decrease of low affinity GABA receptors in the rat cerebral cortex, Brain Res. 341:50-56 (1985).

Cook, D., Kendall, J., Greer, M. and Kramer, R., The effect of acute or chronic ether stress on plasma ACTH concentration in the rat, Endocrin. 93: 1019-1024 (1977).

Corda, M., Blaker, W., Mendelson, W., Guidotti, A. and Costa, E., B-Carbolines enhance shock-induced suppression of drinking in the rat, Proc. Natl. Acad. Sci. USA 80:2072-2078 (1983).

Corda, M., Ferrari, M., Guidotti, A., Kondel, D. and Costa, E., Isolation, purification and partial sequence of a neuropeptide (diazepam-binding inhibitor) precursor of an anxiogenic putative ligand for benzodiazepine recognition site, Neurosci. Lett. 47:319-325 (1984).

Costa, T., Rodbard, D. and Pert, C., Is the benzodiazepine receptor coupled to a chloride ion channel? Nature 277:315-317 (1979).

Crawley, J., Marangos, P., Paul, S., Skolnick, P. and Goodwin, F., Purine-benzodiazepine interaction: Inosine reverses diazepam-induced stimulation of mouse exploratory behavior, Science 211:725-727 (1981).

Dantzer, R. and Perio, A., Behavioral evidence for partial agonist properties of Ro 15-1788, a benzodiazepine receptor antagonist, Eur. J. Pharmacol. 81:655-658 (1982).

deCarvalho, L., Grecksch, G., Chapouthier, G. and Rossier, J., Anxiogenic and non-anxiogenic benzodiazepine antagonists, Nature 301:64-66 (1983).

Dorow, R., Horowski, R., Paschelke, G, Amin, M. and Braestrup, C., Severe anxiety induced by FG 7142, a B-carboline ligand for benzodiazepine receptors, Lancet 9:98-99 (1983).

Ferrero, P., Guidotti, A., Conti-Tronconi, B. and Costa, E., A brain octadecaneuropeptide generated by tryptic digestion of DBI (diazepam binding inhibitor) functions as a proconflict ligand of benzodiazepine recognition sites, Neuropharmacol. 23:1359-1362 (1984).

Ferrero, P., Santi, M., Conti-Tronconi, B., Costa, E. and Guidotti, A., A study of an octadecaneuropeptide derived from diazepam binding inhibitor (DBI): biological activity and presence in rat brain, Proc. Natl. Acad. Sci. USA 83:827-831 (1986).

File, S., Animal anxiety and the effects of benzodiazepines. In: Pharmacology of benzodiazepines. Usdin, E., Skolnick, P., Tallman, J., Greenblatt, D. and Paul, S. (eds.), MacMillan Press, Ltd., London, pp. 355-363 (1982).

File, S., Lister, R. and Nutt, D., The anxiogenic action of benzodiazepine antagonists, Neuropharmacol. 21:1033-1037 (1982).

File, S. and Lister, R., Interactions of ethyl-B-carboline-3-carboxylate and Ro 15-1788 with CGS 8216 in an animal model of anxiety, Neurosci. Letts. 39:91-94 (1983).

File, S. and Pellow, S., The effects of putative anxiogenic compounds, FG 7142, CGS 8216, and Ro 15-1788 on the rat corticosterone response, Physiol. and Behav. 35:587-590 (1985).

Gallager, D., Lakoski, J., Gonsalves, S. and Rauch, S., Chronic benzodiazepine treatment decreases postsynaptic GABA sensitivity, Nature 308:74-76 (1984).

Gavish, M. and Snyder, S., Gamma-aminobutyric acid and benzodiazepine receptors: copurification and characterization, Proc. Natl. Acad. Sci. USA 78:1939-1942 (1981).

Grecksch, G., deCarvalho, L., Venault, P., Chapouthier, G. and Rossier, J., Convulsions induced by submaximal dose of pentylenetetrazole in mice are antagonized by the benzodiazepine antagonist Ro 15-1788, Life Sci. 32:2579-2584 (1983).

Grimm, V. and Hershkowitz, M., The effect of chronic diazepam treatment on discrimination performance and [^{3}H]flunitrazepam binding in the brains of shocked and nonshocked rats, Psychopharmacol. 74:132-136 (1981).

Guidotti, A., Corda, M. and Costa, E., Strategies for the isolation and characterization of an endogenous effector of the benzodiazepine recognition sites. In: Benzodiazepine recognition site ligands: biochemistry and pharmacology. G. Biggio and E. Costa (eds.), Raven Press, New York, pp. 95-103 (1983a).

Guidotti, A., Forchetti, C., Corda, M., Kondel, D., Bennett, C. and Costa, E., Isolation, characterization and purification of homogeneity of an endogenous polypeptide with agonistic action on benzodiazepine receptors, Proc. Natl. Acad. Sci. USA 80:3531-3535 (1983b).

Guillemin, R., Vargo, T., Rossier, J., Minick, S., Ling, N., Rivier, C., Vale, W. and Bloom, F., B-Endorphin and adrenocorticotropin are secreted concomitantly by the pituitary gland, Science 197:1367-1369 (1977).

Haefely, W., Antagonists of benzodiazepines: functional aspects. In: Benzodiazepine recognition site ligands: biochemistry and pharmacology. G. Biggio and E. Costa (eds.), Raven Press, New York, pp. 73-93 (1983).

Havoundjian, H., Paul, S. and Skolnick, P., Rapid, stress induced modification of the benzodiazepine receptor coupled chloride ionophore, Brain Res., in press (1986a).

Havoundjian, H., Paul, S. and Skolnick, P., Acute, stress-induced changes in the benzodiazepine/GABA receptor complex are confined to the chloride ionophore, J. Pharmacol. Exp. Ther., in press (1986b).

Havoundjian, H., Paul, S. and Skolnick, P., The permeability of γ-aminobutyric acid gated chloride channels is described by the binding of a cage convulsant, [^{35}S]t-butylbicyclophosphorothionate, Proc. Natl. Acad. Sci. USA, submitted (1986c).

Havoundjian, H., Cohen, R., Paul, S. and Skolnick, P., Differential sensitivity of "central" and "peripheral" benzodiazepine receptors to phospholipase A$_2$, J. Neurochem. 46:804-811 (1986d).

Higgitt, A., Lader, M. and Fonagy, P., The effects of the benzodiazepine antagonist, Ro 15-1788 on psychophysiological, performance and subjective measures in normal subjects, Psychopharmacol., in press (1986).

Hunkeler, W., Mohler, H., Pieri, L., Polc, P., Bonetti, E., Cumin, R., Schaffner, R. and Haefely, W., Selective antagonists of benzodiazepines, Nature 290:514-516 (1981).

Ikeda, M. and Nagatsu, T., Effect of short term swimming stress and diazepam on 3,4-dihydroxyphenylacetic acid (DOPAC) and 5-hydroxyindoleacetic acid (5-HIAA) levels in the caudate nucleus: an in vivo voltammetric study, Naunyn-Schmiedeberg's Arch. Pharmacol. 331:23-26 (1985).

Ikonomidou, C., Turski, L., Klockgether, T., Schwarz, M. and Sontag, K., Effects of methyl-B-carboline-3-carboxylate, Ro 15-1788, and CGS 8216 on muscle tone in genetically spastic rats, Eur. J. Pharmacol. 113:205-213 (1985).

Insel, T., Ninan, P., Aloi, J., Jimerson, D., Skolnick, P. and Paul, S., A benzodiazepine receptor-mediated model of anxiety. Studies in non-human primates and clinical implications, Arch. Gen. Psychiatry 41:741-750 (1984).

Kajima, M., LaSalle, G. and Rossier, J., The partial benzodiazepine agonist properties of Ro 15-1788 in pentylenetetrazole-induced seizures in cats, Eur. J. Pharmacol. 93:113-115 (1985).

Karobath, M. Supavilai, P. and Borea, P., Distinction of benzodiazepine receptor agonists and inverse agonists by binding studies in vitro. In: Benzodiazepine recognition site ligands: biochemistry and pharmacology. G. Biggio and E. Costa (eds.), Raven Press, New York, pp. 37-45 (1983).

Lane, J., Crenshaw, C., Guerin, G., Cherek, D. and Smith, J., Changes in biogenic amine and benzodiazepine receptors correlated with conditioned emotional response and its reversal by diazepam, Eur. J. Pharmacol. 83:183-190 (1982).

Levine, J., Gordon, N., Jones, R. and Fields, H., The narcotic antagonist naloxone enhances clinical pain, Nature 272:826-827 (1978).

Lippa, A., Klepner, C., Yunger, L., Sano, M., Smith, W. and Beer, B., Relationship between benzodiazepine receptors and experimental anxiety in rats, Pharmacol. Biochem. & Behav. 9:853-856 (1978).

MacDonald, J., Barker, J., Paul, S., Marangos, P. and Skolnick, P., Inosine may be an endogenous ligand for benzodiazepine receptors in cultured spinal neurons, Science 205:715-717 (1979).

Marangos, P., Paul, S., Parma, A., Goodwin, F., Syapin, P. and Skolnick, P., Purinergic inhibition of diazepam binding to rat brain in vitro, Life Sci. 24:851-858 (1979).

Marangos, P., Martino, A., Paul, S. and Skolnick, P., The benzodiazepines and inosine antagonize caffeine-induced seizures, Psycopharmacol. 72 269-274 (1981).

Medina, J., Novas, M., Wolfman, C., Levi de Stein, M. and De Robertis, E., Benzodiazepine receptors in rat cerebral cortex and hippocampus undergo rapid and reversible changes after acute stress, Neurosci. 9:331-335 (1983).

Mohler, H. and Okada, T., Benzodiazepine receptor: demonstration in the central nervous system, Science 198:849-851 (1977).

Nicoll, R. and Wojtowicz, J., The effects of pentobarbital and related compounds on frog motoneurons, Brain Res. 191:225-237 (1980).

Nielsen, M., Braestrup, C. and Squires, R., Evidence for a late evolutionary appearance of brain-specific benzodiazepine receptors: an investigation of 18 vertebrate and 5 invertebrate species, Brain Res. 141:342-346 (1978).

Ninan, P. Insel, T., Cohen, R., Cook, J., Skolnick, P. and Paul, S., Benzodiazepine receptor-mediated experimental "anxiety" in primates, Science
218:1332-1334 (1982).

Pieri, L. and Biry, P., Isoniazid-induced convulsions in rats: effects of
Ro 15-1788 and B-CCE, Eur. J. Pharmacol. 112:355-362 (1985).

Schoch, P., Harling, P., Takacs, B., Stahli, C. and Mohler, H., A GABA/benzo-
diazepine receptors complex from bovine brain: purification, reconstitution, and immunological characterization, J. Receptor Res. 4:189-200
(1984).

Schulz, D. and MacDonald, R., Barbiturate enhancement of GABA-mediated inhibition and activation of chloride ion conductance: correlation with
anticonvulsant and anesthetic actions, Brain Res. 209:177-188 (1981).

Schwartz, R., Thomas, J., Kempner, E., Skolnick, P. and Paul, S., Radiation
inactivation studies of the benzodiazepine/GABA/chloride ionophore
receptor complex, J. Neurochem. 5:2963-2970 (1985).

Scollo-Lavizzari, G. and Steinmann, E., Reversal of hepatic coma by benzodiazepine antagonist (Ro 15-1788), Lancet, June 8, 1985, 1324.

Sigel, E., Stephenson, F., Mamalaki, C. and Barnard, E., A γ-aminobutyric
acid/benzodiazepine receptor complex of bovine cerebral cortex, J. Biol.
Chem. 258:6965-6971 (1983).

Skerritt, J. and MacDonald, R., Benzodiazepine Ro 15-1788: electrophysiological evidence for partial agonist activity, Neurosci. Letts. 43:321-
326 (1983).

Skolnick, P., Syapin, P., Paugh, B., Marangos, P. and Paul, S., Inosine, an
endogenous ligand of the brain benzodiazepine receptor antagonizes PTZ-
induced seizures, Proc. Natl. Acad. Sci. USA 76:1515-1518 (1979).

Skolnick, P., Marangos, P. and Paul, S., Putative endogenous ligands of the
benzodiazepine receptor. In: Anxiolytics. Malick, J., Enna, S., and
Yamamura, H. (eds.), Raven Press, New York, pp. 41-53 (1983).

Slater, P. and Longman, D., Effects of diazepam and muscimol on GABA-mediated
neurotransmission: interaction with inosine and nicotinamide, Life Sci.
25:1963-1967 (1979).

Soubrie, P., Theibot, M., Jobert, A., Montasruc, J., Hery, F. and Hamon, M.,
Decreased convulsant potency of picrotoxin and pentylenetetrazole and
enhanced [^{3}H]flunitrazepam cortical binding following stressful manipulations in rats, Brain Res. 189:505-517 (1980).

Squires, R. and Braestrup, C., Benzodiazepine receptors in rat brain, Nature
266:732-734 (1977).

Squires, R., Casida, J., Richardson, M. and Saederup, E., [^{35}S]t-butylbi-
cyclophosphorothionate binds with high affinity to brain specific sites
coupled to gamma-aminobutyric acid-A and ion recognition sites, Mol.
Pharmacol. 23:326-336 (1983).

Supavilai, P., Mannonen, A., Collins, J. and Karobath, M., Anion dependent
modulation of ^{3}H-muscimol binding and of GABA-stimulated ^{3}H-flunitraze-
pam binding by picrotoxin and related CNS convulsants, Eur. J. Pharmacol. 81:687-691 (1982).

Takeuchi, A. and Takeuchi, N., A study of the actions of picrotoxin on the
inhibitory neuromuscular junction of the crayfish, J. Physiol. 205:377-
391 (1969).

Trullas, R., Havoundjian, H., Paul, S. and Skolnick, P., Environmentally induced modification of benzodiazepine-GABA receptor coupled chloride channel, <u>Psychopharmacol</u>., submitted (1986).

Vogel, J., Beer, B. and Clody, D., A simple and reliable conflict procedure for testing antianxiety agents, <u>Psychopharmacol</u>. 21:1-7 (1971).

Witkin, J. and Barrett, J., Benzodiazepine-like effects of inosine on punished behavior of pigeons, <u>Pharmacol. Biochem. & Behav</u>. 24:121-125 (1986).

Wong, D., Threlkeld, P., Bymaster, F. and Squires, R., Saturable binding of ^{35}S-t-butylbicyclophosphorothionate to the sites linked to the GABA receptor and the interaction with GABAergic agents, <u>Life Sci</u>. 34:853-860 (1984).

INTERACTIONS OF THE ALKYL-ETHER-PHOSPHOLIPID, PLATELET ACTIVATING FACTOR (PAF) WITH PLATELETS, NEURAL CELLS, AND THE PSYCHOTROPIC DRUGS TRIAZOLOBENZODIAZEPINES

E. Kornecki, R. H. Lenox, D. H. Hardwick, J. A. Bergdahl
and Y. H. Ehrlich

The Neuroscience Research Unit, Department of Psychiatry
University of Vermont College of Medicine, Burlington
Vermont 05405

SUMMARY

PAF-acether, a naturally occurring phospholipid, is a potent activator of various biological processes, including platelet aggregation. The mechanisms of action of PAF are largely unknown. We have found that the psychotropic triazolobenzodiazepine drugs, alprazolam and triazolam, potently (IC_{50} < 1μM) inhibit PAF-induced shape change, aggregation and secretion of human platelets. These effects are specific for PAF-activation, since the responses of human platelets to other agonists (ADP, thrombin, epinephrine, collagen, arachidonate and the Ca^{++} ionophore, A23187) are not inhibited by these triazolobenzodiazepines. The action of triazolobenzodiazepines on PAF-induced platelet function has clinical relevance, especially in diseases where enhanced platelet aggregability may lead to thrombosis and atherosclerosis. In addition, the ability of triazolobenzodiazepines to inhibit other PAF-mediated cellular-responses, such as anaphylactic shock or bronchoconstriction, suggests that these drugs may be useful in preventing several known pathophysiological effects of PAF.

The specific antagonism of PAF action by psychotropic drugs also suggests that PAF or PAF-like phospholipids may play a role in neuronal function. This possibility was tested by examining the effects of PAF on neural cells of the clonal line NG108-15, grown in culture in a chemically defined, serum-free medium. Low concentrations of PAF (0.5-2.5μM) induced neurite extension in NG108-15 cells, whereas higher concentrations (>3μM) were cytotoxic. Using NG108-15 cells preloaded with aequorin, it was found that PAF causes an increase in intracellular ionized calcium concentration, which is dependent on the presence of extracellular calcium. These results suggest that PAF-induced Ca^{++} uptake may play a role in neuronal development, and that circulating PAF may contribute to the neuronal degeneration caused by the exposure of neural tissues to blood in situations such as spinal cord injury, trauma, or stroke.

INTRODUCTION

Platelet activating factor (PAF; PAF-acether) is a naturally occurring

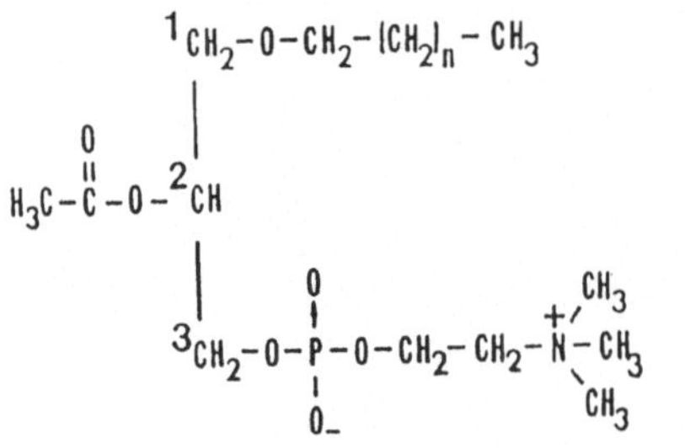

$$^1CH_2-O-CH_2-(CH_2)_n-CH_3$$

$$H_3C-\overset{O}{\overset{\|}{C}}-O-{}^2CH$$

$$^3CH_2-O-\overset{O}{\underset{O_-}{\overset{|}{\underset{|}{P}}}}-O-CH_2-CH_2-\overset{+}{N}\overset{\diagup CH_3}{\underset{\diagdown CH_3}{-CH_3}}$$

FIG. 1: The structure of Platelet-Activating Factor (PAF) as 1-0-alkyl-2-acetyl-sn- glycero-3-phosphocholine.

neutral ether-phospholipid (1-0-alkyl, 2-acetyl-sn-glyceryl-3 phosphoryl-choline) also referred to as acetyl-glyceryl-ether-phosphorylcholine (AGEPC) (Demopoulos et al., 1979; Benveniste et al., 1979; Blank et al., 1979). The structure of PAF is shown in Fig. 1. PAF acts as an extra-cellular messenger in the communication of a variety of cells (for recent reviews, see Lee and Winslow, 1986). PAF is released in vivo during IgE-induced anaphylaxis (Vargaftig et al., 1981; Princkard et al., 1979). PAF is released in vitro by basophils upon immunological challenge, as well as from platelets, mast cells, monocytes, neutrophils, eosinophils, endothelial cells, and alveolar macrophages in response to specific stimuli (Marcus et al., 1981; Alam et al., 1983; Chignard et al., 1979; Lynch et al., 1979; Mencia-Huerta and Benveniste, 1979; Lotner et al., 1980; Betz and Henson, 1980; Clark et al., 1980; Arnoux et al., 1980; Camussi et al., 1977; Camussi et al., 1983; Lee et al., 1984). PAF is a potent mediator of inflammation, possesses anti-hypertensive activity, and induces bronchoconstriction and contraction of smooth muscles (Vargaftig et al., 1981; Pinckard et al., 1979; Blank et al., 1979; Findlay et al., 1981). PAF is one of the most powerful platelet activators known, inducing platelet shape change, aggregation and secretion (Demopoulos et al., 1979; Benveniste et al., 1975; Henson, 1977; McManus et al., 1981; Chesney et al., 1982). The metabolism of PAF has been studied in neutrophils (Mueller et al., 1983; Alonso et al., 1982; Chilton et al., 1984), platelets (Benveniste et al., 1982; Chap et al., 1981) and macrophages (Ninio et al., 1981; Albert and Snyder, 1983) where it is thought to be synthesized from preformed pools of 1-0-alkyl-2-acyl-glyceryl phosphorylcholine. In the proposed cycle, PAF is formed through the action of phospholipase A_2 which removes arachidonic acid from position 2 of the precursor, followed by the action of acetyltransferase which inserts an acetyl group in its place. PAF is then degraded to lyso-PAF by a specific acetylhydrolase. Highest levels of enzymes for the biosynthesis and degradation of PAF are found in the kidney; relatively high levels of these enzymes are also found in brain tissue (Blank et al., 1981).

Benzodiazepines are used clinically as both anxiolytic and hypnotic agents. Derivatives with a triazole ring, called triazolobenzodiazepines, have specialized clinical applications. The primary clinical use of the triazolobenzodiazepine alprazolam is in the treatment of panic disorder with or without agoraphobia (Cohn, 1981; Chouinard et al., 1982; Sheehan et al., 1984). A second triazole derivative of benzodiazepine, triazolam, is prescribed as an effective, short acting sleep medication (Pakes et al., 1981). The structures of alprazolam and triazolam are shown in Fig. 2. Benzodiazepine receptors are present primarily in brain (see Skolnick et al., this volume), but also have been found in platelets (Wang et al., 1980; Benavides et al., 1984; Moingeon et al., 1984). Calcium-dependent protein phosphorylation activity can be inhibited by benzodiazepines (DeLorenzo et al., 1981). Since platelet activation by agonists involves calcium-dependent protein phosphorylation systems (Hathaway and Adelstein, 1979; Nishizuka, 1984), we

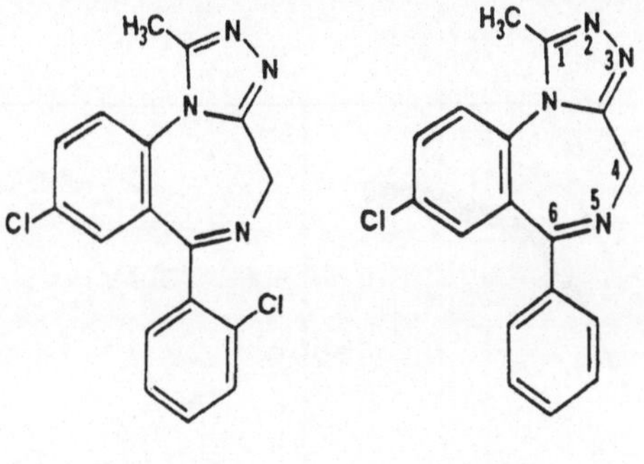

TRIAZOLAM   ALPRAZOLAM

FIG. 2: Structure of alprazolam and triazolam. The chemical formula for
        alprazolam (Xanax) is 8-chloro-1-methyl-6-phenyl-4H-s-triazolo-
        [4,3-a][1,4] benzodiazepine and for triazolam (Halcion) is 8-
        chloro-6-(o-chlorophenyl)-1-methyl-4H-s-triazolo[4,3-a] [1,4]-
        benzodiazepine.

tested whether platelet activation by various agonists could be inhibited
by benzodiazepines.

Inhibition of PAF induced platelet activation by triazolobenzodiazepines

     The benzodiazepines most commonly prescribed, diazepam and chlordiaze-
poxide, were tested at a concentration range of 0.1-40μM and had no detectable effects on either the initial velocity nor on the extent of aggregation of human platelets stimulated by ADP, thrombin, epinephrine, the calcium ionophore A23187, collagen or arachidonate. Similarly, neither alprazolam nor triazolam had significant effects on platelet aggregation induced
by these agonists. In contrast, these triazolobenzodiazepines potently
inhibited the activation of platelets by PAF; benzodiazepines without a triazole ring had very low potency in inhibiting PAF-action (Kornecki et al.,
1984). We have concluded from these results that alprazolam and triazolam
are potent and specific antagonists of PAF-induced cellular responses. Some
of these studies are summarized here, together with select examples of the
most salient findings.

     Our initial studies have determined that addition of alprazolam or
triazolam to human platelet-rich-plasma (PRP) blocked PAF induced platelet
activation (Kornecki et al., 1984). Subsequent experiments were performed
to examine whether the effects of alprazolam and triazolam on platelet function are mediated by a plasma factor, or involve direct interaction with
platelets. This was done using washed platelets obtained from PRP by gel
filtration. In Fig. 3, panels A and C show aggregometer tracings which
demonstrate that PAF-induced shape change (shown as a decrease in light
transmission) and aggregation (shown as an increase in light transmission)
of human gel-filtered platelets (GFP) were completely inhibited by the preincubation of GFP with 10μM alprazolam. In panels B and D of Fig. 3, it
can be seen that secretion from platelet granules, monitored by measuring
ATP release from GFP with a Lumi aggregometer, was completely inhibited by
this concentration of alprazolam. Fifty percent inhibition of PAF-induced
aggregation of GFP occurred at a concentration of alprazolam of 500nM and
the IC_{50} value for inhibition of PAF-induced shape change by alprazolam was
2.3μM. Diazepam, chlordiazepoxide and a preferred ligand for the peripheral
benzodiazepine receptor, Ro5-4864, had no significant effects on PAF-induced
aggregation of gel-filtered platelets when tested at concentrations in which
both alprazolam and triazolam produced complete inhibition.

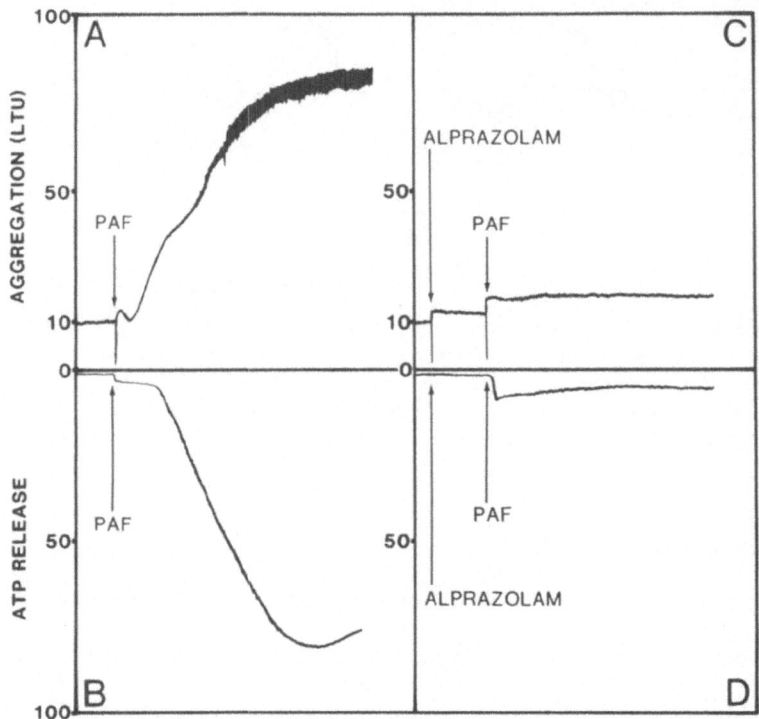

FIG. 3: Inhibition of PAF-induced aggregation and granular ATP release by
alprazolam in washed human platelets. Aliquots (0.45ml) of gel-
filtered platelets were suspended in Tyrode's - albumin-containing
buffer at a final calcium concentration of 0.5mM. The suspensions
were stirred at 1,200 rpm in a ChronoLog Lumi Aggregometer for 1
min at 37°C in the absence (panels A and B) and in the presence
(panels C and D) of 10μM alprazolam. Platelet aggregation is shown
as an increase in light transmission units (LTU). The release of
ATP from platelet granules was monitored by adding 100μl of a luci-
ferin/luceiferase reagent (ChronoLog) into the incubation mixture.
Platelet aggregation and secretion were initiated by the addition
of 51 nanomolar PAF (final concentration).

The inhibition of PAF-induced platelet aggregation by alprazolam and
triazolam could be completely reversed by increasing the concentration of
PAF. The concentration of PAF required for inducing 50% of the maximal initial velocity of aggregation of GFP in a typical experiment is approximately
7nM. This value shifted to 35nM, 150nM and 400nM PAF in the presence of
10μM, 20μM and 40μM alprazolam, respectively (Kornecki, et al., 1984). Such
parallel shifts of the dose-response curve to the right indicate competitive
antagonism.

We found similar results to those described above with washed rabbit
platelets. At a final PAF concentration of 0.049nM, the IC_{50} values for
inhibition of PAF-induced aggregation and shape change by alprazolam were
375nM and 750nM, respectively.

Fibrinogen receptor exposure and fibrinogen binding are necessary for
platelet aggregation to physiological agonists (Kornecki et al., 1983). We
have found that ^{125}I-fibrinogen (Fg) binding to PAF-stimulated platelets was

TABLE 1: IC$_{50}$ Values of Neuroactive Drugs Tested for Inhibition of PAF-Induced Aggregation of Human Platelets

Drugs Tested	IC$_{50}$ Values
Triazolam	0.3μM
Alprazolam	0.5μM
Ro 5-4864	45.0μM
Chlordiazepoxide	67.0μM
Trazodone	110.0μM
Diazepam	130.0μM
Adinazolam	> 500.0μM

Gel-filtered platelets (2 x 10^8/ml) were prepared and assayed as described by Kornecki et al. (1984). The final concentration of PAF was 51nM.

inhibited by alprazolam (200ng ^{125}I-Fg Bound/10^8 platelets in the absence of alprazolam vs 39ng ^{125}I-Fg Bound/10^8 platelets in the presence of 20μM alprazolam). However, this inhibition is not likely to be due to direct blockade of fibrinogen binding sites, since fibrinogen binding and fibrinogen-dependent aggregation of platelets to various agonists other than PAF was not blocked by alprazolam, nor by triazolam.

A selective ligand of the peripheral benzodiazepine receptor, Ro5-4864, was about 100 times less potent than triazolam or alprazolam in inhibiting the action of PAF (Table 1). In addition, Ro5-4864 did not block the inhibitory effect of alprazolam or triazolam on PAF-induced platelet function. Therefore, binding to the platelet benzodiazepine receptor does not appear to be involved in the inhibition of PAF-induced platelet activation by triazolobenzodiazepines, and it is more likely that direct interaction of these drugs with PAF-receptor(s) is the underlying mechanism. Studies of the structure-function relationships have shed light on the structural requirements necessary for inhibiting the action of PAF by triazolobenzodiazepines. Triazolam and alprazolam inhibited the aggregation of washed human platelets stimulated by PAF with IC$_{50}$ values of approximately 300nM and 500nM, respectively. In contrast, benzodiazepines without the triazole ring, diazepam and chlordiazepoxide were far less potent and had IC$_{50}$ values of 130μM and 67μM, respectively. These results indicate that the triazole ring is required for inhibition of PAF action. However, presence of the triazole ring per-se is not sufficient for PAF-antagonism, since we found that trazodone, a phenylpiperazine derivative of triazolopyridine, had an IC$_{50}$ value of 110μM. Adinazolam, a triazolobenzodiazepine which contains a dimethylamino substitution at the C-1 position of the traizole ring, was ineffective as an PAF antagonist (IC$_{50}$>500μM) (Table 1).

Based on all these findings, we conclude that triazolobenzodiazepines with appropriate substitutions can serve as potent antagonist of PAF-induced cellular responses. Such drugs may prove to be useful in the treatment of several clinical disorders resulting from the pathophysiological activity of PAF. Indeed, Darius et al. (1986) reported that PAF-mediated anaphylaxis in guinea pigs could be prevented by alprazolam administration in vivo.

The PAF-stimulated shape change and secretion of platelet granular contents have been associated with increased phosphatidyl inositol breakdown, phosphatidic acid formation and the phosphorylation of endogenous platelet proteins (Lapetina, 1982; Ieyasu et al., 1982). However, these events are activated by other platelet agonists as well. Investigation of the process by which alprazolam and triazolam selectively inhibit PAF-mediated platelet activation may shed light on biochemical mechanisms specific for the action of PAF, as well as the mechanism of action of these triazolobenzodiazepines. Within the context of this volume, it is important to emphasize that the interaction with PAF represents a unique property of alprazolam and triazolam, which is not shared by benzodiazepines without the triazole ring. It can be speculated that among the benzodiazepines, alprazolam and triazolam have found specialized psychotherapeutic uses (Cohn, 1981; Pakes et al., 1981; Sheehan et al., 1984) since their mechanism of action is unique. This unique mechanism may involve interaction with processes mediated by PAF or PAF-like ether phospholipids in the central nervous system. This suggestion is supported by reports (Blank et al., 1981; Snyder, 1986) that the brain contains relatively high levels of enzymes for the biosynthesis and degradation of alkyl ether phospholipids, such as PAF. However, effects of PAF on the function of neuronal cells have never been reported. We have initiated this line of investigation by examining the interactions of PAF with a homogenous population of a cloned neuronal cell line.

Cloned cells of the line designated NG108-15 have been extensively investigated in recent years as a model system in studies of various neuronal functions (Nirenberg et al., 1983; Hamprecht et al., 1985). Under standard growth conditions, NG108-15 cells divide rapidly and exhibit an exponential growth curve. However, when these cells are exposed in culture to agents which induce neuronal differentiation, their growth is arrested, and they develop a multitude of morphological, physiological and biochemical properties characteristic of mature neurons (Nirenberg et al., 1983; Hamprecht et al., 1985; Ehrlich et al., 1986). One of the main morphological manifestations of this differentiation is the extension of long processes called neurites. Differentiation of NG108-15 cells in culture can be quantitated by determining the percentage of cells that, under the influence of a differentiating agent, cease multiplying and extend neurites. Using this criterion, our initial studies have determined that PAF can induce neuronal differentiation.

NG108-15 cells are grown in our laboratory in a DMEM medium supplemented with fetal calf serum (FCS) as described by Nirenberg et al. (1983) except that the FCS concentration used is 5% or 1% instead of 10%. In select experiments, we maintain the cells in a chemically defined, serum-free medium (Ehrlich et al., 1986). We have observed that addition of PAF to the growth medium of cultured NG108-15 cells induced growth arrest and morphological differentiation, expressed as enhanced neurite extension. The differentiating effect of PAF on NG108-15 cells was found to be both concentration- and time-dependent. In a medium containing 1% FCS, maximal differentiation required treatment of the cells with $10\mu M$ PAF for a period of 4-5 days. With lower concentrations of PAF, the rate of differentiation was slower (Kornecki et al., 1986). Moreover, we have found that in serum-free medium, PAF was about 10 times more potent in enhancing neurite extension than with NG108-15 cells grown in serum-containing media. Under these conditions, incubation of NG108-15 cells with as little as 50 nanomolar PAF induced measurable enhancement of neurite extension. As can be seen in Fig. 4, when differentiation was measured by assessing the percentage of cells bearing long neurites, maximal effects were observed after exposing NG108-15 cells in a chemically-defined medium to $4\mu M$ PAF for 3 days.

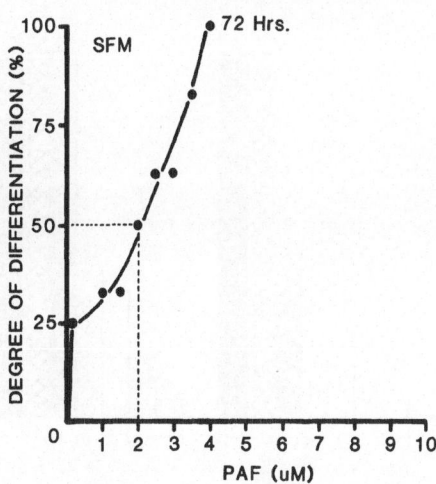

FIG. 4: Differentiation by PAF. PAF-induced neurite extension in NG108-15
cells. Cells were plated in a chemically-defined, serum-free
medium. Twenty-four hours after plating, the PAF concentrations
shown above were added. Degree of differentiation was assessed
after 72 hours exposure by determining the percentage of neurite-
bearing cells.

Increasing the dose of PAF beyond the levels which induce maximal neurite extension revealed that, at relatively high concentrations, PAF is
cytotoxic to NG108-15 cells. The cytotoxic effects of PAF were dose and
time-dependent. In serum containing medium, a differentiating concentration
of PAF (10μM) had no cytotoxic effect on NG108-15 cells. A slightly higher
level, 12.5μM PAF, caused growth arrest and differentiation within the first
24 hours of treatment, but no cell death. After 48 hours exposure of the
cells to 12.5μM PAF in a serum-containing medium, cytotoxicity was in evidence and cell death rapidly progressed over time. At higher PAF concentrations, this process was accelerated (Kornecki et al., 1986). As might be
expected, with cells grown in a chemically defined, serum-free medium, neu-
rotoxicity was evident at lower concentrations of PAF than those observed in
media containing serum. Fig. 5 shows the time- and concentration-dependency
of these neurotoxic effects. With 3.5μM PAF, 50% of the cells lost viability after about 96 hours exposure. With 5μM PAF the same effect was
observed after 48 hours, and no cell survived more than a 24 hour exposure
to 10μM PAF. This finding is consistent with a previous report which
described a cytotoxic effect of PAF on non-neuronal cells, such as human
leukemia cells, polymorphonuclear neutrophils and human skin fibroblasts
(Hoffman, et al., 1984).

The biochemical mechanisms underlying the neurotropic effects of PAF
reported above are currently under investigation. In platelets, PAF induces
an influx of calcium ions which can be measured both by [45]calcium uptake
from the extracellular fluid into the platelet (Lee, T-C., et al., 1983),
and by measuring PAF-induced increase in fluorescence intensity of platelets
preloaded with the calcium binding dye quin 2 (Valone and Johnson, 1985).
Such mechanism may operate also in neurons. In initial studies we have
loaded NG108-15 cells with the calcium probe aequorin, and were able to measure a significant PAF-induced rise in free intracellular Ca^{++}-ions concentrations in these neural cells (Kornecki and Ehrlich, 1986). The PAF-
induced increase in intracellular free Ca^{++} could be completely blocked by
including EDTA instead of Ca^{++} in the extracellular medium, suggesting that
the mechanism involves a PAF-induced influx of extracellular Ca^{++} into the
cells.

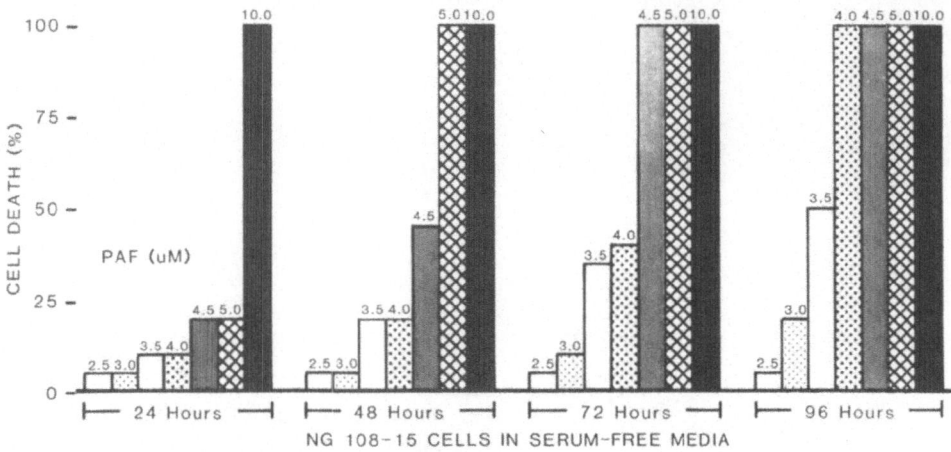

FIG. 5: Cytotoxicity of PAF. Time course and dose-dependency of PAF
effects on the viability of NG108-15 cells. NG108-15 cells were
plated in serum-free medium. Twenty-four hours after plating, the
cells were incubated with various concentrations of PAF ranging
from 3.0µM to 10µM. The viability of the cells was determined
microscopically by trypan blue exclusion at various time points
ranging from 24 to 96 hours after PAF addition.

Previous studies of non-neural cells have shown that increased levels
of intracellular ionized calcium is an early event in processes leading to
either cell differentiation or programmed cell-death, depending on the magnitude of the initial rise (Leonard and Salpeter, 1979). Our findings are
consistent with these reports, and suggest that such series of events may
be induced by PAF in neuronal cells. The effects of PAF on the morphology
and viability of NG108-15 cells may be triggered by increases in intracellular ionized calcium concentrations induced by PAF. Such an initial rise
in calcium may initiate a series of events which lead to either physiological responses or pathological consequences, depending on the PAF concentration to which the neural cells are exposed.

CONCLUSIONS AND FUTURE DIRECTIONS

The studies summarized here have demonstrated that neurotropic drugs,
the triazolobenzodiazepines triazolam and alprazolam, are potent and spe-
ific antagonists of PAF-induced platelet responses. These drugs may there-
ore be of clinical relevance in disorders involving platelet activation such
as in thrombosis and atherosclerosis. Our results, as well as reports by
other investigators (see "Introduction" of this chapter), suggest that
alprazolam and triazolam may be useful for the prevention of various patho-
hysiological effects of PAF (Lee and Winslow, 1986). We have also shown
here that PAF itself can exert specific effects on the function of neural
cells. In the mature brain, PAF or PAF-like phospholipids may be involved
in cell-cell communication. Antagonism of this process by triazolam or
alprazolam may be involved in mechanisms underlying the therapeutic effects
of these drugs. In the developing nervous system, low levels of PAF pro-
uced by neuronal and/or glial cells may play a role in cell differentiation
or maturation. Indeed, neuronal dysfunction has been found in individuals
with Zellweger Syndrome, a genetic disorder in which there is lack of
enzymes needed for the synthesis of ether-phospholipids (Datta et al.,

484

1984). The advanced methodology available today for the determination and
separation of various phospholipids from brain (see Bazan et al., this
volume), combined with the very sensitive bioassay of platelet aggregation
by PAF (sensitivity in the nanomolar range) make it feasible to carry out
experiments that could detect the presence of PAF in brain tissue. In addi-
ion, under certain pathophysiological conditions, neurons may be exposed to
high levels of PAF originating from cells in the circulation. This could
occur whenever cells in the nervous system are exposed to blood. Our finding, that at such levels PAF is cytotoxic to neural cells, suggests the possibility that PAF may be one of the circulating factors responsible for the
irreversible neuronal degeneration which is associated with spinal cord
injury, trauma or stroke. Further studies of the responsiveness of neuronal
cells to PAF could provide new insights into the role of ether-phospholipids
in the regulation of neurophysiological events, and could reveal new information relevant for the treatment of certain pathological states in the nervous system.

REFERENCES

Alam, I., Smith, J. B. and Silver, M. J., Human and rabbit platelets form
platelet-activating factor in response to calcium ionophore, Thromb.
Res. 30:71-79 (1983).

Albert, D. H. and Snyder, F., Biosynthesis of 1-alkyl-2-acetyl-sn-glycero-
3-phosphocholine (platelet activating factor) from 1-alkyl-2-acyl-sn-
glycero-3-phosphocholine by rat alveolar macrophages, J. Biol. Chem.
258:97-102 (1983).

Alonso, F., Gil, M. G., Sanchez-Crespo, M. and Mato, J. M., Activation of
1-O-alkyl-2-lyso-glycero-3-phosphocholine acetyl-CoA transferase during
phagocytosis in human polymorphonuclear leukocytes, J. Biol. Chem. 257:
3376-3378 (1982).

Arnoux, B., Darval, D. and Benveniste, J., Release of platelet-activating
factor (PAF-acether) from alveolar macrophages by the calcium ionophore
A23187 and phagocytosis, Eur. J. Clin. Invest. 10:437-441 (1980).

Benveniste, J., Chignard, M., Le Couedic, J. P. and Vargaftig, B. B., Biosynthesis of platelet-activating factor (PAF-acether). II. Involvement of phospholipase A_2 in the formation of PAF-acether and lyso-PAF-
acether from rabbit platelets, Throm. Res. 25:375-385 (1982).

Benveniste, J., Tence, M., Varenne, P., Bidault, J., Boullet, C. and Polon-
sky, J., Semi-synthese et structure proposee du facteur activant les
plaquettes (PAF): PAF-acether, un alkyl ether analogue de la lysophos-
phatidyl-choline, C. R. Acad. Sci. (D), 289:1037-1040 (1979).

Benveniste, J., LeCouedic, J. P. and Kamoun, P., Aggregation of human platelets by platelet-activating factor, Lancet 1:344-345 (1975).

Benavides, J., Quarteronet, D., Plouin, P-F., Imbault, F., Phan, T., Uzan,
A., Renault, C., Dubroeuco, M-C., Gueremy, C. and LeFur, G., Characterization of peripheral type benzodiazepine binding sites in human and
rat platelets by using [^{3}H]PK 11195. Studies in hypertensive patients,
Biochem. Pharm. 33:2467-2472 (1984).

Betz, S. J. and Henson, P. M., Production and release of platelet-activating factor (PAF): dissociation from degranulation and superoxide production in the human neutrophil, J. Immunol. 125:2756-2763 (1980).

Blank, M. L., Lee, T-C., Fitzgerald, V. and Snyder, F., A specific acetylhydrolase for 1-alkyl-2-acetyl-sn-glycero-3-phosphocholine (a hypotensive and platelet-activating lipid), J. Biol. Chem. 256:175-178 (1981).

Blank, M. L., Snyder, F. Byers, L. W., Brooks, B. and Muirhead, E. E., Anti-
hypertensive activity of an alkyl ether analog of phosphatidylcholine,
Biochem. Biophys. Res. Commun. 90:1194-1200 (1979).

Camussi, G. Aglietta, M., Malavasi, F., Bussolino, F., Piacibello, W., Sana-
vio, F. and Tetta, C., Release of platelet-activating factor from human
endothelial cells. In: Benveniste, J. and Arnoux, B., eds. Platelet

activating factor. Amsterdam: Elsevier, 1983:83-90 (INSERM symposium No. 23) (1983).

Camussi, G., Mencia-Huerta, J. M. and Benveniste, J., Release of platelet-activating factor and histamine. I. Effect of immune complexes, complement, and neutrophils on human and rabbit mastocytes and basophils, Immunology 33:523-534 (1977).

Chap, H., Mauco, G., Simon, M., Benveniste, J. and Douste-Blazy, L., Biosynthetic labelling of platelet activating factor from radioactive acetate by stimulated platelets, Nature 289:312-314 (1981).

Chesny, C. M., Pifer, D. D., Byers, L. W. and Muirhead, E. E., Effect of platelet-activating factor (PAF) on human platelets, Blood 59:583-585 (1982).

Chignard, M., LeCouedic, J. P., Tence, M., Vargaftig, B. B. and Benveniste, J., The role of platelet-activating factor in platelet aggregation, Nature (London) 279:799-800 (1979).

Chilton, F. H., Ellis, J. M., Olson, S. C. and Wykle, R. L., 1-O-alkyl-2-arachidonoyl-sn-glycero-3-phosphocholine. A common source of platelet-activating factor and arachidonate in human polymorphonuclear leukocytes, J. Biol. Chem. 259:12014-12019 (1984).

Chilton, F. H., O'Flaherty, J. T., Ellis, J. M., Swendsen, C. L. and Wykle, R. L., Metabolic fate of platelet-activating factor in neutrophils, J. Biol. Chem. 258:6357-6361 (1983).

Chouinard, G. Annable, L., Fontaine, R. and Solyom, L., Alprazolam in the treatment of generalized anxiety and panic disorders: a double-blind placebo-controlled study, Psychopharmacol. 77:229-233 (1982).

Clark, P. O., Hanahan, D. J. and Pinckard, R. N., Physical and chemical properties of platelet-activating factor obtained from human neutrophils and monocytes and rabbit neutrophils and basophils, Biochem. Biophys. Acta 628:69-75 (1980).

Cohn, J. B., Multicenter double-blind efficacy and safety study comparing alprazolam, diazepam and placebo in clinically anxious patients, J. Clin. Psychiatry 42:347-351 (1981).

Datta, N. S., Wilson, G. N. and Hajra, A. K., Deficiency of enzymes catalyzing the biosynthesis of glycerol-ether lipids in Zellweger syndrome, New Eng. J. Med. 311:1080-1083 (1984).

Darius, H., Lefer, D. J., Smith, J. B. and Lefer, A. M., Role of platelet-activating factor-acether in mediating guinea pig anaphylaxis, Science 232:58-60 (1986).

DeLorenzo, R. J., Burdette, S. and Holderness, J., Benzodiazepine inhibition of the calcium-calmodulin protein kinase system in brain membranes, Science 213:546-549 (1981).

Demopoulos, C. A., Pinckard, R. N. and Hanahan, D. J., Platelet-activating factor: evidence for 1-O-alkyl-2-acetyl-sn-glyceryl-3-phosphorylcholine as the active component (a new class of lipid chemical mediators), J. Biol. Chem. 254:9355-9358 (1979).

Ehrlich, Y. H., Davis, T. B., Bock, E., Kornecki, E. and Lenox, R. H., Ecto-protein kinase activity on the external surface of neural cells, Nature 320:67-70 (1986).

Findlay, S. R., Lichtenstein, L. M., Hanahan, D. J. and Pinckard, R. N., Contraction of guinea pig ileal smooth muscle by acetyl glyceryl ether phosphorylcholine, Amer. J. Physiol. 241:C130-C133 (1981).

Hamprecht, B., Glaser, T., Reiser, G., Bayer, E. and Propst, F., Culture and characteristics of hormone-responsive neuroblastoma X glioma hybrid cells, Methods in Enzymology 109:316-341 (1985).

Hathaway, D. R. and Adelstein, R. S., Human platelet myosin light chain kinase requires the calcium-binding protein calmodulin for activity, Proc. Natl. Acad. Sci. USA 76:1653-1657 (1979).

Henson, P. M., Activation of rabbit platelets by platelet-activating factor derived from IgE-sensitized basophils, J. Clin. Invest. 60:481-490 (1977).

Hoffman, D. R., Hajdu, J. and Snyder, F., Cytotoxicity of platelet activating factor and related alkyl-phospholipid analogs in human leukemia cells, polymorphonuclear neutrophils, and skin fibroblasts, Blood 63: 545-552 (1984).

Ieyasu, H., Takai, Y., Kaibuchi, K., Sawamura, M. and Nishizuka, Y., A role of calcium-activated, phospholipid-dependent protein kinase in platelet-activating factor-induced serotonin release from rabbit platelets, Biochem. Biophys. Res. Commun. 108:1701-1708 (1982).

Johnson, P. C., Ware, J. A., Cliveden, P. B., Smith, M., Dvorak, A. M. and Salzman, E. W., Measurement of ionized calcium in blood platelets with the photoprotein aequorin, J. Biol. Chem. 260:2069-2076 (1985).

Kornecki, E. and Ehrlich, Y. H., Stimulation of calcium uptake in neural cells by the alkyl-ether phospholipid platelet activating factor (PAF), Society for Neuroscience, Vol. 12:1244 (1986).

Kornecki, E., Lenox, R. H., Hardwick, D. H. and Ehrlich, Y. H., A role for platelet activating factor (PAF) in neuronal function: specific inhibition of platelet activation by triazolobenzodiazepines and interactions of PAF with cultured neural cells. In: New Horizons in Platelet Activating Factor Research, (eds. Lee, M. L. and Winslow, C. M.) John Wiley and Sons Ltd. (1986).

Kornecki, E., Tuszynski, G. P. and Niewiarowski, S., Inhibition of fibrinogen receptor-mediated platelet aggregation by heterologous anti-human platelet membrane antibody, J. Biol. Chem. 258:9349-9356 (1983).

Kornecki, E., Ehrlich, Y. H. and Lenox, R. H., Platelet-activating factor induced aggregation of human platelets specifically inhibited by triazolobenzodiazepines, Science 226:1454-1456 (1984).

Lapetina, E. G., Platelet-activating factor stimulates the phosphatidylinositol cycle, J. Biol. Chem. 247:7314-7317 (1982).

Lee, M. L. and Winslow, C. M., (Eds.). New Horizons in Platelet Activating Factor Research, John Wiley & Sons, Ltd., (1986).

Lee, T., Lenihan, D. J., Malone, B., Roddy, L. L. and Wasserman, S. I., Increased biosynthesis of platelet activating factor in activated human eosinophils, J. Biol. Chem. 259:5526-5530 (1984).

Lee, T-C., Malone, B. and Snyder, F., Stimulation of calcium uptake by 1-alkyl-2-acetyl-sn-glycero-3-phosphocholine (Plaatelet-Activating Factor) in rabbit platelets: possible involvement of the lipoxygenase pathway, Archiv. Biochem. Biophys. 223:33-39 (1983).

Leonard, J. P. and Salpeter, M. M., Agonist-induced myopathy at the neuromuscular junction is mediated by calcium, J. Cell Biol. 82:811-819 (1979).

Lotner, G. Z., Lynch, J. M., Betz, S. J. and Henson, P. M., Human neutrophil-derived platelet activating factor, J. Immunol. 124:676-684 (1980).

Lynch, J. M., Lotner, G. Z., Betz, S. J. and Henson, P. M., The release of a platelet-activating factor by stimulated rabbit neutrophils, J. Immunol. 123:1219-1226 (1979).

Marcus, A. J., Safier, L. B., Ullman, H. L., Wong, K. T. H., Broekman, J., Weksler, B. B. and Kaplan, K. L., Effects of acetylglyceryl ether phosphorylcholine on human platelet function in vitro, Blood 58:1027-1030 (1981).

McManus, L. M., Hanahan, D. J. and Pinckard, R. N., Human platelet activation by acetyl glyceryl ether phosphorylcholine, J. Clin. Invest. 67: 903-906 (1981).

Mencia-Huerta, J. M. and Benveniste, J., Platelet-activating factor and macrophages. I. Evidence for the release from rat and mouse peritoneal macrophages and not from mastocytes, Eur. J. Immunol. 9:409-415 (1979).

Moingeon, Ph., Dessaux, J. J., Fellous, R., Alberici, G. F., Bidart, J. M., Motte, Ph. and Bohuon, C., Benzodiazepine receptors on human blood platelets, Life Sci. 35:2003-2009 (1984).

Mueller, H. W., O'Flaherty, J. T. and Wykle, R. L., Biosynthesis of platelet activating factor in rabbit polymorphonuclear-neutrophils, J. Biol. Chem. 258:6213-6218 (1983).

Ninio, E., Mencia-Huerta, J. M., Heymans, F. and Benveniste, J., Biosynthesis of platelet activating factor. I. Evidence for an acetyltranferase activity in murine macrophages, Biochem. Biophys. Acta. 710:23-31 (1982).

Nirenberg, M., Wilson, S., Higashida, H., Rotter, A., Krueger, K., Busis, N., Ray, R., Kenimer, J. G. and Adler, M., Modulation of synapse formation by cyclic adenosine monophosphate, Science 222:793-799 (1983).

Nishizuka, Y., Turnover of inositol phospholipids and signal transduction, Science 225:1365-1370 (1984).

Pakes, G. E., Brogden, R. N., Heel, R. C. and Speight, T. M., Triazolam: a review of its pharmacological properties and therapeutic efficacy in patients with insomnia, Drugs 22:81-110 (1981).

Pinckard, R. N., Farr, R. S. and Hanahan, D. J., Physicochemical and functional identity of rabbit platelet-activating factor (PAF) released in vivo during IgE anaphylaxis with PAF released in vitro from IgE sensitized basophils, J. Immunol. 123:1847-1857 (1979).

Sheehan, D. V., Coleman, J. H., Greenblatt, D. J., Jones, K. J., Levine, P. H., Orsulak, P. J., Peterson, M., Schnildkraut, J. J., Uzogara, E. and Watkins, D., Some biochemical correlates of panic attacks with agoraphobia and their response to a new treatment, J. Clin. Psychopharmacol. 4:66-75 (1984).

Snyder, F., The significance of dual pathways for the biosynthesis of platelet-activating factor alkyl-acetylglycerols and lyso-PAF as immediate precursors. In: New Horizons in Platelet Activating Factor Research (eds. Lee, M. L. and Winslow, C. M.), John Wiley and Sons, Ltd. (1986).

Valone, F. H. and Johnson, B., Modulation of cytoplasmic calcium in human platelets by the phospholipid platelet-activating factor 1-0-alkyl-2-acetyl-sn-glycero-3-phosphorylcholine, J. Immunol. 134:1120-1124 (1985).

Vargaftig, B. B., Chignard, M., Benveniste, J., Lefort, J. and Wal, F., Background and present status of research on platelet-activating factor (PAF-acether), Ann. NY Acad. Sci. 370:119-137 (1981).

Wang, J. K. T., Taniguchi, T. and Spector, S., Properties of [^{3}H]diazepam binding sites on rat blood platelets, Life Sci. 27:1881-1888 (1980).

THE SEROTONIN-NOREPINEPHRINE LINK HYPOTHESIS OF AFFECTIVE

DISORDERS: RECEPTOR-RECEPTOR INTERACTIONS IN BRAIN

Fridolin Sulser and Elaine Sanders-Bush

Departments of Pharmacology and Psychiatry, Vanderbilt
University School of Medicine, Nashville, Tennessee
37232

INTRODUCTION

The morphological organization of central monoamine systems has suggested a functional linkage of noradrenergic and serotonergic neurons ever since
Dahlström and Fuxe (1964) demonstrated, by means of a sensitive fluorescence
method, the neuronal localization of norepinephrine (NE) and serotonin (5HT)
and mapped NE and 5HT containing cell bodies and terminals in the central
nervous system. These and more recent studies show that NE and 5HT neurons
form monosynaptic pathways between the lower brain stem and the cerebral
cortex (Dahlstrom and Fuxe, 1964; Anden et al., 1966; Moore and Bloom, 1979;
Levitt and Moore, 1978; Morrison et al., 1982; Consolazione and Cuello,
1982). With the introduction of immunohistochemical techniques, it became
evident that NE and 5HT containing neurons project to the entire neuraxis
and, within the cerebral cortex, to all six cortical layers, though some
topographical and species differences exist (Lindvall and Bjorklund, 1984;
Levitt et al., 1984). Early work also indicated a high degree of catecholamine innervation of 5HT cell bodies in the raphe nuclei (Dahlstrom and
Fuxe, 1964) and the existence of a 5HT innervation of NE cell bodies in the
locus coeruleus (Pickel et al., 1975). The details of the monoaminergic
pathways in the central nervous system have been authoritatively reviewed
(Jacobowitz, 1978; Moore and Bloom, 1979; Consolazione and Cuello, 1982).
As pointed out by Fuxe et al. (1978), there is little doubt that the 5HT
and NE neurons are linked together neuroanatomically and that they influence one another at various points on the neural axis. Recent experiments
conducted in animals with specific lesions of serotonergic neurons and
psychopharmacological studies with antidepressant drugs have generated new
support for this aminergic link at the level of the beta-adrenoceptor
coupled adenylate cyclase system. The evidence for this biomolecular link
at the level of central beta-adrenoceptor systems and the role of biochemical effector systems activated by NE and 5HT via the corresponding receptors
are discussed below.

The Role of Noradrenergic and Serotonergic Neuronal Input in the Regulation
of Density and Function of Central Noradrenergic Receptor Systems

The evidence for the crucial role of the endogenous agonist, NE, in the
process of the in vivo desensitization of the NE receptor coupled adenylate
cyclase system in brain by antidepressants has been previously reviewed
(Sulser, 1983). Using the terminology of Sibley and Lefkowitz (this mono-

graph), the antidepressant induced desensitization of the beta adrenoceptor coupled adenylate cyclase system in brain seems to be of the homologous type though the fate of beta adrenoceptors that are no longer labelled by [^{3}H]dihydroalprenolol ([^{3}H]-DHA) is not known.

While agonist induced homologous desensitization in peripheral tissues (e.g. frog erthrocytes) involves the uncoupling of beta adrenoceptors from the guanine nucleotide regulatory protein(s) followed by sequestration of the receptors away from their effector system, it is not clear if the in vivo desensitization process in brain also involves a two-step phenomenon. Thus, the use of conventional methods to probe coupling mechanisms following treatment with antidepressants e.g. the determination of GTP induced shifts of agonist competition binding curves, has not revealed unequivocal changes in the interconversion of the high affinity state of the receptor to a form that recognizes NE with lower affinity (O'Donnell and Frazer, 1985; unpublished results from this laboratory). However, when Okada et al. (1986) studied the time course for activation of adenylate cyclase by Gpp(NH)p in cortical membrane preparations, they found that the isoproterenol induced facilitation of the enzyme activation by the guanine nucleotide was absent in preparations from rats treated with desipramine (DMI), even at an early time when no significant changes were observed in the density of beta adrenoceptors. These results seem to indicate that DMI-like drugs can modify beta adrenoceptor-N_s-protein and/or N_s protein catalyst interactions.

In view of the morphological organization of central aminergic systems, it was of interest to study the consequences of selective lesions of 5HT neurons with the neurotoxin 5,7-dihydroxytryptamine (5,7-DHT) on the density and the function of beta adrenoceptors coupled to adenylate cyclase. As can be seen in Table 1, selective lesions of 5HT neurons significantly increase the density of beta adrenoceptors (labelled with [^{3}H]-DHA) in cortex, limbic forebrain and striatum. The increase in beta adrenoceptor density is particularly marked in the striatum which lacks noradrenergic innervation. This change in density confirms previously reported results (Janowsky et al., 1982). Stockmeier et al. (1985) also reported a marked increase in the density of beta adrenoceptors following 5,7-DHT lesions in frontal cortex and hippocampus and showed that serotonergic denervation selectively affects the beta adrenoceptor component of noradrenergic receptor systems with no change in the binding of [^{3}H]-prazosin to alpha$_1$ adrenoceptors and of p-[^{3}H]-amino-clonidine to alpha$_2$-adrenoceptors. In frontal cortex, limbic forebrain and hippocampus, Scatchard analysis of saturation isotherms indicates that the increase in ligand binding to beta adrenoceptors following 5,7-DHT lesions is due to an increase in the B_{max} value without changes in the K_d value. In the striatum on the other hand, the marked increase in the B_{max} value of beta adrenoceptors following lesions is accompanied by a slight but significant increase in the K_d value for [^{3}H]-DHA. Moreover, in agreement with previous results from this and other laboratories (Brunello et al., 1982; Janowsky et al., 1982; Drumbrille-Ross and Tang, 1983; Nimgonakar et al., 1985), the impairment of serotonergic neuronal input to cortex and limbic forebrain made beta adrenoceptors resistant to down-regulation by desipramine (DMI). It is also evident from the data in Table 1 that the availability of the endogenous agonist NE is a prerequisite for down-regulation of beta adrenoceptors by DMI and for the desensitization of the NE sensitive adenylate cyclase. Thus, DMI fails to down-regulate the beta adrenoceptors in the striatum of control animals and also does not cause desensitization of the striatal NE receptor coupled adenylate cyclase (Table 2). These results unequivocally indicate the importance of the availability of both NE and 5HT for the antidepressant induced down-regulation of beta adrenoceptors. In contrast to beta adrenoceptor antagonist binding which is largely unaffected by 5,7-DHT induced lesions, an impaired serotonergic neuronal input markedly alters agonist affinity to beta adrenoceptors (Table 3). Selective lesions with 5,7-DHT or a reduction in the synaptic availa-

490

TABLE 1. Consequences of 5,7-DHT Induced Lesions of Serotonergic Neurons on Regional Alterations of Beta Adrenoceptors by Desipramine (DMI)

	Sham-lesioned			
	Saline		DMI	
	B_{max} fmol/mg protein ± SEM	K_d nM ± SEM	B_{max} fmol/mg protein ± SEM	K_d nM ± SEM
Frontal Cortex	112 ± 4 (5)	1.2 ± 0.2 (5)	81 ± 4[e] (4)	1.5 ± 0.2 (4)
Limbic Forebrain	91 ± 7 (6)	1.5 ± 0.2 (6)	66 ± 8[a] (6)	1.8 ± 0.4 (6)
Striatum	112 ± 8 (9)	1.1 ± 0.1 (9)	95 ± 4 (9)	0.9 ± 0.1 (9)

	5,7-DHT lesioned			
	Saline		DMI	
	B_{max} fmol/mg protein ± SEM	K_d nM ± SEM	B_{max} fmol/mg protein ± SEM	K_d nM ± SEM
Frontal Cortex	172 ± 14[d] (5)	1.7 ± 0.04[b] (5)	143 ± 4 (4)	2.1 ± 0.2 (4)
Limbic Forebrain	143 ± 17[c] (4)	1.5 ± 0.3 (4)	172 ± 19 (4)	2.3 ± 0.4 (4)
Striatum	218 ± 11[f] (9)	1.6 ± 0.1[d] (9)	197 ± 11 (9)	1.4 ± 0.1 (9)

Serotonergic neurons were lesioned by intraventricular injection of 5,7-dihydroxytryptamine (5,7-DHT) as described in Methods. Ten days after lesioning (or sham-lesioning), the animals were treated daily with desipramine (15 mg/kg i.p.) or saline for 3 weeks. The animals were sacrificed 24 hours after the last injection and Scatchard analyses performed as described in Methods. Numbers in parentheses indicate the number of animals. (From Manier et al., 1986).
[a] $p < 0.05$ DMI vs. Saline (sham-lesioned)
[b] $p < 0.05$ 5,7-DHT lesioned vs. sham-lesioned (saline)
[c] $p = 0.01$ 5,7-DHT lesioned vs. sham-lesioned (saline)
[d] $p < 0.005$ 5,7-DHT lesioned vs. sham-lesioned (saline)
[e] $p = 0.001$ DMI vs. saline (sham-lesioned)
[f] $p < 0.001$ 5,7 DHT lesioned vs. sham-lesioned (saline)

TABLE 2. Responsiveness of the Noradrenaline (NA) Sensitive Adenylate
Cyclase System Following Desipramine (DMI) in Frontal Cortex,
Limbic Forebrain and Striatum of the Rat Brain.

	pmol cyclic AMP/mg protein ± SEM 1 week	
	Saline	DMI
Frontal Cortex	71.2 ± 3.9 (10)	49.4 ± 6.4[b] (9)
Limbic Forebrain	139.7 ± 11.2 (9)	77.4 ± 8.1[c] (8)
Striatum	25.5 ± 1.3 (10)	28.8 ± 3.1 (8)

	pmol cyclic AMP/mg protein ± SEM 2 weeks	
	Saline	DMI
Frontal Cortex	112.8 ± 9.4 (5)	48.9 ± 4.2[c] (5)
Limbic Forebrain	174.3 ± 15.0 (5)	111.2 ± 13.3[a] (4)
Striatum	46.9 ± 4.3 (5)	45.7 ± 9.6 (5)

	pmol cyclic AMP/mg protein ± SEM 3 weeks	
	Saline	DMI
Frontal Cortex	117.6 ± 13.0 (5)	68.1 ± 6.2[b]
Limbic Forebrain	211.9 ± 10.8 (5)	157.7 ± 12.6[a]
Striatum	60.1 ± 5.4 (5)	57.8 ± 8.3

Desipramine (DMI) was administered daily (15 mg/kg i.p.) for 1,2, and 3
weeks. The animals were sacrificed by decapitation 24 hours after the last
dose and cyclic AMP responses to 100 M NA determied in slices of the
respective brain areas as described under Methods. The data represent the
stimulated levels of the cyclic nucleotide ± SEM. Basal levels of cyclic
AMP in pmol/mg protein ± SEM were as follows: Cortex, 23.4 ± 2.0 (24);
Limbic Forebrain, 32.1 ± 2.4 (21); Striatum, 10.3 ± 1.0 (23). The numbers
in parentheses indicate the number of animals, each animal sample being
analyzed in duplicate. (From Manier et al., 1986). [a]p<0.02; [b]p<0.01;
[c]p<0.001.

bility of 5HT by inhibition of tryptophan hydroxylase with p-chlorophenylalanine (PCPA) causes a 5- to 6-fold decrease in the affinity of beta adrenoceptors for the beta adrenoceptor agonist, isoproterenol. Interestingly, in
the striatum, which lacks noradrenergic innervation, selective lesions of
5HT neurons by 5,7-DHT exaggerate the decrease in agonist affinity i.e. the
K_d values for isoproterenol are increased as much as 40 fold (Manier et al.,
1986). Whether or not these changes in agonist affinity are sufficient to
account for the unchanged stimulation of adenylate cyclase by a presumed
maximal concentration of NE (Table 3) under conditions of an increased density of beta adrenoceptors remains to be further studied. While studies
from our laboratory show that the neurohormonal responsiveness to both NE
(Table 2) and isoproterenol (Janowsky et al., 1982) is reduced following
DMI to essentially the same degree in tissue preparations from animals with
selective lesions of serotonergic neurons and from control animals, Barbaccia
et al. (1983) reported that imipramine fails to attenuate the NE sensitive
adenylate cyclase following destruction of 5HT terminals with 5,7-DHT. The
reasons for these different results are not known although different drugs
and different analytical methodology were used in the two studies.

To ascertain whether the antidepressant induced change in the density
of beta adrenoceptors and in the sensitivity of the NE receptor coupled adenylate cylase system are reversible, the synaptic availability of 5HT was
acutely reduced by inhibiting tryptophan hydroxylase with PCPA in rats with
a DMI induced down-regulation of beta adrenoceptors. PCPA, given for two
days, nullifies - despite the continuous administration of DMI- the decrease
in the density of beta adrenoceptors, but does not significantly alter the
DMI induced attenuation of the NE sensitive adenylate cylase (Manier et al.,
1984). Since PCPA, unlike the neurotoxin 5,7-DHT, does not alter the concentration of possible cotransmitters, substance P or TRH (Gilbert et al.,
1981), the results with PCPA indicate more definitely the co-requirement of
5HT in the process of in vivo down regulation of central beta adrenoceptors
by DMI-like antidepressants. Brunello et al. (1985) have provided additional
evidence that the decrease in the B_{max} value of [^{3}H]-DHA binding to cortical

TABLE 3. The Effect of Impaired Serotonergic Neuronal Activity on Beta
Adrenoceptor Agonist and Antagonist Affinities

	K_d [^{3}H]-DHA nM ± SEM	IC_{50} Isoproterenol μM ± SEM
Controls (saline)	1.31 ± 0.13	0.43 ± 0.03
5,7-DHT lesioned	1.56 ± 0.08	2.19 ± 0.43*
p-Chlorophenylalanine	1.49 ± 0.14	2.49 ± 0.54*

Serotonergic neurons were lesioned with intraventricular 5,7-dihydroxytryp-
tamine (5,7-DHT) as described by Janowsky et al. (1982). The animals were
sacrificed 17-19 days after lesioning and K_d values for the antagonist [^{3}H]-
dihydroalprenolol (DHA) and IC_{50} values for the agonist isoproterenol determined. P-chlorophenylalanine (200 mg/kg) was given i.p. for 2 days and the
animals were sacrificed 24 hours after the last dose. The concentration of
[^{3}H]-DHA in the competition binding curve was 3.0 nM and that of isoproterenol varied from 0.001 and 100 μM. The IC_{50} value designates the concentration of isoproterenol that displaces 50% of specific [^{3}H]-DHA binding. The
K_d values were determined from saturation isotherms (Scatchard Analysis).
(From Sulser et al., 1986). * p<0.005

membranes of rat brain induced by long term administration of DMI can be prevented by concomitant treatment with PCPA. Nimgaonkar et al., (1985) confirmed that selective lesions of 5HT neurons by 5,7-DHT prevented the down-regulation of beta adrenoceptors by a number of antidepressant treatments including ECT, but they failed to observe similar effects with PCPA. The reason for this negative data is unknown. As do 5,7-DHT induced lesions, PCPA treatment causes a marked decrease in agonist affinity (Table 3) of beta adrenoceptors without alteration of antagonist affinity. Resupplying 5HT by circumventing the inhibition of tryptophan hydroxylase by co-administration of 5-hydroxytryptophan (5HTP) with PCPA converted a DMI resistant to a DMI responsive beta adrenoceptor population (Manier et al., 1984) and shifted the reduced agonist affinity of beta adrenoceptors towards control values (Table 4). These results are of considerable clinical interest as it has been shown almost a decade ago that PCPA can nullify, within two days, the therapeutic efficacy of antidepressants including tricyclics and MAO inhibitors (Shopsin et al., 1975; 1976). The demonstrated co-requirement of 5HT for a down-regulation of beta adrenoceptors by tricyclic antidepressants and the conversion by 5HTP of a DMI resistant to a DMI responsive beta adrenoceptor population may also provide a rationale for the clinical reports indicating that 5HT precursors potentiate the antidepressant response to tricyclic antidepressants and MAO inhibitors (Coppen and Wood, 1982; Van Praag, 1982; Mendlewicz and Youdim, 1980).

Biochemical Effector Systems Which Are Coupled to 5HT Receptors

Studies of the interactions at the cellular level between second messengers formed by activation of NE and 5HT receptors may elucidate the molecular mechanism of the co-requirement of NE and 5HT in the down-regulation of beta-adrenoceptors. We therefore initiated a series of studies of the signal transduction pathways of central 5HT receptors. These studies show that the first step in signal transduction for at least two central 5HT receptors is hydrolysis of inositol phospholipids in the membrane (Conn and Sanders-Bush, 1984; 1985; Conn et al., 1986). Progress in understanding the phosphoinositide hydrolysis signal transducing system has recently exploded and readers are referred to recent reviews for details (Berridge, 1984; Hirasawa and Nishizuka, 1985). The initial event in receptor activated phosphoinositide hydrolysis is thought to be phospholipase C mediated hydrolysis of phosphatidylinositol-4,5-bisphosphate, which leads to the formation of inositol-1,4,5-trisphosphate (IP3) and diacylglycerol (DAG), both of which function as second messengers. IP3 elevates cytosol calcium levels by releasing intracellular storage pools. Calcium ions then serve as a "third messenger" to evoke responses by activating specific enzymes, e.g. calcium/calmodulin dependent protein kinases. DAG activates protein kinase C, an enzyme which regulates cellular events by phosphorylating key proteins. Evidence is emerging that, among other things, protein kinase C is involved in heterologous receptor desensitization (see Sibley and Lefkowitz, this volume). Another important consequence of the rise in DAG is the release of arachidonic acid, which leads to the formation of prostaglandins and leukotrienes.

The effect of 5HT on phosphoinositide hydrolysis in cerebral cortex and choroid plexus in illustrated in Figure 1. For these studies, tissue phosphoinositides are prelabelled by incubation with [³H]-inositol and 5HT induced release of [³H]-inositol-1-phosphate (IP) is then measured, using the method of Berridge et al. (1982). In cortical slices, 5HT induces a two-fold increase in the release of [³H]-IP with half-maximum effect at a concentration of 1-3 μM. In choroid plexus, the phosphoinositide hydrolysis response to 5HT is much larger (6-fold) and 5HT is 20 to 50 times more potent. As outlined below, additional studies of the 5HT phosphoinositide hydrolysis response in the cerebral cortex and choroid plexus suggest that different receptors mediate these effects.

TABLE 4. Role of Serotonin in the Regulation of Number and Function
β-Adrenoceptors by DMI

	[^{3}H]-DHA Binding			
	B_{max} fmol/mg protein±SEM	% Control	K_d nM±SEM	IC_{50} (-)-isoproterenol M ± SEM
A. Normal Control				
Saline	180 ± 7 (12)	100	1.40 ± 0.11	0.40 ± 0.03 (13)
DMI	116 ± 9 (10)[a]	64[a]	1.25 ± 0.14	0.44 ± 0.03 (6)
B. PCPA Treated				
Saline	238 ± 10 (4)[d]	100	1.49 ± 0.14	2.49 ± 0.54 (6)[d]
DMI	216 ± 12 (12)[ns]	91	2.14 ± 0.24[c]	7.48 ± 1.46 (7)[e]
5HTP + DMI	151 ± 3 (7)[b]	63[b]	1.93 ± 0.25	0.85 ± 0.14 (6)[f]

A. Control animals were given DMI (15 mg/kg twice daily for 2 days & 15 mg/kg/day for an additional 2 days. The animals were sacrificed 24 hrs after the last dose of the drug and Scatchard analyses and competition binding curves performed as described under Methods. The IC_{50} value designates the concentration of (-)-isoproterenol that displaces 50% of [^{3}H]-DHA.

B. Animals were given DMI as above. At day 3, PCPA (200 mg/kg/day) or PCPA + 5HTP (100 mg/kg/day) were administered for two days and the animals were sacrificed 24 hours after the last dose of the drugs and the characteristics of beta adrenoceptors determined as described under Methods. Numbers in parentheses indicate the number of animals. (From Manier et al., 1984). a) $p < 0.001$; DMI vs. Saline; b) $p < 0.001$; 5HTP + DMI vs. DMI; c) $p < 0.0 5$; DMI vs. Saline; d) $p < 0.005$; PCPA vs. Saline controls; e) $p < 0.001$; DMI vs. Saline; f) $p < 0.001$; 5HTP + DMI vs. DMI.

Based on pharmacological studies, we and others have concluded that 5HT stimulated phosphoinositide hydrolysis in cerebral cortex is mediated by $5HT_2$ receptor activation (Conn and Sanders-Bush, 1985; 1986a; Kendall and Nahorski, 1985). Perhaps the most convincing evidence is the finding that K_d values of antagonists at the phosphoinositide hydrolysis linked cortical 5HT receptor correlate significantly (r = 0.98) with K_i values at the $5HT_2$ binding sites (Conn and Sanders-Bush, 1986a). Furthermore, studies of the adaptive regulation of 5HT stimulated phosphoinositide hydrolysis in cortex and of the $5HT_2$ binding site are consistent with the conclusion that the $5HT_2$ site is linked to phosphoinositide hydrolysis. For example, chronic administration of the 5HT antagonist, mianserin, paradoxically induces a down-regulation of $5HT_2$ binding sites and a comparable decrease in the maximal 5HT stimulated phosphoinositide hydrolysis response in cerebral cortex (Conn and Sanders-Bush, 1986a). Studies of 5HT stimulated phospho-inositide hydrolysis in platelets (de Chaffoy de Courcelles et al., 1985) and in aorta (Roth et al., 1984; 1985) are consistent with the conclusion that the $5HT_2$ site utilizes phosphoinositide hydrolysis as its transducing

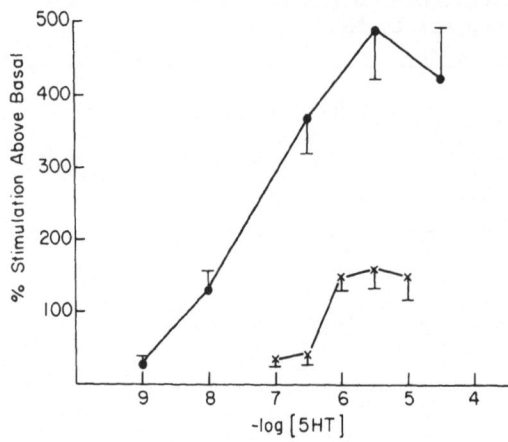

FIG. 1: Effect of increasing concentrations of serotonin upon phosphoinositide hydrolysis in choroid plexus(·--·) and cerebral cortex (X--X).
Serotonin induced release of [^{3}H]-inositol-1-phosphate from phosphoinositides prelabelled with [^{3}H]-inositol was used as a measure
of phosphoinositide hydrolysis. The data are presented as percent
stimulation above basal [^{3}H]-inositol-1-phosphate radioactivity
which was 660 ± 29 cpm in cerebral cortex and 1300 ± 119 cpm in
choroid plexus. Maximum stimulation resulted in 1807 ± 174 cpm
and 7885 ± 832 cpm in cerebral cortex and choroid plexus, respectively. The vertical bars represent standard errors of the means
(N=6). From Conn et al. (1986).

mechanism. Furthermore, in platelets, 5HT evokes calcium fluxes (Affolter
et al., 1982) and stimulates the phosphorylation of two proteins which are
protein kinase C substrates (de Chaffoy de Courcelles et al., 1984). Both
of these effects are blocked by ketanserin, suggesting that 5HT$_2$ receptor
activation leads to the activation of the phosphoinositide hydrolysis signal
cascade.

In the choroid plexus, the effects of 5HT are apparently mediated by
activation of a receptor other than the 5HT$_2$ site, since this binding site
is not present in this tissue (Pazos et al., 1984). Two groups have
recently described a new 5HT binding site, the 5HT$_{1c}$ site, which is highly
localized in the choroid plexus (Pazos et al., 1984; Yagaloff and Hartig,
1985). In Table 5, the pharmacology of the phosphoinositide hydrolysis
response in choroid plexus is compared with the pharmacology of the 5HT$_2$ and
the 5HT-1c binding sites. The K$_i$ value of 5HT at competing for the 5HT$_{1c}$
binding site is similar to its EC$_{50}$ value at the phosphoinositide-linked
receptor in choroid plexus, but not in cerebral cortex. Furthermore, the
rank order and absolute potencies of antagonists at blocking choroidal phosphoinositide hydrolysis and 5HT$_{1c}$ binding are in excellent agreement. We
have therefore proposed that the 5HT$_{1c}$ binding site on choroid plexus epithelial cells is a functional receptor which is linked to phosphoinositide
hydrolysis (Conn et al., 1986). It is possible that this receptor mediates
the effects of 5HT on CSF production (Maeda, 1983).

The phospholipid substrates and the second messenger products of 5HT$_2$

initiated phosphoinositide hydrolysis have not yet been characterized in brain. Phospholipase C-type phosphodiesterases catalyze the hydrolysis of all three of the membrane phosphoinositides. With most phosphoinositide linked receptors, including 5HT stimulated phosphoinositide hydrolysis in insect salivary glands (Berridge et al., 1983; Berridge, 1983), the primary substrate for the activated phospholipase C is phosphatidylinositol-4,5-bis-phosphate leading to the formation of two second messengers, IP_3 and DAG. However, we have been unable to demonstrate 5HT stimulated formation of IP_3 in cerebral cortex, probably because the signal is too weak. Alternatively, perhaps one of the other phosphoinositides is the substrate for the 5HT activated phospholipase C in cerebral cortex. This question has more than theoretical interest, since it would determine whether one or two second messengers were formed. In the choroid plexus, on the other hand, the robust signal has made it possible to study early times and we have been able to demonstrate a $5HT_{1c}$ mediated increase in IP_3 formation, which is detectable within 2.5 minutes and peaks at 5 minutes (Conn and Sanders-Bush, 1986b).

A number of neurotransmitters, including NE, stimulate adenylate cyclase in the choroid plexus. The adenylate cyclase response to NE is robust and it is apparently mediated by activation of beta-adrenoceptors (Nathanson, 1979; Crook et al., 1984). It is not known if the choroidal beta-adrenoceptors are regulated by antidepressants in a manner analogous to those in brain, but if they are, the choroid plexus may provide a useful model system for studying the molecular mechanism of the 5HT/NE interaction.

Molecular Aspects of the 5HT/NE Link in Brain and Its Signal Transduction

The changes reported in agonist properties of beta adrenoceptors following impairment of serotonergic neuronal input are reminiscent of beta adrenoceptors that have been (partially) uncoupled from the guanine nucleotide regulatory protein(s). The rightward shift of competition binding curves for beta agonists by guanine nucleotides can be taken as an index of the capacity of coupling of beta adrenoceptors with the nucleotide regulatory protein(s) (Harden, 1983). It was thus of interest to ascertain whether GTP sensitivity would be altered under conditions of impaired sero-

TABLE 5. Potencies at Serotonin Receptors in Cerebral Cortex and Choroid Plexus[a]

Compound	Phosphoinositide Response		Radioligand binding	
	Cerebral Cortex[b]	Choroid Plexus	$5HT_2$[c]	$5HT_{1c}$
	K_i (nM)		K_i (nM)	
Mianserin	14	12	5.0	5.1
Ketanserin	12	115	3.1	196
Spiperone	2	6200	2.0	4600
	EC_{50} (nM)			
Serotonin	1300	46	6200	94

[a] modified from Conn et al. (1986). [b] from Conn and Sanders-Bush (1986b).
[c] from Conn and Sanders-Bush (1985).

497

tonergic neuronal activity. However, in contrast to the marked GTP induced shift of agonist affinity observed in peripheral systems, and in agreement with results obtained by Hegstrand et al. (1979), the effect of GTP on isoproterenol inhibition of antagonist binding was only marginal or absent in cortical membrane preparations from control rats. In the absence of any demonstration of GTP sensitivity in cortical membrane preparations from control rats, it became impossible to evaluate the predicted lack of GTP sensitivity in membrane preparations from rats with impaired serotonergic neuronal activity. More recently, we have observed moderate GTP sensitivity in cortical beta adrenoceptor preparations by using NE rather than isoproterenol as the displacing agent. When the guanine nucleotide guanyl-5'-imidodiphosphate [Gpp-(NH)p] is present during the binding assay, the competition binding curve for the agonist NE is shifted slightly to the right indicating that the receptor is converted to a form that recognizes NE with lower affinity. Preliminary results derived from competition binding curves in cortical membrane preparations from animals lacking 5HT input (5,7-DHT lesions) indicate low affinity agonist (NE) binding of the receptor that cannot be further shifted by guanine nucleotides (unpublished results). Further experimentation is required, however, to validate these preliminary results.

While the basic phenomena of regulation in receptor number and function (neuronal responsiveness) of the linked 5HT/NE system are now firmly established in the CNS, due to the complexity of brain tissue, the analysis of the molecular mechanisms underlying desensitization and resensitization of adrenergic receptor systems has been accomplished almost entirely in peripheral model systems (see e.g. Sibley and Lefkowitz, this volume). Nevertheless, a fascinating picture of second messengers has emerged in the CNS as a consequence of beta adrenoceptor and 5HT receptor activation by NE and 5HT respectively. Although the role of these second messengers (cyclic AMP, diacyglycerol, inositoltrisphosphate and calcium) in the regulation of central neuronal responsiveness is not established and it has not yet been possible to identify substrate phosphoproteins in brain that reflect the changes in the sensitivity of 5HT/NE receptor systems, in analogy to peripheral model systems as discussed in this symposium, protein kinase mediated protein phosphorylations are most likely the final common pathway. Studies involving the injection of protein kinases or protein kinase inhibitors into neurons have provided direct evidence that activation of protein kinases is an obligatory step in the sequence of events by which extracellular signals produce physiological responses in neurons (Nestler and Greengard, 1983; Nestler et al., 1984; Greengard, this symposium). It remains a challenge for molecular psychopharmacology to elucidate how the various second messenger - protein kinase interactions alter [in concert with changes in phospholipid methylation (Hirata and Axelrod, 1980) and steroid hormone action (Mobley et al., 1983)] neuronal responsiveness and transmembrane signalling of the linked 5HT/NE receptor system.

The Formulation of the "Serotonin/Norepinephrine Link Hypothesis" of Affective Disorders - Conclusions

In this chapter, we have reviewed evidence for a biomolecular link between serotonergic and noradrenergic neuronal systems in brain at the level of the regulation of density and function of beta adrenoceptors that are coupled to adenylate cyclase. Though the final pathways of signal transfer via second messengers produced by NE and 5HT receptor interactions in brain remain to be elucidated, protein phosphorylation via activation of protein kinases is the most likely mechanism for the ultimate regulation of receptor mediated cellular responses (Nestler and Greengard, 1983). Psychopharmacological research strongly supports the notion that the delayed in vivo downregulation of the linked 5HT/NE beta adrenoceptor system by antidepressant treatments reflects a therapeutically relevant biological action. The evidence is as follows: (1) Most (if not all) clinically effective antidepressant treatments cause, upon administration on a clinically relevant time basis, a desensitization of this NE receptor system coupled to adenylate cyclase. (2) Pharmacological interventions which prevent the down-regulation of beta adrenoceptors (e.g. PCPA) have been shown to nullify the therapeutic efficacy of antidepressants. (3) 5HT precursors (e.g. 5-hydroxytryptophan) which convert a DMI resistant to a DMI responsive beta adrenoceptor population have been reported to potentiate the antidepressant effect of tricyclic antidepressants or MAO inhibitors. (4) Supersensitivity of the 5HT/NE beta adrenoceptor system caused by treatment with reserpine or the beta adrenoceptor blocker propranolol, or in patients with adrenocortical insufficiency, has been shown to precipitate depressive reactions in man. Moreover, neurophysiological studies in patients with affective disorders have demonstrated two decades ago that central hyperarousal is a prominent characteristic of both depressive and manic states (Whybrow and Mendels, 1969).

Considering all the evidence - neuroanatomical, neurochemical, neurophysiological, neuropharmacological, clinical - it has been tempting to integrate the various pieces of information into a testable "Serotonin/Norepinephrine Link Hypothesis" of affective disorders (Sulser, 1984), suggesting that the multicomponent beta adrenoceptor system in brain functions as an integrative amplification/adaptation system of a wide variety of vital physiological functions including mood, sleep, arousal, pain, neuroendocrine and central autonomic regulation. Under normal physiological conditions, the system seems to function as a protective adaptive system against excessive oscillations in sensory input. Conversely, an impairment in the regulation of the 5HT/NE receptor-effector system - occurring at any one of the multiple steps involved - is hypothesized to result in maladaptation leading to the precipitation of depressive illness. Successful antidepressant treatments (pharmacotherapy and ECT) would then depend on a successful facilitation of this adaptation (i.e. a successful deamplification of the NE signal). How affective and cognitive mental experiences arise from neurochemical and molecular biological events remains a mystery. It is our belief, however, that an intensive integration of knowledge generated in the multidisciplinary field of neuroscience - neuroanatomy, neurochemistry, neurophysiology, neuropharmacology, neuropsychology, developmental neurobiology, molecular neurobiology - will eventually provide these new and fundamental levels of understanding.

ACKNOWLEDGMENTS

The studies from our laboratories have been supported by USPHS grants MH-29228, MH-34007, MH-26463 and by the Tennessee Department of Mental Health and Mental Retardation.

REFERENCES

Affolter, H., Erne, P., Burgisser, E. and Pletscher, A., Ca++ as a messenger of 5HT$_2$ receptor stimulation in human blood platelets, Naunyn-Schmiedeberg's Arch. Pharmacol, 325:337-342 (1984).

Anden, N. E., Dahlstrom, A., Fuxe, K., Larsson, K., Olson, L. and Ungerstedt, U., Ascending monoamine neurons to the telencephanon and diencephalon, Acta physiol. Scand. 67:313-326 (1966).

Barbaccia, M. L., Brunello, N., Chuang, D. M. and Costa, E., On the mode of action of imipramine: Relationship between serotonergic axon terminal function and down-regulation of beta adrenergic receptors, Neuropharmacology 22:373-383 (1983).

Berridge, M. J., Rapid accumulation of inositol triphosphate reveals that agonists hydrolyse polyphosphoinositides instead of phosphatidylinositol, Biochem. J. 212:849-858 (1983).

Berridge, M. J., Inositol trisphosphate and diacylglycerol as second messengers, Biochem. J., 220:345-360 (1984).

Berridge, M. J., Dawson, C., Downes, C. P., Heslop, J. P. and Irvine, R. K., Changes in the levels of inositol phosphates after agonist-dependent hydrolysis of membrane phosphoinositides, Biochem. J. 212:473-482 (1983).

Berridge, M. J., Downes, P. C. and Hanely, M. R., Lithium amplifies agonist-dependent phosphatidylinositol responses in brain and salivary glands, Biochem. J. 206:587-595 (1982).

Brunello, N., Chuang, D. M., Costa, E., Use of specific brain lesions to study the site of action of antidepressants, Adv. Biosc. 40:141-145 (1982).

Brunello, N., Volterra, A., Cagiano, R., Ianieri, G. C., Cuomo, V. and Racagni, G., Biochemical and behavioral changes in rats after prolonged treatment with desipramine: Interaction with p-chlorophenylalanine, Naunyn-Schmiedeberg's Arch. Pharmacol. 331:20-22 (1985).

Conn, P. J. and Sanders-Bush, E., Selective 5HT$_2$ antagonists inhibit serotonin stimulated phosphatidylinositol metabolism in cerebral cortex, Neuropharmacology 23:993-996 (1984)..

Conn, P. J. and Sanders-Bush, E., Serotonin-stimulated phosphoinositide turnover: mediation by the S$_2$ binding site in rat cerebral cortex but not in subcortical regions, J. Pharmacol. Exp. Ther. 234:195-203 (1985).

Conn, P. J. and Sanders-Bush, E., Regulation of serotonin stimulated phosphoinositide hydrolysis: Relation to the serotonin 5HT$_2$ binding site, J. Neurosci. (submitted) (1986a).

Conn, P. J. and Sanders-Bush, E., Agonist induced phosphoinositide hydrolysis in rat choroid plexus, J. Neurochem. (submitted) (1986b).

Conn, P. J., Sanders-Bush, E., Hoffman, B. J. and Hartig, P. R., A unique serotonin receptor in choroid plexus is linked to phosphoinositide hydrolysis, Proc. Natl. Acad. Sci., in press (1986).

Consolazione, A. and Cuello, A. C., CNS serotonin pathways, In: Biology of serotonergic transmission, N. N. Osborne, ed., pp. 29-61, John Wiley and Sons, Ltd., Baffius Lane, England (1982).

Coppen, A. and Wood, K., 5-Hydroxytryptamine in the pathogenesis of affective disorders, Adv. Biochem. Psychopharmacol. 34:249-258 (1982).

Crook, R. B., Farber, M. B. and Prusiner, S. B., Hormone and neurotransmitters control cyclic AMP metabolism in choroid plexus epithelial cells, J. Neurochem. 42:340-350 (1984).

Dahlström, A. and Fuxe, K., Evidence for the existence of monoamine-containing neurons in the central nervous system I. Demonstration of monoamines in the cell bodies of brain stem neurons, Acta. Physiol. Scand. 62 (Suppl. 232):1-55 (1964).

de Chaffoy de Courcelles, D., Roevens, P. and Van Belle, H., Stimulation by serotonin of 40 KDa and 20 KDa protein phosphorylation in human platelets, FEBS Lett. 171:289-292 (1984).

de Chaffoy de Courcelles, D., Leysen, J. E., de Clerck, F., Van Belle, H. and Janssen, P. A. J., Evidence that phospholipid turnover is the signal transducing system coupled to serotonin-S$_2$ receptor sites, J. Biol. Chem. 260:7603-7608 (1985).

Dumbrille-Ross, A. and Tang, S. W., Noradrenergic and serotonergic input necessary for imipramine induced changes in beta but not S$_2$ receptor densities, Psychiatry Res. 9:207-215 (1983).

Fuxe, K., Hokfelt, T., Agnati, L. F., Johansson, O., Goldstein, M., Perez de la Mora, M., Possami, L., Tapia, R., Teran, L. and Palacios, R., Mapping out central catecholamine neurons: Immunohistochemical studies on catecholamine-synthesizing enzymes. In: Psychopharmacology: A generation of progress, M. A. Lipton, A. DiMascio, K. K. Killam, eds. pp. 67-94, Raven Press, NY, NY (1978).

Gilbert, R. F. T., Bennett, G. W., Marsden, C. A. and Emson, P. C., The effects of 5-hydroxytryptamine depleting drugs on peptides in the ventral spinal cord, Europ. J. Pharmacol. 76:203-210 (1981).

Harden, T. K., Agonist-induced desensitization of the beta adrenergic receptor-linked adenylate cyclase. Pharmacol. Rev. 35:5-32 (1983).

Hegstrand, L. R., Minneman, K. P. and Molinoff, P. B., Multiple effects of guanosine triphosphate on beta adrenergic receptors and adenylate cyclase activity in rat heart, lung and brain, J. Pharmacol. Exp. Ther. 210:215-221 (1979).

Hirasawa, K. and Nishizuka, Y., Phosphatidylinositol turnover in receptor mechanisms and signal transduction, Ann. Rev. Pharmacol, Toxicol. 25: 147-170 (1985).

Hirata, F. and Axelrod, J., Phospholipid methylation and the transmission of biological signals through membranes, Science 209:1082-1090 (1980).

Jacobowitz, D. M., Monoaminergic pathways in the central nervous system, In: Psychopharmacology: A generation of progress, M. A. Lipton, A. DiMascio, K. K. Killam, eds., pp. 119-129, Raven Press, New York, NY (1978).

Janowsky, A., Okada, F., Manier, D., Applegate, C. D. and Sulser, F., Role of serotonergic input in the regulation of the beta-adrenergic receptor-coupled adenylate cyclase system, Science 218:900-901 (1982).

Kendall, D. A. and Nahorski, S. R., 5-hydroxytryptamine stimulated inositol phospholipid hydrolysis in rat cerebral cortex slices: Pharmacological characterization and effects of antidepressants, J. Pharmaco. Exper. Therap. 233:473-479 (1985).

Levitt, P. and Moore, R. Y., Noradrenaline neuron innervation of the neurocortex in the rat, Brain Res. 139:219-232 (1978).

Levitt, P., Rakic, P. and Goldman-Rakic, P. S.. Comparative assessment of monoamine afferents in mammalian cerebral cortex, Neurology and Neurobiology 10:41-59 (1984).

Lindvall, O. and Bjorklund, A., General organization of cortical monoamine systems, Neurology and Neurobiology 10:9-40 (1984).

Maeda, K., Monoaminergic effect on cerebrospinal fluid production, Nihon Univ. J. Med. 25:155-174 (1983).

Manier, D. H., Gillespie, D. D., Steranka, L. R. and Sulser, F., A pivotal role for serotonin in the down-regulation of beta-adrenoceptors by antidepressants: Reversibility of the action of p-chlorophenylalanine (PCPA) by 5-hydroxytryptophan, Experientia 40:1223-1226 (1984).

Manier, D. H., Gillespie, D. D., Sanders-Bush, E. and Sulser, F., The serotonin/noradrenaline-link in brain: I. The role of noradrenaline and serotonin in the regulation of density and function of beta adrenoceptors and its alteration by DMI, Naunyn-Schmiedeberg's Arch. Pharmacol. submitted (1986).

Mendlewicz, J. and Youdim, M. B. H., Antidepressant potentiation of 5-hydroxytryptophan by l-deprenyl in affective illness, J. Affect. Disord. 2:137-146 (1980).

Mobley, P. L., Manier, D. H. and Sulser, F., Norepinephrine-sensitive adenylate cyclase system in rat brain: Role of adrenal corticosteroids, J. Pharmacol. Exp. Ther. 226:71-77 (1983).

Moore, R. Y., Bloom, F. E., Central catecholamine neuron systems: Anatomy and physiology of the norepinephrine and epinephrine systems, Ann. Rev. Neurosci. 2:113-168 (1979).

Morrison, J. H., Foote, S. L., Molliver M. E., Bloom, F. E. and Lidov, H. G. W., Noradrenergic and serotonergic fibers innervate complementary layers in monkey primary visual cortex: An immunohistochemical study, Proc. Natl. Acad. Sci. USA 79:2401-2405 (1982).

Nathanson, J., Beta-adrenergic sensitive adenylate cyclase in secretory cells of choroid plexus, Science 204:843-844 (1979).

Nestler, E. J. and Greengard, P., Protein phosphorylation in brain. Nature 305:583-588 (1983).

Nestler, E. J., Walaas, S. T. and Greengard, P., Neuronal phosproteins: Physiological and clinical implications, Science 225:1557-1364 (1984).

Nimgaonkar, V. L., Goodwin, G. M., Davies, C. L. and Green, A. R., Down-regulation of beta-adrenoceptors in rat cortex by repeated administration of desipramine, electroconvulsive shock and clembuterol requires 5HT neurones but not 5HT, Neuropharmacology 24:279-283 (1985).

O'Donnell, J. M. and Frazer, A., Effects of clenbuterol and tricyclic antidepressants on beta-adrenergic receptor, N-protein coupling in rat cerebral cortex, Fed. Proc. 43:839 (1984).

Okada, F., Tokumitsu, Y. and Ui, M., Desensitization of beta adrenergic receptor coupled adenylate cyclase in cerebral cortex after treatment in vivo of rats with desipramine, J. Neurochem. in press (1986).

Pazos, A., Hoyer, D. and Palacios, J. M., The binding of serotonergic ligand to the porcine choroid plexus: Characterization of a new type of serotonin recognition site, Europ. J. Pharmacol. 106:539-546 (1984).

Pickel, V. M., Joh, T. H., Field, P. M., Becker, C. G. and Reis, D. J., Cellular localization of tyrosine hydroxylase by immunohistochemistry, J. Histochem. Cytochem. 23:1-12 (1975).

Roth, B. L., Nakaki, T., Chuang, D. M., Chernow, B. and Costa, E., Characterization of $5HT_2$ receptors linked to phospholipase C in rat aorta, Fed. Proc. 44:1244 (1985).

Roth, B. L., Nakaki, T., Chuang, D. M. and Costa, E., Aortic recognition sites for serotonin (5HT) are coupled to phospholipase C and modulate phosphatidylinositol turnover, Neuropharmacology 23:1223-1225, (1984).

Shopsin, B., Friedman, E. and Gershon, S., The use of synthesis inhibitors in defining a role for biogenic amines during imipramine treatment in depressed patients, Psychopharmacol. Comm. 1:239-249 (1975).

Shopsin, B., Friedman, E. and Gershon, S., Parachlorophenylalanine reversal of tranylcypromine effects in depressed patients, Arch. Gen. Psych. 33: 811-819 (1976).

Stockmeier, C. A., Martino, A. M. and Kellar, K. J., A strong influence of serotonin axons on beta-adrenergic receptors in rat brain, Science 230: 323-325 (1985).

Sulser, F., Noraderenergic receptor regulation and the action of antidepressants. In "Depression and Antidepressants - Recent Events, pp. 24-36, excerpta Medica, Amsterdam (1983).

Sulser, F., The serotonin-noradrenaline link-hypothesis of affective disorders, in: Psychiatry, vol. 2, Eds. P. Pichot, P. Berner, R. Wolf and K. Thau, Plenum Publishing Corporation, pp. 411-416 (1985).

Sulser, F., Conn, P. J., Zawad, J. S. and Sanders-Bush, E., Molecular aspects of altered transmembrane regulation of the noradrenaline signal by antidepressants, Benzon Symposium on Drug Action, Copenhagen, in press (1986).

Van Praag, H. M., Serotonin precursors in the treatment of depression, Adv. Biochem. Psychopharmacol. 34:259-286 (1982).

Whybrow, P. and Mendels, J., Toward a biology of depression: Some suggestions from neurophysiology, Amer. J. Psychiatry 125:45-54 (1969).

Yagaloff, K. A. and Hartig, P. R., [125I]-LSD binds to a novel serotonergic site on rat choroid plexus epithelial cells, J. Neurosci. 5:3178-3183 (1985).

"SUBSTANCE M", A SEROTONIN MODULATOR CANDIDATE FROM

HUMAN URINE?

K. G. Walton, T. McCorkle, T. Hauser, C. MacLean,
R. K. Wallace, J. Ieni* and L. R. Meyerson*

Neurochemistry Laboratory, Maharishi International
University, Fairfield, Iowa 52556 and *Department of
Chemical Pharmacology, Medical Research Division of
American Cyanamid Company, Ramapo College, Mahwah
New Jersey 07430

INTRODUCTION

Intense interest in serotonin (5-hydroxytryptamine, 5-HT) stems from
the involvement of this biogenic amine in an extraordinary range of physiological functions. (See, for example, the five-volume series edited by
Essman, 1977-1979.) Most tissues and organs are affected by 5-HT, either
through serotonergic neurons or following its release from blood platelets
and other non-neuronal sites. Both platelets and serotonergic neurons have
mechanisms for the active uptake of 5-HT, and the platelet has become a
model for studying aspects of the regulation of 5-HT levels. The discovery
that platelets of endogenously depressed patients have a decreased number
of high-affinity binding sites for [^{3}H]-imipramine (Briley et al., 1980;
Paul et al., 1981) has kindled research on the imipramine binding site, and
this site is now thought to effect a modulation of 5-HT uptake (Wennogle
and Meyerson, 1983, 1985; Meyerson et al., submitted).

Imipramine is a tricyclic antidepressant drug that inhibits 5-HT uptake
in an identical manner in neurons and platelets. (See review by Rehavi et
al., 1984.) Studies of the effects of different agents on the dissociation
rate of imipramine from its binding site have indicated that the imipramine
site is not the 5-HT transporter and that imipramine participates in an allosteric regulation of 5-HT uptake (Wennogle and Meyerson, 1983, 1985; Sette
et al., 1983; Briley, 1985; Meyerson et al., submitted). In a sequence of
steps similar to those involved in the discovery of the opiate peptides,
perhaps a bioassay based on competition with imipramine for its binding
site could lead to discovery of endogenous ligands controlling serotonin
uptake. Such ligands might modulate serotonergic activity through enhancing or reducing the synaptic concentration of 5-HT. This approach has in
fact been initiated in several laboratories.

Using the inhibition of imipramine binding as a bioassay, Barbaccia
et al. (1983) have found a substance in rat brain that inhibits both imipramine binding and 5-HT uptake in a concentration-related manner. The
substance has not been identified, but appears to be a small molecule,
conceivably a derivative of beta-carboline. Some of the beta-carboline
compounds compete effectively with imipramine for binding to the [^{3}H]-

imipramine site and also inhibit 5-HT uptake (Langer et al., 1984). Trace amounts of these compounds have been found in brain and body fluids (Rommelspacher et al., 1979a,b; Bloom, 1982). Other groups have reported both small peptide and non-peptidic species isolated from rat serum and brain that inhibit both imipramine binding and 5-HT uptake. (See Lal et al., 1985.)

More recently, a large peptide has been discovered which inhibits imipramine binding but, contrary to the other candidates for endogenous ligands of the imipramine site, stimulates 5-HT uptake (Abraham and Meyerson, 1985; Abraham, Ieni, and Meyerson, submitted). It is an acidic glycoprotein of molecular weight about 45,000, isolated from plasma, and appears present in sufficient concentration to exert effects on 5-HT uptake in platelets in vivo. Antibodies to this protein have been prepared, and data will soon be available from tests of its presence in brain. Thus, there are at present several possible candidates for endogenous modulators which act at the imipramine binding site.

The research described here gives evidence that a substance from human urine shows the ability to inhibit imipramine binding and 5-HT uptake in platelets. This compound may be identical with a small molecule, "Substance M", as yet unidentified but previously reported to form a colored complex with nitrosonaphthol and to vary either diurnally or in response to a specific type of behavior (Walton et al., 1983). In the present report, three procedures of partially purifying Substance M and the inhibitor of imipramine binding are described, and the possibility that Substance M might be a candidate for endogenous modulator of serotonin uptake is discussed.

Circadian and/or Behaviorally Related Changes in an Unidentified Nitrosonaphthol-positive Substance

Walton et al. (1983), while studying diurnal changes in urinary 5-hydroxyindoles using an adaptation of the nitrosonaphthol assay of Udenfriend et al. (1955), found that a substance other than 5-HIAA was detected with this colorimetric assay and that excretion of the substance appeared to

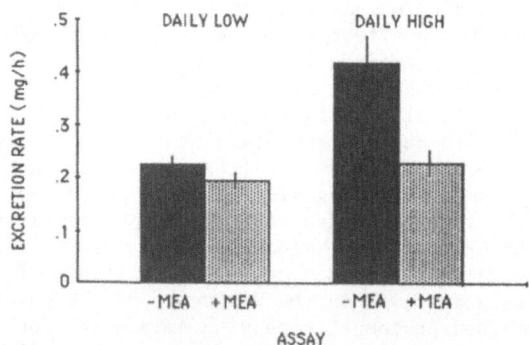

FIG. 1: Differences in the apparent amount of nitrosonaphthol-positive substance by two procedures (-MEA and +MEA). Urine collected in 8 three-h periods per day beginning at 0715 h was analyzed in the absence and presence of mercaptoethylamine (MEA), an amine which increases the specificity for 5-HIAA (Shihabi and Wilson, 1982). 5-HIAA served as standard in both assays. Probabilities that differences between assays in the absence and presence of MEA were due to chance are < 0.01 and < 0.001, respectively, for the daily low and the daily high values (t-test for related measures). Means ± SD for n = 9 days (one subject).

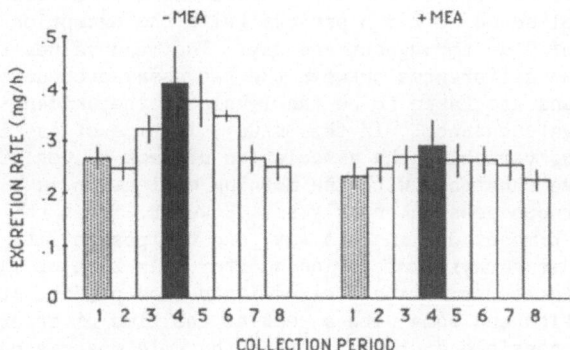

FIG. 2: Differences between values obtained in the presence of MEA (specific for 5-HIAA) and in the absence of MEA (less specific) on
the same set of 3-h urine samples. Urine collected in 8 three-
h periods per day beginning at 0715 h was analyzed by the two
procedures. 5-HIAA was the standard in both assays. Foods high
in 5-HT were omitted from the diet. Grey and black bars represent
the morning and evening periods, respectively, during which the
TM and TM-Sidhi program was practiced. Values are means ± SEM
for 14 consecutive days of collection (one subject). Results from
the two methods were different at the .05 level for periods 3-6
(t-test).

change with either behavior or time of day. The lack of complete specificity of the Udenfriend assay was known (Mustala, 1965), and when a modified
assay employing the addition of mercaptoethylamine (MEA) to improve specificity for 5-hydroxyindoleacetic acid (5-HIAA) was reported (Shihabi and
Wilson, 1982), a routine comparison of the two showed a highly significant
difference (Figure 1). Out of the 3-h urine collections taken over a 24-h
period, those giving the highest and the lowest values in the Udenfriend
assay were also assayed by the new procedure. Not only did the two procedures give different results, but also the difference was much greater at
the high point than at the low point of the day. This suggested that substantial amounts of a nitrosonaphthol-positive substance other than 5-HIAA
was being detected in the less specific assay.

This observation alone might only have provoked a switch to the more
specific assay had it not been for the fact that the difference between the
results obtained with the two assays appeared systematic, i.e. occurred
around a particular time of day (Figure 2). The initial purpose of the
diurnal studies had been to test the possibility that a circadian rhythm in
excretion of 5-HIAA might have been responsible for the reported rise of 5-
HIAA with the afternoon practice of a mental technique for personal development, the Transcendental Meditation (TM) technique (Bujatti and Riederer
1976). Figure 2 shows that consecutive 3-h collections for 14 days in one
subject peaked in the unknown nitrosonaphthol-positive substance at the time
of the afternoon practice of the TM and TM-Sidhi techniques. (TM-Sidhi
refers to specific additional mental procedures added to the usual TM practice. See Dillbeck et al., 1981.) These mental techniques are usually
practiced twice daily, in the early morning and late afternoon. Figure 2
shows an insignificant difference in the two assays at the time of the morning session. Thus, these data from one subject do not permit a definite
conclusion regarding the question of behavioral effect versus circadian
rhythm of this nitrosonaphthol-positive substance. The next study, however, may support the behavioral connection rather than a simple circadian
rhythm.

Figure 3 shows results from a study involving 14 meditating subjects whose urine was collected in 2-3 h periods (with the exception of an overnight collection of 9 h) throughout one day. The mean values represented by the bars are the differences between the two assay procedures (i.e. with and without MEA) and are taken to be the levels of the unidentified nitrosonaphthol-positive substance. In this study, if time of day is not considered a confounding variable, the association of peak values of the nitrosonaphthol-positive substance with the morning meditation is significant, and with the afternoon session, nearly so. However, it is not statistically valid to rule out time of day in this way, and the possibility remains that either: 1) there is a rhythm of period shorter than 24 h or 2) there are two groups of subjects, some with a circadian period peaking at the time of morning meditation and some with a peak at the time of afternoon meditation. Although possible diurnal or other rhythmic changes cannot be excluded, one result does seem clear. The fact that no food was consumed between the previous evening meal and the end of the morning meditation argues against a dietary source of the elevated levels of the substance seen at certain times of day.

Possible Association between the Nitrosonaphthol-positive Substance and Inhibition of Imipramine Binding and Serotonin Uptake

Before carrying out other tests on this unidentified nitrosonaphthol-positive substance, it was necessary to separate it from 5-HIAA. The first attempt (Method I), involving extraction with chloroform:ethyl acetate:amyl alcohol (2:1:1) under slightly basic conditions (pH 8), resulted in significant amounts of the substance in the organic phase. (5-HIAA was not extracted from water or from spiked urine under these conditions.) Such extracts were found to produce a colored complex with nitrosonaphthol which showed an absorbance maximum (visible range) at 540 nm, similar to the 5-HIAA-nitrosonaphthol complex. However, when MEA was added, the absorbance maximum of the complex shifted to 570 nm, rather than the shift to 640 nm

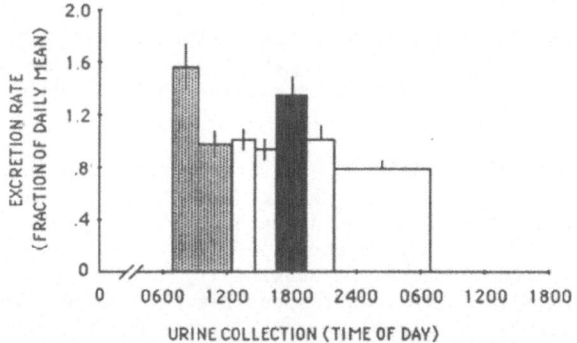

FIG. 3: Possible circadian or behaviorally-related change in a nitrosonaphthol-positive substance other than 5-HIAA. Urine samples from 7 periods of collection, starting at 0700 h, were assayed in the absence and presence of MEA. Differences between the two values represented nitrosonaphthol-positive substance other than 5-HIAA. The gray and black bars represent the morning and afternoon periods respectively during which the TM and TM-Sidhi program was practiced. If time of day were not a confounding variable, period 1 would be significantly different (.05 level) from all other periods (one-way ANOVA). Means ± SEM for n = 14 male subjects.

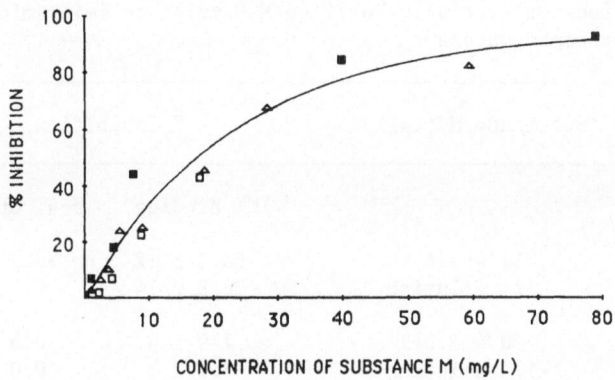

FIG. 4: Inhibition of imipramine binding to platelet membranes by organic
extracts of urine. Binding inhibition by samples extracted from
urine by Method I (see text) was assayed in duplicate at five
dilutions using 2.5 nM tritium-labeled imipramine, and 100-200 μg
of platelet membrane protein per tube. Points represent the averages of two binding assays carried out on separate days. Each
type of symbol denotes urine from one subject.

characteristic of the 5-HIAA complex. Thus, the difference between values
obtained by the two assays (- and + MEA) could be understood, and did
indeed appear due to a second substance in urine. Significant portions of
the substance were back extracted into 0.1 N HCl, and the first tests of
the substance on imipramine binding were carried out with samples dried
down from either the organic solvent or from the back extractions and taken
up in aqueous buffer (Figure 4).

Figure 4 shows the effects of urine extracts prepared by Method I on
the binding of [^{3}H]-imipramine to platelet membranes. As above, the unknown
nitrosonaphthol-positive substance was taken as the difference between the
two assays (using 5-HIAA as standard). (For simplicity, the term "Substance
M" was adopted to signify the nitrosonaphthol-positive substance.) The
relationship between Substance M concentration and percent inhibition of
imipramine binding is exponential, with a correlation coefficient of .917.
This type of curve suggests a binding-saturation, with inhibition of imipramine binding approaching 100%.

Inhibition of imipramine binding as well as 5-HT uptake was tested in
another experiment using urine samples from two groups of five subjects
(Table I). In this experiment Method II of extracting Substance M was used
and is described as follows. Urine adjusted to pH 9-10 was put through C18,
reverse-phase extraction columns ('Baker'-10 SPE) preconditioned with methanol and water. The columns were then washed twice with water followed by
ethyl acetate, and the Substance M was eluted with acetonitrile. Substance
M was assayed colorimetrically, as above. The results shown in Table I
confirm those of Figure 4 for concentration-related inhibition of imipramine
binding, and appear to show slightly less, but also concentration-related,
inhibition of 5-HT uptake.

The large differences in concentration of nitrosonaphthol-positive substance as well as inhibition of imipramine binding and 5-HT uptake between
urine extracts from Group 1 (long-term meditating) and Group 2 (nonmedita-
ting) subjects were unexpected. The results suggest that meditators tend
to have more of these activities than nonmeditators. However, even though
comparison (t-test) of the two groups shows highly significant differences,

TABLE 1: The Effect of Partially Purified M Samples on Imipramine Binding
and Serotonin Uptake[a]

Subjects[b]	Substance M (mg/L)	% Inhibition	
		IMI Binding	5-HT Uptake
Group 1	11.9 ± 6	60.3 ± 12	38.7 ± 12
	1/4 dilution	38.8 ± 14	6.5 ± 1
Group 2	0.4 ± 0.2	7.9 ± 4	0.5 ± 0.3
	1/4 dilution	3.7 ± 2	0.0

[a]Substance M was separated from 24-h urine samples by Method II, which
involved adsorption onto C18 reverse-phase extraction columns (Baker Chemical Co.) followed by washes with water then ethyl acetate and finally
elution with acetonitrile. Concentrations of tritiated imipramine (IMI)
and 5-HT were 2.5 nM and 1 μM, respectively. For 5-HT uptake the numbers
of platelets per tube ranged from 3.4×10^8 to 6.3×10^8. For IMI binding
100-200 μg of platelet membrane protein were used in each tube. Data are
expressed as means ± SEM. [b]The 5 Group 1 subjects were long-term (over 5
years) practitioners of the TM and TM-Sidhi program. The 5 Group 2 subjects had never practiced meditation or any other self-improvement techniques. There was one female in each group.

the subjects were not matched for age, diet, exercise, drinking habits or
other possibly significant variables. To detect a difference due to meditation the studies will have to be extended to larger groups of subjects
matched for these variables. The main point of the data is that inhibition
of both imipramine binding and 5-HT uptake is significant and appears related to the concentration of a nitrosonaphthol-positive substance other than
5-HIAA. (Most of the 5-HIAA was found to elute from the column in the water
wash).

Purification and Assay of Substance M by Liquid Chromatography

Since completing the experiments described above, further progress in
the purification of Substance M has been made (Walton, McCorkle, MacLean,
Hauser, Wallace, Ieni and Meyerson, in preparation). The third method of
purification (Method III) used preparative liquid chromatography of the
organic extract of urine. The urine was extracted at pH 1 with hexane:ethyl
acetate (1:2). After evaporation of the organic solvent, the residue was
dissolved in dimethylsulfoxide and separated on a semipreparative C18
reverse-phase HPLC column (Phenomenex, 25 cm x 10 mm) with a mobile phase
containing 30% methanol and 0.1 M acetic acid in water. Figure 5 shows the
chromatogram of a urine extract from a subject with unusually high M concentration. The M peak was the only one showing significant activity in
the nitrosonaphthol test. This peak was collected and rechromatographed
in the same system. After the second collection from this system, a portion injected onto an analytical column (C18, 5 particle size, 20 cm in
length, E. Merck) showed only one other peak (with peak height 5% of the
major peak) of material absorbing at 280 nm. At 2.5 mg/L (based on the
nitrosonaphthol assay) this sample of Substance M showed a 55% inhibition
of imipramine binding. Most recently, however, chromatography of this
fraction on a size-exclusion column (Waters I-60) suggests that there are
several substances represented under the major peak on the C18 column. In
any case, the enrichment of Substance M, based on the dry weight of nonvo-

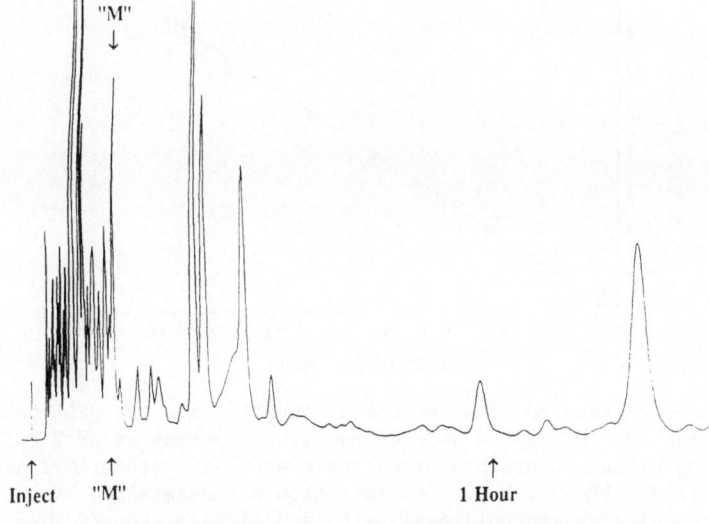

FIG. 5: Chromatogram showing preparative HPLC separation of Subtance M
from an organic extract of urine. A 24-h urine sample containing
47 mg/L of Substance M was extracted at pH 1 with hexane:ethyl
acetate (1:2). After drying the extract was redissolved and
~140 µg were separated using a C18 reverse-phase semipreparative
column. (See text for details.) Substance M ("M") eluted at 12
minutes.

latile constituents, was 800- to 1000-fold using Method III, and recovery
was 40-70%.

HPLC of urine extracts on the C18 analytical column may prove to be a
means of assaying Substance M. Figure 6 shows values obtained from nine
urine extracts using peak height for the HPLC assay and mg/L (based on a
5-HIAA standard) from the colorimetric assay. These assays showed comparable sensitivities, with a limit around 0.4 mg/L. (Greater overall sensitivity could be obtained in either assay by simply evaporating solvent to
give a higher concentration of the Substance M.) Furthermore, the substance has been detected amperometrically, with an apparent optimum at 0.5
V on the working electrode. Using an analytical HPLC system with amperometric detection, greater than a four-hundred fold improvement in sensitivity over the above assays was obtained.

Stability and Possible Chemical Structure of Substance M

The extraction and chromatographic properties of Substance M appear
more compatible with its being a small, nonpolar molecule than a peptide.
The lack of attack by Proteinase K (Type XI, Sigma Chemical Co.), a highly
active, fungal proteinase, supports this conclusion. The compound appears
moderately sensitive to acid (60% loss after 6 h at 25° in 6 N HCl), quite
sensitive to base (50% loss after 48 h at 25° in 0.04 N NaOH) and fairly
sensitive to heat, at least in aqueous solution, (85% loss after 6 h at
90°).

The strongest clue to the structure of the compound may come from the
known structural requirements for the nitrosonaphthol assay. A single free
hydroxyl group attached to an aromatic ring appears to be an absolute
requirement (Mutsala, 1965; Knight et al., 1983). Compounds with multiple

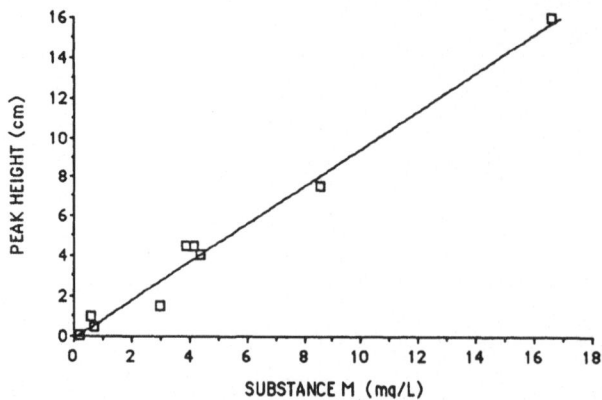

FIG. 6: Simple regression of substance M concentration vs. HPLC peak height
at 280 nm. M samples were extractions of urine at pH 2 and pH 8
using either chloroform or chloroform:ethyl acetate:amyl alcohol
(2:1:1). (R^2 = .983.) Concentration was determined by cholorime-
tric (nitrosonaphthol) assay with 5-HIAA as standard.

hydroxyl groups on the ring will not react. In addition, the compound must
have an appropriate electron-donating group either ortho or para to the
hydroxyl (Knight et al., 1983). This requirement would make the hydroxyl
easily oxidized, a condition which appears to be upheld by the relatively
low voltage required for oxidation of Substance M in the amperometric
detector. These and other restrictions on the structure allow the selec-
tion of compounds likely to be nitrosonaphthol-positive. The number of
such compounds conceivably present in urine appears less than 20. Fifteen
of these are available commercially and, along with 14 other substances with
similar structures not expected to give a positive nitrosonaphthol reaction,
have been compared with Substance M in the colorimetric assay as well as by
GC/MS (Walton et al., in preparation). None seems to be identical to Substance M.

CONCLUSIONS AND DISCUSSION

In brief, these studies appear to support the following conclusions:

1.   The urinary excretion of an unidentified nitrosonaphthol-positive substance, "Substance M", appears to increase in a circadian manner and/or
during the practice of a specific mental technique for self-development.

2.   Urine extracts containing this low-molecular-weight, relatively nonpolar substance inhibit both imipramine binding and serotonin uptake in a
manner related to the concentration of the substance.

3.   Based on total dry weight of the nonvolatile constituents, Substance
M can be enriched 1000-fold through organic extraction followed by liquid
chromatography on C18 reverse-phase columns, and the fractions obtained in
this manner retain high activity in the inhibition of imipramine binding.

4.   The parallel enrichment of nitrosonaphthol-positive and imipramine-
binding-inhibitory activities through three procedures suggests that there
is a single substance responsible for both activities and that this substance is worthy of further study as a potential modulator of serotonin
function.

Although preliminary evidence is available, the following important questions have not been definitively answered. The chemical identification of Substance M will help in providing answers, but until this is achieved, only preliminary conclusions can be reached.

1.   Is it possible that Substance M (or the ligand for the imipramine site, if these should eventually prove to be different entities) is an artifact of the purification procedure?

Artifactual synthesis of M during the purfication procedure is unlikely, since a given sample of urine appeared to have approximately the same amount of M (and of activity in the imipramine competition assay, although these assays were more variable) regardless of which of the three methods of enrichment was used. Since the methods are quite different from one another, in terms of solvents used, in pH and in other variables expected to be critical to most paths of artifactual synthesis, the possibility of an artifact during purification appears unlikely. Also, the amount of nitrosonaphthol-positive substance detected in each urine sample is approximately the same in the unextracted urine as in the total of two extractions with, for example, hexane:ethyl acetate. However, it is conceivable that some reaction goes on in the urine itself while still in the bladder. Demonstration of Substance M in blood or other bodily tissues would help to rule out the latter possibility.

2.   Is it likely that a substance potentially active at imipramine binding sites in both the circulatory and the nervous systems would be excreted in the urine, or could the substance reported here be only a slightly active metabolite of some other functional agent?

There are many metabolites of neurotransmitters and hormones found in urine, and it is conceivable that one or more of these would have some ability to compete with imipramine for binding to specific sites. Although several of these have already been compared with Substance M (in the nitrosonaphthol reaction, by HPLC and by GC/MS) the list is not exhaustive, and the critical compound may have been omitted. On the other hand, some active neurotransmitters and hormones (e.g. 5-HT and the catecholamines) are normally found in urine, and at concentrations two orders of magnitude higher than in plasma. Cell growth factors are also found in normal urine. So it is not unlikely that physiologically active substances capable of regulating 5-HT uptake could be found in the urine.

3.   Is there reason to expect that an endogenous ligand for the imipramine binding site would increase during the practice of Transcendental Meditation or during any other well-defined behavior?

On the basis of what is now known about the physiological site of action of imipramine, the most likely acute effect of an endogenous ligand mimicking its activity would appear to be inhibition of 5-HT uptake. In the CNS, this would be expected to produce an acute rise of serotonergic activity (De Montigny and Blier, 1985). Suggesting an increase in serotonergic activity, levels of 5-HIAA, the major metabolite of serotonin, increase in brain and urine during immobilization and certain other stressors in animals (e.g. Yuwiler, 1979). Furthermore, a recent study with the receptor blocker, metergoline, active on serotonin receptors in the CNS, supports the notion that increased serotonergic activity in man may combat some stressful or anxiety-producing situations (Graeff et al., 1985). Thus the TM technique, which has been reported to decrease the negative effects of stress (Orme-Johnson, 1973) and to reduce the distress associated with pain (Mills and Farrow, 1981), might do so in part by increased serotonergic activity.

Other changes related to the TM practice might also be explained by increased central 5-HT activity. For example, plasma cortisol was found to decrease (Jevning et al., 1978a), and plasma prolactin to increase (Jevning et al., 1978b), acutely with the practice of TM. Both of these hormonal changes could be mediated by increased central serotonergic activity (Krieger, 1978). Finally, in most subjects there is a slowing of breath rate during the practice of TM, and in some subjects complete suspensions of breath occur for periods up to 90 s (Farrow and Hebert, 1982; Badawi et al., 1984; Kesterson and Clinch, 1985). The reduction of respiration during TM is also associated with reduced sensitivity to CO_2 (Kesteron and Clinch, 1985). Animal studies have convincingly demonstrated that increased activity of a central serotonergic system can mediate both a depression of respiration and a reduced sensitivity to CO_2 (Lundberg et al., 1980). In sum, there is ample evidence for an association between the TM practice and increased serotonergic activity. Whether the substance isolated here could be a link in this association remains to be determined.

The effects of behavior on unknown urinary substances is also being studied in other laboratories. For example, a urinary substance which increases in response to environmental stress as well as in high-anxiety states and in lactate-provoked panic has been reported (Clow et al., 1983). This substance competes for binding at the benzodiazepine receptor and may also inhibit monoamine oxidase. Since the urine is replete with active substances and the present as well as other studies suggest that new substances related to behavior may well be discovered there, the over-riding message seems to be to pursue their chemical identification and characterization. This is the goal now being pursued in our laboratories.

ACKNOWLEDGMENTS

Special thanks are due Mary Anne Walton, Steve Winn, Donald Dannemann and Chris Banus as well as Drs. Byron Rigby, Bevan Morris and Neil Paterson for invaluable support and to Margaret Lerom for expert clerical assistance.

REFERENCES

Abraham, K. I., Ieni, J. R. and Meyerson, L. R., Biochim. Biophys. Acta., submitted.

Abraham, K. I. and Meyerson, L. R., Isolation of a plasma acid protein modulator for the human platelet imipramine binding site, Society for Neuroscience Abstracts 11:749 (1985).

Badawi, K., Wallace, R. K., Orme-Johnson, D. and Rouzere, A. M., Electrophysiologic characteristics of respiratory suspension periods occuring during the practice of the Transcendental Meditation program, Psychosomatic Medicine 46:267 (1984).

Barbaccia, M. L., Gandolfi, O., Chuang, D. M. and Costa, E., Modulation of neuronal serotonin uptake by a putative endogenous ligand of imipramine recognition sites, Proc. Natl. Acad. Sci. USA 80:5134 (1983).

Bloom, F. E., Open Questions: a summary of the workshop discussions, in: "Beta-Carbolines and Tetrahydroisoquinolines" (Progress in Clinical and Biological Research; v. 90), F. Bloom, J. Barchas, M. Sandler and Usdin, E., eds., Alan R. Liss, New York (1982).

Briley, M., Imipramine binding: its relationship with serotonin uptake and depression, in: "Neuropharmacology of Serotonin", A. R., Green, ed., Oxford University Press, New York (1985).

Briley, M. S., Langer, S. Z., Raisman, R., Sechter, D. and Zarifian, E., Tritiated imipramine binding sites are decreased in platelets of untreated depressed patients, Science, 209:303 (1980).

Bujatti, M. and Riederer, P., Serotonin, noradrenaline and dopamine metabolites in the Transcendental Meditation technique, J. Neural Transm., 39:257 (1976).

de Montigny, C. and Blier, P., Electrophysiological aspects of serotonin neuropharmacology: implications for antidepressant treatments, in: "Neuropharmacology of Serotonin", A. R. Green, ed., Oxford University Press, New York (1985).

Clow, A., Glover, V., Armando, I. and Sandler, M., New endogenous benzodiazepine receptor ligand in human urine: identity with endogenous monoamine oxidase inhibitor? Life Sciences, 33:735 (1983).

Dillbeck, M. C., Orme-Johnson, D. W. and Wallace, R. K., Frontal EEG coherence, H-reflex recovery, concept learning, and the TM-Sidhi program, Internat. J. Neurosci., 15:151 (1981).

Essman, W. B. (Editor), "Serotonin in Health and Disease", Vols I-V, Spectrum Publications, Inc., New York (1977-1979).

Farrow, J. T. and Hebert, J. R., Breath suspension during the transcendental meditation technique, Psychosomatic Medicine 44:131 (1982).

Graeff, F. G., Zuardi, A. W., Giglio, J. S., Lima Filho, E. C. and Karniol, I. G., Effect of metergoline on human anxiety, Psychopharmacology 86: 334 (1985).

Jevning, R., Wilson, A. F. and Davidson, J. M., Adrenocortical activity during meditation, Hormones and Behavior, 10:54 (1978a).

Jevning, R., Wilson, A. F. and Van der Laan, E. F., Plasma prolactin and growth hormone during meditation, Psychosom. Med. 40:329 (1978b).

Kesterson, J. and Clinch, N., Peripheral and central control mechanisms during respiratory suspensions in Transcendental Meditation as evidenced by latency, hypoxia and RQ change, Soc. Neurosci. Abstr. 11: 1144 (1985).

Knight, J. A., Robertson, G. and Wu, J. T., The chemical basis and specificity of the nitrosonaphthol reaction, Clin. Chem., 29:1969 (1983).

Krieger, D. T., Endocrine processes and serotonin, in: "Serotonin in Health and Disease Vol III: The Central Nervous System," W. B. Essman, ed., Spectrum Publications, Inc., New York (1978).

Lal, H., Laballa, F. and Lane, J., eds., "Endocoids" (Progress in Clinical and Biological Research, v. 192), Alan R. Liss, New York (1985).

Langer, S. Z., Raisman, R., Tahraoui, L., Scatton, B., Niddam, R., Lee, C. R. and Claustre, Y., Substituted tetrahydro-beta-carbolines are possible candidates as endogenous ligand of the 3H-imipramine recognition site, Eur. J. Pharmacol., 98:153 (1984).

Lundberg, D. B. A., Mueller, R. A. and Breese, G. R., An evaluation of the mechanism by which serotonergic activation depresses respiration, J. Pharmacol. and Exp. Ther. 212:397 (1980).

Meyerson, L. R., Ieni, J. R. and Wennogle, L. P., Allosteric interactions between platelet imipramine binding and serotonin transport, J. Neurochem (submitted).

Mills, W. W. and Farrow, J. T., The Transcendental Meditation technique and acute experimental pain, Psychosomatic Medicine 43:157 (1981).

Mustala, O., Specificity of the nitrosonaphthol reaction in the determination of urinary 5-hydroxyindoleacetic acid, Ann. Med. Exp. et Biologie Fenn., 63, Suppl. 8:1 (1965).

Orme-Johnson, D. W., Autonomic stability and Transcendental Meditation, Psychosomatic Medicine 35:341 (1973).

Paul, S. M., Rehavi, M., Skolnick, P., Ballenger, J. C. and Goodwin, F. K., Depressed patients have decreased binding of 3H-imipramine to the platelet serotonin "transporter", Arch. Gen. Psychiat., 38:1315 (1981).

Rehavi, M., Skolnick, P. and Paul, S. M., High-affinity binding sites for tricyclic antidepressants in brain and platelets in: "Brain Receptor Methodologies. Part B. Amino Acids, Peptides, Psychoactive Drugs," P. J. Marangos et al., eds., Academic Press, Inc., New York (1984).

Rommelspacher, H., Honecker, H., Barbey, M. and Meinke, B., 6-Hydroxytetrahydronorharmane (6-hydroxy-tetrahydro-beta-carboline), a new active metabolite of indole-alkylamines in man and rat, Naunyn-Schmiedeberg's Arch, Pharmacol., 310:35 (1979a).

Rommelspacher, H., Strauss, S. and Lindemann, J., Excretion of tetrahydro-harmane and harmane into the urine of man and rat after a load with ethanol, FEBS Letters, 109:209 (1979b).

Sette, M., Briley, M. S. and Langer, S. Z., Complex inhibition of 3H-imipra-mine binding by serotonin and nontricyclic serotonin uptake blockers, J. Neurochem., 40:622 (1983).

Shihabi, Z. K. and Wilson, E. L., Colorimetric assay of urinary 5-hydroxy-3-indoleacetic acid, Clin. Biochem., 15:106 (1982).

Udenfriend, S., Titus, E. and Weissbach, H., The identification of 5-Hydroxy-3-indoleacetic acid in normal urine and a method for its assay, J. Biol. Chem., 216:499 (1955).

Walton, K. G., Francis, D., Lerom, M. and Tourenne, C., Behaviorally-induced alterations in urinary 5-hydroxyindoles, Trans. of the Amer. Soc. Neuro-chem., 14:199 (1983).

Wennogle, L. P. and Meyerson, L. R., Serotonin modulates the dissociation of 3H-imipramine from human platelet recognition sites, Eur. J. Pharmacol. 86:303 (1983).

Wennogle, L. P. and Meyerson, L. M., Serotonin uptake inhibitors differen-tially modulate high affinity imipramine dissociation in human plate-lets, Life Sciences 36:1541 (1985).

Yuwiler, A., Stress and serotonin, in: "Serotonin in Health and Disease Vol. V: Clinical Applications," W. B. Essman, ed., SP Med and Scien-tific Books, New York (1979).

ROLE OF RECEPTOR COUPLING TO PHOSPHOINOSITIDE METABOLISM IN THE
THERAPEUTIC ACTION OF LITHIUM

Robert H. Lenox

Neuroscience Research Unit, Department of Psychiatry
College of Medicine, University of Vermont, Burlington
Vermont 05405

INTRODUCTION

Lithium is one of the most effective psychopharmacological treatments for affective illness and offers a potential avenue for furthering our understanding of the neurobiological components of this disorder. Among the most specific and potent actions of lithium is its ability to not only treat the acute episode of mania but also reduce the frequency and severity of recurrent episodes of mania and depression in bipolar patients and depression in unipolar patients (Bunney and Garland, 1984). Thus, the unique action of lithium appears to be its long-term ability to dampen pathological neurobiological oscillations in individuals vulnerable to profound cyclic affective disturbances.

In order to explain the mechanism of action of lithium in the treatment of affective illness, we must understand the ability of this monovalent cation to affect significant changes in regions of the brain mediating the behavioral symptomatology associated with this clinical disorder. How can we explain its relatively profound activity in the central nervous system compared to the periphery? Wherein lies the physiological selectivity of its action and its long term ability to stabilize mood? Several lines of investigation characterizing a second messenger system that involves phospholipid metabolism have recently served to shed further light on the aforementioned questions.

Lithium and Phosphoinositide Hydrolysis

Over 30 years ago, Hokin and Hokin (1953, 1954) first described that acetylcholine could stimulate inositol lipid turnover in exocrine pancreas. Since that time, evidence has accumulated to demonstrate that phosphoinositide hydrolysis appears to be a common response of cells to activation of a variety of receptors which utilize calcium as an intracellular messenger (Michell, 1975; Berridge, 1981). Current models conceptualize receptor stimulation through coupling to the enzyme phospholipase C which then mediates hydrolysis of the polyphosphoinositides (see Figure 1). This receptor activated hydrolysis generates potent intracellular signals i.e. IP_3 and DAG (Berridge, 1984; Williamson, 1986). Through the mobilization of intracellular calcium, IP_3 can rapidly activate Ca^{++} dependent processes and contribute to the activation of a unique phospholipid dependent protein kinase (PKC) by DAG (Nishizuka, 1984a).

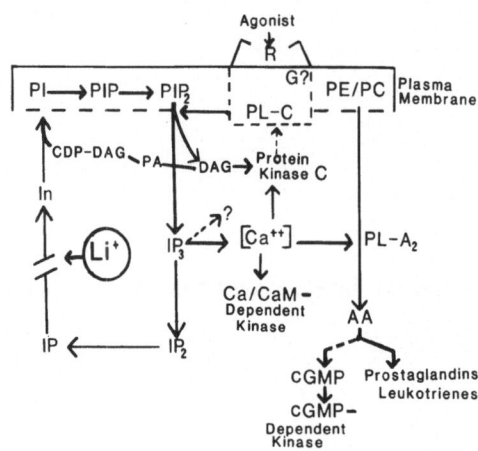

FIG. 1: Receptor coupled hydrolysis of phosphoinositides in the presence
of lithium. Receptor (muscarinic, alpha$_1$ adrenergic, etc.) activation results in phospholipase C (PL-C) mediated hydrolysis of
phosphatidylinositol 4,5 biphosphate (PIP$_2$) resulting in the production of 1,2 diacylglycerol (DAG) and inositol 1,4,5 triphosphate (IP$_3$). DAG is a potent activator of protein kinase C activity. IP$_3$ (1,4,5) has been shown to mobilize nonmitochondrial
Ca^{++} stores and increase intracellular levels of Ca^{++}. Phospholipase A$_2$ stimulates cGMP dependent kinase. IP$_3$ (1,3,4) has also
been shown to appear but its role remains unknown. IP$_3$ (1,4,5) is
metabolized to IP$_2$ and IP and finally to inositol (In) which combines with CDP-DAG to produce phosphatidylinositol (PI). Lithium
is a potent inhibitor of inositol-1-phosphatase. Abbreviations:
PE, phosphatidylenthanolamine; PC, phosphatidylcholine; AA, ara-
chadonic acid; PA, phosphatidic acid; IP, inositol phosphate; G,
guanine nucleotide binding protein possibly coupling the receptor
binding site to PL-C.

During this period of time Allison and Stewart (1971) made their initial observation that lithium decreased the concentration of inositol in
rat cerebral cortex. Additional studies revealed that lithium treated rats
accumulated inositol-1-phosphate in brain and this accumulation (as well
as the inositol reduction) could be inhibited by a muscarinic antagonist
(Allison et al., 1976). Moreover these same investigators demonstrated
that administration of cholinomimetic agents resulted in an increase in
inositol-1-phosphate in rat brain (Allison, 1978). Lithium has since been
shown to significantly affect the activity of this phospholipid second messenger system by acting as a potent inhibitor of inositol-1-phosphatase
which converts inositol-1-phosphate to inositol (Allison et al., 1976;
Hallcher and Sherman, 1980). Since the brain does not have ready access to
free inositol circulating in the plasma, the effects of chronic lithium
might be uniquely apparent in this organ system (Berridge et al., 1982;
Sherman et al., 1985). In particular, studies by Sherman et al., (1981;
1985) have shown that chronic lithium administration in the rat at clinically therapeutic levels results in more than a 10 fold sustained increase
in concentration of inositol-1-phosphate while reducing inositol concentrations in the cerebral cortex. Van Rooijen et al. (1985) recently demonstrated that PIP2 can directly inhibit PIP kinase and thus may indirectly
stimulate synthesis of this phosphoinositide substrate. Thus, relative
concentrations of these phospholipids may be critical to proper regulation
of this second messenger pathway and such dramatic shifts in the concentration of inositol phospholipids in the presence of chronic lithium may

contribute to significantly altered receptor mediator responses in certain regions of the brain.

While the action of chronic lithium on this second messenger system may be particularly significant in the brain, its physiological selectivity may be further explained by the receptor types coupled to the inositol-lipid response system and their regional distribution throughout the brain. Two neurotransmitter systems utilizing this second messenger response that are particularly associated with the neurophysiology of limbic function and implicated in the pathogenosis of affective illness are the muscarinic-cholinergic and adrenergic systems.

Cholinergic/Adrenergic Interaction in Brain

Animal studies have provided anatomical and pharmacological data for a close interaction among central cholinergic/adrenergic pathways (Robinson, 1985). The majority of these studies have involved cell bodies and/or terminal areas involved with limbic function, relevant to behavioral mechanisms and the actions of psychotropic medications. One of the best characterized is the septal hippocampal pathway with cell bodies in the medial septum and nucleus of the diagonal band projecting through the fimbria to the hippocampus where axodendritic synapses are made with pyramidal and granular cells. Stimulation of the medial septum results in enhanced turnover and release of acetylcholine from the hippocampus where it apparently has excitatory neurotransmitter activity, resulting in a reduction of hippocampal concentration of acetylcholine. The presence of muscarinic receptors have been well documented in areas of the hippocampus, with receptor subtype characterization suggesting the presence of both pre and postsynaptic localization (Raiteri et al., 1984; Mash et al., 1985). In parallel, both the septal area and hippocampus receive afferents from noradrenergic cell bodies in the brain-stem. Norepinephrine appears to activate acetylcholine turnover through the septum and amphetamine administration will enhance both hippocampal turnover and release of acetylcholine through noradrenergic terminals in the septal area (Robinson et al., 1978). Analogously, acetylcholine appears to be excitatory in the cortex and administration of amphetamine will result in stimulation of acetylcholine turnover and release in the cortex, presumably through afferent noradrenergic projections from the brain-stem to the Nucleus Basalis (Butcher and Woolf, 1983). Similarly, catecholamine and muscarinic systems are intimately associated in the neostriatum and brain stem where this interaction leads respectively to fundamental alterations in control of motor activity as well as regulation of sleep staging onset through the interaction of cholinergic neurons of the pontine-gigantocellular tegmental field and noradrenergic neurons of the locus ceruleus (Hobson et al., 1975; Robinson, 1985).

Effects of Lithium on Muscarinic/Adrenergic Activity in Brain

Research efforts attempting to examine the effects of lithium on the regulation of catecholamine and/or muscarinic neurotransmission in animal and human studies have been generally fraught with methodological inconsistencies. A number of these include: difference in dose and mode of lithium administration, acute vs chronic schedule, lack of plasma/tissue lithium concentration, use of different agonist/antagonist receptor binding ligands, different brain regions, variable assay procedures for turnover/release of neurotransmitter, etc. While these issues have resulted in problems of interpretation, it remains instructive to review some of the most salient findings briefly.

Studies examining lithium effects on presynaptic regulation of catecholamine metabolism have been generally inconclusive. A number of studies have shown no change in the concentration or rate of synthesis of norepinephrine in the brain following chronic lithium (Schubert, 1973; Poitou and Bohuon, 1975; Goodrich and Gershon, 1983). Ho et al. (1970), however, have reported that chronic lithium did decrease norepinephrine turnover in the hypothalmus while Schildkraut et al., (1969) have reported enhanced turnover of norepinephrine associated with decreased level of norepineprhine and increase in level of deaminated metabolites. Studies examining uptake of norepinephrine from synaptic cleft have reported increases following chronic lithium (Colburn et al., 1967; Kuriyama and Speken, 1970; Cameron and Smith, 1980). Furthermore, Ahulwalia and Singhal (1981) found that enhanced uptake varied and was regionally specific, occuring in the hypothalmus and striatum but reduced in the midbrain and cortex. A recent report does suggest, however, that chronic lithium will enhance potassium evoked release of norepinephrine from synaptosomal preparations from cortex (Ebstein et al., 1985).

Studies of the effects of chronic lithium on noradrenergic receptors in brain have also been inconsistent. Two laboratories have reported increased alpha$_1$ adrenergic receptor binding in cortex (Rosenblatt et al., 1979; Kafka et al., 1982), while Treiser and Kellar (1979) could find no such change. Earlier studies also reported a reduction of beta receptor binding in brain following chronic lithium treatment (Cameron and Smith, 1980; Rosenblatt et al., 1979), but a later study has not substantiated the finding in a number of brain regions (Maggi and Enna, 1980). If studies of lithium's effect on adrenergic receptor binding are equivocal, it is still possible that lithium could affect the regulation of the receptor. There are reports that chronic lithium will prevent reserpine-induced up regulation of the beta receptor (Treiser and Kellar, 1979) and supersensitivity of the noradrenergic stimulated adenylate cyclase response (Hermoni et al., 1980) in brain, without affecting reserpine-induced alpha$_1$ receptor supersensitivity. This must be viewed with caution in light of the evidence for lithium's direct inhibitory action on β adrenergic mediated norepinephrine-stimulated adenylate cyclase in rat brain, albeit at concentrations significantly higher than therapeutic (IC_{50}>5mM). Thus, chronic lithium may result in a reduction of norepinephrine-stimulated adenylate cyclase which may contribute to its possible action as an antidepressant, in as much as classical antidepressants have been shown to down-regulate the beta receptor in brain. Lithium, however, had little effect on the down-regulation of beta receptor following long term treatment with impramine (Rosenblatt et al., 1979). These findings, for the most part, may suggest that chronic lithium reduces both norepinephrine available at the synapse and the potential effectiveness of released norepinephrine at β adrenergic receptor sites.

Chronic lithium administration does seem to increase levels and turnover of acetylcholine in brain with enhanced affinity, uptake of choline and tendency toward increased synthesis and release (Miyauchi et al., 1980; Jope, 1979). Miyauchi et al. (1980) found enhanced levels of acetylcholine in midbrain and cortex with an associated increase in turnover noted in the cortex. Jope (1979) found an increased rate of synthesis of acetylcholine in striatum, hippocampus and cortex of rat brain and enhanced high affinity uptake of choline and its conversion to acetylcholine in synaptosomes of forebrain prepared from similarily treated lithium animals. This effect particularly is noted in the striatum, where enhanced potassium stimulated release of acetylcholine was observed.

Lerer (1985) and Dilsaver (1984) recently reviewed studies examining effects of chronic lithium administration on the muscarinic receptor. These studies have primarily focused upon [^{3}H]QNB binding sites with conflicting results showing a modest increase or no change. Controversy currently exists in the literature as to lithium's effect on muscarinic antagonist induced up-regulation of muscarinic receptors in rat brain. Levy et al. (1982) found that chronic lithium completely antagonized a 23% up-regulation of [^{3}H]QNB binding in whole brain, caused by the administration of atropine in rats for 5 days. On the other hand, Lerer (1985) has examined the effects of chronic lithium on scopolamine-induced up-regulation of [^{3}H]QNB sites in rat brain regions and has reported no effects of lithium on up-regulation in cortex, hippocampus, and striatum; with evidence for increased [^{3}H]QNB binding in the striatum of animals receiving only chronic lithium. However, in a parallel set of animals, chronic lithium enhanced pilocarpine induced hypothermia and catalepsy to the same extent as chronic scopolamine; and the effect of combined chronic lithium-scopolamine was found to be additive. These latter results are consistent with the observations of Samples et al. (1977) reporting that lithium significantly enhanced the toxicity of physostigmine in animals. This study, however, is confounded by extremely high levels of lithium and opposite findings have been observed by the same laboratory in later reports of antagonist properties of lithium on the behavioral effects of cholinomimetic agents (Janowsky et al., 1979). An extension of these studies was carried out by Honchar et al. (1983) wherein acute administration of lithium in rats resulted in seizures and cytopathological changes in limbic regions of the brain upon concomitant administration of cholinomimetic agents. Since these toxic affects were blocked by prior administration of atropine, it is possible that these findings may be attributable to enhanced cholinergic function in the presence of chronic lithium, or in part may be explained by parallel actions of both the cholinomimetic and lithium on a common second messenger system. Thus animal studies of the interaction of lithium and muscarinic-cholinergic activity in the brain are intriguing but remain plagued by inconsistent methodology and the limited techniques available in the past for assessment of the state of muscarinic activity in regions of the central nervous system.

Cholinergic/Adrenergic Balance in Clinical Studies

In a recent review, Janowsky et al. (1985) set forth accumulating evidence supporting a role for adrenergic/cholinergic interactions in the regulation of affect and affective disorders. The authors point out that activation of central cholinergic systems in animals results in "depressant and inhibitory behavioral affects, including lethargy, decreased locomotion and decreased self stimulation, while centrally active sympathomimetics and anticholinergic agents produce behavioral arousal, hyperactivity and increased self stimulation and stereotyped behaviors." Similar findings have been obtained in clinical studies following administration of physostigmine, i.e. tiredness, lethargy, psychomotor retardation and withdrawal. These symptoms were antagonized by the centrally-acting anticholinergic, atropine, and not by the peripheral anticholinergic, methylscopolamine; and they were not produced following administration of the peripheral cholinesterase inhibitor, neostigmine. Furthermore there is ample evidence that the behavioral effects of adrenergic vs cholinergic may be antagonistic (Janowsky et al., 1984).

Similar studies have been extended to patients with affective disorder wherein DFP and physostigmine have been shown to possess antimanic properties (Janowsky et al. 1985; Risch and Janowsky, 1984). Following administration, patients manifested a significant reduction in manic symptomatology and in addition reported feelings of anergia, fatigue, apathy, and paucity of thoughts, often associated with sadness and tearfulness. Similar findings have been observed in groups of depressive patients given the cholinergic agonist arecoline. It is also of interest that a behavioral biphasic response has been observed following administration of cholinomimetic agents to both animals and man. While the acute response is consistently inhibitory, there is a secondary phase within hours, of an apparent excitation period. In animals this is reflected in enhanced locomotor activity and in man an increase in irritability and agitation leading to a "rebounding" of manic symptomatology. This cholinergic mediated activation has been used recently to explain the antidepressant withdrawal syn-

drome (Dilsaver et al., 1983) and may be attributable to a secondary activation of noncholinergic catecholaminergic systems mediating a homeostatic/
compensatory response.

Clinical investigations examining a number of physiological variables
including: noradrenergic metabolites in urine/CSF; pupillary responses to
phenylephrine; amphetamine and AMPT administration; and neuroendocrine studies; reveal an apparent hyperresponsive adrenergic system in bipolar
depressed patients and hyporesponsive system in unipolar depressed (Sitaram
et al., 1984). Studies of patients, in particular bipolar, during both
remission and illness, reveal exaggerated physiological responses i.e. shortening of REM latency, overreactive pupillary and neuroendocrine responses
to the administration of muscarinic agonists. Most recently, a number of
these hyperresponsive muscarinic mediated physiological events have been
observed in first degree relevance of probands with a lifetime diagnosis of
major depressive disorder (Gillin, 1985). These clinical studies would
appear to be consistent with state dependent alterations of noradrenergic
activity and state independent or trait alterations of muscarinic mediated
responses associated with affective illness in some individuals.

"Thus the interaction and reciprocal relationship between cholinomi-
metics and psychostimulants as they affect or relate to each other in man
could suggest an adrenergic/cholinergic continuum model of behavior in
which behavior ranges from adrenergic activation (increased thoughts, cheer-
fulness, talkativeness, emotionality, mania etc) to cholinergic inhibition,
(decreased thoughts, dysphoria, lethargy, anergy, and depression). Clinically such a continuum could determine a normal individual's relative
aggressiveness, talkativeness, and activation and in the manic depressive,
such a balance could determine whether a psychiatric patient was activated
or manic, or motor retarded and depressed." (Janowsky et al., 1985)

In clinical studies, chronic lithium has been reported to attenuate
the behavioral effects of physotigmine and amphetamine (Janowsky et al.,
1985). However, Oppenheim et al. (1979) have found that physostigmine can
cause depressive symptoms in a significant proportion of a small group of
bipolar patients maintained on lithium. Clinical studies by Sitaram et al.
(1984) have also shown that administration of lithium (2 weeks) in normal
subjects can prevent supersensitivity of the phenylephrine (alpha$_1$ adrenergic) mediated pupillary response following treatment with alpha methyl para-
tyrosine. Interpretation of such clinical observations are necessarily
confounded by the limited number of independent studies, absence of repli-
cation, and the small sample populations involved.

Thus altered regulation of central muscarinic activity may play a per-
missive or predisposing role to the mood altering effects of changing catecholaminergic activity. Patients with primary affective disorder through
an integration of genetic, developmental and situational factors, possess
such a vulnerability to cyclic episodes of affective illness, the nature of
which (depression or mania) is determined in part by the relative balance
of muscarinic/catecholaminergic activity. The use of a pharmacological
intervention, such as lithium, may serve to alter the balance of choliner-
gic/adrenergic activity in critical regions of the brain thereby significantly altering such a predisposition to affective episodes. Studies in
our laboratory have been addressing such a mode of action for lithium.

Regulation of Muscarinic/Adrenergic Receptor Mediated Phosphoinositide
Hydrolysis in Brain

Efforts to examine the effects of chronic lithium on muscarinic and
alpha adrenergic mediated neurotransmission in the brain of animals and
man have been limited, not only by the variability of methodology as noted

above, but also by the necessity to utilize indirect means to determine alterations in receptor response. Early studies demonstrating that muscarinic and alpha$_1$ adrenergic receptors in rat brain mediate intracellular calcium mobilization through the second messenger inositol-lipid system has offered a more direct avenue with which to explore the action of lithium on the relative activity of the muscarinic/adrenergic systems of the brain (Berridge et al., 1982; Brown et al., 1984; Gonzales and Crews, 1984; Janowsky et al., 1984; Fisher et al., 1984). Our laboratory initiated a series of investigations to further define characteristics of these two receptor systems. In spite of their common second messenger response system, we have evidence that these two receptor systems appear coupled to inositol phospholipid response independently (Lenox et al., 1987). This offers the possibility for differential regulation by lithium, leading to significant alterations in the balance of muscarinic/alpha adrenergic activity in limbic regions of the brain.

Based upon modifications of the methods described by Berridge et al. (1982), we examined the inositol-1-phosphate response of brain slices exposed to either carbachol or epinephrine in the presence of 10mM lithium. Pharmacological characterization of these agonist responses using selective antagonists i.e. atropine and prazosin were consistent with carbachol acting at a muscarinic receptor site and epinephrine at an alpha$_1$ adrenergic receptor site. (Table 1) When both angonists were added simultaneously, the IP response was additive and only partially blocked by either atropine or prazosin; but completely antagonized by the presence of both antagonists. As shown in Figure 2, the kinetics of the IP response of the two agonists at maximum concentrations in both cortex and striatum differ, i.e. the accumulation of IP following stimulation with carbachol (1mM) is linear for the first 20 minutes then attains a plateau, while epinephrine (30µM) stimulation results in a more prolonged linear response. The EC_{50} for carbachol is 25µM while that for epinephrine is 3uM. Similar observations have been

TABLE 1: Inhibition of Agonist-Stimulated IP Accumulation in Hippocampal Slices by Selected Antagonists

ANTAGONIST (1µM)

AGONIST:	NONE	ATR	PHE	PRA	PRA+ATR	HEX	PRO	YOH
NONE	11.5							
CARBACHOL, 1mM	34.5	10.8 (103)		35.2 (-3)	33.5 (4)			
EPINEPHRINE, 30µM	30.5	31.4 (-5)	14.2 (86)	11.4 (100)			31.5 (-5)	32.3 (-9)
EPINEPHRINE, 30µM PLUS CARBACHOL 1mM	45.6	34.6 (32)		35.0 (31)	13.9 (93)			

Hippocampal slices were incubated as described in Table 2 and responses to agonists in the presence and absence of various antagonists were examined in the presence of 10mM lithium. The values are the means of duplicate incubations and represent the CPM in the IP fraction as a percent of the total CPM in the IP and PI fractions. Values in parentheses are percent inhibition. Abbreviations: ATR, Atropine; PHE, Phentolamine; PRA, Prazosin; HEX, Hexamethonium; PRO, Propranolol; YOH, Yohimbine.

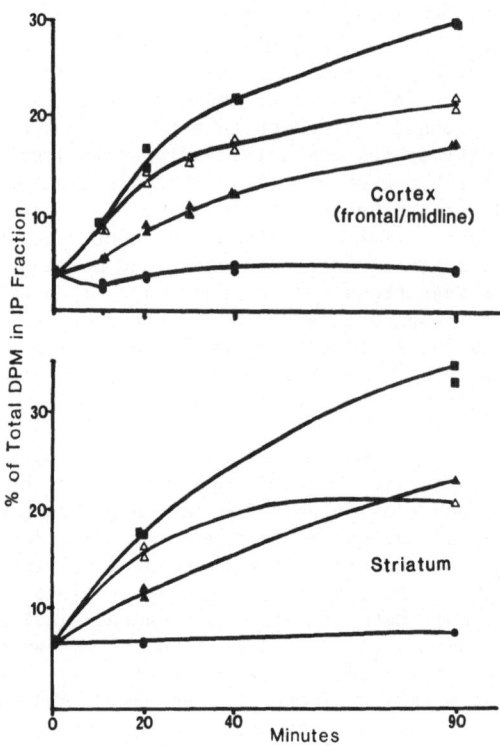

FIG. 2: Comparison of the time course of IP_1 accumulation in cortical and striatal slices following stimulation by carbachol (1mM △), epinephrine (30μM ▲), carbachol plus epinephrine (▓), or no agonists (●). Total DPM is derived from the CPM determined in the inositol (I), phosphoinositide (PI) and IP fractions.

TABLE 2: Agonist Stimulated Accumulation of IP at Different Lithium Concentrations (MEAN ± SEM)

	Epinephrine	Carbachol	Epinephrine & Carbachol
10mM LiCl	11.8 ± 1.0	9.3 ± 0.8	21.1 ± 0.5
60mM LiCl	27.4 ± 1.1	10.6 ± 1.3	35.8 ± 2.3

Hippocampal slices are incubated in Krebs-Ringer bicarbonate buffer for 30 minutes at 37°C under 95% O_2/5% CO_2 prior to being labelled with 0.3μM [^{3}H]-inositol for 1 hour. After washing to remove extracellular label, IP responses to carbachol (1mM), epinephrine (30μM) or carbachol plus epinephrine were determined in the presence of 10mM or 60mM LiCl. The indicated concentrations of the agonists have been shown to produce maximal response in this system. Simultaneous determinations of phosphoinositides (PI) and inositol (I) were carried out using modifications of the method employing anion exchange resin columns described by Berridge (1983). The results shown are from 4 separate slice preparations and each assay was carried out in duplicate. Values represent the net percent of IP accumulation (as percent of total DPM in I + PI + IP fractions) beyond basal levels.

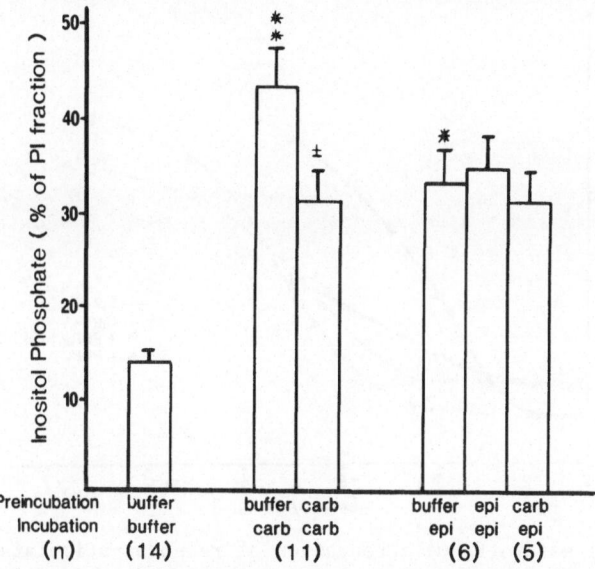

FIG. 3: Effect of preincubation with agonist on agonist-stimulated accumulation of IP. Hippocampal slices were preincubated for 2 hours
at 37°C with 1mM carbachol or 0.1mM epinephrine, or buffer alone.
The slices were then washed in KRB and carbachol (100μM) or epinephrine (10μM) stimulated accumulation of IP over 60-90 minute
period was determined in the presence of 10mM lithium chloride.
Results were expressed as percent IP of the total phosphoinositide fraction. The numbers in parentheses are the number of separate experiments. Bars represent S.E.M. *p<.05 **p<.01 - basal
vs stimulated; ±p<.05 - stimulated vs agonist preincubated.

reported in the hippocampus (Lenox et al., 1987). Addition of both agonists simultaneously appears to result in an additive response at two different concentrations of lithium (see Table 2). It is of interest that we
find that there is a preferential enhancement of the alpha$_1$ adrenergic
response in the presence of 60mM of lithium as compared to the muscarinic
mediated response.

We then carried out a series of studies to examine the possibility
that these two receptor systems might regulate differently in spite of
being coupled to the same second messenger system. Pre-exposure of hippocampal slices to carbachol resulted in a significant decrease in the subsequent IP response to carbachol (see Figure 3). In a parallel series of
experiments there was no evidence for desensitization of the IP response
to epinephrine following preincubation with either carbachol or epinephrine.
In addition, we found no evidence for changes in muscarinic receptor binding
sites as determined by [^{3}H]QNB following a similar preincubation with carbachol, suggesting that the subsensitivity to agonist may result from changes
in coupling of the muscarinic receptor to the inositol-lipid response system
beyond the binding site. Thus, the receptor response determined in the presence of lithium shows desensitization in the muscarinic system under conditions in which a similar change was not observed in the adrenergic receptor.

Studies in a number of different cell systems have demonstrated inhibitory activity of tumor-promoting phorbol esters on receptor mediated activation of inositol-lipid response (Labarca et al., 1984; Cooper et al., 1985;
Corvera and Garcia-Sainz, 1984). The phorbol esters are known to directly
activate protein kinase C which is known to be stimulated by diacylglycerol

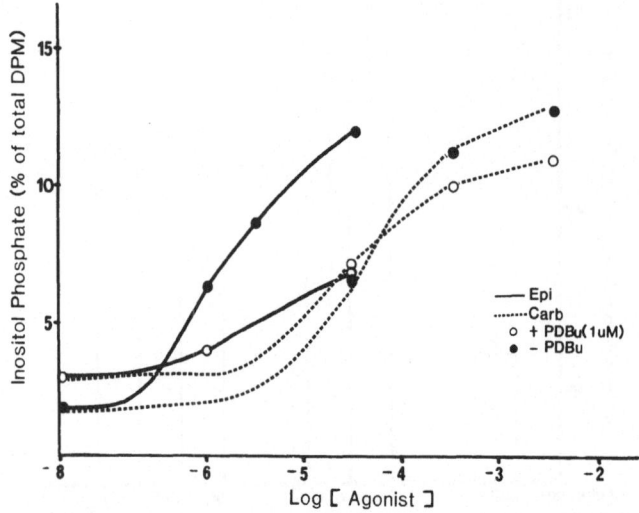

FIG. 4: Phorbol ester (PDBu) attenuation of receptor-stimulated accumulation of IP. Hippocampal slices were incubated for 10 minutes in
the absence (●) or presence (○) of 1µM PDBu prior to exposure
to various concentrations of either epinephrine (———) or carbachol (----) for 60 minutes in the presence of LiCl (10mM). Accumulation of IP was determined as indicated in Table 1 and is
expressed as a percentage of IP of the total DPM observed in the
inositol plus phosphoinositide plus IP fractions.

generated during the inositol-lipid response. It is thought, in fact, that
protein kinase C mediated phosphorylation of the receptor complex may represent a mechanism of for desensitization of receptor mediated responses such
as the alpha₁ adrenergic. Studies in our laboratory have demonstrated what
appears to be a differential sensitivity to the inhibitory action of the
phorbol ester PDBu on the two receptor mediated responses (Figure 4). If
agonist induced receptor desensitization is mediated through an activation
of protein kinase C, it may be that the time sequence of cellular events is
also significantly different in the two receptor systems.

These data clearly indicate that the muscarinic and adrenergic receptor coupled inositol phospholipid systems appear to have different kinetic
response characteristics suggesting that the two receptors act upon independent pools of substrate, enzyme, or coupling proteins, which are regulated differentially and are likely localized to different cell populations.

Studies in both GH₃ pituitary (Drummond and Raeburn, 1984) and pancreatic acinar cells (Downes and Stone, 1986) in culture, in the presence
of lithium, reveal an inhibition of resynthesis of the polyphosphoinositides
and a reduction of free inositol, which results in an accumulation of the
glycerol backbone, 1,2 DAG. The net increase in DAG would certainly be
consistent with an enhanced activation of the protein C kinase system.
Activation of protein kinase C appears to represent a site wherein long
term regulatory changes in receptor mediated function may occur, and become
amplified through the phosphorylation of key target proteins in the cell
(Nishizuka, 1984b; 1986). Moreover, its distribution in brain appears to
be primarily associated with neurons and particularly to the intrinsic neurons of the hippocampus and cortex (Worley et al., 1986). Stimulation of
protein kinase C activity in a number of cell systems results in a shift
of enzyme activity from cytosol to plasma membrane (Wolf et al., 1985).

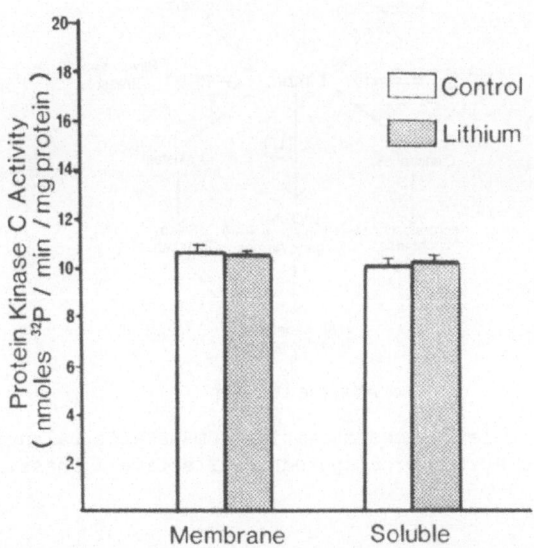

FIG. 5: Protein kinase C activity in soluble and membrane fractions of hippocampus following chronic lithium. Protein kinase C activity was determined using the phosphorylation of histone (IIIs) in accordance with procedures outlined (Lenox et al., 1986). Specific activity is measured as the difference between activity in the presence and absence of 100 nM PDBu and 5μM phosphotidylserine (PS). In the absence of PDBu/PS, protein kinase activity in the soluble and membrane fraction at 0.1μM free calcium was only 3% of net protein kinase C activity.

We have initiated a series of investigations to explore the possibility that chronic lithium may alter the balance of muscarinic vs adrenergic receptor activity by altering the protein kinase C activity and/or its site of action in the brain. Initial studies of protein kinase C activity in the hippocampus of animals treated with lithium for a period of three weeks revealed no evidence for altered specific activity of the enzyme in the crude cytosol or membrane fraction (Figure 5). However, it is entirely possible that chronic effects of lithium on protein kinase C activity may not be manifested by a sustained alteration in the relative activity of protein kinase C in the cell, but rather by long term changes in the phosphorylation state of specific protein substrates.

In as much as the assay determining distribution of protein kinase C activity utilizes an exogenous protein (histone III) to determine phosphorylation activity, we have recently initiated additional studies in our laboratory to examine possible endogenous substrates for protein kinase C activity in the presence of chronic lithium. In light of the substantive alterations in the inositol-lipid pathway in the brain in the presence of chronic lithium at clinically relevant therapeutic levels; long term alterations in cellular events, such as receptor response, may be mediated by alterations in the phosphorylation state of as-yet unknown protein substrates. Studies by Gispen (1986) and others (Jolles et al., 1980) have described a phosphoprotein in neuronal membranes, B-50, which is a substrate for protein kinase C. The phosphorylated form of B-50 inhibits PIP kinase activity and may represent a form of feedback inhibition for receptors coupled to phosphoinositide hydrolysis. Furthermore, $ACTH_{1-24}$ which inhibits the phosphorylation of B-50 can antagonize phorbol-mediated inhibition of muscarinic-stimulated inositol phosphate accumulation in hippocampal slices

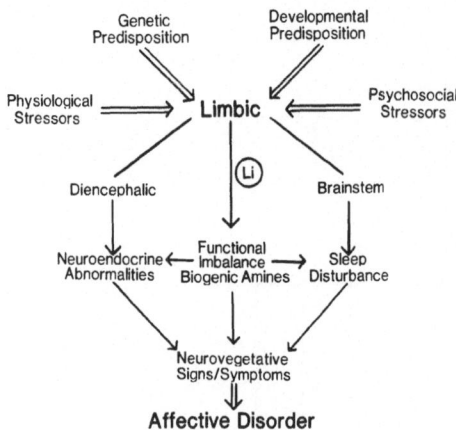

FIG. 6: Lithium's role in reestablishing homeostatic balance in an unstable limbic system predisposed to affective illness.

(Schrama et al., 1986). It is also of interest that the 48kd phosphoprotein substrate of protein kinase C in rat hippocampus described by (Akers et al., 1986) appears to be identical to the B-50 protein and has been implicated in the mediation of long term potentiation (LTP), thought to be a unique feature of the hippocampal complex and function of the limbic system. Chronic lithium may alter the dynamic state of phosphorylation of such specific proteins responsible for modulating receptor activity in various regions of the brain.

     A constellation of factors: genetic, developmental, physiological, and psychosocial; all contribute to the eventual manifestation of an affective episode at a particular point in time (Figure 6). Dysfunction associated with the limbic system and its midbrain and hypothalamic connections can result in the neurovegetative signs/symptoms (sleep disturbance, anorexia, autonomic changes, endocrine changes, mood disturbance, psychomotor agitation, etc.) manifested as depression and/or mania. As described herein, it appears that cholinergic and adrenergic neurotransmission may uniquely be involved in maintaining normal physiological functioning of this region of the brain. In particular, there is evidence that up-regulation of the muscarinic receptor system may represent a predisposing factor in the central nervous system, making critical limbic regions of the brain more vulnerable to fluctuations in catecholamine activity which are outside of homeostatic control. Lithium induced alterations in inositol lipid metabolism may differentially alter muscarinic/adrenergic balance in specific regions of the limbic system, resulting in long term adaptive regulation of receptor mediated activity and the restabilization of mood and behavior.

ACKNOWLEDGMENTS

     The author would like to acknowledge the contributions of Drs. Ellis, Hendley and Ehrlich, the fine technical assistance of Mrs. Ann Wood, and the typing support of Miss GiGi Beach. Studies described herein were supported in part by NIA PHS R01-AG05214 and NIMH PHS R01-41571-01.

REFERENCES

Ahluwalia, P and Singhal, R. L., Monamine uptake into synaptosomes from various regions of rat brain following lithium administration and withdrawal, Neuropharmacology 20:483-487 (1981).

Akers, R. F., Lovinger, D. M., Colley, P. A., Linden, D. J. and Routtenberg, A., Translocation of PKC activity may mediate hippocampal long term potentiation, Science 231:587-589 (1986).

Allison, J. H., Lithium and brain myo-inositol metabolism, in: Cyclitols and phosphoinositides, pp. 507, Wells, W. W. and Eisenberg, F., Jr. (eds.), Academic Press, New York (1978).

Allison, J. H. and Blisner, M. E., Inhibition of the effect of lithium on brain inositol by atropine and scopolamine, Biochem. Biophys. Res. Commun. 68:1332-1338 (1976).

Allison, J. H., Blisner, M. E., Holland, W. H., Hipps, P. P. and Sherman, W. R., Increased brain myo-inositol 1-phosphate in lithium-treated rats, Biochem. Biophys. Res. Commun. 71:664-670 (1976).

Allison, J. H. and Stewart, M. A., Reduced brain inositol in lithium-treated rats, Nature (London) New. Biol. 233:267-268 (1971).

Berridge, M. J., Phosphatidyinositol hydrolysis: A multifunctional transducing system, Mol. Cell Endocrinol. 24:115-140 (1981).

Berridge, M. J., Inositol triphosphate and diacylglycerol as second messengers, Biochem. J. 220:345-360 (1984).

Berridge, M. J., Downes, C. P. and Hanley, M. R., Lithium amplifies agonist-dependent phosphatidylinositol responses in brain and salivary glands, Biochem. J. 206:587-595 (1982).

Berridge, M. J., Rapid accumulation of inositol trisphosphate reveals that agonists hydrolyse polyphosphoinositides instead of phosphatidylinositol, Biochem. J. 212:849-858 (1983).

Brown, E., Kendall, D. A. and Nahorski, S. R., Inositol phospholipid hydrolysis in rat cerebral coutical slices: I. receptor characterisation, J. Neurochem. 42:1379-1387.

Bunney, W. E. and Garland, B. L., Lithium and its possible modes of action in: Neurobiology of Mood Disorders, pp. 731-743, Post, R. M. and Ballenger, J. C. (eds.), Williams & Wilkins, Baltimore (1984).

Butcher, L. L. and Woolf, N. J., Cholinergic systems: Synopsis of anatomy and overview of physiology and pathology, in: The Biological Substrates of Alzheimer's Disease, pp. 73-86, Scheibel, A. B., Wechsler, A. F. and Brazier, M. A. B. (eds.), Academic Press, New York (1986).

Cameron, O. G. and Smith, C. B., Comparison of acute and chronic lithium treatment on [^{3}H]-norepinephrine uptake by rat brain slices, Psychopharmacology 67:81-85 (1980).

Colburn, R. W., Goodwin, F. K., Bunney, W. E., Jr. and Davis, J. M., Effect of lithium on the uptake of noradrenaline by synaptosomes, Nature (London) 215:1395-1397 (1967).

Cooper, R. H., Coll, K. E. and Williamson, J. R., Differential effects of phorbol ester on phenylephrine and vasopressin-induced Ca^{2+} mobilization in isolated hepatocytes, J. Biol. Chem. 260:3281-3288 (1985).

Corvera, S. and Garcia-Sainz, J. A., Phorbol esters inhibit alpha adrenergic stimulation of glycogenolysis isolated rat hepatocytes, Biochem. Biophys. Res. Comm. 119:1128-1133 (1984).

Dilsaver, S. C., Lithium's effects on muscarinic receptor binding parameters: a relationship to therapeutic efficacy, Biol. Psychiatry 19:1551-1565 (1984).

Dilsaver, S. C., Konfol, Z., Sackellars, J. C. and Greden, J. F., Antidepressant withdrawal syndromes: evidence supporting the cholinergic overdrive hypothesis, J. Clin. Psychopharm. 3:157-164 (1983).

Downes, C. P. and Stone, M. A., Lithium-induced reduction in intracellular inositol supply in cholinergically stimulated parotid gland, Biochem. J. 234:199-204 (1986).

Drummond, A. H. and Raeburn, C. A., The interaction of lithium with thyrotropin-releasing hormone-stimulated lipid metabolism in GH pituitary tumour cells, Biochem. J. 224:129-136 (1984).

Ebstein, R. P., Lerer, B., Shlaufman, M. and Belmaker, R. H., ECS and noradrenaline release, in: ECT: Basic mechanisms, pp. 62-66, Lerer, B., Weiner, R. D. and Belmaker, R. H. (eds.), John Libbey and Company Limited, London (1984).

Fisher, S. K., Figueiredo, J. C. and Bartus, R. T., Differential stimulation of inositol phospholipid turnover in brain by analogs of oxotremorine, J. Neurochem. 43:1171-1179 (1984).

Gillin, C., Muscarinic supersensitivity in affective illness, Proceedings of the American Psychiatric Association, 138:124 (1985).

Gispen, W. H., Phosphoprotein B-50 and phosphoinositides in brain synaptic plasma membranes: a possible feedback relationship, Biochem. Soc. Transact. 14:163 (1986).

Gonzales, R. A. and Crews, F. T., Characterization of the cholinergic stimulation of phosphoinositide hydrolysis in rat brain slices, J. Neurosci., 4:3120-3127 (1984).

Goodnick, P. and Gershon, S., Lithium, in: Handbook of Neurochemistry, pp. 001.9, Lajtha, A. (ed.), Plenum Press, New York (1983).

Hallcher, L. M. and Sherman, W. R., The effects of lithium ion and other agents on the activity of myo-inositol-1-phosphatase from bovine brain, J. Biol. Chem. 255:10896-10901 (1980).

Hermoni, M., Lerer, B., Ebstein, R. P. and Belmaker, R. H., Chronic lithium prevents reserpine-induced supersensitivity of adenylate cyclase, J. Pharm. Pharmacol. 32:510-511 (1980).

Ho, A. K. S., Loh, H. H., Graves, F., Hitzemann, R. J. and Gershon, S., The effect of prolonged lithium treatment on the synthesis rate and turnover of monoamines in brain regions of rats, Eur. J. Pharmacol. 10: 72-78 (1970).

Hobson, J. A., McCartey, R. W. and Wyzinski, P. N., Sleep cycle oscillation: Reciprical discharge by two brain stem neuronal groups, Science 189:55-58 (1975).

Hokin, M. R. and Hokin, L. E., Enzyme secretion and the incorporation of ^{32}P into phospholipids of pancreas slices, J. Biol. Chem. 203:967-977 (1953).

Hokin, M. R. and Hokin, L. E., Effects of acetylcholine on phospholipids in the pancreas, J. Biol. Chem. 209:549-558 (1954).

Honchar, M. P., Olney, J. W. and Sherman, W. R., Systemic cholinergic agents induce seizures and brain damage in lithium-treated rats, Science 220:323-325 (1983).

Janowsky, A., Labarca, R. and Paul, S. M., Characterization of neurotransmitter receptor-mediated phosphotidylinositol hydrolysis in the rat hippocampus, Life Sci. 35:1953-1961 (1984).

Janowsky, D. S., El-Yousef, M. K., Davis, J. M. and Sekerke, H. J., A cholinergic-adrenergic hypothesis of mania and depression, Lancet ii:632-635 (1972).

Janowsky, D. S., Abrams, A. A., Groom, G. P., Judd, L. L. and Cloptin, P., Lithium antagonizes cholinergic behavioral effects in rodents, Psychopharmacology, 63:147-150 (1979).

Janowsky, D. S., Risch, S. C., Judd, L. L., Huey, L. Y. and Parker, D. C., Brain cholinergic system and the pathogenesis of affective disorders, in: Central Cholinergic Mechanisms and Adaptive Dysfunctions, pp. 309-333, Singh, M. D., Warburton, D. M., Lal, H. and Mason, B. (eds.), Plenum Press, New York (1985).

Jolles, J., Zwiers, H., Van Dongen, C. J., Schotman, P., Wirtz, K. W. A. and Gispen, W. H., Modulation of brain polyphosphoinositide metabolism by ACTH sensitive protein phosphorylation, Nature 286:623 (1980).

Jope, R. S., Effects of lithium treatment in vitro and in vivo on acetylcholine metabolism in rat brain, J. Neurochem. 33:487-495 (1979).

Kafka, M. S., Wirz-Justice, A., Naber, D., Marangos, P. J., O'Donohue, T. L. and Wehr, T. A., The effect of lithium on circadian neurotransmitter receptor rhythms, Neuropsychobiology 8:41-50 (1982).

Kuriyama, K. and Speken, R., Effect of lithium on content and uptake of norepinephrine and 5-hydroxytryptamine in mouse brain synaptosomes and mitochondria, Life Sci. 9:1213-1220 (1970).

Labarca, R., Janowsky, A., Patel, J. and Paul, S. M., Phorbol esters inhibit agonist-induced [3H] inositol-1-phosphate accumulation in rat hippocampal slices, Biochem. Biophys. Res. Comm., 123:703-709 (1984).

Lenox, R. H., Hendley, D. D. and Ellis, J., Receptor coupled hydrolysis of phosphoinositides in hippocampal slices: Muscarinic vs adrenergic, New York Acad. Sci. 736 (1987).

Lenox, R. H., Meyers, S., Hendley, D., Ellis, J. and Ehrlich, Y. H., Effects of chronic lithium on protein kinase C activity in rat brain, Society for Neuroscience Abstracts, 12:566 (1986).

Lerer, B., Studies on the role of brain cholinergic systems in the therapeutic mechanisms and adverse effects of ECT and lithium, Biol. Psychiat. 20:20-40 (1985).

Lerer, B. and Stanley, M., Effect of chronic lithium on cholinergically mediated responses and [3H]QNB binding in rat brain, Brain Res. 334: 211-219 (1985).

Levy, A., Zohar, J. and Belmaker, R. H., The effect of chronic lithium pretreatment on rat brain muscarinic receptor regulation, Neuropharmacology 21:1199-1201 (1982).

Maggi, A. and Enna, S. J., Regional alterations in rat brain neurotransmitter systems following chronic lithium treatment, J. Neurochem 34: 888-889 (1980).

Mash, D. C., Flynn, D. D. and Potter, L. T., Loss of M2 muscarine receptors in the cerebral cortex in Alzheimer's disease and experimental cholinergic denervation, Science 228:1115-1117 (1985).

Michell, R. H., Inositol phospholipids and cell surface receptor function, Biochem. Biophys. Acta. 415:81-147 (1975).

Miyauchi, T., Oikawa, S. and Kitada, Y., Effects of lithium chloride on the cholinergic system in different brain regions in mice, Biochem. Pharmacol. 29:654-657 (1980).

Nishizuka, Y., The role of protein kinase C in cell surface signal transduction and tumour promotion, Nature 308:693-698 (1984a).

Nishizuka, Y., Turnover of inositol phospholipids and signal transduction, Science 225:1365-1370 (1984b).

Nishizuka, Y., Studies and perspectives of protein kinase C, Science 233: 305-312 (1986).

Oppenheim, G., Ebstein, R. P. and Belmaker, R., Effect of lithium on the physostigmine-induced behavioral syndrome and plasma cyclic GMP, J. Psychiat. Res. 15:133-138 (1979).

Poitou, P. and Bohuon, C., Catecholamine metabolism in the rat brain after short and long term lithium administration, J. Neurochem. 25:535-537 (1975).

Raiteri, M., Leardi, R. and Marchi, M., Heterogeneity of presynaptic muscarinic receptors regulating neurotransmitter release in the rat brain, J. Pharmacol. and Exp. Ther. 228:209-214 (1984).

Risch, S. G. and Janowsky, D. S., Cholinergic-adrenergic balance in affective illness, in: Neurobiology of Mood Disorders pp. 652-663, Post R. M. and Ballenger, J. C. (eds.), Williams and Wilkins, Baltimore (1984).

Robinson, S. E., Cholinergic pathways in the brain, in: Central Cholinergic Mechanisms and Adaptive Dysfunctions, pp. 37-61, Singh, M. M., Warburton, D. M., Lal, H., Mason, B. (eds.), Plenum Press, New York (1985).

Robinson, S. E., Cheney, D. L. and Costa, E., Effect of nomifensine and other antidepressant drugs on acetylcholine turnover in various regions of rat brain, Naunyn. Schmeid. Arch. Pharmacol. 304:263-269 (1978).

Rosenblatt, J. E., Pert, C. B., Tallman, J. F., Pert, A. and Bunney, W. E.,

Jr., The effect of imipramine and lithium on alpha and beta receptor binding in rat brain, Brain Res. 160:186-191 (1979).

Samples, J., Janowsky, D. S., Pechnick, R. and Judd, L. L., Lethal effects of physostigmine plus lithium in rats, Psychopharmacology 52:307-309 (1977).

Schildkraut, J. J., Logue, M. A. and Dodge, G. A., The effect of lithium salts on the turnover and metabolism of norepinephrine in rat brain, Psychopharmacology 14:135-141 (1969).

Schrama, L. H., De Graan, P. N. E., Eichberg, J. and Gispen, W. H., Feedback control of inositol phospholipid response in rat brain is sensitive to ACTH, Eur. J. Pharmacol. 121:403 (1986).

Schubert, J., Effect of chronic lithium treatment on monoamine metabolism in the rat brain, Psychopharmacology 32:301-311 (1973).

Sherman, W. R., Leavitt, A. L., Honchar, M. P., Hallcher, L. M. and Phillips, B. E., Evidence that lithium alters phosphoinositide metabolism: chronic administration elevates primarily d-myo-inositol-1-phosphate in cerebral cortex of the rat, J. Neurochem. 36:1947-1951 (1981).

Sherman, W. R., Munsell, L. Y., Gish, B. G. and Honchar, M. P., Effects of systemically administered lithium on phosphoinositide metabolism in rat brain, kidney, and testis, J. Neurochem. 44:798-807 (1985).

Sitaram, N., Gillin, J. C. and Bunney, W. E., Cholinergic and catecholaminergic receptor sensitivity in affective illness: Strategy and theory, in: Neurobiology of Mood Disorders, pp. 629-651, Post, R. M. and Ballenger, J. C. (ed.), Williams and Wilkins, Baltimore (1984).

Treiser, S. and Kellar, K. J., Lithium effects on adrenergic receptor supersensitivity in rat brain, Eur. J. Pharmacol. 58:85-86 (1979).

Van Rooijen, L. A. A., Rossowska, M. and Bazan, N. G., Inhibition of phosphatidylinositol-4-phosphate kinase by its product phosphatidylinositol-4,5-bisphosphate, Biochem. Biophys. Res. Commun. 126:150 (1985).

Williamson, J. R., Role of inositol lipid breakdown in the generation of intracellular signals, Hypertension 8:140-156 (1986).

Wolf, M., Levine, H. III, May, S. W., Cuatrecasas, P. and Sahyoun, J., A model for intracellular translocation of protein kinase C involving synergism between Ca^{++} and phorbol esters, Nature 317:546-549 (1985).

Worley, P. F., Baraban, J. M. and Snyder, S. H., Heterogeneous localization of protein kinase C in rat brain: Autoradiographic analysis of phorbol ester receptor binding, J. Neurosci. 6(1):199-207 (1986).

AGONIST-STIMULATION OF CEREBRAL PHOSPHOINOSITIDE TURNOVER

FOLLOWING LONG-TERM TREATMENT WITH ANTIDEPRESSANTS

Pamela D. Butler and Amiram I. Barkai

New York State Psychiatric Institute and Department
of Psychiatry, College of Physicians and Surgeons
Columbia University, 722 West 168th Street
New York, NY 10032

ABSTRACT

Receptor-mediated stimulation of the formation of inositol phosphates
(IP) in cerebral tissue may serve as a useful tool for studying long-term
changes in the function of serotonin-2 (5-HT2), alpha-1-adrenergic (a1),
and muscarinic-cholinergic (musc) receptors. In this study we have evaluated the effects of chronic treatment with various antidepressants on
receptor-mediated formation of IP in rat brain. Imipramine (IMI: 10 mg/
kg/day; 14 days), Bupropion (BUPR: 40 mg/kg/day; 14 days), Lithium (Li:
0.5% in diet; 7 days) and electroshock treatment (EST: 20-30 mA/day; 7
days) were investigated. Cross-chopped slices of cerebral cortex from control and treated rats were prelabelled with myo-^{3}H-inositol in HEPES buffer
containing 11.1 mM LiCl. Accumulation of IP was measured in the presence
and absence of serotonin (5-HT, 10 uM), norepinepherine (NE, 5 uM), and
carbamylcholine (CCH, 100 uM). Values for agonist-stimulated IP formation
in control rats were: 5-HT=123 ± 5%; NE=268 ± 16%; CCh=205 ± 21% of the
basal level. The IP response to 5-HT was significantly lower following
BUPR and higher following EST. Responses to NE and CCH were significantly
lower following BUPR treatment but were not affected by the other antidepressant treatments. These observations are consistent with results of
receptor-binding studies indicating up-regulation of 5-HT2 receptors by
EST but are not consistent with studies showing down-regulation of 5-HT2
receptors by IMI and a lack of effect on 5-HT2 receptors by BUPR. Our
results are not supportive of the notion, based mainly on [^{3}H]prazosin
binding studies, that a1 receptors are up-regulated by EST as well as by
different antidepressant drugs.

The role of phospholipids, particularly phosphatidylinositols, in the
transmission of signals across neuronal membranes has been investigated
extensively in recent years. It has been shown that a number of neurotransmitters and hormones may stimulate the phospholipase C (PLC)-catalyzed
hydrolysis of phosphoinositides, and that such stimulation is receptor-
mediated (for reviews see: Abdel-Latif, 1982; Berridge, 1984; Fisher et
al., 1984b; Hawthorne & Pickard, 1979; Michell, 1975; Nishizuka, 1984a).
In the cerebral cortex, the enzymic hydrolysis of phosphoinositides has
been shown to be coupled to serotonin-2 (5-HT2) receptors (Brown et al.,
1984; Conn & Sanders-Bush, 1984; 1985; Kendall & Nahorski, 1985), alpha-1-
adrenergic (a1) receptors (Brown et al., 1984; Gonzales & Crews, 1985;

Schoepp et al., 1984), and muscarinic cholinergic (musc) receptors (Brown et al., 1984; Fisher et al., 1984a; Gonzales & Crews, 1984; 1985). Recent work indicates that the receptor-mediated effects on the metabolism of phosphatidylinositol (PI) and its phosphorylated derivatives phosphatidy-linositol-4-phosphate (PIP) and phosphatidylinositol-4,5-bisphosphate (PIP2) primarily involves the hydrolysis of PIP2 (Agranoff et al., 1983; Berridge, 1983; Martin, 1983). The products of PIP2 hydrolysis are diacylglycerol (DAG) and inositol-1,4,5-triphosphate (IP3) which may both act as second messengers. DAG activates protein kinase C and also serves as a source for arachidonic acid and the subsequent synthesis of prostaglandins, throm-boxanes, and leukotrienes (Bell et al., 1979; Nishizuka, 1984b; Okazaki et al., 1981), whereas IP^3 acts as a second messenger by stimulating Ca^2 release from intracellular vesicles (Berridge & Irvine, 1984; Streb et al., 1983).

Until recently it was difficult to study agonist-stimulated hydrolysis of phosphoinositides because of their rapid resynthesis via the phosphati-dylinositol turnover cycle (for reviews see Fisher et al., 1984b; Michell, 1975). However, the finding that lithium (Li) blocks the degradation of myo-inositol 1-phosphate (Allison et al., 1976; Sherman et al., 1981) allowed agonist-stimulated formation of labelled inositol phosphates (IP) to be measured in brain slices incubated in the presence of Li (Berridge et al., 1982). The amount of $[^3H]IP$ accumulation after preincubation with $[^3H]$inositol reflects the metabolism of phosphoinositides which participate in receptor coupled activation of the Ca^{2+}-dependent responses of the cell. Thus, receptor-mediated stimulation of the formation of IP in cerebral tis-sue may serve as a useful tool for studying possible changes in the func-tion of 5-HT2, a1, and musc receptors induced by a variety of manipulations, including long-term antidepressant treatment.

Antidepressants were originally thought to exert their therapeutic effect by enhancing noradrenergic neurotransmission (Schildkraut, 1965; Schildkraut & Kety, 1967). This notion was based on the observation that both monoamineoxidase- (MAO) inhibitor and tricyclic antidepressants increased the availability of NE in the brain (Schildkraut, 1965; Bunney & Davis, 1965). But, while such NE accumulation occurs after acute admi-nistration (Iversen & Mackay, 1979), several weeks are required for anti-depressants to exert their therapeutic effect (Oswald et al., 1972). More recently, it has been found that chronic, but not acute, antidepressant treatment produces changes in both serotonergic and noradrenergic recep-tors. The decrease in β-adrenergic receptor density after chronic treat-ment with antidepressant drugs and EST (Banerjee et al., 1979; Pandey et al., 1979b; Wolfe et al., 1977) has received much attention. This and the finding that chronic antidepressant treatment decreases the sensitivity of the β-adrenergic receptor-coupled adenylate cyclase system to NE and to the β-adrenergic agonist isoproterenol (for reviews see Charney et al., 1981; Sulser, 1979; Sulser et al., 1978) has led to a hypothesis opposite to that originally proposed, namely that antidepressants work by decreasing noradrenergic transmission (Sulser et al., 1983; Sulser et al., 1978). However, changes involving 5-HT2 and a1 receptors may also be important for the therapeutic effect of antidepressants. Following chronic treatment with antidepressant drugs, there is a decreased density of 5-HT2 receptors (Fuxe et al., 1982; Kellar et al., 1981b; Kendall et al., 1982; Peroutka & Snyder, 1980), while chronic electroshock treatment (EST) produces an increased density of 5-HT2 receptors (Green et al., 1983a,b; Kellar et al., 1981b; Vetulani et al., 1981). Chronic antidepressant treatment failed to produce changes in a1 receptors when $[^3H]$WB-4101 was used as a ligand (Bergstrom & Kellar, 1979; Peroutka & Snyder, 1980; Rosenblatt et al., 1979; Tang et al., 1981). However, recent studies using the more specific a1 ligand $[^3H]$prazosin have found an increased density of a1 receptors following chronic treatment with a number of antidepressant drugs and EST (Vetulani,

1983; Vetulani & Antkiewicz-Michaluk, 1985). These recent observations led
Vetulani and Antkiewicz-Michaluk (1985) to suggest that since many different
antidepressant drugs and EST produce up-regulation of a1-adrenoceptors in
conjunction with down-regulation of β- and a2-adrenoceptors, these changes
may play an important role in their therapeutic action.

While a number of antidepressant drugs produce anticholinergic effects
(Danielsson et al., 1985; Snyder & Yamamura, 1977), they generally fail to
alter the number or affinity of musc receptors (see Charney et al., 1981).
On the other hand, studies examining the effects of chronic EST have yielded
controversial results. Some studies report little change in [^{3}H]QNB binding
in rat cortex or hippocampus (Deakin et al., 1981; Kellar et al., 1981a)
while others have found significant down-regulation of musc receptors in
these areas following chronic EST (Dashieff et al., 1982; Lerer et al.,
1983). However, there are very few studies bearing directly on the possibility that antidepressant drugs or EST may alter receptor-coupled phosphoinositide metabolism.

The goal of the present experiments was to examine whether long-term
treatment with antidepressants, which are expected to produce changes in
5-HT$_2$, a1, and perhaps musc receptors, would produce parallel changes in
5-HT-, NE-, or CCH-stimulated accumulation of [^{3}H]IP in cerebral tissue.
This approach would provide information regarding whether or not a change
in the function of these receptors had taken place during chronic treatment
with antidepressants.

A recent report by Kendall and Nahorski (1985) showed that chronic
treatment with imipramine (IMI) and iprindole did indeed decrease 5-HT-,
but not NE-, or CCH-stimulated [^{3}H]IP accumulation in the rat cortex. However, Janowsky et al. (1985) found that in the hippocampus, IMI-treated
animals showed increased 5-HT-, NE-, and CCH-stimulated accumulation of
[^{3}H]IP. Because of this apparent discrepancy we decided to include IMI in
the present studies.

The effects of the novel antidepressant bupropion (BUPR) were also
studied. BUPR was of interest because, while it is clinically effective
(Halaris et al., 1981; Maxwell et al., 1981), its mechanism of action is
presently obscure. It is neither a tricyclic nor an MAO inhibitor (Ferris
et al., 1981; Soroko et al., 1977), and it appears to be free of anticholinergic or cardiovascular effects (Cooper et al., 1984). Some studies have
found a down-regulation of β receptors in rat and mouse brain cortex following chronic BUPR (Gandolfi et al., 1983; Pandey et al., 1979a; Perumal
et al., 1986; Sellinger-Barnett et al., 1980). However, others have failed
to find any changes in β-adrenergic, 5-HT$_2$, or a2 receptors in rat cortex
(Ferris & Beaman, 1983) or 5-HT$_2$ receptors in mouse cortex (Perumal et al.,
1986). Further, the receptor changes are produced only by relatively high
doses of BUPR. It has been suggested that B.W.306U, a metabolite of BUPR,
may be responsible for its therapeutic effects. The accumulation of B.W.
306U in human plasma is 20-100 times greater than that of the parent compound (Cooper et al., 1984) and like many other antidepressants (Iversen &
Mackay, 1979), it inhibits NE uptake and is twice as active as BUPR in
doing so (Perumal et al., 1986). In addition, it decreases the density of β
receptors in limbic forebrain but not frontal cortex of the mouse (Perumal
et al., 1986). Since the mechanism of action of BUPR is unclear, it was
of interest to see if its effects on agonist-stimulated IP formation would
be similar to those of other antidepressant treatments.

EST represents a unique antidepressant treatment which, unlike psychoactive drugs, does not intervene chemically with the physiologic mechanisms
involved in the therapeutic effect. While repeated EST retains a central
role in the treatment of certain psychiatric disorders, it is not clearly

understood how it exerts its therapeutic effects. The bulk of research on the biochemical and physiologic mechanisms of this treatment have been discussed comprehensively in a recent symposium (Malitz & Sackeim, 1986). Like antidepressant drugs, repeated EST produces changes in the responsiveness of cerebral monoamine receptors which generally parallel the time course of its therapeutic effects. As noted above, the long-term effects of EST on adrenoreceptors are generally similar to those observed after treatment with antidepressant drugs, but in contrast to antidepressant drugs EST has been shown to up-regulate 5-HT$_2$ receptors. Kellar and Stockmeier (1986) have suggested that the opposite effects of EST and antidepressant drugs on 5-HT$_2$ receptors may be related to their therapeutic effects in the treatment of depressions which have different underlying pathologies. Certain kinds of depression may be treated successfully with EST whereas treatment with antidepressant drugs is ineffective and might even exacerbate the illness. Since opposite changes in 5-HT$_2$ receptors are seen after long-term EST and IMI treatment, it was of particular interest to examine whether these treatments would also produce opposite effects on 5-HT$_2$ stimulated [^{3}H]IP accumulation. To our knowledge, there have been no reports describing the effects of long-term EST on agonist-stimulated formation of [^{3}H]IP.

The effects of chronic Li were also investigated in the present study. While it is unclear how Li exerts its therapeutic effects, it is known that chronic Li produces changes in cerebral neurotransmitter systems and this could also affect agonist-stimulated accumulation of [^{3}H]IP. Li appears to increase 5-HT activity (Broderick & Lynch, 1982; Furukawa et al., 1979; Meltzer, 1984; Sangdee & Franz, 1980) inhibit 5-HT uptake (Ahluwalia & Singhal, 1981), and down-regulates 5-HT receptors in the rat brain (Maggi & Enna, 1980; Treiser & Kellar, 1980). Li also appears to increase Ach synthesis and release (Jope, 1979). It may also increase cholinergic-mediated catalepsy and hypothermia (Lerer & Stanley, 1985). However, there are conflicting results with regard to muscarinic-cholinergic receptor changes following chronic Li treatment in rats. Li treatment has been found to produce no changes in [^{3}H]QNB binding in rat cortex, hippocampus, or striatum (Maggi & Enna, 1980), to increase muscarinic binding in the striatum but not hippocampus, cortex, or hypothalamus (Lerer & Stanley, 1985), or to increase binding in the whole brain (Levy et al., 1982) and forebrain (Kafka et al., 1982). A study of the receptor-mediated stimulation of [^{3}H]IP formation in Li-treated rats was therefore initiated to address the question of whether LI treatment may have an effect on signal transduction in the cholinergic, adrenergic, and serotonergic systems.

METHODS

Reagants, Drugs, and Chemicals

myo (2-^{3}H) Inositol dissolved in ethanol was obtained from American Radiolabeled Chemicals, Inc. (St. Louis, MO). BUPR was the gift of Burroughs Wellcome Laboratories and kindly provided by Dr. Ray Suckow. Lithium (0.5%) in rat chow was prepared by Ralston Purina Co., (Richmond, IND). IMI, serotonin, norepinepherine, carbamylcholine, ammonium formate, formic acid, and the chemicals used in making the buffer solution were obtained from Sigma Chemical Co., (St. Louis, MO). IMI and BUPR were delivered via osmotic pumps (model 2ML2) which were obtained from Alza Corp. (Palo Alto, CA). Dowex resin (AG 1-X8, 100-200 mesh in the formate form) was purchased from BioRad Laboratories (Richmond, CA). Other reagants and solvents were purchased from commerical sources.

The buffer solution was made according to Fisher et al. (1984a). Buffer A contained: 142 mM NaCl, 5.6 mM KCl, 2.2 mM CaCl$_2$, 3.6 mM NaHCO3, 1 mM Mg-Cl$_2$, 5.6 mM D-glucose, and 30 mM 4-(2-hydroxyethyl)-1-piperazineethanesulfonic acid (HEPES-N+) buffer (PH=7.4).

Subjects and Experimental Design

Male, Sprague-Dawley rats (250-300 g) were used. Food and water were provided ad libitum. Animals were divided into four groups. Each experimental animal in a group was paired with a control animal on the basis of weight. The experimental animals in Groups 1 (n=7) and 2 (n=6) received 10 mg/kg IMI or 40 mg/kg BUPR respectively for 14 days via subcutaneously implanted Alzet pumps. The matched controls for each group received sham operations. The pumps were removed under either anesthesia 24 hrs before the animals were sacrificed by decapitation. Experimental animals in Group 3 (n=8) received EST (20-30 mA; 1 sec) once a day for seven days and were sacrificed 24 hrs after the end of treatment. The experimental animals in Group 4 (n=8) were maintained on a diet consisting of 0.5% Li in chow for 7 days. This diet was continued until the time of sacrifice. In each group, the drug-treated animal and its matched control were sacrificed and brain slices incubated within 15 min of each other.

Measurement of [^{3}H]IP Accumulation

The [^{3}H] myo-inositol was purified the day before each experiment. The ethanol solvent was evaporated from a 100 uCi aliquot of [^{3}H]inositol solution under a stream of N_2 and then 0.5 ml H_2O was added. The aqueous [^{3}H]inositol solution was then washed by passing it through a Dowex column three times, adding 1 ml of H_2O each time (J. Figueiredo, personal communication). A 1.6 ml aliquot of a concentrated solution (10 x) of Buffer A containing 11.1 mM LiCl was added to each 100 uCi solution of [^{3}H]inositol in H_2O. The volume was brought up to 16 ml with H_2O to produce the desired concentration of 2.5 uCi [^{3}H]inositol/0.4ml Buffer A containing 11.1 mM LiCl.

Agonist-induced stimulation of the formation of inositol phosphates was measured according to the procedure of Fisher et al. (1984a). Briefly, following decapitation, the brain was rapidly removed and the cerebral cortex was dissected free from underlying structures and cross chopped (350 um x 350 um) on a McIlwain tissue slicer. The slices were placed in a vial and washed three times in warm Buffer A and then incubated in excess Buffer A for 30 min at 37°C with gentle shaking. Vials were gassed (95% 02: 5% C02) and capped between each wash and just prior to the 30 min incubation. After the 30 min incubation, the excess buffer was aspirated and 50 ul aliquots of tissue slices were added to vials containing 2.5 uCi [^{3}H]inositol/0.4 ml Buffer A with 11.1 mM LiCl. Each vial was gassed again, capped and preincubated for 60 min. After the 60 min preincubation, a 50 ul aliquot of 5-HT, NE, or CCH was added. Each vial was gassed, capped, and incubated for another 60 min. The final concentrations of the agonists used were: 5-HT=10 uM; NE=5 uM; CCH=100 uM. Reactions were terminated by adding 1.5 ml choloform/methanol (1:2 by volume). The vials were let stand for 15 min and then vortexed for 10 sec. After the addition of 1 ml chloroform and 0.5 ml H_2O, the vials were vortexed and centrifuged and a 1.5 ml aliquot of the upper (aqueous) phase was removed. Dowex resin (0.5 ml of a 50% weight/volume solution) was added to the aliquot of the upper phase and H_2O was added to bring the volume up to 3 ml. The vials were vortexed and centrifuged and the supernatant was aspirated. The resin was then washed four times with 5 mM myo-inositol. One ml of a 1M ammonium formate/0.1M formic acid solution was used to elute the labeled inositol phosphates from the Dowex resin. Each vial was vortexed and centrifuged and a 0.8 ml aliquot of the supernatant in 10 ml hydrofluor was used for counting.

Determination of Drug Levels

Plasma and brain tissue from the IMI, BUPR, and Li treated rats were obtained at the time of sacrifice. The output of the Alzet pumps were measured in vitro to ensure that they were working properly. After removal from the animal, each pump was placed in 20 ml of saline and put in a 37°C H_2O bath for 24 hrs. The concentration of IMI or BUPR in the saline was then determined.

IMI and its metabolites in brain, plasma, and saline were determined by a modification of the HPLC method of Suckow and Cooper (1981). BUPR and its metabolites in brain, plasma, and saline were determined by HPLC according to Cooper et al., (1984). Li was measured in brain and plasma by atomic absorption spectrometry.

Data Analysis and Presentation

The basal level of [^{3}H]IP accumulation in antidepressant-treated and control animals was measured as cpm of [^{3}H]IP in slices incubated in the absence of any agonist. Agonist-stimulated [^{3}H]IP accumulation in antidepressant-treated animals and their matched controls was expressed as percent of the basal level of [^{3}H]IP accumulation. Matched pair t tests were performed to compare the responses between experimental and control animals in each treatment group. The ratio between values for treated and control animals was expressed as relative activity for both basal levels and agonist-stimulated [^{3}H]IP accumulation.

RESULTS

Agonist-induced changes in [^{3}H]IP accumulation in control animals

Dose response curves for agonist-stimulated [^{3}H]IP formation were obtained with varying concentrations of 5-HT, NE, and CCH (Fig. 1). The maximal response for 5-HT was obtained with a concentration of 10^{-4} M. This maximal 5-HT response represented approximately 170% of the basal level of [^{3}H]IP accumulation. Maximal responses for NE and CCH were obtained with

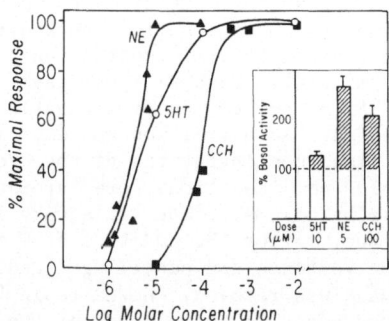

FIG. 1: Dose-Response Curve for 5-HT-, NE-, and CCH-Stimulated Accumulation of [^{3}H]IP. Dose-response curves for agonist-stimulated [^{3}H]IP formation were obtained with varying concentrations of 5-HT, NE, and CCH. Based on these curves, final concentrations of 10 uM 5-HT, 5 uM NE, and 100 uM CCH were used in subsequent experiments. The mean responses elicited by these concentrations in control animals, expressed as % of the basal [^{3}H]IP accumulation, were 123 ± 5%, 268 ± 16%, and 205 ± 21% for 5-HT, NE, and CCH respectively (insert).

536

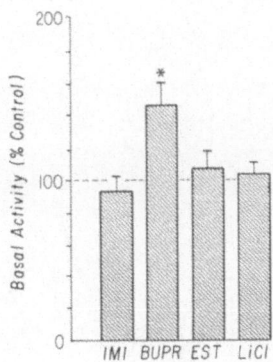

FIG. 2: Basal Accumulation of [^{3}H]IP Following Chronic Treatment with IMI,
BUPR, EST, and Li. Each value for [^{3}H]IP accumulation in experimental animals was expressed % of the value for [^{3}H]IP accumulation
obtained in the same experiment for a matched control animal. Long-
term treatment with IMI, EST, and Li had no significant effect on
the basal level of [^{3}H]IP formation in cortical slices. However,
chronic treatment with BUPR produced a significant, 42% elevation
of [^{3}H]IP accumulation compared to untreated controls.

concentrations of 10^{-5} M and 5×10^{-4} M respectively. The maximal NE
response represented approximately 400% whereas the maximal CCH response
was approximately 350% of the basal level of [^{3}H]IP accumulation. The concentrations of 5-HT, NE, and CCH used in subsequent experiments to stimulate
[^{3}H]IP formation under various experimental conditions were 10^{-5} M for 5-HT,
5×10^{-6}M for NE, and 10^{-4}M for CCH. The mean responses elicited by these
concentrations in control animals, expressed as percent of the basal [^{3}H]IP
accumulation, were $123 \pm 5\%$, $268 \pm 16\%$, and $205 \pm 21\%$ for 5-HT, NE, and CCH
respectively (Fig. 1 insert).

Basal levels of [^{3}H]IP accumulation following long-term treatment with
antidepressants

The basal level of [^{3}H]IP accumulation in each animal subjected to
long-term antidepressant treatment was measured as cpm of [^{3}H] in 0.8 ml of
the ammonium formate/formic acid eluate of the Dowex resin. Each value for
[^{3}H]IP accumulation in experimental animals was expressed as percent of the
value of [^{3}H]IP accumulation obtained in the same experiment for a matched
control animal. Long-term treatment with IMI, EST, or LI had no significant
effect on the basal level of [^{3}H]IP formation in cortical slices. However,
chronic treatment with the novel antidepressant drug BUPR resulted in a
significant, 42% elevation of [^{3}H]IP accumulation compared to untreated
control animals ($t(5)=2.86$, $p<0.05$, Fig. 2).

Agonist-induced alterations in [^{3}H]IP accumulation following chronic antidepressant treatment

The [^{3}H]IP accumulation to a given agonist was measured in pairs of
animals in which one member of the pair received antidepressant treatment
and the other served as an untreated control. Pairs of control animals
were also studied to rule out possible differences in [^{3}H]IP accumulation
due to different time of sacrifice or other unpredicted variables in a
given experiment. The accumulation of [^{3}H]IP was measured for each rat in
the presence and absence of the agonist and the value obtained in the presence of agonist was expressed as percent of the basal level. Values for

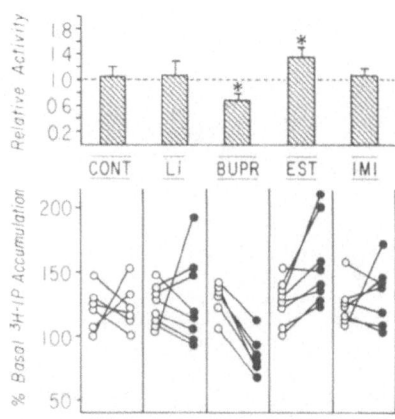

FIG. 3: 5-HT-Stimulated [³H]IP Accumulation Following Chronic IMI, BUPR,
EST, and Li Treatment. Values of [³H]IP accumulation were compared between pairs of control and experimental animals in
response to 5-HT and expressed as % of basal accumulation (Fig.
3, bottom: 0=control animals; ● =experimental animals). Fig. 3
(top) shows the ratio between values for treated and control animals expressed as relative activity. Data for NE and CCH were
analyzed in the same manner.

percent of the basal [³H]IP accumulation were compared between treated animals and their matched controls and the ratio between values for treated
and control animals expressed as relative activity. Relative activity
values were obtained for the various agonists studied in each treatment
group. As expected, the mean relative activity value for 5-HT-stimulated
[³H]IP accumulation obtained when control rats were compared with each
other was not significantly different from 1.0 (Fig. 3). An example of the
results obtained for individual animals tested for 5-HT-stimulated [³H]IP
release after different treatments with antidepressants is shown in Fig. 3
(bottom). Mean values for the relative activity in the same experiments
are shown in Fig. 3 (top).

5-HT-stimulated [³H]IP accumulation: The accumulation of [³H]IP in
the presence of 10 uM 5-HT was significantly lower in animals treated chronically with BUPR (t(5)=8.19, p<0.001) relative to matched controls. The
relative activity value for the BUPR group was 0.69 ± 0.09 (Figs. 3,4).
The response of [³H]IP accumulation to 5-HT was, however, significantly
higher in EST-treated rats (t(7)=4.17, p<0.01) compared to their matched
controls (Figs. 3,4). The mean relative activity for this group was 1.31
± 0.11. The responses to 5-HT in Li- and IMI-treated rats were not significantly different from their matched controls (Figs. 3,4).

NE-stimulated [³H]IP accumulation: The response to 5 uM NE was not
appreciably altered in animals treated with chronic IMI, EST, or Li compared
to their matched controls. However, animals treated with BUPR showed significantly lower values for [³H]IP accumulation in the presence of 5 uM NE
than their matched controls (t(5)=2.87, p<0.05). The relative activity for
the BUPR-treated group was 0.75 ± 0.07 (Fig. 4).

CCH-stimulated [³H]IP accumulation: There were no significant differences in [³H]IP accumulation in the presence of 100 uM CCH between IMI,
EST, or Li treated rats and their matched controls. Significantly lower
responses to CCH were seen, however, in BUPR-treated rats (t(5)=3.04,

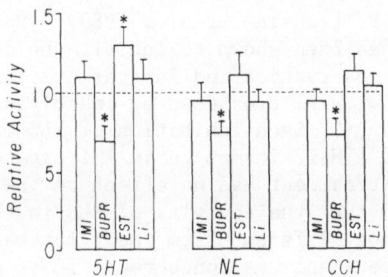

FIG. 4: Agonist-Stimulated [3H]IP Accumulation Following Chronic IMI,
BUPR, EST, and Li Treatment. Chronic BUPR produced a significant
decrease and EST produced a significant increase in 5-HT-stimu-
lated [3H]IP accumulation. The relative activities for the BUPR
and EST groups were .70 ± 0.09 and 1.31 ± 0.11 respectively.
Chronic BUPR also produced a significant decrease in the NE- and
CCH-stimulated [3H]IP accumulation. The relative activities in
the BUPR-treated animals for NE and CCH were .75 ± 0.07 and .73
± 0.1 respectively. EST, IMI, and Li failed to significantly
affect NE- and CCH-stimulated [3H]IP accumulation.

$p < 0.05$). The relative activity of the BUPR-treated group was 0.73 ± 0.1
(Fig. 4).

Drug Levels

Chronic treatment with IMI and BUPR was continued until 24 hrs before
sacrifice. IMI levels in plasma and brain obtained at the time of sacrifice were 1.63 + .39 ng/ml and 1.09 + .14 ug/g respectively. BUPR levels
in plasma and brain obtained at the time of sacrifice ranged from trace
amounts to being undetectable. The IMI and BUPR pumping rates in vitro
were in good agreement with the pre-calculated values required to maintain
the appropriate dose. Li treatment was continued until the time of sacrifice. The mean Li concentration in plasma obtained at this time was 0.42 +
0.09 mEq/liter.

DISCUSSION

The values obtained in the present experiments for agonist-stimulated
accumulation of [3H]IP in cerebral cortical slices of control rats were
generally in agreement with those obtained by other investigators; NE and
CCH produced a large and 5-HT produced a relatively small (max=170% of
basal) stimulation of [3H]IP accumulation (Brown et al., 1984; Schoepp et
al., 1984). However, Kendall and Nahorski (1985) found a biphasic curve
for 5-HT-stimulated [3H]IP accumulation in which very high doses (5-10 mM)
produced a 300% increase over basal activity.

Long-term administration of IMI, Li, and EST had no significant effect
on the basal level of [3H]IP accumulation in cerebral tissue while chronic
BUPR treatment significantly increased the basal level of [3H]IP accumulation. The lack of effect of chronic treatment with IMI on phosphoinositide
metabolism deserves attention since cationic amphiphilic drugs, including
IMI, are known to produce profound changes in phospholipid metabolism when
administered over an extended period of time (Hauser & Pappu, 1982). IMI
was shown to produce phospholipidosis in animals upon prolonged administration (Lüllmann-Rauch, 1975). In most cases of drug-induced phospholipidosis it has been shown that there is a disproportionate effect on acidic

phospholipids, including PI (Fauster et al., 1983). Desipramine (DMI), a major metabolite of IMI has been shown to inhibit the degradation of PI in human fibroblasts. This observation led Fauster et al. (1983) to suggest that DMI could interfere with PI breakdown by inhibition of the PLC which may act preferentially on PI. Such inhibition of PLC might be expected to reduce the basal level of [^{3}H]IP formation in IMI-treated animals. The present finding that IMI-treatment had no effect on the basal level of [^{3}H]-IP accumulation indicates that the activity of PLC in the cerebral cortex of IMI-treated rats was not different from their matched, untreated controls at the time the experiment was conducted. It is possible that inhibition of PI breakdown requires the presence of a drug-phospholipid complex (Matsuzawa & Hostetler, 1980), a condition that might not have been met in the present experiments since the drug was present only in trace amounts after the 24 hr washout period. Alternatively, the inhibition of phospholipid degradation by amphiphilic drugs might be confined to lysosomes (Brindley & Bowley, 1975) whereas [^{3}H]IP release in the present experiments may have occurred predominantly following PIP$_2$ breakdown in the plasma membrane. It is also possible that the IMI dose used in the present study was not high enough to produce changes in cerebral phospholipid metabolism. Relatively high doses of cationic amphiphilic drugs are required to produce such changes (Lullman et al., 1973). However, long-term treatment with BUPR resulted in a significant increase in the basal level of [^{3}H]IP release. These results indicate that, in contrast to other amphiphilic drugs, BUPR might stimulate PLC activity. While the functional significance of this BUPR effect is not known, it indicates that treatment with BUPR over an extended period of time may enhance basal phosphoinositide turnover.

The failure of long-term administration of Li or EST to produce changes in the basal level of [^{3}H]IP accumulation in cerebral tissue indicates that these treatments have no effect on the basal level of phosphoinositide turnover. While the effects of EST on phosphoinositide metabolism have not been studied extensively, it is known that Li inhibits the phosphatase which catalyses the breakdown of IP (Hallcher & Sherman, 1980). The present results suggest that the inhibition of this phosphatase in vivo may have no compensatory effect on the basal level of PLC activity that governs the turnover of polyphosphoinositides. Alternatively, the present failure to detect an increase in basal levels of [^{3}H]IP accumulation following chronic Li treatment could be related to the Li dose. Sherman et al., (1981) found an increase in myo-inositol-1-phosphate (MIP) in rat cortical tissue following chronic (9 day) Li treatment using a dose of Li (3.6-7.2 mEq/kg of body weight) which would produce a higher level of plasma Li than that found in the present experiment (.42 mEq/liter).

The 5-HT-stimulated accumulation of [^{3}H]IP was significantly decreased in animals treated with BUPR but increased in animals receiving repeated EST. Chronic IMI and Li treatment did not affect 5-HT-stimulated [^{3}H]IP accumulation. The changes in 5-HT-stimulated [^{3}H]IP accumulation following long-term EST treatment are consistent with the up-regulation of 5-HT$_2$ receptors produced by this treatment. Interpretation of the decrease in 5-HT-stimulated [^{3}H]IP accumulation in BUPR-treated animals is, however, more difficult; unlike other antidepressant drugs, BUPR fails to down-regulate 5-HT$_2$ receptors (Ferris & Beaman, 1983; Perumal et al., 1986). However, it is possible that a decreased [^{3}H]IP response to 5-HT in BUPR-treated rats may be produced by a mechanism which is uncoupled to 5-HT$_2$ receptors. For example, BUPR may exert its effect directly on second messenger systems; Ferris and Beaman (1983) reported a decrease in NE-stimulated adenylate cyclase activity which was not accompanied by a change in β-receptor density in BUPR-treated animals.

While chronic IMI treatment did not significantly affect 5-HT-stimulated [^{3}H]IP accumulation in rat cortical tissue in the present experiments,

Kendall and Nahorski (1985) did find a significant decrease in 5-HT-stimulated [3H]IP accumulation in rat cortex following chronic IMI treatment. However, they found that the decrease in Bmax for [3H]ketanserin binding was not of the same magnitude as the decrease in 5-HT-stimulated [3H]IP accumulation following chronic IMI treatment and suggested that the effect of antidepressants on phospholipid metabolism may occur at a point beyond the receptor. It is also possible that a subset of 5-HT$_2$ receptor sites is involved in the [3H]IP response to 5-HT. Although the ability of ketanserin to inhibit 5-HT-stimulated [3H]IP accumulation in rat cortex correlates to some extent with [3H]ketanserin binding to the 5-HT$_2$ site, the correlation for other antagonists is low in terms of their affinities at 5-HT$_2$ sites in [3H]ketanserin binding assays and their ability to inhibit 5-HT-stimulated [3H]IP accumulation (Conn & Sanders-Bush, 1985; Kendall & Nahorski, 1985). Finally, Janowsky et al., (1985) found that chronic IMI treatment increased, rather than decreased, 5-HT-stimulated [3H]IP accumulation in hippocampal slices. This also shows that, at present, it is difficult to make simple correlations between alterations in receptor binding and phosphoinositide turnover following chronic antidepressant treatment. 5-HT-stimulated [3H]-IP accumulation in the hippocampus appears to be associated with 5-HT$_1$ receptors rather than 5-HT$_2$ receptors (Janowsky et al., 1984) and while chronic IMI treatment increases 5-HT-stimulated [3H]IP accumulation in the hippocampus, antidepressants fail to affect 5-HT$_1$ receptor density in this structure (Fuxe et al., 1983).

Chronic treatment with Li had no significant effect on 5-HT-stimulated [3H]IP release. This finding suggests that increased 5-HT function during Li treatment (Meltzer, 1984) probably does not involve 5-HT$_2$ receptors. However, further experiments are warranted because animals with higher plasma Li levels showed a tendency to have increased [3H]IP accumulation in response to 5-HT.

The NE-stimulated accumulation of [3H]IP was decreased significantly in animals treated with BUPR but was unaffected in animals treated with IMI, EST, or Li. These results are not consistent with recent reports indicating that chronic administration of IMI, EST, and several other anti-depressants up-regulate a1 receptors (Vetulani & Antkiewicz-Michaluk, 1985). It is possible, however, that the increase in a1 adrenoreceptor density seen in [3H]prazosin binding experiments following long-term treatment with anti-depressants is uncoupled to the [3H]IP response. Another possibility is that, as with 5-HT$_2$ receptors, a subpopulation of a1 receptors may be involved in NE-stimulated [3H]IP accumulation.

The CCH-stimulated accumulation of [3H]IP was decreased significantly in animals chronically treated with BUPR, but was unaffected following chronic treatment with IMI, EST, and Li. This parallels the finding that anti-depressants and Li generally fail to up- or down-regulate cortical musc receptors.

Unlike the other antidepressant treatments, BUPR significantly decreased 5-HT-, NE-, and CCH-stimulated [3H]IP accumulation while increasing basal [3H]IP accumulation. The finding that [3H]IP accumulation in the presence of 5-HT, NE, and CCH is decreased, indicates that in BUPR-treated rats, the agonist may initiate cellular events which act to inhibit [3H]IP formation. It is possible that BUPR may exert its action at a point beyond the receptor. It may create favorable conditions for the production of an endogenous substance which inhibits PLC. Such a substance may be derived from DAG whose production would be stimulated in the presence of the ago-nist. The inhibitory substance itself could be a product of the arachidonic acid cascade which may exert negative feedback on PLC and thus inhibit PIP2 breakdown. The increase seen in the basal level could be explained by a lack of such negative feedback until the presumed inhibitor reaches a critical level, as could occur in the presence of agonists.

It is important to note that infusion pumps were used in these studies rather than the more commonly intraperitoneal injections. While we are not aware of any studies looking at receptor changes following chronic IMI infusion with pumps, studies with DMI show that the pumps and injections produce similar receptor changes (Dooley et al., 1983; Williams et al., 1983).

In conclusion, chronic treatment with the novel antidepressant BUPR significantly increased the basal level of [^{3}H]IP accumulation. Thus, treatment with BUPR over an extended period of time may enhance basal levels of phosphoinositide turnover. In contrast, chronic IMI, EST, and Li treatment failed to significantly affect the basal level of [^{3}H]IP accumulation.

Chronic treatment with BUPR significantly decreased and chronic EST significantly increased 5-HT-stimulated [^{3}H]IP accumulation. Chronic IMI and Li treatment had no appreciable effect. The EST results are consistent with up-regulation of 5-HT$_2$ receptors by EST. The effects of the two antidepressant drugs are not consistent with 5-HT$_2$ receptor changes since IMI down-regulates 5-HT$_2$ and BUPR fails to affect 5-HT$_2$ receptors.

NE-stimulated accumulation of [^{3}H]IP was decreased in BUPR-treated animals but no change was found in IMI-, EST-, or Li-treated animals. Recent studies indicate that there is an up-regulation of al receptors following treatment with IMI, EST, and a number of other antidepressants, but not BUPR. Thus, changes in al receptor density following EST and IMI do not appear to influence the al receptor population that is coupled to the [^{3}H]IP response.

CCH-stimulated [^{3}H]IP accumulated was decreased in BUPR-treated animals but was unaffected in IMI-, EST-, and Li-treated animals. This is consistent with the finding that there are no clear musc receptor changes following these treatments. However, the finding that BUPR decreased 5-HT-, NE-, and CCH-stimulated [^{3}H]IP accumulation indicates that it may initiate cellular events which act to inhibit [^{3}H]IP formation. For instance, BUPR may create favorable conditions for the agonist-stimulated production of an endogenous substance which would inhibit PLC.

ACKNOWLEDGMENTS

The studies reported here were supported in part by NIMH grant 33690 to Dr. A. Barkai. The authors are grateful to Dr. Ray Suckow for performing the drug determinations and to Dr. Basalingappa Hungund for his assistance. The valuable suggestions of Dr. Stephen K. Fisher during the initial stages of this work are also gratefully acknowledged.

REFERENCES

Abdel-Latif, A. A., Metabolism of phosphoinositides. In: Handbook of Neurochemistry, Second Edition (Lajtha, A., ed), Plenum Press, N. Y., 91-131 (1982).

Agranoff, B. W., Murthy, P. and Seguin, E. B., Thrombin-induced phosphodiesteratic cleavage of phosphatidylinositol bisophosphate in human platelets, J. Biol. Chem., 258:2076-2078 (1983).

Ahluwalia, P. and Singhal, R. L., Monoamine uptake into synaptosomes from various regions of rat brain following lithium administration and withdrawal, Neuropharmacology, 20:483-487 (1981).

Allison, J. H., Blisner, M. E., Holland, W. H., Hipps, P. P. and Sherman, W. R., Increased brain myo-inositol-1-phosphate in lithium treated rats, Biochem. Biophys. Res. Commun., 71:664-670 (1976).

Banerjee, S. P., Kung, L. S., Riggi, S. J. and Chanda, S. K., Development of B-adrenergic receptor subsensitivity by antidepressants, Nature, 268: 455-456 (1977).

Bell, R. L., Kennerly, D. A., Stanford, N. and Majerus, P. W., Diglyceride lipase: A pathway for arachidonate release from human platelets, Proc. Natl. Acad. Sci. USA, 76:3238-3241 (1971).

Bergstrom, D. A. and Kellar, K. J., Effect of electroconvulsive shock on monoaminergic receptor binding sites in rat brain, Nature, 278:464-466 (1979).

Berridge, M. J., Rapid accumulation of inositol triphosphate reveals that agonists hydrolyse polyphosphoinositides instead of phosphatidylinositol, Biochem. J., 212:849-858 (1983).

Berridge, M. J., Inositol triphosphate and diacylglycerol as second messengers, Biochem. J., 220:345-360 (1984).

Berridge, M. J., Downes, C. P. and Hanley, M. R., Lithium amplifies agonist-dependent phosphatidylinositol responses in brain and salivary glands, Biochem. J., 206:587-595 (1982).

Berridge, M. J. and Irvine, R. F., Inositol triphosphate, a novel second messenger in cellular signal transduction, Nature (London), 312:315-321 (1984).

Brindley, D. N. and Bowley, M., Drugs affecting the synthesis of glycerides and phospholipids in rat liver, Biochem. J., 148:461-469 (1975).

Broderick, P. and Lynch, V., Behavioral and biochemical changes induced by lithium and L-tryptophan in muricidal rats, Neuropharmacology, 21:671-679 (1982).

Brown, E., Kendall, D. A. and Nahorski, S. R., Inositol phospholipid hydrolysis in rat cerebral cortical slices: I. receptor characterization, J. Neurochem., 42:1379-1387 (1984).

Bunney Jr., W. E. and Davis, J. M., Norepinepherine in depressive reactions: a review, Arch. Gen. Psychiatry, 13:483-494 (1965).

Charney, D. S., Menkes, D. B., Phil, M. and Heninger, G. R., Receptor sensitivity and the mechanism of action of antidepressant treatment, Arch. Gen. Psychiatry., 38:1160-1180 (1981).

Conn, P. J. and Sanders-Bush, E., Selective 5-HT$_2$ antagonists inhibit serotonin stimulated phosphatidylinositol metabolism in cerebral cortex, Neuropharmacology, 23:993-996 (1984).

Conn, P. J. and Sanders-Bush, E., Serotonin-stimulated phosphoinositide turnover: mediation by the S2 binding site in rat cerebral cortex but not in subcortical regions, J. Pharm. Exp. Ther., 234:195-203 (1985).

Cooper, T. B., Suckow, R. F. and Glassman, A., Determination of bupropion and its major basic metabolites in plasma by liquid chromatography with dual wavelength in UV detection, J. Pharm. Sci., 73:1104-1107 (1984).

Danielsson, E., Peterson, L-L., Grundin, R., Ogren, S.-O. and Bartfai, T., Anticholinergic potency of psychoactive drugs in human and rat cerebral cortex and striatum, Life Sci., 36:1451-1457 (1985).

Dashieff, R. M., Savage, D. D. and McNamara, J. O., Seizures down-regulate muscarinic cholinergic receptors in hippocampal formation, Brain Res., 235:327-334 (1982).

Deakin, J. F. W., Owen, F., Cross, A. J. and Dashwood, M. J., Studies on possible mechanisms of action of electroconvulsive therapy: Effects of repeated electrically induced seizures on rat brain receptors for monoamines and other neurotransmitters, Psychopharmacology, 73:345-349 (1981).

Dooley, D. J., Hauser, K. L. and Bittinger, H., Differential decrease of the central beta-adrenergic receptor in the rat after subchronic infusion of desipramine and clenbuterol, Neurochem. Int., 5:333-338 (1983).

Fauster, R., Honegger, U. and Weismann, U., Inhibition of phospholipid degradation and changes of the phospholipid pattern by desipramine in cultured human fibroblasts, Biochem. Pharm., 32:1737-1744 (1983).

543

Ferris, R. M. and Beaman, O. J., Burpropion: a new antidepressant drug, the mechanism of action of which is not associated with down-regulation of post-synaptic B-adrenergic, serotonergic ($5HT_2$, alpha$_2$ adrenergic, imipramine and dopaminergic receptors in brain, Neuropharmacology, 22: 1257-1267 (1983).

Ferris, R. M., White, H. L., Cooper, B. R., Maxwell, R. A., Tang, F. L. M., Beaman, O. J. and Russell, A., Some neurochemical properties of a new antidepressant bupropion hydrochloride (Wellbrutin), Drug Devel. Res., 1:21-35 (1981).

Fisher, S. K., Figueiredo, J. C. and Bartus, R. T., Differential stimulation of inositol phospholipid turnover in brain by analogs of oxotremorine, J. Neurochem., 43:1171-1179 (1984a).

Fisher, S. K., Van Rooijen, L. A. A. and Agranoff, B. W., Renewed interest in the polyphosphoinositides, Trends Bio. Sci., 53-56 (1984b).

Furukawa, T., Yamada, K., Kohno, Y. and Nagasaki, N., Brain serotonin metabolism with relation to the head twitches elicited by lithium in combination with reserpine in mice, Pharmacol. Biochem. Behav., 10:547-549 (1979).

Fuxe, K., Ogren, S-O., Agnati, L. F., Andersson, K. and Eneroth, P., Effects of subchronic antidepressant drug treatment on central serotonergic mechanisms in the male rat. In: Typical and Atypical Antidepressants: Molecular Mechanisms (Costa, E. and Racagni, G., eds). Raven Press, New York, 91-107 (1982).

Fuxe, K., Ogren, S-O., Agnati, L. F., Benfenati, F., Fredholm, B., Andersson, K., Zini, I. and Eneroth, P., Chronic antidepressant treatment and central 5-HT synapses, Neuropharm, 22:389-400 (1983).

Gandolfi, O., Barbaccia, M. L., Chuang, D. M. and Costa, E., Daily bupropion injections for three weeks attenuate the NE-stimulation of adenylate cyclase and the number of B-adrenergic recognition sites in rat frontal cortex, Neuropharmacology, 22:927-929 (1983).

Gonzales, R. A. and Crews, F. T., Characterization of the cholinergic stimulation of phosphoinositide hydrolysis in rat brain slices, J. Neurosci., 4:3120-3127 (1984).

Gonzales, R. A. and Crews, F. T., Cholinergic- and adrenergic-stimulated inositide hydrolysis in brain: interaction, regional distribution, and coupling mechanisms, J. of Neurochem., 45:1076-1084 (1985).

Green, A. R., Heal, D. J., Johnson, P., Laurence, B. E. and Nimgaonkar, V. L., Antidepressant treatments: effects in rodents on dose-response curves of 5-hydroxytryptamine- and dopamine-mediated behaviours and $5-HT_2$ receptor number in frontal cortex, Br. J. Pharmacol., 80:377-385 (1983a).

Green, A. R., Johnson, P. and Nimgaonkar, V. L., Increased $5-HT_2$ receptor number in brain as a probable explanation for the enhanced 5-hydroxytryptamine-mediated behavior following repeated electroconvulsive shock administration to rats, Br. J. Pharmacol., 80:173-177 (1983b).

Halaris, A. E., Stern, W. and Truox-Horto, M. S., Clinical efficacy of the antidepressant bupropion (Wellbutrin), Psychopharm. Bull., 17:140 (1981).

Hallcher, L. M. and Sherman, W. R., The effects of lithium ion and other agents on the activity of myo-inositol-1-phosphate from bovine brain, J. Bio. Chem., 255:10896-10901 (1980).

Hauser, G. and Pappu, A. S., Effects of propranolol and other cationic amphiphilic drugs on phospholipid metabolism. In: phospholipids in the Nervous System (Horrocks, L. A., Amsel, G. B. and Porcellati, G. eds) Raven Press, New York, 283-300 (1982).

Hawthorne, J. N. and Pickard, M. R., Phospholipids in synaptic function, J. Neurochem., 32:4-14 (1979).

Iversen, L. L. and Mackay, A. V. P., Pharmacodynamics of anti-depressants and antimanic drugs. In: Psychopharmacology of Affective Disorders (Paykel, E. S. and Coppen, A. eds). New York, Oxford University Press Inc., 60-90 (1979).

Janowsky, A., Labarca, R. and Paul, S., Characterization of neurotransmitter receptor-mediated phosphatidylinositol hydrolysis in the rat hippocampus, Life Sciences, 35:1953-1961 (1984).

Janowsky, A., Labarca, R. and Paul, S. M., Neurotransmitter receptor-mediated myo-inositol-1-phosphate accumulation in hippocampal slices. In: Inositol and Phosphoinositides: Metabolism and Regulation (Bleasdale, J. E., Eichberg, J., and Hauser, G. eds). Humana Press, Clifton, N.J., 83-90 (1985).

Jope, R. S., Effects of lithium treatment in vitro and in vivo on acetylcholine metabolism in rat brain, J. Neurochem., 33:487-495 (1979).

Kafka, M., Wirz-Justice, A., Naber, D., Marangos, P., O'Donohue, T. and Wehr, T., Effect of lithium on circadian neurotransmitter receptor rhythms, Neuropsychobio., 8:41-50 (1982).

Kellar, K. J., Cascio, C. S., Bergstrom, D. A., Butler, J. A. and Iadorola, P., Electroconvulsive shock and reserpine: Effects on B-adrenergic receptors in rat brain, J. Neurochem., 37:830-836 (1981a).

Kellar, K. J., Cascio, C. S., Butler, J. A. and Kurtzke, R. N., Differential effects of electroconvulsive shock and antidepressant drugs on serotonin-2 receptors in rat brain, Eur. J. Pharmacol., 69:515-518 (1981b).

Kellar, K. J. and Stockmeier, C. A., Effects of ECS and serotonin axon lesions on B-adrenergic and serotonin-2 receptors in rat brain. In: ECT: Clinical and Basic Research Issues (Malitz, S. and Sackeim, H. eds) Ann. N.Y. Acad. Sciences, vol. 452 (1986).

Kendall, D. A., Duman, R., Slopis, J. and Enna, S. J., Influence of adrenocorticotropin hormone and yohimbine on antidepressant-induced declines in rat brain neurotransmitter receptor binding and function, J. Pharmacol. Exp. Ther., 22:566-571 (1982).

Kendall, D. A. and Nahorski, S. R., 5-Hydroxytryptamine-stimulated inositol phospholipid hydrolysis in rat cerebral cortex slices: pharmacological characterization and effects of antidepressants, J. Pharm. Exp. Ther., 233:473-479 (1985).

Lerer, B. and Stanley, M., Effect of chronic lithium on cholinergically mediated responses and [^{3}H]QNB binding in rat brain, Br. Res., 344:211-219 (1985).

Lerer, B., Stanley, M., Demetriou, S. and Gershon, S., Effect of electroconvulsive shock on muscarinic cholinergic receptors in rat cerebral cortex and hippocampus, J. Neurochem., 6:1680-1683 (1983).

Levy, A., Zohar, J. and Belmaker, R. H., The effect of chronic lithium pretreatment on rat brain muscarinic receptor regulation, Neuropharm., 21:1199-1201 (1982).

Lullman, H., Lullman-Rauch, R. and Wassermann, O., Drug-induced phospholipidosis, Germ. Med. 3:128-135 (1973).

Lullman-Rauch, R., Lipodosislike renal changes in rats treated with chlorphentermine or with tricyclic antidepressants, Virchows Arch. B. Cell Pathol., 18:51-60 (1975).

Maggi, A. and Enna, S. J., Regional alterations in rat brain neurotransmitter systems following chronic lithium treatment, J. Neurochem., 34:888-892 (1980).

Malitz, S. and Sackheim, H. eds., ECT: Clinical and Basic Research Issues, Ann. N.Y. Acad. Sciences, Vol. 452 (1986).

Martin, T. F. J., Thyrotropin-releasing hormone rapidly activates the phosphodiesteratic hydrolysis of polyphosphoinositides in GH3 pituitary cells, J. Biol. Chem., 258:14816-14822 (1983).

Matsuzawa, Y. and Hostetler, K. Y., Inhibition of lysosomal phospholipase A and phospholipase C by chloroquine and 4,4'-Bis(diethylaminoethoxy)a,B-diethyldiphenylethance, J. Bio. Chem., 255:5190-5194 (1980).

Maxwell, R. A., Mehta, N. B., Tucker, W. E., Schroeder, D. H. and Stern, W. C., Bupropion. In: Pharmacological and Biochemical Properties of Drug Substances. Vol. 3, (M. E. Goldberg ed.) American Pharmaceutical Association Academy of Pharmaceutical Sciences, Washington, D. C. 1-55 (1981).

545

Meltzer, H. Y., Serotonergic function in the affective disorders: The effects of antidepressants and lithium on the 5-hydroxytryptophan-induced increase in serum cortisol. In: Presynaptic Modulation of Postsynaptic Receptors in Mental Diseases, (Salama, A. I. ed.), <u>Annals of the N. Y. Academy of Sciences</u>, vol. 430, 115-137 (1984).

Michell, R. H., Inositol phospholipids and cell surface receptor function, <u>Biochim. Biophys. Acta</u>, 415:81-147 (1975).

Nishizuka, Y., Turnover of inositol phospholipids and signal transduction, <u>Science</u>, 225:1365-1370 (1984a).

Nishizuka, Y., The role of protein kinase C in cell surface signal transduction and tumor promotion, <u>Nature</u>, 308:693-698 (1984b).

Okazaki, T., Sagawa, N., Okita, J. R., Bleasdale, J. E., MacDonald, P. C. and Johnston, J. M., Diacylglycerol metabolism and arachidonic acid release in human fetal membranes and decidual vera, <u>J. Biol. Chem.</u>, 256:7316-7321 (1981).

Oswald, I., Brezinova, V. and Dunleavy, D. L. F., On the slowness of action of tricyclic antidepressant drugs, <u>Br. J. Psychiatry</u>, 120:673-677 (1972).

Pandey, G. N., Heinze, W. B., Brown, B. D. and Davis, J. M., Effects of antidepressants on B-adrenergic receptor sensitivity in rat brain, <u>Fedn. Proc. Fedn. Am. Socs. Exp. Biol.</u>, 38:592 (1979a).

Pandey, G. N., Heinze, W. B., Brown, B. D. and Davis, J. M., Electroconvulsive shock treatment decreases B-adrenergic receptor sensitivity in rat brain, <u>Nature</u>, 280:234-235 (1979b).

Peroutka, S. J. and Snyder, S. H., Long-term antidepressant treatment decreases spiroperidol-labelled serotonin receptor binding, <u>Science</u>, 210:88-90 (1980).

Perumal, A. S., Smith, T. M., Suckow, R. F. and Cooper, T. B., Bupropion and BW306U on B-receptors in mouse and guinea-pig brain, <u>Trans. Am. Soc. Neurochem.</u>, 17:268 (1986).

Rosenblatt, J. E., Pert, C. B., Tallman, J. F., Pert, A. and Bunney Jr., W. E., The effects of imipramine and lithium on a- and B-receptor binding in rat brain, <u>Br. Res.</u>, 160:186-191 (1979).

Sangdee, C. and Franz, D. N., Lithium enhancement of central 5-HT transmission induced by 5-HT precursors, <u>Biol. Psychiatry</u>, 15:59-75 (1980).

Schoepp, D. D., Knepper, S. M. and Rutledge, C. O., Norepinepherine stimulation of phosphoinositide hydrolysis in rat cerebral cortex is associated with the alpha$_1$-adrenoceptor, <u>J. Neurochem.</u>, 43:1758-1761 (1984).

Schildkraut, J. J., The catecholamine hypothesis of depression: A review of supporting evidence, <u>Am. J. Psychiat.</u>, 122:509-522 (1965).

Schildkraut, J. J. and Kety, S. S., Biogenic amines and emotion, <u>Science</u>, 156:21-30 (1967).

Sellinger-Barnett, M. M., Mendels, J. and Frazer, A., The effect of psychoactive drugs on B-adrenergic receptor binding sites in rat brain, <u>Neuropharmacology</u>, 19:447-454 (1980).

Sherman, W. R., Leavitt, A. L., Honchar, M. P., Hallcher, L. M. and Phillips, B. E., Evidence that lithium alters phosphoinositide metabolism: Chronic administration elevates primarily D-myo-inositol-1-phosphate in cerebral cortex of the rat, <u>J. Neurochem.</u>, 36:1947-1951 (1981).

Snyder, S. H. and Yamamura, H. I., Antidepressants and the muscarinic acetylcholine receptor, <u>Arch. Gen. Psychiatry</u>, 34:236-239 (1977).

Soroko, F. E., Mehta, N. B., Maxwell, R. A., Ferris, R. M. and Schroeder, D. H., Buproprion hydrochloride ((+) a-t-butylamino-3-chloropropiophenone HCl): a novel antidepressant agent, <u>J. Pharm. Pharmac.</u>, 29:767-770 (1977).

Streb, H., Irvine, R. F., Berridge, M. J. and Schulz, I., Release of Ca^{2+} from a nonmitochondrial intracellular store in pancreatic acinar cells by inositol-1,4,5-triphosphate, <u>Nature (London)</u>, 306:67-69 (1983).

Suckow, R. F. and Cooper, T. B., Simultaneous determination of imipramine and desipramine and their 2-hydroxy metabolites in plasma by ion-pair reversed-phase high performance liquid chromatography with amperometric detection, J. Pharm. Sci., 70:257-261 (1981).

Sulser, F., New perspectives on the mode of action of antidepressant drugs, Trends Pharmacol. Sci., 1:92-94 (1979).

Sulser, F., Janowsky, A. J., Okada, F., Manier, D. H. and Mobley, P. L., Regulation of recognition and action function of the norepinepherine (NE) receptor-coupled adenylate cyclase system in brain: implications for the therapy of depression, Neuropharmacology, 22:425-431 (1983).

Sulser, F., Vetulani, J. and Mobley, P. L., Mode of action of antidepressant drugs, Biochem. Pharmacol., 27:257-261 (1978).

Tang, S. W., Seeman, P. and Kwang, S., Differential effects of chronic desipramine and amitriptyline treatment on rat brain adrenergic and serotonergic receptors, Psychiatry Res., 4:129-138 (1981).

Treiser, S. and Kellar, K. J., Lithium effects on serotonin receptors in rat brain, Eur. J. Pharmacol., 64:83-185 (1980).

Vetulani, J., Changes in responsiveness of central aminergic structures after chronic ECS. In: ECT: Basic Mechanisms (Lerer, B., Weiner, R. D., and Belmaker, R. H., eds.), Libbey, London, 33-45 (1983).

Vetulani, J. and Antkiewicz-Michaluk, L., Alpha-adrenergic receptor changes during antidepressant treatment. Acta Pharmacologica et Toxicologica, 56, Suppl 1:55-65 (1985).

Vetulani, J., Lebrecht, U. and Pilc, A., Enhancement of responsiveness of the central serotonergic system and serotonin-2 receptor density in rat frontal cortex by electroconvulsive shock treatment, Eur. J. Pharmacol., 76:81-85 (1981).

Williams, M., Risley, E. A. and Robinson, J. L., Chronic in vivo treatment with desmethylimipramine and mianserin does not alter adenosine A-1 radioligand binding in rat cortex, Neurosci. Lett., 35:47-51 (1983).

Wolfe, B. B., Harden, T. K., Sporn, J. R. and Molinoff, P. B., Presynaptic modulation of Beta adrenergic receptors in rat cerebral cortex after treatment with antidepressants, J. Pharmacol. Exp. Ther., 207:446-457 (1977).

A POSSIBLE ROLE FOR THYROTROPIN RELEASING HORMONE (TRH) IN

ANTIDEPRESSANT TREATMENT

Albert Sattin

R. L. Roudebush V. A. Medical Center, Indianapolis
and Institute of Psychiatric Research, Indiana
University Medical Center, Indianapolis, IN 46233

INTRODUCTION

TRH is a tripeptide which was discovered in the context of hypotha-
lamic-thyroid function but which is now known to be widely distributed
throughout the CNS (Jackson, 1982). The majority of TRH in the CNS is
extrahypothalamic and its regional concentrations are highly variant (Kubek
et al., 1977).

There are three bases for raising the question of a role for TRH in
antidepressant treatment and in affective function generally. The first is
psychopharmacological. Attempts were made in the early 1970's to treat
depression with oral and parenteral TRH. The results were characterized by
occasional transient euphoria but no lasting antidepressant effects (Prange
et al., 1972; Kastin et al., 1972; Itil et al., 1975; Breese et al., 1981).
However, in retrospect, this peptide is rapidly metabolized in the blood and
its penetration through the blood-brain barrier is low (Breese et al., 1981;
Loosen and Prange, 1980). These factors would make the native tripeptide a
poor candidate for clinical trials even if it had inherent antidepressant
effects.

The second basis for raising this question is neuroendocrinological.
A biological subgroup of Major Depressive Disorder shows a reduction of the
increase in circulating TSH following intravenous injection of 200-500 ug
of TRH. In some instances this "blunted response" to TRH is normalized
after successful antidepressant treatment (Kirkegaard, 1981; Loosen, 1985).
The relationship, if any, of this neuroendocrine phenomenon to possible CNS
dysfunctions in depression remains unclear.

The third basis for raising this question is neuropharmacological.
Considerable evidence supports interactive or modulatory roles for TRH upon
catecholamines and ACh (Yarbrough, 1979), both of which are thought to be
involved in the pathophysiology of depressive disorders (Sachar and Baron,
1979). In addition, TRH has been co-localized with 5-HT in some of the
raphe neurons that project into the spinal cord (Johansson et al., 1971).
Thus, TRH could be involved in the modulation of 5-HT function. Abnormalities of 5-HT function have also been implicated in depressive disorders
(Asberg et al., 1976; van Praag, 1977). Although the neuropharmacology of
TRH is largely undeveloped, the involvement of this tripeptide with amine
systems subserving affective function is already apparent.

Our approach to investigation of the role of TRH has utilized an animal model of antidepressant treatment, i.e., the repeated administration of seizures to rats. Electroconvulsive treatment (ECT) is recognized as the most efficacious of all presently available medical treatments for Major Depressive Disorder (Rose, 1985; Frankel, 1978). Even though this treatment is generally reserved for cases that do not remit following adequate trials of two or more drugs, most patients recover following 6-12 seizures administered at intervals of 2-4 days. As with any biological treatment of depression, there is a measurable placebo effect but genuine ECT produces a significant additional antidepressant effect (Freeman et al., 1978; Lamborn and Gill, 1978; Johnstone et al., 1980; West, 1981; Brandon et al., 1981). The duration of the antidepressant effect is highly variable and, as with antidepressant drug treatment, reflects the natural course of the illness.

In the rat model of electroconvulsive shock (ECS) we asked the following questions: 1) Is the content of TRH altered by an ECS paradigm that mimics clinical ECT?, 2) In what regions does this change occur?, 3) Are the changes related to the number of ECS given?, 4) Do changes occur following a subconvulsive stimulus?, 5) How long do the changes persist? Seizure-dependent changes were indeed found and the answers to these and related questions will be reviewed.

METHODS

Sprague-Dawley male rats (150-180 gm) were obtained from Harlan Industries (Indianapolis). After allowing a week for subsidence of the transportation stress tonic-clonic seizures were induced with 35-60 millicoulombs (60 Hz, a.c.). Sham ECS rats received identical handling but no current. Subconvulsive shock was induced with 5 mA x 0.4s (= 2.0 millicoulombs). Subconvulsively shocked rats leaped, squealed and displayed piloerection but showed no behavioral seizure. (Using similar currents, an electrocortical seizure discharge was previously shown to be present in the high current and absent in the subconvulsively shocked rats; Perumal and Barkai, 1982). Following decapitation, brains were immediately removed, rinsed in ice-cold saline and dissected under low-power magnification on a moistened, ice-chilled wax plate as previously described (Balcom et al., 1975). After frozen storage of individually weighed brain regions the tissues were extracted with acetic acid and TRH was assayed by an RIA method sensitive to 4 pg. Further details of the extraction, assay and statistical analysis of results have been previously given (Kubek and Sattin, 1984; Kubek et al., 1985).

RESULTS

In our initial experiments the brain was divided into five regions two days after a series of five ECS given on alternate days. The two-day interval following the final seizure was adopted in order to avoid possible transient post-ictal changes. Of the five regions observed (anterior cortex, A.Ctx; hippocampus, HC; striatum, ST; thalamus + midbrain and hypothalamus, HY) the noteworthy change occurred in HC where TRH level increased five-fold from 6.18 to 30.37 pg/mg wet weight (sham ECS vs. ECS). In these initial experiments an increase was also noted in ST but in all subsequent observations ECS induced changes in ST have been equivocal or absent. It is important to note that no change occurred in HY despite the fact that the basal level of TRH is highest of all the regions studied (Kubek and Sattin, 1984; Kubek et al., 1985). Thus, it was initially established that repeated ECS induces a long-lasting (two day) increase in TRH that is highly selective in regard to brain region.

We next expanded the brain dissection to include the amygdala-pyriform cortex (AY-PYR), septum, ventral striatum (olfactory tubercle + n. accumbens), thalamus, midbrain and brain stem. The large increase in HC was confirmed (two days following five ECS), 5.96 vs. 14.83 pg/mg protein, and significant increases were seen in AY-PYR (10.87 vs. 21.86) and whole Ctx (0.82 vs. 1.44) but none of the other regions was affected (Kubek et al., 1985). Subsequently AY and PYR were processed separately revealing significant changes in both regions, 28.62 vs. 46.01 and 5.15 vs. 28.05 pg/mg protein, respectively (Kubek et al., 1985). The increase in PYR was over five-fold.

Subconvulsive shock (SCS) was given to rats in parallel with ECS and sham ECS groups. Analysis of whole CTX plus the nine subcortical regions showed no change in any region two days after five SCS given on alternate days except in ST where a small but significant increase was observed following SCS but not following ECS (Kubek et al., 1985).

Subsequently sham ECS rats plus three experimental groups were given one, three and five ECS, then sacrificed two days after the last ECS or sham ECS. In PYR, HC and AY a plateau of increase in TRH was reached after ECS x 3, i.e., TRH levels were not significantly greater after ECS x 5. After ECS x 1 a smaller but significant increase occurred in PYR, a non-significant increase in HC and no increase in AY.˙ In ST a small increase after ECS x 3 was no longer present after ECS x 5 (Kubek et al., 1985).

The ECS x 3 paradigm was then adopted for a study of the longer-term duration of the ECS-induced increase in TRH. The results, to be published elsewhere, showed that substantial increases in TRH persisted for longer than two days in PYR and HC. In PYR the increases were 702%, 270% and 49% at 2, 6 and 12 days, respectively following ECS x 3. Significant increases persisted for six days in PYR and HC after ECS x 1 (Sattin et al., 1985).

In a preliminary experiment two groups of six rats were injected daily (i.p.) with saline or desipramine (10 mg/kg) and sacrificed 24 h. after the last of 7 daily injections. Assay of TRH from extracts of HC showed no significant change in TRH resulting from the drug treatment (M. Kubek and A. Sattin, unpublished).

DISCUSSION

The results show that ECS given to rats in a paradigm that mimics clinical ECT induces a substantial augmentation of the content of TRH in specific limbic regions and smaller but significant elevations in CTX. The largest percent increase is seen in PYR, followed by HC and AY. Regional specificity of this effect is shown by the absence of any change in TRH in 7 other subcortical regions two days after ECS x 5 (Kubek et al., 1985). Lighton et al. (1984) have also studied the effects of 5 ECS (given on alternate days) on TRH in rat brain. However, their observations were made 24 h after the last ECS at which time 40% reductions were reported in N. accumbens and lumbar spinal cord. In our hands no alterations in TRH levels could be seen in N. accumbens (combined with olf. tubercule) 48 h after ECS x 5 (Kubek et al., 1985).

Subconvulsive shock failed to alter TRH in the three limbic regions (PYR, HC and AY) where ECS x 5 induced large increases (Kubek et al., 1985). This suggests that the observed increases of TRH in these regions is seizure-dependent. Additional support for this idea comes from analogous observations made with electrically and chemically kindled rats. In fully (Stage 5) electrically kindled rats (daily, 500 uA, 0.1 mSec, Ay placement) TRH increases in PYR, HC and AY were similar to those observed following ECS x 3-5. In partially kindled rats (Stage 2-3) a smaller elevation was noted in PYR only (Walczak et al., 1983). Analysis of PYR and AY from rats fully kindled following daily subconvulsive pentylenetetrazole (20 mg/kg) showed similar increases in TRH (Kubek and Morzorati, 1985). This also constitutes additional evidence that the limbic changes in TRH are seizure dependent and not a consequence of electrical stimulation per se. In this connection it may be noted that therapeutic seizures in humans can also be produced non-electrically (Small and Small, 1972).

In the PYR the ECS-induced increase in TRH was consistently the largest relative to the sham controls. The increased TRH content in this region was also the most prolonged, lasting for at least 12 days following ECS x 3 (Sattin et al., 1985). It is interesting to note that in clinical ECT an antidepressant effect of two weeks duration is highly predictable. There is a small but significant subgroup of patients who do not obtain a sustained remission but do show this two week-long antidepressant effect (Barton et al., 1973). Thus, the temporal pattern of our results to date with TRH does correlate with a known clinical human phenomenon. It is conceivable that significant increases occur in limbic TRH in humans undergoing ECT and that the onset of the antidepressant effect is correlated with that increase. In the usual more favorable outcome the two week-long increase in TRH might be associated with other changes that result in the more prolonged remission but in the treatment-resistant cases the antidepressant effect subsides when the limbic TRH returns to the pre-ECT level. As stated in the Introduction TRH is known to interact with the biogenic amines and through these interactions other longer-term changes of antidepressant significance might occur. This theory differs significantly from the older idea that exogenously administered TRH might have antidepressant effects (3-6) since it is based upon in situ changes in endogenous TRH in specific brain regions in a pattern that would appear to be very difficult to simulate through an exogenous pharmacological approach.

However, any further speculation about interaction with other transmitters is premature since the physio-pharmacological significance of these elevated TRH levels remains unknown. The seizure-dependent increases might have resulted from increased synthesis or storage of TRH or decreased release of TRH. We still lack basic information about the neuroanatomical source of the TRH in the regions examined and it is still unknown whether this peptide is physiologically released like a conventional neurotransmitter. Finally, the temporal association of this TRH phenomenon with human antidepressant treatment is not necessarily causal. Similar changes of smaller magnitude and involving other brain regions have been reported for opiate peptide which have also. been suggested to have antidepressant relevance (Hong et al., 1979).

Our preliminary negative observation about the effect of chronic desipramine on the TRH level in HC requires further confirmation. If more adequate and lengthy exposures to tricyclic and other chemical antidepressants prove to have no effect on limbic content of TRH then the TRH phenomenon would join a very limited number of neuropharmacological changes that are associated only with the seizure model of antidepressant treatment and this might provide additional opportunities to establish the neuropharmacological basis for the greater clinical efficacy of ECT vis-a-vis the drug treatments. The other long-lasting changes specifically associated with ECS have been reviewed by Lerer and Sitaram (1983). In addition to the increased met-enkephalin they include up-regulation of 5-HT$_2$ receptors and down-regulation of muscarinic receptors.

The data reviewed here are also relevant to the regulation of seizure activity in the brain. As stated above, kindled seizures produce regional TRH changes in the forebrain similar to those seen following ECS, yet in contrast to kindled seizures which mimic epilepsy, ECS is anti-epileptic

(Babington and Wedeking, 1975; Post et al., 1981) and some limited clinical observation supports this in humans also (Fink, 1979). In both animals and humans treatment with DN-1417, a metabolically stable analogue of TRH has demonstrated antiepileptic effects (Morimoto et al., 1983; Matsuishi et al., 1983).

Thus the rat forebrain responds to infrequently repeated single seizures with dramatic and prolonged increases of the TRH content of hippocampus and pyriform cortex and significant increases in TRH in whole cortex and amygdala. The molecular mechanisms responsible for this phenomenon are unknown and this is a challenge to further study. In the animal model and possibly in humans this phenomenon might be associated with both antidepressant and antiepileptic effects. Further data are needed to clarify these hypotheses and considerably more work will be required to elucidate the functional significance of this phenomenon and in particular its interactions with the biogenic amine systems.

ACKNOWLEDGMENTS

Supported by the Medical Research Service of the Veterans Administration.

REFERENCES

Asberg, M., Thoren, P., Traskman, L., Bertilsson, L. and Ringberger, V., Serotonin depression - a biochemical subgroup within the affective disorders?, Science 191:478-480 (1976).

Babington, R. G. and Wedeking, P. W., Blockade of tardive seizures in rats by electroconvulsive shock, Brain Res. 88:141-144 (1975).

Balcom, G. J., Lenox, R. H. and Meyerhoff, J., Regional γ-aminobutyric acid levels in rat brain determined after microwave fixation, J. Neurochem. 24:609-613 (1975).

Barton, J. L., Mehta, S. and Snaith, R. P., The prophylactic value of extra ECT in depressive illness, Acta Psychiat. Scand. 49:386-392 (1973).

Brandon, S., Cowley, P., McDonald, C., Neville, P., Palmer, R. and Wellstood-Eason, S., Electroconvulsive therapy: Results in depressive illness from the Leicestershire trial, Br. Med. J. 288:22-25 (1984).

Breese, G. R., Mueller, R. A., Mailman, R. B. and Frye, G. P., Effects of TRH on central nervous system function. In Lombardini, J. B. and Kenny, A. D., Eds. The Role of Peptides and Amino Acids as Neurotransmitters, Alan R. Liss, N. Y., pp. 99-116 (1981).

Fink, M., Convulsive Therapy: Theory and Practice, Raven Press, NY, p. 46 (1979).

Frankel, F. H. (Ed.), Report No. 14 of The Amer. Psychiat. Assn. Task Force on Convulsive Therapy, APA, Washington, DC. (1978).

Freeman, C. P. L., Basson, J. V. and Crighton, A., Double-blind controlled trial of electroconvulsive therapy (ECT) and simulated ECT in depressive illness, Lancet i:738-740 (1978).

Hong, J. S., Gillin, J. C., Yang, H.-Y. T. and Costa, E., Repeated electroconvulsive shocks and the brain content of endorphins, Brain Res. 177:273-278 (1979).

Itil, T. M., Patterson, C. D., Polvan, N., Bigelow, A. and Bergey, B., Clinical and CNS effects of oral and I. V. thyrotropin-releasing hormone in depressed patients, Diseases of Nervous Syst. 36:529-536 (1975).

Jackson, I. M. D., Thyrotropin-Releasing Hormone, New Eng. J. Med., 306:145-155 (1982).

Johansson, O., Hokfelt, T., Pernow, B., Jeffcoate, S. L., White, N., Steinbusch, H. W. M., Verhofstad, A. A. J., Emson, P. C. and Spindel, E., Immunohistochemical support for three putative transmitters in one neuron: coexistence of 5-hydroxytryptamine, substance P- and thyrotropin releasing hormone-like immunoactivity in medullary neurons projecting to the spinal cord, Neurosci. 6:1857-1881 (1981).

Johnstone, E. C., Deakin, J. F. W., Lawler, P., Frith, C. D., Stevens, M., McPherson, K. and Crow, T. J., The Northwick Park electroconvulsive therapy trial, Lancet ii:1317-1320 (1980).

Kastin, A. J., Ehrensing, R. H., Schalch, D. S. and Anderson, M. S., Improvement in mental depression with decreased thyrotropin response after administration of thyrotropin-releasing hormone, Lancet ii:740-742 (1972).

Kirkegaard, C., The thyrotropin response to thyrotropin-releasing hormone in endogenous depression, Psychoneuroendocrinology 6:189-212 (1981).

Kubek, M. J. and Morzorati, S. L., Effect of pentylenetetrazol (PTZ) kindled seizures on TRH levels in specific areas of the rat brain, Abstr. Soc. for Neurosci. 11:884 (1985).

Kubek, M. J., Meyerhoff, J. L., Hill, T. G., Norton, J. A. and Sattin, A., Effects of subconvulsive and repeated electroconvulsive shock on thyrotropin-releasing hormone in rat brain, Life Sci. 36:315-320 (1985).

Kubek, M. J. and Sattin, A., Effect of electroconvulsive shock on the content of thyrotropin-releasing hormone in rat brain, Life Sci. 34:1149-1152 (1984).

Kubek, M. J., Lorincz, M. A. and Wilber, J. F., The identification of thyrotropin releasing hormone (TRH) in hypothalamic and extrahypothalamic loci of the human nervous system, Brain Res. 126:196-200 (1977).

Lamborn, J. and Gill, D., A controlled comparison of simulated and real ECT., Br. J. Psychiatry 133:514-519 (1978).

Lerer, B. and Sitaram, N., Clinical strategies for evaluating ECT mechanisms - pharmacological, biochemical and psychophysiological approaches, Progr. Neuro-Psychopharmacol. Biol. Psychiat. 7:309-333 (1983).

Lighton, C., Marsden, C. A., Bennett, G. W., Minchin, M. and Green, A. R., Decrease in levels of thyrotropin-releasing hormone (TRH) in the N. accumbens and lumbar spinal cord following repeated electroconvulsive shock, Neuropharmacol. 23:963-966 (1984).

Loosen, P. T., The TRH-induced TSH response in psychiatric patients: A possible neuroendocrine marker, Psychoneuroendocrinology 10:237-260 (1985).

Loosen, P. T. and Prange, A. J., Jr., Thyrotropin releasing hormone (TRH): A useful tool for psychoneuroendocrine investigation, Psychoneuroendocrinology 5:63-80 (1980).

Matsuishi, T., Yano, E., Inanaga, K., Terasawa, K., Ishihara, O., Shiotsuki, Y., Katafuchi, Y., Aoki, N. and Yamashita, F., A pilot study on the anticonvulsive effects of a thyrotropin-releasing hormone analog in intractable epilepsy, Brain & Development 5:421-428 (1983).

Morimoto, K., Moriwake, T., Akiyama, T., Sato, M. and Otsuki, S., Anticonvulsant effect of a novel TRH analogue (DN-1417) on amygdaloid kindled seizure, Brain and Nerve 35:189-195 (1983).

Perumal, A. S. and Barkai, A. I., Beta-adrenergic receptor binding in different regions of rat brain after various intensities of electroshock, Neurosci. Res. 7:289-296 (1982).

Post, R. M., Putnam, F. W. and Contel, N. R., Electroconvulsive shock inhibits amygdala kindling, Abstr. Soc. for Neurosci. 7:587 (1981).

Prange, A. J., Jr., Wilson, I. C., Lara, P. D., Alltop, L. B. and Breese, G. R., Effects of thyrotopin-releasing hormone in depression, Lancet ii:999-1002 (1972).

Rose, R. M. (Chairman), Consensus Development Conference Statement: Electroconvulsive Therapy, Vol. 5, No. 11, Office of Med. Appl. of Res., NIH, Bethesda, MD, USA, Chapt. 1. (1985).

Sachar, E. J. and Baron, M., The biology of affective disorders, Ann. Rev. Neurosci. 2:505-518 (1979).

Sattin, A., Hill, T. G., Meyerhoff, J. L. and Kubek, M. J., Thyrotropin releasing hormone (TRH) remains elevated in limbic regions for 1-2 weeks following electroconvulsive shock (ECS), <u>Abstr. Soc. for Neurosci</u>. 11:622 (1985).

Small, J. G. and Small, I. F., Clinical results: Indoklon vs. ECT, <u>Semin. Psychiat</u>. 4:13-26 (1972).

van Praag, H. M., New evidence of serotonin-deficient depression, <u>Neuropsychobiol</u>. 3:56-63 (1977).

Walczak, D., Meyerhoff, J., Bates, V. E., Lynch, T. and Kubek, M. J., Effect of partial and fully generalized kindled seizures on thyrotropin releasing hormone levels in specific cortical and subcortical regions of rat brain, <u>Abstr. Soc. for Neurosci</u>. 9:485 (1983).

West, E. D., Electroconvulsive therapy in depression: A double-blind controlled trial, <u>Br. Med. J</u>. 282:355-357 (1981).

Yarbrough, G. G., On the neuropharmacology of thyrotropin releasing hormone (TRH), <u>Progr. Neurobiol</u>. 12:291-312 (1979).

INDEX

A23187, 21, 242
AAGTP, 126-131
Acetylcholine, 346, 368, 517
Acetylsalicylic acid, 237
Acidic amino acids, 238
ACTH, see Adrenocorticotropin
Actin, 444
Active avoidance behavior, 393
Action potential, 33
Acyl CoA synthetase, 50, 60, 61
Adenylate cyclase, 95, 123-131, 331-
        341, 489, 518
  and microtubule disrupting drugs
  NaF activated, 124
Adrenal glomerulosa cells, 19-20
Adrenal medulla, 367-373
Adrenocorticotropin, 21, 25-26, 393,
        398, 401-404
  -induced excessive grooming, 401-403
    phorbol ester effects, 402
ADP-ribosylation, 95, 337
  cholera toxin catalyzes, 95
  inhibition by thrombin, 96
  pertussis toxin induced, 95
Affective illness, 515, 519
Aldosterone secretion, 19-21
  angiotensin II-induced, 21-22
Alpha-adrenergic receptors, 223, 232,
        521
  and 5HT release, 232-233
Alprazolam, 479
Alzheimer's disease, 427
Amiloride, 379
2-Amino-5-phosphono-valeric acid, 204
Amygdala, 361
Amyloid deposits, 428, 434-435
Anaphylactic shock, 477
Angiotensin II, 21, 25
Anticonvulsant compounds, 413-415
Antidepressant, 124, 129, 549-553
  tricyclic, 129, 503
  treatment, 549-553
Anxiety, 460, 469
Aplysia, 276
2-APV, see 2-Amino-5-phosphono-valeric
        acid

Arachidonic acid, 27, 46, 237, 324,
        532
Associative learning, 279, see also
        Classical conditioning
Astrocytoma cells, 255
Atherosclerosis, 484
Atropine, 52, 69
Axonal window, 170
Axoplasmic transport, 170

B50, 82, 320, 393-404, 420-421
  biochemical characterization, 394-
        395
  kinase C activation, 397-398
  phosphorylation, 393-404
  phosphoinositide breakdown, 393
B50-kinase, 75
BAY K 8644, 24, 162
  stimulation of Ca^{++} influx rate,
        24
Benzodiazepines, 359, 364, 414, 459,
        478-479
Benzodiazepine receptor complex,
        459-471
Beta-adrenoceptor, 489-499
Beta-pleated conformation, 431
Bicuculline, 52
Bipolar depression, 46
Biochemical memory, 28
Blood-brain barrier, 213, 215
  transport, 215
Boltzman probabilities, 103
8-bromo cAMP, 223, 233
Buproprion, 531, 533

Ca^{++}, see calcium
CA, 299
Ca^{++} channel, 11, 14, 83, 161, 275
  antagonists, 83, 161
  kinetics, 11, 14
Calcium, 19, 21-22, 160, 207
  -calmodulin dependent protein
        kinase, 126, 141, 276, 282,
        285, 409-421
  current, 286
  cycling, 21-22

557

223, 254, 532

K$^+$ channel, 275
Kainate, 204
Kainic acid, 345-353
K$^+$ current, 276-286, 412
  calcium dependent, 286
Ketanserin, 541
Kinase C see Protein kinase C
Kindling, 357, 418-420

Laminin, 446
Large neutral amino acids, 213
Laser microprobe mass analyzer, 432
L-aspartate, 204
Leukotrienes, 61, 237, 532
Leupeptin, 304
Li$^+$, see Lithium
Light adaptation, 117
  coupled-sequence model, 116
Limbic, 521, 526, 551
  system, 521, 526
  regions, 551
Linoleic acid, 243
Linolenic acid, 48
Lipid peroxidation, 63
Lipases, 51
Lipomodulin, 58
Lipoxygenase, 62, 237
  pathway, 62
Lithium, 55, 73, 380, 515-526
  therapeutic action of, 515-526
LNAA, see Large neutral amino acids
Long term potentiation, 194-196, 298,
      300, 313, 526
LTP, see Long term potentiation

mACh R-subtypes, 84-86
Macroscopic current, 36
Mammalian motor neurones, 33
Mania, 515
MAPs, see Microtubule associated pro-
      teins
Melatonin, 226
Membrane depolarization, 46
Memory storage, 285
Methylation, 380
Methoxy-hydroxy-phenylethylglycol sul-
      fate, 213
MHPG-SO4, see Methoxy-hydroxy-
      phenylethylglycol sulfate
Microinjection, 202
Microtubules, 295, 430
Microtubule associated proteins, 295,
      302, 430
Microwave irradiation, 46, 60
Midazolam, 364
Molecular adaptation, 187
Molecular geometries, 101
Molecular mechanics, 102
Monoamine oxidase, 532

Mouse spinal cord cells, 33
mRNA, 201
Muscarinic receptors, 53-55, 69,
      81-88, 518, 521
  inositide cycle, 81-88
  and phospholipase C activation,
      128
Myelination, 452
Myo-inositol, 70-72

Na$^+$ channel, 275
N-CAM, see Neuronal cell adhesion
      molecule
Nerve growth factor, 443
Nerve transection, 453
N-ethylmaleimide, 85
NFTs, see Neurofibrillary tangles
Neurite extension, 482-483
Neurofibrillary tangles, 428-437
  aluminum-induced, 432
Neurofilaments, 168-183
  crosslinked, 168
  translocation, 168
Neurofilament proteins, 167-182,
      430, 446
  axonal transport of, 177-178
  perikaryal accumulation, 169
  phosphorylation, 169-182
  polyclonal antibodies, 174
Neuronal degeneration, 477, 485
Neuronal development, 477
Neuronal differentiation, 482
Neurogenerative diseases, 427-438
Neuronal cell adhesion molecules,
      187, 192, 196
Neurotransmitter release, 147, 156,
      237
Neuronotrophic factors, 443
NFPs, see Neurofilament proteins
NG108-15 neural cells, 190-192, 482
Nifedepine, 162
Nigral-striatal pathway, 385
NMDA, see N-methyl-d-aspartate
N-methyl-d-aspartate, 204, 300
Norepinephrine, 223, 227, 331, 489-
      499
  and melatonin synthesis, 227
  mediated release of 5HT, 223-234
  second messenger involvement,
      233-234
  uptake, 196
Nucleus basalis of Meynert, 428

OAG, see 1-oleoyl-2-acetylglycerol
Oleic acid, 47, 324-325
1-oleoyl-2-acetylglycerol, 276, 286
Oxotremorine, 338

P96, 157-165
Paired helical filaments, 429-431
Palmitic acid, 47, 243

Paramecium, 375-382
Parkinson's disease, 435
Patch-clamp, 33
P-chloro-phenylalanine, 493-494
PDAc, see Phorbol 10,11 diacetate
PDH, see Pyruvate dehydrogenase
Pentylene tetrazole, 358
Periaqueductal gray, 402
Perforant path, 315
Pertussis toxin, 337
Phentolamine, 521
Phenylephrine, 223, 232
Phenytoin, 414
PHFs, see Paired helical filaments
Phorbol 10,11 diacetate, 101-105
Phorbol esters, 75, 223, 286, 320,
    523
  PMA, 223, 233
  TPA, 320
Phosphatidic acid, 52
Phosphatidylcholine, 216
Phosphatidylinositol, 69, 77, 291,
    482, 531
  turnover, 77-78
Phosphatidylinositol-4 phosphate, 19,
    52
  kinase, 82
Phosphatidylinositol-4,5 bis phos-
    phate, 19, 52, 69, 398
Phosphatidylserine, 320
Phosphodiesterase, 107
  calmodulin dependent, 116
  light dependent regulation of reti-
    nal rod, 107
  light activated, 110-112
Phosphoinositide metabolism, 69-78,
    195, 393, 494, 531-542
Phospholipase A , 57, 60
Phospholipase C, 95, 531
  induced degradation of inosotides,
    96-97
  muscarinic receptor mediated activa-
    tion, 128
Phospholipids, 237, 515
Phosphorylation, 275, 295, 524
Photoreceptor outer segments, 49
Photoreceptor cell biology, 108-109
Physostigmine, 519
PI, see Phosphatidylinositol
Pick's disease, 437
PIP, see Phosphatidylinosotol-4 phos-
    phate
PIP kinase, 398, 516, 525
PIP_2, see Phosphatidylinositol-4,5
    bis phosphate
Pineal glands, 223
Pinealocytes, 225
PKC, see Protein kinase C
Platelet activating factor, 477-485
Platelet aggregation, 477, 480, 485
Polyphosphoinositides, 51, 82, 398

Polyunsaturated fatty acids, 45
Postsynaptic densities, 295
Postsynaptic density protein, 413
Potassium channels, 33-41
  activation of voltage gated, 33
  delayed rectifier, 40
Prazosin, 223, 531, 532
Presynaptic membranes, 45
Presynaptic depolarization, 12
Presynaptic terminal, 4, 135
Proenkephalin mRNA, 385, 389
Progressive supranuclear palsy, 437
Prostaglandins, 59, 61, 237, 245,
    532
Protachykinin mRNA, 385
Protein dephosphorylation, 155-165
Protein F1, 313-327, 393, see also
    B50, GAP43
  phosphorylation and LTP, 314-319
  substrate for protein kinase C,
    319-320
Protein kinase C, 51, 69, 82, 101,
    155, 162-163, 194, 234,
    291, 302, 313-327, 397-398,
    523, 525
  B50 phosphorylation, 397-398
  calpain effect on, 302
  membrane and cytosolic activity,
    322
  translocation and F1 phosphoryla-
    tion, 325-327
Protein phosphatases, 314
Protein phosphorylation, 22, 135,
    155-165, 187, 314, 411
  neurofilament proteins, 167-182
  temporal patterns of, 22
  in synaptosomes, 155-165
Protein IIIa, 155
PSDs, see Postsynaptic densities
Psychopharmacological treatments,
    515
Pyruvate dehydrogenase, 157, 292

Quantal content of synaptic poten-
    tial, 13
Quin 2, 379

Receptor cell, 375
Receptors, 188, 201-207, 253-268,
    459-471
  Benzodiazepine, 459-471
  Beta-adrenergic, 253-268
  excitatory amino acids, 202
    subtypes, 207
  GABA, 459
  glutamate, 207
    NMDA-type, 322
  Kappa opiod, 345
  noradrenergic, 532
  serotonergic, 532
Reinnervation, 449